Environmental Engineering and Waste Management

Vineet Kumar • Sartaj Ahmad Bhat
Sunil Kumar • Pradeep Verma
Editors

Environmental Engineering and Waste Management

Recent Trends and Perspectives

Springer

Editors
Vineet Kumar
Department of Microbiology
School of Life Sciences
Central University of Rajasthan
Ajmer, Rajasthan, India

Sunil Kumar
Waste Re-processing Division
CSIR-National Environmental
Engineering Research Institute
(CSIR-NEERI)
Nagpur, Maharashtra, India

Sartaj Ahmad Bhat
River Basin Research Center
Gifu University
Gifu, Japan

Pradeep Verma
Department of Microbiology
School of Life Sciences
Central University of Rajasthan
Ajmer, Rajasthan, India

ISBN 978-3-031-58440-4 ISBN 978-3-031-58441-1 (eBook)
https://doi.org/10.1007/978-3-031-58441-1

This Springer imprint is published by the registered company Springer Nature Switzerland AG
The registered company address is: Gewerbestrasse 11, 6330 Cham, Switzerland

This book;
Dedicated to my late Grandparents for their enduring legacy and the values they instilled in me and to my family for their abundant support, patience, understanding, love, encouragement, blessings, and ongoing education have been the cornerstone of this endeavor. Without the unwavering support of my family, this book would not have been possible.

—Vineet Kumar

Dedicated to my family, with special acknowledgment to my wife. Without her unwavering support, this project would not have been possible.

—Sartaj Ahmad Bhat

Dedicated to our students, teachers, and mentors, from whom I continue to learn and my family for their unwavering support, blessings, motivation, and love.

—Sunil Kumar

Dedicated to my family members and colleagues for their enduring patience and understanding, providing me with the time and inspiration needed for research and editing.

—Pradeep Verma

Preface

In this era of rapid urbanization, industrialization, and population growth, waste generation has reached unprecedented levels. Simultaneously, the impact of human activities on the environment has become more pronounced. In a world that generates an ever-increasing amount of waste, and our planet faces an array of complex ecological, climatic, and sustainability challenges, the need for innovative solutions in environmental engineering and waste management has never been more critical. Recent advances in these areas have provided innovative solutions to address these issues, offering hope for a more sustainable future. For example, the concept of a circular economy, where waste is minimized, resources are conserved, and products are designed for reuse and recycling, has gained traction as a sustainable approach to waste management.

This book, *Environmental Engineering and Waste Management: Recent Trends and Perspectives*, seeks to address these challenges head-on by exploring the latest advancements and insights in the field. The book comprises four parts, encompassing a total of 24 chapters. The chapters cover a wide range of topics, including but not limited to wastewater treatment, solid waste management, air quality, environmental monitoring, sustainable practices, and emerging technologies. By presenting diverse perspectives and insights, this book provides a holistic view of the field and offers practical solutions to some of our most pressing environmental challenges. These chapters are authored by leading experts from around the world, each contributing valuable insights in their respective fields. The content spans the latest research and developments in various facets of environmental engineering and waste management technologies, all geared towards ecological restoration and ensuring public health safety.

Within this volume, esteemed authors provide critical reviews of the foremost research and development in this dynamic field. Additionally, they explore various topics related to both fundamental and advanced knowledge in waste management technologies, all with a focus on fostering environmental sustainability.

Part I Waste and Impact comprises three chapters.

Chapter 1 offers a thorough examination of emerging contaminants in tannery sludge, covering their environmental impact, health hazards, management approaches, challenges, and future perspectives.

Chapter 2 provides a concise overview of methods for collecting, transporting, and segregating municipal solid waste (MSW). It analyzes the impact of these approaches on sustainable management and environmental protection. The authors explore how technological advancements enhance the efficiency of MSW processes. The chapter concludes by addressing challenges in MSW management and offering recommendations to enhance existing strategies.

Chapter 3 emphasizes wastewater sludge valorization methods for resource recovery, focusing on obtaining economically attractive end products aligned with a critical circular economy approach. The chapter also covers various methods and recent technological developments in this context.

Part II Waste Recycling, Reuse, and Resource Recovery comprises nine chapters.

Chapter 4 delves into utilizing waste sewage sludge as a fertilizer, presenting it as an effective option for nutrient recycling and waste disposal. The chapter examines the impact of sewage sludge-derived fertilizer on soil fertility, crop quality, and potential environmental hazards. Additionally, it highlights challenges associated with using sewage sludge as fertilizer, addressing future directions and potential solutions for mitigating concerns related to high contaminant contents that may affect ecology and social acceptance.

Chapter 5 offers an overview of agri-food waste management, discussing treatment strategies, innovations in agri-food supply systems, and challenges in valorizing agri-food wastes/by-products. Highlighting pre-treatment procedures, the chapter emphasizes their role in maximizing the potential use of phytochemicals for biofuels, livestock feed additives, and natural food additives. This approach contributes to a more sustainable and circular economy, addressing environmental concerns and recovering valuable phytochemicals.

Chapter 6 delivers comprehensive information on the management of construction and demolition waste (CDW), highlighting its significance as a crucial environmental issue. The chapter addresses bottlenecks, regulations, and policy frameworks, presenting examples and global best practices. Key challenges in effective CDW management, such as insufficient awareness among stakeholders and inadequate infrastructure, are discussed as major impediments.

Chapter 7 focuses on the treatment of petroleum oil sludge generated in oil refinery industries. The chapter explores various steps and methods, evaluating their advantages and disadvantages while revisiting their mechanisms. The study aims to enhance the scientific understanding of petroleum oil sludge treatment technologies, paving the way for more efficient and environmentally friendly practices in the petroleum industry.

Chapter 8 offers insights into the impact of the COVID-19 pandemic on waste generation, highlighting the increase in single-use plastic waste, challenges in waste disposal, and the necessity for sustainable waste management strategies. It examines the implications on the global waste management cycle, diverse government responses, and the crucial role of effective healthcare waste management systems.

The chapter concludes with recommendations for future policies and practices, advocating a holistic approach involving behavioral changes, institutional shifts, and innovative technological solutions to address the escalating issue of single-use plastics.

Chapter 9 centers on the strategies, trends, and future outlook of the cement industry in addressing the net-zero challenge. The author asserts that while the cement industry may not decrease clinker production in the coming decades, it aims to store emitted CO_2 from calcination in the form of carbonates (mineralization). The ultimate goal is for concrete, derived from cement, to absorb and store all the CO_2 emitted through constant use and over-recycling, effectively closing the emissions cycle.

Chapter 10 highlights the production of extremozymes through an innovative approach for environmentally friendly biocatalyst synthesis. This method involves waste management of agriculture and food, aiming to reduce environmental risks and enhance economic feasibility, creating a "win-win situation" in the process.

Chapter 11 provides a detailed examination of the textile industry, focusing on its processes responsible for microfiber (microplastic) generation. The chapter explores factors influencing microfiber production in textile processes and discusses potential measures to address this issue. Additionally, the chapter touches upon future research needs in this area.

Chapter 12 emphasizes the use of alternative approaches for the management and treatment of agri-food wastes, including composting, biochar production, utilizing crop residues for cultivating edible mushrooms, generating energy sources, organic fertilizers, animal feed, bioactive compounds, etc. The chapter advocates for these practices over improper and unregulated dumping and disposal, underscoring the significance of maximizing the valorization of agri-food wastes.

Part III Environmental Technologies comprises seven chapters.

Chapter 13 offers an overview of the current state of legacy waste dumpsite management in India, detailing the workflow process of LFBM operation. The chapter also identifies various challenges that must be addressed for the effective implementation of LFBM. It predicts that in the coming years, LFBM, as a reclamation approach to legacy waste dumpsites, will gain traction nationwide with funding from SBM 2.0.

Chapter 14 offers an overview of microbial recycling for various organic wastes to achieve a circular waste management system, addressing greenhouse gas emissions, nutrient cycling, and resource recovery. It explores the pivotal role of microorganisms in decomposing and transforming organic wastes into valuable resources. The chapter also delves into opportunities, challenges, and future perspectives in this field, drawing insights from recent scientific research.

Chapter 15 underscores the significance of conventional and advanced high-throughput molecular techniques in achieving comprehensive microbial profiling. These techniques play a crucial role in understanding the entire microbial community's function in solid waste treatment, ultimately enhancing the efficiency and effectiveness of waste treatment processes.

Chapter 16 provides insights about composting and vermitechnology in managing organic waste.

Chapter 17 concentrates on the challenges associated with the field-scale applicability of phytoremediation and offers solutions to maximize its effectiveness in large-scale applications. The chapter provides a summary of successes and challenges encountered in implementing phytoremediation strategies on a larger scale. It also delves into information on both conventional and improved phytoremediation strategies, covering their principles, mechanisms, possibilities, advantages, and limitations.

Chapter 18 provides a concise exploration of the detailed mechanisms and pathways of cellulose degradation by cross-linked endonuclease. Through various case studies, the chapter elucidates the applicability of endonuclease from diverse microorganisms. Additionally, it discusses the utilization of the cross-linking immobilization technique for the endocellulase enzyme and its application in the bioconversion of agro-industrial waste.

Chapter 19 underscores the elimination of toxic pollutants through the synergistic effects of environmentally friendly nanoparticles and microbial technologies. It explores the applications of these technologies in various fields. The chapter also provides a brief overview of the advantages and challenges associated with using specific biological agents in green nanomaterial synthesis, while outlining a roadmap for future research.

Part IV Recent Trends and Development comprises five chapters.

Chapter 20 offers critical insights into various aspects of integrated environmental technologies in waste management, encompassing natural resource recycling, development, modeling, characteristics, and the benefits of different smart integrated waste management technologies. The chapter emphasizes the integration of these technologies into conventional waste management systems as the key to environmental sustainability, fostering a better and safer future for the next generation.

Chapter 21 provides a concise overview of state-of-the-art technologies for wastewater treatment in pulp and paper manufacturing industries. It emphasizes the integration of physicochemical and biological processes for treating effluents from pulp and paper mills. The chapter explores how biological treatment processes can be integrated with techniques like adsorption, advanced oxidation processes, and membrane filtration for the effective treatment of chlorinated phenolic compounds and adsorbable organic halides (AOX) found in industry effluents. Additionally, the chapter highlights the recovery of value-added byproducts as a futuristic approach for the sustainable management of paper and pulp industry effluent.

Chapter 22 places a strong emphasis on comprehending the role of circular waste management in diminishing greenhouse gas emissions (GHGs). It highlights the connection of sources, advanced approaches, technological innovations, and policies geared towards GHG reduction. The chapter underscores the significance of adopting circular approaches and explores strategies, technological innovations, as well as policies and regulations in circular waste management. Additionally, it addresses challenges and barriers faced during the implementation of these initiatives.

Chapter 23 outlines the substantial benefits of implementing Industry 4.0 (I4.0) in waste management, including cost reduction, enhanced efficiency, waste elimination, and reduced labor risks. The chapter also addresses barriers, challenges, and limitations associated with the practical application of I4.0 technologies. It explores proposals for overcoming these challenges and adopting I4.0 as an effective tool for waste management.

Chapter 24 offers a robust theoretical framework for life cycle and cost estimation in solid waste management. It emphasizes the need to consider both financial and economic aspects before implementing a waste management method. The chapter's detailed analysis of various studies, methodologies, and recommended indicators is valuable for strategic planning and decision-making in the creation of sustainable municipal solid waste management practices.

This book serves as a compendium of cutting-edge research and practices in environmental engineering and waste management. It brings together contributions from esteemed experts and scholars who are at the forefront of their respective fields.

As editors, we hope that this book will serve as a valuable resource for researchers, practitioners, policymakers, and students in the fields of environmental engineering and waste management. In closing, we invite readers to explore the diverse chapters within this book, each offering a unique perspective on the multifaceted field of environmental engineering and waste management. Together, we can work towards a more sustainable and resilient future for our planet.

Ajmer, Rajasthan, India — Vineet Kumar
Gifu, Japan — Sartaj Ahmad Bhat
Nagpur, Maharashtra, India — Sunil Kumar
Ajmer, Rajasthan, India — Pradeep Verma

Acknowledgments

The realization of this book, *Environmental Engineering and Waste Management: Recent Trends and Perspectives*, has been a collaborative endeavor marked by the contributions and support of numerous individuals and institutions. We extend our heartfelt appreciation to all those who have been instrumental in bringing this project to fruition.

First and foremost, we express our sincere gratitude to the dedicated authors whose expertise, research, and commitment have enriched the content of this book. Your invaluable contributions have illuminated various facets of environmental engineering and waste management, making this publication a comprehensive and enlightening resource.

We extend our thanks to the diligent reviewers whose rigorous evaluations and constructive feedback have helped uphold the quality and scholarly integrity of the chapters presented in this volume. Your expertise and dedication to excellence have played a vital role in shaping the final product.

We would also like to thank Aaron Schiller, Associate Editor, Geography & Sustainability at Springer Nature, New York, USA, for giving us the opportunity to accomplish this book project and share the knowledge with the academic and scientific fraternity. We thank for his guidance and support throughout the publication process. His expertise and suggestions have been invaluable.

We wish to acknowledge the team at Springer Nature, particularly Ms. Tina Silva, Production Editor at Springer Nature for their professionalism, guidance, and support throughout the publishing process. Your commitment to advancing scientific knowledge and facilitating scholarly communication is deeply appreciated.

The editors extend their sincere thanks to Gulsunnoorfurr Adamshah K, Production Project Manager at Straive and her dedicated team for their exceptional work in transforming over 650 pages of text and art manuscript into the remarkable learning tool now presented before you. The publishing group's unwavering support and helpful contributions have played a pivotal role in the realization of this project.

Furthermore, we are grateful to the academic institutions, research organizations, and universities that have provided the necessary resources, platforms, and environments conducive to the research and expertise featured in this book.

We extend our heartfelt thanks to our families, friends, and loved ones for their unwavering support, understanding, and encouragement during the editorial journey. Your patience and belief in our endeavors have been a source of strength. Any success that we have achieved or will achieve in the future would not be possible without the love and moral support of our beloved families.

In anticipation of any potential errors in the text, we extend our sincere apologies. Your understanding is appreciated, and we welcome feedback on the content and style of *Environmental Engineering and Waste Management: Recent Trends and Perspectives* from students, professionals, environmentalists, policymakers, and readers. All comments will be thoughtfully considered for inclusion in any subsequent editions.

In conclusion, this book stands as a testament to the collaborative spirit of the scientific and research community, as well as the shared commitment to addressing environmental challenges through innovation and knowledge dissemination. We hope that the insights presented in these pages will continue to inspire positive change and contribute to a more sustainable and resilient future.

Vineet Kumar
Sartaj Ahmad Bhat
Sunil Kumar
Pradeep Verma

Contents

Contributors

Saravanan Alamelu Department of Biochemistry and Biotechnology, Faculty of Science, Annamalai University, Chidambaram, Tamil Nadu, India

M. V. Angelicola Planta Piloto de Procesos Industriales Microbiológicos (PROIMI-CONICET), Tucumán, Argentina

Gayatri Barabde Department of Environmental Science, The Institute of Science, Dr. Homi Bhabha State University, Mumbai, India

Department of Analytical Chemistry, The Institute of Science, Dr. Homi Bhabha State University, Mumbai, India

Ankita Bharshankh Academy of Scientific and Innovative Research (AcSIR), CSIR-Human Resource Development Centre (CSIR-HRDC), Ghaziabad, India

Wastewater Technology Division, CSIR-National Environmental Engineering Research Institute, Nagpur, Maharashtra, India

Rima Biswas Academy of Scientific and Innovative Research (AcSIR), CSIR-Human Resource Development Centre (CSIR-HRDC), Ghaziabad, India

Wastewater Technology Division, CSIR-National Environmental Engineering Research Institute, Nagpur, Maharashtra, India

Akshay Botle Department of Environmental Science, The Institute of Science, Dr. Homi Bhabha State University, Mumbai, India

Sankar Chakma Department of Chemical Engineering, Indian Institute of Science Education and Research Bhopal, Bhopal, India

Niti Chawla Department of Biotechnology, Chaudhary Bansi Lal University, Bhiwani, Haryana, India

Suchismita Das Department of Life Science and Bioinformatics, Assam University, Silchar, Assam, India

Dileep Dasari Department of Biological Science and Technology, National Pingtung University of Science and Technology, Pingtung, Taiwan

Ana Laura De la Colina Martínez Ciencia de Materiales por la Facultad de Química, Universidad Autónoma del Estado de México, Toluca, Estado de México, Mexico

Hemen Deka Ecology and Environmental Remediation Laboratory, Department of Botany, Gauhati University, Guwahati, India

David Joaquín Delgado Hernández Facultad de Ingeniería, Universidad Autónoma del Estado de México, Toluca, Estado de México, Mexico

Debasis Dhak Nanomaterials Research Lab, Department of Chemistry, Sidho-Kanho-Birsha University, Purulia, West Bengal, India

Vishal Diwan Division of Environmental Monitoring and Exposure Assessment (Water & Soil), ICMR – National Institute for Research in Environmental Health, Bhopal, India

Laura Patricia Domínguez-Jaimes Centro de Investigación y Desarrollo Tecnológico en Electroquímica, Querétaro, Mexico

P. M. Fernández Planta Piloto de Procesos Industriales Microbiológicos (PROIMI-CONICET), Tucumán, Argentina

Facultad de Ciencias Exactas y Naturales (FACEN), Universidad Nacional de Catamarca (UNCA), Catamarca, Argentina

Facutlad de Bioquímica, Química y Farmacia, Universidad Nacional de Tucumán (UNT), Tucumán, Argentina

B. Garolera Planta Piloto de Procesos Industriales Microbiológicos (PROIMI-CONICET), Tucumán, Argentina

Arghya Ghosh Department of Civil Engineering, Indian Institute of Technology Guwahati, North Guwahati, Assam, India

Lalita Gupta Department of Zoology, Chaudhary Bansi Lal University, Bhiwani, Haryana, India

Md. Rakib Hasan Department of Applied Chemistry and Chemical Engineering, Faculty of Engineering and Technology, University of Dhaka, Dhaka, Bangladesh

Mehedi Hasan Department of Applied Chemistry and Chemical Engineering, Faculty of Engineering and Technology, University of Dhaka, Dhaka, Bangladesh

Mihir Herlekar Department of Environmental Science, The Institute of Science, Dr. Homi Bhabha State University, Mumbai, India

Nivedha Jayaseelan Department of Biochemistry and Biotechnology, Faculty of Science, Annamalai University, Chidambaram, Tamil Nadu, India

Charu Juneja Wastewater Technology Division, CSIR-National Environmental Engineering Research Institute (CSIR-NEERI), Nagpur, India

Academy of Scientific and Innovative Research (AcSIR), Ghaziabad, India

Sathish-Kumar Kamaraj Instituto Politécnico Nacional (IPN)- Centro de Investigación en Ciencia Aplicada y Tecnología Avanzada (CICATA), Altamira, Tamps, Mexico

Arshiya Kariapper Department of Civil Engineering, University of Moratuwa, Katubedda, Moratuwa, Sri Lanka

Shno Karimi Department of Soil Science, Faculty of Agriculture, Isfahan University of Technology, Isfahan, Iran

Suresh A. Kartha Department of Civil Engineering, Indian Institute of Technology Guwahati, North Guwahati, Assam, India

Shiromi Karunaratne Department of Civil Engineering, Sri Lanka Institute of Information Technology, Malabe, Sri Lanka

Gulzar Alam Khan Department of Commerce, Assam University, Silchar, India

Ratul Ahmed Khan Department of Applied Chemistry and Chemical Engineering, Faculty of Engineering and Technology, University of Dhaka, Dhaka, Bangladesh

Julekha Khatun Nanomaterials Research Lab, Department of Chemistry, Sidho-Kanho-Birsha University, Purulia, West Bengal, India

Pinki Koiri Department of Life Science and Bioinformatics, Assam University, Silchar, Assam, India

Harish Krishnan Department of Biochemistry and Biotechnology, Faculty of Science, Annamalai University, Chidambaram, Tamil Nadu, India

Pawan Kumar Department of Biotechnology, School of Life Sciences, Mahatma Gandhi Central University, Motihari, Bihar, India

Sanjeev Kumar Department of Biotechnology, Chaudhary Bansi Lal University, Bhiwani, Haryana, India

Babli Kumari Department of Biotechnology, School of Life Sciences, Mahatma Gandhi Central University, Motihari, Bihar, India

Taimima Mahbub Department of Applied Chemistry and Chemical Engineering, Faculty of Engineering and Technology, University of Dhaka, Dhaka, Bangladesh

Amrita Ghosh Majumdar School of Biotechnology, Kalinga Institute of Industrial Technology (KIIT), Deemed to be University, Bhubaneswar, India

Ashootosh Mandpe Department of Civil Engineering, Indian Institute of Technology Indore, Indore, India

Vivek Manyapu Department of Biological Science and Technology, National Pingtung University of Science and Technology, Neipu, Pingtung, Taiwan

Department of Tropical Agriculture and International Cooperation, National Pingtung University of Science and Technology, Neipu, Pingtung, Taiwan

Mouna Mechri National Institute of Field Crops, Bousalem, Tunisia

Sandhya Mishra Applied Phycology and Biotechnology Division, CSIR-Central Salt and Marine Chemicals Research Institute, Bhavnagar, Gujarat, India

Priti S. Mohanty School of Biotechnology, Kalinga Institute of Industrial Technology (KIIT), Deemed to be University, Bhubaneswar, India

School of Chemical Technology, Kalinga Institute of Industrial Technology (KIIT), Deemed to be University, Bhubaneswar, India

Arnab Mukherjee Nanomaterials Research Lab, Department of Chemistry, Sidho-Kanho-Birsha University, Purulia, West Bengal, India

Khondaker Nafisa Nawar Department of Applied Chemistry and Chemical Engineering, Faculty of Engineering and Technology, University of Dhaka, Dhaka, Bangladesh

Priti Pal Shri Ramswaroop Memorial College of Engineering & Management, Lucknow, India

Sukdeb Pal Wastewater Technology Division, CSIR-National Environmental Engineering Research Institute (CSIR-NEERI), Nagpur, India

Academy of Scientific and Innovative Research (AcSIR), Ghaziabad, India

Deepak Panchal Wastewater Technology Division, CSIR-National Environmental Engineering Research Institute (CSIR-NEERI), Nagpur, India

Academy of Scientific and Innovative Research (AcSIR), Ghaziabad, India

Brijesh Pandey Department of Biotechnology, School of Life Sciences, Mahatma Gandhi Central University, Motihari, Bihar, India

Biswajit Pany School of Biotechnology, Kalinga Institute of Industrial Technology (KIIT), Deemed to be University, Bhubaneswar, India

School of Chemical Technology, Kalinga Institute of Industrial Technology (KIIT), Deemed to be University, Bhubaneswar, India

Sankha Subhra Parua School of Chemical Technology, Kalinga Institute of Industrial Technology (KIIT), Deemed to be University, Bhubaneswar, India

Vishwa Patel Applied Phycology and Biotechnology Division, CSIR-Central Salt and Marine Chemicals Research Institute, Bhavnagar, Gujarat, India

Shailesh Kumar Patidar Department of Environmental Science, Central University of Rajasthan, Kishangarh, Rajasthan, India

W. J. Pech-Rodríguez Department of Mechatronics, Polytechnic University of Victoria, Tamaulipas, Mexico

Anand Prakash Department of Biotechnology, School of Life Sciences, Mahatma Gandhi Central University, Motihari, Bihar, India

Om Prakash Department of Biotechnology, School of Life Sciences, Mahatma Gandhi Central University, Motihari, Bihar, India

Om Prakash Wastewater Technology Division, CSIR-National Environmental Engineering Research Institute (CSIR-NEERI), Nagpur, India

Briska Jifrina Premnath Department of Biochemistry and Biotechnology, Faculty of Science, Annamalai University, Chidambaram, Tamil Nadu, India

Bablesh Ranawat Department of Earth Sciences, Indian Institute of Technology, Mumbai, Maharashtra, India

Taslim Ur Rashid Department of Applied Chemistry and Chemical Engineering, Faculty of Engineering and Technology, University of Dhaka, Dhaka, Bangladesh

Taqi Raza Department of Biosystems Engineering & Soil Science, University of Tennessee, Knoxville, TN, USA

E. Rocha-Rangel Department of Master Engineering program, Polytechnic University of Victoria, Tamaulipas, Mexico

Nihat Ege Sahin Department of Biological and Chemical Engineering – Process and Materials Engineering, Aarhus University, Aarhus, Denmark

Rojalin Sahu Academy of Scientific and Innovative Research (AcSIR), CSIR-Human Resource Development Centre (CSIR-HRDC), Ghaziabad, India

Wastewater Technology Division, CSIR-National Environmental Engineering Research Institute, Nagpur, Maharashtra, India

Sayli Salgaonkar Department of Environmental Science, The Institute of Science, Dr. Homi Bhabha State University, Mumbai, India

Miguel A. Sandoval Instituto Tecnológico Superior de Guanajuato, Tecnológico Nacional de México, Guanajuato, Mexico

Prakash Kumar Sarangi College of Agriculture, Central Agricultural University, Imphal, India

Shradhanjali Sethi Academy of Scientific and Innovative Research (AcSIR), CSIR-Human Resource Development Centre (CSIR-HRDC), Ghaziabad, India

Wastewater Technology Division, CSIR-National Environmental Engineering Research Institute, Nagpur, Maharashtra, India

Freny Shah Applied Phycology and Biotechnology Division, CSIR-Central Salt and Marine Chemicals Research Institute, Bhavnagar, Gujarat, India

Abhishek Sharma Wastewater Technology Division, CSIR-National Environmental Engineering Research Institute (CSIR-NEERI), Nagpur, India

Academy of Scientific and Innovative Research (AcSIR), Ghaziabad, India

Parag Shil Department of Commerce, Assam University, Silchar, India

Douglas J. H. Shyu Department of Biological Science and Technology, National Pingtung University of Science and Technology, Pingtung, Taiwan

Satyabrata Si School of Biotechnology, Kalinga Institute of Industrial Technology (KIIT), Deemed to be University, Bhubaneswar, India

School of Chemical Technology, Kalinga Institute of Industrial Technology (KIIT), Deemed to be University, Bhubaneswar, India

Akhilesh Kumar Singh Department of Biotechnology, School of Life Sciences, Mahatma Gandhi Central University, Motihari, Bihar, India

Maharudra Pratap Singh Department of Biotechnology, School of Life Sciences, Mahatma Gandhi Central University, Motihari, Bihar, India

Shubhra Singh Department of Tropical Agriculture and International Cooperation, National Pingtung University of Science and Technology, Pingtung, Taiwan

Department of Biological Science and Technology, National Pingtung University of Science and Technology, Pingtung, Taiwan

Surya Singh Division of Environmental Monitoring and Exposure Assessment (Water & Soil), ICMR – National Institute for Research in Environmental Health, Bhopal, India

W. James Singha Ecology and Environmental Remediation Laboratory, Department of Botany, Gauhati University, Guwahati, India

Manoj Kumar Srinivasan Department of Biochemistry and Biotechnology, Faculty of Science, Annamalai University, Chidambaram, Tamil Nadu, India

G. G. Suarez-Velázquez Universidad Politécnica de Altamira, Departamento de Ingeniería en energía, Altamira, Tamaulipas, Mexico

Redwan Arifin Tashfiq Department of Applied Chemistry and Chemical Engineering, Faculty of Engineering and Technology, University of Dhaka, Dhaka, Bangladesh

Rohan Mani Thomas Applied Phycology and Biotechnology Division, CSIR-Central Salt and Marine Chemicals Research Institute, Bhavnagar, Gujarat, India

Rahul Tiwari CSIR-National Environmental and Engineering Research Institute (CSIR-NEERI), Nagpur, India

Purusottam Tripathy Wastewater Technology Division, CSIR-National Environmental Engineering Research Institute (CSIR-NEERI), Nagpur, India

Kamalesh Balakumar Venkatesan Department of Biochemistry and Biotechnology, Faculty of Science, Annamalai University, Chidambaram, Tamil Nadu, India

S. C. Viñarta Planta Piloto de Procesos Industriales Microbiológicos (PROIMI-CONICET), Tucumán, Argentina

Facultad de Ciencias Exactas y Naturales (FACEN), Universidad Nacional de Catamarca (UNCA), Catamarca, Argentina

Junpei Yamanaka Graduate School of Pharmaceutical Sciences, Nagoya City University, Nagoya, Japan

About the Editors

Vineet Kumar is currently a National Postdoctoral Fellow in the Department of Microbiology, School of Life Sciences at the Central University of Rajasthan, Rajasthan, India. He got his M.Sc. (2008) and M.Phil. (2012) in Microbiology at Chaudhary Charan Singh University, Meerut, Uttar Pradesh (U.P.), India. Subsequently, he earned his Ph.D. (2018) in Environmental Microbiology from Babasaheb Bhimrao Ambedkar (A Central) University, Lucknow, U.P., India, focusing on the treatment of distillery effluent through two-step treatment system towards environmental sustainability. Dr. Kumar's research work has been mainly dedicated to wastewater treatment and solid waste management, focused on the valorization of agro-industrial wastes for bioenergy production and the development of sustainable approaches to mitigate the environmental impact of refractory pollutants discharged in wastewater and sludge. As a postdoctoral fellow, he works on role of biofilm-forming microbial community in in situ and ex situ remediation of pulp-paper mill waste. In the last few years, his research has been conducted in collaboration with some of the leading international environmental microbiology and biotechnology research groups across the world (Brazil, Mexico, China, and USA, among others) with diverse expertise. He has published more than 50 articles in peer-reviewed international journals of repute, 24 books, and 52 book chapters, on various aspects of science and engineering, with more than 2900 citations and *h*-index of 32 (Google Scholar). Dr. Kumar has been serving as a guest editor and reviewer in more than 70 prestigious international journals. He has served on the editorial board of various reputed journals. Dr. Kumar is a recipient of several prestigious fellowships, such as the Rajiv Gandhi National Fellowship by the University Grants Commission, India, and the National Postdoctoral Fellowship by the Science and Engineering Research Board (SERB), Department of Science and Technology, Government of India. He has presented several papers relevant to his research areas at national and international conferences. He is an active member of numerous scientific societies including the Microbiology Society (UK), the Indian Science Congress Association (India), the Association of Microbiologists of India (India), etc. He serves as the President of the Society for Green Environment, India (website: www.sgeindia.org). He can be reached at drvineet.micro@gmail.com; vineet.way18@gmail.com .

Sartaj Ahmad Bhat is a JSPS Postdoctoral Researcher at the River Basin Research Center, Gifu University, Japan. He received his Ph.D. in Environmental Sciences from Guru Nanak Dev University, Amritsar, India, in 2017. His research interests focus on vermicomposting treatment of various solid wastes, especially for investigations on fate and behavior of emerging pollutants during biological treatment of organic wastes. He has published more than 70 papers in peer-reviewed journals and edited over 15 books published by Elsevier Science, CRC Press, Royal Society of Chemistry, International Water Association Publishing and Springer Nature. Dr. Bhat is serving as an Associate/Academic Editor and Editorial Board Member/Advisory Board Member of more than 20 journals published by Frontiers, Springer, Elsevier, PLOS, Wiley, and De Gruyter. Dr. Bhat is a recipient of several prestigious awards such as the JSPS Postdoctoral Fellowship to pursue research at River Basin Research Center, Gifu University, Japan; the Basic Scientific Research Fellowship (BSR JRF, SRF) by the University Grants Commission (UGC) India; the DST-SERB National Postdoctoral Fellowship at CSIR-NEERI, Nagpur, India; and Swachhta Saarthi Fellowship by the Government of India. Dr. Bhat has also received the 2020 Outstanding Reviewer Award by the *International Journal of Environmental Research and Public Health*, MDPI, and Top Peer Reviewer 2019 Award in Environment and Ecology by Web of Science and has more than 800 verified reviews and 80 editor records to his credit.

Sunil Kumar is a well-rounded researcher with more than 24 years of experience in leading, supervising, and undertaking research in the broader field of Environmental Engineering and Science with a focus on solid and hazardous waste management. Dr. Kumar is a graduate in Environmental Engineering and Management from the Indian Institute of Technology, Kharagpur, India. He completed his Ph.D. in Environmental Engineering from Jadavpur University, Kolkata, India. His primary area of expertise is solid waste management (municipal solid waste, electronic waste, etc.) over a wide range of environmental topics including contaminated sites, EIA, and wastewater treatment. His contributions in these fields led to a citation of 20000, h-index of 70, and i10-index of 286 (Google Scholar). His contributions since inception at CSIR-National Environmental Engineering Research Institute (NEERI), India, in 2000 include 275 refereed publications, 8 books, and 40 book chapters, 10 edited volumes, and numerous project reports to various governmental bodies and private, local, and international academic/research bodies. He is the Associate Editor of peer-reviewed journals of international repute, i.e., *Environmental Chemistry Letters*, *International Journal of Environmental Science and Technology*, and *ASCE Journal of Hazardous, Toxic and Radioactive Waste*. He also served as Editorial Board of *Bioresource Technology*, Elsevier. He has completed many research projects as PI with 20 (13 awarded) Ph.D. and 22 M.Phil./M.Tech. thesis/dissertations. Dr. Kumar was awarded the most prestigious award Alexander von Humboldt-Stiftung Jean-Paul-Str.12 D-53173 Bonn, Germany, as a Senior Researcher for developing a global network and excellence for more advanced research and technology innovation.

Pradeep Verma is a Professor in the Department of Microbiology, School of Life Sciences at Central University of Rajasthan, Rajasthan, India. He is a well-rounded researcher with more than 22 years of experience in leading, supervising, and undertaking research in the broader field of bioprocess and bioenergy production from lignocellulosic waste with a focus on waste management. He earned his PhD in Microbiology from Sardar Patel University Gujarat, India in 2002. His research area of expertise involves microbial diversity, bioremediation, bioprocess development, and lignocellulosic and algal biomass-based biorefinery. He has more than 92 research and review articles in peer-reviewed international journals and contributed to 64 book chapters in different edited books with citations of more than 6750, and *h*-index of 43. He has also edited sixteen books by international publishers such as Springer, Taylor & Francis, CRC Press, and Elsevier. He also holds 12 international patents in the field of microwave-assisted biomass pretreatment and bio-butanol production. He is a guest editor to several journals such as *Biomass Conversion and Biorefinery* (Springer), *Frontier in Nanotechnology* (Frontiers), and *International Journal of Environmental Research and Public Health* (MDPI). He is also an editorial board member for the *Journal Current Nanomedicine* (Bentham Sciences). He is acting as reviewers for more than 70 journals in different publication houses such as Springer, Elsevier, RSC, ACS, Nature, Frontiers, MDPI, etc. He has also been honored with various prestigious fellowship and awards, such as JSPS Post-Doctoral Fellowship and Ron-Cockcroft Award by Swedish Society, and UNESCO Fellow, ASCR, Prague. He has been awarded with Fellow of Mycological Society of India (MSI-2020), Prof. P.C. Jain Memorial Award, Mycological Society of India 2020, and Fellow of Biotech Research Society, India (2021). He is a member of various national and international societies/ academies.

Part I
Waste and Impact

Chapter 1
Emerging Contaminants of Tannery Sludge and Their Environmental Impact and Health Hazards

Julekha Khatun, Arnab Mukherjee, and Debasis Dhak

1 Introduction

The rapid growth of urbanization and industrialization has increased the release of toxic wastewater into water bodies (Raman et al., 2021; Kumar & Verma, 2023). Around 80% of the wastewater from various industries and municipalities is released into water bodies without first undergoing any proper treatment (Yang et al., 2020a). According to the World Health Organization (WHO), ~25% of human health diseases are triggered by environmental pollution (soil, air, and water pollution) (Mukherjee et al., 2023). According to WHO (2018), globally, ~844 million people lack safe drinking water (Mukherjee et al., 2022), and this water scarcity has been addressed in the United Nations (UN) Sustainable Developmental Goal (SDG) 6: clean water and sanitation. According to the World Health Organization, about 80% of diseases are in one way or the other linked to infected water and several nations are unable to meet their standards for secure ingesting water (Khan et al., 2013). It is estimated that ~10^9 people will face sheer water shortage by 2025 (Kirandeep et al., 2021). There are many factors contributing to this hassle, but the most widespread one is the dramatic increase in the number of industries over the last few years given the role that they play in economic development. The tanning industry is a promising sector that generates an annual turnover of ~€7.8 billion by producing ~224 million square meters of finished leather across Europe, thus contributing ~16% of the total global production (Pietrelli et al., 2020). China is the largest producer of leather in the world with an annual average production of 6170 million square feet of leather, followed by Brazil with 2360 million square feet, Russia with 1652 million square feet, and India with 1560 million square feet (Saira & Shanthakumar, 2023). India

J. Khatun · A. Mukherjee · D. Dhak (✉)
Nanomaterials Research Lab, Department of Chemistry, Sidho-Kanho-Birsha University, Purulia, West Bengal, India
e-mail: debasis.chem@skbu.ac.in

V. Kumar et al. (eds.), *Environmental Engineering and Waste Management*,
https://doi.org/10.1007/978-3-031-58441-1_1

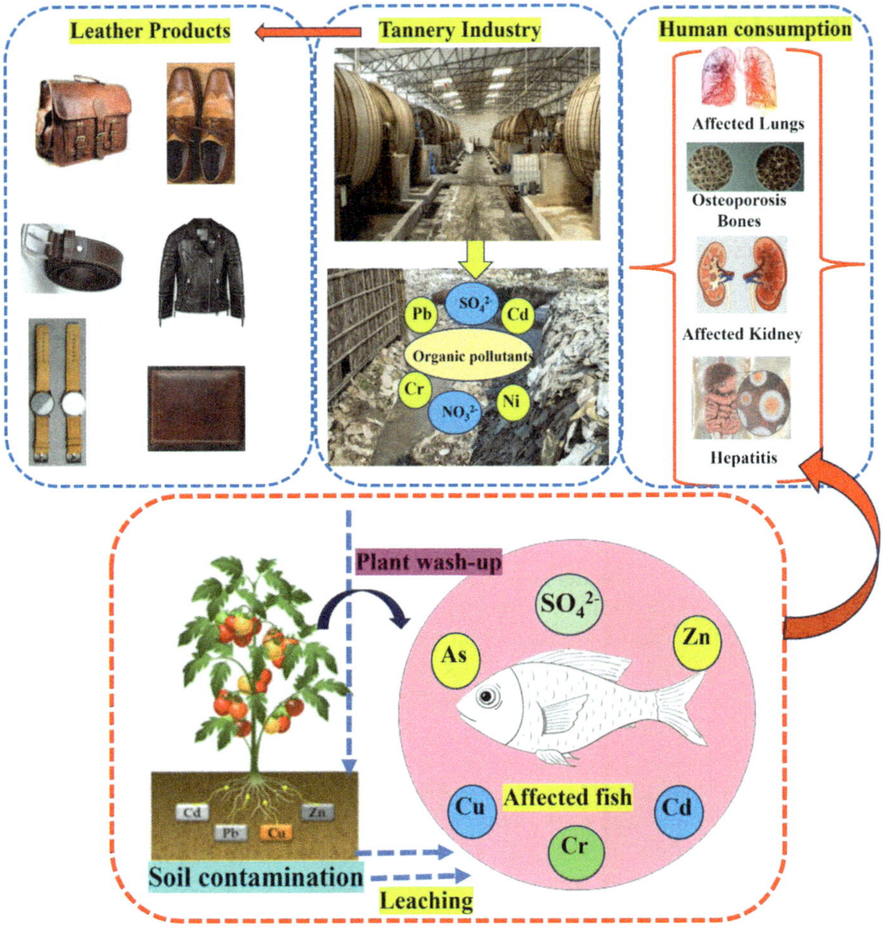

Fig. 1.1 The whole process of environmental pollution with the possible diseases associated with various tannery products

also exported ~3.68 billion US dollars' worth of leather products in 2020–2021, and, in 2022, it became the largest inventory of cattle population in the world (Hashem et al., 2022). Raw materials for the leather industry are imported from various slaughterhouses (Kumar & Thakur, 2020; Panneerselvam et al., 2022). The whole process of environmental pollution, with the possible diseases associated with various tannery products, is depicted in Fig. 1.1.

2 The Tanning Process

The leather industry is an ancient manufacturing sector that produces processed leather goods, such as footwear, bags, garments, etc. (Sivaram & Barik, 2019). It is a growing industry with a global market value of ~271.21 billion dollars in 2021 and has seen a continuous growth in the last 5 years (Saira & Shanthakumar, 2023). In this industry, tanning is used to transform animal skins into leather using a series of mechanical and chemical processes to obtain the final product that has the ability to resist microbial degradation, heat, and moisture stress (Alibardi & Cossu, 2016). Raw materials are imported from various slaughterhouses (Panneerselvam et al., 2022), and, then, a complex technique is performed to create leather using the tanning process that involves the following two steps:

The Preparatory Stage The different procedures that go into making leather are depicted in Fig. 1.2a. In the first process, which is known as preparatory or beam house operations, animal hides are washed to remove blood and grime. This is followed by soaking these hides in water to soften them and remove the leftover salts (Dixit et al., 2015). By fleshing, fatty acids are also eliminated. Liming is used to swell the hides so that the tanning products used can penetrate and remove all unwanted hair more effectively. To chemically dissolve the hair and epidermis, an alkaline medium of sulfur and lime is used. This involves a sequence of drums containing lime liquors (calcium hydroxide) and sharpening agents, although oxidative, thermal, and chemical procedures may also be used (Dixit et al., 2015). Liming results in the production of an effluent that contains a significant amount of lime, organic debris, and sodium sulfide. The hides are treated with enzymes after being neutralized by ammonium salts to eliminate hair residues and break down proteins. As a result, the effluent generates a significant amount of ammonium waste. Following these steps, the hides are pickled to prepare them for tanning. The pH value of the hides is managed by sulfuric acid and brine solution (Dixit et al., 2015).

Tanning Tanning transforms raw hides or skin into a stable, flexible material that does not putrefy and is appropriate for a wide range of end uses (Krishnamoorthy et al., 2012). Salts are used to stop the swelling of the skin. The collagen fibers found in the skins react with chromium (Cr), tannins, alum, etc. during the process. Numerous substances, including alum, formaldehyde, syntans, heavy oils, and glutaraldehyde, are employed as tanning agents. Per tonne of hides, more than 300 kg of chemicals are utilized in the tanning process. In general, two tanning procedures are utilized, depending on the eventual application of the tanned leather (Dixit et al., 2015; China et al., 2020).

(i) Chrome tanning that includes chromium salts
(ii) Natural tanning that includes polyphenolic biomolecules obtained from bark and leaves

Vegetable/natural tanning, also considered a "green tanning" method, is eco-friendly as it uses extracts from renewable sources that provide considerable

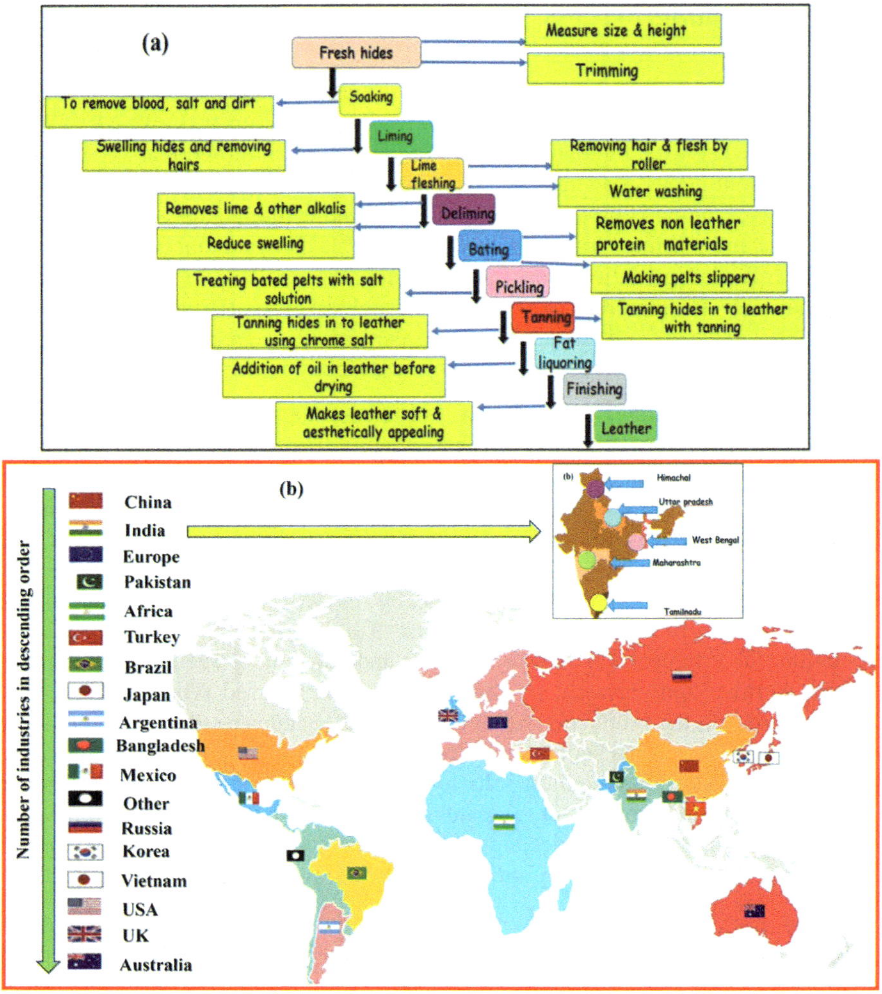

Fig. 1.2 (**a**) Processes involved in leather manufacturing. (**b**) Worldwide tanneries

benefits, e.g., comfort, dimensional stability, etc., thus causing less environmental pollution (China et al., 2021; Das et al., 2022). Green tanning agents contain tannin, non-tannin, gum, oligomeric and polymeric flavan-3-ol compounds, etc. based on plant sources and produce full, compact, and embossable leather (Das et al., 2022) with properties such as water permeability, solidness, good stability, high strength, and the ability to mold (Zhang et al., 2021). Recently, most commercially green tanning agents have been derived from mimosa, quebracho, sumac, tara, valonia, divi-divi, oak, and chestnut plants (Pisitsak et al., 2016). The *Mimosa tenuifora* (mimosa tannin) is available in South America and Eastern Africa over Asia and is used in Bangladeshi tanneries which are imported from Brazil, and Eastern Africa (Arne Witt & van Wilgen, 2018). The worldwide tanneries are shown in Fig. 1.2b. The process of vegetable tanning usually takes 3 weeks so that the dye penetrates the leather and is then washed with sodium bicarbonate or sulfuric acid. Then, before being set out to dry, lignosulfate, corn sugar, and oils may be added to the leather for mechanical strength generation. The use of various vegetable-driven tanning chemicals shows good tanning efficiency and has made the process a green one (Mohammed et al., 2020). However, for large-scale production of tanned leathers, this green process is not efficient, and, as a result, the chrome tanning process has been widely accepted by industries around the globe.

Chrome tanning has quickly become the most common method of tanning, in which chromium salts, which form highly stable transverse bonds with the collagen structure, are mainly used as the tanning agent to produce leather with excellent elastic, tensile, thermal, stable, and water resistant/repellent properties (China et al., 2020). On average, the overall tanning process requires ~6–8 hours, and only ~50–60% of the chromium is anchored to the leather, whereas the rest is disposed of as waste (Pietrelli et al., 2020). Besides the chromium salts, various azo dyes (acid orange) are used to make leather with vibrant colors (Alibardi & Cossu, 2016). The tanning process consumes >100 L of water and about 2 kg of chemicals per square meter of the final leather product, and it requires >1 ton of oil equivalent per 1000 m^2 of the final product as energy supply. Thus, the tanning industry has high environmental impacts along with causing air and water pollution due to the tannery sludge (TS) and effluents (proteins, hair, lime, salts, acids, tannins, dyes, oils) (Alibardi & Cossu, 2016).

3 Typical Features of Waste Produced

Tannery wastes include both solid and liquid wastes, such as splits, shavings, and trimmings, which contain minerals, fats, wastewater, and sludge (Saikia et al., 2017). Nearly 850 kg of solid wastes are generated in the leather processing industry for every 1000 kg of raw hide, and only 150 kg is transformed into leather (Kanagaraj et al., 2006). Skin trimmings (5–7%), fleshing wastes (50–60%), keratin wastes, buffing wastes (35–40%), and chrome shaving wastes are the most common solid wastes generated by the leather industry (Kanagaraj et al., 2006). Solid wastes generated in the preparation of hides and skins are as follows: 80% beam house; 1% finishing; and 19% tanning (Saha et al., 2021). Here, proteins are the main component.

The waste produced by tannery companies is characterized by a vivid color, a large biological oxygen demand (BOD), a high concentration of dissolved salts, and a high pH. It also contains a mixture of organic and inorganic contaminants. Tanning agents are used to transform hides into leather; as a result, effluents with high turbidity, color, and an unpleasant odor are produced (Buljan & Kral, 2011; Dargo & Ayalew, 2014). Chromium and chlorinated phenols were discovered in tannery effluents, according to studies related to the tannery industry (Saha et al., 2021). According to Kotaś and Stasicka (2000), chromium is an inorganic contaminant that is most frequently present in the trivalent and hexavalent oxidation states. Chlorinated phenols, which are organic pollutants connected to the tanning industry, are hazardous to animal cellular components that may come into contact with these molecules (Liknaw et al., 2017). According to Mwinyihija (2010), some significant contaminants that warrant attention include cadmium (Cd), antimony (Sb), cobalt (Co), copper (Cu), lead (Pb), zinc (Zn), nickel (Ni), selenium (Se), pesticide residues, barium (Ba), mercury (hg), arsenic (As), polychlorinated biphenyls, and formaldehyde resins. According to the complete industrial wastewater tannery process, effluents are by far the worst polluters (Shen, 1995). Tannery wastes can be divided into the following categories:

3.1 Liquid Waste

One of the essential requirements of the tanning industry is water, which is wasted on a tremendous scale during the manufacturing process. Therefore, the majority of tanneries are situated close to river banks since it takes about 30–40 liters of water to turn 1 kg of raw hide into a finished leather product. Tannery effluents are a serious environmental concern since they are largely responsible for compromising the standards of the water system by releasing hazardous compounds into it. Tannery waste comprises proteins and carbohydrates as biodegradable organic matter, which causes dissolved oxygen loss in aquatic environments due to microorganism decomposition (Saha et al., 2021). The amount of dissolved oxygen is greatly reduced, which harms aquatic organisms and increases anaerobic function, resulting in the production of noxious gases and harming the nutrient profile of aquatic organisms (Saha et al., 2021). The direct release of pollutants into streams and rivers is especially concerning because of their impact on the marine environment and the maintenance of viable water supplies during a time of near-global water scarcity. High Cr concentrations in water and wastewater have an inactivating effect on soil microorganisms and enzymes (Hutton & Shafahi, 2019). Chrome tannins in discharged effluents hinder decomposition and hence cripple the land areas they come into contact with. Because of the need to remove "non-collagenous proteins and the use of ammonium salts," deliming, unhairing/liming, and bating generate a significant amount of nitrogen-rich wastewater. Besides Cr, nitrogen as a primary tannery effluent ingredient is of concern due to its tendency to cause eutrophication and hence damage aquatic systems. Eutrophication can reduce biodiversity among fish species while increasing the density of aquatic plants (i.e., algae), all of which compete for resources. The leather business generates 1.3 million pounds of sludge and 80 billion gallons of wastewater each year, and this figure is expected to rise. The high concentrations of hazardous chemicals in tannery effluents contribute to an increase in biochemical oxygen demand (BOD), total dissolved solids (TDSs), chemical oxygen demand (COD), total suspended solids (TSSs), nitrogen (N),

Table 1.1 Solid and liquid wastes associated with the tanning process

Process	Solid waste	Liquid waste
Trimming	Trimming waste	--
Soaking	Salts	BOD, TDSs, COD, blood, alkali, dung, dirt
Liming	--	BOD, TDSs, COD, alkali, H_2S gas
Unhairing	Keratinous waste	Alkali, TDSs, organics, fats
Fleshing	Fleshing waste	Alkali, fats
Deliming	--	Salt, ammonia
Bating	--	Organics, proteoglycans, lime
Pickling	--	Salt, acid
Tanning	--	Cr, BOD, salts, COD
Samming and shaving	Splitting waste, shaving dust	Splitting waste, shaving dust
Neutralization	--	Salt, Cr
Re-tanning	--	Salts, Cr, phenol
Dyeing	--	Dye
Fat liquoring and fixing	--	Acid, oil, fat

and Cr. The sludge produced by tannery factories has an impact on agricultural activities and degrades groundwater in the surrounding area. The liquid and solid waste produced during the tanning process contains a significant number of organic and inorganic pollutants (Tables 1.1 and 1.2). The discharged compounds' chemical profile, including their names, formulas, structures, and toxicity levels—specifically the acute oral toxicity dose (LD_{50}) in rats, measured in milligrams per kilogram—and the human organs they target, is detailed in Table 1.2.

3.2 *Solid Waste*

The amount of solid waste generated is dependent on a variety of criteria, including the type of leather manufactured, the process used, and the source of the hides and skins, according to the European Commission (Ashraf et al., 2021). Solid waste is extremely difficult to eliminate, with the majority of these being biological. Solid contamination is caused by chemicals in the effluents that have leached into adjacent soils as a result of inappropriate disposal (Hutton & Shafahi, 2019). A significant amount of chromium-containing trash is produced during the chrome tanning process (Saha et al., 2021). In general, these wastes include 1–3% Cr and 90% collagen (Saha et al., 2021). Concerning chemical substances are mostly trivalent chromium (Cr^{3+}) and hexavalent chromium (Cr^{6+}). After leaching into soils, the creation of Cr^{6+} from Cr^{3+} can cause "severe allergic contact dermatitis" and damage ground waterways (Hutton & Shafahi, 2019). Trivalent chromium is now used in 80–90% of leather because of its time and cost advantages (Hutton & Shafahi, 2019). The industry is regarded as "one of the most polluting industries in the world," with an

Table 1.2 A list of some hazardous chemicals used in the tanning industry

Chemical	Structure/formula	Use	Toxicity (LD_{50} in rats, oral milligram per kilogram)	Target organ	References
Chromium	Cr	Dyeing	3250	Kidneys, CNS, hematopoietic system	Hossain et al. (2023)
Arsenic	As	Preservation	763	Liver, kidneys, skin, lungs, lymphatic system and carcinogen	Athulya et al. (2023)
Formaldehyde	HCHO	Leather finishing	100	Eyes, lungs, and carcinogen	Hassan et al. (2023)
Dibutyltin	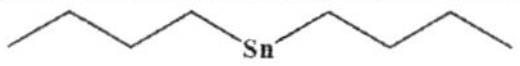	As a catalyst	175	Gastrointestinal tract, liver, and carcinogen	Xiong et al. (2023)
Azo dyes (orange II)	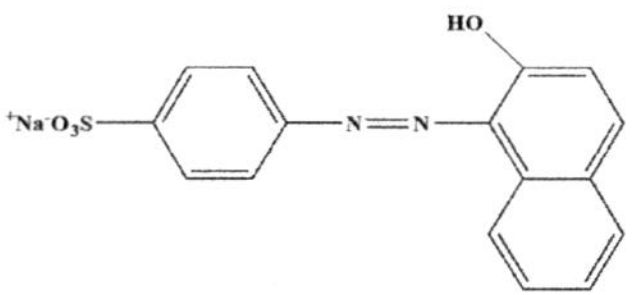	Dyeing	3418	Blood, liver, testes, and carcinogen	Van et al. (2023)
Methylisothiazolinone	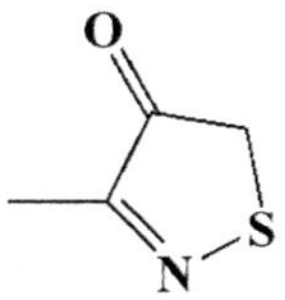	Biocide, microbiological protection	1800	Skin, eyes, and carcinogen	Singh et al. (2023)
Cobalt dichloride	$CoCl_2$		80		

		Leather dyeing and finishing as well as found in tanned leather		Lungs, liver, kidneys, heart, skin	Xiao et al. (2023)
Sodium dichromate	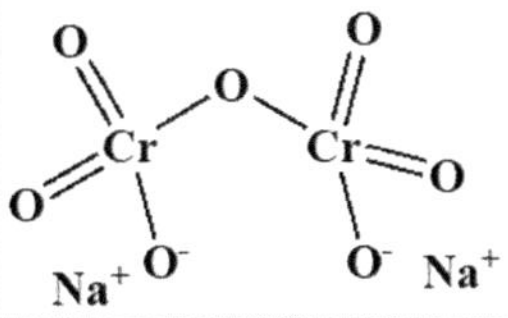	The principal raw material is used in the production of chrome tanning materials for the leather industry like chrome tanning salts	–	Blood, kidneys, heart, lungs, eyes, and carcinogen	Arti and Mehra (2023)
Nonylphenol	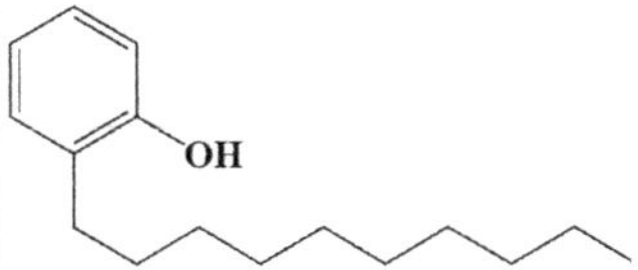	Finishing	1475	Blood, lungs, eyes, skin, CNS, kidneys	Korpe et al. (2023)
Dibutyl phthalate	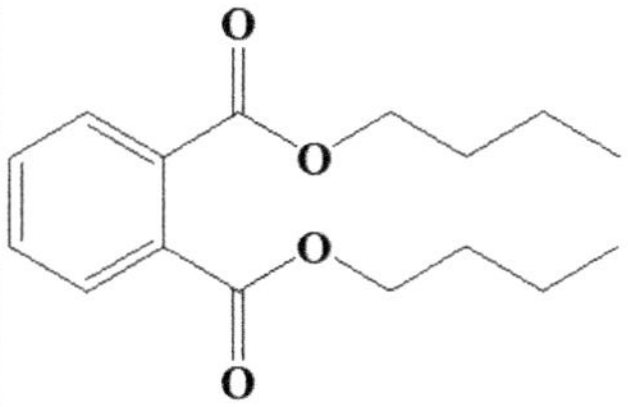	Used as a phthalate plasticizer in the artificial leather industry	7499	Eyes, lungs, gastrointestinal tract, testes	Devi et al. (2023)
Benzyl butyl phthalate	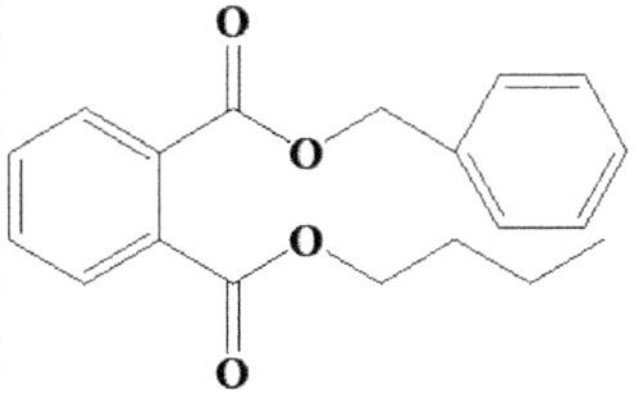	Used in the process for the production of a microporous artificial leather coating/water vapor-permeable sheet materials	2330	Eyes, lungs, liver, reproductive system	Kumar et al. (2023)

(continued)

Table 1.2 (continued)

Chemical	Structure/formula	Use	Toxicity (LD_{50} in rats, oral milligram per kilogram)	Target organ	References
Bis(2-ethylhexyl) phthalate	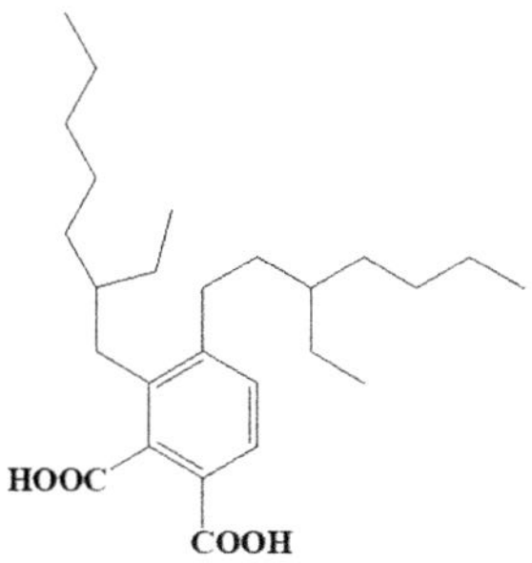	Used as plasticizers in the processing of shoe soles and in artificial leather manufacturing	30,000	Liver and testes	Ningthoujam et al. (2023)

CNS central nervous system

estimated 40% of worldwide Cr contamination (Hutton & Shafahi, 2019). The following categories of solid waste are produced at tanneries:

(a) Waste non-collagenous proteins
(b) Collagen with a tan
(c) Unbronzed collagen

Every year, the world's tanneries consume 8-9 MT of skin and hide, generating 1.4 MT of solid waste. Consumers who work in the glue, textile, gelatine, and artificial leather industries receive some of the solid waste or by-products. In the form of effluent tannery sludge (TS), flashings, showings, and brimming, about 35% of the original protein material is lost. The leather industry in developing nations is suffering a lot of solid waste problems, and several tanneries have closed for not satisfying the BOD and TDS requirements (Kanagaraj et al., 2006). The solid wastes produced during the various steps of the process are listed in Table 1.2. Their physico-chemical characteristics, their chemical profile including their chemical name, formula, structure, and toxicity level (LD_{50} in rats, oral milligram per kilogram), and the target organs of our body are described in Table 1.1.

3.3 *Gaseous Waste*

Gaseous air pollutants are released by the tanning industry during leather processing, particularly during unhairing and liming, deliming, pickling, and finishing activities when ammonia (NH_3), hydrogen sulfide (H_2S), chlorine (Cl_2), and volatile organic compounds (VOCs) are emitted (Hashem & Jor, 2015). Tanning activities also generate unpleasant odors caused by the decomposition of protein-rich waste materials as well as by the presence of ammonia, sulfide, and other volatile organic compounds (Kanagaraj et al., 2006). The poisonous H_2S gas has a deleterious influence on the environment due to the photochemical reaction process; it increases the emission of the greenhouse gas methane. In the atmosphere, the gaseous form of NH_3 combines with available acids to generate their corresponding salts, which induce cloudiness and eventually return to the earth's surface via wet or dry deposition, affecting aquatic life (Hashem & Jor, 2015). Workers in tanning firms, who are directly exposed to the harmful chlorine gas, suffer from a variety of health complications (Hashem & Jor, 2015).

4 Effect of Waste on the Environment

We are currently experiencing a terrible ecological crisis as a result of pollution from the tannery industry. The tannery business is contaminating the environment as a result of poor management practices, and the situation is getting worse since roughly 95% of the tanneries around the world have been constructed haphazardly

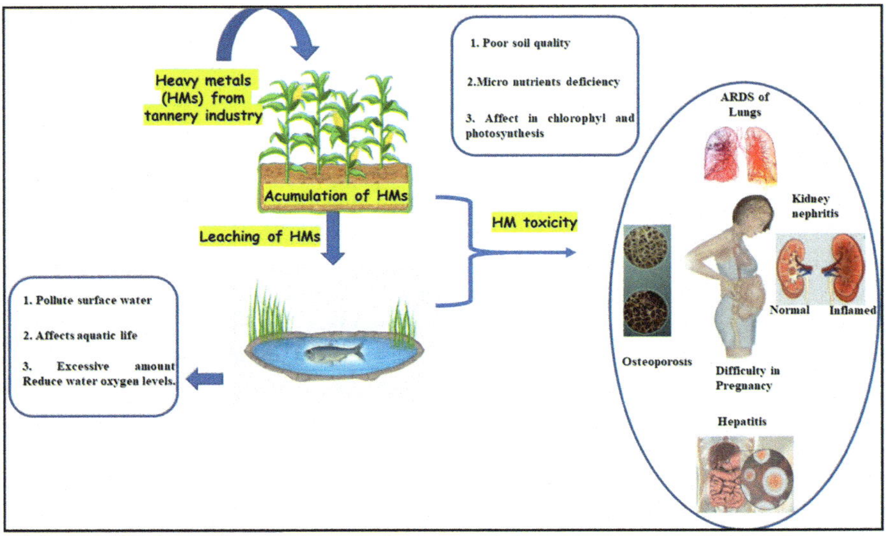

Fig. 1.3 The toxic effects of heavy metals (HMs) on the soil system, aquatic system, and human health

(Wicaksana & Rachman, 2018c). Due to their characteristics, including discoloration, hazardous chemical contents, and high oxygen consumption, wastes produced by tanneries are the main environmental contaminants and can harm both land and water (Song et al., 2000).

Chrome tanning is a popular tanning procedure because it saves time and delivers a product with excellent hydrothermal stability, softness, and better dyeing characteristics (Hutton & Shafahi, 2019). Trivalent Cr salts are used to tan around 80–90% of leather (Hutton & Shafahi, 2019). However, Cr exerts carcinogenic, mutagenic, and teratogenic effects on humans, animals, numerous plants, and microbes that live in water settings (Hutton & Shafahi, 2019). These effects could include dizziness, headache, irritation of the eyes and lungs, skin poisoning of the liver, kidneys, or nervous system, or even passing out from a lack of oxygen (Ashraf et al., 2021). Due to the growth of tanneries, Cr pollution is on the rise throughout the world. This is particularly evident in nations like India, South Africa, Pakistan, Burkina Faso, Latin America, Sudan, and Ethiopia, where many tanneries operate and wastes containing toxic Cr^{6+} and phenolics are released from each of them (Felsner & Kiruthu, 1996). Along with this, several harmful heavy metals (Cr, Cd, Pb, Ni, etc.) are also emitted from tanneries. All these heavy metals mix with the soil and also leach into the groundwater and pollute it, as shown in Fig. 1.3. Some of the toxicities of the elements are discussed below:

4.1 Toxicity Associated with Chromium

Chromium is a micronutrient crucial for regulating the metabolism of mammalian carbohydrates, lipids, and proteins, making it an essential part of both animal and human diets (Ashraf et al., 2021). Animals and humans both have altered lipid and glucose metabolisms as a result of their nutritional deficits. The glucose tolerance factor complex (GFC) contains chromium. Chromium has no known beneficial impacts on either plants or microbes (Nies, 1999). However, increased Cr levels are always dangerous because the degree of toxicity depends on the oxidation state of the metal (Mukherjee et al., 2022). According to Parveen et al. (Ram Bharose & Singh, 2017), Cr^{6+} is exceedingly hazardous to all life-forms, is mutagenic to microorganisms, people, and animals, and is carcinogenic Cr^{6+} exposure can cause cytotoxicity, genotoxicity, and carcinogenicity of Cr^{6+} -containing substances, DNA mutations and chromosomal damage at various stages, and oxidative protein alterations (Ayele & Godeto, 2021). When ingested in large doses, Cr^{6+} can result in death in both humans and animals (Mukherjee et al., 2022). The ease with which Cr^{6+} diffuses transversely the cell membrane in both eukaryotic and prokaryotic species, followed by its subsequent reduction in cells, resulting in free radicals, that can directly cause DNA changes and adverse effects (Arslan et al., 1987). Chromium compounds generated naturally and/or as a result of various anthropogenic operations are being continually discharged into the soil, where various flora develop. These compounds are most typically found in soil in the form of $HCrO_4^-$ and CrO_4^{2-}, which are easily absorbed by plants and swiftly flow down into the deep layers of the soil, supporting surface water and groundwater pollution (Ayele & Godeto, 2021). The detrimental effects of Cr^{6+} on algae, bacteria, and plants have been extensively researched (Cervantes et al., 2001; Ahemad 2014). Due to toxicity concerns in the USA, the total Cr concentration is restricted to 0.1 mg L^{-1} in drinking water, with 5 mg L^{-1} leaching from solids in the toxicity trait, particularly in soils that have been heavily contaminated with chromium (~5000 mg kg^{-1} of soil) due to tannery activities. Hexavalent chromium exposure can change the shape of microorganisms, such as both Gram-negative and Gram-positive bacteria, cause metabolic changes, and enable growth alterations by modifying their nucleic acid structure, disturbing their cell membranes, causing functional damage, limiting their enzyme activity and oxidative phosphorylation, and causing lipid peroxidation and osmotic equilibrium (Ayele & Godeto, 2021).

4.2 Toxicity Associated with Lead

Lead is regarded as a toxicant with documented toxicological effects, which is unsafe for use in the workplace. According to reports, the blood lead (BPb) threshold levels for neurotoxicity in adults are 18 g dL^{-1} and 5 g dL^{-1} in children (Khatun et al., 2022). However, no "safe" amount of lead exposure has been identified, and there does not seem to be any threshold level of lead that is necessary or beneficial to the body. However, the primary factor contributing to lead's persistent existence in the environment is the fact that it is not biodegradable. Adults absorb 35–50% of Pb in their drinking water, but children may absorb up to 50%. Lead absorption is influenced by age and physiological state. The kidney absorbs the most lead inside the human body, accompanied by the liver and other tender tissues, which include the brain and heart (Casas & Sordo, 2006).

4.3 *Toxicity Associated with Cadmium*

Cadmium (Cd) is an environmental and industrial toxin that harms the liver, lungs, pancreas, testicles, kidneys, and placenta in humans and other mammals. Acute respiratory distress syndrome (ARDS) has been reported in workers exposed to cadmium-containing vapors. The circulatory system is a significant conduit for Cd distribution, and blood vessels are regarded as the primary organs of cadmium poisoning. Cadmium absorbed through inhalation enters the bloodstream in the form of cadmium–cysteine complexes. Cadmium is a lung and gastrointestinal irritant that can be fatal if inhaled or swallowed (Tchounwou et al., 2012). Abdominal pain, nausea, salivation, burning sensation, muscle cramps, shock, vomiting, loss of consciousness, convulsions, and vertigo are typical, often occurring 15–30 minutes after severe ingestion (Tchounwou et al., 2012). Cadmium toxicity give rise to itai-itai disease, where bone and renal system both are affected (Nogawa et al., 2004). Cd poisoning can lead to cancer. There are preceding reviews of most renal cancers in humans (Kolonel, 1976). More specifically, the following harmful effects of tannery runoffs can be studied:

4.4 *The Soil System*

We are all aware of the significance of soil in the environment: it is essential for the formation and growth of both animal and plant life due to its role as a growth substrate. However, the inorganic and organic impurities formed during the tanning process, which are discharged along with effluents, cause harm to this vital source of life. Pollutants significantly influence the soil close to tanneries, resulting in heavy metal pollution in agricultural soil, as seen in Jajmau, Kanpur, India (Ashraf et al., 2021). Continuous sludge irrigation causes accumulation of heavy metals, such as zinc, cadmium, chromium, manganese, lead, and several others, in the soil, resulting in heavy metal release into soil solution or groundwater available for plant consumption due to a decrease in soil capacity to hold heavy metals. Soil pollution with heavy metals or micronutrients in phytotoxic concentrations harms both human and plant health (Ashraf et al., 2021). Because fields are often irrigated using chromium sulfate-rich wastewater, the quantity of hazardous chromium in the soil increases

(Babyshakila & Usha, 2009). At high concentrations, Cr is carcinogenic, mutagenic, toxic, and teratogenic, and it exists as an extremely dangerous anion. When sodium chloride (NaCl) is employed as a raw material in the tanning process, chlorides and nitrates are generated as the end products (Patel et al., 2021). The presence of sodium carbonate, sodium chloride, sodium bicarbonate, and calcium chloride created during the tanning process causes alkalization of the soil (Liknaw et al., 2017).

Because of their high toxicity, chromium-containing tannery solid wastes classed as shavings are of great concern amid the numerous waste pollutants. The usage of Cr in tanning operations increases the degree of chromium mobility and transfer, which substantially surpasses the rate of natural cycle progressions, resulting in significant issues in the tanning industry (Taylor et al., 1990). While the tanning industry is environmentally vital as the primary consumer of meat industry waste, it is also seen as a resource collector and polluter. A single metric tonne of raw hide produces 200 kg of end leather commodities (3 kg Cr), 250 kg of non-tanned solid trash, 200 kg of tanned waste (3 kg Cr), and 50,000 kg of wastewater (5 kg Cr) (Hüffner, 2003; Huffer & Taeger, 2004). Thus, about 20% of the feedstock is converted into leather, and liquid and solid wastes contain more than 60% of chromium. Annually, with an output potential of 0.7 million tonnes of hides and skins, ~2500 units are scattered throughout India, with the primary concentration areas being Tamil Nadu, Uttar Pradesh, and West Bengal (Garg et al., 2012). The use of basic chromium sulfate (BCS) has increased from 25,000 to 40,000 tonnes per year, with chromium tanning being used by 80% of tanners (Garg et al., 2012). Tannery runoff and waste sludge soil pollution is essentially different from water or air pollution since heavy metal deposition in the soil is purportedly slower than in other biosphere compartments. Heavy metal elimination from polluted soils is tough due to its diverse content and substances.

4.5 The Water System

Tanning necessitates a large volume of water to provide an aqueous medium for chemicals to permeate the hide's three-dimensional (3D) skin (Hutton & Shafahi, 2019). Effluents generated during tannery operations emit color and degrade water quality, making them a major environmental problem (Pepper et al., 1996).

The tanning and pre-tanning operations increase the concentrations of BOD, COD, TDSs, SS, chlorides, and also sulfates in tannery effluents (Dixit et al., 2015; Hutton & Shafahi, 2019). Because of the decrease in oxygen levels, anaerobic activity begins in the water system, resulting in the liberation of poisonous gases into the water system, which is damaging to aquatic species' survival (Dixit et al., 2015; Hutton & Shafahi, 2019). The sulfide generated throughout the tanning procedure harms fish and other aquatic species.

4.6 The Atmospheric System

Contaminated air comprises particulate matter, polycyclic aromatic hydrocarbons (PAHs), carbon monoxide, ground-level ozone, ammonia, heavy metals, sulfur dioxide, benzene, nitrogen dioxide, and hydrogen sulfide, all of which are readily present in wastes generated from the tannery industry during unhearing, deliming, liming, and bating methods and mix with air to cause air pollution (Ashraf et al., 2021).

Another dangerous chemical that is being discharged into the environment is Cr^{6+}. Anthropogenic activities are responsible for 60–70% of the chromium in the atmosphere (Ashraf et al., 2021). Because chromium is carcinogenic, its retention in the lungs poses serious health hazards.

4.7 Vegetable and Plant Growth

When tannery water is used as a source of irrigation, crop development and production suffer greatly (Joyia et al., 2021; Faiz et al., 2010). Furthermore, it causes the buildup of several heavy hazardous metals throughout the food chain at various tropical levels, as shown in Fig. 1.4. Tannery wastewater used to irrigate crops causes phytotoxicity in plants and creates salinity pressure in plants. It affects numerous metabolic activities, which leads to a reduction in vegetative development and a loss of reproductive growth in plants over time. Other than that, it influences activities such as respiration, photosynthesis, mitosis, and shortened germ sprouting as well as enhances the number of reactive oxygen species in the system (Faiz et al., 2010). Heavy metal accumulation is controlled by several parameters, including plant species, redox, pH, dissolved oxygen, temperature, cation exchange capacity, root secretion, and its bioavailability and element content in waste.

Higher quantities of dissolved salts in tannery effluents could also be a barrier to crop development (Joyia et al., 2021). According to research, the increase in effluent discharge from tanneries reduces all of the factors important for vegetative growth and productivity (Joyia et al., 2021). Another study found that tannery wastewater had the same negative effect on the growth of other crops such as soybeans, wheat, and maize (Hewitt & Keller, 2003). More than 80% of effluent concentrations are detrimental to plants' vegetative and reproductive stages Faiz et al. (2010).

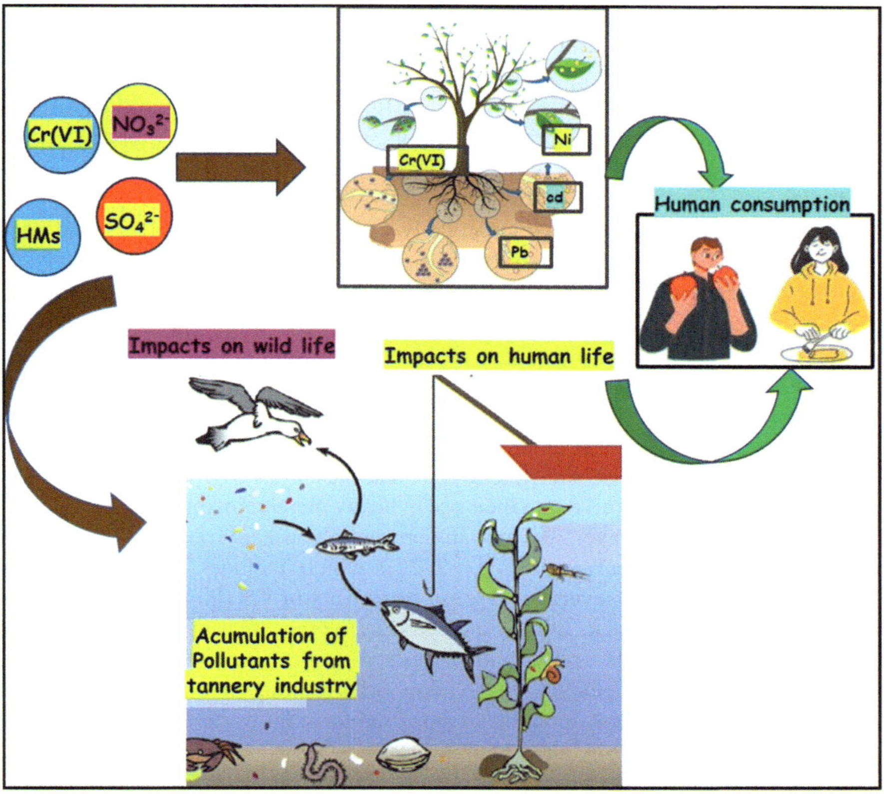

Fig. 1.4 Wastes from the tannery industry entering the food chain

5 Management of the Emerging Contaminants in the Polluted Environment

A noticeable increase in solid waste generation has been observed with the development of industries, such as paper mills, textiles, and, especially, the tannery industry that saw a rise from ~4.1 million tons to ~5.1 million tons from 2020 to 2021 (Araujo et al., 2023b). The solid waste generated during the tanning process is known as tannery sludge (TS), which is a major solid waste composed of organic and inorganic compounds with a high amount of chromium (Araujo et al., 2020). TS also contains a high proportion of sulfur and heavy metals/metalloids, e.g., arsenic, nickel, cobalt, copper, zinc, iron, systems, pigments, azo dyes, etc., used in the tanning process (Rossi et al., 2023). TS has been classified as a hazardous waste and also causes secondary pollution, which poses substantial challenges to the tanning industry and the environment (Malaiškienė et al., 2019). Only ~50–60% of chromium salts are utilized for the tanning process, and the residual part is discharged as wastewater with a high chromium content (Yang et al., 2020b). Hexavalent chromium is 10–100 times more toxic than trivalent chromium, which has strong mobility and higher biological activity, leading to carcinogenicity and bioaccumulation in humans (Singh et al., 2020). Generally, TS in landfills (Alibardi & Cossu, 2016) has several drawbacks, e.g., soil pollution, the transmission of hexavalent chromium into the food chain, etc., as untreated TS is highly toxic and may also cause secondary pollution (Yu et al., 2021). Meanwhile, some sludge treatment technologies are growing parallelly where the biomass of TS is converted into biochar (BC) material through pyrolysis (Hossain et al., 2020). The pyrolyzed sludge materials produce high-value fuel materials and low-priced BC for pollutant removal (Zheng et al., 2021). However, as TS contains various heavy metals and metalloids, it cannot be used for the synthesis of BC materials directly, which is the major drawback. Pietrelli et al. (2020) have developed a plasma pyro-gasification method to prepare a BC material using TS for the removal of heavy metals like lead, cadmium, zinc, etc. Ghezzi et al. (2023) and Ruan et al. (2023) have recently reported on the synthesis of different BC materials from TS. The production of various BC materials from TS and their use in heavy metal remediation may lead to a circular economy. However, the researchers mainly use pyrolysis techniques to produce BC from TS (Wang et al., 2022), which includes high temperature (>900 °C) heating, and several acid–base treatments of TS (Pietrelli et al., 2020), which is again a drawback. Some other processes other than pyrolysis have been developed/applied at the pilot and industrial scales for TS treatment, e.g., composting, anaerobic digestion, vitrification, brick making (Rossi et al., 2023), etc.

Thermal plasma gasification/vitrification is used for solid waste management where TS is converted into nontoxic slags with lower emissions of air pollutants (Zhou et al., 2023). Simultaneously, due to the application of high temperatures, the organic components present in the TS (dioxins etc.) are destroyed (Pei et al., 2020). This process is also helpful in converting TS into high-strength bricks by adding $CaO–SiO_2–Al_2O_3$ using the thermal plasma technology in TS waste treatment (Zhou et al., 2023). Vitrification is a useful process to treat TS effectively, but using a high temperature of ~1230 °C (Ma et al., 2021) makes this a less efficient and sustainable treatment process. On the other hand, composting is a biological method in which organic wastes are mainly decomposed under thermophilic conditions to eliminate pathogens (Araujo et al., 2023a). As this process involves biodegrading and recycling of the organic compounds present in TS (Ayilara et al., 2020), it has been adopted to treat cattle manure, sewage sludge, and industrial wastes (Araujo et al., 2020). During the last decade, composting has been used to recycle TS before its application in soil as an organic fertilizer (Yang et al., 2023). de Sousa et al. (2017) observed increased soil fertility and higher plant growth on applying composted TS. Laila et al. have reported that sunflower plants can take up toxic metals from TS-filled land (chromium: 540 mg kg^{-1} > cadmium 330 mg kg^{-1} > lead 285 mg kg^{-1}); these toxic metals were found in the sunflower seeds below the detection limit using atomic absorption spectrophotometry (Laila et al., 2023). Composted tannery sludge (CTS) application can improve soil fertility and crop yield (Araujo et al., 2020). A higher amount of organic matter and nutrients, e.g., N, P, S, K, etc., are present in the

CTS and also reported in a higher cowpea and maize plant yield after 6 years of CTS application. The rhizosphere is a microbial hotspot of soil, is present in the roots of legume plants (Sousa et al., 2020), and usually releases organic compounds that promote fertilization naturally (Araujo et al., 2023b). Heavy metal-containing TS-filled soils take up heavy metals (mainly chromium) and accumulate in the roots and shoots (Sousa et al., 2020), which lowers the microbe interaction. Dotaniya et al. (2022) reported treating TS with *Brassica juncea* for better performance. Use of TS as an agricultural fertilizer is a green way to treat them but uptake of various heavy metals from them by the plants, their toxicity and also the inhibition of them in soil is also reported agricultural fertilizer is heavy metals their toxicity (Araujo et al., 2023b). Thus, more research is needed for developing a proper methodology of treating heavy metal-containing TS and treating TS in a sustainable way. This may lead to fulfilling the waste-to-wealth agenda and help achieve United Nations Sustainable Development Goals 11 (sustainable cities and communities), 12 (responsible consumption and production), 6 (clean water and sanitation), 14 (life below water), 15 (life on land), etc.

6 Challenges and Future Perspectives

In this technologically advanced era, the use of industrial products belonging to daily life is directly associated with population expansion. Besides, the drainage of various toxic chemicals or effluents containing heavy metals, especially Cr^{6+}, Pb^{2+}, Ni^{2+}, Cd^{2+}, etc., causes various crises, e.g., environmental pollution, lethal diseases, habitat loss, etc. Among various industries, the tannery industry belongs to the top 10 listed pollution-causing industries, which affect ~32 million people's health worldwide. However, a huge amount of water and energy are required in the tanning procedure, which is also a drawback of this industry. In India, ~3000 tanneries generate ~50 million L of liquid waste along with ~305 million kg of solid waste per annum. About 2000–3200 tons of elemental Cr is also discharged from different tanneries along with other toxic heavy metals, such as Cd^{2+}, Pb^{2+}, Ni^{2+}, etc., in India annually. The sludge coming from physical-chemical and biological treatment units from tannery industries containing toxic heavy metals [Cr(VI), Pb^{2+}, Ni^{2+}, Cd^{2+}], proteins, animal hairs, etc. when exposed to soil or any agricultural land the toxic materials such as heavy metals, tannery dyes combine with the crops and enter into the food chain. In present times, improperly treated tannery sludge is directly disposed of into landfills in many developed/developing countries, which causes soil pollution, thus leading to an ecological imbalance.

Regarding this issue, the United Nations (UN) made a treaty in May 2001 in Stockholm, Sweden, to address pollution via micropollutants. The reduction/elimination of the various toxic chemicals released from various industries was the main concern of this treaty. The European Union (EU) has also made a Water Framework Directive regulated by Environmental Quality Standards (EQS) to control water pollution. The EQS had made a standard/permissible limit of various elements present in natural waters where 45 prior toxic substances, e.g., Cd^{2+}, Pb^{2+}, Hg^{2+}, Ni^{2+}, etc., were enlisted. Although global communities like the UN, EU, WHO, etc. have come up with prominent treaties, the number of newly produced chemicals across industries is also increasing exponentially. Most of the tanneries lack a common effluent treatment plant (CETP) and generate sludge-enriched hazardous solid waste that is transported to landfills. A huge variety of man-made chemicals are being regularly designed for industrial or agricultural use. These newly arrived chemicals are not only being consumed by the society but are also frequently exposed to the environment without undergoing prior investigation of their adverse effects on the environment. Thus, new research should be conducted with newer chemicals used in tanneries and newer drug discovery should be initiated by the scientific community to combat the various diseases caused by tannery sludge pollution.

7 Conclusions

This chapter highlights the environmental impact and challenges associated with the leather industry. From start to finish, the tanning process generates many wastes that damage water, air, and soil, propelling it to the top of the list of the world's most polluting industries. Tannery wastes include solid wastes like shavings, splits, and trimmings that contain minerals and fats as well as liquid wastes such as wastewater and sludge that are high in organic compounds. In recent years, the management of solid waste generated by tanneries has become a major challenge as it causes severe harm to both mankind and the environment. At present, more green alternatives are in demand owing to increased consumer awareness, and the strict rules imposed by global authorities demand the use of leather products that cause the least environmental harm.

References

Ahemad, M. (2014). Bacterial mechanisms for Cr(VI) resistance and reduction: An overview and recent advances. *Folia Microbiology (Praha), 59*, 321–332. https://doi.org/10.1007/s12223-014-0304-8

Alibardi, L., & Cossu, R. (2016). Pre-treatment of tannery sludge for sustainable landfilling. *Waste Management, 52*, 202–211. https://doi.org/10.1016/j.wasman.2016.04.008

Araujo, A. S. F., de Melo, W. J., Araujo, F. F., & Van den Brink, P. J. (2020). Long-term effect of composted tannery sludge on soil chemical and biological parameters. *Environmental Science and Pollution Research, 27*(33), 41885–41892. https://doi.org/10.1007/s11356-020-10173-9

Araujo, A. S. F., Jia, X., Miranda, A. R. L., de Araujo Pereira, A. P., Melo, V. M. M., Rocha, S. M. B., et al. (2023a). Changes in the bacterial rare biosphere after permanent application of composted tannery sludge in a tropical soil. *Chemosphere, 313*, 137487. https://doi.org/10.1016/j.chemosphere.2022.137487

Araujo, A. S. F., Miranda, A. R. L., de Araujo Pereira, A. P., de Melo, W. J., Melo, V. M. M., Ventura, S. H., et al. (2023b). Microbial communities in the rhizosphere of maize and cowpea respond differently to chromium contamination. *Chemosphere, 313*, 137417. https://doi.org/10.1016/j.chemosphere.2022.137417

Arne Witt, T. B., & van Wilgen, B. W. (2018). An assessment of the distribution and potential ecological impacts of invasive alien plant species in eastern Africa. *Transactions of the Royal Society of South Africa, 73*(3), 217–236. https://doi.org/10.1080/0035919X.2018.1529003

Arslan, P., Beltrame, M., & Tomasi, A. (1987). Intracellular chromium reduction. *Biochimica et Biophysica Acta (BBA) - Molecular Cell Research, 931*. https://doi.org/10.1016/0167-4889(87)90044-9

Arti, & Mehra, R. (2023). Analysis of heavy metals and toxicity level in the tannery effluent and the environs. *Environmental Monitoring and Assessment, 195*(5), 554. https://doi.org/10.1007/s10661-023-11154-4

Ashraf, R. S., Abid, Z., Shahid, M., Rehman, Z. U., Muhammad, G., Altaf, M., & Raza, M. A. (2021). *Methods for the treatment of wastewaters containing dyes and pigments*. https://doi.org/10.1007/978-3-030-52395-4_17

Athulya, R., Nandini, J., Bhoi, T. K., & Sundararaj, R. (2023). Recent advances of nanotechnology in wood protection: A comprehensive review. *Wood Material Science & Engineering, 0*, 1–12. https://doi.org/10.1080/17480272.2023.2239800

Ayele, A., & Godeto, Y. G. (2021). Bioremediation of chromium by microorganisms and its mechanisms related to functional groups. *Journal of Chemistry, 2021*, 1. https://doi.org/10.1155/2021/7694157

Ayilara, M. S., Olanrewaju, O. S., Babalola, O. O., & Odeyemi, O. (2020). Waste management through composting: Challenges and potentials. *Sustainability, 12*(11). https://doi.org/10.3390/su12114456

Babyshakila, P., & Usha, K. (2009). Effect of diluted effluent on soil properties and plant growth. *Advanced Studies in Biology, 1*, 391–398.

Buljan, J., & Kral, I. (2011). *Introduction to treatment of tannery effluents*. UNIDO.

Casas, J. A. S., & Sordo, J. A. (2006). *Leader to Leader* (pp. 158–228). https://doi.org/10.1016/B978-0-444-52945-9.X5000-3

Cervantes, C., Campos-García, J., Devars, S., Gutiérrez-Corona, F., Loza-Tavera, H., Torres-Guzmán, J. C., & Moreno-Sánchez, R. (2001). Interactions of chromium with microorganisms and plants. *FEMS Microbiology Review, 25*, 335–347. https://doi.org/10.1016/S0168-6445(01)00057-2

China, C. R., Maguta, M. M., Nyandoro, S. S., Hilonga, A., Kanth, S. V., & Njau, K. N. (2020). Alternative tanning technologies and their suitability in curbing environmental pollution from the leather industry: A comprehensive review. *Chemosphere, 254*, 126804. https://doi.org/10.1016/j.chemosphere.2020.126804

China, C. R., Nyandoro, S. S., Munissi, J. J. E., Maguta, M. M., Meyer, M., & Schroepfer, M. (2021). Tanning capacity of Tessmannia burttii extracts: The potential eco-friendly tanning agents for the leather industry. *Journal of Leather Science and Engineering, 3*(1), 13. https://doi.org/10.1186/s42825-021-00055-2

Dargo, H., & Ayalew, A. (2014). *Tannery waste water treatment: A review. International Journal of Emerging Trends Science and Technology.*

Das, R. K., Mizan, A., Zohra, F. T., Ahmed, S., Ahmed, K. S., & Hossain, H. (2022). Extraction of a novel tanning agent from indigenous plant bark and its application in leather processing. *Journal of Leather Science and Engineering, 4*(1), 18. https://doi.org/10.1186/s42825-022-00092-5

de Sousa, R. S., Santos, V. M., de Melo, W. J., Nunes, L. A. P. L., van den Brink, P. J., & Araújo, A. S. F. (2017). Time-dependent effect of composted tannery sludge on the chemical and microbial properties of soil. *Ecotoxicology, 26*(10), 1366–1377. https://doi.org/10.1007/s10646-017-1861-9

Devi, A., Verma, M., Saratale, G. D., Saratale, R. G., Ferreira, L. F. R., Mulla, S. I., & Bharagava, R. N. (2023). Microalgae: A green eco-friendly agents for bioremediation of tannery wastewater with simultaneous production of value-added products. *Chemosphere, 336*, 139192. https://doi.org/10.1016/j.chemosphere.2023.139192

Dixit, S., Yadav, A., Dwivedi, P. D., & Das, M. (2015). Toxic hazards of leather industry and technologies to combat threat: A review. *Journal of Cleaner Production, 87*(C), 39–49. https://doi.org/10.1016/j.jclepro.2014.10.017

Dotaniya, M. L., Rajendiran, S., Saurabh, K., Saha, J. K., Dotaniya, C. K., & Patra, A. K. (2022). Immobilization of chromium bioavailability through application of organic waste to Indian mustard (Brassica juncea) under chromium-contaminated Indian soils. *Environmental Monitoring and Assessment, 195*(1), 31. https://doi.org/10.1007/s10661-022-10625-4

Faiz, H., Ahmad, S., Athar, M., et al (2010). Effect of tannery effluents on seed germination and growth of twosunflower cultivars. *African Journal of Biotechnology* 9:5113–5120

Felsner, G., & Kiruthu, S. (1996). Status of the leather industry in eight African countries. *PT, 2.*

Garg, S. K., Tripathi, M., & Srinath, T. (2012). *Strategies for chromium bioremediation of tannery effluent.* https://doi.org/10.1007/978-1-4614-2329-4_2

Ghezzi, L., Mugnaioli, E., Perchiazzi, N., Duce, C., Pelosi, C., Zamponi, E., et al. (2023). Hexavalent chromium release over time from a pyrolyzed Cr-bearing tannery sludge. *Scientific Reports, 13*(1), 16283. https://doi.org/10.1038/s41598-023-43579-9

Hashem, M. A., & Jor, A. (2015). Gaseous air pollutants and its environmental effect-emitted from the tanning industry at Hazaribagh, Bangladesh. *American Journal of Engineering Research (AJER), 04*(05), 138–144.

Hashem, M. A., Payel, S., Mim, S., Hasan, M. A., Nur-A-Tomal, M. S., Rahman, M. A., & Sarker, M. I. (2022). Chromium adsorption on surface activated biochar made from tannery liming sludge: A waste-to-wealth approach. *Water Science and Engineering, 15*(4), 328–336. https://doi.org/10.1016/j.wse.2022.09.001

Hassan, M. M., Harris, J., Busfield, J. J. C., & Bilotti, E. (2023). A review of the green chemistry approaches to leather tanning in imparting sustainable leather manufacturing. *Green Chemistry, 25*(19), 7441–7469. https://doi.org/10.1039/D3GC02948D

Hewitt, F., & Keller, T. (2003). Water pollution impacts on bean and cereals. *Journal of Agronomy, 11*(3), 175–185.

Hossain, M. Z., Bahar, M. M., Sarkar, B., Donne, S. W., Ok, Y. S., Palansooriya, K. N., et al. (2020). Biochar and its importance on nutrient dynamics in soil and plant. *Biochar, 2*(4), 379–420. https://doi.org/10.1007/s42773-020-00065-z

Hossain, M. A., Sultana, R., Moktadir, M. A., & Hossain, M. A. (2023). A novel bio-adsorbent development from tannery solid waste derived biodegradable keratin for the removal of hazardous chromium: A cleaner and circular economy approach. *Journal of Cleaner Production, 413*, 137471. https://doi.org/10.1016/j.jclepro.2023.137471

Huffer, S., & Taeger, T. (2004). Sustainable leather manufacturing a topic with growing importance. *Journal of the American Leather Chemists Association, 99*(10).

Hüffner, S. (2003). *Photoelectron spectroscopy: Principles and applications* (Vol. 3, 3rd ed.). Springer.

Hutton, M., & Shafahi, M. (2019). Water pollution caused by leather industry: A review. In *ASME 2019 13th international conference on energy sustainability, ES 2019, collocated with the ASME 2019 heat transfer summer conference*, (December). https://doi.org/10.1115/ES2019-3949.

Joyia, F. A., Ashraf, M. Y., Shafiq, F., Anwar, S., Nisa, Z., Khaliq, B., & Malik, A. (2021). Phytotoxic effects of varying concentrations of leather tannery effluents on cotton and brinjal. *Agricultural Water Management, 246*, 106707. https://doi.org/10.1016/j.agwat.2020.106707

Kanagaraj, J., Velappan, K. C., Chandra Babu, N. K., & Sadulla, S. (2006). Solid wastes generation in the leather industry and its utilization for cleaner environment - A review. *Journal of Scientific and Industrial Research, 65*(7), 541–548. https://doi.org/10.1002/chin.200649273

Khan, N., Hussain, S. T., Saboor, A., Jamila, N., Kim, K. S. (2013). Physicochemical investigation of the drinking water sources from Mardan, Khyber Pakhtunkhwa, Pakistan. *International Journal of Physical Sciences, 8*(33), 1661–1671.

Khatun, J., Intekhab, A., & Dhak, D. (2022). Effect of uncontrolled fertilization and heavy metal toxicity associated with arsenic(As), lead(Pb) and cadmium (Cd), and possible remediation. *Toxicology, 477*(July), 153274. https://doi.org/10.1016/j.tox.2022.153274

Kirandeep, S., Sharma, A., Sahoo, S. C., Kumar, G., Mehta, S. K., & Kataria, R. (2021). Synthesis and characterization of 1D-Co/Zn MOFs having potential for efficient dye adsorption from wastewater. *Journal of Molecular Structure, 1226*, 129327. https://doi.org/10.1016/j.molstruc.2020.129327

Kolonel, L. N. (1976). Association of cadmium with renal cancer. *Cancer, 37*(4), 1782–1787. https://doi.org/10.1002/1097-0142(197604)37:4<1782::AID-CNCR2820370424>3.0.CO;2-F

Korpe, S., Rao, P. V., & Sonawane, S. H. (2023). Performance evaluation of hydrodynamic cavitation in combination with AOPs for degradation of tannery wastewater. *Journal of Environmental Chemical Engineering, 11*(3), 109731. https://doi.org/10.1016/j.jece.2023.109731

Kotaś, J., & Stasicka, Z. (2000). Chromium occurrence in the environment and methods of its speciation. *Environmental Pollution, 107*(3), 263–283. https://doi.org/10.1016/S0269-7491(99)00168-2

Krishnamoorthy, G., Sadulla, S., Sehgal, P. K., & Mandal, A. B. (2012). Green chemistry approaches to leather tanning process for making chrome-free leather by unnatural amino acids. *Journal of Hazardous Materials, 215–216*, 173–182. https://doi.org/10.1016/j.jhazmat.2012.02.046

Kumar, V., & Thakur, I. S. (2020). Extraction of lipids and production of biodiesel from secondary tannery sludge by in situ transesterification. *Bioresource Technology Reports, 11*, 100446. https://doi.org/10.1016/j.biteb.2020.100446

Kumar, V., & Verma, P. (2023). A critical review on environmental risk and toxic hazards of refractory pollutants discharged in chlorolignin waste of pulp and paper mills and their remediation approaches for environmental safety. *Environmental Research, 236*, 116728. https://doi.org/10.1016/j.envres.2023.116728

Kumar, A., Nighojkar, A., Varma, P., Prakash, N. J., Kandasubramanian, B., Zimmermann, K., & Dixit, F. (2023). Algal mediated intervention for the retrieval of emerging pollutants from aqueous media. *Journal of Hazardous Materials, 455*, 131568. https://doi.org/10.1016/j.jhazmat.2023.131568

Laila, U., Nazir, A., Bareen, F., & Shafiq, M. (2023). Role of composted tannery solid waste and its autochthonous microbes in enhancing phytoextraction of toxic metals and stress abatement in sunflower. *International Journal of Phytoremediation, 25*(2), 229–239. https://doi.org/10.1080/15226514.2022.2070597

Liknaw, G., Tekalign, T., & Guya, K. (2017). Impacts of tannery effluent on environments and human health: A review article. *International Institute for Science, Technology and Education E-Journals, 54*, 58–67.

Ma, W., Shi, W., Shi, Y., Chen, D., Liu, B., Chu, C., et al. (2021). Plasma vitrification and heavy metals solidification of MSW and sewage sludge incineration fly ash. *Journal of Hazardous Materials, 408*, 124809. https://doi.org/10.1016/j.jhazmat.2020.124809

Malaiškienė, J., Kizinievič, O., & Kizinievič, V. (2019). A study on tannery sludge as a raw material for cement mortar. *Materials, 12*(9). https://doi.org/10.3390/ma12091562

Mohammed, S. A., Naisini, A., Madhan, B., & Demessie, B. A. (2020). Rumex abyssinicus (mekmeko): A newer alternative for leather manufacture. *Environmental Progress \& Sustainable Energy, 39*(6), e13453. https://doi.org/10.1002/ep.13453

Mukherjee, A., Kundu, S., Chatterjee, D., & Dhak, D. (2022). A critical review on photoreduction using metal–organic frameworks: Kinetics, pH and mechanistic Studies and anthropogenic/natural sources of Cr(VI). *Chemistry Africa, 5*(6), 1783–1796. https://doi.org/10.1007/s42250-021-00295-z

Mukherjee, A., Dhak, P., Hazra, V., Goswami, N., & Dhak, D. (2023). Synthesis of mesoporous Fe/Al/La trimetallic oxide for photodegradation of various water-soluble dyes: Kinetic,

mechanistic, and pH studies. *Environmental Research, 217*, 114862. https://doi.org/10.1016/j.envres.2022.114862
Mwinyihija, M. (2010). Main Pollutants and Environmental Impacts of the Tanning Industry. *In, 2010*. https://doi.org/10.1007/978-1-4419-6266-9_2
Nies, D. H. (1999). Microbial heavy-metal resistance. *Applied Microbiology and Biotechnology, 51*, 730–750.
Ningthoujam, R., Satiraphan, M., Sompongchaiyakul, P., Bureekul, S., Luadnakrob, P., & Pinyakong, O. (2023). Bacterial community shifts in a di-(2-ethylhexyl) phthalate-degrading enriched consortium and the isolation and characterization of degraders predicted through network analyses. *Chemosphere, 310*, 136730. https://doi.org/10.1016/j.chemosphere.2022.136730
Nogawa, K., Kobayashi, E., Okubo, Y., & Suwazono, Y. (2004). Environmental cadmium exposure, adverse effects and preventive measures in Japan. *Biometals, 17*(5), 581–587. https://doi.org/10.1023/B:BIOM.0000045742.81440.9c
Panneerselvam, M., Rangaraj, A., & Monisha, K. M. (2022). Experimental study on solid waste management in slaughter house. *Materials Today: Proceedings, 68*, 1324–1329. https://doi.org/10.1016/j.matpr.2022.06.358
Patel, N., Shahane, S., Chauhan, D., Rai, D., Khan, M. Z. A., & Bhunia, B., et al. (2021). Environmental impact and treatment of tannery waste. In Inamuddin, M. I. Ahamed, & E. Lichtfouse, (Eds.), *Water pollut remediat org pollut* [Internet] (pp. 577–595). Cham: Springer International Publishing. Available from: https://doi.org/10.1007/978-3-030-52395-4_16
Pei, S.-L., Chen, T.-L., Pan, S.-Y., Yang, Y.-L., Sun, Z.-H., & Li, Y.-J. (2020). Addressing environmental sustainability of plasma vitrification technology for stabilization of municipal solid waste incineration fly ash. *Journal of Hazardous Materials, 398*, 122959. https://doi.org/10.1016/j.jhazmat.2020.122959
Pepper, I. L., Gerba, C. P., & Brusseau, M. L. (1996). Pollution science. Academic Press
Pietrelli, L., Ferro, S., Reverberi, A. P., & Vocciante, M. (2020). Removal and recovery of heavy metals from tannery sludge subjected to plasma pyro-gasification process. *Journal of Cleaner Production, 273*, 123166. https://doi.org/10.1016/j.jclepro.2020.123166
Pisitsak, P., Hutakamol, J., Jeenapak, S., Wanmanee, P., Nuammaiphum, J., & Thongcharoen, R. (2016). Natural dyeing of cotton with Xylocarpus granatum bark extract: Dyeing, fastness, and ultraviolet protection properties. *Fibers and Polymers, 17*(4), 560–568. https://doi.org/10.1007/s12221-016-5702-x
Ram Bharose, S. P., & Singh, D. (2017). Assessment of physico-chemical properties of tannery waste water and its impact on fresh water quality. *International Journal of Current Microbiology and Applied Sciences, 6*(4), 1879–1887. https://doi.org/10.20546/ijcmas.2017.604.224
Raman, C. D., Sellappa, K., & Mkandawire, M. (2021). Facile one step green synthesis of iron nanoparticles using grape leaves extract: Textile dye decolorization and wastewater treatment. *Water Science and Technology, 83*(9), 2242–2258. https://doi.org/10.2166/wst.2021.140
Rossi, D., Cappello, M., Antognoli, M., Brunazzi, E., & Seggiani, M. (2023). Pyrolyzed tannery sludge as adsorbent of volatile organic compounds from tannery air emissions. *Chemical Engineering Journal, 454*, 140320. https://doi.org/10.1016/j.cej.2022.140320
Ruan, R., Wu, H., Yu, C., Zhao, C., Zhou, D., Shi, X., et al. (2023). Impacts of magnetic biochar from reed straw on anaerobic digestion of pigment sludge: Biomethane production and the transformation of heavy metals speciation. *Process Biochemistry, 125*, 96–103. https://doi.org/10.1016/j.procbio.2022.12.004
Saha, B., Ahammed, F., Azam, B., Saha, B., Ahammed, F., & Azam, B. (2021). Texti le & leather review probable ways of tannery ' s solid and liquid waste management in Bangladesh—An overview probable ways of tannery ' s solid and liquid waste management in Bangladesh—An overview, *6257*(June), 76–95. https://doi.org/10.31881/TLR
Saikia, P., Goswami, T., Dutta, D., Dutta, N. K., Sengupta, P., & Neog, D. (2017). Development of a flexible composite from leather industry waste and evaluation of their physico-chemical properties. *Clean Technologies and Environmental Policy, 19*(8), 2171–2178. https://doi.org/10.1007/s10098-017-1396-z

Saira, G. C., & Shanthakumar, S. (2023). Zero waste discharge in tannery industries—An achievable reality? A recent review. *Journal of Environmental Management, 335*, 117508. https://doi.org/10.1016/j.jenvman.2023.117508

Shen, T. T. (1995). *Industrial pollution prevention* (pp. 15–35). Springer.

Singh, P., Itankar, N., & Patil, Y. (2020). Biomanagement of hexavalent chromium: Current trends and promising perspectives. *Journal of Environmental Management, 279*, 111547. https://doi.org/10.1016/j.jenvman.2020.111547

Singh, K., Kumari, M., & Prasad, K. S. (2023). Tannery effluents: Current practices, environmental consequences, human health risks, and treatment options. *CLEAN—Soil, Air, Water, 51*(3), 2200303. https://doi.org/10.1002/clen.202200303

Sivaram, N. M., & Barik, D. (2019). Chapter 5 - Toxic waste from leather industries. In D. Barik (Ed.), *Energy from toxic organic waste for heat and power generation* (pp. 55–67). Woodhead Publishing. https://doi.org/10.1016/B978-0-08-102528-4.00005-5

Song, Z., Williams, C. J., & Rgj, E. (2000). Sedimentation of tannery wastewater. *Water Research, 34*, 2171.

Sousa, R. M. S., Mendes, L. W., Antunes, J. E. L., de Souza Oliveira, L. M., de Cassia Batista Sousa, A. M., Gomes, R. L. F., et al. (2020). Diversity and structure of bacterial community in rhizosphere of lima bean. *Applied Soil Ecology, 150*, 103490. https://doi.org/10.1016/j.apsoil.2019.103490

Spb, K., Megharaj, M., Juhasz, A. L., Sethunathan, N., & Naidu, R. (2003). Chromium microorganism.

Taylor, M. M., Diefendorf, E. J., & Na, G. C. (1990). Enzymic treatment of chrome shavings. *The Journal of the American Leather Chemists Association (USA).*

Tchounwou, P. B., Yedjou, C. G., Patlolla, A. K., & Sutton, D. J. (2012). Molecular, clinical and environmental toxicicology volume 3: Environmental toxicology. *Molecular, Clinical and Environmental Toxicology, 101*. https://doi.org/10.1007/978-3-7643-8340-4

Van, H. T., Hoang, V. H., Luu, C., & Vi, L. (2023). Enhancing acid orange II degradation in ozonation processes with CaFe 2 O 4 nanoparticles as a heterogeneous catalyst. *RSC Advances, 13*, 28753–28766. https://doi.org/10.1039/d3ra04553f

Waalkes, M. P., Rehm, S., Bare, R. M., Poirier, L. A., Riggs, C. W., Devor, D. E., et al. (1989). Cadmium carcinogenesis in male Wistar [Crl:(WI)BR] rats: Dose-response analysis of effects of zinc on tumor induction in the prostate, in the testes, and at the injection site. *Cancer Research, 49*(15), 4282–4288.

Wang, X., Wei-Chung Chang, V., Li, Z., Song, Y., Li, C., & Wang, Y. (2022). Co-pyrolysis of sewage sludge and food waste digestate to synergistically improve biochar characteristics and heavy metals immobilization. *Waste Management, 141*, 231–239. https://doi.org/10.1016/j.wasman.2022.02.001

Xiao, G., Xie, Q., He, Y., Huang, X., Richardson, J. J., Dai, M., et al. (2023). Synergistic adsorption and in situ catalytic conversion of SO2 by transformed bimetal-phenolic functionalized biomass. *Environmental Science & Technology, 57*(34), 12911–12921. https://doi.org/10.1021/acs.est.3c03827

Xiong, W., Chen, B., Zhang, H., Peng, J., Pan, X., Guo, M., et al. (2023). A bio-based waterborne polyurethane with high toughness, superior wear resistance, and water resistance enabled by sorbitol monooleate. *Progress in Organic Coatings, 185*, 107895. https://doi.org/10.1016/j.porgcoat.2023.107895

Yang, Y., Li, X., Zhou, C., Xiong, W., Zeng, G., Huang, D., et al. (2020a). Recent advances in application of graphitic carbon nitride-based catalysts for degrading organic contaminants in water through advanced oxidation processes beyond photocatalysis: A critical review. *Water Research, 184*, 116200. https://doi.org/10.1016/j.watres.2020.116200

Yang, Y., Ma, H., Chen, X., Zhu, C., & Li, X. (2020b). Effect of incineration temperature on chromium speciation in real chromium-rich tannery sludge under air atmosphere. *Environmental Research, 183*, 109159. https://doi.org/10.1016/j.envres.2020.109159

Yang, K., Wu, Y., Lan, M., Li, X., & Wang, X. (2023). The feasibility evaluation of nitrogen recovery from tannery sludge leachate combined with phosphogypsum leachate by the struvite precipitation. *Water, Air, & Soil Pollution, 234*(4), 260. https://doi.org/10.1007/s11270-023-06258-z

Yu, S., Zhang, H., Lü, F., Shao, L., & He, P. (2021). Flow analysis of major and trace elements in residues from large-scale sewage sludge incineration. *Journal of Environmental Sciences, 102*, 99–109. https://doi.org/10.1016/j.jes.2020.09.023

Zhang, Z., Liu, Y., Wang, J., Xie, T., Sun, L., & Li, Z. (2021). A chrome-free combination tanning strategy: Based on silicic acid and plant tannin. *Journal of Leather Science and Engineering, 3*(1), 15. https://doi.org/10.1186/s42825-021-00058-z

Zheng, C., Yang, Z., Si, M., Zhu, F., Yang, W., Zhao, F., & Shi, Y. (2021). Application of biochars in the remediation of chromium contamination: Fabrication, mechanisms, and interfering species. *Journal of Hazardous Materials, 407*, 124376. https://doi.org/10.1016/j.jhazmat.2020.124376

Zhou, J., Zhai, Z., Wu, J., Yang, W., & Liu, J. (2023). Thermal plasma vitrification treatment of oil-based drill cuttings: Product characterization and harmless transformation. *Journal of Environmental Management, 331*, 117285. https://doi.org/10.1016/j.jenvman.2023.117285

Chapter 2
Municipal Solid Waste Collection, Transportation, and Segregation

Khondaker Nafisa Nawar, Taimima Mahbub, Redwan Arifin Tashfiq, and Taslim Ur Rashid

1 Introduction

The increasing global population, intensifying industrialization, and rapid urbanization have all directly contributed to the rapid increase in municipal solid waste (MSW) production, which has become a pressing concern for municipalities worldwide. Adapting to consumer culture, over-purchasing leading to food waste, the rapid obsolescence of electrical devices, and excessive use of plastic packaging are some of the common unsustainable practices that have contributed to the rapid increase in waste generation. MSW is a complex matrix of diverse materials, such as paper, food scraps, metals, glass, wood, batteries, etc., disposed of by households, offices, schools, parks, and other institutions. About 2 billion tons of MSW are produced annually worldwide, and it is estimated that this number will increase by 70% to a staggering 3.4 billion metric tons a year by 2050 (Valavanidis, 2023). While the volume of waste is increasing day by day, measures taken to effectively manage and reduce waste have been ineffective in protecting the environment and public health, especially in underdeveloped countries (Adnan et al., 2020). The accumulation of all sorts of wastes impacts life on land and water not only for the current generation but also for the future ones. All the elements of the environment—air, water, and soil—have become polluted to the extent that they now pose a significant health hazard to living beings. To avoid further degradation of the environment, particularly due to the over-concentration of waste, waste management is a necessity (Kumar & Thakur, 2020; Kumar et al., 2021). For effective results, the existing policies of waste management need to be revised and

Khondaker Nafisa Nawar and Taimima Mahbub contributed equally with all other contributors.

K. N. Nawar · T. Mahbub · R. A. Tashfiq · T. U. Rashid (✉)
Department of Applied Chemistry and Chemical Engineering, Faculty of Engineering and Technology, University of Dhaka, Dhaka, Bangladesh
e-mail: taslim@du.ac.bd

V. Kumar et al. (eds.), *Environmental Engineering and Waste Management*,
https://doi.org/10.1007/978-3-031-58441-1_2

improvised. Changes need to be made at the grassroots level so that further focus on the three basic steps—collection, transportation, and segregation—can bring about impactful changes in municipal waste management.

The management of MSW entails a range of actions that include every functional step from creation to ultimate disposal. Generation, collection, transportation, segregation, and disposal are the functional elements of MSW management systems (Yadav et al., 2016). Collection and transportation are the first steps practiced for the sustainable management of MSW, where waste is systemically collected from its point of generation and transported to processing facilities or disposal sites (Yadav & Karmakar, 2020). The majority of the total budget for MSW management is spent on collection and transportation systems, ranging from 80–90% in low-income countries to 50–80% in middle-income countries (Das & Bhattacharyya, 2015). However, lower-income countries still often have insufficient collection systems, whereas developed countries have higher collection rates. Waste management can be improved by raising awareness and encouraging residents to adopt sustainable waste management practices as well as by bringing about changes in the current management infrastructure (Nepal et al., 2023). Waste collection and transportation systems can be optimized by incorporating geographic information system (GIS) technologies that enable efficient waste collection by generating an optimal route, which is the shortest possible path for the transportation of MSW, thus reducing both travel time (Lella et al., 2017) and transportation costs (Das & Bhattacharyya, 2015). The collected waste consists of materials that can be recycled, organic wastes that can be composted, and other materials that can be processed for energy recovery. Municipal solid waste (MSW) is a lucrative source for the production of biofuels, biofuel carriers, and biochemicals using energy-efficient technologies (Naveenkumar et al., 2023). The recovery of valuable materials makes segregation the most crucial of all functional components of MSW management, as it reduces the volume of waste in landfills (Oberlin, 2013). Waste segregation is carried out in households where waste is created or at materials recovery facilities (MRFs). Sorting waste in households enhances the efficacy of waste segregation at MRFs because it takes less time and less energy to separate presorted waste (Kochar, 2021). Adoption of sustainable waste management practices involving effective collection, transportation, and segregation of waste facilitates the transition toward a circular economy (Wang et al., 2021), reduces environmental pollution (Ayodele et al., 2018), and enhances the potential for resource recovery from waste (Oberlin, 2013).

In this book chapter, we dissect and analyze the amalgamation of the collection, transportation, and segregation of MSW and their impact on its sustainable management as well as environmental protection. The sources and composition are first discussed, and, then, a detailed overview of the methods of MSW management is presented. We further look into how technological advancements can improve the functioning and mechanisms of MSW collection, transportation, and segregation. In the end, the challenges that need to be resolved for better management of MSW are addressed and some recommendations are made to improve the existing strategies.

2 Municipal Solid Waste

Municipal solid waste (MSW) is defined as waste gathered by a municipality or disposed of at a particular disposal site. It includes residential, institutional, commercial wastes, etc. (Daniel Hoornweg and Perinaz Bhada-Tata, 2012). Among the types of materials collected, paper, metals, plastics, glass, and organics are the most common. Determining the source, classification, and comprehensive characterization of MSW is the initial step for efficient waste management systems (Babatunde et al., 2013).

2.1 Sources

Solid waste characterization is a chief parameter that affects environmental impact assessment. MSW characterization has been observed to depend on social structure and income levels (Ozcan et al., 2016). Although several sources within a community produce MSW, presently, the rise in the amount of wastes produced by urban families stands out in particular (Gu et al., 2015). Independent of the size of the population, the urban division includes the sources of solid wastes found in populated areas. Both residential (dwellings) and nonresidential (commercial, institutional/services, construction/demolition, and special) are considered the two subdivisions of the urban division. However, the urban, industrial, and rural divisions are each treated as a separate class in the source categorization of MSW. Seven source classes can be derived from these distinctions (Buenrostro et al., 2001). Residential waste mainly includes garbage produced by homes and apartments. It covers four varied forms of wastes: bulky waste, household hazardous waste (HHW), garden waste, and waste from multifamily and single-family dwellings (Christensen et al., 2011). This type of waste includes packaging materials, paper, glass bottles, plastic containers, food waste, etc. Commercial waste can also be considered a part of municipality waste, consisting of packaging materials, office paper, cardboard, food waste, and other waste produced during commercial operations. Commercial waste is mainly produced by businesses, offices, retail stores, hotels, and restaurants (Das, 2021). The findings of a study of commercial activities in a tourism city in Vietnam revealed that 35.1% of municipal waste came from commercial sources, with restaurants being the largest contributor (74.5%) (Phu et al., 2019). Institutional and service waste includes garbage produced by public and private workplaces, educational institutions, museums, libraries, archaeological sites, and entertainment venues like stadiums and movie theaters. The service sector is frequently viewed as having little impact on pollution and waste production. However, the extensive service industry operations can result in the production of enormous amounts of waste (Chandrappa et al., 2012).

2.2 Composition

Simply put, MSW is composed of wastes produced in houses and other places, which have similar characteristics and compositions (Urbańska et al., 2023). Continuous advancements in technology, industrial expansion, rising consumer demands, and the age of heightened consumerism are driving the sustained increase in diverse products and services, thus leading to a rise in waste generation. Based on its composition, MSW can be broadly divided into biodegradable and nonbiodegradable waste. Biodegradable waste is composed of decayable organic materials, such as fruits, vegetables, food products, etc. Agri-food waste is organic material thrown away during food production, like leftover crops or spoiled produce (Pereira et al., 2023). It also includes animal and plant by-products, such as bones and fat. This waste contains useful substances that can benefit both animals and humans if proper waste management techniques are implemented. It would be greatly profitable if ways are found to use these leftovers wisely. Dry wastes like glass, metals, leather, textiles, and household wastes are not biodegradable (Bharadwaj et al., 2015).

2.2.1 Physical Composition

Different types of materials are found in the waste stream collected from municipal areas. In Europe, the most significant plastic packaging waste stream is known as DKR-350 (Genuino et al., 2023). This mixture comprises plastics left after the easily recyclable ones are sorted out. DKR-350 is a diverse blend of various plastic types, including polyethylene (PE), polypropylene (PP), polystyrene (PS), polyethylene terephthalate (PET), and polyvinyl chloride (PVC). It also includes multilayer flexible materials from plastic, aluminum, and paper layers as well as tangled items. The increasing global demand for plastic products persists due to their significant benefits, including affordability, flexibility, strength, resistance to corrosion, and low heat and electrical conductivity (Chen & Bai, 2023). However, recycling polyvinyl chloride (PVC) from municipal solid waste (MSW) is challenging and often leads to landfilling due to the harmful impact that PVC has on recycling processes. Various harmful microorganisms, such as fecal coliform bacteria, salmonellae, protozoan parasites, and hepatitis B virus, are frequently present in municipal solid waste (Han et al., 2023). These pathogens can spread to different hosts. Workers who collect and handle municipal solid waste are especially vulnerable to these infectious agents. Typical data on the composition of MSWs and their parent substances are depicted in Table 2.1.

Table 2.1 Composition of MSWs and their parent substances

Type of waste material	Parent substances	Remarks
Paper and cardboard	Newspapers, magazines, books, office and writing paper, cardboard boxes, packaging materials, tissue paper	Common recyclable materials are widely found in households and businesses, and they can be processed to create new paper products and packaging
Plastics	Bottles, containers, toys, plastic bags, hangers	These present a significant environmental challenge due to their nonbiodegradable nature and abundance in waste streams
Glass	Bulbs, glassware, windows, mirrors	Recycling glass reduces the need for raw materials and conserves energy
Metals	Food cans, utensils and cutlery (spoons, forks, knives), hardwire tools (screws, nails)	These can be recycled repeatedly without losing quality, making them environmentally beneficial materials to recycle
Textiles	Discarded clothing and fabrics	These can be reused or recycled to minimize their impact on landfills
Construction waste	Concrete, bricks	Recycling or reusing these materials can significantly reduce waste in the construction industry
Organics	Food scraps, yard trimmings, wood waste	These can be composted for nutrient-rich soils while diverting them from landfills, where they would produce harmful greenhouse gas emissions

2.2.2 Parameters Affecting the Physical Composition of Waste

The composition of MSW depends upon various factors such as the topography of the area, seasonal variation, food habits of residents, income, type of job, etc. (Thitame et al., 2010). Topography plays an important role, as challenges can be encountered in waste management in mountainous or coastal areas compared to flat areas. The dietary habits of residents are important determinants, with neighborhoods favoring fresh produce generating more organic waste, whereas areas eating more packaged foods generate more plastic packaging waste. This means that low-income families generally produce more organic waste. Understanding these factors is critical to designing sustainable and community-involved waste management strategies (Adeleke et al., 2021).

2.2.2.1 Topography

The factors concerning topography include cultural practices, climate, economic conditions, population density, life standard and consumption patterns, waste

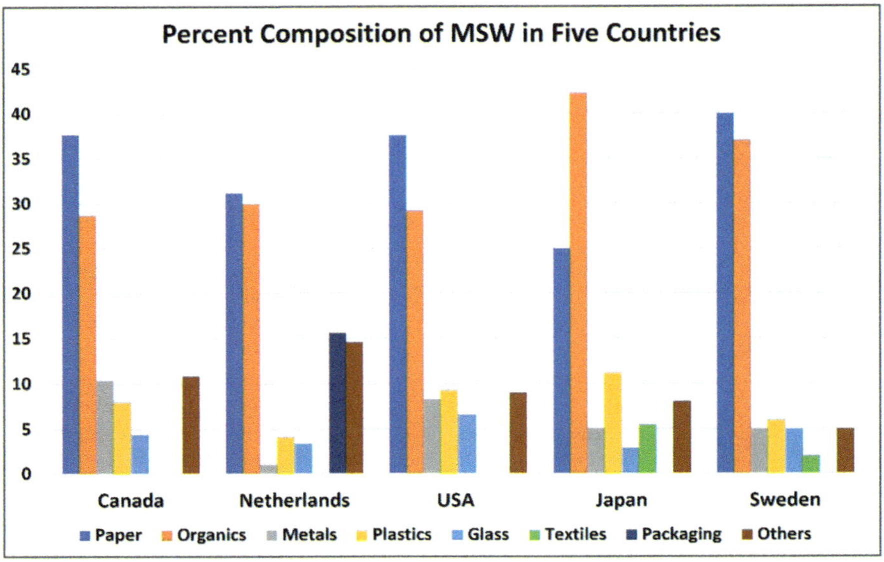

Fig. 2.1 Composition of municipal solid waste (MSW) components (paper, organics, metals, plastics, glass, textiles, packaging, etc.) in five countries. (Data collected from Sakai et al., 1996)

management infrastructure, and regulatory frameworks specific to each region. According to a study of MSW compositions in 297 cities of China from 2002 to 2017, climatic conditions contributed between 20.1% and 37.6% to the overall composition of MSW (Zhao et al., 2023). An increase in population by one person increases MSW disposed of by approximately 1.1×10^{-4} kg/person/day in urban areas, compared to 2.1×10^{-3} kg/person/day in rural areas (Chen, 2010). Urban areas often have a larger population density, resulting in reduced service costs for MSW collection. On the contrary, some rural communities do not have access to garbage collection services—thus, many households resort to open dumping of their domestic waste. Hence, for each unit change in population density in the rural areas, the change for the MSW disposed is larger (Chen, 2010). Urbanized regions have a higher percentage of packaging waste, electronic waste, and construction and demolition waste compared to developing ones. The variation in the composition of MSW in five different countries is displayed in Fig. 2.1.

2.2.2.2 Season and Climate

The existing body of literature extensively explores studies that have measured and described waste patterns during various seasons. These studies aim to understand the variations in the amount of waste produced and the composition of different types of physical waste streams throughout different times of the year (Oluwatobi Adeleke et al., 2023). The amount of organic waste often rises in the warmer seasons. The chief cause of this is an increase in fresh produce consumption, which leads to an increase in food and yard waste. Moreover, the composition of waste might be affected by seasonal agricultural operations. Crop residues and agricultural trash may be produced during harvest times. The percentage of leather and textile wastes in total MSW is projected to rise in winter as more people use clothing and textiles to remain warm (Adeleke et al., 2021). The high temperatures and increasing humidity during summer and monsoon are the cause of high PET content in wastes, since more people drink beverages (juices, water, flavored drinks, etc.) in plastic containers during these times. Ash makes up 60% of the MSW produced in the winter in Ulan Bator, Mongolia, where coal is utilized for house heating, compared to just 20% in the summer (Hoornweg & Bhada-Tata, 2012a, b). Kamran et al. (2015) analyzed waste in Lahore, Pakistan. According to their findings, during the spring season, the majority of the waste generated comprised of food and yard waste, whereas, in winter, the predominant types were plastic and textile wastes.

2.2.2.3 Income

The amount and composition of MSW is mostly influenced by household size and income (Richardson & Havlicek, 1974). Industrialized cities tend to dump more garbage that contains more recyclable materials and electronics, whereas industrializing cities tend to discard less waste but have higher biodegradable fractions (Dangi et al., 2011). According to a study conducted in Pakistan, the percentage of paper and cardboard is the highest in the high-income group, then in the middle-income group, and, finally, in the low-income group (Zia et al., 2017). The justification for this would be that households with greater incomes may purchase more items in cardboard packaging than would those with lower incomes, including toys, electrical equipment, tableware, and processed food items.

2.3 Impact on the Environment

MSW can be a threat to the environment unless it is properly managed. Despite past research on efficient ways to stop environmental contamination caused by MSW, difficulties still persist, necessitating more recent data and fresh tactics (Vinti et al., 2021). There are four criteria of MSW that contribute to pollution, as defined by Gumus (2009).

(i) *Corrosive*: This describes solid waste products having corrosive qualities that have the potential to harm the environment, machinery, or infrastructure. Batteries containing corrosive electrolytes, specific cleaning products, automotive fluids (such as battery acid or brake fluid), various industrial chemicals, and specific forms of electronic wastes (which may contain corrosive materials like lead or mercury) are examples of MSW of this class.

(ii) *Ignitable*: This class includes wastes that can create fires under the right conditions, e.g., waste oils and solvents.

(iii) *Reactive*: Reactive solid waste refers to unstable waste items that can undergo hazardous chemical reactions and harm infrastructure, the environment, or human health. They may cause explosions and release toxic fumes when heated.

(iv) *Toxic:* This includes waste that can be lethal or hazardous if consumed or absorbed.

The primary health concerns arising from improper MSW management come from the breeding of disease-spreading animals, especially flies and rats (Southwood, 1984). Environmental contamination in the surrounding area is frequently caused by garbage decomposing into its component substances (Kwarteng & Kumah, 2022). Developing countries are especially affected by this issue. The world's poorest countries have very few landfills that comply with the environmental standards recognized in wealthier countries, and, because of financial limitations, it is unlikely that many sites will be thoroughly analyzed before usage in the future. Again, this situation is made worse by the problems caused by growing urbanization.

2.3.1 Air Pollution

Uncollected and untreated wastes are a big challenge, since they contribute to climate change and global warming by emission of greenhouse gases (GHGs), including methane. The general public may be exposed to toxic compounds, causing respiratory issues and skin diseases. Moreover, these wastes can act as the breeding grounds for rodents and pests. Anaerobic decomposition of biodegradable waste provides approximately 18% and 2.9% of global methane and greenhouse gas (GHG) emissions, respectively (IPCC, 2014). Methane is also responsible for fires and explosions (Kumar et al., 2017). Emissions from solid waste management (SWM) in developing countries are increasing due to rapid economic expansion and rising living standards (Cao et al., 2018).

The carbon footprint of food waste is estimated to be 3.3 billion tons of CO_2 that is equivalent to the GHGs discharged into the atmosphere per year (Paritosh et al., 2017). Apart from inorganic criteria air pollutants such as carbon monoxide (CO), nitrogen oxides (NOx), and sulfur dioxide (SO_2), the presence of polymers, poly bags, and organic matter suggests that organic pollutants like polycyclic aromatic hydrocarbons (PAHs) and volatile organic compounds (VOCs) also play an undeniable role (Okedere et al., 2019). The People's Republic of China's Atmospheric Pollution Prevention and Control Law outlaws burning in the open, and, yet, unlawful MSW burning persists (Cao & Wang, 2017). Because open burning frequently takes place in inadequately ventilated areas or at temperatures too low to ensure complete combustion, dangerous compounds such carbon monoxide (CO), heavy metals, and particulate matter are released (Estrellan & Iino, 2010). Urban open burning locations are frequently situated close to heavily inhabited regions, and this proximity and direct exposure pose serious health risks to the locals.

2.3.2 Water Pollution

Irregular waste collection is directly linked to water pollution. In 2010, 192 coastal countries produced 275 million metric tons of plastic debris and up to 12.7 million metric tons, which is an alarming 4.4%, reached ocean ecosystems (Jambeck et al., 2015). Plastic debris gathers and stagnates water, creating a breeding ground for mosquitos, thus increasing the danger of dengue and malaria (Kumar et al., 2017). Garbage trucks frequently run uncovered due to improper management, and, hence, exhaust fumes and dust from collection and transportation contribute to pollution and significant health issues (Alam & Ahmade, 2013). Paper and pulp waste is produced in industries through various stages, as different by-products and waste materials are generated during washing, bleaching, and pulping operations (Gupta & Shukla, 2020). Paper and pulp waste primarily affects aquatic life due to its high organic content and release of harmful inorganic chemicals (Zacharewski et al., 1995).

2.3.3 Heavy Metal Contamination

The term "heavy metal" is used to describe a group of metals and metalloids with an atomic density > 5.0 g/cm^3, possessing toxic or poisonous effects even at extremely low concentrations (Duffus, 2001). Heavy metals are a major public health problem due to their toxicity and tendency to accumulate in humans and other living creatures (Hutton & Symon, 1986). Their persistence in aquatic habitats is also a big challenge (Tomno et al., 2020). Pollution sources in metropolitan areas include industrial activity, transportation emissions, and poor waste disposal, all of which release heavy metals into the environment (Li et al., 2008). These metals can then build in urban soils as a result of air deposition or runoff from contaminated surfaces. Heavy metals can exert toxicity through various pathways like oxidative stress, neurotoxicity, nephrotoxicity, and cellular dysfunction (Sieghardt et al., 2005). According to studies, children were found to be more susceptible to heavy metal contamination than were adults (Ihedioha et al., 2017). Heavy metal deposition in soil can gradually poison the environment over time, creating a long-term hazardous concern. These metals slowly seep into water sources, and crops and plants that grow in such soil absorb and assimilate these toxins into their bodies. As a result, polluted crops and plants enter the food chain, posing a risk to human health (Thongyuan et al., 2021). The two major activities that have led to elevated levels of heavy metals in soils are waste disposal and agriculture (Modaihsh et al., 2004). Until the 1980s, many people focused so much on the means of reducing expenses that they paid little attention to solid waste management (Zurbrugg, 2003). More specifically, in underdeveloped nations, solid waste management has been hampered by a lack of public and private finances to support existing initiatives and the corrupt management structures (Gupta et al., 1998) and also mainly due to the fact that municipal solid waste management is influenced by a complex interrelationship of politics, law, the environment, and economics (Kumar et al., 2005).

3 Solid Waste Collection

The problem of "waste siege" is characterized by the pattern of continual growth in the amount of MSW in several emerging nations. Making sure that as much MSW is gathered for planned harmless treatment and disposal is one strategy to successfully address the issue of the continuously growing amount of MSW (Leal Filho et al., 2016). Variations exist in the type of vehicle used for collection of waste, primarily depending on the type of collection container and the measurement of the road (Chiplunkar et al., 1981).

In low-income countries, waste collection is often infrequent, inefficient, and impacted by a large percentage of biodegradable and inert waste (Giusti, 2009). The total amount collected is under 50%. In nations with moderate incomes, enhanced transportation and collection services exist in residential areas due to major infrastructural investments and adoption of modern technology; 50% and 80% of collections are made overall. In nations with high income, where transfer stations, trucks with heavy machinery, and compactors are widespread, $a > 90\%$ rate of collection is evident.

3.1 Types of Collection Services

The activities involved in waste collection will vary with respect to population density, urbanization, economic conditions, cultural customs, regulation, infrastructure, and seasonal considerations. The most common types of collection services utilized now are (i) commingled (unseparated) waste collection and (ii) Source-separated waste collection (Tanskanen & Kaila, 2001).

Waste products that have not been divided into distinct categories or streams are referred to as commingled (unseparated) wastes (Kinsella & Gleason, 2003). Usually, these wastes are collected in a single bin or container that has been combined with other waste. Mixed recyclables like paper, plastics, glass, and metals that are gathered together for sorting and processing at a later time are examples of commingled trash. Common names for this kind of waste collection include "single-stream recycling" and "commingled recycling." Commingled waste is hauled to a materials recovery facility (MRF) where it is processed and sorted to separate various commodities for reuse.

Source-separated wastes, on the other hand, are wastes that have been segregated at the point of generation based on their unique properties or composition (Sukholthaman & Sharp, 2016). This means that various waste materials are separated and gathered into unique containers or bins. Recyclable trash streams, including paper, cardboard, plastic bottles, and metal cans, as well as organic garbage like food scraps and yard trimmings are typical examples of source-separated waste streams. By encouraging the segregation of various materials, source-separated waste collection makes it simpler to recycle and manage them efficiently (Zeng et al., 2018). Waste source separation has a number of advantages. Since the materials are already sorted according to their kind, it makes recycling and recovery procedures more effective and decreases the need for sorting at recycling facilities (De Feo et al., 2016). Source separation aids in reducing contamination and preserving recyclable materials' quality. It also enables the targeted treatment or elimination of particular waste streams, such as the composting of organic waste separately from other waste kinds. However, both source-separated and commingled waste management strategies have certain benefits and drawbacks. The decision between them is frequently influenced by regional recycling programs, infrastructure, and waste management practices (Tchobanoglous & Kreith, 2002). Certain localities or municipalities may use a combination of source-separated rubbish collection and commingled waste collection systems due to their particular needs and available resources.

The five most typical residential collection methods for low-rise detached homes are curb, alley, setout-setback, setout, and backyard carry (Singh et al., 2014). When using curb service, it is the homeowner's responsibility to place the empty containers at the curb on collection day and to bring the full containers back to the location where they were kept until the next collection event. This is frequently utilized in residential areas where residents put their trash cans close to the edge of the roadway, allowing collection trucks to easily access them. Alley storage of solid waste containers is widespread in areas where alleyways constitute a fundamental feature of the design of a city or a particular residential neighborhood. This is frequently observed in urban settings where collection vehicles can approach the waste from the back of the properties through narrow streets or alleyways. In a setout-setback service, extra crew assist the collection crew in loading the collection vehicle by setting out containers from the homeowner's property and setting them back once they have been emptied. Setout service is quite similar to setout-setback service, with the exception that the homeowner is in charge of bringing the containers back to their storage place (Olukanni et al., 2016).

In manual curb collection, wastes are directly transported from the containers to the collection vehicle, employing collection vehicles with low loading heights. In some instances, collection vans have additional containers that are used to empty the waste. Mechanical techniques are used to mechanically empty the auxiliary containers into the collection vehicle. Small satellite vehicles are used in yet another variation. Waste is emptied into a huge container that a satellite vehicle carries; then, the satellite car is transported to the collection car, and the container is mechanically emptied. The container utilized for on-site garbage storage must be carried to the curb or another appropriate area for collection when mechanical self-loading collection vehicles are used. As previously mentioned, either the homeowner or an employee of a collection agency may bring the containers to the curb and return them to their usual location. In most cases, mechanical collection vehicles are employed in conjunction with huge containers.

According to their mode of operation, collection systems are divided into two classes: (i) hauled container systems (HCSs) and (ii) stationary container systems (SCSs). A hauled container system (HCS) is a collecting method in which the waste storage containers are transported to the processing, transfer, or disposal locations, emptied, and then brought back to their original position or another place (Sharholy et al., 2007). Because relatively large containers are employed, HCSs are preferred, as their rate of waste generation is high (Lee & Lin, 2007). On the other hand, except when they are transported to the curb or another area to be emptied, waste storage containers in stationary container systems (SCSs) remain at the point of generation. Compared to SCSs, HCSs offer more planning freedom for container placement, deter animal raiding, and maintain a clean environment (Kumar & Goel, 2009).

Globally, one of the biggest challenges of management of MSW is the accessibility to waste management services, particularly in developing countries. Three billion people are reportedly not served by coordinated systems but rather by informal systems where residents dispose of waste in an unchecked manner (Sandhu et al., 2017). Advanced collection systems are now being created by integrating the collecting systems' drop-off locations, bins, and door-to-door (curbside) (Calabrò & Komilis, 2019). Drop-off facilities (also known as ecological islands) require garbage producers to transport MSW to a fixed or mobile collecting system in accordance with the municipal regulations (Calabrò & Komilis, 2019). Bin collection (also known as curbside collection in certain papers) is based on the placement of containers (bins of various sizes and types) in the streets. Customers can store their waste in the containers, which are emptied on a regular basis by specialized trucks (usually compactors). Door-to-door/curbside collection systems differ from traditional bin collection systems, in that street bins are replaced by smaller containers managed by households, shops, or offices and are placed for collection at the boundary of the property, following a predefined schedule that specifies which waste fraction is collected. In another type of collection system, called the pneumatic system, MSW is collected and transported through pipes by the use of pressurized air (Yadav & Karmakar, 2020). Even though it can manage up to 80% of the MSW, it is unsuitable for large waste items, liquids or hazardous materials, and waste electrical equipment (Punkkinen et al., 2012). ECOFIL is a smart urban waste management system that has collection points distributed just like bins distributed in the streets; however, each bin is composed of a camera and a computerized system for user identification and waste classification prior to disposal (Rossi et al., 2022). Each bin can be collected and stored by cloud systems via the Global System for Mobile Communications (GSM) or the General Packet Radio Service (GPRS), and, once the bin is at least 90% full, it sends information to the transportation center to collect the waste. Multiple route management methods could be used to calculate the requisite routes (Abdullah et al., 2019). Waste collection bins serve as a link between sources and transfer stations or treatment plants (Khwaja, 2010).

The colors assigned to waste bins are usually part of a standardized waste management system that is intended to make waste sorting and recycling easier for the public (Leeabai et al., 2021). The blue container is used to recycle items such as paper, plastics, glass, and metals. Organic waste, such as food scraps and yard debris, can be composted in the green or brown bin. The black or gray container is for nonrecyclable general rubbish, whereas the yellow or red bin is for hazardous waste such as batteries or chemicals. Although some regions use different colors or have additional bins for specific trash types, the usage of color-coded garbage bins attempts to improve waste separation and recycling processes, making it more convenient and environmentally friendly. Table 2.2 shows the different types of bins (differentiated by their colors) generally used in various regions across the world.

3.2 Geographic Information Systems (GISs) for Primary and Secondary Collection

A geographic information system, GIS for short, allows users to manipulate and examine geographic information, which can consist of both visual and textual data (Lella et al., 2017). The collection crew can use the navigational information provided by network analysis to more easily and efficiently collect waste from the generators by reducing travel time and facilitating more efficient waste collection from disposing areas and dustbin locations.

In a study conducted in Ipoh city, Malaysia, AutoCAD data were initially converted to the shapefile format with the help of MapInfo Professional (Malakahmad et al., 2014). During the conversion, every layer, including the road, drainage, elevation, and contour layers, was taken into consideration. Then, the coordinating system was standardized using the projected coordinate. The locations of geographic features, images, and observations, including Global Positioning System (GPS) locations, are described using a reference tool called the coordinate system within a shared geographic framework (ESRI, 2011). Distance criteria and collecting time by the vehicle in concern (independent of traffic delays) were taken into account and developed in order to optimize the route (Rasdorf et al., 2002). The GIS helped gather information on possible routes, and the optimal routes for solid waste collection were chosen after considering the restrictions imposed by the state of the roads and the terrain. The routes were selected so as to reduce the amount of resources utilized in the collection, in addition to the time and length of the route itself. A useful extension in this context, the ArcGIS Network Analyst, is a powerful tool that enables network-based spatial analysis, easing tasks like finding the best route, directions of travel, nearest facilities, and service area analysis (Karadimas et al., 2007). An ideal route for the collection of large trash items in the research

Table 2.2 Collection of different types of MSW components in different types of bins generally used in various regions over the world (Zhang et al., 2022)

Region	Types of wastes	MSW components	Types of bins	References
New York (USA)	Paper	Newspapers, magazines, cardboard, office paper, and packaging materials	Green bins	Aphale et al. (2015)
	Metal, glass, or plastic cartons	Soda cans, food cans, spray paint bottles, plastic bottles, caps, and food containers	Blue bins	
	Mixed waste	–	Black bins	
Berlin (Germany)	Paper	Old newspapers, magazines, cards	Blue wheelie bins	Schulze (2003)
	Organic waste	Fruits and vegetables, spoilt foods, leftovers, coffee grounds, eggshells, yard clippings	Brown wheelie bins	
	Glass	White/colored glass (derived from bottles, jars, windows, and mirrors)	White recycling bins, green recycling bins	
Shanghai (China)	Residual waste (dry waste)	Rechargeable batteries, oil paints, pesticides, plastic films, textiles, rubber, mercury products	Red bins	Chen et al. (2020)
	Recyclable waste	Paper, glass, metals, plastics, textiles	Blue bins	
	Hazardous waste	Batteries, cleaning chemicals, fluorescent light bulbs, electronic waste (e-waste), pesticides, solvents	Yellow bins	
Greece	Recyclable waste	Paper, glass, metals, plastics, textiles	Blue and green bins	Keramitsoglou and Tsagarakis (2018)
Thailand	Recyclable waste	Paper, glass, metals, plastics, textiles	Yellow bins	Schloss et al. (2018)
	Compostable waste	Food scraps, yard waste, paper products	Green bins	

region is developed in the current work utilizing Network Analyst. Based on two criteria, Network Analyst resolves the routing problem: distance criteria and time criteria (Bhambulkar & Khedikar, 2011). In distance criteria, only the location of the huge waste items is taken into account when generating the route. In this situation, the amount of traffic on the roadways is not taken into account. For time criteria, each road segment's overall travel time should be taken into account when the sum of the vehicle's total runtime and the time spent collecting bulky waste is the route's total trip time. The length of each road and the vehicle's speed on each one are taken into account when calculating the runtime of the vehicle. The time required for the truck to pick up these things from all loading locations would be considered the collection time for heavy waste materials. In fact, the GIS has been used to simulate a number of waste management applications, such as landfill and transfer station placement, thus improving the efficiency of collection and transportation and predicting the amount of garbage in a given locality (Khan & Samadder, 2014). Spatial modeling tools and GISs can be applied to optimize journey duration and length, fuel use, and environmental emissions (Chalkias & Lasaridi, 2009).

3.3 Hazardous Waste Collection

Recently, interest in waste management and collection has been growing. The use of hazardous wastes also needs more attention due to the serious risks they pose to our health and the ecosystem. However, the literature still lacks strategies specifically devoted to optimization of such a critical activity (Dotoli & Epicoco, 2017). Household hazardous materials such as different coolants, cleaners, insecticides, oil paints, cells, and waste medicines may typically have the following characteristics: toxic, caustic, combustible, and reactive (Wessel, 2004).

A survey found that 93% of those interviewed had used at least one pesticide product in their home in the previous year, and nearly 95% said that they had disposed of the containers in regular trash cans. Similar findings were found in the USA, where a pertinent poll showed that 88% of those questioned said that they disposed of pesticide containers with regular trash (Inglezakis & Moustakas, 2015). Different types of collection methods are implemented in different countries, some of which involve the citizens. These include having locations where they can leave their waste, recycling facilities (sites for public convenience), stores that buy back or take back waste, waste bins, and vans placed in public areas on a specific date (Bernstad et al., 2011). Again, the curbside collection option includes special collecting boxes collected as part of ordinary rubbish collection or on a regular basis by a special collection car, mobile collection sporadically throughout the year, and on-demand, call-in collection systems.

In Germany, two alternative types of collection schemes are available for the collection of paints, oils, other chemicals, and batteries: recycling centers and temporary recycling centers (trucks/vehicles that collect frequently from specific locations) (Óskarsson et al., 2022). In Ireland, the most used method of collecting household hazardous waste (HHW) is through private mobile collection services contracted by the local government. In California, HHW is collected through curbside, door-to-door, mobile, temporary, and periodic facilities, load checks, and special events, including cleanup days and electronic waste collection events, and permanent facilities have collected more than half of all HHW (Lim-Wavde et al., 2017). For complex and easily recyclable or recoverable equipment like cameras, computers, and copiers, some manufacturers have product return policies available. For instance, Xerox no longer sells copiers directly to businesses. Instead, it rents copiers to them. Models that have been used before are sent back to the manufacturer, and they are made with numerous interchangeable parts and are easy to demanufacture. This ensures that businesses have an inherent incentive to conserve resources and create environmentally friendly products as it is selling services rather than hardware. The goal of Japan's e-waste regulations is to separate e-waste from the MSW stream by making manufacturing companies and relevant local intermediaries (Widmer et al., 2005) voluntarily help in recycling goods.

The "Home Appliance Recycling Law" in Japan, which was passed in 1998 and became fully operative in 2001, requires manufacturers or local intermediaries to recycle televisions, refrigerators, washing machines, and air conditioners (Kahhat et al., 2008). Additionally, consumers will be required to pay an end-of-life fee to help defray some of the costs associated with recycling and shipping. Consumers must pay between US$ 23 and US$ 46 ($1 = JPY 107) for the recycling charge and an extra US$ 4–US$ 19 ($1 = JPY 107) for the product's transportation to authorized collection sites.

Although the designs for each HHW collection fixed facilities require unique considerations depending upon the site itself, they can always take advantage of the preexisting, commonly applied concepts (Tchobanoglous & Kreith, 2002). These include keeping adequate space to accommodate any further needs, storing low-volume reactive HHW in cabinets designed for hazardous waste, having extra capacity for spot ventilation during liquid bulking/lab packing, maintaining covered and heated areas for waste handling and storage, improving the movement and entry of traffic, allotting sanitary and aerated spaces for workers, adopting a straightforward process flow, having a waste exchange area, installing flammable gas monitors in bulk storage locations, and having secondary containment with chemical-resistant coatings to ensure safety during waste handling.

3.4 The Economics of Waste Collection

The economics of waste collection in this modern day and age hinges upon the initial points of waste deposits (containers), the waste collection service itself, the transfer stations, and the final transportation to the disposal sites. The cost-effectiveness of waste collection remains one of the core components of the final cost of the entire integrated waste management system. The productivity of every individual within the collection route determines labor efficiency. Parameters for measuring such include crew size, service time, travel time, and time at the disposal site (Worrell et al., 2016). Generally, a single-crew, mechanically loaded collection system is preferred owing to the reduction of job-related injuries and overall cost efficiency. However, this is not always the case, since in large cities with extreme vehicle congestion and intense competition for street space, crew are ideally larger (about four people/vehicle) with manual collection. Again, in communities with adequate space and conveniently placed storage containers for waste, the ideal crew size is one person/vehicle and the vehicle is loaded mechanically. Reduction of municipal waste collection frequency can reduce vehicular needs by 20–40%, thereby increasing savings in costs of labor, investment, and repair ("Collection efficiency: Strategies for success. United states environmental protection agency. Available at: https://archive.Epa.Gov/epawaste/nonhaz/municipal/web/pdf/k99007.Pdf. Last accessed: August 03, 2023"). This will also allow expansion into recyclable/yard trimming collection without fleet/staffing increases. Territorial expansion negatively affects the return on assets ratio owing to the increased costs associated with waste management activities (Bartolacci et al., 2018). Moreover, the separate waste collection ratio positively impacts the return on assets ratio of waste management because separate waste collection products can be sold as secondary raw materials, thus increasing revenue and decreasing costs. Revenues from resale, energy generation, and public fees can pay for waste collection and disposal and improve the capacity to maintain and expand public services.

Although they are inherent qualities of municipalities and outside the direct impact of economic policies, higher wage expenses and population growth are key variables that cause a noticeable increase in MSW collection costs (Voltes-Dorta et al., 2014). According to research, cooperative small towns have reduced garbage collection costs, although whether a production is public or private does not always affect expenses (Bel & Mur, 2009). A larger degree of dispersion within a municipal region has a good impact on total cost expenses, but population density does not always have a significant relationship with municipal costs (Fernández-Aracil et al., 2018).

4 The Transportation System

The management of MSW involves multidisciplinary activities of generation, source separation, storage, collection, transfer and transport, processing and recovery, and disposal (Rada et al., 2013). To establish an efficient material-cycle society, the need for a cost-effective, integrated MSW management system is essential. Problems with waste collection and transportation rank among the most challenging operational issues in developing an integrated waste management system (Nuortio et al., 2006). To lower the expense of collection and transportation, municipalities have been encouraged to create new plans for the transportation and collection of solid waste (Das & Bhattacharyya, 2015).

4.1 Types of Vehicles Used

MSW is transported using a variety of vehicles. These vehicles are designed to gather, transport, and get rid of different types of MSWs produced industrially, commercially, and residentially. Each of them has its own benefits and drawbacks. Most of these vehicles are equipped with a hydraulic system to lift up waste containers and a compactor mechanism to reduce the volume of waste collected, thus maximizing the capacity of the vehicle. The first step in ensuring efficient waste transportation is choosing the right collecting vehicle for the task. The most commonly used vehicles are listed in Table 2.3.

4.2 The Transportation Route

The transportation of MSW can be divided into two steps. Residential, industrial, and commercial MSWs are collected and transferred to the nearest transfer stations. Each transfer station is composed of a number of bins that have the same capacity. Then, vehicles that start from the depot move through transfer stations following a scheduled route, collect MSW, and finish at landfills, waste-to-energy (WtE) facilities, recycling facilities, or composting sites (Nguyen-Trong et al., 2017). This road network is shown in Fig. 2.2. For each trip, a vehicle will go to its last stop when it reaches the last transfer station of the trip or, if full of waste, it will go to the landfill for unloading and then return to the next station in the plan. These cases can be treated as an open vehicle routing problem (OVRP) (Li et al., 2007). This is a version of the classical vehicle routing problem (VRP). A VRP is the case when vehicles are not required to return to the depot after servicing (Laporte, 1992). This waste transportation problem can be converted into a VRP by ensuring that the overall volume of waste on a trip does not go beyond the limit of the vehicle. In more complex scenarios with multiple vehicles or total waste volume exceeding capacity, it can be divided into several classic VRPs (Nguyen-Trong et al., 2017). However, most of the developing countries contain the same depot and landfilling spots or the distance between them is known.

4.3 Optimization of The Transportation System

A significant challenge in designing waste management systems is the efficient management of MSW transportation. Due to high collection and transportation costs, existing systems face considerable expenses. The cost of collection and transportation for the current MSW management systems is 80–90% of the overall costs in low-income countries, and 50–80% of the overall costs in middle-income countries (Das & Bhattacharyya, 2015). The dispersed and diverse distribution of waste sources throughout cities further increases this cost. Adapting a plan that puts emphasis on the shortest and most direct waste transportation routes can help solve this problem while also reducing costs and increasing the efficiency. For identifying the optimal waste transportation routes, a GIS can be implemented. A GIS is capable of storing, retrieving, and analyzing a large amount of data as well as visualizing output with response time (Malakahmad et al., 2014). The suggested method lacks consideration for the dynamic and temporal aspects of urban life and traffic. They

Table 2.3 Description and usage of commonly used MSW transportation vehicles

Vehicle type	Description	Usage	Disadvantages
Rear loader	This is a waste collection vehicle with a hydraulic compactor at the rear	Residential and commercial waste collection	Not efficient in larger commercial areas
Front loader	This has front hydraulic forks. Using slots on the sides, the forks raise a trash can. After that, they flip the bin over so that the lid will open and the contents will be dumped into the truck's hopper	Commercial and industrial waste transportation	Not suitable for smaller residential neighborhoods with narrow streets and tight turns
Side loader	This vehicle has a mechanical arm or robotic system on the side, capable of lifting waste containers and emptying them into the storage compartment	Transportation of waste in densely populated areas	Not efficient in larger commercial areas
Roll-off truck	This is a rugged and highly versatile vehicle that transports open-topped waste containers to the local landfill or recycling center	Construction and bulky waste transportation	Loading and unloading need to be done manually
Recycling truck	This has been specifically designed for recyclable materials	Collection and transportation of recyclable materials	Causes traffic hazard
Compactor	This is equipped with compacting mechanisms that compress the waste to maximize the load capacity	Residential and commercial waste disposal	Lacks regulation
Road sweeper	In the traditional road sweeper design, two spinning brushes are positioned on either side of the vehicle to disturb the trash, which is subsequently picked up by a cylindrical brush that is positioned further behind. After that, any trash is dumped into a storage hopper so that it can be disposed of	Cleaning of streets and public areas	Requires an extremely flat pavement surface
Dump truck	This collects and transports waste to the transfer stations or straight to the landfills or recycling facilities. It generally relies on a hydraulic system to unload waste from the truck	Landfill waste disposal	Slipping or crushing can occur
Bulk pneumatic tanker	This tanker is vacuum-sealed to prevent moisture, dirt, or pathogen contamination	Transportation of items like food, chemicals, or even polluted soil	Cannot collect bulky waste
Grapple truck	This is a vehicle with a grapple loader fitted to its frame used to load and occasionally move bulky waste	Construction and waste container transport	Has less volume capacity

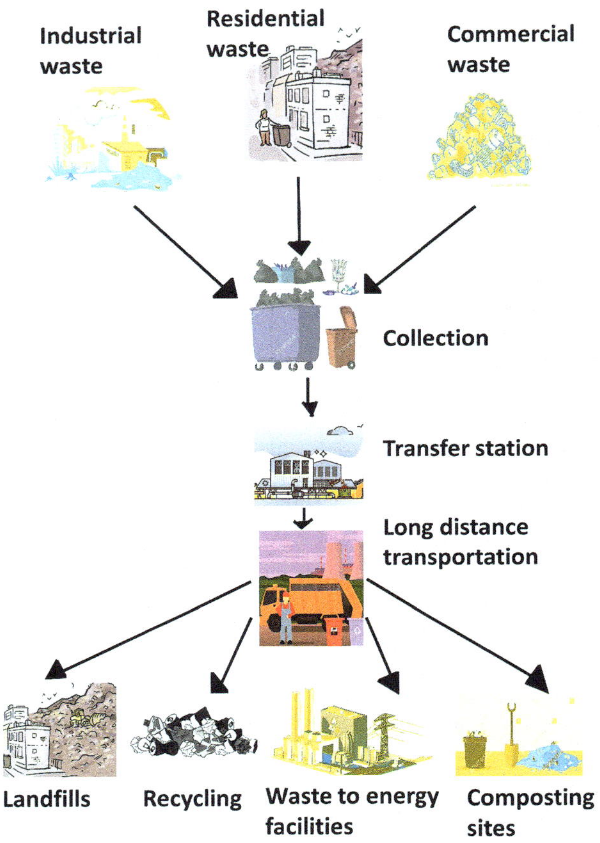

Fig. 2.2 The road network of municipal solid waste (MSW) transportation (Das & Bhattacharyya, 2015)

only offer solutions based on static conditions, ignoring the constantly changing weather traffic flow influenced by various factors like timing, river behavior, and the state of the weather (Das & Bhattacharyya, 2015). A potent method for overcoming the limitations of a GIS is an agent-based model (ABM) (Nguyen-Trong et al., 2017). Five types of agents can be modeled in an ABM.

4.3.1 The Road Agent

The real road is represented by this agent. As real data are obtained from GIS files, the location, length, and other geometric parameters are all present in this type of agent. Each agent represents a short segment of the road (Nguyen et al., 2015). Position is the most important of these attributes. An additional attribute of this agent is the possibility of the occurrence of landslides. A road agent usually does not show any behavior.

4.3.2 The Transport Agent

Transport includes various kinds of vehicles, each with distinct attributes, behaviors, and mobility. As a result, modeling them as agents is required due to their impact on the efficiency of MSW collection and transportation. The attributes of this agent are transport name, length, width, maximum speed, current speed, acceleration, and deceleration factor (Nguyen et al., 2015). Start time and start position destination are also included. A transport agent exhibits behavior that can be acceleration or deceleration by the agent, emergency calling, stopping, and finding a different path.

4.3.3 The Garbage Agent

A garbage agent is a representation of the GIS-generated transfer stations. This agent shows an increase in the quantity of waste over time. The amount of waste will increase according to the ratio of waste generation if it is not during the collection time. The attributes and behaviors of a garbage agent are waste generation rate, position, and volume (Nguyen-Trong et al., 2017).

4.3.4 The Garbage Truck Agent

A garbage truck agent represents vehicles used to transport waste from transfer stations to their final destination. The attributes of this agent are storage capacity at full state, current state, and the list of transfer stations that the agent must visit (Nguyen-Trong et al., 2017). The movement behaviors of a garbage truck agent are almost the same as those of a transport agent. However, the main behaviors of this agent are collection, cleaning, and unloading.

4.3.5 The Depot Agent

Both garbage and garbage truck agents are overseen by this agent, who will initiate a plan for each garbage truck agent at the starting of a trip. The attributes of this agent are position, number of garbage trucks, collection times, and control of garbage truck agents (Nguyen-Trong et al., 2017).

5 Segregation

After the waste is collected from different sources, it is transported to disposal sites or waste processing facilities where the waste is segregated. The majority of developed nations presort waste before transporting it to facilities for material recovery (Zhang et al., 2022). However, source separation of waste is not as well-practiced in underdeveloped nations. MSW is a matrix of different types of wastes that have the potential for a variety of applications, such as composting, recycling, and energy recovery (Zhang et al., 2022). For optimal utilization of its potential, MSW must be divided into distinct fractions since each has a separate set of uses. The most basic step of waste segregation is to separate wet waste, which is usually perishable organic waste, from dry waste (Abdel-Shafy & Mansour, 2018). This ensures that the reusable materials are not contaminated and thus enhances the potential for resource recovery. Segregation of waste is the most crucial step of sustainable waste management as it diverts potentially valuable materials in the waste stream from ending up at landfills (Wan et al., 2019). The first step in the reuse, reduce, and recycle (RRR) strategy of waste is segregation, which gives waste a new lease on life (Kihila et al., 2021).

5.1 Waste Categorization: Disposable Versus Recyclable

As municipal solid waste is waste disposed of by households and other institutions, it is composed of a variety of organic and inorganic substances. Its complex matrix includes food waste, plastics, metals, electrical equipment, hazardous waste, paper waste, bottles, clothing items, and other miscellaneous items (Funk et al., 2013). To ensure proper treatment of these waste materials, they must be separated on the basis of their potential for recycling and disposal. This method of classifying and separating waste based on its characteristics is known as waste categorization. Based on the economic value and reusability of waste materials, it can be categorized into disposable or recyclable waste. Further discussion on disposable and recyclable waste is presented in the next section of this chapter.

5.1.1 Disposables

Wastes that cannot be reused or do not have any economic value are disposable or nonrecyclable. Food scraps, disposable dishware (paper plates, cups), Styrofoam, batteries, soiled paper or cardboard (pizza boxes, coffee cups, used tissues, etc.), hygiene products (razors, sanitary napkins, etc.), and disposable diapers are some examples of nonrecyclable wastes. Unscientific disposal of these wastes can have a detrimental effect on the environment, such as evolution of methane gas (Stafford et al., 1980), blockage of sewer pipes, consumption of plastics by animals (Bharadwaj et al., 2015), heavy metal contamination (Prechthai et al., 2008), etc. Disposable wastes should be segregated and disposed of properly to minimize their negative environmental impact and reduce health hazards.

5.1.2 Recyclables

These are nonbiodegradable wastes that have economic value because of their recycling potential. They are generally discarded but can be recovered and reused; therefore, their energy values can also be harnessed (Bharadwaj et al., 2015). Plastic materials (plastic bottles, containers, boxes), paper and cardboard, glass containers (glass bottles, jars), metals (cans, aluminum foils), electronic wastes (laptops and computers), etc. are materials that have recycling potential. Recycling helps to significantly reduce the amount of waste that ends up in landfills (Agunwamba, 2003), reduces environmental pollution, and saves energy (Birara & Kassahun, 2018). Figure 2.3 illustrates different types of disposable and recyclable solid wastes.

About 75% of the total waste generated worldwide ends up in landfills, whereas 19% is recycled and 11% is used for energy generation (Kaur et al., 2021). Waste categorization and segregation promote recycling, reduce volumes of waste in landfills, and facilitate proper waste management (Stoeva & Alriksson, 2017). It also allows safe disposal of nonrecyclable waste. Some disposal techniques include incineration with energy recovery, composting or anaerobic digestion for biodegradable waste, etc. (Khan et al., 2016). Proper segregation and recycling can reduce the disposal of about 65% of waste generated in landfills (Donnini Mancini et al., 2007).

5.2 Source Segregation of Waste

In terms of waste management, incremental steps make a substantial impact. Although it is difficult to segregate recyclable materials and implement waste management strategies for large amounts of waste, sorting out waste into small

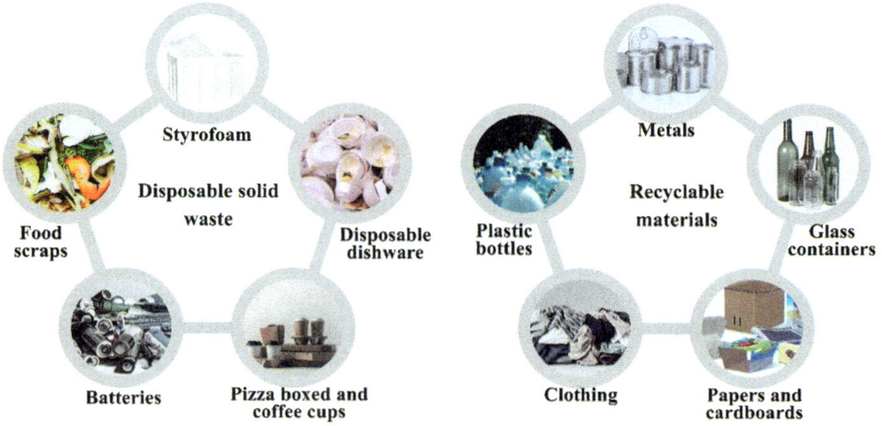

Fig. 2.3 Categorization of municipal solid waste (MSW) into disposable solid waste and recyclable materials

batches according to particular waste types makes the task much more accessible. This sorting of trash at the root level has emerged as the inventive fix and is termed "source waste segregation." Source segregation means separating functional or recyclable materials from the waste stream at its origin (houses, commercial buildings, public and private institutions). MSW presents a valuable opportunity for recycling (Surroop & Mauthoor, 2012), composting (Farrell & Jones, 2009), and energy generation (Chakraborty et al., 2013). Waste segregation is vital for properly utilizing the potential of MSW.

Waste is separated at the collection stage and/or at materials recovery facilities (MRFs). Waste is mechanically or manually sorted at MRFs since no available machines can effectively divide all fractions of the waste stream (Rousta & Bolton, 2019). When the waste is not source-separated, recyclables are often contaminated with other wastes, making them even more challenging to isolate and requiring additional operating costs (Rousta & Dahlén, 2015). For effective recycling and waste management, starting from the source and laying the groundwork is first necessary. The mechanism of source segregation follows an easy pattern so that it is well-practiced from the micro levels, such as households, commercial buildings, etc. It simply calls for having multiple colored or labeled trash cans instead of one for all types of wastes (Rousta & Bolton, 2019). The labels that the trash cans will include will not be limited to paper waste, organic waste, plastic waste, metal waste, etc. After the labeling is done, the rest of the job is a self-explanatory process. However, inspection and maintenance must be conducted regularly for a notable result. This way, waste management can begin at the foundation.

5.2.1 Factors Influencing Source Segregation

To avail effective results from source separation for a greener environment, the public, municipal authorities, and government must together come forward and actively contribute to the agenda (Zhang et al., 2022). Although the resources, planning, and initiatives of the government contribute significantly to creating the vision of an eco-friendly society, it is the public who can bring that vision into reality. Understanding the public's attitude and the factors contributing to their waste-sorting behavior is essential for more robust MSW management. These factors include environmental concerns, recycling habits, moral norms, social norms, convenience and effort for segregation, economic factors, and many more (Knickmeyer, 2020). Chen and Lee (2020) divided the factors affecting people's waste-sorting behavior into external and internal stimuli. Policy regulation is an external stimulus that determines mentality, whereas internal motivation, like awareness of consequences, is strongly related to people's attitudes. These factors can be considered while making policies regarding waste separation to create a more positive attitude and trigger a pro-environmental behavioral change in citizens. However, factors creating a negative attitude toward waste management, such as fines, charges, sorting rules, and improper recycling procedures, should be avoided during policy-making (Wu et al., 2021). The concept of the Internet of Things (IoT)-enabled accountability based on accountability theory, established by Wang et al. (2021), can also be adopted by governments and waste management organizations to find solutions to sustainable waste management. Municipalities play a crucial role in household waste management, and their proactive initiatives and authority can ensure the smooth functioning of the project. Their measures include supplying necessary resources, improving the existing disposal capacity, creating awareness-based campaigns, and providing adequate waste collection and disposal systems (Ko et al., 2020; Srivastava, 2016). Successful waste management results from the joint effort of municipalities and the general public. This cooperative strategy creates the foundation for a cleaner and more environmentally friendly community by encouraging a culture of waste segregation and recycling.

5.2.2 Disparity Regarding Source Separation in Developed and Developing Countries

Developed countries have greater economic prosperity and a higher standard of living, leading to increased consumption patterns and more waste generation (Kaza et al., 2018). The waste composition data show that high-income countries generate less food waste and more recyclable waste like plastics, paper, cardboard, etc. (Ali & Sion, 2014). By practicing source segregation, they have achieved high recycling rates and better waste management systems (Bassi et al., 2017). Among Asian countries, Japan has one of the highest recycling rates and a remarkable waste management practice. In Japan, municipalities consider the public's opinion when forming waste-sorting policies, and 92% have source-separated collection programs (Matsumoto, 2011). The government of South Korea adopted the volume-based waste fee (VWF) system in response to the increased concern of the South Korean public about the environmental implications of MSW (Seo et al., 2004). The VWF system has effectively reduced waste and improved household recycling (Seejeen Park & Lah, 2015). German residents' household waste management behavior has significantly contributed to the country's overall recycling rates, where 70% of the MSW is recycled (Mintz et al., 2019). Sweden has dramatically increased its recovery rates by encouraging systematic development in roadside collection infrastructure and by considering economic incentives and source-separated food waste collection (Takahashi, 2020). Other developed countries like Norway, Austria, and Italy have also adopted source waste separation in their waste management policy. Establishing laws, engagement of municipal authorities, and public awareness greatly influence the success of the source waste management of the developed nation.

Source waste segregation is not as widely practiced in developing countries. A study in the city of Lubumbashi in Congo identified insufficient knowledge regarding the classification of waste, a high percentage of wet waste in the waste stream, leading to contamination of dry recyclable waste, and a shortage of compatible facilities for segregated waste as the main challenges in waste segregation at the source (Mukemo et al., 2020). Lack of facilities, inadequate policy enforcement, and lack of understanding and strategies for its promotion are among the main obstacles to household waste management faced by developing countries (Kihila et al., 2021). However, by implementing successful public awareness programs by local governments and placing recycling bins close to residential areas, source separation may be made more effective in developing countries (Sarbassov et al., 2019).

5.2.3 Benefits of Source Segregation

Waste segregation is essential for waste reduction and proper management of municipal solid wastes. The general public can act as front runners and contribute to this agenda by sorting out the trash at home or public places, such as parks and commercial buildings, into categories such as paper, plastics, glass, organic waste, etc. This requires less time and makes sorting trash easier at materials recovery facilities (Kochar, 2021). By taking on the primary role, the public can help enhance the efficacy of waste segregation, thus making them more conscious of producing less waste. Waste segregation at the point of origin lowers the amount of waste that must be handled, thus increasing the effectiveness of waste collection and disposal. The reduced volume of waste eases the handling and processing of waste and reduces operational costs (Oberlin, 2013). Sorting waste at the source preserves the quality of recyclable products, enhancing the potential for recycling (Medina, 2009). Separating recyclables, organic wastes for composting, nonrecyclables and non-compostable wastes for incineration and energy recovery, and other types of wastes for further treatment significantly reduces the volume of wastes in landfills. Countries that practice waste segregation at source points have high recycling rates and more sustainable waste management (Bassi et al., 2017). The increased recycling rates owing to source segregation contribute the most to reducing environmental pollution created by municipal waste. Source waste separation facilitates the transition from an unviable linear economy of "take–make–dispose," to a more sustainable circular economy of "take–make–use." (Elisha, 2020; Wang et al., 2021). In a linear economy, all factors of production and consumption face the ultimate fate of disposal, thus wasting valuable resources and losing the unlimited potential for reuse and recycling of waste. The circular economy system utilizes waste as a valuable resource by practicing and preaching the "reduce, reuse, and recycle" principle. Segregation of waste at the source is a more sustainable approach to waste management as it provides economic benefits, reduces environmental pollution, and utilizes the proper potential of waste.

5.3 *Waste Segregation Equipment*

Efficient waste management requires systematically implementing strategies and technologies for waste segregation and sorting. Waste segregation is carried out at the source or at materials recovery facilities. Setting up well-equipped MRFs requires significant capital investments. However, in the long run, it significantly reduces labor costs by minimizing manual labor, making the segregation process more cost-effective and creating safer conditions for workers working at MRFs (Kochar, 2021). There is currently no single waste-sorting equipment available that can effectively separate all fractions of waste from a waste stream, as different types of wastes require different sorting techniques (Rousta & Bolton, 2019). The waste management equipment sector has seen revolutionary inventions since the concept of waste segregation was introduced. These machines are non-sensor-based and sort waste directly based on their material properties or are employed with sensors for more selective separation of recyclable materials (Gundupalli et al., 2017).

5.3.1 Direct Sorting Equipment

These machines are usually not sensor-based, and they sort out trash directly based on their material properties, such as magnetic properties, electrical conductivity, and density using magnetic field, eddy current, and gravity, respectively (Gundupalli et al., 2017). These machines usually require presorted wastes. Only materials with specific properties can be separated with direct waste sorting. So, materials recovery facilities typically contain more than one type of sorting equipment. Table 2.4 lists various kinds of conventional machines that separate municipal solid waste.

5.3.2 Sensor-Based Waste Sorting

Sensor-based waste sorting is a more advanced waste-sorting technique that separates trash based on material properties beyond the scope of traditional mechanical waste sorting. While conventional waste-sorting techniques utilize material properties like electrical conductivity, density, size, magnetic properties, etc. to sort waste, sensor-based devices can sort trash based on color, shapes, texture features, etc. (Huang et al., 2010). The non-contacting and nondestructive sensor-based sorting is faster, more accurate, and more efficient in terms of separating and recovering recyclable materials (Zhao & Li, 2022). Unlike traditional techniques, sensor-based sorting is more efficient with higher purity and requires low operational cost due to reduced manual labor (Rahman et al., 2014). Optical sensors, laser-induced breakdown spectroscopy (LIBS), X-ray technology, and spectral imaging are technologies used in automated sensor-based waste sorting, which are depicted in Fig. 2.4.

(i) Optical Sensor-Based Sorting

Camera-based sensors are used in optical sorting to sort trash based on visual characteristics like color, shape, and texture (Huang et al., 2010). Camera scanners are placed on top of a conveyor belt carrying waste particles. Industrial-grade scan cameras can capture three-dimensional (3D) features of a single waste particle, such

Table 2.4 Summary of non-sensor-based equipment for sorting municipal solid waste

Type of machine	Basis for separation	Equipment description	Working principle	Limitations	References
Magnetic drum separator	Based on ferromagnetic properties; ferrous fractions from non-ferrous waste	Consists of a drum containing a magnetic assembly placed on half of its circumference and a conveyor belt containing waste	A conveyor belt supplies the mixture of waste The strong magnetic reflux attracts and holds the ferromagnetic materials to the revolving drum	Cannot separate materials that have no ferromagnetic properties	Kelland et al. (1974)
Eddy current separator	Based on nonmagnetic properties. Non-ferrous materials from nonmagnetic materials	Consists of a rotating drum lined with neodymium (NdFeB) magnets placed inside a conveyor belt	A thin layer of the waste mixture is carried via a conveyor belt The electrically conductive metal portions that do not have magnetic properties are repelled by external magnetic reflux	Not suitable for heat-conductive materials	Balakrishnan et al. (2016), Bonifazi and Serranti (2012)
Air classifier	Based on gravity; light fractions of the waste stream, e.g., paper, Styrofoam, fibers, etc.	Consists of a throat section through which a fan draws air. The feed hopper introduces the waste into the throat. A sliding gate valve and a Venturi orifice meter regulate airflow	A high-pressure air jet released through a compressed air nozzle carries lighter particles to a longer distance, separating them from heavier particles	Can only separate materials with similar densities	Worrell and Vesilind (1979)
Tribo-electrostatic separator	Based on electric charge; plastic materials	Consists of a charging section where plastic particles are charged. The charged particles are separated in the separation chamber and collected into separate bins	Shredded plastics get charged due to friction electrification The charged waste is placed under an electric field to separate them	At high humidity (77%), the recovery rate is extremely low (24.8%)	Silveira et al. (2018)

Screw press sorter	Wet waste from organic waste	Consists of a porous cylindrical drum with an inclined feed end	Feed materials spiral down a rotating drum The organic waste squeezes through the screen slits and passes through it, and the hard and dry waste is collected through the other end	Requires maintenance as screen slits can get clogged	Gundupalli et al. (2017)
Hydrocyclone separator	Based on density, wettability, and shape factors; polyethylene, PVC (polyvinyl chloride), etc.	Consists of a cyclone unit for separation, an inlet for slurry intake, and an outlet at the top and bottom for collecting segregated products	The feed is tangentially fed to the inlet, causing the formation of a vortex Materials heavier than the fluid move outward, collide with the container wall, and fall Lighter materials are carried to the outlet by the central vortex	Reduced efficiency in separating fine particles	Serranti and Bonifazi (2019)

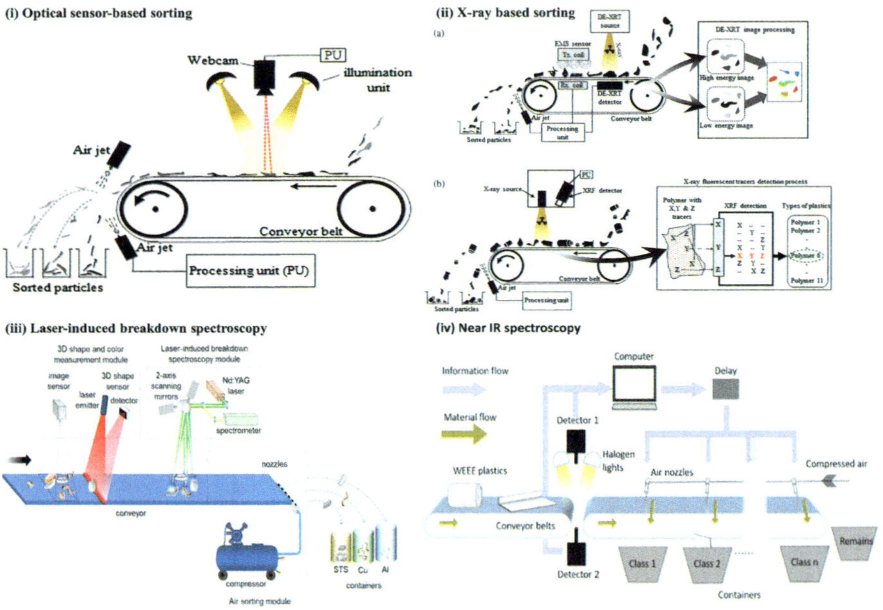

Fig. 2.4 A schematic diagram of different sensor-based technologies for waste sorting. (i) Optical sensor-based sorting using a camera sensor to separate materials using texture information. (Reprinted with permission from Gundupalli et al. (2017). Copyright © 2016 Elsevier Ltd. (ii) Two types of X-ray-based sorting: (ii)(a) DE-XRT-based sorting technique and (ii)(b) XRF-based sorting technique. Reprinted with permission from Gundupalli et al. (2017). Copyright © 2016 Elsevier Ltd. (iii) Laser-induced breakdown spectroscopy for identification and separation of scrap metals. Reprinted with permission from Seongyun Park et al. (2022). Copyright © 2021 Springer Nature. (iv) Near-IR spectroscopy technique for sorting plastic waste. Reprinted with permission from Wu et al. (2020). Copyright © 2020 Elsevier Ltd)

as shape parameters, particle size, actual positions, and color. A computer processes and extracts the 3D features, and a mechanical sorting system uses the data to separate the different types of trash. Optical sensors are highly efficient and have a recovery rate ranging from 85% to 95% (Gundupalli et al., 2017). However, these sensors are less effective in separating materials with similar properties and appearance, e.g., aluminum and zinc (Kutila et al., 2005).

(ii) X-Ray-Based Sorting

When X-rays pass through a material, some of its energy is absorbed and the rest can be transmitted through a detector to measure its intensity. Denser materials will absorb more energy than will lighter materials (Ying et al., 2022). X-ray transmission (XRT)-based sorting uses this principle to sort out waste based on density (Kutila et al., 2005). X-ray sorting can be classified into dual-energy X-ray transmission (DE-XRT) and X-ray fluorescence (XRF). XRF technology has more widespread use in sorting plastic and wood waste. XRT-based sorting is fast and has a recovery rate of 90–98% (Gundupalli et al., 2017).

(iii) Laser-Induced Breakdown Spectroscopy (LIBS)

The LIBS approach was created to quickly and accurately identify several categories of plastic polymers and scrap metals using spectral ratio analysis (Gondal & Siddiqui, 2007). The LIBS technique utilizes an Nd:YAG laser to produce plasma emission, which is captured by a spectrometer and sent to a processing unit to analyze the different types of waste particles (Gondal & Siddiqui, 2007). The advantages of LIBS are its fast speed, less intrusive nature, and the fact that it requires little to no sample preparation (Liu et al., 2018). However, despite being highly efficient at classifying metals and polymers, LIBS sorting is more costly than other sorting methods since it uses multiple spectrometers (Jull et al., 2018).

(iv) Near-Infrared (NIR) Spectroscopy

Near-IR spectroscopy is a spectral imaging technology that identifies different types of wastes using spectral reflectance measurement and image processing technologies (Bonifazi & Serranti, 2006). This technique enables fast, nondestructive, and accurate materials detection without surface preparations (Zhu et al., 2019). NIR reflectance spectroscopy is a fast and efficient technique for identifying different types of plastics. However, it is ineffective in identifying dark-colored plastics (Ruj et al., 2015).

Automated waste segregation is essential for efficient waste management and recycling, as it reduces manual labor, increases efficiency, and lowers costs. While materials recovery facilities commonly practice automatic segregation, integrating IoT-based waste segregators and smart bins can automate source segregation with minimal effort, further reducing human intervention and optimizing the waste management process.

5.4 *Recycling, Composting, and Energy Recovery of Segregated Waste*

Sustainable MSW management utilizes waste as a valuable resource instead of viewing it as a disposable burden. Waste segregation is crucial in waste management because it enhances the potential for recycling and recovering valuable materials from waste and enables efficient reuse, recycling, and recovery (Kihila et al., 2021). The treatment options for MSW include recycling, composting energy generation and incineration, and landfilling. Recycling is recovering valuable materials that were otherwise destined for landfills and reusing them to reduce waste and conserve resources. Recyclable MSW includes plastics, paper, cardboard, metals, etc. The efficient separation of recyclable materials from nonrecyclables prevents contamination of recoverable materials, reduces the volume of waste in landfills, and increases the efficacy of material recovery (Kochar, 2021). Reusing and recycling waste reduce environmental pollution and energy consumption and help conserve natural resources. Recycling 1 ton of office paper can save 18 trees, 27,300 liters of water, and 2450 liters of fuel (Lamma, 2021). The organic fraction of MSW can also be reused for composting and making biodegradable materials and bioplastics (Kaur et al., 2021). MSW compost contains high levels of nitrogen, which is an essential nutrient for the growth of plants and crops (Shiralipour et al., 1992). Although recycling and composting of MSW are more sustainable practices of waste management, landfilling and incineration are the two most widely used methods of treatment of waste (Fei et al., 2018). Incineration has a higher energy recovery rate (20.5% energy recovery), but it has negative environmental impacts due to the large amounts of fly ash and exhaust gas production (Fei et al., 2018). Mechanical biological treatment (MBT) is a more environmentally friendly approach to waste energy recovery, with a recovery rate of 38.5% when connected to a biogas treatment unit (Fei et al., 2018). MBT facilities need significant financial investments, thus making them economically unfeasible for underdeveloped nations with inadequate resources for waste treatment.

6 Challenges, Opportunities, and Future Perspectives

According to Ferronato et al., deficiencies in public awareness, low compliance by users for selective collection, geography, financial situation of people, and unwillingness to pay for MSW management (MSWM) facilities are the biggest challenges in MSWM (Ferronato et al., 2019). Proper management of municipal solid waste (MSW) is hindered by the high cost of collection and transportation. This challenge is particularly prevalent in developing countries, where the expense of collection alone can make up 60–80% of a municipality's total budget (Singh et al., 2014). Thus, it can get difficult to expect significant participation from the general public in waste management programs. The price of recycled materials fluctuating greatly is also a big issue. Other challenges include insufficient infrastructure, lack of skilled workers, few managerial commitments, weak cooperation among stakeholders, lack of regular monitoring and evaluation, scarcity of land for solid waste management activities, inefficient technical aspects of MSWM systems, and weak enforcement of solid waste management policy and legal frameworks (Hirpe & Yeom, 2021). As a consequence, there are irregularities in MSW collection, necessitating additional expenditure in waiting time on the side of homes for the collection vehicle. Timely garbage collection is important to ensure proper treatment and disposal of MSW as failure to collect waste on time results in improper dumping and burning of waste at the source. Furthermore, in search of reusable and recyclable materials, waste pickers might throw away trash elsewhere, thus creating more issues for sustainable solid waste management (Hazra & Goel, 2009).

Around three billion people, mostly in developing countries, depend on informal networks and local open dumps to dispose of their waste without any regulation

(Calabrò & Satira, 2020). Recent developments in the field of MSWM have vouched for the potential of waste in the circular economy since the huge amount of raw materials and energy present in MSW are crucial for the successful application of this economic theory (van Ewijk & Stegemann, 2020). Therefore, future waste management techniques should prioritize waste reduction at the source, reusing things after refurbishment, recycling materials, and recovering energy through combustion while minimizing landfilling. In fact, recycling waste materials has the potential to create a whole new industrial sector (Kaushal et al., 2012). The Gujarat government in India is already considering constructing a "Recycling Industry Park." Recycling companies will benefit from a cheap and consistent supply of input materials, in addition to being free of the unpredictability of price swings and constraints in the availability of virgin raw materials. The existing waste management costs can be decreased by designing scientific waste management systems with a focus on "RRR." Residual waste volume after material recovery/recycling can be considerably decreased, thus lowering treatment and disposal expenses. According to the United Nation Environment Programme (UNEP) studies, using the integrated solid garbage management strategy, residual garbage that needs to be disposed of can be readily reduced to 30–40%. Earnings from recovered materials and resources can help reduce waste management budget requirements. Since 2000, new technologies such as geographic information systems (GISs) and other advanced optimization software have been used to optimize the collection and transportation route distances (Sanjeevi & Shahabudeen, 2016). Future waste management is projected to undergo a dramatic transformation as a result of newly developed technologies like the Internet of Things (IoT), artificial intelligence (AI), and blockchain (Farrokhi-Asl et al., 2020).

7 Conclusions

This chapter focuses on garbage collection, segregation, and management, which are undeniably the key components of creating sustainable and ecologically conscious communities. Effective waste collection systems ensure that waste is efficiently collected from its source, reducing litter and potential health problems. Proper segregation strategies allow different waste streams to be separated, thus facilitating recycling and improving resource conservation. In fact, waste management projects that include innovative technologies and community involvement pave the way for a cleaner and greener future. However, challenges persist, and further efforts are required to combat them. Governments, companies, and individuals must all work together to implement evidence-based waste management plans and policies. The authors firmly believe that sustained research and innovation will be critical in bringing about a positive change and ensuring a more sustainable future. Only then will our planet be able to thrive for generations to come.

References

Abdel-Shafy, H. I., & Mansour, M. S. (2018). Solid waste issue: Sources, composition, disposal, recycling, and valorization. *Egyptian Journal of Petroleum, 27*(4), 1275–1290.

Abdullah, N., Alwesabi, O. A., & Abdullah, R. (2019). *Iot-based smart waste management system in a smart city.* Paper presented at the Recent Trends in Data Science and Soft Computing: Proceedings of the 3rd International Conference of Reliable Information and Communication Technology (IRICT 2018).

Adeleke, O., Akinlabi, S., Jen, T., & Dunmade, I. (2021). *An overview of factors affecting the rate of generation and physical composition of municipal solid waste.* Paper presented at the IOP Conference Series: Materials Science and Engineering.

Adeleke, O., Akinlabi, S., Jen, T.-C., & Dunmade, I. (2023). A machine learning approach for investigating the impact of seasonal variation on physical composition of municipal solid waste. *Journal of Reliable Intelligent Environments, 9*(2), 99–118.

Adnan, M., Jha, A., Kumar, S., & Adnan, M. (2020). Municipal solid waste management and its impact: A review. *International Journal of Advanced Research in Engineering and Technology, 11*(5), 685–693.

Agunwamba, J. (2003). Analysis of scavengers' activities and recycling in some cities of Nigeria. *Environmental Management, 32*, 116–127.

Alam, P., & Ahmade, K. (2013). Impact of solid waste on health and the environment. *International Journal of Sustainable Development and Green Economics (IJSDGE), 2*(1), 165–168.

Ali, N., & Sion, H. (2014). *Solid waste management in asian countries: A review of solid waste minimisation (3'r) towards low carbon.* Paper presented at the IOP Conference Series: Earth and Environmental Science.

Aphale, O., Thyberg, K. L., & Tonjes, D. J. (2015). Differences in waste generation, waste composition, and source separation across three waste districts in a New York suburb. *Resources, Conservation and Recycling, 99*, 19–28.

Ayodele, T., Alao, M., & Ogunjuyigbe, A. (2018). Recyclable resources from municipal solid waste: Assessment of its energy, economic and environmental benefits in Nigeria. *Resources, Conservation and Recycling, 134*, 165–173.

Babatunde, B., Vincent-Akpu, I., Woke, G., Atarhinyo, E., Aharanwa, U., Green, A., & Isaac-Joe, O. (2013). Comparative analysis of municipal solid waste (MSW) composition in three local government areas in rivers state, Nigeria. *African Journal of Environmental Science and Technology, 7*(9), 874–881.

Balakrishnan, K., Swathy, R., & T D, S. (2016). Automatic waste segregator and monitoring system. *Journal of Microcontroller Engineering and Applications, 3*, 1–7.

Bartolacci, F., Paolini, A., Quaranta, A. G., & Soverchia, M. (2018). Assessing factors that influence waste management financial sustainability. *Waste Management, 79*, 571–579.

Bassi, S. A., Christensen, T. H., & Damgaard, A. (2017). Environmental performance of household waste management in europe-an example of 7 countries. *Waste Management, 69*, 545–557.

Bel, G., & Mur, M. (2009). Intermunicipal cooperation, privatization and waste management costs: Evidence from rural municipalities. *Waste Management, 29*(10), 2772–2778.

Bernstad, A., la Cour Jansen, J., & Aspegren, H. (2011). Property-close source separation of hazardous waste and waste electrical and electronic equipment–a Swedish case study. *Waste Management, 31*(3), 536–543.

Bhambulkar, A., & Khedikar, I. (2011). Municipal solid waste (msw) collection route for laxmi nagar by geographical information system (GIS). *International Journal of Advanced Engineering Technology, 2*(4), 1–6.

Bharadwaj, A., Yadav, D., & Varshney, S. (2015). Non-biodegradable waste–its impact & safe disposal. *International Journal of Advanced Technology and Engineering Science, 3*(1).

Birara, E., & Kassahun, T. (2018). Assessment of solid waste management practices in bahir dar city, Ethiopia. *Pollution, 4*(2), 251–261.

Bonifazi, G., & Serranti, S. (2006). Imaging spectroscopy based strategies for ceramic glass contaminants removal in glass recycling. *Waste Management, 26*(6), 627–639.

Bonifazi, G., & Serranti, S. (2012). Recycling technologies. In R. A. Meyers (Ed.), *Encyclopedia of sustainability science and technology* (pp. 8794–8848). Springer.

Buenrostro, O., Bocco, G., & Cram, S. (2001). Classification of sources of municipal solid wastes in developing countries. *Resources, Conservation and Recycling, 32*(1), 29–41.

Calabrò, P. S., & Komilis, D. (2019). A standardized inspection methodology to evaluate municipal solid waste collection performance. *Journal of Environmental Management, 246*, 184–191.

Calabrò, P., & Satira, A. (2020). Recent advancements toward resilient and sustainable municipal solid waste collection systems. *Current Opinion in Green and Sustainable Chemistry, 26*, 100375.

Cao, R., & Wang, H. (2017). Research on the pollution hazard of municipal solid waste in China and its prevention and control legal countermeasures. *Nature Environment & Pollution Technology, 16*(2).

Cao, S., Xu, D., & Liu, S. (2018). A study of the relationships between the characteristics of the village population structure and rural residential solid waste collection services: Evidence from China. *International Journal of Environmental Research and Public Health, 15*(11), 2352.

Chakraborty, M., Sharma, C., Pandey, J., & Gupta, P. K. (2013). Assessment of energy generation potentials of msw in delhi under different technological options. *Energy Conversion and Management, 75*, 249–255.

Chalkias, C., & Lasaridi, K. (2009). A gis based model for the optimisation of municipal solid waste collection: The case study of nikea, athens, Greece. *Technology, 1*, 11–15.

Chandrappa, R., Das, D. B., Chandrappa, R., & Das, D. B. (2012). Wastes from industrial and commercial activities. In *Solid waste management: Principles and practice* (pp. 217–247). Springer.

Chen, C.-C. (2010). Spatial inequality in municipal solid waste disposal across regions in developing countries. *International Journal of Environmental Science & Technology, 7*, 447–456.

Chen, X., & Bai, X. (2023). A single-step upcycling of pvc-containing municipal solid waste compositions for greener chemicals and clean solids as fuel or oil absorbent. *Journal of the Energy Institute, 101405.*

Chen, B., & Lee, J. (2020). Household waste separation intention and the importance of public policy. *International Trade, Politics and Development, 4*(1), 61–79.

Chen, S., Huang, J., Xiao, T., Gao, J., Bai, J., Luo, W., & Dong, B. (2020). Carbon emissions under different domestic waste treatment modes induced by garbage classification: Case study in pilot communities in shanghai, China. *Science of the Total Environment, 717*, 137193.

Chiplunkar, A. V., Mehndiratta, S., & Khanna, P. (1981). Optimization on refuse collection systems. *Journal of the Environmental Engineering Division, 107*(6), 1203–1210.

Christensen, T. H., Fruergaard, T., & Matsufuji, Y. (2011). Residential waste. In *Solid waste technology and management* (pp. 85–96). Wiley.

Collection efficiency: Strategies for success. *United States environmental protection agency*. Available at: https://archive.Epa.Gov/epawaste/nonhaz/municipal/web/pdf/k99007.Pdf. Last accessed: 03 August 2023.

Dangi, M. B., Pretz, C. R., Urynowicz, M. A., Gerow, K. G., & Reddy, J. (2011). Municipal solid waste generation in Kathmandu, Nepal. *Journal of Environmental Management, 92*(1), 240–249.

Das, D. B. (2021). *Recycling and processing of waste: Needs of the time*. Available at SSRN 3777738.

Das, S., & Bhattacharyya, B. K. (2015). Optimization of municipal solid waste collection and transportation routes. *Waste Management, 43*, 9–18.

De Feo, G., Ferrara, C., Iuliano, C., & Grosso, A. (2016). Lca of the collection, transportation, treatment and disposal of source separated municipal waste: A southern Italy case study. *Sustainability, 8*(11), 1084.

Donnini Mancini, S., Rodrigues Nogueira, A., Akira Kagohara, D., Saide Schwartzman, J. A., & de Mattos, T. (2007). Recycling potential of urban solid waste destined for sanitary landfills: The case of indaiatuba, sp, Brazil. *Waste Management & Research, 25*(6), 517–523.

Dotoli, M., & Epicoco, N. (2017). A vehicle routing technique for hazardous waste collection. *IFAC-PapersOnLine, 50*(1), 9694–9699.

Duffus, J. H. (2001). "Heavy metals"–a meaningless term. *Chemistry International--Newsmagazine for IUPAC, 23*(6), 163–167.

Elisha, O. D. (2020). Moving beyond take-make-dispose to take-make-use for sustainable economy. *International Journal of Scientific Research and Education, 13*(3), 497–516.

ESRI. (2011). *Geographic information systems and environmental health: Incorporating esri technology and services*. Accessible from: chrome-extension://efaidnbmnnnibpcajpcglclefindmkaj/https://www.esri.com/content/dam/esrisites/sitecore-archive/Files/Pdfs/library/whitepapers/pdfs/gis_and_env_health.pdf. Last Accessed: August 08, 2023.

Estrellan, C. R., & Iino, F. (2010). Toxic emissions from open burning. *Chemosphere, 80*(3), 193–207.

Farrell, M., & Jones, D. (2009). Critical evaluation of municipal solid waste composting and potential compost markets. *Bioresource Technology, 100*(19), 4301–4310.

Farrokhi-Asl, H., Makui, A., Ghousi, R., & Rabbani, M. (2020). Developing a hazardous waste management system with consideration of health, safety, and environment. *Computers & Electrical Engineering, 82*, 106553.

Fei, F., Wen, Z., Huang, S., & De Clercq, D. (2018). Mechanical biological treatment of municipal solid waste: Energy efficiency, environmental impact and economic feasibility analysis. *Journal of Cleaner Production, 178*, 731–739.

Fernández-Aracil, P., Ortuño-Padilla, A., & Melgarejo-Moreno, J. (2018). Factors related to municipal costs of waste collection service in Spain. *Journal of Cleaner Production, 175*, 553–560.

Ferronato, N., Ragazzi, M., Portillo, M. A. G., Lizarazu, E. G. G., Viotti, P., & Torretta, V. (2019). How to improve recycling rate in developing big cities: An integrated approach for assessing municipal solid waste collection and treatment scenarios. *Environmental Development, 29*, 94–110.

Funk, K., Milford, J., & Simpkins, T. (2013). *Waste not, want not: Analyzing the economic and environmental viability of waste-to-energy (wte) technology for site-specific optimization of renewable energy options*. National Renewable Energy Lab.(NREL), Golden, CO (United States). https://doi.org/10.2172/1068584

Genuino, H. C., Ruiz, M. P., Heeres, H. J., & Kersten, S. R. (2023). Pyrolysis of mixed plastic waste: Predicting the product yields. *Waste Management, 156*, 208–215.

Giusti, L. (2009). A review of waste management practices and their impact on human health. *Waste Management, 29*(8), 2227–2239.

Gondal, M. A., & Siddiqui, M. N. (2007). Identification of different kinds of plastics using laser-induced breakdown spectroscopy for waste management. *Journal of Environmental Science and Health Part A, 42*(13), 1989–1997.

Gu, B., Wang, H., Chen, Z., Jiang, S., Zhu, W., Liu, M., et al. (2015). Characterization, quantification and management of household solid waste: A case study in China. *Resources, Conservation and Recycling, 98*, 67–75.

Gumus, A. T. (2009). Evaluation of hazardous waste transportation firms by using a two step fuzzy-ahp and topsis methodology. *Expert Systems with Applications, 36*(2), 4067–4074.

Gundupalli, S. P., Hait, S., & Thakur, A. (2017). A review on automated sorting of source-separated municipal solid waste for recycling. *Waste Management, 60*, 56–74.

Gupta, G. K., & Shukla, P. (2020). Insights into the resources generation from pulp and paper industry wastes: Challenges, perspectives and innovations. *Bioresource Technology, 297*, 122496.

Gupta, S., Mohan, K., Prasad, R., Gupta, S., & Kansal, A. (1998). Solid waste management in India: Options and opportunities. *Resources, Conservation and Recycling, 24*(2), 137–154.

Han, J., He, S., Shao, W., Wang, C., Qiao, L., Zhang, J., & Yang, L. (2023). Municipal solid waste, an overlooked route of transmission for the severe acute respiratory syndrome coronavirus 2: A review. *Environmental Chemistry Letters, 21*(1), 81–95.

Hazra, T., & Goel, S. (2009). Solid waste management in kolkata, India: Practices and challenges. *Waste Management, 29*(1), 470–478.

Hirpe, L., & Yeom, C. (2021). Municipal solid waste management policies, practices, and challenges in Ethiopia: A systematic review. *Sustainability, 13*(20), 11241.

Hoornweg, D., & Bhada-Tata, P. (2012a). *What a waste: A global review of solid waste management.* World Bank Group. United States of America. Accessible from https://policycommons.net/artifacts/1513146/what-a-waste/2185670/. Last Accessed: August 08, 2023

Hoornweg, D., & Bhada-Tata, P. (2012b). *What a waste: Waste management around the world* (pp. 9–15). World Bank.

Huang, J., Pretz, T., & Bian, Z. (2010). *Intelligent solid waste processing using optical sensor based sorting technology.* Paper presented at the 2010 3rd international congress on image and signal processing.

Hutton, M., & Symon, C. (1986). The quantities of cadmium, lead, mercury and arsenic entering the UK environment from human activities. *Science of the Total Environment, 57*, 129–150.

Ihedioha, J., Ukoha, P., & Ekere, N. (2017). Ecological and human health risk assessment of heavy metal contamination in soil of a municipal solid waste dump in uyo, Nigeria. *Environmental Geochemistry and Health, 39*, 497–515.

Inglezakis, V. J., & Moustakas, K. (2015). Household hazardous waste management: A review. *Journal of Environmental Management, 150*, 310–321.

IPCC. (2014). *Ar5 climate change 2014: Mitigation of climate change.* IPCC.

Jambeck, J. R., Geyer, R., Wilcox, C., Siegler, T. R., Perryman, M., Andrady, A., et al. (2015). Plastic waste inputs from land into the ocean. *Science, 347*(6223), 768–771.

Jull, H., Bier, J., Künnemeyer, R., & Schaare, P. (2018). Classification of recyclables using laser-induced breakdown spectroscopy for waste management. *Spectroscopy Letters, 51*(6), 257–265.

Kahhat, R., Kim, J., Xu, M., Allenby, B., Williams, E., & Zhang, P. (2008). Exploring e-waste management systems in the United States. *Resources, Conservation and Recycling, 52*(7), 955–964.

Kamran, A., Chaudhry, M. N., & Batool, S. A. (2015). Effects of socio-economic status and seasonal variation on municipal solid waste composition: A baseline study for future planning and development. *Environmental Sciences Europe, 27*, 1–8.

Karadimas, N. V., Kolokathi, M., Defteraiou, G., & Loumos, V. (2007). *Municipal waste collection of large items optimized with arc gis network analyst.* Paper presented at the Proceedings 21st European conference on modelling and simulation.

Kaur, P., Kaur, G. J., Routray, W., Rahimi, J., Nair, G. R., & Singh, A. (2021). Recent advances in utilization of municipal solid waste for production of bioproducts: A bibliometric analysis. *Case Studies in Chemical and Environmental Engineering, 4*, 100164.

Kaushal, R. K., Varghese, G. K., & Chabukdhara, M. (2012). Municipal solid waste management in India-current state and future challenges: A review. *International Journal of Engineering Science and Technology, 4*(4), 1473–1489.

Kaza, S., Yao, L., Bhada-Tata, P., & Van Woerden, F. (2018). *What a waste 2.0: A global snapshot of solid waste management to 2050.* World Bank Publications.

Kelland, D., Kolm, H., deLatour, C., Maxwell, E., & Oberteuffer, J. (1974). High gradient magnetic separation: An industrial application of magnetism. In *Superconducting machines and devices: Large systems applications* (pp. 581–594). Springer.

Keramitsoglou, K. M., & Tsagarakis, K. P. (2018). Public participation in designing the recycling bins to encourage recycling. *Sustainability, 10*(4), 1240.

Khan, D., & Samadder, S. R. (2014). Municipal solid waste management using geographical information system aided methods: A mini review. *Waste Management & Research, 32*(11), 1049–1062.

Khan, M. N., Luna, I. Z., Islam, M. M., Sharmeen, S., Salem, K. S., Rashid, T. U., et al. (2016). Cellulase in waste management applications. In *New and future developments in microbial biotechnology and bioengineering* (pp. 237–256). Elsevier.

Khwaja, R. (2010). *Report of the committee to evolve road map on management of wastes in India. Ministry of Environment and Forest (MoEF), India (Accessed 20 August 2016).* https://www.moef.nic.in/sites/default/files/Roadmap-Mgmt-Waste.pdf

Kihila, J. M., Wernsted, K., & Kaseva, M. (2021). Waste segregation and potential for recycling-a case study in dar es salaam city, Tanzania. *Sustainable Environment, 7*(1), 1935532.

Kinsella, S., & Gleason, G. (2003). *Single stream: An investigation into the interaction between single stream recycling collection systems and recycled paper manufacturing*. Conservatree.

Knickmeyer, D. (2020). Social factors influencing household waste separation: A literature review on good practices to improve the recycling performance of urban areas. *Journal of Cleaner Production, 245*, 118605.

Ko, S., Kim, W., Shin, S.-C., & Shin, J. (2020). The economic value of sustainable recycling and waste management policies: The case of a waste management crisis in South Korea. *Waste Management, 104*, 220–227.

Kochar, Q. (2021). The increasing the efficiency of material recovery facility through source separation of residential solid waste, the case of Duhok city. *Sustainable Future and Technology Development, 3*(3).

Kumar, K. N., & Goel, S. (2009). Characterization of municipal solid waste (msw) and a proposed management plan for Kharagpur, West Bengal, India. *Resources, Conservation and Recycling, 53*(3), 166–174.

Kumar, V., & Thakur, I. S. (2020). Extraction of lipids and production of biodiesel from secondary tannery sludge by in situ transesterification. *Bioresource Technology Reports, 11*, 100446. https://doi.org/10.1016/j.biteb.2020.100446

Kumar, V., Bee, D. J., Shirodkar, P. S., Tumkor, S., Bettig, B. P., & Sutherland, J. W. (2005). *Towards sustainable "product and material flow" cycles: Identifying barriers to achieving product multi-use and zero waste.* Paper presented at the ASME International Mechanical Engineering Congress and Exposition.

Kumar, S., Smith, S. R., Fowler, G., Velis, C., Kumar, S. J., Arya, S., et al. (2017). Challenges and opportunities associated with waste management in India. *Royal Society Open Science, 4*(3), 160764.

Kumar, V., Srivastava, S., & Thakur, I. S. (2021). Enhanced recovery of polyhydroxyalkanoates from secondary wastewater sludge of sewage treatment plant: Analysis and process parameters optimization. *Bioresource Technology Reports, 15*, 100783.

Kutila, M., Viitanen, J., & Vattulainen, A. (2005). *Scrap metal sorting with colour vision and inductive sensor array.* Paper presented at the International Conference on Computational Intelligence for Modelling, Control and Automation and International Conference on Intelligent Agents, Web Technologies and Internet Commerce (CIMCA-IAWTIC'06).

Kwarteng, P., & Kumah, V. A. (2022). Solid waste disposal practices on the environment. *Research Journal of Ecology and Environmental Sciences*, 26–45.

Lamma, O. A. (2021). The impact of recycling in preserving the environment. *IJAR, 7*(11), 297–302.

Laporte, G. (1992). The vehicle routing problem: An overview of exact and approximate algorithms. *European Journal of Operational Research, 59*(3), 345–358.

Leal Filho, W., Brandli, L., Moora, H., Kruopienė, J., & Stenmarck, Å. (2016). Benchmarking approaches and methods in the field of urban waste management. *Journal of Cleaner Production, 112*, 4377–4386.

Lee, C., & Lin, S. D. (2007). *Handbook of environmental engineering calculations*. McGraw-Hill Education.

Leeabai, N., Areeprasert, C., Khaobang, C., Viriyapanitchakij, N., Bussa, B., Dilinazi, D., & Takahashi, F. (2021). The effects of color preference and noticeability of trash bins on waste collection performance and waste-sorting behaviors. *Waste Management, 121*, 153–163.

Lella, J., Mandla, V. R., & Zhu, X. (2017). Solid waste collection/transport optimization and vegetation land cover estimation using geographic information system (gis): A case study of a proposed smart-city. *Sustainable Cities and Society, 35*, 336–349.

Li, F., Golden, B., & Wasil, E. (2007). The open vehicle routing problem: Algorithms, large-scale test problems, and computational results. *Computers & Operations Research, 34*(10), 2918–2930.

Li, C., Lu, F., Zhang, Y., Liu, T., & Hou, W. (2008). Spatial distribution characteristics of heavy metals in street dust in Shenyang city. *Ecology and Environment, 17*(2), 560–564.

Lim-Wavde, K., Kauffman, R. J., & Dawson, G. S. (2017). Household informedness and policy analytics for the collection and recycling of household hazardous waste in california. *Resources, Conservation and Recycling, 120*, 88–107.

Liu, K., Tian, D., Li, C., Li, Y., Yang, G., & Ding, Y. (2018). A review of laser-induced breakdown spectroscopy for plastic analysis. *TrAC Trends in Analytical Chemistry, 110*. https://doi.org/10.1016/j.trac.2018.11.025

Malakahmad, A., Bakri, P. M., Mokhtar, M. R. M., & Khalil, N. (2014). Solid waste collection routes optimization via gis techniques in ipoh city, Malaysia. *Procedia Engineering, 77*, 20–27.

Matsumoto, S. (2011). Waste separation at home: Are japanese municipal curbside recycling policies efficient? *Resources, Conservation and Recycling, 55*(3), 325–334.

Medina, M. (2009). Municipal solid waste management in third world cities: Lessons learned and a proposal for improvement. *Human Settlement Development, 3*, 264.

Mintz, K. K., Henn, L., Park, J., & Kurman, J. (2019). What predicts household waste management behaviors? Culture and type of behavior as moderators. *Resources, Conservation and Recycling, 145*, 11–18.

Modaihsh, A., Ai-Swailem, M., & Mahjoub, M. (2004). Heavy metals content of commercial inorganic fertilizers used in the Kingdom of Saudi Arabia. *Journal of Agricultural and Marine Sciences [JAMS], 9*(1), 21–25.

Mukemo, A., Mukemo, D., Kalenga, M., Mishika, P., Mukuku, O., & Malonga, F. (2020). Household waste management in lubumbashi, democratic republic of congo. *The Open Journal of Public Health, 2*(1), 1010.

Naveenkumar, R., Iyyappan, J., Pravin, R., Kadry, S., Han, J., Sindhu, R., et al. (2023). A strategic review on sustainable approaches in municipal solid waste management and energy recovery: Role of artificial intelligence, economic stability and life cycle assessment. *Bioresource Technology, 129044*.

Nepal, M., Karki Nepal, A., Khadayat, M. S., Rai, R. K., Shyamsundar, P., & Somanathan, E. (2023). Low-cost strategies to improve municipal solid waste management in developing countries: Experimental evidence from Nepal. *Environmental and Resource Economics, 84*(3), 729–752.

Nguyen, M. H., Ho, T. V., Nguyen, T. K., & Do, M. D. (2015). Modeling and simulation of the effects of landslide on circulation of transports on the mountain roads. *International Journal of Advanced Computer Science and Applications, 6*(8).

Nguyen-Trong, K., Nguyen-Thi-Ngoc, A., Nguyen-Ngoc, D., & Dinh-Thi-Hai, V. (2017). Optimization of municipal solid waste transportation by integrating gis analysis, equation-based, and agent-based model. *Waste Management, 59*, 14–22.

Nuortio, T., Kytöjoki, J., Niska, H., & Bräysy, O. (2006). Improved route planning and scheduling of waste collection and transport. *Expert Systems with Applications, 30*(2), 223–232.

Oberlin, A. S. (2013). Resource recovery potential: A case study of household waste in kinondoni municipality, dar es salaam. *TaJONAS: Tanzania Journal of Natural and Applied Sciences, 4*(1), 563–574.

Okedere, O., Olalekan, A., Fakinle, B., Elehinafe, F., Odunlami, O., & Sonibare, J. (2019). Urban air pollution from the open burning of municipal solid waste. *Environmental Quality Management, 28*(4), 67–74.

Olukanni, D., Adeleke, J., & Aremu, D. (2016). A review of local factors affecting solid waste collection in Nigeria. *Pollution, 2*(3), 339–356.

Óskarsson, G. K., Agnarsson, S., & Davíðsdóttir, B. (2022). Waste management in Iceland: Challenges and costs related to achieving the eu municipal solid waste targets. *Waste Management, 151*, 131–141.

Ozcan, H. K., Guvenc, S. Y., Guvenc, L., & Demir, G. (2016). Municipal solid waste characterization according to different income levels: A case study. *Sustainability, 8*(10), 1044.

Paritosh, K., Kushwaha, S. K., Yadav, M., Pareek, N., Chawade, A., & Vivekanand, V. (2017). Food waste to energy: An overview of sustainable approaches for food waste management and nutrient recycling. *BioMed Research International, 2017.*

Park, S., & Lah, T. (2015). Analyzing the success of the volume-based waste fee system in South Korea. *Waste Management, 43*, 533–538.

Park, S., Lee, J., Kwon, E., Kim, D., Shin, S., Jeong, S., & Park, K. (2022). 3d sensing system for laser-induced breakdown spectroscopy-based metal scrap identification. *International Journal of Precision Engineering and Manufacturing-Green Technology*, 1–13.

Pereira, J. A., Berenguer, C. V., & Câmara, J. S. (2023). Delving into agri-food waste composition for antibacterial phytochemicals. *Metabolites, 13*(5), 634.

Phu, S. T. P., Takeshi, F., Giang, H. M., & Dinh, P. V. (2019). An analysis of the commercial waste characterisation in a tourism city in Vietnam. *International Journal of Environment and Waste Management, 23*(3), 319–335.

Prechthai, T., Parkpian, P., & Visvanathan, C. (2008). Assessment of heavy metal contamination and its mobilization from municipal solid waste open dumping site. *Journal of Hazardous Materials, 156*(1–3), 86–94.

Punkkinen, H., Merta, E., Teerioja, N., Moliis, K., & Kuvaja, E. (2012). Environmental sustainability comparison of a hypothetical pneumatic waste collection system and a door-to-door system. *Waste Management, 32*(10), 1775–1781.

Rada, E. C., Ragazzi, M., & Fedrizzi, P. (2013). Web-gis oriented systems viability for municipal solid waste selective collection optimization in developed and transient economies. *Waste Management, 33*(4), 785–792.

Rahman, M., Hussain, A., & Basri, H. (2014). A critical review on waste paper sorting techniques. *International journal of Environmental Science and Technology, 11*, 551–564.

Rasdorf, W., Janisch, A., Robson, F., & Tilley, C. (2002). Testing and evaluating alternative algorithms for location referencing system route generation. *Journal of Computing in Civil Engineering, 16*(3), 194–205.

Richardson, R. A., & Havlicek, J. (1974). An analysis of seasonal household waste generation. *Journal of Agricultural and Applied Economics, 6*(2), 143–155.

Rossi, M., Papetti, A., & Germani, M. (2022). A comparison of different waste collection methods: Environmental impacts and occupational risks. *Journal of Cleaner Production, 368*, 133145.

Rousta, K., & Bolton, K. (2019). Sorting household waste at the source. In *Sustainable resource recovery and zero waste approaches* (pp. 105–114). Elsevier.

Rousta, K., & Dahlén, L. (2015). Source separation of household waste: Technology and social aspects. In *Resource recovery to approach zero municipal waste*. CRC Press.

Ruj, B., Pandey, V., Jash, P., & Srivastava, V. (2015). Sorting of plastic waste for effective recycling. *International Journal of Applied Science & Engineering Research, 4*(4), 564–571.

Sakai, S., Sawell, S., Chandler, A., Eighmy, T., Kosson, D., Vehlow, J., et al. (1996). World terends in municipal solid waste management. *Waste Management, 16*(5–6), 341–350.

Sandhu, K., Burton, P., & Dedekorkut-Howes, A. (2017). Between hype and veracity; privatization of municipal solid waste management and its impacts on the informal waste sector. *Waste Management, 59*, 545–556.

Sanjeevi, V., & Shahabudeen, P. (2016). Optimal routing for efficient municipal solid waste transportation by using arcgis application in chennai, India. *Waste Management & Research, 34*(1), 11–21.

Sarbassov, Y., Sagalova, T., Tursunov, O., Venetis, C., Xenarios, S., & Inglezakis, V. (2019). Survey on household solid waste sorting at source in developing economies: A case study of nur-sultan city in Kazakhstan. *Sustainability, 11*(22), 6496.

Schloss, K. B., Lessard, L., Walmsley, C. S., & Foley, K. (2018). Color inference in visual communication: The meaning of colors in recycling. *Cognitive Research: Principles and Implications, 3*(1), 1–17.

Schulze, C. (2003). *Municipal waste management in Berlin. Berlin senate department for urban development and the environment communication*. Accessible from. https://www.berlin.de/sen/uvk/en/. Last Accessed: August 08, 2023.

Seo, S., Aramaki, T., Hwang, Y., & Hanaki, K. (2004). Environmental impact of solid waste treatment methods in Korea. *Journal of Environmental Engineering, 130*(1), 81–89.

Serranti, S., & Bonifazi, G. (2019). Techniques for separation of plastic wastes. In *Use of recycled plastics in eco-efficient concrete* (pp. 9–37). Elsevier.

Sharholy, M., Ahmad, K., Vaishya, R. C., & Gupta, R. D. (2007). Municipal solid waste characteristics and management in allahabad, India. *Waste Management, 27*(4), 490–496.

Shiralipour, A., McConnell, D. B., & Smith, W. H. (1992). Uses and benefits of msw compost: A review and an assessment. *Biomass and Bioenergy, 3*(3–4), 267–279.

Sieghardt, M., Mursch-Radlgruber, E., Paoletti, E., Couenberg, E., Dimitrakopoulus, A., Rego, F., et al. (2005). The abiotic urban environment: Impact of urban growing conditions on urban vegetation. In *Urban forests and trees: A reference book* (pp. 281–323). Springer.

Silveira, A., Cella, M., Tanabe, E., & Bertuol, D. (2018). Application of tribo-electrostatic separation in the recycling of plastic wastes. *Process Safety and Environmental Protection, 114*, 219–228.

Singh, G. K., Gupta, K., & Chaudhary, S. (2014). Solid waste management: Its sources, collection, transportation and recycling. *International Journal of Environmental Science and Development, 5*(4), 347.

Southwood, R. (1984). *Tenth report–tackling pollution–experience and prospects (London, royal commission on environmental pollution)*.

Srivastava, R. (2016). Waste management: Developed and developing countries. *International Journal of Science and Research, 5*(3), 202–203.

Stafford, D. A., Hawkes, D. L., & Horton, R. (1980). Methane production from waste organic matter. CRC Press. 285 p.

Stoeva, K., & Alriksson, S. (2017). Influence of recycling programmes on waste separation behaviour. *Waste Management, 68*, 732–741.

Sukholthaman, P., & Sharp, A. (2016). A system dynamics model to evaluate effects of source separation of municipal solid waste management: A case of bangkok, Thailand. *Waste Management, 52*, 50–61.

Surroop, D., & Mauthoor, S. (2012). Assessment of recycling potential of municipal solid waste in Mauritius. *International Journal of Environmental Technology and Management, 15*(3–6), 400–416.

Takahashi, W. (2020). Economic rationalism or administrative rationalism? Curbside collection systems in Sweden and Japan. *Journal of Cleaner Production, 242*, 118288.

Tanskanen, J.-H., & Kaila, J. (2001). Comparison of methods used in the collection of source-separated household waste. *Waste Management & Research, 19*(6), 486–497.

Tchobanoglous, G., & Kreith, F. (2002). *Handbook of solid waste management*. McGraw-Hill Education.

Thitame, S. N., Pondhe, G., & Meshram, D. (2010). Characterisation and composition of municipal solid waste (msw) generated in sangamner city, district ahmednagar, maharashtra, India. *Environmental Monitoring and Assessment, 170*, 1–5.

Thongyuan, S., Khantamoon, T., Aendo, P., Binot, A., & Tulayakul, P. (2021). Ecological and health risk assessment, carcinogenic and non-carcinogenic effects of heavy metals contamination in the soil from municipal solid waste landfill in central, Thailand. *Human and Ecological Risk Assessment: An International Journal, 27*(4), 876–897.

Tomno, R. M., Nzeve, J. K., Mailu, S. N., Shitanda, D., & Waswa, F. (2020). Heavy metal contamination of water, soil and vegetables in urban streams in machakos municipality, Kenya. *Scientific African, 9*, e00539.

Urbańska, W., Janda, A., Osial, M., & Słowikowski, M. (2023). Sustainable municipal waste management during the covid-19 pandemic—a case study of Poland. *Resources, 12*(7), 76.
Athanasios - Valavanidis (2023). *Global municipal solid waste (msw) in crisis. Two billion tonnes of msw every year, a worrying worldwide environmental problem* (Vol. 1, pp. 1–28). Accessible from: https://www.researchgate.net/profile/Athanasios-Valavanidis/publication/367221073_Global_Municipal_Solid_Waste_MSW_in_Crisis_Two_billion_tonnes_of_MSW_every_year_a_worrying_worldwide_environmental_problem/links/63c7c552d9fb5967c2e5656b/Global-Municipal-Solid-Waste-MSW-in-Crisis-Two-billion-tonnes-of-MSW-every-year-a-worrying-worldwide-environmental-problem.pdf. Last Accessed: August 08, 2023.
van Ewijk, S., & Stegemann, J. A. (2020). Recognising waste use potential to achieve a circular economy. *Waste Management, 105*, 1–7.
Vinti, G., Bauza, V., Clasen, T., Medlicott, K., Tudor, T., Zurbrügg, C., & Vaccari, M. (2021). Municipal solid waste management and adverse health outcomes: A systematic review. *International Journal of Environmental Research and Public Health, 18*(8), 4331.
Voltes-Dorta, A., Jiménez, J. L., & Suárez-Alemán, A. (2014). An initial investigation into the impact of tourism on local budgets: A comparative analysis of spanish municipalities. *Tourism Management, 45*, 124–133.
Wan, C., Shen, G. Q., & Choi, S. (2019). Waste management strategies for sustainable development. In *Encyclopedia of sustainability in higher education* (pp. 2020–2028). Springer.
Wang, B., Farooque, M., Zhong, R. Y., Zhang, A., & Liu, Y. (2021). Internet of things (iot)-enabled accountability in source separation of household waste for a circular economy in China. *Journal of Cleaner Production, 300*, 126773.
Wessel, J. (2004). *Household hazardous materials programs, a report to the iowa general assembly*. Iowa Department of Natural Resources, Energy and Waste Management Bureau.
Widmer, R., Oswald-Krapf, H., Sinha-Khetriwal, D., Schnellmann, M., & Böni, H. (2005). Global perspectives on e-waste. *Environmental Impact Assessment Review, 25*(5), 436–458.
Worrell, W. A., & Vesilind, P. A. (1979). Testing and evaluation of air classifier performance. *Resource Recovery and Conservation, 4*(3), 247–259.
Worrell, W. A., Vesilind, P. A., & Ludwig, C. (2016). *Solid waste engineering: A global perspective*. Cengage Learning.
Wu, X., Li, J., Yao, L., & Xu, Z. (2020). Auto-sorting commonly recovered plastics from waste household appliances and electronics using near-infrared spectroscopy. *Journal of Cleaner Production, 246*, 118732.
Wu, Z., Zhang, Y., Chen, Q., & Wang, H. (2021). Attitude of chinese public towards municipal solid waste sorting policy: A text mining study. *Science of the Total Environment, 756*, 142674.
Yadav, V., & Karmakar, S. (2020). Sustainable collection and transportation of municipal solid waste in urban centers. *Sustainable Cities and Society, 53*, 101937.
Yadav, V., Karmakar, S., Dikshit, A., & Vanjari, S. (2016). A facility location model for msw management systems under uncertainty: A case study of nashik city, India. *Procedia Environmental Sciences, 35*, 90–100.
Ying, X., Barlow, N. J., & Tatiparthi, A. (2022). Micro-ct and volumetric imaging in developmental toxicology. In *Reproductive and developmental toxicology* (pp. 1261–1285). Elsevier.
Zacharewski, T. R., Berhane, K., Gillesby, B. E., & Burnison, B. K. (1995). Detection of estrogen- and dioxin-like activity in pulp and paper mill black liquor and effluent using in vitro recombinant receptor/reporter gene assays. *Environmental Science & Technology, 29*(8), 2140–2146.
Zeng, C., Li, H., Xia, F., Niu, D., & Zhao, Y. (2018). Source-separated collection of rural solid waste in China. In *Source separation and recycling: Implementation and benefits for a circular economy* (pp. 151–174). Springer.
Zhang, X., Liu, C., Chen, Y., Zheng, G., & Chen, Y. (2022). Source separation, transportation, pretreatment, and valorization of municipal solid waste: A critical review. *Environment, Development and Sustainability, 24*, 1–43.

Zhao, Y., & Li, J. (2022). Sensor-based technologies in effective solid waste sorting: Successful applications, sensor combination, and future directions. *Environmental Science & Technology, 56*(24), 17531–17544.

Zhao, Q., Tang, W., Han, M., Cui, W., Zhu, L., Xie, H., et al. (2023). Estimation of reduced greenhouse gas emission from municipal solid waste incineration with electricity recovery in prefecture-and county-level cities of China. *Science of the Total Environment, 875*, 162654.

Zhu, S., Chen, H., Wang, M., Guo, X., Lei, Y., & Jin, G. (2019). Plastic solid waste identification system based on near infrared spectroscopy in combination with support vector machine. *Advanced Industrial and Engineering Polymer Research, 2*(2), 77–81.

Zia, A., Batool, S. A., Chauhdry, M. N., & Munir, S. (2017). Influence of income level and seasons on quantity and composition of municipal solid waste: A case study of the capital city of Pakistan. *Sustainability, 9*(9), 1568.

Zurbrugg, C. (2003). *Solid waste management in developing countries*. SWM introductory text on https://www.sanicon.net, 5.

Chapter 3
Industrial Wastewater Sludge Management: Milestones and Challenges

Miguel A. Sandoval and Laura Patricia Domínguez-Jaimes

1 Introduction

Nowadays, several regions worldwide are experiencing water shortage and stress as a consequence of overexploitation of water bodies, inadequate distribution of freshwater, population and industrialization growth, and augmenting pollution of the natural environment due to out-of-date wastewater treatment facilities. The last issue has led to a doubtful supply of clean water, which not only affects casual human activities but also restricts socioeconomic and industrial activities. Several industries largely depend on water to carry out their respective processes. For example, the textile industry consumes more than 2500 L of water to dye and print a single cotton T-shirt. Similarly, the pulp and paper, tannery, distillery, pharmaceutical, mining, and petrochemical industries require substantial amounts of water for cooling, solvent extraction, and chemical reactions (Kumar et al., 2020, 2021a, b, 2022; Jit et al., 2023; Kumar & Verma, 2023). In this vein, the main objective of wastewater treatment plants (WWTPs) is the removal of organic/inorganic contaminants from polluted industry effluents. To improve the biodegradability of such recalcitrant wastewater, various treatment methods have been established, including biological, physicochemical, and advanced oxidation processes. Once wastewater is treated, clear water and large quantities of sewage sludge (SS) are obtained. However, sludge is a complex mixture that can only be disposed of in landfills or on land with prior treatment (Jimenez et al., 2004; Chandra & Kumar, 2017). SS usually

M. A. Sandoval (✉)
Instituto Tecnológico Superior de Guanajuato, Tecnológico Nacional de México, Guanajuato, Mexico
e-mail: miguel.sl@guanajuato.tecnm.mx; miguel.sandoval@usach.cl

L. P. Domínguez-Jaimes
Centro de Investigación y Desarrollo Tecnológico en Electroquímica, Parque Tecnológico s/n Sanfandila, Querétaro, Mexico

V. Kumar et al. (eds.), *Environmental Engineering and Waste Management*,
https://doi.org/10.1007/978-3-031-58441-1_3

needs high costs for appropriate handling, dumping, and treatment—frequently around 30–40% of the capital expenses and approximately 50% of the operating budgets of whole WWTPs (Kumar & Thakur, 2020; Abdel Wahaab et al., 2020).

Sludge is primarily generated during the chemical coagulation and microbiological phases of treatment process. This process may involve aerobic and anaerobic digestion, alkaline stabilization, thermal drying, disinfection, oxidation, and stabilization (Awasthi et al., 2023). This sludge is called biosolids, also called active sludge, which can be considered unusable materials after sludge processing because they may contain an innumerable number of contaminants, including nutrients, organic compounds, and pathogens, which cause hygiene concerns and foul odors (Cheng et al., 2022; Raheem et al., 2018; Tyagi & Lo, 2013). In addition, emerging pollutants such as personal care products, pharmaceuticals, pesticides, microplastics, or heavy metals may be present. Individuals residing close to the deposited sludge could come into contact with its contaminants through direct pathways, including hand-to-mouth ingestion, inhalation, and skin contact. Indirect routes, such as dietary consumption, also present a possibility for exposure. Furthermore, the pollutants in these sludges might seep into nearby soils, surface water, and groundwater or become airborne, potentially harming human health.

Several technologies are being explored for using SS as raw materials for renewable energy and materials as well as for nutrient recovery in agriculture. These technologies include dark fermentation for hydrogen production, bioelectrochemical systems and fermentation for bioplastic production, and integration of thermochemical processes, such as pyrolysis, gasification, and hydrothermal carbonization (HTC), for energy production and biomaterial absorbents. However, anaerobic digestion is expected to be the best process for sludge recovery because it is an environmentally friendly process (Elalami et al., 2019).

Anaerobic digestion processes have been used on an industrial scale to efficiently remove and add value to sludge. It is widely used to treat SS either alone or with other substrates. This process allows for the production of biogas containing sufficient methane content to be classified as a renewable energy source. At the same time, the digestate can be used as a fertilizer because it is rich in nutrients (Jiang et al., 2022). The main advantages and disadvantages of the traditional processes are summarized in Table 3.1, and they will be deeply discussed in the following sections.

The quantity and quality of sludge are both affected by some factors, mainly the kind of industrial wastewater and the treatment method used. Currently, industrial wastewater systems are switching from traditional methods to those where resources can be recovered. This is the principle behind the shift from a linear economy to a circular economy (CE) approach, which consequently contributes to sustainable development (Iqbal et al., 2022).

The concept of CE has emerged as a promising avenue to convert waste into valuable products. An example of this is the conversion of sludge into biochar through pyrolysis—a sustainable and encouraging approach. This method aligns with CE principles and offers potential agricultural, energy, and environmental applications. In addition, by-products like biofuels, macronutrients such as nitrogen

Table 3.1 Sludge recovery and management processes: advantages, disadvantages, and recommendations

Process	Advantages	Disadvantages	Recommendations
Combustion	Significant reduction of sludge volumes Energy recovery Destruction of pathogens Mature technology	High energy consumption Drying/dewatering is required	Management of ashes (concentrated in heavy metals) Gaseous emission treatment
Gasification	Sludge volume reduction Energy production (syngas)	Complexity of the technology High energy consumption Drying/dewatering is required	Optimization of operating conditions Gaseous emission treatment
Anaerobic digestion	Reduction of sludge volume Energy production (biogas) Low energy consumption Digestate used as organic fertilizer mature technology	Low degradability of some sludge, which may require pretreatments (additional costs) Digestate management strategy required	Optimization of biogas-upgrading techniques to reduce the process costs Optimization of pretreatments conditions
Composting	Reduction of sludge volume Stabilization of sludge Low energy consumption Cost-effective on large scales Mature technology	Dewatering of sludge may be needed	Co-composting of sludge with other organic wastes may enhance the compost quality
Pyrolysis and hydrothermal carbonization	Sludge volume reduction Energy production (syngas and bio-oils) Biochar generation	Complexity of the technology High energy consumption and investment costs Drying requirement	Gaseous emission treatment Co-pyrolysis with more carbonaceous materials
Bioelectrochemical system	Reduction of chemical oxygen demand and ammonia in sludge Electrical energy generation	Expensive scale-up Low power generated	Coupling with other processes such as anaerobic digestion to enhance optimization of electrode material selection

(continued)

Table 3.1 (continued)

Process	Advantages	Disadvantages	Recommendations
	Hydrogen production		
Polyhydroxyalkanoate (PHA) production	Reduction of sludge Bioplastic production	Expensive technology Waste disposal costs	Pretreatment of the feed-stock Development of low-cost extraction methods
Dark fermentation	Low energy consumption Biohydrogen production	Low yields of biohydrogen Pretreatments of feed-stock Sustainable management of the residues	Optimization of pretreatment conditions PHA-rich effluents must be valorized

From Elalami et al. (2019)

and phosphorous (agricultural use), and carboxylic acids such as maleic, acetic, oxamic, fumaric, oxalic, and glyoxylic acids, among others with a significant market value can be obtained (dos Santos et al., 2023). Concurrently, the success of the Sustainable Development Goals (SDGs) of the United Nations, especially those marked 6, 11, and 12, is based on the transition toward circularity.

This chapter aims to emphasize and summarize emerging research on resource recovery from different wastewater treatments and the critical role of circular economy approaches. Here, several methods and recent technological developments are included. It is aimed at engineers, technologists, and researchers who are motivated to address the challenges of and capitalize on the practical insights into managing industrial wastewater. Furthermore, future perspectives are discussed.

2 Wastewater Treatment Technologies

Wastewater is produced through several industrial processes as well as from domestic and public activities. Therefore, a combination of organic/inorganic compounds, germs, and pathogens is commonly present. Moreover, wastewater usually contains contaminants of emerging concern (CECs) that must be degraded/mineralized to avoid severe damage to the aquatic environment and human health. Research is currently focused on eliminating these pollutants, and the conventional and alternative wastewater treatment technologies are analyzed in this section.

2.1 Typical Treatment Processes

Traditional approaches to wastewater treatment involve integrating physical, chemical, and biological procedures to eliminate solids, organic substances, and, occasionally, nutrients from wastewater. These methods mainly rely on removing the organic load and nonorganic solids, turbidity, and disinfection (chlorine treatment). However, they are insufficient to deal with the removal or elimination of CECs.

The usual stages, in general terms, used in conventional wastewater treatment plants are the following (Koul et al., 2022):

(a) Preliminary treatment, where almost all large and gross particles floating in the effluent are separated by screeners. At this stage, around 25% of the organic load is eliminated and disposed of in a landfill. This stage is important to prevent the accumulation of excess sludge in pipes or channels.
(b) Primary treatment, where the remaining suspended particles and sedimented organic/inorganic matter are eliminated (~60%). Moreover, biochemical oxygen demand (BOD) is reduced together with grease and oil. The common methods used at this stage are chemical coagulation and flocculation.
(c) Secondary treatment, where residual and biodegradable pollutants are deleted by biological treatment in an aeration tank using microorganisms to degrade "waste" into inorganic by-products.
(d) Tertiary treatment, or the purification stage, where bacteria and viruses are eliminated from treated water by adding disinfectant agents such as chlorine to reach optimal parameters for reusing or recycling in industrial applications, for irrigation company purposes, or for discharging into natural water reservoirs, according to quality standards.

2.2 *Electrochemical Treatment Technologies*

Electrochemical treatment technologies show various advantages over conventional wastewater treatment techniques (Clematis et al., 2023). These advantages include environmental compatibility, quick adaptation to wastewater changes, energy efficiency, simple operational management, selectivity, amenability to automation, and cost-effectiveness. Electrochemical techniques can accelerate the speed and effectiveness of interactions at environmental interfaces by enabling controlled electron transfer in specific directions. Many electrochemical advanced oxidation processes (EAOPs), such as anodic oxidation, electrocoagulation, electrodialysis, and Fenton-based processes, are used to treat different effluents (Yadav & Kamsonlian, 2022).

The conditions for wastewater electrochemical treatment depend on the physicochemical characteristics of the matrix as well as the volume of the polluted water for treatment. In addition, the type of contaminant present in the wastewater will influence the electrochemical treatment to be chosen.

It is important to highlight that a combination of processes is usually recommended to achieve the permissible standard limits before discharge into municipal sewer systems. Similarly, electrochemical processes merged with renewable energy with in situ chemical production produce excellent results.

Electrochemical technologies have an optimistic future owing to their characteristics but, most importantly, their environmental sustainability based on the non-requirement of any additional chemicals.

3 Wastewater Sewage Sludge

Due to municipal WWTP operations, large amounts of SS are generated. This poses a significant concern in numerous countries because of its growing volume and the consequences linked to its improper disposal (Kumar et al., 2021c). SS constitutes the solid residual that is predominantly derived from the secondary sedimentation tank, oxidation ditch, or aeration tank following the completion of biological treatment on municipal wastewater. Sludge from WWTPs may be considered a sink of different concentrated pollutants such as microplastics (fibers and fragments) (Franco et al., 2023) and phenolic compounds (Martínez-Gallardo et al., 2022). The type of contaminant present in sludge differs on the basis of the attributes of the wastewater that enters WWTPs, such as wastewater from the food industry or a municipal sewer, as well as on the treatment capacity and technology.

3.1 Sludge Treatment and Disposal Practices

Before disposal, the typical methods for sludge treatment and stabilization involve reducing moisture content and stabilizing organic matter through digestion, composting, or thermal treatment (Fytili & Zabaniotou, 2008). The procedures of sludge drying, reliant on heat and electricity, are characterized by high energy consumption and limited economic viability. Beyond energy efficiency concerns, a significant challenge in sludge drying lies in the inconsistent quality of the resulting raw material. Furthermore, as mentioned earlier, sludge can be susceptible to contamination by various organic/inorganic pollutants and microorganisms. SS treatment usually consists of some stages such as anaerobic/aerobic digestion, chemical stabilization, composting, drying, and incineration (Amann et al., 2022).

The most common method of managing wastewater sludge is storing it in simulated ponds until it naturally vanishes during the warm phase (summer). However, this method causes pollution of the surrounding areas through lixiviation. There are several biotreatments for bioremediation and valorization of industrial sludge. For instance, Martinez-Gallardo et al. assessed land farming, phytoremediation, composting, and vermicomposting techniques to treat olive mill wastewater sludge. These methods convert the contaminated matrix into a suitable substrate for plant growth or a product with potential for agricultural applications (Martínez-Gallardo et al., 2022). The results showed that land farming and phytoremediation achieved the cleaning of soils, whereas composting/vermicomposting produced high-quality products of agricultural interest in the framework of the circular economy. It is important to highlight that adding earthworms to composting processes improves the motivation of usable biodiversity due to a rational metabolic outline of the microbial population. Wang et al. reviewed wastewater and sludge treatment using potassium permanganate ($KMnO_4$) as a friendly environmentally oxidant. This process decreases dangerous gas traces and enhances sludge dewatering, followed by resource reclamation owing to its powerful oxidation (Yufen Wang et al., 2023a, b). The authors have briefed the processes of improving $KMnO_4$ oxidation and examined the latest studies of potassium permanganate in wastewater and sludge treatment procedures. They mainly concluded that $KMnO_4$ could be stimulated by bisulfite ions (HSO_3^-), radiation (i.e., ultraviolet (UV)/solar light), carbon-based catalysts (i.e., biochar), and ferrous ions (Fe^{2+}). Regarding sludge dewaterability improvement, manganese ions (Mn^{7+}) could fragment highly hydrated extracellular polymeric substances (EPSs), thus altering the configuration of sludge flocs and expediting the liberation of both bound and intracellular water within the sludge. Subsequently, the disintegrated sludge is re-aggregated by in situ formed manganese dioxide (MnO_2) or Fe^{3+}, leading to even more effective dewatering. Additionally, the disruption of EPSs augments the biodegradability of the sludge, offering significant substrates for the accumulation of short-chain fatty acids or the production of methane. These conclusions validate that this process provides a solution for wastewater cleansing, risk modification, and sludge reusing. Figure 3.1 showcases sludge management data for the leading five European producers.

Related to advanced oxidation processes, Fenton sludge may be a potential source of iron that could have positive effects (augmenting methane production,

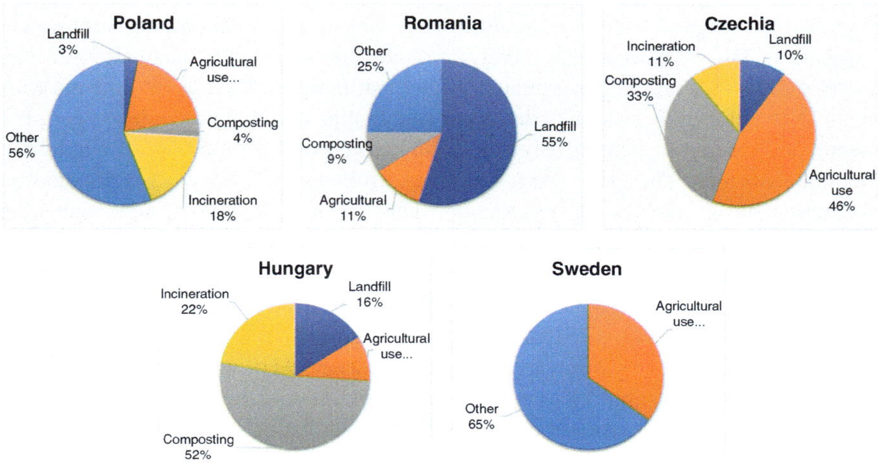

Fig. 3.1 (Landfill, agricultural use, composting, incineration, other). Sludge management in the top five European producers according to the most recent data from Eurostat, 2022

biodegradability, and chemical oxygen demand (COD) removal efficiency) on biological treatment as a consequence of the stimulation in extracellular respiration by enhancement of iron reduction bacteria (Shi et al., 2017; Wang et al., 2022).

Proper treatment of industrial sludge leads to a decrease in greenhouse gas emissions and facilitates the control of the carbon cycle, thus eliminating the need for external carbon sources employed in nitrogen removal processes.

4 The Circular Economy Approach

Sludge valorization is a route of reusing, recycling, and converting residues into more beneficial and valued products. Several industries produce substantial amounts of wastewater during production processes. Treating these effluents results in the production of substantial amounts of sludge, marked by elevated levels of organic matter, nutrients, pathogens, and other contaminants. Inadequate disposal of this sludge could lead to environmental repercussions (Ávila-Pozo et al., 2021). Research has explored utilization of sludge derived from the activated sludge process for various purposes, including as a nutrient source for energy crops, in walnut plantations, for biodiesel production, for biogas generation, in tile manufacturing, for promotion of crop growth, and for production of stabilized compounds for agricultural applications (Agabo-García et al., 2022; Ferreira et al., 2018; Martins et al., 2022; Ozdemir et al., 2020; Pirmoradi et al., 2021), and from other pollutants owing to its greater adsorption capacity (Yan et al., 2014). In this vein, activated sludge and anaerobic digestion, the two most widely used conventional methods to treat sewage and industrial wastewater, are biological processes. The former process involves the following steps: aeration, settling, return, and discharge. This treatment effectively removes organic pollutants and nutrients from wastewater, producing a treated effluent suitable for release into the environment. The latter process breaks down organic matter in the absence of oxygen. It is commonly used for treating organic solid waste, sewage sludge, and industrial organic wastes. The key stages are as follows: hydrolysis, acidogenesis, methanogenesis, and biogas production. The last two steps are important when circular economy is considered owing to the conversion of organic acids and volatile fatty acids into methane gas, which may be collected and used as renewable energy. Moreover, the digested sludge can be used as a fertilizer, as commented below, thus reducing the volume of waste that needs to be disposed of in landfills. Both activated sludge and anaerobic digestion are important components of wastewater and organic waste treatment, contributing to environmental protection and resource recovery.

In Fig. 3.2, a visual representation highlights the distinctions between linear and circular economy models.

The recovery of metallic coagulants can be achieved through sludge acidification, but this approach is both intricate and expensive, particularly for heavily contaminated sludge (Chakraborty et al., 2020). Madeira et al. (2023) studied the feasibility of reusing sludge from the instant one-step lime precipitation stage to treat

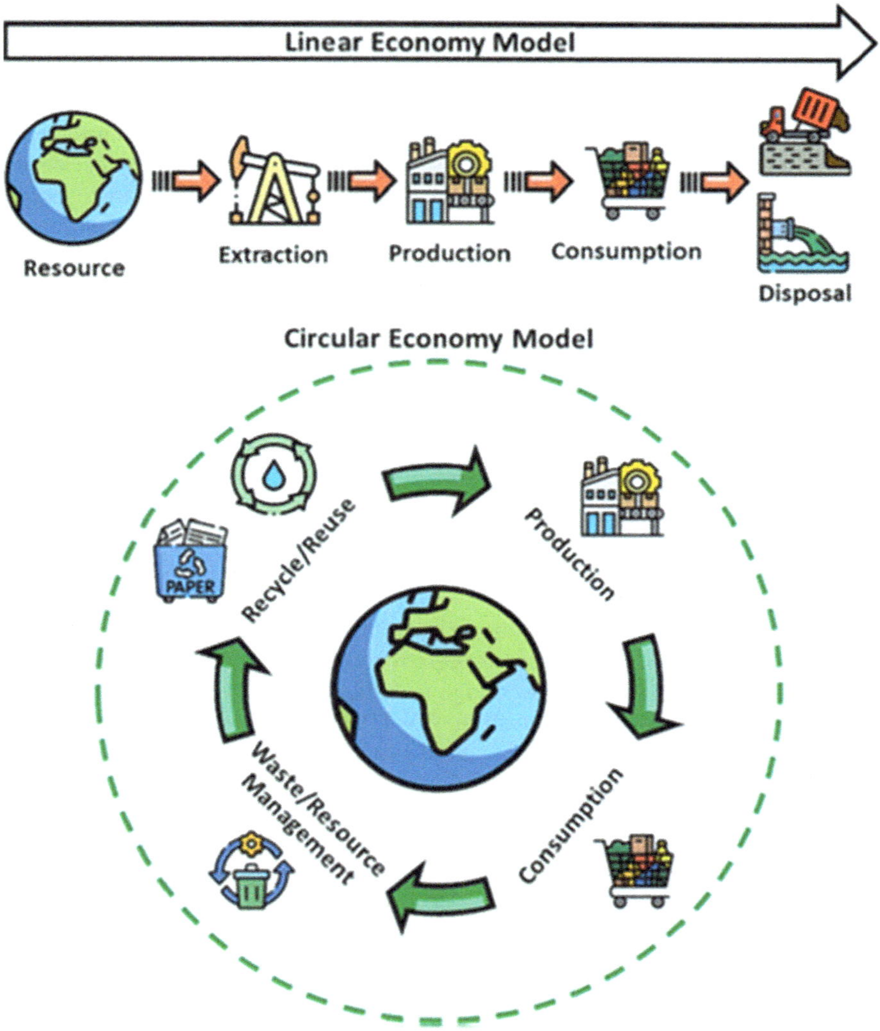

Fig. 3.2 A comparative scheme of the linear and circular economy models. (From Mannina et al. (2022))

slaughterhouse wastewater. The authors assessed four different techniques of sludge reuse, concluding that the increase in phosphorus levels is attributed to the limit of reusability. The utilization of lime sludge produced during the water treatment softening process has been examined for its potential as a reagent, both directly and indirectly, to counteract acidity in industrial wastewater from food processing (van Leeuwen et al., 2011), enhancing the settlement of sludge flocs avoiding the use pH adjustment reagents for phosphorous removal (Bal Krishna et al., 2017).

In the chemical coagulation technique, aluminum and iron salts are frequently used as the coagulating agents in WWTPs; therefore, waste (alum/iron) sludge is generated in large volumes. Some attempts have been made to reuse this type of sludge, such as resources for building and construction or land-based usage (Erdogmus et al., 2021; Shamaki et al., 2021; Wolff et al., 2015; Zhao et al., 2016). In addition, aluminum and iron have an affinity for phosphorus species and other contaminants, even emerging contaminants, owing to their large specific surface area and porosity (Hu et al., 2011). In this line, Kang et al. studied aluminum sludge in liquid form as a primary coagulant to treat wastewater from animal farms. In the treatment of actual wastewater from an animal farm, aluminum-based water treatment sludge was employed as an alternative to conventional chemicals as a coagulant in the coagulation–flocculation process. The authors' findings suggest that this sludge can be effectively reused in the treatment process, thus achieving significant removal efficiencies cost-efficiently (Kang et al., 2022).

5 Resource Recovery from Wastewater Treatment Plants (WWTPs)

The waste by-products generated during water treatment present a dual aspect: they can pose challenges or serve as prospects for procuring valuable raw materials. Within the European Union, raw sludge and substances like fats, oil, and grease sourced from municipal and industrial WWTPs are prohibited from disposal in natural or managed settings. Hence, these materials need to undergo processing, including stabilization and drying, before they can be repurposed as fertilizers, construction materials, or alternative fuel sources suitable for co-incineration in high-temperature furnaces, such as those in power plants or concrete production facilities (Bratina et al., 2016). The economic attractiveness of various by-products depends on the presence or absence of pollutants. As highlighted by Bratina et al. (2016), SS that contains coliform bacteria, viruses, helminth eggs, or limited amounts of heavy metals may not be suitable for use as a fertilizer; however, it can still serve as a viable fuel source. This section resumes, under the CE approach, valuable resources that can be recovered from biosolids/sludge after wastewater treatment. Figure 3.3 illustrates the exploration of circular economy approaches in wastewater management.

5.1 *Waste to Energy (WtE)*

Most SS derived from WWTPs is disposed of in landfills, incinerated, or repurposed for agricultural use as a soil amendment. Effective management and reutilization strategies need to be enhanced to minimize the waste volume destined for disposal.

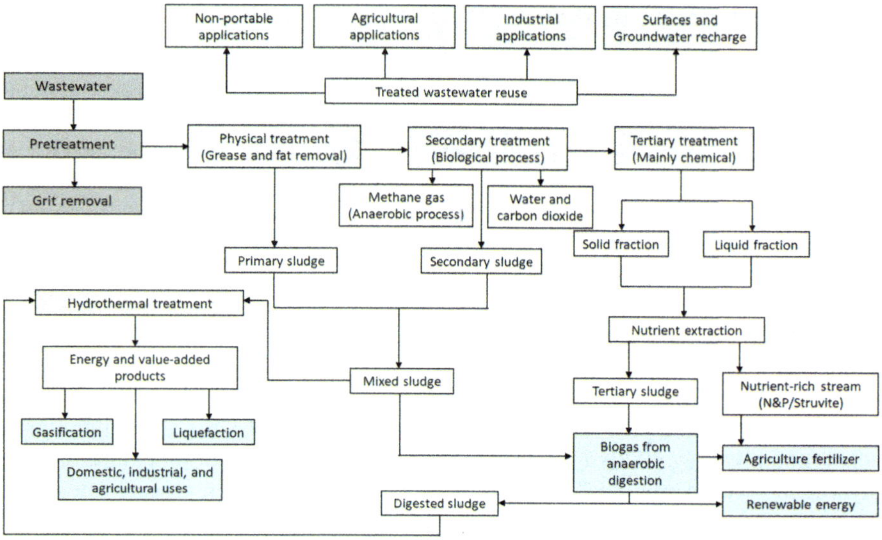

Fig. 3.3 Resource recovery and potential reuse routes (circular economy) between conventional wastewater treatment processes and emerging alternatives in water, energy, and value-added products. (Adapted from Semaha et al. (2023))

This involves converting the sludge into heat and electricity or generating advanced fuels in solid, liquid, and gaseous forms (waste to energy, WtE) (Dong et al., 2019). Thermal valorization seems to be an evident approach to generating renewable energy by pyrolysis (thermal degradation of organic substances at temperatures ranging from moderate to high, reaching up to 973.15 K under anaerobic conditions). The process can be carried out at either a slow or a high heating rate. The former is implemented to reduce waste volume, thereby maximizing biochar production.

Conversely, the latter is employed to produce liquid bio-oil with higher efficiency (Kan et al., 2016); gasification (at higher temperatures in an oxygen-deficient environment, up to 1173.15 K to obtain a fuel in a liquid state (syngas)) (Migliaccio et al., 2021) or combustion/incineration (at the highest temperatures, >1173.15 K in the presence of oxygen) is used to produce heat, which can be used for eliminating harmful substances, thus reducing sludge volume. Besides, the conversion of carbonaceous materials into mainly carbon dioxide and water and substantial heat generation are also achieved (Makarichi et al., 2018). When choosing each technology, it is crucial to consider three vital factors: energy provision, sustainability, and geographical placement (Ronda et al., 2023).

Numerous alternatives exist for material valorization, including repurposing of materials for construction purposes, low-cost adsorbents, or material recovery, as commented below.

5.2 Construction Materials

The dependence on plastics derived from petroleum-based sources has considerably risen in present years. Unfortunately, abundant plastics negatively affect coastal and marine ecosystems worldwide. New biodegradable materials, such as polyhydroxyalkanoates (PHAs), are produced via the bacterial fermentation of carbon-based feedstock under nutrient-limited growth conditions. These "green" plastics are environmentally friendly owing to low landfill requirements, compared with petroleum-based plastics (Goh et al., 2022). A primary drawback of PHA production is its reliance on costly, pure carbon sources, such as glucose (Aslan et al., 2016). The latter concern could be addressed by utilizing wastewater sludge as the feedstock for PHA production. However, the demand for bioplastics is currently limited due to their comparatively higher market price when compared to petrochemical-based polymers. This factor makes bioplastics less competitive globally (Crutchik et al., 2020). Goh et al. optimized the conceptual design of the carbon–hydrogen–oxygen symbiosis network (CHOSYN), focusing on process integration to improve the economic and environmental sustainability of the PHA production process. Their analysis led to the recommendation of expanding the CHOSYN framework to include nitrogen when the feedstock possesses high protein content.

Although specific conventional techniques employed in crafting road and highway pavement blends exhibit efficiency and longevity, they come at the cost of substantial raw material usage, heightened energy consumption, emission of harmful gases into the environment, and logistical safety concerns (Amoori Kadhim et al., 2022). In this line, new developments for making road pavements as well as enhancing mechanical and durability properties have been studied (Shanbara et al., 2021). An example is incorporating cementitious fillers comprising ordinary Portland cement, ground-granulated blast furnace slag, and fly ash within cold mix asphalt. Several benefits of implementing these novel technologies include enhancing mechanical properties, decreasing curing time and temperature, eliminating carbon dioxide emissions during production, promoting the formation of more rigid hydration products, strengthening the interaction between the asphalt matrix and aggregates, and minimizing the presence of air voids (Lu et al., 2021). Al Nageim et al. investigated the performance of conventional cold mix asphalt using both ordinary Portland cement and secondary cementitious material using fly and tail ash produced from burning waste sludge. The latter is used as a replacement of regular limestone fillers (Al Nageim et al., 2023). The results showed that the mechanical and durability features, quantified through the indirect tensile stiffness modulus values, were significantly higher than those performed using the conventional method. The same results were observed in rutting (permanent deformation) resistance. However, these results must be corroborated through practical road section trials involving on-site exposure to traffic and environmental conditions.

5.3 Activated Carbon/Adsorbents

SS rich in moisture is a by-product of wastewater treatment, typically encompassing various CECs as well as heavy metals, microplastics, poly- and perfluoroalkyl substances (PFASs), polybrominated diphenyl ethers (PBDEs), and polycyclic aromatic hydrocarbons (PAHs). The latter is the leading cause of why it is not helpful as a fertilizer (migration of contaminants). Therefore, this sludge contributes to significant environmental and economic issues for governments due to excessive operational investing and advanced methods (Li et al., 2019). Most traditional processes in WWTPs, i.e., coagulation, flocculation, filtration, and ultraviolet disinfection, are not designed to eliminate CECs, which efficiently accumulate in the generated sludge.

Producing activated carbon from SS is a novel advancement studied as a comprehensive and sustainable CE strategy. Activated carbon may capture CECs and heavy metals from WWTP effluents (Li et al., 2020) and stabilize during sludge pyrolysis. After that, the by-products, such as bio-oil, can be used for energy recovery. Figure 3.4 provides an illustrative overview of the various technologies employed in the retrieval of material resources and energy from wastewater sludge.

Biochar, derived from the pyrolysis of organic biomass, holds significant promise as a versatile material that is applicable in agriculture, the environment, and energy sectors. It effectively removes diverse mineral and organic pollutants from wastewater, including inorganic/organic compounds, heavy metals, and CECs. Wang et al. studied a pilot-scale process to reclaim carbon from primary sludge using acidogenic fermentation for 5 days. Then, it was employed for denitrification within an industrial WWTP specializing in wool processing (Wang et al., 2023a, b).

In actual wastewater, the presence of persistent pollutants competes with dissolved organic matter, inorganic nutrients, and other micropollutants for

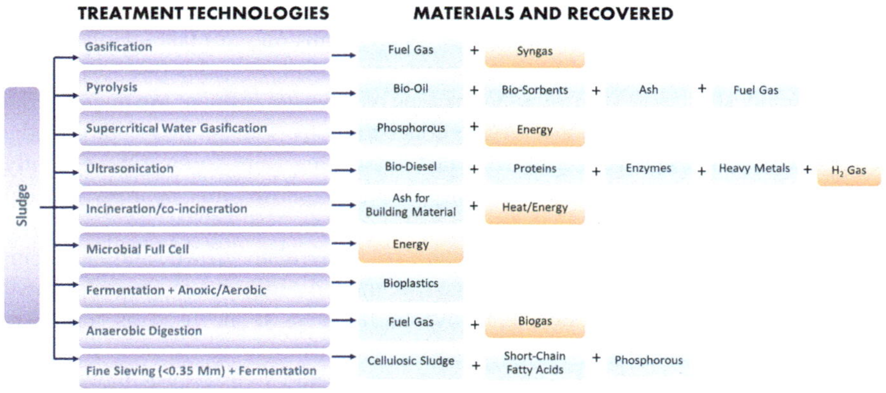

Fig. 3.4 Technologies used for the recovery of material resources and energy from wastewater sludge. (From Gherghel et al. (2019))

adsorption sites. This can result in pore blockage or sluggish pore diffusion, reducing removal efficiency (Shimabuku et al., 2017). Mohamed et al. discussed and evaluated the efficiency of using commercial and municipal sludge-based (secondary treatment, WWTP) activated carbon to remove some emerging contaminants (PFASs, PBDEs, and PAHs) in both warm and cold seasons. Their results showed a percentage removal of PAHs and alkyl-PAHs of 88–100% in the effluents from warm season (Mohamed et al., 2023). Therefore, biosolids can remove organic pollutants, thus diminishing their harmful effects on the aquatic environment. Jellali et al. addressed the sustainable and CE aspects related to industrial sludge. Their focus was on transforming raw industrial sludge into valuable biochar, which could be utilized as an adsorbent for emerging pollutants, particularly pharmaceuticals, in synthetic wastewater. This transformation improved the structural, textural, and surface chemical properties, enhancing adsorption capacities. Additionally, the authors employed biosolids for the removal of amoxicillin (Jellali et al., 2023).

Rumky et al. explored the possibility of valorizing sludge by converting it in municipal WWTPs, located in Finland, into sludge-based adsorbents to assess the recovery of rare earth elements from low-concentration aqueous solutions. Hydrochloric acid and Fenton reagents were used as a priori treatment to improve the surface area and adsorption production. The prepared beads were analyzed to recover, among others, cerium (Ce), lanthanum (La), europium (Eu), scandium (Sc), praseodymium (Pr), gadolinium (Gd), terbium (Tb), yttrium (Y), dysprosium (Dy), samarium (Sm), erbium (Er), holmium (Ho), ytterbium (Yb), and thulium (Tm). The sludge demonstrated a stronger affinity for Sm^{3+} ions with a maximum adsorption capacity of 4.16 mg g^{-1} (Rumky et al., 2022). This confirms that sludge can be used as an attractive adsorbent and as an alternative in reuse strategies.

5.4 Nutrient Recovery

Phosphorus holds crucial significance for all life-forms on our planet. Nonetheless, being a nonrenewable resource, its demand encounters obstacles due to potential future supply shortages. According to the type of industry, effluent treatment sludge contains high phosphorus and nitrogen concentrations. Phosphorus within sludge comprises a combination of organic and precipitated inorganic forms. Consequently, waste sludge has emerged as a potent biological phosphorus source. Through pyrolysis, specific by-products are generated, resulting in nutrient-rich biochar that serves as an advantageous soil conditioner, fostering plant growth and enhancing crop production. This element and nitrogen are the most important nutrients (growth of living organisms) that may be recovered from SS to be used as fertilizers. Moreover, these resources are useful for distinct industrial processes.

Soil has a bigger potential to moderate climate change. In this sense, bio-sludge is an attractive alternative in the fertilizer structure. However, recycling can become complicated due to the coexisting contaminants and economic viability. Zhang et al. developed a comprehensive review of viable methods for reclaiming phosphorus from wastewater and SS. Vivianite ($Fe_3(PO_4)_2{\cdot}8H_2O$) is a phosphorous value-added product obtained after convenient process operation. The authors conducted a comprehensive review of both qualitative and quantitative techniques for detecting vivianite, encompassing ion exchange, electrodialysis, capacitive deionization, membrane bioreactor, anaerobic fermentation, chemical precipitation, electrochemical crystallization, biomineralization, and anaerobic digestion. Furthermore, the mechanisms and common applications of each method were extensively explored. The authors concluded that vivianite production offers notable benefits compared to conventional phosphorus recovery approaches, including minimal chemical expenses, elevated recovery efficiency, and the generation of value-added products (Zhang et al., 2022).

Through chemical or electrocoagulation processes, inorganic flocculants like iron and aluminum salts can facilitate reactions in which phosphate interacts with these elements to yield various minerals, including variscite ($AlPO_4{\cdot}2H_2O$), vivianite ($Fe_3(PO_4)_2{\cdot}8H_2O$), and strengite ($FePO_4{\cdot}2H_2O$). Additionally, during the coking process, the release of calcium results in an increase of calcium ions in coking wastewater. The transformation of Fe-P and Al-P to Ca-P can occur through pyrolysis (Ren et al., 2021). Furthermore, phosphorus recovery from coking sludge was achieved through pyrolysis using a standardized measurement testing protocol. The outcomes indicated that phosphoric substances within coking sludge underwent enrichment and transformation, leading to the formation of apatite phosphorus compounds such as $Ca_5(PO_4)_3OH$, $Ca_5(PO_4)_3Cl$, and $Ca_3(PO_4)_2$ (Ban et al., 2023). This finding confirms the feasibility of converting sludge into bioavailable products, specifically phosphorus-rich biochar. Notably, the levels of heavy metals in the biochar remained well below the national standard limits. This study underscores the transformation of hazardous waste, exemplified by coking sludge, into a resource with a reduced environmental impact.

Kaljunen et al. investigated the adsorptive loading of phosphorus from chemical wastewater sludge onto various biosolid materials, aiming to utilize them sustainably as fertilizer compounds. They assessed four biosolids (biochar, sludge char, lignin, and humus extract from black liquor (paper manufacturing process)) (Kaljunen et al., 2022). Their results showed that sludge char had the highest P-content under optimal conditions of acidic pH and an adsorbent concentration between 0.5 and 0.61 g L^{-1} for all biomaterials. The maximum adsorption capacity values achieved are around 400–500 mg g^{-1} with dependency on temperature. Phosphorus adsorbed on the biosolids was particularly PO_4^{2-}–bound with iron and calcium cations. Fe-containing sludge improves phosphorous adsorption, thus converting the biosolid into a valuable fertilizer. Binding phosphorous with a carbon source results in a route to recycle phosphorous, improving nutrient preservation and soil health. Pellets comprising these nutrients, including potassium, have been produced (Khan et al., 2008).

5.5 Energy Recovery

The utilization of renewable energy sources has surged due to society's heightened demand, driven primarily by concerns over environmental impacts caused by hydrocarbon-containing materials. In this section, some methods of analyzing the generation of environmentally friendly renewable energy from waste residues are described.

Biofuels sourced from third-generation feedstocks, including wastewater, exhibit greater sustainability. Microalgal biomass is harnessed to generate diverse products, encompassing biodiesel, bioethanol, biohydrogen, biogas, bio-oils, bioplastics, biofertilizers, biochar, and biopesticides. Rodrigues et al. explored the utilization of biomass derived from domestic and industrial effluent treatment facilities alongside the bio-oils generated through the fast pyrolysis technique. The authors performed a complete characterization of the raw materials. Copious compounds based on oxygen, sulfur, and nitrogen (a mixture of biomolecules such as proteins, lipids, and nucleic acids) were observed in bio-oils, making them inappropriate for use as renewable fuels given that highly reactive gases like nitrogen and sulfur oxides could be released during combustion. Nevertheless, functionalized alkyl chains showed that bio-oils might be a basis for high value-added by-products (fertilizers, surfactants, and nitrogen solvents) (Rodrigues et al., 2023). The hydrothermal carbonization (HTC) process can potentially enhance SS's characteristics, transforming it into a solid fuel that closely resembles coal (Fakkaew et al., 2018). This technique enables the utilization of biomass feedstock without the need for a preliminary drying treatment, making it particularly well-suited for biomass with a high moisture content. The HTC process yields three distinct product components: (i) solid hydro-char, which finds applications in energy generation, soil enrichment for nutrient recovery and carbon sequestration, removal of heavy metals and pesticides from water, and enhancement of composting and fermentation outcomes (Sun et al., 2020); (ii) a liquid product that can serve as a source for nutrient recovery; and (iii) a gas primarily composed of CO_2 (Aragón-Briceño et al., 2021). In HTC, the operational factors encompass a solid-to-liquid ratio, temperature, and reaction time.

It is well-established that elevated temperatures result in increased carbonization levels, consequently yielding higher quantities of liquid and gaseous products while reducing the formation of solid products (Khoo et al., 2020). Paiboonudomkarn et al. examined the viability of SS derived from brewery wastewater treatment as a biomass feedstock for the HTC process. They investigated the influence of the solid-to-liquid ratio and feedwater pH on the characteristics of the resulting solid fuel. The outcomes indicated that pH levels had the potential to reduce ash content and elevate fixed carbon content (Paiboonudomkarn et al., 2023).

Another significant global industry is pulp and paper manufacturing. In addition to utilizing diverse wood and plant materials, this sector extensively employs water and chemicals throughout various production stages, generating substantial volumes of wastewater (Faubert et al., 2016). A mixed sludge is generated following the treatment procedures, comprising constituents such as wood fibers, papermaking fillers, pitch, and lignin by-products. This sludge contributes to leaching pollution and the production of greenhouse gases. Consequently, incineration of high-water-content sludges becomes imperative. Nevertheless, efficient mechanical and thermal dewatering, as in direct incineration, accounts for a significant portion of operational expenses. Incorporating the biorefinery concept is a promising alternative solution to addressing this drawback (dos Santos et al., 2023). Hämäläinen et al. explored the potential of HTC in managing pulp and paper industry sludge and enhancing its transformation into value-added products. The authors conducted lab-scale HTC experiments on mechanically dewatered mixed sludge, varying parameters such as total solids content, temperature, and residence time (Hämäläinen et al., 2022). The energy content augmented up to 20.5 MJ kg^{-1} (initial value 14.9 MJ kg-1). Related to methane, the generated filtrates had 3–5-fold higher volumetric methane production.

The "sludge-to-energy" approach emerges as a favorable strategy offering numerous advantages. These encompass a significant enhancement in sludge stabilization and reduction in sludge volume. Moreover, this approach facilitates organic matter conversion into biogas while yielding various value-added products such as soil conditioners and fertilizers (Abdel Wahaab et al., 2020). Finally, the study of processes that promote the removal of nitrogen, sulfur, and oxygen atoms from molecules to obtain bio-oils that can be used as alternative fuels should be of interest.

5.6 *Reuse of Treated Wastewater*

As widely recognized, water stands as the fundamental necessity for supporting life in all its manifestations on Earth. The processes of industrialization and urbanization constitute the origins of water pollution, which subsequently gives rise to health hazards affecting all living beings. Given the escalating public consciousness and the stringent enforcement of environmental regulations, a requirement for harnessing wastewater—a renewable and reclaimable resource—has emerged. This resource holds potential as a source of drinking water, energy, nutrients, and other crucial applications (Angelakis et al., 2018). Kumar and Sharma investigated the CE process for treating paper production wastewater. This treated water was assessed to uncover its appropriateness for reuse in farming. The authors observed that reclaimed water was appropriate for germination of mung beans. Furthermore, they deduced that sludge can serve as a compelling supplement not only for fostering plant growth but also for energy recovery and diverse industrial applications (Kumar & Sharma, 2022).

6 Facing Challenges

Implementing sludge management in WWTPs has enabled overcoming the general challenges regarding pollution reduction, ecosystem protection, and public health welfare. However, addressing specific wastewater treatment processes to drive innovation and technological advancements remains challenging. It is necessary to enhance the joint performance of WWTP treatment trains by making processes more economical and ecological while also enabling the adoption and commercialization of advanced technologies such as EAOPs. Despite using activated sludge technology to redirect carbon into energy, maximizing energy recovery from the organic load remains a challenge, such as with biogas production. Finally, lack of government support can hinder addressing wastewater issues in industries and rural communities. Rural areas often face challenges in constructing and operating wastewater treatment plants.

7 Future Perspectives

The future perspectives of wastewater management aim to integrate the circular economy concept, where wastewater would be reused for simpler processes or even for the original process that generated it. Other future perspectives are directed toward:

- Advanced treatment technologies, such as membrane bioreactors, electrocoagulation, and constructed wetlands, can effectively remove complex contaminants and reach higher purification standards in industrial wastewater treatment.
- Solar-powered photocatalytic processes and membrane-based separation methods are projected effective and sustainable alternatives. These methods have shown promising results in providing efficient and environmentally friendly solutions.
- Decentralized wastewater treatment systems can be used in areas without a centralized WWTP infrastructure. The integration of green infrastructure, such as constructed wetlands and green roofs, can help reduce sludge volume and improve wastewater quality.
- Pretreatment of sewage sludge with various additives during co-composting improves waste treatment efficiency.
- Policies to regulate micropollutant discharge and improve wastewater treatment for environmental sustainability should be developed.

In the near future, wastewater management will require technological innovation, regulatory compliance, and individual and collective action to reduce water usage and minimize the environmental impact of wastewater discharge.

8 Conclusions

This chapter shows that there has been an evolution in scientific research dedicated to the administration/reuse of SS generated after wastewater treatment. Nevertheless, this research remains insufficient in various regions worldwide due to inadequate financial resources and constrained areas. In most WWTPs, the absence of sludge stabilization facilities further exacerbates the situation. In addition, wastewater treatment facilities need better technologies for correctly eliminating CECs. Moreover, research on finding strategies that allow mitigate environmental impacts and understand the potential of recycling/reuse benefits is necessary.

The process typically involves initial steps of sludge dewatering and thickening in the present landscape of sludge treatment and disposal. Subsequently, the treated sludge is commonly traded to external vendors for its utilization in agricultural applications. Another avenue pursued is the stabilization of sludge, accomplished through one of several technical approaches: mesophilic anaerobic digestion, co-digestion of sludge with other organic waste streams under anaerobic conditions, windrow composting, and storage in lagoons.

To resume, several important topics deserve attention beyond research, including (i) the study of CECs at different water matrices; (ii) the assessment of disinfection effectiveness and the study of the toxicity of SS; (iii) advancements in technology, which should prioritize solutions that are not only straightforward to implement but also cost-effective while effectively mitigating the creation of undesirable hazardous by-products with a primary focus on wastewater pretreatment; (iv) the evaluation of the possibility of decreasing sludge and energy requirements by CE and 3Rs (reducing waste by reusing and recycling resources and by-products) approaches; and (v) an increase in university–industry–government interactions (technicians' education).

Finally, the presence of microplastics is barely being studied, so research on this type of contaminant should be rapidly addressed to reduce the contamination in water and soil. Furthermore, waste management must be executed to optimize productivity and maximize the overall yield of resource recovery. Achieving this goal requires the establishment of suitable infrastructures, substantial investments in research, and effective project leadership to facilitate well-integrated processes for biofuel production and value-added material production. It is crucial to emphasize that creating highly value-added products through electrochemical methods presents many advantages spanning energy efficiency, economic viability, and environmental considerations.

References

Abdel Wahaab, R., Mahmoud, M., & van Lier, J. B. (2020). Toward achieving sustainable management of municipal wastewater sludge in Egypt: The current status and future prospective. *Renewable and Sustainable Energy Reviews, 127*, 109880. https://doi.org/10.1016/j.rser.2020.109880

Agabo-García, C., Solera, R., & Perez, M. (2022). Anaerobic sequential batch reactor for CO-DIGESTION of slaughterhouse residues: Wastewater and activated sludge. *Energy, 255*, 124575. https://doi.org/10.1016/j.energy.2022.124575

Al Nageim, H., Dulaimi, A., Al-Busaltan, S., Kadhim, M. A., Al-Khuzaie, A., Seton, L., Croft, J., & Drake, J. (2023). The development of an eco-friendly cold mix asphalt using wastewater sludge ash. *Journal of Environmental Management, 329*, 117015. https://doi.org/10.1016/j.jenvman.2022.117015

Amann, A., Weber, N., Krampe, J., Rechberger, H., Peer, S., Zessner, M., & Zoboli, O. (2022). Systematic data-driven exploration of Austrian wastewater and sludge treatment - implications for phosphorus governance, costs and environment. *Science of the Total Environment, 846*, 157401. https://doi.org/10.1016/j.scitotenv.2022.157401

Amoori Kadhim, M., Al-Busaltan, S., Dulaimi, A., Sadique, M., Al Nageim, H., Al-Kafaji, M., & Al-Yasari, R. (2022). Developing a sustainable, post treated, half warm mix asphalt for structural surface layer. *Construction and Building Materials, 342*, 127926. https://doi.org/10.1016/j.conbuildmat.2022.127926

Angelakis, A. N., Asano, T., Bahri, A., Jimenez, B. E., & Tchobanoglous, G. (2018). Water reuse: From ancient to modern times and the future. *Frontiers in Environmental Science, 6*, 26. https://doi.org/10.3389/fenvs.2018.00026

Aragón-Briceño, C. I., Pozarlik, A. K., Bramer, E. A., Niedzwiecki, L., Pawlak-Kruczek, H., & Brem, G. (2021). Hydrothermal carbonization of wet biomass from nitrogen and phosphorus approach: A review. *Renewable Energy, 171*, 401–415. https://doi.org/10.1016/j.renene.2021.02.109

Aslan, A. K. H. N., Ali, M. D. M., Morad, N. A., & Tamunaidu, P. (2016). Polyhydroxyalkanoates production from waste biomass. *IOP Conference Series: Earth and Environmental Science, 36*(1), 12040. https://doi.org/10.1088/1755-1315/36/1/012040

Ávila-Pozo, P., Parrado, J., Caballero, P., Díaz-López, M., Bastida, F., & Tejada, M. (2021). Use of Slaughterhouse Sludge in the bioremediation of an Oxyfluorfen-Polluted soil. *International Journal of Environmental Research, 15*(4), 723–731. https://doi.org/10.1007/s41742-021-00351-z

Awasthi, M. K., Ganeshan, P., Gohil, N., Kumar, V., Singh, V., Rajendran, K., Harirchi, S., Solanki, M. K., Sindhu, R., Binod, P., Zhang, Z., & Taherzadeh, M. J. (2023). Advanced approaches for resource recovery from wastewater and activated sludge: A review. *Bioresource Technology, 384*, 129250. https://doi.org/10.1016/j.biortech.2023.129250

Bal Krishna, K. C., Niaz, M. R., Sarker, D. C., & Jansen, T. (2017). Phosphorous removal from aqueous solution can be enhanced through the calcination of lime sludge. *Journal of Environmental Management, 200*, 359–365. https://doi.org/10.1016/j.jenvman.2017.06.003

Ban, Z., Wei, T., Ke, X., Chen, A., Guan, X., Chen, Y., Qiu, G., Wei, C., Wu, H., Li, F., Peng, Y., & Zhao, W. (2023). Fate and distribution of phosphorus in coking wastewater treatment: From sludge to its derived biochar. *Science of the Total Environment, 881*, 163384. https://doi.org/10.1016/j.scitotenv.2023.163384

Bratina, B., Šorgo, A., Kramberger, J., Ajdnik, U., Zemljič, L. F., Ekart, J., & Šafarič, R. (2016). From municipal/industrial wastewater sludge and FOG to fertilizer: A proposal for economic sustainable sludge management. *Journal of Environmental Management, 183*, 1009–1025. https://doi.org/10.1016/j.jenvman.2016.09.063

Chakraborty, T., Balusani, D., Smith, S., Santoro, D., Walton, J., Nakhla, G., & Ray, M. B. (2020). Reusability of recovered iron coagulant from primary municipal sludge and its impact on chemically enhanced primary treatment. *Separation and Purification Technology, 231*, 115894. https://doi.org/10.1016/j.seppur.2019.115894

Chandra, R., & Kumar, V. (2017). Detection of androgenic-mutagenic compounds and potential autochthonous bacterial communities during in-situ bioremediation of post methanated distillery sludge. *Frontiers in Microbiology, 8*, 87. https://doi.org/10.3389/fmicb.2017.00887

Cheng, X., Wei, C., Ke, X., Pan, J., Wei, G., Chen, Y., Wei, C., Li, F., & Preis, S. (2022). Nationwide review of heavy metals in municipal sludge wastewater treatment plants in China: Sources, composition, accumulation and risk assessment. *Journal of Hazardous Materials, 437*, 129267. https://doi.org/10.1016/j.jhazmat.2022.129267

Clematis, D., Delucchi, M., & Panizza, M. (2023). Electrochemical technologies for wastewater treatment at pilot plant scale. *Current Opinion in Electrochemistry, 37*(1), 101172. https://doi.org/10.1016/j.coelec.2022.101172

Crutchik, D., Franchi, O., Caminos, L., Jeison, D., Belmonte, M., Pedrouso, A., del Rio, A., Mosquera-Corral, A., & Campos, J. L. (2020). Polyhydroxyalkanoates (PHAs) production: A feasible economic option for the treatment of sewage sludge in municipal wastewater treatment plants? *Water, 12*(4). https://doi.org/10.3390/w12041118

Dong, J., Tang, Y., Nzihou, A., & Chi, Y. (2019). Key factors influencing the environmental performance of pyrolysis, gasification and incineration waste-to-energy technologies. *Energy Conversion and Management, 196*, 497–512. https://doi.org/10.1016/j.enconman.2019.06.016

dos Santos, E. V., Martínez-Huitle, C. A., & Rodrigo, M. A. (2023). The electro-refinery in organics: A new arising concept for valorization of wastes. *Current Opinion in Electrochemistry, 39*, 101267. https://doi.org/10.1016/j.coelec.2023.101267

Elalami, D., Carrere, H., Monlau, F., Abdelouahdi, K., Oukarroum, A., & Barakat, A. (2019). Pretreatment and co-digestion of wastewater sludge for biogas production: Recent research advances and trends. *Renewable and Sustainable Energy Reviews, 114*, 109287. https://doi.org/10.1016/j.rser.2019.109287

Erdogmus, E., Harja, M., Gencel, O., Sutcu, M., & Yaras, A. (2021). New construction materials synthesized from water treatment sludge and fired clay brick wastes. *Journal of Building Engineering, 42*, 102471. https://doi.org/10.1016/j.jobe.2021.102471

Eurostat - Your key to European statistics. (2022). www document, URL: https://ec.europa.eu/eurostat/databrowser/view/ten00030/default/table?lang=en

Fakkaew, K., Koottatep, T., & Polprasert, C. (2018). Faecal sludge treatment and utilization by hydrothermal carbonization. *Journal of Environmental Management, 216*, 421–426. https://doi.org/10.1016/j.jenvman.2017.09.031

Faubert, P., Barnabé, S., Bouchard, S., Côté, R., & Villeneuve, C. (2016). Pulp and paper mill sludge management practices: What are the challenges to assess the impacts on greenhouse gas emissions? *Resources, Conservation and Recycling, 108*, 107–133. https://doi.org/10.1016/j.resconrec.2016.01.007

Ferreira, A., Marques, P., Ribeiro, B., Assemany, P., de Mendonça, H. V., Barata, A., Oliveira, A. C., Reis, A., Pinheiro, H. M., & Gouveia, L. (2018). Combining biotechnology with circular bioeconomy: From poultry, swine, cattle, brewery, dairy and urban wastewaters to biohydrogen. *Environmental Research, 164*, 32–38. https://doi.org/10.1016/j.envres.2018.02.007

Franco, A. A., Martín-García, A. P., Egea-Corbacho, A., Arellano, J. M., Albendín, G., Rodríguez-Barroso, R., Quiroga, J. M., & Coello, M. D. (2023). Assessment and accumulation of microplastics in sewage sludge at wastewater treatment plants located in Cádiz, Spain. *Environmental Pollution, 317*, 120689. https://doi.org/10.1016/j.envpol.2022.120689

Fytili, D., & Zabaniotou, A. (2008). Utilization of sewage sludge in EU application of old and new methods—A review. *Renewable and Sustainable Energy Reviews, 12*(1), 116–140. https://doi.org/10.1016/j.rser.2006.05.014

Gherghel, A., Teodosiu, C., & De Gisi, S. (2019). A review on wastewater sludge valorisation and its challenges in the context of circular economy. *Journal of Cleaner Production, 228*, 244–263. https://doi.org/10.1016/j.jclepro.2019.04.240

Goh, Q. H., Farouk, A. A., & Chew, I. M. L. (2022). Optimizing the bioplastic chemical building block with wastewater sludge as the feedstock using carbon-hydrogen-oxygen framework. *Resources, Conservation and Recycling, 176*, 105920. https://doi.org/10.1016/j.resconrec.2021.105920

Hämäläinen, A., Kokko, M., Kinnunen, V., Hilli, T., & Rintala, J. (2022). Hydrothermal carbonization of pulp and paper industry wastewater treatment sludges - characterization and potential use of hydrochars and filtrates. *Bioresource Technology, 355*, 127258. https://doi.org/10.1016/j.biortech.2022.127258

Hu, Y. S., Zhao, Y. Q., & Sorohan, B. (2011). Removal of glyphosate from aqueous environment by adsorption using water industrial residual. *Desalination, 271*(1), 150–156. https://doi.org/10.1016/j.desal.2010.12.014

Iqbal, A., Zan, F., Siddiqui, M. A., Nizamuddin, S., & Chen, G. (2022). Integrated treatment of food waste with wastewater and sewage sludge: Energy and carbon footprint analysis with economic implications. *Science of the Total Environment, 825*, 154052. https://doi.org/10.1016/j.scitotenv.2022.154052

Jellali, S., Khiari, B., Al-Harrasi, M., Charabi, Y., Al-Sabahi, J., Al-Abri, M., Usman, M., Al-Raeesi, A., & Jeguirim, M. (2023). Industrial sludge conversion into biochar and reuse in the context of circular economy: Impact of pre-modification processes on pharmaceuticals removal from aqueous solutions. *Sustainable Chemistry and Pharmacy, 33*, 101114. https://doi.org/10.1016/j.scp.2023.101114

Jiang, X., Lyu, Q., Bi, L., Liu, Y., Xie, Y., Ji, G., Huan, C., Xu, L., & Yan, Z. (2022). Improvement of sewage sludge anaerobic digestion through synergistic effect combined trace elements enhancer with enzyme pretreatment and microbial community response. *Chemosphere, 286*, 131356. https://doi.org/10.1016/j.chemosphere.2021.131356

Jimenez, B., Barrios, J. A., Mendez, J. M., & Diaz, J. (2004). Sustainable sludge management in developing countries. *Water Science and Technology: A Journal of the International Association on Water Pollution Research, 49*(10), 251–258.

Jit, B., Chakraborty, A., & Sehgal, R. (2023). A systematic review of industrial wastewater management: Evaluating challenges and enablers. *Journal of Environmental Management, 348*, 119230. https://doi.org/10.1016/j.jenvman.2023.119230

Kaljunen, J. U., Yazdani, R., Al-Juboori, R. A., Zborowski, C., Meinander, K., & Mikola, A. (2022). Adsorptive behavior of phosphorus onto recycled waste biosolids after being acid leached from wastewater sludge. *Chemical Engineering Journal Advances, 11*, 100329. https://doi.org/10.1016/j.ceja.2022.100329

Kan, T., Strezov, V., & Evans, T. J. (2016). Lignocellulosic biomass pyrolysis: A review of product properties and effects of pyrolysis parameters. *Renewable and Sustainable Energy Reviews, 57*, 1126–1140. https://doi.org/10.1016/j.rser.2015.12.185

Kang, C., Zhao, Y., Tang, C., & Addo-Bankas, O. (2022). Use of aluminum-based water treatment sludge as coagulant for animal farm wastewater treatment. *Journal of Water Process Engineering, 46*, 102645. https://doi.org/10.1016/j.jwpe.2022.102645

Khan, M. A., Kim, K.-W., Mingzhi, W., Lim, B.-K., Lee, W.-H., & Lee, J.-Y. (2008). Nutrient-impregnated charcoal: An environmentally friendly slow-release fertilizer. *The Environmentalist, 28*(3), 231–235. https://doi.org/10.1007/s10669-007-9133-5

Khoo, C. G., Lam, M. K., Mohamed, A. R., & Lee, K. T. (2020). Hydrochar production from high-ash low-lipid microalgal biomass via hydrothermal carbonization: Effects of operational parameters and products characterization. *Environmental Research, 188*, 109828. https://doi.org/10.1016/j.envres.2020.109828

Koul, B., Yadav, D., Singh, S., Kumar, M., & Song, M. (2022). Insights into the Domestic Wastewater Treatment (DWWT) regimes: A review. *Water (Switzerland), 14*(21). https://doi.org/10.3390/w14213542

Kumar, D., & Sharma, C. (2022). Paper industry wastewater treatment by electrocoagulation and aspect of sludge management. *Journal of Cleaner Production, 360*(131970), 1–10. https://doi.org/10.1016/j.jclepro.2022.131970

Kumar, V., & Thakur, I. S. (2020). Extraction of lipids and production of biodiesel from secondary tannery sludge by in situ transesterification. *Bioresource Technology Reports, 11*, 100446. https://doi.org/10.1016/j.biteb.2020.100446

Kumar, V., & Verma, P. (2023). A critical review on environmental risk and toxic hazards of refractory pollutants discharged in chlorolignin waste of pulp and paper mills and their remediation approaches for environmental safety. *Environmental Research, 236*, 116728. https://doi.org/10.1016/j.envres.2023.116728

Kumar, V., Thakur, I. S., & Shah, M. P. (2020). Bioremediation approaches for pulp and paper industry wastewater treatment: Recent advances and challenges. In M. P. Shah (Ed.), *Microbial bioremediation & biodegradation*. Springer. https://doi.org/10.1007/978-981-15-1812-6_1

Kumar, V., Shahi, S. K., Ferreira, L. F. R., Bilal, M., Biswas, J. K., & Bulgariu, L. (2021a). Detection and characterization of refractory organic and inorganic pollutants discharged in biomethanated distillery effluent and their phytotoxicity, cytotoxicity, and genotoxicity assessment using *Phaseolus aureus* L. and *Allium cepa* L. *Environmental Research, 201*, 111551. https://doi.org/10.1016/j.envres.2021.111551

Kumar, V., Ferreira, L. F. R., Sonkar, M., & Singh, J. (2021b). Phytoextraction of heavy metals and ultrastructural changes of *Ricinus communis* L. grown on complex organometallic sludge discharged from alcohol distillery. *Environmental Technology & Innovation, 22*, 101382. https://doi.org/10.1016/j.eti.2021.101382

Kumar, V., Srivastava, S., & Thakur, I. S. (2021c). Enhanced recovery of polyhydroxyalkanoates from secondary wastewater sludge of sewage treatment plant: Analysis and process parameters optimization. *Bioresource Technology Reports, 15*, 100783. https://doi.org/10.1016/j.biteb.2021.100783

Kumar, V., Ameen, F., Islam, M. A., Agrawal, S., Motghare, A., Dey, A., Shah, M. P., Américo-Pinheiro, J. H. P., Singh, S., & Ramamurthy, P. C. (2022). Evaluation of cytotoxicity and genotoxicity effects of refractory pollutants of untreated and biomethanated distillery effluent using *Allium cepa*. *Environmental Pollution, 300*, 118975. https://doi.org/10.1016/j.envpol.2022.118975

Li, L. Y., Gong, X., & Abida, O. (2019). Waste-to-resources: Exploratory surface modification of sludge-based activated carbon by nitric acid for heavy metal adsorption. *Waste Management, 87*, 375–386. https://doi.org/10.1016/j.wasman.2019.02.019

Li, F., Chen, J., Hu, X., He, F., Bean, E., Tsang, D. C. W., Ok, Y. S., & Gao, B. (2020). Applications of carbonaceous adsorbents in the remediation of polycyclic aromatic hydrocarbon-contaminated sediments: A review. *Journal of Cleaner Production, 255*, 120263. https://doi.org/10.1016/j.jclepro.2020.120263

Lu, D., Wang, Y., Leng, Z., & Zhong, J. (2021). Influence of ternary blended cementitious fillers in a cold mix asphalt mixture. *Journal of Cleaner Production, 318*, 128421. https://doi.org/10.1016/j.jclepro.2021.128421

Madeira, L., Ribau, M., Almeida, A., & Santos, T. (2023). Reuse of lime sludge from immediate one-step lime precipitation process as a coagulant (aid) in slaughterhouse wastewater treatment. *Journal of Environmental Management, 342*, 118278. https://doi.org/10.1016/j.jenvman.2023.118278

Makarichi, L., Jutidamrongphan, W., & Techato, K. (2018). The evolution of waste-to-energy incineration: A review. *Renewable and Sustainable Energy Reviews, 91*, 812–821. https://doi.org/10.1016/j.rser.2018.04.088

Mannina, G., Gulhan, H., & Ni, B. J. (2022). Water reuse from wastewater treatment: The transition towards circular economy in the water sector. *Bioresource Technology, 363*, 127951. https://doi.org/10.1016/j.biortech.2022.127951

Martínez-Gallardo, M. R., Jurado, M. M., López-González, J. A., Toribio, A., Suárez-Estrella, F., Sáez, J. A., Moral, R., Andreu-Rodríguez, F. J., & López, M. J. (2022). Biorecovery of olive mill wastewater sludge from evaporation ponds. *Journal of Environmental Management, 319*, 115647. https://doi.org/10.1016/j.jenvman.2022.115647

Martins, G. A., Corrêa, L. B., Guidoni, L. L. C., Lucia, T., Gerber, M. D., Silva, F. M. R., & Corrêa, E. K. (2022). Toxicity and physicochemical parameters of composts including distinct residues from agribusiness and slaughterhouse sludge. *Waste Management, 138*, 75–82. https://doi.org/10.1016/j.wasman.2021.11.032

Migliaccio, R., Brachi, P., Montagnaro, F., Papa, S., Tavano, A., Montesarchio, P., Ruoppolo, G., & Urciuolo, M. (2021). Sewage sludge gasification in a fluidized bed: Experimental investigation and modeling. *Industrial & Engineering Chemistry Research, 60*(13), 5034–5047. https://doi.org/10.1021/acs.iecr.1c00084

Mohamed, B. A., Hamid, H., Montoya-Bautista, C. V., & Li, L. Y. (2023). Circular economy in wastewater treatment plants: Treatment of contaminants of emerging concerns (CECs) in effluent using sludge-based activated carbon. *Journal of Cleaner Production, 389*, 136095. https://doi.org/10.1016/j.jclepro.2023.136095

Ozdemir, S., Yetilmezsoy, K., Nuhoglu, N. N., Dede, O. H., & Turp, S. M. (2020). Effects of poultry abattoir sludge amendment on feedstock composition, energy content, and combustion emissions of giant reed (Arundo donax L.). *Journal of King Saud University - Science, 32*(1), 149–155. https://doi.org/10.1016/j.jksus.2018.04.002

Paiboonudomkarn, S., Wantala, K., Lubphoo, Y., & Khunphonoi, R. (2023). Conversion of sewage sludge from industrial wastewater treatment to solid fuel through hydrothermal carbonization process. *Materials Today: Proceedings, 75*, 85–90. https://doi.org/10.1016/j.matpr.2022.11.107

Pirmoradi, N., Ghaneian, M. T., Ehrampoush, M. H., Salmani, M. H., & Hatami, B. (2021). The conversion of poultry slaughterhouse wastewater sludge into biodiesel: Process modeling and optimization. *Renewable Energy, 178*, 1236–1249. https://doi.org/10.1016/j.renene.2021.07.016

Raheem, A., Sikarwar, V. S., He, J., Dastyar, W., Dionysiou, D. D., Wang, W., & Zhao, M. (2018). Opportunities and challenges in sustainable treatment and resource reuse of sewage sludge: A review. *Chemical Engineering Journal, 337*, 616–641. https://doi.org/10.1016/j.cej.2017.12.149

Ren, Z., Zeng, W., Liu, H., Jia, Y., Peng, X., & Peng, Y. (2021). Enhanced bioavailability of phosphorus in sewage sludge through pyrolysis aided by calcined clam shell powder. *Process Safety and Environmental Protection, 156*, 121–133. https://doi.org/10.1016/j.psep.2021.10.008

Rodrigues, J. L., Camp, J. D. M., Wisniewski, A., Hantao, L. W., Eberlin, M. N., & Santos, J. M. (2023). Chemical evaluation of pyrolysis oils from domestic and industrial effluent treatment station sludges with perspective to produce value-added products. *Waste Management, 168*, 202–210. https://doi.org/10.1016/j.wasman.2023.06.004

Ronda, A., Haro, P., & Gómez-Barea, A. (2023). Sustainability assessment of alternative waste-to-energy technologies for the management of sewage sludge. *Waste Management, 159*, 52–62. https://doi.org/10.1016/j.wasman.2023.01.025

Rumky, J., Deb, A., Ramasamy, D. L., Sillanpää, M., Häkkinen, A., & Repo, E. (2022). Utilization of sludge-based alginate beads for the application of rare earth elements (REEs) recovery from wastewater: A waste to resource approach. *Journal of Cleaner Production, 362*, 132496. https://doi.org/10.1016/j.jclepro.2022.132496

Semaha, P., Lei, Z., Yuan, T., Zhang, Z., & Shimizu, K. (2023). Transition of biological wastewater treatment from flocculent activated sludge to granular sludge systems towards circular economy. *Bioresource Technology Reports, 21*, 101294. https://doi.org/10.1016/j.biteb.2022.101294

Shamaki, M., Adu-Amankwah, S., & Black, L. (2021). Reuse of UK alum water treatment sludge in cement-based materials. *Construction and Building Materials, 275*, 122047. https://doi.org/10.1016/j.conbuildmat.2020.122047

Shanbara, H. K., Dulaimi, A., Al-Mansoori, T., Al-Busaltan, S., Herez, M., Sadique, M., & Abdel-Wahed, T. (2021). The future of eco-friendly cold mix asphalt. *Renewable and Sustainable Energy Reviews, 149*, 111318. https://doi.org/10.1016/j.rser.2021.111318

Shi, Y., Huang, J., Zeng, G., Gu, Y., Chen, Y., Hu, Y., Tang, B., Zhou, J., Yang, Y., & Shi, L. (2017). Exploiting extracellular polymeric substances (EPS) controlling strategies for performance enhancement of biological wastewater treatments: An overview. *Chemosphere, 180*, 396–411. https://doi.org/10.1016/j.chemosphere.2017.04.042

Shimabuku, K. K., Kennedy, A. M., Mulhern, R. E., & Summers, R. S. (2017). Evaluating activated carbon adsorption of dissolved organic matter and micropollutants using fluorescence spectroscopy. *Environmental Science & Technology, 51*(5), 2676–2684. https://doi.org/10.1021/acs.est.6b04911

Sun, X., Atiyeh, H. K., Li, M., & Chen, Y. (2020). Biochar facilitated bioprocessing and biorefinery for productions of biofuel and chemicals: A review. *Bioresource Technology, 295*, 122252. https://doi.org/10.1016/j.biortech.2019.122252

Tyagi, V. K., & Lo, S.-L. (2013). Sludge: A waste or renewable source for energy and resources recovery? *Renewable and Sustainable Energy Reviews, 25*, 708–728. https://doi.org/10.1016/j.rser.2013.05.029

van Leeuwen, J., White, D., Baker, R. J., & Jones, C. H. (2011). Reuse of water treatment residuals from lime softening, Part I: Applications for the reuse of lime sludge from water softening. *Land Contamination & Reclamation*. https://doi.org/10.2462/09670513.1012

Wang, Y., Wang, H., Jin, H., Zhou, X., & Chen, H. (2022). Application of Fenton sludge coupled hydrolysis acidification in pretreatment of wastewater containing PVA: Performance and mechanisms. *Journal of Environmental Management, 304*, 114305. https://doi.org/10.1016/j.jenvman.2021.114305

Wang, Q., Li, X., Liu, W., Zhai, S., Xu, Q., Huan, C., Nie, S., Ouyang, Q., Wang, H., & Wang, A. (2023a). Carbon source recovery from waste sludge reduces greenhouse gas emissions in a pilot-scale industrial wastewater treatment plant. *Environmental Science and Ecotechnology, 14*, 100235. https://doi.org/10.1016/j.ese.2022.100235

Wang, Y., Zheng, K., Guo, H., Tian, L., He, Y., Wang, X., Zhu, T., Sun, P., & Liu, Y. (2023b). Potassium permanganate-based advanced oxidation processes for wastewater decontamination and sludge treatment: A review. *Chemical Engineering Journal, 452*, 139529. https://doi.org/10.1016/j.cej.2022.139529

Wolff, E., Schwabe, W. K., & Conceição, S. V. (2015). Utilization of water treatment plant sludge in structural ceramics. *Journal of Cleaner Production, 96*, 282–289. https://doi.org/10.1016/j.jclepro.2014.06.018

Yadav, S., & Kamsonlian, S. (2022). A review of electrochemical methods for treatment of wastewater. *Materials Today: Proceedings, 78*, 36–39. https://doi.org/10.1016/j.matpr.2022.11.036

Yan, Y., Sun, X., Ma, F., Li, J., Shen, J., Han, W., Liu, X., & Wang, L. (2014). Removal of phosphate from etching wastewater by calcined alkaline residue: Batch and column studies. *Journal of the Taiwan Institute of Chemical Engineers, 45*(4), 1709–1716. https://doi.org/10.1016/j.jtice.2013.12.023

Zhang, J., Chen, Z., Liu, Y., Wei, W., & Ni, B. J. (2022). Phosphorus recovery from wastewater and sewage sludge as vivianite. *Journal of Cleaner Production, 370*, 133439. https://doi.org/10.1016/j.jclepro.2022.133439

Zhao, Y., Ren, B., O'Brien, A., & O'Toole, S. (2016). Using alum sludge for clay brick: An Irish investigation. *International Journal of Environmental Studies, 73*, 719–730. https://api.semanticscholar.org/CorpusID:101633231

Part II
Waste Recycling, Reuse, and Resource Recovery

Chapter 4
Sewage Sludge: Is It a Sustainable Fertilizer or a Source of Contaminants?

Mehedi Hasan, Md. Rakib Hasan, Ratul Ahmed Khan, and Taslim Ur Rashid

1 Introduction

In recent years, management of sewage sludge has become a major concerning issue in municipal wastewater treatment plant (WWTP), since it is the principal byproduct of these plants. And municipal wastewater treatment produces large amounts of sewage sludge everyday all around the world. The annual production of sewage sludge in the world now exceeds 45 million tons of dry solid (Ferrentino et al., 2023). But the production of sewage sludge is increasing day by day since the volume of waste water generation from domestic, industrial, and commercial sources is increasing day by day due to rapid urbanization, population, and improved living conditions (Kumar et al., 2021). Statistically saying, global waste water production is anticipated to rise by 24% by 2030 and 51% by 2050 relative to current data (Qadir et al., 2020).

Sewage sludge may contain organic materials, inorganic materials, microorganisms (pathogenic bacteria, viruses, fungi, and parasite eggs), nutrients (nitrogen, potassium, and phosphorus) and heavy metals (Dume et al., 2022; Rusănescu et al., 2022). Most of the sewage sludge disposed by incineration, landfilling, and 80% of wastewater is released into waterways without being treated (Mukawa et al., 2022), which cause significant environmental pollution and disrupt ecology since it contains pathogens and heavy metals. Various treatment methods including composting, anaerobic digestion, chemical treatments, and thermal treatments have been developed to remove pathogens and heavy metals from sludge in order to use it for different purposes such as fertilizer to save environment from this detrimental

M. Hasan · M. R. Hasan · R. A. Khan · T. U. Rashid (✉)
Department of Applied Chemistry and Chemical Engineering, Faculty of Engineering and Technology, University of Dhaka, Dhaka, Bangladesh
e-mail: taslim@du.ac.bd

V. Kumar et al. (eds.), *Environmental Engineering and Waste Management*,
https://doi.org/10.1007/978-3-031-58441-1_4

pollution (Fijalkowski et al., 2017; Khan et al., 2016; Menéndez et al., 2004; Smoliński et al., 2019).

The following nutrients found in sludges are helpful for fertilization: organic matter, fertilizing elements (nitrogen, potassium, and phosphorus), and microelements (iron, manganese, copper, and molybdenum) required for plants (Rusănescu et al., 2022). It improves soil fertility and physical, chemical, and biological properties of soil (Hussein, 2009). Besides using as fertilizer, it has some other important uses such as biogas that can be produced from sewage sludge to produce electricity (Mukawa et al., 2022). *Sewage sludge* ash (SSA) has potential usage as additive in cement (Goswami et al., 2022). Since the sludge has high calorific value, it is also used as a potential source of renewable energy after stabilizing it by anaerobic digestion (Menéndez et al., 2004; Smoliński et al., 2019). Many wastewater treatment plants (WWTPs) utilize its dewatered waste (5% moisture) by composting under anaerobic condition along with green wastes or other bulking agents to transform it into product suitable for use as fertilizer or fuel (Fijalkowski et al., 2017).

Sewage sludge is a type of "biological cocktail" that contains a variety of different organisms, including pathogens and saprophytes (Fijalkowski et al., 2017), which cause diseases making it harmful to humans, animals, and plants. Heavy metals are the most prevalent inorganic pollutants in sewage sludge. Lead (Pb), cadmium (Cd), nickel (Ni), chromium (Cr), copper (Cu), and zinc (Zn) are common metals detected in sewage sludge (Usman et al., 2012b). One of the issues with these metals is that they cannot be destroyed, unlike other contaminants. As a result, they enter into the food chain by agricultural use and can be consumed and digested by plants and animals, causing heavy metals to bio-accumulate which increases health risk to animals and plants (Sinha & Paul, 2013). Hydrogen peroxide with phosphoric acid at an 8% concentration demonstrated effective heavy metal removal rates that were comparable to those of 1 N hydrochloric acid (Yoshizaki & Tomida, 2000). Other widely used methods such as chemical precipitation, membrane separation, flotation, adsorption, ion exchange separation, and electrochemical deposition can also be employed (Akter et al., 2021; Barakat, 2011; Biswas et al., 2021, 2020). For pathogen inactivation, disinfection techniques like chlorination, ozonation, and ultraviolet (UV) radiation are typically necessary (Deepnarain et al., 2020).

Sewage sludge fertilizer presents a valuable opportunity to sustainably manage waste sewage sludge while also has application in intensive farming as it can improve soil health and condition by supplying essential nutrient contents such as nitrogen, phosphorus, and potassium offering a substitute for synthetic fertilizers (Muter et al., 2022b). However, concerns for risk remain as it contains potential contaminants which impact dangerously on human and environmental wellbeing if not properly treated and managed. These persistent contaminants including a variety of inorganic and organic chemicals, pharmaceuticals, pathogens, and heavy metals (Seleiman et al., 2020) that present in sewage sludge may accumulate over time and leach into ground and surface water, endangering ecosystems (Seleiman et al., 2020). Sewage sludge pollutants also pose risks to farm workers and enter into the food chain (Suthar & Singh, 2009).

While from a waste management and agricultural point of view, sewage sludge fertilizer offers sustainability and benefit, sufficient treatment processes should be implemented to mitigate contaminant risks to human, animal, and environmental health. Strict governance with adequate management skill is required to forecast and observe long-term impacts of sewage sludge fertilizer application and to take preventative steps to limit contamination level below threshold limit for environment. With such controlled regulation and monitoring, sewage sludge fertilizer may help to erase waste disposal problems and to advance us toward a circular bio-economy. This chapter will discuss about characterization and comparative study about sewage sludge about its nutritional and pollutant contents with their possible pathway to pollute environment. Some widely used methodologies that effectively convert the waste sewage sludge to valuable resources such as energy and fertilizer are illustrated and analyzed. In addition, this chapter also demonstrates the suitability and feasibility of using sewage sludge as fertilizer. Future direction along with the challenges associated with its high contaminant contents that could potentially disturb ecology and social acceptance is also considered, and possible solutions are outlined.

2 Sewage Sludge

Sewage sludge (biosolid) is a mud-like semi-liquid or semi-solid substance composed of organic as well as inorganic components which is a heterogeneous material created as a byproduct of a wastewater treatment plant (WWTP) (Gray, 2010). Basically, sediments that are collected on the bottom of either the primary or secondary clarifier of a wastewater treatment plant are collectively called sewage sludge. Other than gravitational settling and microbiological transformation, different chemicals might also be added to these clarifiers to increase sedimentation which leads to the formation of the chemical waste. Raw sludge, which contains generally 97–98% water, can undergo different treatments of its own and account for about 65% of the operating cost of a typical WWTP. So, finding out possible applications and upcycling sewage sludge are probably the ways to justify the management cost. Again, any beneficial form of sewage sludge after treatment is referred to as biosolids (Mathney & Dunn, 2011).

2.1 Characteristics of Sewage Sludge

Sludge characteristics can greatly vary depending on the type of discharge from a WWTP. Unless there are separate drainage systems for municipal wastewater, industrial wastewater, and rainwater runoffs, it is very difficult to generalize sewage sludge characteristics (Feng et al., 2015). Not only industry-wise wastewater, even locality-wise municipal wastewater can also give variable compositions of organic and inorganic materials in sewage sludge from a typical WWTP (Dai et al., 2006). Besides, the different treatment processes of sewage sludge give rise to different characteristics of it (Spellman, 2013). However, some physicochemical accounts of raw, activated, or treated sludge are given in the literature.

Typical activated sewage sludge can contain 75% free water which can be removed by simple gravitational settling, 20% floc water trapped within the floc, 2% capillary water adhered to individual particles, 2.5% particle water chemically bound to individual particles, and the rest 0.5% of solids (Spellman, 2013). Raw (untreated) sewage sludge typically consists of 2.0–8.0% total dry solids (TSs), 60–80% of TS comprises volatile solids (VSs), grease and lipids, proteins, nitrogen, phosphate, potassium, cellulose, iron, silica, and organic acids (mg/L as HAc) and alkalinity (mg/L as $CaCO_3$) (Tchobanoglus et al., 2003). Concentration of solids in sewage sludge as colloid, suspension, or dissolved form gives rise to viscous yield stress and causes to show non-Newtonian fluid characteristics which is very important to know for sludge handling (Forster, 2002).

2.1.1 Major Nutrients Content

The result of sewage sludge analysis shows that major nutrients were present in very high concentrations and ranges, e.g., available nitrogen (0.46–0.63%), phosphorous (0.0044–0.0060%), potassium (0.029–0.041%), and total nitrogen (1.54–1.92%), phosphorous (0.61–0.92%), potassium (0.35–0.43) (Dubey et al., 2006). Nitrogen can be found in sewage sludge in mainly two distinct forms such as organic (amino acids, proteins, nucleic acids, etc.) and inorganic (ammonium, nitrate, and nitrite). For phosphorous, it can also be classified into two parts—organic (phospholipids, nucleic acids, and phosphoproteins) and inorganic (orthophosphate, polyphosphates, and phosphonates). Potassium is not as abundant in sewage sludge as nitrogen and phosphorus; it is still an important nutrient for plant growth and can contribute to the fertilizing value of the sludge when used as a soil amendment. It is present in the form of amino acids, nucleotides, etc.

2.1.2 Heavy Metals and Pathogens

Sewage sludge frequently contains elevated levels of heavy metals like lead, cadmium, and chromium originated from industrial discharge, infrastructure, and environmental contamination. The heavy metals found in sewage sludge from wastewater treatment plant of different countries are listed in Table 4.1. The major metal types found in sludge include those that are soluble, precipitate, co-precipitate in metal oxides, adsorbates, and linked with biological residues (Lester, 1983). The distribution of metals among the various forms varies considerably depending on the chemical characteristics of each metal and those of the sludge, which are a result of the physical and chemical constraints imposed by the specific sludge treatment procedure (Gould & Genetelli, 1978).

Table 4.1 Heavy metal content of sewage sludge found from wastewater treatment plant of region such as Europe, India, and Pakistan

Heavy metals	Range in the European Union member states (mg/kg in dry sewage sludge)	Range in India (mg/kg in dry sewage sludge)	Range in Pakistan (mg/kg in dry sewage sludge)
As	–	8–23	–
Fe	–	–	250
Mn	–	–	210
Cu	39–361	–	202
Zn	142–2000	–	640
Cd	0.4–3.8	2–9	
Cr	16–275	66–1098	
Hg	0.3–3	7–32	
Ni	9–90	12–596	
Pb	13–221	26–154	
References	Andersen (2001)	Dubey et al. (2006)	Jamil et al. (2006)

Sludge from sewage treatment plants contains a variety of pathogens and saprophytes, heterotrophic and chemautotrophic bacteria, saprophytes, and pathogens which make up the majority of the population and are the most diverse. Sewage sludge contains a complex mix of microbes, mainly bacteria that break down organic waste. All of these complex mixes of microbes make sludge a "biological cocktail" (Fijalkowski et al., 2017).

2.1.3 Organic Matter Content

Organic matter content depends on the source of the wastewater the treatment process used and the characteristics of the sludge itself. Organic matters comprise carbohydrates, proteins, organic acids, humus, etc. There are some undissolved organic components in the sewage sludge which are known as particulate material consisting of pulp fiber (lignin and cellulose), food waste, fecal matter, etc. The dewatered sludge contains organic matter (50–70%), minerals (30–50%), nitrogen (3–4%), phosphorus (1–3%), and other micronutrients. Also, dewatered sewage sludge contains 33.4% protein, 6.6% fat, and 3.3% carbohydrate on an organic basis (Inoue et al., 1996).

2.1.4 pH and Salinity

The sludge pH is almost neutral at approximately 7.1 ± 0.2 (Wei et al., 2019). This is because the organic matters and microbial activity in the sludge do not significantly impact pH, but during digestion and stabilization, the pH can increase as NH_3 is released from the breakdown of protein and other nitrogen-based compounds. Lime treatment during treatment significantly raises the pH to above neutrality.

2.2 Overview of Sewage Sludge Applications

Sewage sludge is an unavoidable, unwanted byproduct of wastewater treatment. Historically different approaches for sewage sludge application have been tried, and some of these have been already discarded by local and international standards. Nowadays, all the researches for sewage sludge applications are social, economic, and environmental sustainability centric (Donatello & Cheeseman, 2013). Thermal conversion processes like hydrothermal conversion, pyrolysis, incineration etc., are considered major sustainable approaches to recovering valuable resources from sewage sludge (Fan et al., 2022).

Incineration can be the most direct way to reclaim heat energy from sewage sludge, given that sufficient drying of the sludge is done to maintain the process temperature. Various heat-intensive industries like the cement industry or even a WWTP to dry the sewage sludge itself can count on incineration, granted that sufficient emission control is applied (Hurynovich et al., 2021; Lin et al., 2018; Tytła & Widziewicz-Rzońca, 2021). Besides, incinerated sewage sludge ash has found some use in land reclamation (Lin et al., 2018), ceramic tile (Fan et al., 2022), sintered bricks, cement manufacture (Donatello & Cheeseman, 2013), etc.

Pyrolysis of sewage sludge, like other lignocellulosic materials, can be done from the perspective of the production of solid adsorbents, syngas, and liquid fuels. While syngas and liquid fuels can be an alternative heat source, solid adsorbents from sewage sludge pyrolysis can be used to capture contaminants like H_2S, NO_x, metals, dyes, phenols, etc. (Fonts et al., 2012). Even, owing to high nitrogen and phosphorus content, sewage sludge pyrolysis product has the potential to be used as fertilizer (Frišták et al., 2018). Hydrothermal treatment is an alternative approach to pyrolysis and can also be done from the perspective of syngas, biocrude, and hydro-char production. One of the major advantages of hydrothermal treatment over pyrolysis is the unnecessity of drying the sewage sludge since water itself is a reactant in the process (Fan et al., 2022).

The application of sewage sludge as fertilizer is greatly backed up by its contents of essential nutrients such as nitrogen (N), phosphorus (P), and potassium (K), along with other micronutrients. Furthermore, the organic materials present in sludge can enhance soil fertility, promote beneficial microbial activity, increase water-holding capacity, and improve overall soil physical, biological, and chemical properties. More than 80% of mined phosphate rock is utilized for fertilizer, accounting for roughly half of global fertilizer consumption. The remainder of the P is derived from recycled organic waste, such as animal manure and sewage sludge (Ohtake & Tsuneda, 2018).

3 Environmental Impact of Sewage Sludge

The increasing globalization has led to the production of numerous sources of pollutants, which can have a substantial impact on the environment. Sewage sludge obtained from wastewater treatment can contain a large quantity of heavy metals and disease-containing pathogens which arise a significant environmental concern and can have a significant impact on biodiversity. The extent of environment pollution can vary depending on how it is managed and treated. Unsuitable disposal can lead to serious damage in ecosystem by interacting with water body, soil, and atmosphere. While the application of sewage sludge can have positive impacts on soil fertility and plant growth, it is important to consider the potential risks and ensure safe and sustainable use.

3.1 Water Pollution

Untreated sewage sludge is the major contributor for pollution in aquatic ecology (Owa, 2013). The biggest issue produced by putting raw sewage into water is that it promotes eutrophication (Fig. 4.1), the excessive nutrient richness in a lake or other

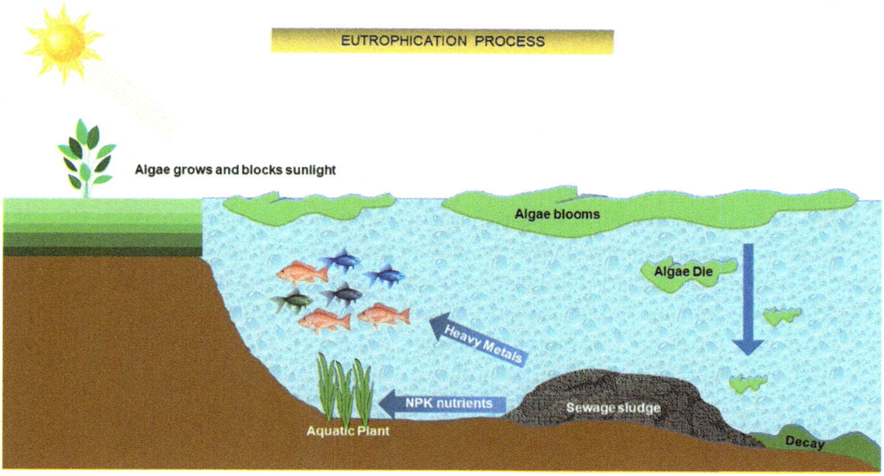

Fig. 4.1 Illustration of nutrients' pathway from sewage sludge to aquatic life. Nutrients from sewage sludge thrown into the water are helping the growth of aquatic plant algae which come to the surface of the water, form dense mats on the water surface, and thus block the sunlight from penetrating into deep water layer. This occurrence reduces the rate of photosynthesis which consequently reduces the production of oxygen. Some O_2 may be consumed during their decomposition process of algae blooms. Consequently, the total amount of dissolved oxygen in water declines significantly. All these eventually disrupt the ecology in aquatic system

body of water, usually caused by run-off from the land, resulting in dense plant growth (Chislock et al., 2013). If sewage sludge, without proper treatment, is directly thrown into water bodies like rivers, canals, lakes, or seas, it can affect the plants and animals that are part of the aquatic ecosystem; significant amounts of nutrients including nitrogen, phosphorus, potassium, and others nutrients are introduced into the water ecosystem that increases the aquatic plant and algae growth by increasing nutrient uptake which can eventually have detrimental effects such as algae bloom. Algae blooms are resulted from the consequence of eutrophication, which stimulates the rapid growth and accumulation of microscopic algae, cyanobacteria, and water-based phytoplankton (Sarkar, 2018). This imparts tastes and color to the water (Owa, 2013) and results in mat formation in the surface of water. Research about Australia's stagnant Murray–Darling River shows that a toxic algae bloom occurs on an about 600 km long segment (Wurtsbaugh et al., 2019) that hinders sunlight to penetrate, consequently reduces photosynthesis. Bacteria absorb oxygen as algal blooms rot and degrade, breaking down the organic materials (Van Berg, 1993), and total dissolved oxygen reduces. Hypoxia or anoxia, conditions marked by low oxygen levels in the water, may result from this (Van Berg, 1993). Thus, the ecology in the aquatic system is largely disrupted (Fig. 4.1).

Heavy metals including zinc, cadmium, mercury, lead, and copper are extremely dangerous to the aquatic ecology and its biota. Because fish are particularly vulnerable to heavy metal contamination, they accumulate in their tissues and endanger fish health (Authman, 2015). Heavy metal contamination is thought to be a poisonous factor for fish health. These metals can have a significant impact on a fish's vital functions and reproduction, as well as damage the immune system and cause pathological alterations. Scientists are particularly interested in their harmful effects in aquatic systems due to the health dangers they provide to live species and humans via the food chain.

The ability of sewage sludge to absorb various pollutants from different sources like pesticides, pharmaceuticals, and organic residues including micro-plastics from wastewater is responsible for potential environmental and health risks (Styszko et al., 2022). It may contain dangerous contaminants like germ and microorganisms such as pathogenic bacteria, fungi, viruses, and worm's eggs (Usman et al., 2012a). These germs can build up in aquatic organisms' tissues and cause serious health issues or even death. In developing countries, about 14,000 people die daily because of tainted drinking water which is polluted by dumping untreated sewage sludge (Owa, 2013).

3.2 *Soil Contamination*

Landfilling is one of the most common methods for disposing sewage sludge, but it can lead to the contamination of agricultural land with heavy metals. Sewage sludge can contain high levels of heavy metals, such as zinc, nickel, lead, cadmium, and mercury, which can be harmful to human health and the environment (Tytła &

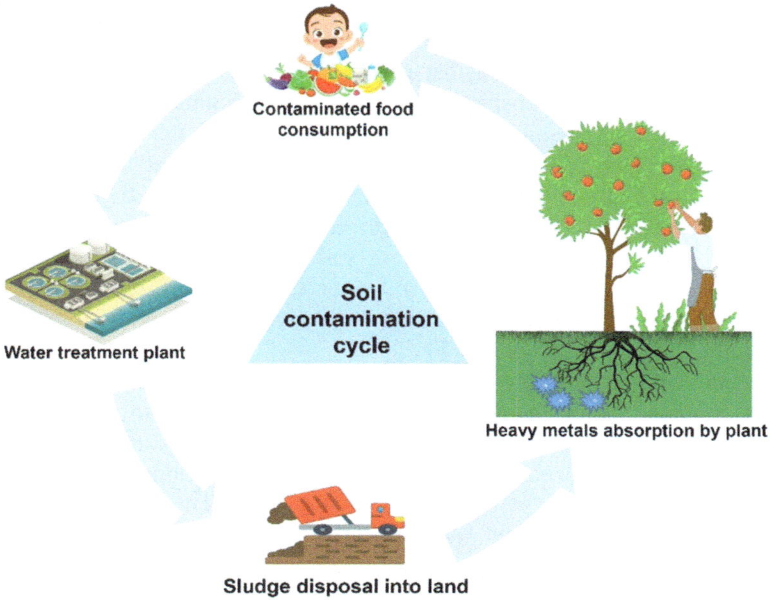

Fig. 4.2 Illustration of a cycle about inclusion of heavy metal in food chain due to soil contamination by sewage sludge dumping directly onto land without proper treatment. Toxic heavy metals permeate through the soil which is then taken up by plant root and subsequently consumed by animal or human. A portion of the heavy metals causes diseases in humans and plants. Other segment is expelled from living organism through excretion, subsequently finding its way back again into wastewater treatment facilities. Sewage sludge obtained from these treatment plants becomes tainted with heavy metals which again joins the cycle if again dumped onto land without proper treatment

Widziewicz-Rzońca, 2021). When sewage sludge is landfilled, these heavy metals can leach into the surrounding soil and water, contaminating agricultural land and groundwater. Heavy metal contamination can also disrupt the food chain, making it difficult for animals to find safe and uncontaminated food sources (Ngodhe, 2018). Dumping sewage sludge in agricultural land without proper treatment can lead to the inclusion of heavy metals in the food cycle, which can have significant environmental impacts (Fig. 4.2). Heavy metal absorption through plant roots is the primary route by which heavy metals enter the food chain, and their subsequent congestion up the food chain poses a serious hazard to animal and human health.

The ecological effects of heavy metal contamination can be significant. Due to bioaccumulation, certain metals may accumulate in the tissues of plants and animals and then enter human tissues, posing serious health risks to people. The food chain can be impacted by the metal that is present in the soil and plant system due to bioaccumulation. Extensive exposure to high levels of heavy metal is recognized to represent a grave health hazard, including significant damage to the vital organs, impairing neurological system, creating reproductive problems, and a risk of cancer that can be fatal (Ngodhe, 2018).

It is crucial to keep in mind that metals reach soils via a variety of sources, including air deposition, phosphate fertilizers, and agricultural chemicals, in large amounts when considering metal inputs from sewage sludge (Ngodhe, 2018). Heavy metals noticeably alter the physical, chemical, and biological properties of soil, including pH, the amount of organic matter present, and the activity of microbial communities. These metals interfere with crucial microbial processes and thus reduce the quantity as well as activity of soil microorganisms. All these factors lead to pose detrimental effects on the soil biota (Chen et al., 2010). The presence of heavy metals causes the reduction in both the biomass and diversity of the bacterial communities within soil (Ngodhe, 2018).

Another major contaminant for environment is microplastic, which is also present in sewage sludge [somewhere around 18–41 particles per gram (Corradini et al., 2019)] and subsequently pollutes the soil. Microplastics are tiny plastic particles smaller than 5 mm in size. They can be extracted from various sources, including small monomer molecules fossil (Zhang & Zhang, 2020). The functional connection between microbial activity and water-stabilized aggregates in soil is changed by microplastics. Furthermore, the presence of microplastics can alter bulk density and water holding capacity of soil (De Souza MacHado et al., 2018). The presence of microplastics in soil along with the other pollutants they adsorb, like heavy metals, antibiotics, and persistent organic compounds, can significantly threaten soil quality and soil organisms. The combined impact of microplastics and the pollutants they carry has the potential to cause serious harm to soil health and the organisms that live within the soil. The microplastics can act as a vector, transporting harmful pollutants into the soil environment and impacting both soil quality and soil organisms in negative ways. The combined effect of microplastics with adsorbed pollutants can pose serious harm to soil safety and soil organisms. It is important to address the issue of microplastic pollution in order to protect soil ecosystems and maintain soil health (Koyuncuoğlu & Erden, 2021).

3.3 Air Pollution

Anaerobic digestion occurs in solid wastes like sewage sludge which dumped in landfills. Due to this, a mixture of gases and odor are produced (Fig. 4.3) (Liu et al., 2016; Sonibare et al., 2019). Greenhouse gas like methane is emitted during anaerobic digestion of landfill wastes (Baldasano & Soriano, 2000). Thus, landfilling with sewage sludge could result in global warming. The surrounding atmosphere becomes particularly unpleasant because of the odor and other gas that landfilling produces. Besides volatile organic components (VOCs) are also released from landfilling of sewage sludge (Nair et al., 2019) .Volatile organic compounds can cause irritation in skin, throat, nose, eyes as well as other body part. Inhaling VOCs

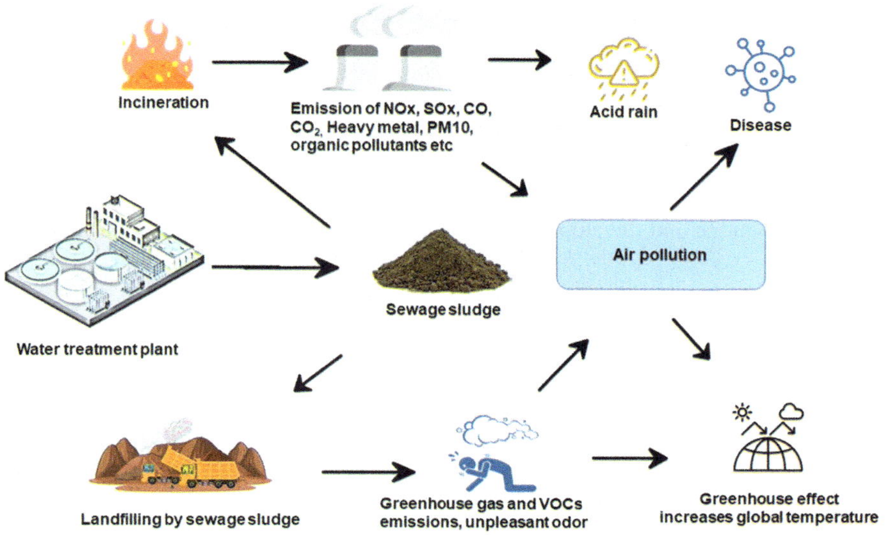

Fig. 4.3 The illustration depicts the route to air pollution due to two popular disposal method of sewage sludge such as landfilling and incineration. Landfilling of sewage sludge results in production of methane gas due to anaerobic digestion which triggers the greenhouse effect resulting elevation of global temperature. Another widely used method is incineration which releases toxic gas like oxides of nitrogen (NOx), oxides of sulfur (SOx), carbon dioxide (CO_2) as well as heavy metals, particulate matters which pollutes air and poses dangerous health issue

at high concentrations for a prolonged period of time causes lung functions to degrade over time, which ultimately leads to cancer (Soni et al., 2018).

Incineration is an efficient and universal approach for disposal of sewage sludge which involves reducing volume by combustion of sludge. During this process, organic pollutants and pathogens are killed. The energy released can be recovered (Liang et al., 2021). But this process has harmful effects on air, as it pollutes air by emission of NO_x, SO_2, particulate matters, heavy metals, methane, dioxins and furans, carbon monoxide (CO), carbon dioxide (CO_2), polychlorinated dibenzo-p-dioxins, polychlorinated dibenzofuran (PCDD/Fs), etc. during this process (Moon, 2019) (Fig. 4.3).

Wastewater treatment plants are acknowledged as a major source of greenhouse gas emissions as they produce sewage sludge emitting carbon dioxide (CO_2), methane (CH_4), and nitrous oxide (N_2O) in their incineration operations (Campos et al., 2016). The incineration of sewage sludge has led to the accumulation of greenhouse gas emissions, endangering humans, and the ecosystem through global climate change (Fig. 4.3). The effects of particulate matter on the environment and people are also concerning. During this incineration process, particulate matter less than 10 μm, PM_{10}, is produced. The current air pollution control devices (APCDs) cannot effectively capture particles with a diameter of less than 2.5 μm ($PM_{2.5}$).

PM_{10} is enriched with heavy metals and has the potential to enter the lungs and can linger in the atmosphere for a very long time (Ye et al., 2012). Short-term and long-term exposure to these numerous contaminants might cause health concerns. AHSMOG (Adventist Health and Smog) study found that PM_{10} causes respiratory deaths and lung cancer mortality. Realistic exposure may shorten life expectancy by 1–2 years compared to other risk factors (Landrigan, 2017). To reduce the number of deaths and make the air safer, we must stop polluting air by uncontrolled incineration of sewage sludge and develop alternative methods or modify incineration method.

4 Sewage Sludge Treatment for Fertilizer Production

Composition of sewage sludge shows that it contains a large proportion of three primary plant nutrients (N, P, and K) as well as organic matter, calcium, magnesium, etc. (Dubey et al., 2006). So instead of incineration or throwing it on the soil and water, we can recycle it and use it as fertilizer at relatively low cost which simultaneously solves the problems associated with sewage sludge disposal. But this practice may lead to some other problem if it is not pretreated properly. The reason for the above problems is the presence of heavy metals, pathogens, and other toxicants in sewage sludge-derived fertilizer (Usman et al., 2012b). To use sewage sludge as a fertilizer, we must solve the aforementioned problems that arise in both plant and soil by getting rid of these heavy metals, pathogens, and other toxicants present in sewage sludge. In recent years, various methods such as composting, anaerobic digestion, and thermochemical treatment had been developed to ensure the safe and uncontaminated use of sewage sludge as fertilizer.

4.1 Composting

Composting is a natural process where biological oxidative transformation takes place that is almost similar to the decomposition of organic matter that occurs in the soil (De Bertoldi et al., 1983). The name compost comes from the Latin word "compositum" which means mixture. In this process, a mixture of substrates being broken down by a microbial community that is made up of diverse populations (Insam & de Bertoldi, 2007). The sewage sludge mixture here provides food and habitat for naturally occurring microbes. The microbes through bio-oxidation then break down the mixture, digesting and decomposing it over time into a nutrient-rich fertilizer that can be used to nourish and feed plants (De Bertoldi et al., 1983; Pan et al., 2012). The composting process takes place in a solid-state environment and under aerobic circumstances (Insam & de Bertoldi, 2007). Sewage sludge undergoes physical and chemical alterations during this process, resulting in the development of a stable humified final product. Composting reduces the bioavailability of both organic pollutants and heavy metals (Kästner & Mahro, 1996; Wong & Selvam,

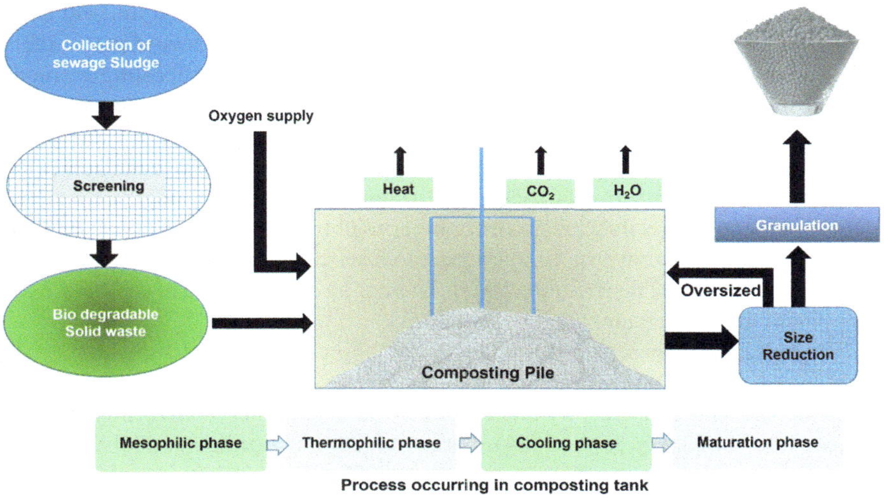

Fig. 4.4 Visual outlines of the process flow for environmentally friendly composting method of waste sewage sludge to produce useable fertilizer by undergoing four distinct phases

2006). Along with a mixture of other acceptable biodegradable wastes and additives, composting is a fruitful treatment process for the modification of sewage sludge to produce fertilizer (Anwar et al., 2015; Stylianou et al., 2008).

The degradation process of materials during composting follows typical biochemical pathways of any other natural degradation process. The substrates involved in composting are primarily biogenic and primarily obtained from plants, animals, or microorganisms. Cellulose, suberin, cutin, starch, glycogen, laminarin, paramylon, dextran, etc. are the major nutrient components present in substrates for composting process (Insam & de Bertoldi, 2007). Therefore, the ratio of sludge to bulking agents should be selected properly in accordance with the quality of the sludge in order to obtain compost of high attribute (Insam & de Bertoldi, 2007). Sewage sludge contains a large amount of moisture which is very inconvenient for composting. So, initially dry bulking agents are added to sewage sludge to absorb excess moisture content, to increase porosity for effective aeration, to support the structure, to allow adequate gas transfer, and to avoid over-packing (Yañez et al., 2009). Composting cannot be done in a continuous process. The batch process of composting generally comprises four phases. Each batch has its distinct biological, physical, and chemical profile (Insam & de Bertoldi, 2007). Figure 4.4 illustrates the process flow for environmentally friendly composting method of waste sewage sludge to produce useable fertilizer by undergoing four distinct phases.

The first phase is the mesophilic phase which is occurred at temperature range from 25 °C to 40 °C. During the starting phase, fungi, actinobacteria, and bacteria which are known as primary decomposers break down the easily degradable components like sugar protein of substrate materials. If the composting process is mechanically undisturbed, other organisms like compost worms, mites, and millipedes may also develop and assist as catalysts, especially in vermicomposting. The case of vermicomposting depends on various factors such as composting methods, type of waste, and composting environment. Hence, this action can be negligible in different scenarios (Insam & de Bertoldi, 2007). Majorly the substrate is reduced due to the degradation of sugar and proteins by the action of mesophilic organisms (Zeng et al., 2011). The quantity of mesophilic organisms in the starting material is a thousand times greater than the number of thermophilic organisms. Despite this, due to the metabolic activity of primary decomposers, the temperature rises during composting (Insam & de Bertoldi, 2007). There is also an increase in carbon dioxide along with the temperature (Zeng et al., 2011).

The mesophilic phase is followed by thermophilic phase. The temperature of the composting pile increases from 40 °C to approximately 70 °C, and the mesophiles are died off and replaced by thermophiles having higher adaptability in high temperature (Sundberg et al., 2004). Large numbers of pathogenic individuals are degraded during this time (Novinscak et al., 2008). Up to 62 °C temperature, the degradation process is running rapidly. But when the temperature rises to 65 °C, the fungal growth largely inhibited. The temperature in the compost pile may climb and approach 80 °C even after most microorganisms have been destroyed above 65 °C. It is more likely that abiotic exothermic reactions, rather than microbial activity, are to blame for this final temperature rise. These nonbiological temperature-increasing activities might involve actinobacterial enzymes, which are stable at high temperatures. For hygienization, the thermophilic phase is crucial. It leads to the eradication of human and plant pathogens, and weed seeds and insect larvae are also killed. In addition to the high temperatures present during this phase, the presence of a specialized flora dominated by actinobacteria produces antibiotics which further kill harmful microbes (Insam & de Bertoldi, 2007).

The third phase, known as the cooling phase or second mesophilic phase, begins with a gradual drop in compost pile temperature. This is because during the completion of second phase, all the substrates are used up and the thermophilic organism's activity comes to an end. At this point, mesophilic organisms start to recolonize as they may be originated from survived spores which were somehow protected or may enter the system from the environment. The second mesophilic phase is known as maturation phase. Here, the composition of microbial populations completely changes where proportion fungi is higher than that of bacteria. Compound in composting does not undergo any further degradation process (Insam & de Bertoldi, 2007). The prepared composts are ready for the land application after some necessary post treatment which varies on the basis of compost quality. Composting has gained widespread favor in recent years as an environmentally friendly, technologically simple, and cost-effective method of biowaste disposal. By recycling organic waste into nutrient-rich compost, composting helps reduce greenhouse gas emissions, lessens the strain on landfills, and fosters healthier soils and plant growth. Its straightforward process can be implemented on various scales, from individual households to large-scale composting facilities, making it accessible to diverse communities and industries seeking sustainable waste management solutions (Al-Rumaihi et al., 2020; Hamidpour et al., 2012; Soudejani et al., 2019).

4.2 Anaerobic Digestion

Anaerobic digestion is a biological process in which microorganisms decompose organic matter, such as sewage sludge, manure, and food waste, in the absence of oxygen (Fig. 4.5). This results in the generation of methane and carbon dioxide gases, along with a wet organic residue known as digestate. Anaerobic digestion is a proven method for treating sewage sludge, efficiently handling its high water content without the need for prior pretreatment (Hanum et al., 2019). Microbial communities that are present in organic waste are crucial to the composting process. Different kinds of organisms that can interact with one another in the environment where they coexist make form a microbial community. Within 4 to 6 weeks, the microbes reduce the organic waste to its most basic constituents. These constituents are subsequently combined to form compost, a humus-like structure (Palaniveloo et al., 2020). There are four main steps that take place during the anaerobic digestion process; these are hydrolysis, acidogenesis, acetogenesis, and methanogenesis, all carried out by a diverse microbial community in the absence of oxygen.

The first phase is called hydrolysis where extracellular enzymes are produced by microorganisms such as cellulase, amylase, protease, and lipase which facilitate hydrolysis. This results in the decomposition of organic materials into soluble substances. Proteins are broken down into individual amino acids, lipids are converted into long- and short-chain fatty acids, starch is transformed into glucose, and

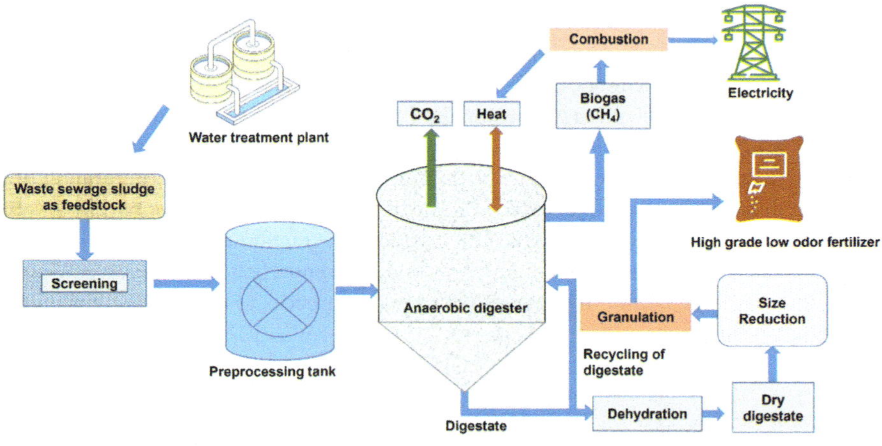

Fig. 4.5 This illustration shows the anaerobic digestion process to produce high grade odorless fertilizer. Additionally, CH_4 is formed which can be used as an energy source to produce electricity

carbohydrates are broken down into simple sugars. This breakdown of complex organic compounds into their smaller soluble components is essential for nutrient recycling and providing the necessary building blocks for various biological processes (Chen & Neibling, 2014). As hydrolysis is the slowest phase in anaerobic digestion, it is frequently referred to as the rate-determining step since it significantly affects how long the feedstock remains in the digester. To increase the reaction rate, pretreatment techniques for anaerobic digestion like thermal treatment is necessary (Jain et al., 1992).

After hydrolysis, acidogens, a subclass of fermentative bacteria, continue to break down the tiny molecules, including amino acids, fatty acids, and sugars. This phase is called acidogenesis. A combination of volatile fatty acids (VFAs), such as acetic acid, propionic acid, and butyric acid, are produced by the conversion of these tiny molecules by acidogens. The process at this point may also produce small byproducts like hydrogen, carbon dioxide, and additional acetic acid. The total anaerobic digestion process begins with this crucial step, which breaks down organic matter and prepares the way for succeeding steps. When complex organic matter is being digested in liquid phase, acidogenesis is typically the stage that happens the fastest (Chen & Neibling, 2014). Acetogenesis is the third phase where acetate, hydrogen, and carbon dioxide are formed from the decomposition of short-chain volatile fatty acids that were formed in the acidogenesis phase. All these three processes are important because materials that are produced up to the final phase will be used in the last phase of anaerobic digestion.

Fourth or final step is methanogenesis in which methanogenic microorganisms in the digester consume the accessible intermediates produced in acidogenesis and acetogenesis (acetate, hydrogen, and carbon dioxide) to produce methane (Chen & Neibling, 2014). Methane, the primary byproduct of the entire anaerobic digestion process, is produced by these microbes via two main pathways (D'Silva et al., 2021). Acetate is transformed into methane and carbon dioxide in the first step, called acetoclastic methanogenesis (Stams et al., 2019). The second mechanism, hydrogenotrophic methanogenesis, uses carbon dioxide and hydrogen to create water and methane (Wagner et al., 2018). Carbon dioxide is also produced as the second most abundant gas (Al-mashhadani et al., 2016). The digestate produced from anaerobic digestion contains high amount on nutrient. Hence, this can be used as high-grade fertilizer after some post treatment such as size reduction, granulation etc. We have seen that anaerobic digestion not only recycles nutrients in the environment but also contributes to the creation of greenhouse gases such as CO_2 and CH_4 (Fig. 4.5). This inconvenience should be comprehended and solved during designing and maintaining the anaerobic digestion process to establish environmental sustainability.

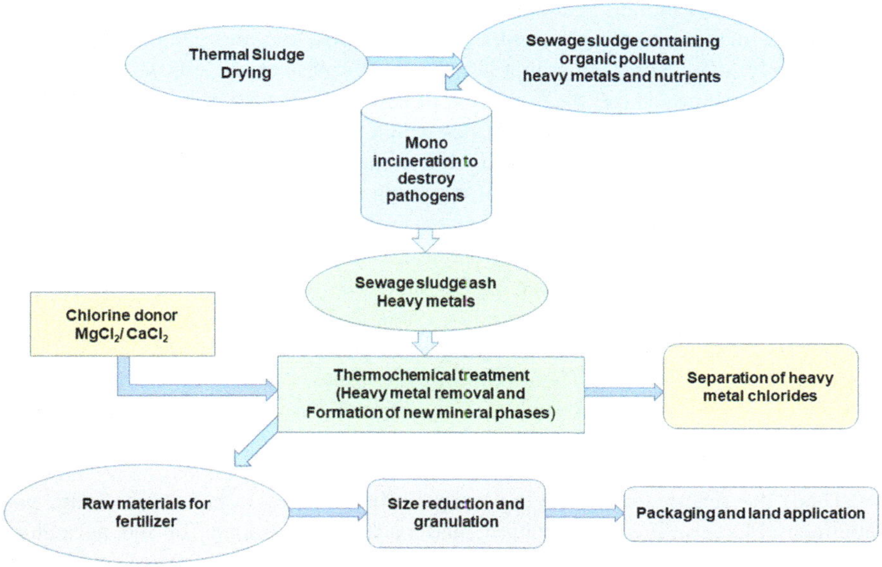

Fig. 4.6 This illustration shows the flowchart of producing fertilizer from sewage sludge waste. This process is initiated with thermal drying of sewage sludge and subsequent mono-incineration to destroy organic pollutants and pathogens. The ash from mono-incineration is further treated with chlorine donor chemicals to separate heavy metals. Thus, the raw material for fertilizer is obtained which eventually undergoes some post-treatment process before being applied to land as soil conditioner

4.3 Thermochemical Treatment

A study conducted in Germany showed that sewage sludge recovery could potentially provide around 19,000 tons of phosphorus per year which could cover up to 13% of total phosphorous used in the agriculture in this country (Krüger & Adam, 2015). Thermochemical approach should be taken to obtain that amount of recovery for a promising method for recycling phosphorus and producing highly plant-available P-fertilizers from sewage sludge which could potentially replace traditional rock phosphate source. At primary stage, this method involves thermal drying of sewage sludge where water is vaporized to reduce moisture content (Fig. 4.6). During this process, high energy is consumed which can be supplied by the plant itself if the plant has an anaerobic digestion segment where biogas is produced (Furness et al., 2000). Mono-incineration sewage sludge is another step in primary stage. During this step, sewage sludge is incinerated at high temperatures to destroy organic pollutants. The process of mono-incineration for sewage sludge treatment can produce various flue gases such as nitrogen oxides (NOx), sulfur dioxide (SO_2), and volatile organic compounds (VOCs). These pollutants can have adverse effects on air quality and human health. Therefore, incineration process requires complex off-gas treatment to remove heavy metals, acid gases, and nitrogen oxides. Incineration is only financially viable on a large scale (Furness et al., 2000).

After the primary stage, the condition of sewage is further enhanced by thermochemical treatment (Fig. 4.6) of the ashes which is obtained from the primary stage. This process involves subjecting to high temperatures and usage of chemicals (chlorine donor) for the purpose of breaking down the organic and inorganic constituents of sewage sludge to convert the phosphorous into mineral phases. Raw ashes from mono-incineration are mixed with a chlorine donor and water. The chlorine donor can be either $MgCl_2$ or $CaCl_2$ which is dissolved in distilled water to make a solution. The ash mixture contains 30% water. The concentration of chorine used ranges from 50 to 200 g Cl/kg ash, and the temperature is maintained between 750 and 1050 °C. The mixture of raw ashes, chlorine donor, and water is intensively mixed and then subjected to thermal treatment in a rotary furnace. Retention periods range from 20 to 120 min with evenly circulated air (Adam et al., 2009).

Then, the reactor is cooled down and solid product is collected. Some post treatment like grinding, granulation, and packaging is done for the agriculture application (Fig. 4.6). During this procedure, heavy metals are removed from the ash and the bioavailability of phosphorous is improved by producing mineral phase such as farringtonite, stanfieldite, and chlorapatite which are more easily taken up by plants (Adam et al., 2009). Through this process, the phosphorous in the waste material can be changed into a form that is more readily available to plants, making it a valuable source of fertilizer. The ashes leftover after incinerating sewage sludge typically contain high levels of phosphorus, around 20% P_2O_5 by weight. This means the ashes represent a potential source of raw material for producing phosphate fertilizers. The high phosphorus content in sewage sludge ashes makes them suitable feedstocks to manufacture phosphorus-containing fertilizers (Adam et al., 2009; Steckenmesser et al., 2022). Table 4.2 represents a comparative study outlining the key differences between composting, anaerobic digestion, and thermochemical treatment methods for making fertilizer from sewage sludge.

5 Sewage Sludge as a Biological Source for Sustainable Fertilizer Option

5.1 Impact on Soil Quality and Fertility

Having enriched with essential plant nutrients, well-treated sewage sludge can improve physical, chemical, and microbiological properties of the soil. But directly applying raw sewage sludge can have harmful effects due to containing heavy metals and pathogens. This adverse effects can be minimized by undergoing some proper treatment processes to remove harmful matters before application. Application of

Table 4.2 Comparative study outlining the key differences between composting, anaerobic digestion, and thermochemical treatment methods for making fertilizer from sewage sludge

Method	Composting	Anaerobic digestion	Thermochemical treatment
Decomposition technique	Aerobic (with oxygen)	Anaerobic (without oxygen)	High-temperature chemical reactions
Condition	The ideal C:N ratio for a compost pile is roughly 25–30 parts carbon to 1 part nitrogen.	Optimal ratio of C:N is between 20 and 30 for the anaerobic digestion process.	N/A
Temperature	40–80 °C	35–37 °C	750–1050 °C
Time	Several months to a year	Weeks to several months	Varies, but typically faster than composting
Residue disposal	Minimal residue as almost all of the compost can be used as soil conditioner	Minimal residue. Maximum amount of digestate can be used as fertilizer.	Ash residue is produced from this process
Environmental Impact	Low greenhouse gas emissions if managed correctly	Low greenhouse gas emissions, and energy recovery	Moderate to high energy use and emissions
End product	Compost, a stable organic material	Biogas (methane) and digestate	Ash-based fertilizer or synthetic gas

sewage sludge can improve soil's physical characteristics including improved water infiltration, increased total porosity, and texture. This high porosity has greater potential to hold water, so increased water holding capacity, decreased surface slide and erosion of water, decreased nutrients penetration resistance, decreased soil bulk density which is better for agriculture, and increased permeability (Angin & Yaganoglu, 2011; Usman et al., 2012b).

There is statistically strong evidence that applying sewage sludge to soil improves its chemical properties significantly. pH can be maintained in desired range. The application of sewage sludge (pH ~6.65) helps to maintain pH of soil in the suitable range for most of the plants to grow properly, and most of the nutrients are available for plant at these pH range (Singh & Agrawal, 2008). Sewage sludge increases soil electrical conductivity which is the measure of total soluble salt concentration and cation exchange capacity. It also increases calcium (Ca) and magnesium (Mg) content in the soil and production of humic acid and carbonic acid. Concentrations of some metals such as Fe, Mn, Cu etc. are also increased but not above the level what a plant cannot tolerate. But applying at excessive rate on soil, these metal concentrations in soil exceed the permissible level which is harmful for trees. So sewage sludge must be applied at required level (Angin & Yaganoglu, 2011; Usman et al., 2012b). After adding sludge, soil enzymatic activity often increases quickly but decreases after 28 days, returning to their baseline levels (Mattana et al., 2014). So regular application at requited rate is necessary to maintain soil quality. Fertility is the major factor that affects crop production. Microorganisms such as fungi, bacteria, actinomycetes, algae, and enzymes largely impact on soil fertility. Treated sewage sludge increases microbial activity in soil. Furthermore, it promotes protease and urease activity, increases nitrogenous organic matter breakdown, releases ammonium ion for plant uptake, and thus enhances soil fertility (Napora & Grobelak, 2014).

5.2 Adaptability for Agriculture Application

5.2.1 Suitability for Crop Husbandry

Sewage sludge, after necessary treatment process, can perform as a valuable organic fertilizer for agricultural application. The agricultural application of these fertilizers has becoming popular in both industrialized and developing nations. However, there have been diverse crop reactions to the application of sewage sludge as fertilizer. In a study on wheat, grain, and straw found that yield rises significantly when application rate of sewage sludge fertilizer increases up to 40 tons per hectare (Gulzar et al., 2022). The yield of lettuce, head circle, plant height, and leaf number are increased by adding sewage sludge fertilizer. This practice also improved the proportion of organic matter, total nitrogen, exchangeable phosphorous, and potassium in top soil (Sönmez & Bozkurt, 2006). In Philippine, sewage sludge application increased production of highland rice after both sewage sludge applied alone and in combination with other synthetic conventional fertilizer (Maguiat, 1998).

5.2.2 Support for Pasture and Forestry Management

In pasture management, cultivating nutritious forage grasses and legumes ensures a sustainable and continuous food source for cattle. Sewage sludge, with its substantial organic matter content (ranging from 40% to 60%), is employed as a fertilizer or soil conditioner to enhance dairy pasture growth. Sewage sludge-treated fertilizer has a positive impact on soil structure, and also it improves aeration and water retention capacity of soil, ultimately contributing to the long-term availability of nourishing feed for cattle while promoting sustainability (Ross & Sim, 1991; Usman et al., 2012b). A good diet for cows must include several nutrient elements to continue growing and producing milk. Treated sewage sludge can serve as a valuable reservoir of diverse macro- and micronutrients which are necessary for plant growth, containing elements such as nitrogen, phosphorus, iron, calcium, and magnesium (Usman et al., 2012b). Adding sewage sludge fertilizer to a pasture used for cow feeding thus enhances milk quality and boosts output.

Sewage sludge is applied in forestry in several countries around the world (Oberg & Mason-Renton, 2018). It improves physical, chemical, and biological properties of the soil and fertility status for creating more favorable conditions for establishment of vegetation (Alonso et al., 2002). The organic matter content of the soil increased by 12.79–80.80% as a result of the application of processed sewage sludge, along with the biomass of the grassland and the intensity of the grass's color (X. Zhang et al., 2020). Forest applications have received increasing attention because of the greater capacity of forest ecosystems to retain sludge constituents. The application of sewage sludge in forestry is considered a viable option.

The prolonged application of sewage sludge potentially may result in metal accumulation which poses a threat to health and ecology. The excessive usage of sewage sludge will also pollute forest land and add an excessive amount of nutrients to the soil, endangering the ecology and directly harming the trees. The growth of weeds might also be accelerated by the excessive usage of sludge. For forestry applications, it is essential to comprehend and establish the precise range and rate at which sewage sludge fertilizer can be applied alongside conventional fertilizer (Usman et al., 2012b; Zhang et al., 2020).

5.2.3 Green House and Nursery Application

Peat moss is widely used as a potting medium in horticulture since the 1950s. Due to increasing peat prices, and the need to recycle organic wastes for environmental concerns, peat moss can be combined with organic waste such as sewage sludge containing high levels of nutrients to formulate potting soil. The sludge should be treated in aerobic composting under thermophilic condition (Raviv, 2009).

A greenhouse pot experiment aimed to assess the physical and chemical properties of five different growing media (De Lucia et al., 2013). These growing media were formulated by blending sewage sludge-based compost and peat in various ratios to determine the potential utility of sewage sludge compost for plant growth and cultivation. The sewage sludge compost ratios used were 25%, 40%, 55%, and 70% (v/v), which also included sphagnum peat and inert materials. The results presented in the study indicate that mixing peat with sewage sludge compost leads to a notable improvement in plant nursery. By incorporating sewage sludge compost into the growing media, the nutrient content becomes more enriched, thereby positively impacting plant growth and development compared to conventional growing medium used in plant nurseries.

5.3 Sustainability and Feasibility

As a result of climate change, almost two-thirds of the world's population is predicted to experience water scarcity by 2025. To provide food for the estimated 9 billion people on the planet, there is a need to increase food production globally by at least 50% by 2050 (Ungureanu et al., 2020). The worldwide water crisis and a scarcity of freshwater lead to the need for massive amount wastewater treatment to meet agricultural, household, and industrial requirements. This treatment process eventually produces large amount of sewage sludge which requires large disposal facility (Gulzar et al., 2022). Over the years, a steady rise is observed in sewage sludge production. For example, in the European Union's 27 member states, the annual dry mass of sewage sludge increased from ten million tons in 2010 to 13.5 million tons in 2020 (Havukainen et al., 2022).

As the conventional method of waste disposal method like landfilling and incineration is not environment friendly, an alternative approach that involves utilizing the nutrient contents of sewage sludge as raw material for producing fertilizer is an excellent choice for promoting sustainability. With all these organic matter and essential nutrients, sewage sludge can serve as a low-cost soil amendment and can potentially replace the traditional synthetic fertilizer (Sugurbekova et al., 2023). The production of fertilizer from sewage sludge has an advantage in terms of raw material supply. Traditional fertilizers that rely on mined or synthesized resources and frequently suffer from crisis for raw material eventually disrupts the supply chain. On the contrary, the raw materials for sewage sludge fertilizer are comparatively more available and accessible. Because with populations continue to grow and more wastewater treatment facilities are being established, the generation of sewage sludge is likely to increase, ensuring a relatively stable, steady, and continuous supply (Chojnacka et al., 2023).

The recovery and utilization of sewage sludge contribute to circular economy, a model aiming at closed loop of manufacturing and consuming goods that includes sharing, reusing, repairing, and recycling already manufactured goods as long as possible allowing for the efficient and sustainable management of waste sludge, reducing environmental impacts and promoting resource conservation (Geissdoerfer et al., 2017). This strategy is associated with the zero-waste strategy since it involves to develop methods for reducing trash generation and encourage the growth of the bio-economy, which offers intelligent waste management (Buta et al., 2021). Utilizing sewage sludge as a resource offers new opportunities for businesses, jobs, and innovation which will promote this model. Products made from sewage sludge can be sold, generating income which makes it a substantial waste management strategy. It offers a solution to the issue of disposing of a large amount of waste as well as contributes to provide more sustainable resources for agriculture and food production.

The usage of sewage sludge fertilizer might reduce the demand for synthetic fertilizers, which are typically made from petroleum products and can harm the environment. With the increase in petroleum products globally, the price of synthetic fertilizer significantly increases. Fertilizer derived from waste sewage sludge requires raw materials which are low cost and easily available. It does not pollute the air or water if treated properly. That makes it a sustainable resource for food production since it includes significant amounts of vital nutrients. Sewage sludge fertilizer used in agriculture helps to recycle nutrients like N, P, and K as well as organic matters that would otherwise be lost. The amount of garbage produced in densely populated places can be decreased by composting sewage sludge which can eventually serve as a reliable fertilizer substitute for nitrogen and phosphorus (Usman et al., 2012b).

6 Challenge and Future Directions

Due to producing enormous volumes of sewage sludge every year containing a variety of chemicals, disposal has become a challenging task which is the main concern in sewage sludge management. The potential health dangers and environment pollution can be exterminated by using sewage sludge as fertilizer due to its rich nutrient content (Gianico et al., 2021). But a range of pollutants, including heavy metals, harmful microbes, and pathogens, can be introduced into agricultural areas if sewage sludge is used as fertilizer (Laturnus et al., 2007). As fertilizer derived from sewage sludge contains pathogens, endotoxins, allergens, and toxicants (Council, 2002), it causes adverse effects on animals and human health when the pollutants penetrate the food chain or groundwater through erosion, runoff, and later enters food chain. There is a possibility that significant amounts of hazardous compounds with mutagenic and carcinogenic properties will be introduced into the soils as well (Oleszczuk, 2006; Muter et al., 2022a). So, despite the obvious benefits of using sewage sludge as fertilizer, the challenge regarding this practice should be addressed.

Regular and comprehensive testing of sewage sludge that is being used as fertilizer is important to trace and identify contaminants such as heavy metals, organic compounds, and pathogens that potentially enter into the bioecology. In case of finding any abnormalities in ecology and animal health, investigation for targeted contaminants will be essential. Sludge containing responsible contaminant for the negative impact should be further analyzed and determined the suitability of that source of sludge. Modification of process in wastewater treatment plant could solve the problem by subjecting it to further chemical or biological treatment. It is critical and necessary to develop new effective onsite sludge disinfection technology that potentially helps to reduce risks associated with the use of sewage sludge fertilizer.

Proper management of sludge is also necessary to minimize risks and maximize the benefits of sludge recovery and utilization. The result associated with each batch of sewage sludge fertilizer is unique due to site-specific parameters such as the applied method, volume of fertilizer used per unit land, soil types, crops, and season. The type of fertilizer may also vary for a specific treatment depending on several factors like source of raw wastewater, the mix of macroelements, microelements, other chemicals, and contaminants present in wastewater. All these specific factors must be taken into account individually to analyze the quality of sewage sludge produced. Interdisciplinary activity such as planning, testing, and storing are required to establish the safe use of sewage (Muter et al., 2022a).

Detailed and empirical analysis centered on sustainability, environment, and animal health; social acceptability is required to ensure the effective use of these fertilizers. Monitoring groundwater quality from application fields at regular intervals after starting the practice of using treated sewage sludge as fertilizer can detect any early signs of contamination. Thus, contaminant levels could then be addressed before leading to any major damage to environment and ecology. Crops with excessive contaminant levels could enter the food chain and disrupt ecology. Regular monitoring of crops fertilized with sewage sludge is important to ensure that contaminant levels remain below safety thresholds. Nutrient mapping is an efficient strategy that can be conducted to analyze soil nutrient levels and crop requirements across different agricultural fields of different regions. This strategy allows us to direct soil-specific application of sewage sludge at rates which is required for a specific crop in a specific land field. This process should involve tracing down the rate and ratio at which essential nutrient content is delivered from fertilizer to crops. By matching the nutrient demands of the crops and delivery rates, we can ensure to avoid excess nutrients that could lead to runoff and pollute soil and water. It is essential to follow necessary protocol strictly in order to reduce any potentially harmful components that could present in sewage sludge fertilizer.

Another technique could be followed to get rid of heavy metal by separating heavy metal from nutrient content during water treatment process. Various methods such as chemical precipitation, ion exchange, membrane filtration processes, adsorption, photocatalysis processes, and nanotechnology, activated carbon technology can be utilized for heavy metal separation and recovery (Azimi et al., 2017; Rashid et al., 2017). The recovered heavy metal can even be reused as a catalyst for different chemical processes (Das & Poater, 2021). Therefore, detailed study, experiment, and analysis are needed to make any change in plant structure for heavy metal and other contaminants removal to minimize risks associated with pollution and maximize the benefits of energy recovery by utilization of sewage sludge as effective fertilizer.

It is important to consider the acceptance of sewage sludge-derived fertilizer to the consumers such as farmer and other personnel related to agriculture and cultivation because full acceptance is likely to be a real challenge even with these benefits regarding the use of these fertilizer and as a source of organic matter. The stigma around its origins may prove difficult to overcome for many consumers. However, as synthetic fertilizers become more expensive and demand for fertilizer increases rapidly with the ever-growing population to feed, sewage sludge fertilizer may gradually become more accepted over time due its low cost and easily available raw materials. Comprehensive outreach program, public meetings, workshops, demonstration, and question answer session should be organized to educate consumers about the safety and sustainability of using sewage sludge as fertilizer. Also, more production plant should be constructed with implementing modern technology to increase the yield, production, and availability of sewage sludge fertilizer. Feasibility should be checked to ensure the continuous raw material supply and easy transportation by identifying suitable locations for new production plant considering the proximity to wastewater treatment facilities.

7 Conclusions

Sewage sludge has the potential to be a sustainable source of fertilizer for farming applications. If sewage sludge is treated properly, it can provide essential nutrients for crops, which is less costly and available than synthetic fertilizer. However, concern arises as sewage sludge may also contain harmful contaminants like heavy metals, organic and inorganic pollutants, as well as pathogens. If it is not managed properly, it could lead to substantial damage to the environment and human health. Therefore, responsible practices, transparent communication, and thorough investigation concerning the food safety, soil health, and groundwater composition are needed to investigate and evaluate the long-term impacts of sewage sludge fertilizer when it is applied for cultivation and land application. Strict guidelines, policies as well as monitoring programs should be conducted on regular intervals to ensure that the quantity of pollutants present in sewage sludge does not go over the threshold limit. We have to make sure that our environment and food system are not being tainted by heavy metals and pathogens. A precautionary approach and further scientific study of sewage sludge may help to meet agricultural demands in a more sustainable way. However, to maximize the sustainability and advantages of sewage sludge fertilizer while reducing environmental and health hazards, a combination of optimized treatment technologies, precise application methods, soil and crop monitoring, and regulatory oversight will be pivotal. The effective and reliable use of sewage sludge fertilizer can be ensured with proper management and consideration of every essential factors.

References

Adam, C., Peplinski, B., Michaelis, M., Kley, G., & Simon, F. G. (2009). Thermochemical treatment of sewage sludge ashes for phosphorus recovery. *Waste Management, 29*(3), 1122–1128. https://doi.org/10.1016/j.wasman.2008.09.011

Akter, M., Bhattacharjee, M., Dhar, A. K., Rahman, F. B. A., Haque, S., Rashid, T. U., & Kabir, S. F. (2021). Cellulose-based hydrogels for wastewater treatment: A concise review. *Gels, 7*(1), 30.

Al-mashhadani, M. K. H., Wilkinson, S. J., & Zimmerman, W. B. (2016). Carbon dioxide rich microbubble acceleration of biogas production in anaerobic digestion. *Chemical Engineering Science, 156*, 24–35. https://doi.org/10.1016/j.ces.2016.09.011

Alonso, E., Alvarez, A., Mochon, M. C., Sanchez, J. J., & Rodriguez, M. (2002). Heavy metal extractable forms in sludge from wastewater treatment plants. *Chemosphere, 47*(7), 765–775.

Al-Rumaihi, A., McKay, G., Mackey, H. R., & Al-Ansari, T. (2020). Environmental impact assessment of food waste management using two composting techniques. *Sustainability (Switzerland), 12*(4). https://doi.org/10.3390/su12041595

Andersen, A. (2001). *Disposal and recycling routes for sewage sludge—part 3: Scientific and technical report*. European commision, DG Environment.

Angin, I., & Yaganoglu, A. V. (2011). Effects of sewage sludge application on some physical and chemical properties of a soil affected by wind erosion. *Journal of Agricultural Science and Technology, 13*(5), 757–768. https://jast.modares.ac.ir/article-23-882-en.html

Anwar, Z., Irshad, M., Fareed, I., & Saleem, A. (2015). Characterization and recycling of organic waste after co-composting – A review. *Journal of Agricultural Science, 7*(4). https://doi.org/10.5539/jas.v7n4p68

Authman, M. M. N. (2015). Use of fish as bio-indicator of the effects of heavy metals pollution. *Journal of Aquaculture Research & Development, 06*(04). https://doi.org/10.4172/2155-9546.1000328

Azimi, A., Azari, A., Rezakazemi, M., & Ansarpour, M. (2017). Removal of heavy metals from industrial wastewaters: A review. *ChemBioEng Reviews, 4*(1), 37–59.

Baldasano, J. M., & Soriano, C. (2000). Emission of greenhouse gases from anaerobic digestion processes: Comparison with other municipal solid waste treatments. *Water Science and Technology, 41*(3), 275–282. https://doi.org/10.2166/wst.2000.0081

Barakat, M. A. (2011). New trends in removing heavy metals from industrial wastewater. *Arabian Journal of Chemistry, 4*, 361–377.

Biswas, S., Rashid, T. U., Debnath, T., Haque, P., & Rahman, M. M. (2020). Application of chitosan-clay biocomposite beads for removal of heavy metal and dye from industrial effluent. *Journal of Composites Science, 4*(1), 16.

Biswas, S., Fatema, J., Debnath, T., & Rashid, T. U. (2021). Chitosan–clay composites for wastewater treatment: A state-of-the-art review. *ACS ES&T Water, 1*(5), 1055–1085.

Buta, M., Hubeny, J., Zieliński, W., Harnisz, M., & Korzeniewska, E. (2021). Sewage sludge in agriculture–the effects of selected chemical pollutants and emerging genetic resistance determinants on the quality of soil and crops–a review. *Ecotoxicology and Environmental Safety, 214*, 112070–112070.

Campos, J. L., Valenzuela-Heredia, D., Pedrouso, A., Val del Río, A., Belmonte, M., & Mosquera-Corral, A. (2016). Greenhouse gases emissions from wastewater treatment plants: Minimization, treatment, and prevention. *Journal of Chemistry, 2016*, 3796352–3796352. https://doi.org/10.1155/2016/3796352

Chen, L., & Neibling, H. (2014). *Anaerobic digestion basics* (p. 6). University of Idaho extension.

Chen, C. J., Huang, J. W., & Hsiao, Y. C. (2010). Knowledge management and innovativeness: The role of organizational climate and structure. *International Journal of Manpower, 31*(8), 848–870.

Chislock, M. F., Doster, E., Zitomer, R. A., & Wilson, A. E. (2013). Eutrophication: Causes, consequences, and controls in aquatic ecosystems. *Nature Education Knowledge, 4*(4), 10–10.

Chojnacka, K., Skrzypczak, D., Szopa, D., Izydorczyk, G., Moustakas, K., & Witek-Krowiak, A. (2023). Management of biological sewage sludge: Fertilizer nitrogen recovery as the solution to fertilizer crisis. *Journal of Environmental Management, 326*, 116602–116602.

Corradini, F., Meza, P., Eguiluz, R., Casado, F., Huerta-Lwanga, E., & Geissen, V. (2019). Evidence of microplastic accumulation in agricultural soils from sewage sludge disposal. *Science of the Total Environment, 671*, 411–420. https://doi.org/10.1016/j.scitotenv.2019.03.368

Council, N. R. (2002). *Biosolids applied to land: Advancing standards and practices*. National Academies Press.

D'Silva, T. C., Isha, A., Chandra, R., Vijay, V. K., Subbarao, P. M. V., Kumar, R., et al. (2021). Enhancing methane production in anaerobic digestion through hydrogen assisted pathways—A state-of-the-art review. *Renewable and Sustainable Energy Reviews, 151*, 111536–111536. https://doi.org/10.1016/j.rser.2021.111536

Dai, J.-Y., Lin, C., Zhao, J.-F., & Naz, M. A. (2006). Characteristics of sewage sludge and distribution of heavy metal in plants with amendment of sewage sludge. *Journal of Environmental Sciences, 18*(6), 1094–1095.

Das, T. K., & Poater, A. (2021). Review on the use of heavy metal deposits from water treatment waste towards catalytic chemical syntheses. *International Journal of Molecular Sciences, 22*(24), 13383–13383.

De Bertoldi, M. D., Vallini, G., & Pera, A. (1983). The biology of composting: A review. *Waste Management & Research, 1*(1), 157–176.

De Lucia, B., Cristiano, G., Vecchietti, L., Rea, E., & Russo, G. (2013). Nursery growing media: Agronomic and environmental quality assessment of sewage sludge-based compost. *Applied and Environmental Soil Science, 2013*. https://doi.org/10.1155/2013/565139

De Souza MacHado, A. A., Lau, C. W., Till, J., Kloas, W., Lehmann, A., Becker, R., & Rillig, M. C. (2018). Impacts of microplastics on the soil biophysical environment. *Environmental Science and Technology, 52*(17), 9656–9665. https://doi.org/10.1021/acs.est.8b02212

Deepnarain, N., Nasr, M., Amoah, I. D., Enitan-Folami, A. M., Reddy, P., Stenström, T. A., et al. (2020). Impact of sludge bulking on receiving environment using quantitative microbial risk assessment (qmra)-based management for full-scale wastewater treatment plants. *Journal of Environmental Management, 267*, 110660–110660. https://doi.org/10.1016/j.jenvman.2020.110660

Donatello, S., & Cheeseman, C. R. (2013). Recycling and recovery routes for incinerated sewage sludge ash (issa): A review. *Waste Management, 33*, 2328–2340.

Dubey, S., Yadav, R., Joshi, P., Chaturvedi, R., Goyel, B., Yadav, R., & Minhas, P. (2006). *Agricultural uses of sewage sludge and water and their impact on soil water and environmental health in haryana, India.* Paper presented at the Abstract of 18th World Congress of Soil Science.

Dume, B., Hanc, A., Svehla, P., Michal, P., Chane, A. D., & Nigussie, A. (2022). Vermicomposting technology as a process able to reduce the content of potentially toxic elements in sewage sludge. *Agronomy, 12*(9), 2049.

Fan, Y., Hornung, U., & Dahmen, N. (2022). Hydrothermal liquefaction of sewage sludge for biofuel application: A review on fundamentals, current challenges and strategies. *Biomass and Bioenergy, 165*. https://doi.org/10.1016/j.biombioe.2022.106570

Feng, L., Luo, J., & Chen, Y. (2015). *Dilemma of sewage sludge treatment and disposal in China* (Vol. 49, pp. 4781–4782). American Chemical Society.

Ferrentino, R., Langone, M., Fiori, L., & Andreottola, G. (2023). Full-scale sewage sludge reduction technologies: A review with a focus on energy consumption. *Water (Switzerland), 15*, 615. MDPI.

Fijalkowski, K., Rorat, A., Grobelak, A., & Kacprzak, M. J. (2017). The presence of contaminations in sewage sludge – The current situation. *Journal of Environmental Management, 203*, 1126–1136.

Fonts, I., Gea, G., Azuara, M., Ábrego, J., & Arauzo, J. (2012). Sewage sludge pyrolysis for liquid production: A review. *Renewable and Sustainable Energy Reviews, 16*, 2781–2805.

Forster, C. F. (2002). The rheological and physico-chemical characteristics of sewage sludges. *Enzyme and Microbial Technology, 30*(3), 340–345.

Frišták, V., Pipíška, M., & Soja, G. (2018). Pyrolysis treatment of sewage sludge: A promising way to produce phosphorus fertilizer. *Journal of Cleaner Production, 172*, 1772–1778. https://doi.org/10.1016/j.jclepro.2017.12.015

Furness, D., Hoggett, L., & Judd, S. (2000). Thermochemical treatment of sewage sludge. *Water and Environment Journal, 14*(1), 57–65.

Geissdoerfer, M., Savaget, P., Bocken, N. M. P., & Hultink, E. J. (2017). The circular economy—A new sustainability paradigm? *Journal of Cleaner Production, 143*, 757–768. https://doi.org/10.1016/j.jclepro.2016.12.048

Gianico, A., Braguglia, C. M., Gallipoli, A., Montecchio, D., & Mininni, G. (2021). Land application of biosolids in europe: Possibilities, con-straints and future perspectives. *Water, 13*(1), 103.

Goswami, S., Shukla, D. K., & Singh, P. K. (2022). Sustainable transformation of sewage sludge ash and waste industrial additive into green cement blend. *Structural Concrete, 23*(4), 2337–2351.

Gould, M. S., & Genetelli, E. I. (1978). Heavy metal complexation behavior in anaerobically digested sludges. *Water Research, 12*(8), 505–512.

Gray, N. F. (2010). *Water technology electronic resource: An introduction for environmental scientists and engineers/n.F.* IWA Pub.

Gulzar, S., Gul, S., Patra, A., Mohapatra, K. K., Jatav, H. S., & Rajput, V. D. (2022). Socio economic aspects of sewage sludge use in agriculture. In V. D. Rajput, A. N. Yadav, H. S. Jatav, S. K. Singh, & T. Minkina (Eds.), *Sustainable management and utilization of sewage sludge* (pp. 281–296). Springer International Publishing.

Hamidpour, M., Afyuni, M., Khadivi, E., Zorpas, A., & Inglezakis, V. (2012). Composted municipal waste effect on chosen properties of calcareous soil. *International Agrophysics, 26*(4), 365–374. https://doi.org/10.2478/v10247-012-0051-5

Hanum, F., Yuan, L. C., Kamahara, H., Aziz, H. A., Atsuta, Y., Yamada, T., & Daimon, H. (2019). Treatment of sewage sludge using anaerobic digestion in Malaysia: Current state and challenges. *Frontiers in Energy Research, 7*. Frontiers Media S.A.

Havukainen, J., Saud, A., Astrup, T. F., Peltola, P., & Horttanainen, M. (2022). Environmental performance of dewatered sewage sludge digestate utilization based on life cycle assessment. *Waste Management, 137*, 210–221.

Hurynovich, A., Kwietniewski, M., & Romanovski, V. (2021). Evaluation of the possibility of utilization of sewage sludge from a wastewater treatment plant—Case study. *Desalination and Water Treatment, 227*, 16–25. https://doi.org/10.5004/dwt.2021.27199

Hussein, A. H. A. (2009). Impact of sewage sludge as organic manure on some soil properties, growth, yield and nutrient contents of cucumber crop. *Journal of Applied Sciences, 9*(8), 1401–1411.

Inoue, S., Sawayama, S., Ogi, T., & Yokoyama, S.-Y. (1996). Organic composition of liquidized sewage sludge. *Biomass and Bioenergy, 10*(1), 37–40.

Insam, H., & de Bertoldi, M. (2007). Chapter 3: Microbiology of the composting process. In L. F. Diaz, M. de Bertoldi, W. Bidlingmaier, & E. Stentiford (Eds.), *Waste management series* (Vol. 8, pp. 25–48). Elsevier.

Jain, S., Lala, A. K., Bhatia, S. K., & Kudchadker, A. P. (1992). Modelling of hydrolysis controlled anaerobic digestion. *Journal of Chemical Technology & Biotechnology, 53*(4), 337–344. https://doi.org/10.1002/jctb.280530404

Jamil, M., Qasim, M., & Umar, M. (2006). Utilization of sewage sludge as organic fertilizer in sustainable agriculture. *Journal of Applied Science, 6*(3), 531–535.

Kästner, M., & Mahro, B. (1996). Microbial degradation of polycyclic aromatic hydrocarbons in soils affected by the organic matrix of compost. *Applied Microbiology and Biotechnology, 44*(5), 668–675. https://doi.org/10.1007/BF00172501

Khan, M. N., Luna, I. Z., Islam, M. M., Sharmeen, S., Salem, K. S., Rashid, T. U., et al. (2016). Cellulase in waste management applications. In *New and future developments in microbial biotechnology and bioengineering* (pp. 237–256). Elsevier.

Koyuncuoğlu, P., & Erden, G. (2021). Sampling, pre-treatment, and identification methods of microplastics in sewage sludge and their effects in agricultural soils: A review. *Environmental Monitoring and Assessment, 193*(4), 175–175. https://doi.org/10.1007/s10661-021-08943-0

Krüger, O., & Adam, C. (2015). Recovery potential of german sewage sludge ash. *Waste Management, 45*, 400–406. https://doi.org/10.1016/j.wasman.2015.01.025

Kumar, V., Srivastava, S., & Thakur, I. S. (2021). Enhanced recovery of polyhydroxyalkanoates from secondary wastewater sludge of sewage treatment plant: Analysis and process parameters optimization. *Bioresource Technology Reports, 15*, 100783. https://doi.org/10.1016/j.biteb.2021.100783

Landrigan, P. J. (2017). Air pollution and health. *The Lancet Public Health, 2*, e4–e5. Elsevier Ltd.

Laturnus, F., Von Arnold, K., & Grøn, C. (2007). Organic contaminants from sewage sludge applied to agricultural soils. False alarm regarding possible problems for food safety? (8 pp). *Environmental Science and Pollution Research, 14*, 53–60.

Lester, J. N. (1983). Significance and behaviour of heavy metals in waste water treatment processes i. Sewage treatment and effluent discharge. *Science of the Total Environment, 30*, 1–44.

Liang, Y., Xu, D., Feng, P., Hao, B., Guo, Y., & Wang, S. (2021). Municipal sewage sludge incineration and its air pollution control. *Journal of Cleaner Production, 295*, 126456. https://doi.org/10.1016/j.jclepro.2021.126456

Lin, W. Y., Ng, W. C., Wong, B. S. E., Teo, S. L. M., Sivananthan, G. D. O., Baeg, G. H., et al. (2018). Evaluation of sewage sludge incineration ash as a potential land reclamation material. *Journal of Hazardous Materials, 357*, 63–72. https://doi.org/10.1016/j.jhazmat.2018.05.047

Liu, Y., Lu, W., Guo, H., Ming, Z., Wang, C., Xu, S., et al. (2016). Aromatic compound emissions from municipal solid waste landfill: Emission factors and their impact on air pollution. *Atmospheric Environment, 139*, 205–213. https://doi.org/10.1016/j.atmosenv.2016.05.043

Maguiat, I. (1998). Sewage sludge: Turning an environmental pollutant into an agricultural resource. https://agris.fao.org/search/en/providers/122430/records/64722b0977fd37171a735ee4. Last visited: August 08, 2023.

Mathney, J. M. J., & Dunn, M. L. (2011). *A critical review of the u.S. Epa's risk assessment for the land application of sewage sludge* (Vol. 21, pp. 43–56). SAGE Publications Ltd.

Mattana, S., Petrovičová, B., Landi, L., Gelsomino, A., Cortés, P., Ortiz, O., & Renella, G. (2014). Sewage sludge processing determines its impact on soil microbial community structure and function. *Applied Soil Ecology, 75*, 150–161. https://doi.org/10.1016/j.apsoil.2013.11.007

Menéndez, J. A., Domínguez, A., Inguanzo, M., & Pis, J. J. (2004). Microwave pyrolysis of sewage sludge: Analysis of the gas fraction. *Journal of Analytical and Applied Pyrolysis, 71*(2), 657–667. https://doi.org/10.1016/j.jaap.2003.09.003

Moon, D. (2019). Pollution and health impacts of waste-to-energy incineration-emissions from emissions from. https://policycommons.net/artifacts/2621214/pollution-and-health-impacts-of-waste-to-energy-incineration/3643886/. Last visited: August 08, 2023.

Mukawa, J., Pająk, T., Rzepecki, T., & Banaś, M. (2022). Energy potential of biogas from sewage sludge after thermal hydrolysis and digestion. *Energies, 15*(14). https://doi.org/10.3390/en15145255

Muter, O., Dubova, L., Kassien, O., Cakane, J., & Alsina, I. (2022a). Application of the sewage sludge in agriculture: Soil fertility, techno-economic, and life-cycle assessment. In *Hazardous waste management*. IntechOpen.

Muter, O., Dubova, L., Kassien, O., Cakane, J., & Alsina, I. (2022b). Application of the sewage sludge in agriculture: Soil fertility, technoeconomic, and life-cycle assessment. In R. B. Jeyakumar, K. Sankarapandian, & Y. K. Ravi (Eds.), *Hazardous waste management*. IntechOpen.

Nair, A. T., Senthilnathan, J., & Nagendra, S. M. S. (2019). Emerging perspectives on voc emissions from landfill sites: Impact on tropospheric chemistry and local air quality. *Process Safety and Environmental Protection, 121*, 143–154. https://doi.org/10.1016/j.psep.2018.10.026

Napora, A., & Grobelak, A. (2014). Sewage sludge influence on microbiological and biochemical soil activity. *Engineering and Environmental Protection, 17*, 619–630.

Ngodhe, S. O. (2018). Effects of sludge on the concentration of heavy metals in soil and plants in obunga slum, Kisumu county, Kenya. *International Journal of Environmental Sciences & Natural Resources, 15*(2). https://doi.org/10.19080/ijesnr.2018.15.555907

Novinscak, A., Surette, C., Allain, C., & Filion, M. (2008). Application of molecular technologies to monitor the microbial content of biosolids and composted biosolids. *Water Science and Technology, 57*(4), 471–477. https://doi.org/10.2166/wst.2008.019

Oberg, G., & Mason-Renton, S. A. (2018). On the limitation of evidence-based policy: Regulatory narratives and land application of biosolids/sewage sludge in bc, Canada and Sweden. *Environmental Science & Policy, 84*, 88–96. https://doi.org/10.1016/j.envsci.2018.03.006

Ohtake, H., & Tsuneda, S. (2018). *Phosphorus recovery and recycling*. Springer.

Oleszczuk, P. (2006). Characterization of polish sewage sludges with respect to fertility and suitability for land application. *Journal of Environmental Science and Health, Part A, 41*(7), 1197–1215.

Owa, F. D. (2013). Water pollution: Sources, effects, control and management. *Mediterranean Journal of Social Sciences, 4*(8), 65–68. https://doi.org/10.5901/mjss.2013.v4n8p65

Palaniveloo, K., Amran, M. A., Norhashim, N. A., Mohamad-Fauzi, N., Peng-Hui, F., Hui-Wen, L., et al. (2020). Food waste composting and microbial community structure profiling. *Processes, 8*, 1–30. MDPI AG.

Pan, I., Dam, B., & Sen, S. K. (2012). Composting of common organic wastes using microbial inoculants. *3 Biotech, 2*(2), 127–134. https://doi.org/10.1007/s13205-011-0033-5

Qadir, M., Drechsel, P., Jiménez Cisneros, B., Kim, Y., Pramanik, A., Mehta, P., & Olaniyan, O. (2020). Global and regional potential of wastewater as a water, nutrient and energy source. *Natural Resources Forum, 44*(1), 40–51. https://doi.org/10.1111/1477-8947.12187

Rashid, T. U., Islam, M. S., Sharmeen, S., Biswas, S., Zaman, A., Khan, M. N., et al. (2017). Applications of chitosan derivatives in wastewater treatment. In *Handbook of composites from renewable materials* (pp. 471–517). Wiley.

Raviv, M. (2009). *The future of composts as ingredients of growing media.* Paper presented at the International Symposium on Growing Media and Composting 891.

Ross, A., & Sim, L.-L. (1991). *Guidelines for the use of sewage sludge on agricultural land.* NSW Agriculture.

Rusănescu, C. O., Rusănescu, M., Voicu, G., Paraschiv, G., Biriş, S. Ş., & Popescu, I. N. (2022). The recovery of vermicompost sewage sludge in agriculture. *Agronomy, 12*(11), 2653.

Sarkar, S. K. (2018). *Marine algal bloom: Characteristics, causes and climate change impacts* (Vol. 4). Springer.

Seleiman, M. F., Santanen, A., & Mäkelä, P. S. A. (2020). Recycling sludge on cropland as fertilizer—advantages and risks. *Resources, Conservation and Recycling, 155*, 104647. https://doi.org/10.1016/j.resconrec.2019.104647

Singh, R. P., & Agrawal, M. (2008). Potential benefits and risks of land application of sewage sludge. *Waste Management, 28*(2), 347–358. https://doi.org/10.1016/j.wasman.2006.12.010

Sinha, S. N., & Paul, D. (2013). *Impact of sewage water on seed germination and vigour index of cicer arietinum l. And pisum sativum l.* Retrieved from http://www.cibtech.org/jfav.htm

Smoliński, A., Karwot, J., Bondaruk, J., & Bak, A. (2019). The bioconversion of sewage sludge to bio-fuel: The environmental and economic benefits. *Materials, 12*(15). https://doi.org/10.3390/ma12152417

Soni, V., Singh, P., Shree, V., & Goel, V. (2018). Effects of vocs on human health. In N. Sharma, A. K. Agarwal, P. Eastwood, T. Gupta, & A. P. Singh (Eds.), *Air pollution and control* (pp. 119–142). Springer Singapore.

Sonibare, O. O., Adeniran, J. A., & Bello, I. S. (2019). Landfill air and odour emissions from an integrated waste management facility. *Journal of Environmental Health Science and Engineering, 17*, 13–28.

Sönmez, F., & Bozkurt, M. A. (2006). Lettuce grown on calcareous soils benefit from sewage sludge. *Acta Agriculturae Scandinavica Section B: Soil and Plant Science, 56*(1), 17–24. https://doi.org/10.1080/09064710510005813

Soudejani, H. T., Kazemian, H., Inglezakis, V. J., & Zorpas, A. A. (2019). Application of zeolites in organic waste composting: A review. *Biocatalysis and Agricultural Biotechnology, 22*, 101396–101396. https://doi.org/10.1016/j.bcab.2019.101396

Spellman, F. R. (2013). *Handbook of water and wastewater treatment plant operations* (3rd ed.). Lewis Publishers.

Stams, A. J. M., Teusink, B., & Sousa, D. Z. (2019). Ecophysiology of acetoclastic methanogens. In *Biogenesis of hydrocarbons* (pp. 1–14). Springer International Publishing.

Steckenmesser, D., Vogel, C., Herzel, H., Félix, R., Adam, C., & Steffens, D. (2022). Thermal treatment of sewage sludge for phosphorus fertilizer production: A model experiment. *Journal of Plant Nutrition, 45*(8), 1123–1133. https://doi.org/10.1080/01904167.2021.1994595

Stylianou, M. A., Inglezakis, V. J., Moustakas, K. G., & Loizidou, M. D. (2008). Improvement of the quality of sewage sludge compost by adding natural clinoptilolite. *Desalination, 224*(1), 240–249. https://doi.org/10.1016/j.desal.2007.06.009

Styszko, K., Durak, J., Kończak, B., Głodniok, M., & Borgulat, A. (2022). The impact of sewage sludge processing on the safety of its use. *Scientific Reports, 12*(1). https://doi.org/10.1038/s41598-022-16354-5

Sugurbekova, G., Nagyzbekkyzy, E., Sarsenova, A., Danlybayeva, G., Anuarbekova, S., Kudaibergenova, R., et al. (2023). Sewage sludge management and application in the form of sustainable fertilizer. *Sustainability (Switzerland), 15*, 6112. MDPI.

Sundberg, C., Smårs, S., & Jönsson, H. (2004). Low ph as an inhibiting factor in the transition from mesophilic to thermophilic phase in composting. *Bioresource Technology, 95*(2), 145–150. https://doi.org/10.1016/j.biortech.2004.01.016

Suthar, S., & Singh, S. (2009). Bioconcentrations of metals (fe, cu, zn, pb) in earthworms (eisenia fetida), inoculated in municipal sewage sludge: Do earthworms pose a possible risk of terrestrial food chain contamination? *Environmental Toxicology, 24*(1), 25–32. https://doi.org/10.1002/tox.20388

Tchobanoglus, G., Burton, F., & Stensel, H. D. (2003). Wastewater engineering: Treatment and reuse. *American Water Works Association Journal, 95*(5), 201.

Tytła, M., & Widziewicz-Rzońca, K. (2021). Heavy metals in municipal sewage sludge—a brief characteristic of potential threats and methods used to assess the ecological risk. *Environment, Earth and Ecology, 5*(1), 18–25. https://doi.org/10.24051/eee/134276

Ungureanu, N., Vlăduţ, V., & Voicu, G. (2020). Water scarcity and wastewater reuse in crop irrigation. *Sustainability (Switzerland), 12*, 1–19. MDPI.

Usman, K., Khan, S., Ghulam, S., Khan, M. U., Khan, N., Khan, M. A., & Khalil, S. K. (2012a). Sewage sludge: An important biological resource for sustainable agriculture and its environmental implications. *American Journal of Plant Sciences, 3*(12).

Usman, K., Khan, S., Ghulam, S., Khan, M. U., Khan, N., Khan, M. A., & Khalil, S. K. (2012b). Sewage sludge: An important biological resource for sustainable agriculture and its environmental implications. *American Journal of Plant Sciences, 03*(12), 1708–1721. https://doi.org/10.4236/ajps.2012.312209

Van Berg, J. J. D. (1993). Effects of sewage sludge disposal. *Land Degradation & Development, 4*(4), 407–413.

Wagner, T., Watanabe, T., & Shima, S. (2018). Hydrogenotrophic methanogenesis. In A. J. M. Stams & D. Sousa (Eds.), *Biogenesis of hydrocarbons* (pp. 1–29). Springer International Publishing.

Wei, H., Tang, Y., Li, A., & Yang, H. (2019). Insights into the effects of acidification on sewage sludge dewaterability through ph repeated adjustment. *Chemosphere, 227*, 269–276.

Wong, J. W. C., & Selvam, A. (2006). Speciation of heavy metals during co-composting of sewage sludge with lime. *Chemosphere, 63*(6), 980–986. https://doi.org/10.1016/j.chemosphere.2005.08.045

Wurtsbaugh, W. A., Paerl, H. W., & Dodds, W. K. (2019). Nutrients, eutrophication and harmful algal blooms along the freshwater to marine continuum. *Wiley Interdisciplinary Reviews: Water, 6*(5), e1373.

Yañez, R., Alonso, J. L., & Díaz, M. J. (2009). Influence of bulking agent on sewage sludge composting process. *Bioresource Technology, 100*(23), 5827–5833.

Ye, W., Han, J., Qin, L., Li, Y., Masami, F., & Yao, H. (2012). Emission characteristics of pm 10 during sewage sludge combustion. *Aerosol and Air Quality Research, 12*(3), 416–421. https://doi.org/10.4209/aaqr.2011.10.0164

Yoshizaki, S., & Tomida, T. (2000). Principle and process of heavy metal removal from sewage sludge. *Environmental Science & Technology, 34*(8), 1572–1575.

Zeng, J., Singh, D., & Chen, S. (2011). Thermal decomposition kinetics of wheat straw treated by phanerochaete chrysosporium. *International Biodeterioration & Biodegradation, 65*(3), 410–414. https://doi.org/10.1016/j.ibiod.2011.01.004

Zhang, C., & Zhang, D. (2020). Microplastics. In W. Cui, S. Fu, & Z. Hu (Eds.), *Encyclopedia of ocean engineering* (pp. 1–9). Springer Singapore.

Zhang, X., Zha, T., Zhu, J., Guo, X., & Liu, Y. (2020). Loading capacity of sewage sludge for forestry application in chinese provincial capital cities. *Sustainability (Switzerland), 12*(18). https://doi.org/10.3390/su12187551

Chapter 5
Sustainable Solutions: Navigating Agrifood Waste with Eco-smart Approaches for a Greener Environment

Maharudra Pratap Singh, Pawan Kumar, Priti Pal, Babli Kumari, Om Prakash, Akhilesh Kumar Singh, Prakash Kumar Sarangi, Brijesh Pandey, and Anand Prakash

1 Introduction

World's population increases exponentially, resulting in the exploitation of natural resources to fulfill the energy and food demands which results in the depletion of resources based on fossil fuels concurrently. Thus, an alternative approach must be needed to complete the need of the population, which brings the agrifood waste in the light. Because agrifood sector produces significant amounts of waste on farm and during processing, this impacts the environment, society, and economy (Paritosh et al., 2017; Castro et al., 2020). Estimation says that about one-third to half the global food production is wasted, accounting for 1.3 billion tons of food annually. On a compressional basis, developed countries have high agrifood waste generation rate than that of developing countries; this is because developed countries have high losses at the consumer level which account for 56% of global agrifood waste, and developing counties losses at the post-harvest and processing level. To overcome this, a comprehensive and integrated global plan is a need to utilize the agrifood waste to provide food and energy security for the limited resources and changing climate, without affecting the ecosystem and biodiversity (Ishangulyyev et al., 2019; Leong & Chang, 2022). Agrifood waste management accounts for an important

M. P. Singh · P. Kumar · B. Kumari · O. Prakash · A. K. Singh · B. Pandey · A. Prakash (✉)
Department of Biotechnology, School of Life Sciences, Mahatma Gandhi Central University, Motihari, Bihar, India
e-mail: anandprakash@mgcub.ac.in

P. Pal
Shri Ramswaroop Memorial College of Engineering & Management, Lucknow, India

P. K. Sarangi
College of Agriculture, Central Agricultural University, Imphal, India

V. Kumar et al. (eds.), *Environmental Engineering and Waste Management*,
https://doi.org/10.1007/978-3-031-58441-1_5

challenge for the upcoming decades because recycling of organic matter from microbes, plants, and animals can contribute toward a sustainable future. In addition, the food industry also utilizes water during agrifood processing result to generate wastewater. To reduce the environmental impact of wastewater discharged from the food industry, a water recirculation system, such as treating and reusing wastewater must be implemented which leads to a more eco-friendly and sustainable future (Carmona-Cabello et al., 2018; Esteban & Ladero, 2018). Globally, 1.3–1.6 billion tons of food is wasted annually, accounting for one-third of food production and over one-fourth of agricultural output. Agrifood waste impacts society, health, environment, and economy thus gaining the attention of both society and governments because agrifood waste reduction can help in reducing hunger, increasing food security, and also helps in reducing climate change by lowering greenhouse gas emissions. Efforts are made globally to reduce agrifood waste by emphasizing sustainable and innovative approaches, controlled supply chains, and increasing awareness (Amicarelli & Bux, 2020; Freitas et al., 2021). Agrifood waste is a major issue that arises both on and off the farm. On-farm waste can result from crop spoilage, inefficient harvesting techniques, and overproduction, while off-farm waste occurs during processing, transportation, storage, and retail. With an average moisture level of 82.5%w/w, the dry content of food waste is made up of 51.2% carbon, 38.1% oxygen, 7.2% hydrogen, 2.8% nitrogen, and 0.7% w/w sulfur. This waste not only compromises safety but also has an adverse effect on the environment by releasing CO_2, emitting methane, and causing eutrophication. Food waste disposal emits around 4.4 billion metric tons of CO_2, accounting for approximately 6% w/w of worldwide CO_2 emissions (Makanjuola et al., 2020; Tsegaye et al., 2021). European Union (EU) generates 130 million tons of agrifood waste annually, with 39% from households, 24% coming from primary production, 23% from processing and manufacturing, 9% from the hospitality industry, and 5% from retail. Since these streams are frequently made up of a single biomass resource, up to 52% can be easily recycled from them. Fruits and vegetables account for almost 50% of total food losses in the EU, highlighting the significant impact of these items on the overall agrifood waste problem. Reducing waste in this category could significantly decrease total food losses. Addressing agrifood waste at every stage of the supply chain is crucial for a more sustainable and efficient food system. Around 26% of total agrifood waste is generated by the drink industries followed by 21% dairy industry, 14.8% fruit and vegetable industry, and 12.9% cereal industry, making drink industries a primary resource of waste production (Caldeira et al., 2019; Gómez-García et al., 2021). Agrifood wastes comprise liquid wastes like whey and olive mill wastewaters; solid wastes like fruit pomace and pericarp; and waste biomass from plant material such as leaves and stalks after pruning. By including energy and other resources necessary for plant development and output, these losses may become greater. Waste biomass disposal can exacerbate environmental issues such as rising emissions of greenhouse gases and soil degradation. Hence, in order to reduce waste and mitigate the overall environmental effect of the agrifood industry, it is critical to utilize sustainable ways for managing and utilizing these plant components (Esteban & Ladero, 2018; Squillaci et al., 2021). Agrifood waste materials are a valuable supply of phytochemicals, with numerous applications. Typical solutions include incineration, composting, livestock feed, extracting chemicals with additional value, and biotransformation utilizing microorganisms. Using waste properly can result in byproducts that have promising uses in a number of industries, including materials, biopolymers, biofuels, and components for food and cosmetics. As a result, higher-value income is produced and waste streams have a smaller negative environmental impact. Waste that is not landfilled or burned supports sustainable waste management and reduces greenhouse gas emissions. New technologies and procedures for transforming agrifood waste into high-value products have the potential to generate additional employment possibilities and contribute to a circular economy (Ben-Othman et al., 2020; Socas-Rodríguez et al., 2021). Waste materials contain organic carbon, fostering the development of a circular bio-based economy. Agrifood waste contains many different organic compounds, which include carbohydrates, biophenols (phenolic acids, flavonoids, and tannins), proteins, lipids, essential oils (terpenoids, hydrocarbons, lactones, and polyacetylenes), and pigments (anthocyanins, carotenoids, betalains, and chlorophyll), which range from a few mg to a few g per kg and have commercial values. The presence of these compounds presents an opportunity for recovery and commercialization. The extraction and purification processes can increase the value of the final product, making it economically viable to utilize these waste materials. Additionally, utilizing these compounds contributes to a sustainable and circular economy by reducing waste and maximizing resource efficiency. They can be used to recover phytochemicals and biofuels because they are cheap, abundant, renewable resources. It is critical to ensure the safety of the end products obtained from this procedure, especially when different solvents are used to ensure that these items are safe, strict safety studies are necessary (Carmona-Cabello et al., 2018; Rodríguez García & Raghavan, 2022). This chapter emphasizes the applicability of agrifood waste valorization in addressing environmental problems and recovering valuable phytochemicals. Pre-treatment procedures maximize the potential to utilize the phytochemicals in biofuels, livestock feed additives, and natural food additives, so contributing to a more sustainable and circular economy.

2 Agrifood Waste Generation and Their Types

Urbanization and the global population have contributed to a rise in waste from agrifood sources as a result of poor waste management practices. The majority of agrifood waste is generated in the course of production, harvesting, transporting, manufacturing, and consumption activities. This waste has a negative impact on the environment, increases greenhouse gas emissions and water pollution, and misses opportunities to solve food insecurity and hunger (Singh et al., 2022). Agrifood processing industries generate large volumes of organic waste, including crop residues, agri-industrial waste, and fruit and vegetable waste. This waste includes cereals, legumes, veggies, fruits, dairy products, meat, eggs, and seafood,

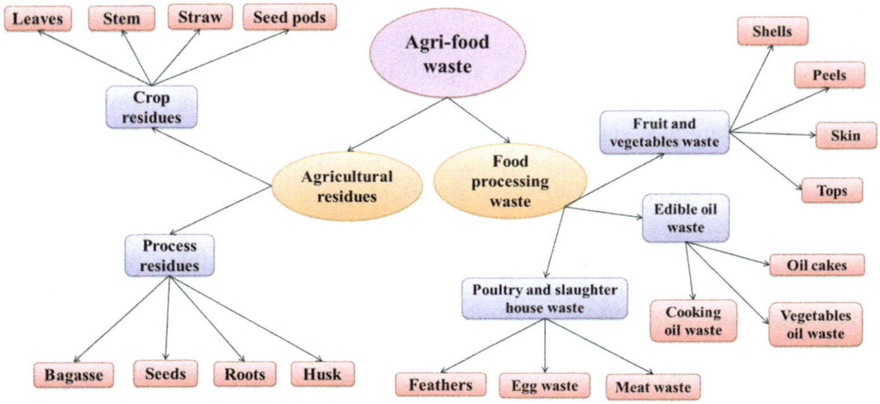

Fig. 5.1 Overview of different types of agrifood waste

contributing to waste management problems (Fig. 5.1) (Haldar et al., 2022). To reduce waste and ensure efficient resource use, it is crucial to improve waste management procedures and support sustainable production and consumption behaviors (Dhanker et al., 2022). Agrifood waste can be converted into substrates for the production of enzymes and other bio-products, which can help ease environmental and waste management concerns. New technology and alternative approaches should be implemented to successfully utilize nonconsumable sections, resulting in appropriate waste management as well as sustainable growth. For sustainable development, emphasis should be placed on implementing the latest technologies and alternative approaches (Yaashikaa et al., 2022).

2.1 *Food Processing Waste*

The food processing and packaging sectors provide a wide range of food products; yet unprocessed food, leftovers, spoilage, and contamination are inevitable events. Food waste, which includes starch-based vegetable trimmings, fruit peels, dairy, meat, fish, wasted cereals, and pulp debris, can be used as low-cost starting materials to create primary and secondary metabolites with commercial values. Nutrient loss and environmental pollution are frequent outcomes of traditional disposal techniques such as landfilling, incineration, composting, and anaerobic digestion. As a result, there is an increasing interest in developing novel and environmentally friendly methods for using food waste, such as bioconversion techniques or biorefineries, to maximize value while limiting environmental harm (Ravindran et al., 2018).

2.1.1 Fruit and Vegetable Waste (FVW)

Fruits and vegetables (FVW) are among the most commonly used horticultural crops. Due to the nutrients and health-improving elements they contain, they are eaten raw, lightly processed, and processed. Producing and processing of horticulture commodities, particularly fruits and vegetables, have greatly increased in response to the expanding population and changing dietary preferences to meet the growing demands (Sharma et al., 2020a, b). A significant amount of fruit and vegetable global production includes citrus (124.73 MMT), bananas (114.08 MMT), apples (84.63 MMT), grapes (74.49 MMT), mangoes, mangosteen, guavas (45.22 MMT), pineapples (25.43 MMT), potatoes (3820.00 MMT), tomatoes (171.00 MMT), cabbages and other brassicas (71.77 MMT), carrots and turnips (38.83 MMT), peas (17.42 MMT), and cauliflowers, broccolis (24.17 MMT). These vital food commodities, however, experience enormous losses and waste as a result of high production and growth, improper handling techniques, and poor infrastructure. According to UN estimates, at least one-third of the 1.3 billion metric tons produced worldwide each year are lost or discarded, with horticulture products suffering the greatest losses, up to 60% (Sagar et al., 2018). FVW is produced as a result of microbial attack, discoloration, and metabolic reactions. Landfilling and incineration of FVW are unsustainable because of greenhouse gas emissions and high construction and operating costs. Anaerobic decomposition of FVW into biofertilizers can reduce pollution while increasing soil nutrition. In developing countries, FVW is transformed into animal feed to alleviate feed scarcity, reduce food waste, and provide a sustainable solution for the livestock industry. By completing the nutritional cycle and lowering reliance on imported feed ingredients, this method supports the circular economy (Sahoo et al., 2021).

2.1.2 Edible Oil Waste

Cooking or Edible oil waste is generated by the edible oil industry during a variety of processing steps, including neutralization, degumming, deodorization, bleaching, and hydrolytic/oxidative rancidity. If improperly managed, this waste poses threats to the environment and human health. The edible oil industry must put in place efficient waste management techniques to reduce adverse environmental effects and assure safe disposal or recycling (Okino-Delgado et al., 2017). Waste cooking oil, a dark-colored liquid obtained by deep-frying, is unfit for human consumption due to its fatty acids, aldehydes, ketones, and other components. Industrial waste from edible oils can, however, be co-fermented with food waste to serve as a less expensive substrate for microbial lipid production. This method offers a sustainable way to produce important microbial lipids that can be used in the production of biofuels, cosmetics, and pharmaceuticals while reducing the impact on the environment (Sahoo et al., 2021).

2.1.3 Kitchen Waste

Significant kitchen waste (KW) is produced by households, restaurants, public catering facilities, and hotels as a result of urbanization, economic development, and rapid population expansion. KW comprises prepared foods, leftovers, meat, shells, eggshells, oil, and grease (Sharma et al., 2020a, b). As it contains a wealth of bioactive chemicals, organic waste (KW) decomposes quickly. KW residues are desirable substrates for the microbial synthesis of enzymes because they include salts, lignin, lipids, proteins, and carbohydrates (Sharma et al., 2022). With the use of solid-state fermentation, *Rhizopus oryzae* may co-produce α-amylase and protease from moistened bread waste at a low cost. Via enzymatic reactions, this technique enhances the food supply chain and food security by reducing waste and transforming waste into value-added items. Ultimately, this improves food security by resulting in a more efficient and sustainable food supply chain (Benabda et al., 2019).

2.1.4 Poultry, Slaughterhouse, and Egg Processing Waste

The meat, poultry, egg, and associated food processing sectors produce a large quantity of waste. Animal waste has a variety of valuable elements that, when recycled properly, can be used as fertilizer for crops, animal feed, and energy production. If slaughterhouse waste, like skin, hair, feathers, hooves, horns, and deboning leftovers, is not managed, it poses health and environmental risks. The risk can be minimized, and sustainability is supported by utilizing it as a raw resource in commercial operations. This method generates goods such as fertilizer, animal feed, biofuels, and pharmaceuticals while minimizing the environmental impact (Singh et al., 2014). Chicken feather waste is a useful source of protein for a variety of industrial uses, and bacteria like *Bacillus* sp. *FPF-1, Brevibacillus* sp., *Chryseobacterium* sp. *FPF-8,* and *Nnolim-K2* have excellent keratinase-producing capacity. These bacteria efficiently degrade keratin into smaller peptides and amino acids, making them useful in the manufacturing of animal feed and biofertilizers. Furthermore, the use of chicken feather waste for the synthesis of keratinase helps lessen environmental damage brought on by inappropriate disposal (Nnolim et al., 2020).

2.2 *Agricultural Waste (AW)*

Agricultural waste (AW) is generated in millions of tons each year, causing environmental and ecosystem damage. Governments, environmentalists, and other stakeholders strive to turn AW into valuable goods by utilizing practical, sustainable, and cost-effective techniques. Effective management and use can support ecological sustainability and the security of the energy supply. As part of AW, bran, bagasse, husks, leaves, peels, seeds, stems, and stalks are used to make fertilizers, animal feed, and soil conditioners. These AW components can also be used to make biofuel, bioplastics, and other sustainable materials. Developing new ways to extract value from AW, not only lessen its negative environmental impact but also contribute to a more circular and sustainable economy (Yaashikaa et al., 2022). Agri-waste is a source of lignocellulosic biomass (LCB), which is predominantly made up of carbonaceous materials like cellulose and other important parts. In order to manufacture second-generation biofuels and other important commodities, lignocellulosic biorefineries, which are being studied, may use these polymers as potential substrates. The conversion of LCB into biofuels requires enzyme-mediated saccharification of complicated polysaccharides into straightforward fermentable sugars utilizing cellulolytic enzymes (Sharma et al., 2020a, b).

3 Agrifood Waste Management Systems

Agrifood waste management systems are critical for tackling the environmental and economic difficulties connected with waste disposal during agricultural and food production methods. These systems strive to reduce waste, encourage sustainability, and improve resource efficiency. Agrifood waste management systems typically consist of the following important elements and techniques (Koul et al., 2022):

3.1 Waste Prevention and Reduction

Reduce waste production at the source as much as possible is the first step toward waste prevention and reduction. Agricultural procedures should be improved to limit waste creation, such as precision farming, irrigation optimization, and integrated pest management, which will cut down on crop losses and output waste. Post-harvest handling and supply chain management involve improved storage, handling, and coordination to minimize losses and waste at various stages of the supply chain (Lin et al., 2009).

3.2 Waste Sorting and Separation

Waste sorting and separation are critical for efficient waste treatment. This entails separating organic waste such as crop residue, discarded food, and animal waste from inorganic waste such as plastics, packaging, and chemical containers. Recycling and composting can reduce waste sent to landfills and produce useful resources like organic fertilizer and animal feed. Composting organic waste yields nutrient-rich compost, enhancing the fertility and structure of the soil. Anaerobic digestion transforms organic waste into biogas and digestate, which are suitable for use as renewable energy sources and nutrient-rich fertilizers in agriculture, effectively closing the nutrient loop (Zhang et al., 2022).

3.3 Biomass-to-Energy Conversion

Agrifood waste can be transformed into biofuels, charcoal, or syngas using conversion processes like gasification and pyrolysis. These materials can be utilized to generate energy, in industrial operations, or as a feedstock for bio-based products. Technologies for biomass conversion offer various environmental advantages in addition to their potential for energy generation. These methods aid in reducing methane emissions and waste management by keeping agrifood waste out of landfills. Also, by converting biomass into biochar, soil fertility, and carbon sequestration can be improved, boosting sustainable agricultural practices (Kumar Sarangi et al., 2023).

3.4 Policy and Regulations

Governments have a significant impact on the promotion and enforcement of waste management laws, the encouragement of sustainable behaviors, and the funding of research and development for waste reduction solutions. Governments and regulatory organizations are essential in developing waste management regulations, creating standards, and offering incentives to promote ethical waste handling practices. These rules can encourage ethical waste management and aid in the adoption of green technologies. In addition, governments can work with businesses and communities to raise awareness about the value of recycling and waste reduction (Salimi, 2021).

3.5 Public Awareness and Education

Increasing awareness about waste reduction, recycling, and responsible consumption among farmers, food producers, and consumers can induce behavioral changes and encourage participation in waste management efforts. By educating them about the environmental impact of waste and sustainable strategies, they can make an informed choice to minimize waste generation. A more integrated approach to waste management can be developed through collaboration among stakeholders, leading to more efficient and long-lasting solutions (Zhang et al., 2022).

Agrifood waste management systems necessitate collaboration and engagement from stakeholders such as farmers, food processors, merchants, waste management firms, and policymakers. These methods lessen the impact on the environment, maximize resource use, and develop a circular, sustainable agrifood system. Innovative solutions, such as waste reduction methods, composting, recycling, and alternative applications, can help offset negative effects, offer economic possibilities, and contribute to a more resilient and responsible management system (Kumar Sarangi et al., 2023).

4 Agrifood Waste Utilization Routes

As the population increases, agrifood production, as well as consumption rate, also increases to cope with the food demand. Research suggests that agrifood manufacturing accounts for 30–40% of global loss of food production per year due to the current emission-intensive, multifaceted, and inefficient manufacturing process. Agrifood waste is a major concern due to rapid resource consumption and poor waste management methods like landfill, incineration, and composting which pollute the environment and cause high costs for consumers, producers, and public bodies. Thus, upcycling agrifood waste is crucial, as it has a global value of $700 billion annually. This involves innovative repurposing, reducing environmental impact, creating economic opportunities, and addressing food insecurity by providing nutritious and affordable options (Castro et al., 2020; Leong & Chang, 2022). Due to inadequate methods of transforming carbon-based resources into value-added products, agrifood waste contributes 7% of global emissions of greenhouse gases. Adopting a circular economy can help meet the rising need for key nutrients and also help to minimize greenhouse gas emissions, save resources, and promote long-term growth. Agrifood management can adopt circular economy strategies to increase resistance to climate change and provide new economic opportunities (Makanjuola et al., 2020; Ben-Othman et al., 2020). A circular economy reduces the need for raw materials by reusing products and components to maximize resource yields. Such an approach extends the longevity of agricultural byproducts by evaluating all stages of agrifood manufacturing and distribution, including waste valorization processes. Unavoidable waste is being transformed into useful chemicals, nutrients, and fuels for industry. Agrifood waste valorization techniques offer value-added products extraction and producing biofuels utilizing pyrolysis (Paritosh et al., 2017; Castro et al., 2020).

4.1 Extraction Techniques for Value-Added Products from Agrifood Wastes

Green extraction techniques, such as supercritical fluid extraction, subcritical water extraction, pulsed electric fields, ultrasound-assisted extraction, microwave-assisted extraction, enzyme-assisted extraction, and pressurized liquid extraction, are drawing attention for their cost-effectiveness, environmental friendliness, and emphasis on safety (Table 5.1). Six green chemistry principles are used in these processes: using non-denatured, biodegradable extracts; using water or green solvents; reducing energy input; co-product production from waste; and minimal unit operations. These methods are critical for extracting value-added products from agrifood waste (More et al., 2022).

4.1.1 Supercritical Fluid Extraction (SFE)

Supercritical fluid extraction is a green approach that employs carbon dioxide (CO_2) to recover bioactive compounds from agrifood waste. It has a low critical temperature and pressure, allowing for moderate operation between 40–60 °C and 200–400 bars. By rapid fluid diffusion across solids, this method allows for solvent physicochemical changes, improves selectivity, and increases yields (Fornereto Soldan et al., 2021). CO_2 is thermodynamically stable, nontoxic, noncarcinogenic, nonmutagenic, and nonflammable. Its low polarity, meanwhile, is a drawback. Little quantities of co-solvents or modifiers, which boost solute solubility and increase extraction yield, can be employed to reduce this effect (Torres-Valenzuela et al., 2019). SFEs can be used in a variety of fields, including manufacturing and the pharmaceutical industry for the extract of phenolic substances, antioxidants, saponins, and carotenoids. The extraction procedure employs extraction times and flux ranging from 30 to 180 min and 1.7 to 133 g/min, as well as pressures, temperatures, and co-solvents ranging from 30 to 40 bar, 33 to 230 C, and 5–15% v/v (de Andrade Lima et al., 2018).

4.1.2 Subcritical Water Extraction (SCWE)

Subcritical water extraction is an environmentally friendly, economically feasible alternative to solvent extraction, using water as an extractant, making it non-flammable, renewable, and economically viable (Munir et al., 2018). SCWE involves heating feedstock in the aqueous phase at sub-critical temperature (150–320 °C) and pressure (20–150 bar), enhancing mass transfer and extractability of water-soluble bioactive compounds. This is due to the hydrolysis of bonds between phenolic compounds and the agrifood waste matrix, enhancing the properties of water change (Benito-román et al., 2020). SCWE enhances the mass transfer ratio for extracting bioactive compounds from agrifood waste, including stilbenes,

Table 5.1 An overview of the use of various green extraction techniques for the extraction of value-added products from agrifood waste

Extraction method	Agrifood waste	Target compound	References
Supercritical fluid extraction	Agave salmiana bagasse	Antioxidants and saponins	Santos-Zea et al. (2019)
	Avocado peel and seeds	Catechin and quercetin	Restrepo-Serna et al. (2022)
	Capsicum annuum waste	Phenolics, flavonoids, fatty acids, and carotenoids	Kostrzewa et al. (2022)
	Carrot peels	Carotenoid	de Andrade Lima et al. (2018)
	Grape seeds	Triacylglycerols	de Souza et al. (2020)
	Mandarin peel	Limonene, hesperidin	Phucharoenrak et al. (2022)
	Mango peel	Carotenoids	Villacís-Chiriboga et al. (2021)
	Melon seeds	Phytosterols	Ekinci and Gürü (2019)
	Pomegranate seed	Seed oil	Tan et al. (2021)
Subcritical water extraction	Citrus peel	Hesperidin, narirutin	Hwang et al. (2021)
	Grape pomace	Phenolic compounds	Pedras et al. (2020)
	Kiwifruit peel	Phenolic compounds	Guthrie et al. (2020)
	Onion peel and skin	Phenolic compounds	Benito-Román et al. (2020)
	Peach palm	Phenolic compounds, Sugars	Giombelli et al. (2020)
	Date pits	Phenolics, Antioxidants, and Dietary fibers	Li et al. (2020)
Pulsed electric fields	Apple pomace	Phenolic compounds	De la Peña-Armada and Mateos-Aparicio (2022)
	Banana peels	Phenolic compounds	Hendrawan et al. (2019)
	Jackfruit waste	Pectin polysaccharide	Lal et al. (2021)
	Lemon peels	Phenolic compounds	Peiró et al. (2019)
	Potato peels	Phenolic compounds	Frontuto et al. (2019)
	Olive pomace	Phenolic compounds	Andreou et al. (2020)
	Apple pomace	Phenolic compounds	Pollini et al. (2021)

(continued)

Table 5.1 (continued)

Extraction method	Agrifood waste	Target compound	References
Ultrasound-assisted extraction	Beet leaves	Bioactive compounds	Nutter et al. (2021)
	Grape pomace	Phenolic compounds	da Rocha and Noreña (2020)
	Kiwi peel	Flavonoids	Giordano et al. (2021)
	Orange peels	Bioactive compounds	Boukroufa et al. (2015)
	Tomato peel	Lycopene	Munde et al. (2017)
Microwave-assisted extraction	Coffee pulp	Phenolic compounds, flavonoids, chlorogenic acid, and caffeine	Tran et al. (2022)
	Peach waste	Phenolic compounds and anthocyanins	Theocharidou et al. (2023)
	Eggplant peel	Phenolic compounds, flavonoids, anthocyanins	Doulabi et al. (2020)
	Lemon peel waste	Essential oil (limonene, β-pinene, and γ-terpinene) and pigment	Martínez-Abad et al. (2020)
	Perillae folium leaves	Flavonoids	Shang et al. (2022a, b)
	Spent onion skins	Flavonoids (quercetin, kaempferol, luteolin, and quercetin-3-O-β-D-glucoside)	Shang et al. (2022a, b)
Enzyme-assisted extraction	Apricot pulp	Polysaccharides (sodium glycocholate and sodium taurocholate)	Xu et al. (2022)
	Tomato peel and seed	Carotenoids and Lycopene-rich oleoresins	Lombardelli et al. (2020)
	Beetroot cell wall	Betalains	Lombardelli et al. (2021)
	Sugar beet leaves	Protein	Akyuz and Ersus (2021)
	Raspberry pomace	Phytosterols and Polyphenols	Saad et al. (2019)
Pressurized liquid extraction	Pomegranate peel and carpelar	Phenolic compounds (α, β punicalagin, and ellagic acid)	García et al. (2021)
	Beetroot leaves and stems	Phenolic compounds (ferulic acid, vitexin, and sinapaldehyde)	Battistella Lasta et al. (2019)
	Olive pomace	Phenolic compounds (phenolic alcohols, secoiridoids, flavonoids, and lignans)	Cea Pavez et al. (2019)
	Pomegranate peel and seed	Protein and Phenolic compounds (phenolic acids, flavonoids, and hydrolysable tannins)	García et al. (2021)

phenolic compounds, and protein hydrolysates, by reducing water viscosity, surface tension, and diffusivity (Rodrigues et al., 2021).

4.1.3 Pulsed Electric Fields (PEFs)

The pulsed electric field is a non-thermal agrifood processing method using high-intensity pulses causing membrane electroporation to extract bioactive compounds (Arshad et al., 2021). PEF treatment efficiency depends on factors like electric field strength, energy input, treatment time, and temperature. Moderate electric field intensity (0.5–10 kV/cm) and relatively low energy input (1–10 kJ/kg) have favorable effects on plant cell membrane permeabilization, enabling high recovery yields of intracellular compounds from agrifood wastes (Pataro et al., 2020). PFE treatments offer low solvent consumption, shorter treatment times, energy efficiency, continuous operability, ease of scale-up, non-destructive nature, and high selectivity but rely on medium composition and equipment cost (Carpentieri et al., 2021).

4.1.4 Ultrasound-Assisted Extraction (UAE)

In comparison to traditional extraction techniques, ultrasound-assisted extraction is a cost-effective, environmentally sustainable method that uses the usual laboratory equipment and requires less solvent (Ranjha et al., 2021). Drying and solid–liquid separation are drawbacks of the UAE process. It is dependent on cavitation, which is brought on by ultrasound (20 kHz–100 MHz frequency) compression and expansion cycles. As a result of the bubbles that acoustic waves produce, particles are dislodged and the diffusion of bioactive compounds into the solvent is improved. The energy released by bubble implosions affects temperature, pressure, and the rate of mass transfer. Sample parameters like consistency, rheology, and particle mobility all have an impact on UAE extraction efficiency (Panzella et al., 2020).

4.1.5 Microwave-Assisted Extraction (MAE)

Microwave-assisted extraction is a low-cost, environmentally friendly approach that increases productivity, decreases extraction time, and uses less solvent. It makes use of electromagnetic radiations between 300 MHz and 300 GHz to transform energy into thermal energy, raising the extraction temperature and mass transfer rate (Wen et al., 2020). The heating effect of microwaves is dependent on the dielectric properties of the solvent, and MAE breaks down hydrogen bonds, causing dipole rotation, ion migration, and component dissolution (Saini et al., 2019). Microwave power, frequency, irradiation period, sample matrix particle size, solvent composition, extraction temperature, pressure, and cycles all have an impact on MAE extraction. The choice of solvent depends on solubility, dielectric constant, and dissipation factors. Increased dielectric constants and dissipation factors result in improved extraction, which can be accomplished by combining water with solvents such as ethanol or methanol. MAE is frequently used to extract bioactive chemicals, particularly from plant sources (Wen et al., 2020).

4.1.6 Enzyme-Assisted Extraction (EAE)

Enzyme-assisted extraction is a safe, efficient, and gentle method using water as solvent rather than organic chemicals. EAE catalyzes cell wall hydrolysis and binding to target components, while pectinases, cellulases, and hemicellulases speed up recovery by rupturing cell membranes and walls (Nadar et al., 2018). EAE is affected by a number of factors, including enzyme concentration, composition, water-to-solid ratio, enzyme-to-substrate ratio, pH, extraction temperature, and hydrolysis duration. Careful selection is required for effective chemical extraction under certain conditions. Additional important influencing factors are the matrix-to-solvent ratio and particle size. The inability of EAE to hydrolyze matrix cell walls limits phytochemical and bioactive component extraction yields (Wen et al., 2020).

4.1.7 Pressurized Liquid Extraction (PLE)

Pressurized liquid extraction is a cost-efficient green extraction method that increases solvent pressures, improving solubility and diffusion rates. It reduces surface tension and solvent viscosity and can be improved by adjusting solvent, temperature, pressure, and time. PLE is often combined with chromatographic and spectral techniques like UPLC-MS/MS for component identification, characterization, quantification, and antioxidant activity evaluation (Dobroslavi'c et al., 2022).

4.2 Biofuel Production by Pyrolysis

Pyrolysis is the anaerobic decomposition of organic material that produces lower temperatures and fewer pollutants. It yields three fractions: bio-oils, charcoal, and biogas, which are all useful as gaseous energy sources, liquid fuels, and bio-fertilizers. The temperature and heating rate can be adjusted to control the process (Chen et al., 2014). The products have great potential value because bio-char can be used as a fertilizer or burned as energy, while bio-oil can be converted into a biofuel or used to generate power. Rice straw, wheat straw, sugarcane bagasse, walnut shells, and olive husks are just a few examples of agrifood crop wastes that could be used as pyrolysis feedstock. Three different pyrolysis temperatures (300 °C, 400 °C, and 500 °C) and particle sizes were applied to the biomass, and the results showed that the higher the reaction temperature, the more the gas was produced. Potential uses for the bio-oil fraction include medications, fertilizers, and transportation. Due to its high oxygen content, bio-oil for transportation requires upgrading to remove organic compounds such as acids, ketones, esters, phenols, and hydrocarbons, which can cause instability and corrosivity (Gong et al., 2021). Microbial fuel cells have employed bio-char as a catalyst, while wastewater treatment methods have used it as an adsorbent. The digestate utilized as a feedstock, electrolyte, and bacterium supply for MFCs was designed simply, with two graphite felt electrodes. Smaller fuel cells and a higher surface area-to-volume ratio for electrochemical processes were both used to increase energy efficiency. MFCs are stacked parallel to maximize power production and meet specialized energy generating requirements. This might offer a different method for producing localized electricity from agrifood waste, which might be advantageous for use or sale locally (Islam et al., 2023). Pyrolysis minimizes waste volume and disposal costs while also adding value by producing biofuel, reducing energy and fossil fuel use, and lowering the carbon footprint. With the help of this circular, sustainable process, low-value waste materials can be reused rather than discarded and converted into energy and byproducts (Chen et al., 2014).

5 Strategies for Producing Renewable Value-Added Products from Agrifood Waste

Agrifood wastes, including peels, seeds, shells, and leaves, contain various nutrients, bioactive molecules like anthocyanins, carotenoids, and phenols, and secondary metabolites like alkaloids, polyphenols, and terpenes. These renewable resources are utilized in the pharmaceutical, cosmetic, and food industries for various purposes (Tsegaye et al., 2021). A number of inventive systems are being invented for this purpose, one of which treats agrifood waste as biorefineries, including all streams exiting the target system for value-added products, with the goal of minimizing residues. Additional approaches include using waste as an energy source for the production of biofuels, as nanomaterials for food packaging, and as a nutrient source for both human and animal food (Paini et al., 2021). Energy-intensive waste conversion technologies are causing waste streams. Newer technologies are super-critical CO_2 extraction, enzyme-activated extraction, pulsed electric field, high-pressure treatment, and ultra-sonication produce minimal waste. Food processing must re-evaluate to retain nutrients and create value-added products, ultimately aiming for circular integration (Samborska et al., 2019).

5.1 As a Source of Pigments and Essential Oils

Agrifood wastes are a rich source of organic compounds, namely carbohydrates, bio-phenols, proteins, lipids, essential oils, and pigments. Given the high content of these wastes, they can be used as a culture medium for microorganism growth

(Papaioannou et al., 2022). As an alternative to chemical synthesis techniques, agrifood waste has potential for the generation of bioactive chemicals such as aroma compounds, essential oils, and pigments. By utilizing agrifood waste, we can reduce our reliance on chemical synthesis methods that often have negative environmental impacts. Furthermore, by reusing waste materials and transforming them into valuable resources for a number of industries, this approach promotes sustainability (Hadj Saadoun et al., 2021).

5.1.1 Pigments

Agrifood waste contains natural food pigments like anthocyanins, betalains, carotenoids, and chlorophylls, providing bioactive compounds and natural colorants. Tiwari et al. (2019) obtained high carotenoid (82.66 g/g) and beta-carotene (78.37 g/g) yields from carrot pomace by ultrasonically extracting it with flaxseed oil. Also, in order to extract carotenoids from pomegranate peels using ultrasonic method with sunflower oil and soy oil as green solvents, 0.6715 mg and 0.6134 mg of carotenoid per 100 g of dry peels is yielded, respectively. Moreover, the process removed betalains from red beetroot waste while operating at 44 kHz for 30 min at 30 °C. In order to extract pigment from agrifood waste, supercritical CO_2 was also implemented to extract beta-carotene from tomato by-products (28.4–58.8%) and pumpkin seeds (18.50 mg/100 g sample) (Kehili et al., 2017). Ghada et al. (2020) used HPLC-DAD-ESI/MS to extract anthocyanins from passion fruit epicarp, yielding 3.4 mg/g dried epicarp and 9 mg/g extract. Mitrea et al. (2020) investigated the physicochemical properties of polyvinyl alcohol biofilms enhanced with eco-friendly polyols and carotenoid-rich extracts. They created beta-carotene pigments from tomato peels and seeds, substituting synthetic colorants with natural pigments derived from agrifood waste and employing biodegradable polymers in bioplastic industry packaging applications. When quercetin and its derivatives were extracted from onion waste using a green technique, it was discovered that cyanidin 3-O-glucoside was the primary colorant, while quercetin and its derivatives were the predominant polyphenols. On the other hand, yogurt matrix uses cyanidin 3-O-glucoside as a stable and natural colorant (Mourtzinos et al., 2018). Agrifood wastes can be used for microbial pigment synthesis, providing nontoxic, high-quality, biodegradable, eco-friendly, and season-independent pigment synthesis. Microorganisms such as bacteria and fungi produce a wide range of pigments (Kaur et al., 2022). Kantifedaki et al. (2018) extracted pigments from orange peels using the fungal strains *Monascus purpureus* and *Penicillium purpurogenum*, yielding 9 AU/g, 0.95 AU/mL, and 0.58 AU/mL for yellow, orange, and red pigments, respectively. The solid-state fermentation of beta-carotene by *Planococcus* sp. (47.13 1.9 mg/g) and the hydrolysate of sugarcane bagasse utilized by *Monascus ruber* to produce red pigments (18.71 AU) demonstrate the potential of sugarcane bagasse as a source of pigments. This approach has issues with product stability and safety because microorganisms can produce mycotoxin and natural pigments can fade over time (Kaur et al., 2022).

5.1.2 Essential Oils

Essential oils are concentrated plant extracts obtained through distillation or cold pressing from flowers, leaves, stems, bark, or roots. For centuries, they have been utilized in cosmetics, aromatherapy, and natural remedies. Residual aromatic biomass, which is high in secondary metabolites, is extracted from wasted fruits and vegetables and used to make health-promoting additives and nutraceuticals (Zaccardelli et al., 2021). da Silva and Jorge (2016) extracted essential oils from citrus fruit waste using chloroform, methanol, and water. These oils contain various compounds, including limonene, geranial, linalool, citronellol, neral, humulene, α-phellandrene, and ocimene. Another important source of essential oils is citrus leaves. Boluda-Aguilar and López-Gómez (2013) used residual lemon peel for the isolation of D-limonene, with the yield of 6 L per 1000 kg. Additionally, Grape seeds also contain essential oils like linoleic, oleic, palmitic, and α-linoleic acids. Kovalcik et al. (2020) extracted essential oils at 4.2 g per 100 g of seeds using n-hexane.

5.2 *Agrifood Waste as a Source of Bioactive Compounds*

Agrifood wastes contain bioactive components that can be used in the biomedical, cosmetics, and food industries as food supplements and functionalized materials, such as bioactive peptides, carotenoids, dietary fibers, and phenolic compounds (Fig. 5.2) (Panzella et al., 2020).

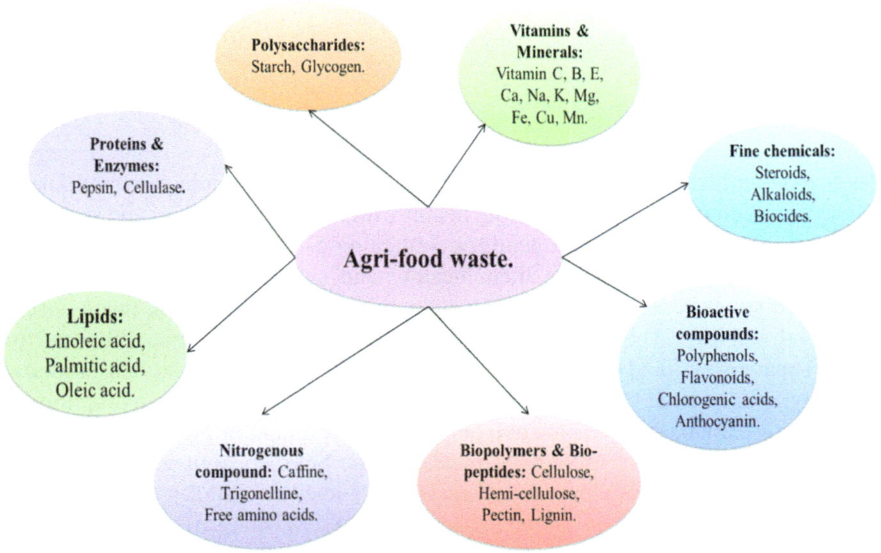

Fig. 5.2 Overview of agrifood waste as a source of various bioactive molecules

5.2.1 Phenolic Compounds

Phenolic compounds are secondary metabolites synthesized through the phenylpropanoid and shikimic acid pathways. These substances, which can be found in plant tissues and agrifood waste, have antibacterial, anti-inflammatory, antioxidant, and anti-proliferative qualities that make them valuable components in a variety of industries (Albuquerque et al., 2021). Dietary polyphenols are crucial for human health because they lower the risk of chronic diseases like cancer, cardiovascular disease, and inflammation. Oxidative stress caused by reactive oxygen and nitrogen species is associated with several disorders. Polyphenols are the antioxidants in food, neutralizing free radicals, serving as direct radical scavengers, and metal chelators of transition metals such as Fe^{2+}. Furthermore, they prevent oxidation brought on by active hydroxyl radicals and slow down Fenton processes (Di Lorenzo et al., 2021). Grape and wine waste, which contains a variety of phenolic compounds like lignin, tannins, phenolic acids, hydroxytyrosol, flavonoids, and anthocyanins, generate nine million tons of waste annually. Byproducts from the orange and lemon juice industries, such as citrus peels, seeds, and pulp, produce 15 million tons of waste each year that are rich in flavonoids and hydroxycinnamic acids (Mourtzinos & Goula, 2019). In agrifood waste, fruit byproducts contain phenolic compounds, whereas vegetable byproducts contain anthocyanins, chlorogenic acids kaempferol, naringenin, quercetin, and rutin (Panzella et al., 2020). Byproducts from the processing of cereals such as corn, wheat, rice, and barley contain phenolic compounds such as caffeic acids, catechin chlorogenic, ferulic, hydroxybenzoic, protocatechuic, sinapic, syringic, vanillic, vanillin, and others. Additional lignocellulosic agrifood wastes include used coffee grounds, which serve as a reservoir of chlorogenic and protocatechuic acids (Farcas et al., 2021).

5.2.2 Bioactive Peptides

Bioactive peptides are small amino acids produced during protein enzymatic proteolysis and food processing, promoting health with high tissue affinity and specificity (Daliri et al., 2017). Bioactive peptides containing 5–16 amino acid residues have antioxidant properties, making them healthy and safe compounds. They have a simplified structure, a low molecular weight, a low cost, high action, quick absorption, and stability. These peptides function as lipid peroxidation inhibitors, free radical scavengers, and transition metal ion chelators. Their antioxidant properties are heavily influenced by their composition, structure, sequence, and hydrophobicity (Sarmadi & Ismail, 2010). Bioactive peptides can be found in protein-rich agrifood waste like pumpkin seeds, rapeseed meal, sesame bran, coconut pulp, tomato and apple pomace, and potato and orange peels (Gorguc et al., 2020).

5.2.3 Carotenoids

Carotenoids are a class of pigmented compounds produced by microorganisms and plants that are classified as xanthophyll and carotenoids. Carotenoids are phytochemicals that are supposed to be responsible for the beneficial characteristics of vegetables and fruits, which help in the prevention of human diseases such as cardiovascular disease, cancer, and other chronic diseases. Certain carotenoids also contribute to dietary vitamin A in addition to their powerful antioxidant activity. Vitamin A is produced via the conversion of carotenoids such as α and β-carotene, along with xanthophylls like β-cryptoxanthin (Maoka, 2020). Vitamin A is a necessary anti-inflammatory vitamin that is vital for human growth and disease prevention. It enhances immune function and increases resistance to infectious diseases by promoting and regulating the adaptive immune system and innate immune system. Vitamin A is required for maintaining vision, development, and the integrity of the epithelium and mucus. Nonetheless, vitamin A insufficiency is a significant dietary problem in children because it impairs the innate immune response, decreases keratinized epithelial cells' resistance to pathogens, and increases the risk of respiratory tract infections, diarrhea, and other diseases (Huang et al., 2018). Diseases can be efficiently prevented by developing functional foods from agrifood waste high in provitamin A components. Fruit and vegetables contain carotenoids such as lycopene, making them a possible source of carotenoids. Tomato pulp, skins, and seeds are substantial agrifood waste worldwide, including high quantities of lycopene and low levels of carotene, α-, β-, γ-, and ε-carotenes, phytoene, phytofluene, neurosporene, and lutein. Carotenoids are found in large quantities in fruit industry byproducts like carrots, peanut butter fruit, durian, hog plum, banana peels, and kwai muk (Ben-Othman et al., 2020).

5.2.4 Dietary Fibers

Dietary fibers are complex carbohydrates and lignin, which are not digested by humans but contribute to energy production via fermentation by intestinal bacteria. They are classified as insoluble or soluble, with water-insoluble fibers such as cellulose and hemicellulose and soluble fibers like β-glucans, gums, fructans, mucilage, and pectin. Fiber-rich diets enhance human health and lower the occurrence of disease (Dhingra et al., 2012). Dairy fiber helps prevent and treat chronic conditions such as colon cancer, diabetes, metabolic syndrome, and cardiovascular disease. It increases dietary bulk, adsorbs and sequesters cholesterol, reduces hepatic absorption, and enhances excretion via bile and fecal lipids. Soluble fibers, which are fermentable by gut bacteria, produce short-chain fatty acids, which are quickly absorbed by the large intestine. As a result of this absorption, the liver produces less cholesterol, which lowers blood cholesterol levels and increases sodium and water absorption into the colon mucosa (Guan et al., 2021). Insoluble fibers shorten transit times, enhance fecal bulk, and ease constipation. These fibers eliminate toxins, mutagens, and carcinogens, preventing negative effects. Dietary fibers are intricately woven substances that are abundant in many fruits, vegetables, and their byproducts. Apple pomace is a significant source of dietary fiber, with 47.3 g per 100 g of apple pomace (Skinner et al., 2018). Citrus shells and pomace are rich sources of dietary fiber, with mango peels having high levels (51.2–78.4%). Dietary fibers can also be found in pineapple pomace and peel, which have 45% and 76%, respectively. Banana peels contain 50% dry matter fibers, while artichoke leaves, cabbage leaves, and carrot pomace are also rich in fiber content (Pop et al., 2021). Various methods are used to extract dietary fibers from agrifood waste, including chemical, enzymatic, microwave-assisted methods, and ultrasound-assisted. Combining two different methods increases extraction yield, particularly for soluble fibers (Moczkowska et al., 2019).

5.3 Agrifood Waste as a Source of Biomass in Biodiesel and Bioethanol Production

Although fossil fuels are a common source of energy, the limited reserves and unsustainable usage of these resources make them a barrier to economic growth. As a cost- and environmentally friendly alternative replacement for fossil fuels, biofuels are becoming more prevalent. These fuels reduce hazardous emissions while adding to the pollution issue (Karmee & Lin, 2014). Agrifood wastes can be used to produce biofuels like biodiesel and bioethanol since they are a valuable source of lipids, carbohydrates, amino acids, and phosphates. Biodiesel is a fatty acid methyl ester created by trans-esterifying fats, greases, and plant oils such as soybean, rapeseed, and canola. The technique utilizes a base-catalyzed trans-esterification process at low temperatures and pressures in which a triglyceride combines with an alcohol in the presence of a strong alkaline catalyst to produce biodiesel and crude glycerol (Hafid et al., 2021). An alternative to fossil diesel, biodiesel can be used for heating, energy production, transportation, and agriculture. Although its fuel qualities are dependent on the feedstock, alcohol, and catalyst, it has an energy content of about 45 MJ/kg or about 90% of the heating value of diesel obtained from fossil fuels (Isah & Ozbay, 2020). Bioethanol is produced from agrifood waste used in place of gasoline in automotive engines, eliminating environmental and health risks. It involves the microbial fermentation of carbohydrates derived from crops high in sugar and grain starch. Nearly 50% of agrifood waste can be transformed into bioethanol. Bioethanol may be produced from cereal crops such as wheat, barley, rice, and maize with yields of 291, 310, 280, and 290 L/ton, respectively. Fruit residues like apple, orange, grape, and date yields of 241, 171, and 68 L/ton, respectively. Vegetables like carrots, sugar beet, sugarcane, and potato yield of 228, 260–375, 260–280, and 211 L/ton, respectively (Sánchez-Borrego et al., 2021). However, industrial production from agrifood waste is largely dependent on seasonal availability and geographical locations, hydrolysis effectiveness, lipid and carbohydrate content, fermentation, and trans-esterification processes. To create fuels from agrifood waste on an industrial scale, low-cost, greener, and advanced technologies are required (Isah & Ozbay, 2020).

6 Applications, Challenges, and Solutions Related to Valorizing Agrifood Wastes

Agrifood wastes can be easily managed for biodegradation and applications as food additives, therapeutics, and environmental applications because they contain bioactive components like antibiotics, pigments, and pigments (Hadj Saadoun et al., 2021). Agrifood wastes contain phytochemicals such as phenolics, carotenoids, and tocopherols, which make them appealing to the food, pharmaceutical, and cosmetic industries due to their antioxidant, therapeutic, and dietary benefits (Fig. 5.3) (Osorio et al., 2021). Agrifood waste can be used as a source for the production of aroma compounds, which can then be used to create ecologically acceptable food additives that can be used in a variety of industrial applications. In fermentation operations, this method requires fewer solvents for carbohydrate, protein, and lipid extraction and its breakdown of lipids, proteins, and carbohydrates contributes to the creation of the primary metabolites required for producing aroma molecules (Hadj Saadoun et al., 2021). Furthermore, it has been demonstrated that the inclusion of precursors results in the creation of specific molecules. For example, *Saccharomyces cerevisiae* produced isoamyl acetate with a distinct banana aroma after adding leucine and valine to agrifood waste, and L-phenylalanine from the yeasts *Kluyveromyces marxianus* and *Pichia kudriavzevii* was utilized to create the volatile chemicals 2-phenylethanol and 2-phenyl acetate, which have a rose-like

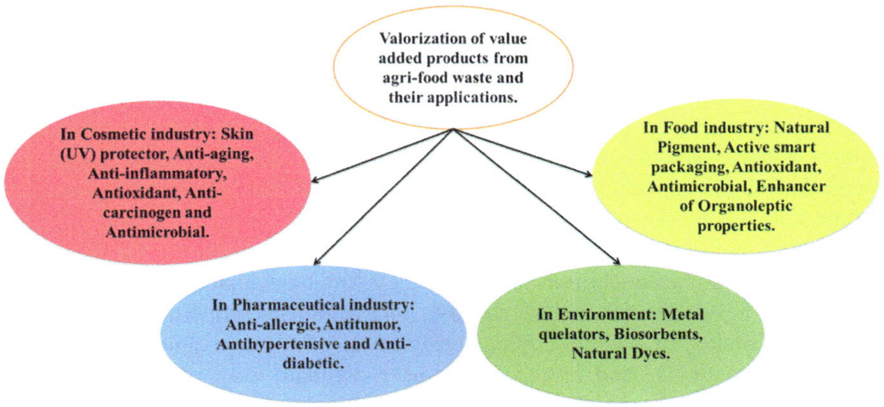

Fig. 5.3 Overview of the application of value-added products valorized from agrifood waste

aroma (Martinez et al., 2018). Following the fermentation operations, the final aroma compounds are produced using bioseparation procedures such as compound extraction, compound purification, and compound recovery. Moreover, enzymes such as cellulose, glucosidase, lipase, and protease are frequently administered to the substrate throughout the aroma production process, where they act as precursors of aroma compounds (Hadj Saadoun et al., 2021). Agrifood waste can be utilized to extract pigments that can be used as additives and nutrients in the food industry. In addition to their use as food coloring, microbial pigments have intriguing biological effects, including anti-oxidant, anti-microbial, anti-mutagenic, anti-cancer, immunoregulation, anti-inflammatory, anti-proliferative, and immunosuppressive capabilities. Natural components, particularly bioactive chemicals, are gaining popularity in the cosmetic and pharmaceutical industries. Agrifood waste contains nutrients and bioactive chemicals with antimicrobial, anti-inflammatory, and immunomodulatory activities, such as flavonoids and anthocyanins. Moreover, these substances have antioxidant properties that can combat free radicals, which makes them great therapeutic sources (Osorio et al., 2021). In the pharmaceutical and cosmetic industries, volatile fragrance molecules from microbial activities, such as ε-pinene, 6-pentyl-α-pyrone, ethyl acetate, isoamyl acetate, and vanillin, can improve or alter product scents. Pigments, which are vitamin precursors, are utilized in biopharmaceuticals and cosmetic items to help prevent cancer, chronic diseases, and cataracts (Lopes & Ligabue-Braun, 2021). To avoid dangerous chemical additions and stop oxidation processes, the food industry is putting more and more emphasis on natural antioxidants. Because of its antioxidants property phytochemical substances derived from agrifood waste, including pulp, seeds, skins, and pomace, have attracted a lot of attention. Antioxidants are used in cosmetics such as creams, powders, and oils to slow cell aging and prevent oxidation. Phenolic compounds such as resveratrol, curcumin, flavan-3-ols, and procyanidins are used as moisturizers and antioxidants in cosmetic products to replace preservatives such as formaldehyde and parabens, which may be harmful to human metabolism (Matwiejczuk et al., 2020). Synthetic antimicrobial chemicals are used to treat respiratory disorders, infections, and conditions; however, they can have contraindications and side effects. To address these difficulties, medical research has concentrated on bioactive compounds derived from agrifood waste. These residues, which include peels, seeds, shells, and leaves, are high in phytochemical substances, which increase bioavailability and the potential for novel medications. Citrus fruit byproducts include phenolic compounds that have antiviral, anti-inflammatory, and antimicrobial activities (Castrica et al., 2019).

Agrifood wastes offer bioactive substances that are valuable in the food, cosmetics, and pharmaceutical industries. Yet their low bioavailability and bioactivity make them unsuitable for widespread use. The beneficial properties of these substances are limited by their solubility, sensitivity to oxygen, enzymes, salts, humidity, pH fluctuations, and instabilities under standard processing and storage conditions. To avoid these restrictions and enhance the stability of bioactive compounds derived from agrifood wastes, a nano-technological approach: encapsulation has been recently developed. The encapsulation technique helps in the improvement of properties like bioavailability, bioactivity, solubility, stability, etc. of bioactive compounds, which enables extended shelf life and controlled release, resulting in novel and useful products for the food and pharmaceutical industries. Encapsulation techniques and wall materials improve the solubility and stability of bioactive compounds from agrifood waste. The methodology to be employed is determined by the entrapment technique and wall material, which may be a solution, dispersion, or emulsion (Marcillo-Parra et al., 2021). Phytochemicals can be trapped in capsules made of carbohydrates, proteins, lipids, and surfactants using a number of methods to improve compound management, including coacervation, crystallization, molecular encapsulation, extrusion, freeze drying, spray drying, and so on. Commonly used wall components in encapsulation methods are gum arabic, alginate, pectin, maltodextrins, cyclodextrins, casein, whey protein, gelatin, soy protein, hydrogenated vegetable oils, and phospholipids. Encapsulation helps to isolate chemicals while also enhancing controlled release, stability, bioavailability, and water solubility. In addition to offering antibacterial, anti-inflammatory, antioxidant, and enzymatic inhibitory effects, encapsulated bioactive chemicals improve food stability, oxidation resistance, microbial proliferation, enrichment, fortification, and coloring. Common encapsulation techniques for extracting bioactive chemicals from agrifood waste involve conservation, spray-drying, and freeze-drying. Controlled releases of beneficial chemicals and byproduct recovery are accomplished via freeze-drying and spray-drying. Coacervation provides high carrying capacity, a lower temperature, better thermal stability, and controlled active material release. This simple process uses little agitation, non-toxic solvents, and no special tools (Ozkan et al., 2019). Santos et al. (2019) used maltodextrin spray-drying to encapsulate the anthocyanins they isolated from blackberry pomace. In 7 days, the microcapsules were tested for pH stability, color fluctuation, antioxidant activity, and kinetic breakdown, which resulted in efficient samples' color preservation and decreased anthocyanin degradation under high pH. Aliakbarian et al. (2018) stabilized phenolic chemicals in olive pomace using spray-drying microcapsules, achieving 85–92% encapsulation efficiency, 65–82% yield, and high polyphenol content (38–52 mg gallic acid/g dw) and antioxidant activity (230–487 g Trolox/g dw). de Araujo et al. (2020) investigated using both maltodextrin and gelatin for the encapsulation of orange peel essential oil, discovering good yields, stability, and the preservation of antimicrobial and antioxidant properties. This proposes a possible substitute for naturally produced foods, delivering functional foods and health advantages by carefully controlling the supply of phytochemicals to specific locations. Šeregelj et al. (2020) used sodium alginate to encapsulate carotenoids from carrot waste. Yogurt was fortified with the encapsulated carotenoids, keeping its physicochemical and microbiological qualities while adding β-carotene. The antioxidant activity of the yogurt improved from 4.21 to 9.36 mol Trolox equivalents/180 g. Gheonea et al. (2021) produced lycopene microparticles from tomato peels via complicated coacervation and freeze-drying. The powder showed a 2.15 mMol Trolox/g dw antioxidant activity and an encapsulation efficiency of 83.6%. It also contained 27.34 mg of lycopene per gram of dry weight. The powder was utilized to functionalize dressing samples, resulting in improved antioxidant activity. Packaging used in the food industry produces waste and residue. Incorporating encapsulated bioactive substances into biodegradable films, such as phenolics, anthocyanins, and essential oils, offers promising active packaging that guards food goods against microbial and oxidative deterioration (Ozkan et al., 2019). Alves et al. (2018) created chitosan films with grape seed extract and carvacrol microcapsules to improve the shelf life of refrigerated salmon. After 7 days of storage, the salmon's shelf life was increased by the carvacrol microcapsules' improved resistance to microbes. Encapsulation strategies have potential, but they are limited in industrial applications due to low extraction yield, expensive purification procedures, and the need for multiple procedures to purify bioactive compounds from agrifood waste (Marcillo-Parra et al., 2021).

7 Conclusions and Future Perspectives

Along with the ongoing exponential growth in human population, the negative effects of climate change on agriculture have resulted in declines in agrifood production and, consequently, nutritional losses. To meet this need, it is now urgently necessary to valorize agrifood waste. Agrifood wastes, which are primarily of plant origin, pose serious societal problems, including a large negative effect on the global economy and environmental issues connected to pollution and also the release of greenhouse gases (GHGs). Agrifood waste contains valuable chemicals like amino acids, bioactive compounds, enzymes, lipids, minerals, polysaccharides, secondary metabolites, vitamins, etc. that have anti-microbial, antioxidant, and anti-inflammatory properties, which allow for various applications in the food, health, pharmaceutical, cosmetic, and environmental sectors, creating value-added and beneficial products from recyclable organic resources. Green analytical chemistry techniques like EAE, MAE, PEF, PLE, SFE, SCWE, and UAE offer diverse applications for producing value-added and marketable products. These techniques efficiently extract and recover valuable compounds, making them economically viable in various fields. Combining two or more techniques provides a highly promising approach regarding rapid and effective extraction, resulting in greater yields and improved products. However, the majority of the aforementioned compounds has poor solubility and is sensitive to standard manufacturing and storage conditions, making them difficult to directly incorporate into many food products. In recent times, encapsulation techniques have been utilized to promote the bioavailability, bioactivity, solubility, and stability of bioactive compounds derived from agrifood wastes, preserve their nutritional value from the adverse environmental conditions of food processing, and improve their shelf life. By isolating components of interest and releasing substances under regulated conditions, these methods can produce novel and useful products, particularly in the food and pharmaceutical industries. Encapsulation methods use protective coverings to enhance the delivery and release of bioactive compounds from agrifood waste. As a result, functional components like probiotics and microencapsulated antioxidants are created and can be used in both food and drug formulations. In the agrifood industry, sustainable waste management techniques increase product efficacy and quality while reducing waste and promoting sustainable resource use. These technologies boost the yield of bioactive component extraction and cut expenses, creating new commercial opportunities and furthering the idea of the circular economy. Applying these methods leads toward a more resilient and sustainable agrifood industry by maximizing resource usage and lowering the carbon emissions associated with waste management. However, encapsulation techniques in the industrial sector suffer restrictions due to their low extraction yields and high technological expenses in purification procedures. Continuous research for the development of novel extraction procedures can result in long-term strategies for utilizing valuable compounds from agrifood waste.

References

Akyuz, A., & Ersus, S. (2021). Optimization of enzyme assisted extraction of protein from the sugar beet (*Beta vulgaris L.*) leaves for alternative plant protein concentrate production. *Food Chemistry, 335*, 127673.

Albuquerque, B. R., Heleno, S. A., Oliveira, M. B. P. P., Barros, L., & Ferreira, I. C. F. R. (2021). Phenolic compounds: Current industrial applications, limitations and future challenges. *Food & Function, 12*, 14–29.

Aliakbarian, B., Sampaio, F. C., de Faria, J. T., Pitangui, C. G., Lovaglio, F., Casazza, A. A., Converti, A., & Perego, P. (2018). Optimization of spray drying microencapsulation of olive pomace polyphenols using response surface methodology and artificial neural network. *Lwt-Food Science and Technology, 93*, 220–228.

Alves, V. L. C. D., Rico, B. P. M., Cruz, R. M. S., Vicente, A. A., Khmelinski, I., & Vieira, M. C. (2018). Preparation and characterization of a chitosan film with grape seed extract-carvacrol microcapsules and its effect on the shelf-life of refrigerated Salmon (*Salmo salar*). *LWT-Food Science and Technology, 89*, 525–534.

Amicarelli, V., & Bux, C. (2020). Food waste measurement toward a fair, healthy and environmental-friendly food system: A critical review. *British Food Journal, 123*, 2907–2935.

Andreou, V., Psarianos, M., Dimopoulos, G., Tsimogiannis, D., & Taoukis, P. (2020). Effect of pulsed electric fields and high pressure on improved recovery of high-added-value compounds from olive pomace. *Journal of Food Science, 85*, 1500–1512.

Arshad, R. N., Abdul-Malek, Z., Roobab, U., Munir, M. A., Naderipour, A., Qureshi, M. I., Bekhit, A. E., Liu, Z. W., & Aadil, R. M. (2021). Pulsed electric field: A potential alternative towards a sustainable food processing. *Trends in Food Science and Technology, 111*, 43–54.

Battistella Lasta, H. F., Lentz, L., Gonçalves Rodrigues, L. G., Mezzomo, N., Vitali, L., & Salvador Ferreira, S. R. (2019). Pressurized liquid extraction applied for the recovery of phenolic compounds from beetroot waste. *Biocatalysis and Agricultural Biotechnology, 21*, 101353.

Benabda, O., M'hir, S., Kasmi, M., Mnif, W., & Hamdi, M. (2019). Optimization of protease and amylase production by *Rhizopus oryzae* cultivated on bread waste using solid-state fermentation. *Journal of Chemistry, 3738181*.

Benito-román, Ó., Blanco, B., Sanz, M. T., & Beltrán, S. (2020). Subcritical water extraction of phenolic compounds from onion skin wastes (*Allium cepa* cv. Horcal): Effect of temperature and solvent properties. *Antioxid, 9*(12), 1233.

Ben-Othman, S., Jõudu, I., Bhat, R., Beatriz, M., Oliveira, P., & Alves, R. C. (2020). Bioactives from agri-food wastes: Present insights and future challenges. *Molecules, 25*, 510.

Boluda-Aguilar, M., & Lopez-Gomez, A. (2013). Production of bioethanol by fermentation of lemon (*Citrus limon L.*) peel wastes pretreated with steam explosion. *Industrial Crops and Products, 41*, 188–197.
Boukroufa, M., Boutekedjiret, C., Petigny, L., Rakotomanomana, N., & Chemat, F. (2015). Bio-refinery of orange peels waste: A new concept based on integrated green and solvent free extraction processes using ultrasound and microwave techniques to obtain essential oil, polyphenols and pectin. *Ultrasonics Sonochemistry, 24*, 72–79.
Caldeira, C., De Laurentiis, V., Corrado, S., van Holsteijn, F., & Sala, S. (2019). Quantification of food waste per product group along the food supply chain in the European Union: A mass flow analysis. *Resources, Conservation and Recycling, 149*, 479–488.
Carmona-Cabello, M., Garcia, I. L., Leiva-Candia, D., & Dorado, P. (2018). Valorization of food waste based on its composition through the concept of biorefinery. *Current Opinion in Green Sustainable Chemistry, 14*, 67–79.
Carpentieri, S., Soltanipour, F., Ferrari, G., Pataro, G., & Donsì, F. (2021). Emerging green techniques for the extraction of antioxidants from agri-food by-products as promising ingredients for the food industry. *Antioxid, 10*, 1417.
Castrica, M., Rebucci, R., Giromini, C., Tretola, M., Cattaneo, D., & Baldi, A. (2019). Total phenolic content and antioxidant capacity of agri-food waste and by-products. *Italian Journal of Animal Science, 18*, 336–341.
Castro, L. A., Lizi, J. M., Chagas, E. G. L. D., Carvalho, R. A., & Vanin, F. M. (2020). From orange juice by-product in the food industry to a functional ingredient: Application in the circular economy. *Food, 9*(5), 593.
Cea Pavez, I., Lozano-Sánchez, J., Borrás-Linares, I., Nuñez, H., Robert, P., & Segura-Carretero, A. (2019). Obtaining an extract rich in phenolic compounds from olive pomace by pressurized liquid extraction. *Molecules, 24*, 3108.
Chen, D., Yin, L., Wang, H., & He, P. (2014). Pyrolysis technologies for municipal solid waste: A review. *Waste Management, 34*(12), 2466–2486.
da Rocha, C. B., & Noreña, C. P. Z. (2020). Microwave-assisted extraction and ultrasound-assisted extraction of bioactive compounds from grape pomace. *International Journal of Food Engineering, 16*, 20190191.
da Silva, A. C., & Jorge, N. (2016). Bioactive compounds of oils extracted from fruits seeds obtained from agroindustrial waste. *European Journal of Lipid Science and Technology, 119*, 1600024.
Daliri, E. B. M., Oh, D. H., & Lee, B. H. (2017). Bioactive peptides. *Foods, 6*, 32.
de Andrade, L. M., Charalampopoulos, D., & Chatzifragkou, A. (2018). Optimisation and modelling of supercritical CO_2 extraction process of carotenoids from carrot peels. *Journal of Supercritical Fluids, 133*, 94–102.
de Araujo, J. S. F., de Souza, E. L., Oliveira, J. R., Gomes, A. C. A., Kotzebue, L. R. V., da Silva Agostini, D. L., de Oliveira, D. L. V., Mazzetto, S. E., da Silva, A. L., & Cavalcanti, M. T. (2020). Microencapsulation of sweet orange essential oil (*Citrus aurantium var. dulcis*) by liophylization using maltodextrin and maltodextrin/gelatin mixtures: Preparation, characterization, antimicrobial and antioxidant activities. *International Journal of Biological Macromolecules, 143*, 991–999.
De la Peña-Armada, R., & Mateos-Aparicio, I. (2022). Sustainable approaches using green technologies for apple by-product valorisation as a new perspective into the history of the apple. *Molecules, 27*(20), 6937.
de Souza, R. D., Machado, B. A., Barreto, G. D., Leal, I. L., Anjos, J. P., & Umsza-Guez, M. A. (2020). Effect of experimental parameters on the extraction of grape seed oil obtained by low pressure and supercritical fluid extraction. *Molecules, 25*(7), 1634.
Dhanker, R., Rawat, S., Chandna, V., Das, S., Sharma, A., Kumar, V., & Kumar, R. (2022). Recovery of silver nanoparticles and management of food wastes: Obstacles and opportunities. *Environmental Advances, 9*, 100303.

Dhingra, D., Michael, M., Rajput, H., & Patil, R. T. (2012). Dietary fibre in foods: A review. *Journal of Food Science and Technology, 49*, 255–266.
Di Lorenzo, C., Colombo, F., Biella, S., Stockley, C., & Restani, P. (2021). Polyphenols and human health: The role of bioavailability. *Nutrients, 13*(1), 273.
Dobroslavi'c, E., Elez Garofuli'c, I., Šeparovi'c, J., Zori'c, Z., Pedisi'c, S., & Dragovi'c-Uzelac, V. (2022). Pressurized liquid extraction as a novel technique for the isolation of *Laurus nobilis* L. Leaf Polyphenols. *Molecules, 27*, 5099.
Doulabi, M., Golmakani, M. T., & Ansari, S. (2020). Evaluation and optimization of microwave-assisted extraction of bioactive compounds from eggplant peel by-product. *Journal of Food Processing & Preservation, 44*, e14853.
Ekinci, M. S., & Gürü, M. (2019). Extraction of phytosterols from melon (*Cucumis melo*) seeds by supercritical CO_2 as a clean technology. *Green Processing and Synthesis, 8*, 677–682.
Esteban, J., & Ladero, M. (2018). Food waste as a source of value-added chemicals and materials: A biorefinery perspective. *International Journal of Food Science & Technology, 53*, 1095–1108.
Farcas, A., Dretcanu, G., Pop, T. D., Enaru, B., Socaci, S., & Diaconeasa, Z. (2021). Cereal processing by-products as rich sources of phenolic compounds and their potential bioactivities. *Nutrients, 13*, 3934.
Fornereto Soldan, A. C., Arvelos, S., Watanabe, É. O., & Hori, C. E. (2021). Supercritical fluid extraction of oleoresin from *Capsicum annuum* industrial waste. *Journal of Cleaner Production, 297*, 126593.
Freitas, L. C., Barbosa, J. R., da Costa, A., Bezerra, F., Pinto, R., & de Carvalho Junior, R. N. (2021). From waste to sustainable industry: How can agro-industrial wastes help in the development of new products? *Resources, Conservation and Recycling, 169*, 105466.
Frontuto, D., Carullo, D., Harrison, S. M., Brunton, N. P., Ferrari, G., Lyng, J. G., & Pataro, G. (2019). Optimization of pulsed electric fields-assisted extraction of polyphenols from potato peels using response surface methodology. *Food and Bioprocess Technology, 12*, 1708–1720.
García, P., Fredes, C., Cea, I., Lozano-Sánchez, J., Leyva-Jiménez, F. J., Robert, P., Vergara, C., & Jimenez, P. (2021). Recovery of bioactive compounds from pomegranate (*Punica granatum L.*) peel using pressurized liquid extraction. *Food, 10*(2), 203.
Ghada, B., Pereira, E., Pinela, J., Prieto, M. A., Pereira, C., Calhelha, R. C., Stojković, D. S., Soković, M. D., Zaghdoudi, K., Barros, L., & Ferreira, I. C. (2020). Recovery of anthocyanins from passion fruit epicarp for food colorants: Extraction process optimization and evaluation of bioactive properties. *Molecules, 25*, 3203.
Gheonea, I., Aprodu, I., Circiumaru, A., Rapeanu, G., Bahrim, G. E., & Stanciuc, N. (2021). Microencapsulation of lycopene from tomatoes peels by complex coacervation and freeze-drying: Evidences on phytochemical profile, stability and food applications. *Journal of Food Engineering, 288*, 110166.
Giombelli, C., Iwassa, I. J., da Silva, C., & Bolanho Barros, B. C. (2020). Valorization of peach palm by-product through subcritical water extraction of soluble sugars and phenolic compounds. *Journal of Supercritical Fluids, 165*, 104985.
Giordano, M., Pinela, J., Dias, M. I., Calhelha, R. C., Stojkovi'c, D., Sokovi'c, M., Tavares, D., Cánepa, A. L., Ferreira, I. C. F. R., Caleja, C., & Barros, L. (2021). Ultrasound-assisted extraction of flavonoids from Kiwi peel: Process optimization and bioactivity assessment. *Applied Sciences, 11*, 6416.
Gómez-García, R., Campos, D. A., Aguilar, C. N., Madureira, A. R., & Pintado, M. (2021). Valorisation of food agro-industrial by-products: From the past to the present and perspectives. *Journal of Environmental Management, 299*, 113571.
Gong, C., Singh, A., Singh, P., & Singh, A. (2021). Anaerobic digestion of agri-food wastes for generating biofuels. *Indian Journal of Microbiology, 61*, 427–440.
Gorguc, A., Gencdag, E., & Yilmaz, F. M. (2020). Bioactive peptides derived from plant origin by-products: Biological activities and techno-functional utilizations in food developments-A review. *Food Research International, 136*, 109504.

Guan, Z. W., Yu, E. Z., & Feng, Q. (2021). Soluble dietary fiber, one of the most important nutrients for the gut microbiota. *Molecules, 26*, 6802.

Guthrie, F., Wang, Y., Neeve, N., Quek, S. Y., Mohammadi, K., & Baroutian, S. (2020). Recovery of phenolic antioxidants from green kiwifruit peel using subcritical water extraction. *Food and Bioproducts Processing, 122*, 136–144.

Hadj Saadoun, J., Bertani, G., Levante, A., Vezzosi, F., Ricci, A., Bernini, V., & Lazzi, C. (2021). Fermentation of agri-food waste: A promising route for the production of aroma compounds. *Food, 10*, 707.

Hafid, H. S., Omar, F. N., Abdul Rahman, N. A., & Wakisaka, M. (2021). Innovative conversion of food waste into biofuel in integrated waste management system. *Critical Reviews in Environmental Science and Technology, 52*, 3453–3492.

Haldar, D., Shabbirahmed, A. M., Singhania, R. R., Chen, C. W., Dong, C. D., Ponnusamy, V. K., & Patel, A. K. (2022). Understanding the management of household food waste and its engineering for sustainable valorization-a state-of-the-art review. *Bioresource Technology, 358*, 127390.

Hendrawan, Y., Larasati, R., Wibisono, Y., Umam, C., Sutan, S. M., & Choviya Hawa, L. (2019). Extraction of phenol and antioxidant compounds from Kepok Banana skin with PEF pre-treatment. *IOP Conference Series: Earth and Environmental Science, 305*, 012065.

Huang, Z., Liu, Y., Qi, G., Brand, D., & Zheng, S. G. (2018). Role of Vitamin A in the immune system. *Journal of Clinical Medicine, 7*, 258.

Hwang, H. J., Kim, H. J., Ko, M. J., & Chung, M. S. (2021). Recovery of hesperidin and narirutin from waste citrus unshiu peel using subcritical water extraction aided by pulsed electric field treatment. *Food Science and Biotechnology, 30*, 217–226.

Isah, S., & Ozbay, G. (2020). Valorization of food loss and wastes: Feedstocks for biofuels and valuable chemicals. *Frontiers in Sustainable Food Systems, 4*, 82.

Ishangulyyev, R., Kim, S., & Lee, S. H. (2019). Understanding food loss and waste-why are we losing and wasting food? *Food, 8*(8), 297.

Islam, M. A., Karim, A., Ethiraj, B., Raihan, T., Khan, M. M. R., Kadier, A., Al Nadhari, S., Al-Masri, A. A., & Ameen, F. (2023). Interspecies microbial interactions in bioelectrochemical system and biodegradation: A state of the art review. *Science of the Total Environment, 891*, 164623.

Kantifedaki, A., Kachrimanidou, V., Mallouchos, A., Papanikolaou, S., & Koutinas, A. A. (2018). Orange processing waste valorisation for the production of bio-based pigments using the fungal strains *Monascus purpureus* and *Penicillium purpurogenum*. *Journal of Cleaner Production, 185*, 882–890.

Karmee, S. K., & Lin, C. S. K. (2014). Valorisation of food waste to biofuel: Current trends and technological challenges. *Sustainable Chemical Processes, 2*, 22.

Kaur, P., Singh, S., Ghoshal, G., Ramamurthy, P. C., Parihar, P., Singh, J., & Singh, A. (2022). Valorization of agri-food industry waste for the production of microbial pigments: An eco-friendly approach. In S. K. Nayak, B. Baliyarsingh, I. Mannazzu, A. Singh, & B. B. Mishra (Eds.), *Advances in agricultural and industrial microbiology* (pp. 137–167). Springer.

Kehili, M., Kammlott, M., Choura, S., Zammel, A., Zetzl, C., Smirnova, I., Allouche, N., & Sayadi, S. (2017). Supercritical CO2 extraction and antioxidant activity of lycopene and β-carotene-enriched oleoresin from tomato (*Lycopersicum esculentum L.*) peels by-product of a Tunisian industry. *Food and Bioproducts Processing, 102*, 340–349.

Kostrzewa, D., Dobrzyńska-Inger, A., Mazurek, B., & Kostrzewa, M. (2022). Pilot-scale optimization of supercritical CO_2 extraction of dry Paprika *Capsicum annuum*: Influence of operational conditions and storage on extract composition. *Molecules, 27*(7), 2090.

Koul, B., Yakoob, M., & Shah, M. P. (2022). Agricultural waste management strategies for environmental sustainability. *Environmental Research, 206*, 112285.

Kovalcik, A., Pernicova, I., Obruca, S., Szotkowski, M., Enev, V., Kalina, M., & Marova, I. (2020). Grape winery waste as a promising feedstock for the production of polyhydroxyalkanoates and other value-added products. *Food and Bioproducts Processing, 124*, 1–10.

Kumar Sarangi, P., Subudhi, S., Bhatia, L., Saha, K., Mudgil, D., Prasad Shadangi, K., Srivastava, R. K., Pattnaik, B., & Arya, R. K. (2023). Utilization of agricultural waste biomass and recycling toward circular bioeconomy. *Environmental Science and Pollution Research International, 30*(4), 8526–8539.

Lal, A. M. N., Prince, M. V., Kothakota, A., Pandiselvam, R., Thirumdas, R., Mahanti, N. K., & Sreeja, R. (2021). Pulsed electric field combined with microwave-assisted extraction of pectin polysaccharide from jackfruit waste. *Innovative Food Science and Emerging Technologies, 74*, 102844.

Leong, Y. K., & Chang, J. S. (2022). Valorization of fruit wastes for circular bioeconomy: Current advances, challenges, and opportunities. *Bioresource Technology, 359*, 127459.

Li, B., Akram, M., Al-Zuhair, S., Elnajjar, E., & Munir, M. T. (2020). Subcritical water extraction of phenolics, antioxidants and dietary fibres from waste date pits. *Journal of Environmental Chemical Engineering, 8*, 104490.

Lin, A. Y., Huang, S. T., & Wahlqvist, M. L. (2009). Waste management to improve food safety and security for health advancement. *Asia Pacific Journal of Clinical Nutrition, 18*(4), 538–545.

Lombardelli, C., Liburdi, K., Benucci, I., & Esti, M. (2020). Tailored and synergistic enzyme-assisted extraction of carotenoid-containing chromoplasts from tomatoes. *Food and Bioproducts Processing, 121*, 43–53.

Lombardelli, C., Benucci, I., Mazzocchi, C., & Esti, M. (2021). A novel process for the recovery of betalains from unsold red beets by low-temperature enzyme-assisted extraction. *Food, 10*, 236.

Lopes, F. C., & Ligabue-Braun, R. (2021). Agro-industrial residues: Eco-friendly and inexpensive substrates for microbial pigments production. *Frontiers in Sustainable Food Systems, 5*, 589414.

Makanjuola, O., Arowosola, T., & Du, C. (2020). The utilization of food waste: Challenges and opportunities. *Journal of Food Chemistry and Nanotechnology, 6*, 182–188.

Maoka, T. (2020). Carotenoids as natural functional pigments. *Journal of Natural Medicines, 74*, 1–16.

Marcillo-Parra, V., Tupuna-Yerovi, D. S., Gonzalez, Z., & Ruales, J. (2021). Encapsulation of bioactive compounds from fruit and vegetable by-products for food application-A review. *Trends in Food Science and Technology, 116*, 11–23.

Martinez, O., Sanchez, A., Font, X., & Barrena, R. (2018). Enhancing the bioproduction of value-added aroma compounds via solid-state fermentation of sugarcane bagasse and sugar beet molasses: Operational strategies and scaling-up of the process. *Bioresource Technology, 263*, 136–144.

Martínez-Abad, A., Ramos, M., Hamzaoui, M., Kohnen, S., Jiménez, A., & Garrigós, M. C. (2020). Optimisation of sequential microwave-assisted extraction of essential oil and pigment from lemon peels waste. *Food, 9*(10), 1493.

Matwiejczuk, N., Galicka, A., & Brzoska, M. M. (2020). Review of the safety of application of cosmetic products containing parabens. *Journal of Applied Toxicology, 40*, 176–210.

Mitrea, L., Călinoiu, L. F., Martău, G. A., Szabo, K., Teleky, B. E., Muresan, V., Rusu, A. V., Socol, C. T., & Vodnar, D. C. (2020). Poly (vinyl alcohol)-based biofilms plasticized with polyols and colored with pigments extracted from tomato by-products. *Polymers, 12*, 532.

Moczkowska, M., Karp, S., Niu, Y., & Kurek, M. A. (2019). Enzymatic, enzymatic-ultrasonic and alkaline extraction of soluble dietary fibre from flaxseed–A physicochemical approach. *Food Hydrocolloids, 90*, 105–112.

More, P. R., Jambrak, A. R., & Arya, S. S. (2022). Green, environment-friendly and sustainable techniques for extraction of food bioactive compounds and waste valorization. *Trends in Food Science and Technology, 128*, 296–315.

Mourtzinos, I., & Goula, A. M. (2019). Polyphenols in agricultural byproducts and food waste. In R. W. Ronald (Ed.), *Polyphenols in plants* (2nd ed., pp. 23–44). Academic Press.

Mourtzinos, I., Prodromidis, P., Grigorakis, S., Makris, D. P., Biliaderis, C. G., & Moschakis, T. (2018). Natural food colourants derived from onion wastes: Application in a yoghurt product. *Electrophoresis, 39*, 1975–1983.

Munde, P. J., Muley, A. B., Ladole, M. R., Pawar, A. V., Talib, M. I., & Parate, V. R. (2017). Optimization of pectinase-assisted and tri-solvent-mediated extraction and recovery of lycopene from waste tomato peels. *3 Biotech, 7*(3), 206.

Munir, M. T., Kheirkhah, H., Baroutian, S., Quek, S. Y., & Young, B. R. (2018). Subcritical water extraction of bioactive compounds from waste onion skin. *Journal of Cleaner Production, 183*, 487–494.

Nadar, S. S., Rao, P., & Rathod, V. K. (2018). Enzyme assisted extraction of biomolecules as an approach to novel extraction technology: A review. *Food Research International, 108*, 309–330.

Nnolim, N. E., Okoh, A. I., & Nwodo, U. U. (2020). Bacillus Sp. FPF-1 produced keratinase with high potential for chicken feather degradation. *Molecules, 25*, 1505.

Nutter, J., Fernandez, M. V., Jagus, R. J., & Agüero, M. V. (2021). Development of an aqueous ultrasound-assisted extraction process of bioactive compounds from beet leaves: A proposal for reducing losses and increasing biomass utilization. *Journal of the Science of Food and Agriculture, 101*, 1989–1997.

Okino-Delgado, C. H., Prado, D. Z., Facanali, R., Marques, M. M. O., Nascimento, A. S., Fernandes, C. J. D. C., Zambuzzi, W. F., & Fleuri, L. F. (2017). Bioremediation of cooking oil waste using lipases from wastes. *PLoS One, 12*, e0186246.

Osorio, L. L., Flórez-López, E., & Grande-Tovar, C. D. (2021). The potential of selected agri-food loss and waste to contribute to a circular economy: Applications in the food, cosmetic and pharmaceutical industries. *Molecules, 26*, 515.

Ozkan, G., Franco, P., De Marco, I., Xiao, J., & Capanoglu, E. (2019). A review of microencapsulation methods for food antioxidants: Principles, advantages, drawbacks and applications. *Food Chemistry, 272*, 494–506.

Paini, J., Benedetti, V., Ail, S. S., Castaldi, M. J., Baratieri, M., & Patuzzi, F. (2021). Valorization of wastes from the food production industry: A review towards an integrated agri-food processing biorefinery. *Waste and Biomass Valorization, 13*(1), 31–50.

Panzella, L., Moccia, F., Nasti, R., Marzorati, S., Verotta, L., & Napolitano, A. (2020). Bioactive phenolic compounds from agri-food wastes: An update on green and sustainable extraction methodologies. *Frontiers in Nutrition, 7*, 60.

Papaioannou, E. H., Mazzei, R., Bazzarelli, F., Piacentini, E., Giannakopoulos, V., Roberts, M. R., & Giorno, L. (2022). Agri-food industry waste as resource of chemicals: The role of membrane technology in their sustainable recycling. *Sustainability, 14*, 1483.

Paritosh, K., Kushwaha, S. K., Yadav, M., Pareek, N., Chawade, A., & Vivekanand, V. (2017). Food waste to energy: An overview of sustainable approaches for food waste management and nutrient recycling. *Biomed Research International*, 2370927.

Pataro, G., Carullo, D., Falcone, M., & Ferrari, G. (2020). Recovery of lycopene from industrially derived tomato processing by-products by pulsed electric fields-assisted extraction. *Innovative Food Science and Emerging Technologies, 63*, 102369.

Pedras, B. M., Regalin, G., Sá-Nogueira, I., Simões, P., Paiva, A., & Barreiros, S. (2020). Fractionation of red wine grape pomace by subcritical water extraction/hydrolysis. *Journal of Supercritical Fluids, 160*, 104793.

Peiró, S., Luengo, E., Segovia, F., Raso, J., & Almajano, M. P. (2019). Improving polyphenol extraction from lemon residues by pulsed electric fields. *Waste and Biomass Valorization, 10*, 889–897.

Phucharoenrak, P., Muangnoi, C., & Trachootham, D. (2022). A green extraction method to achieve the highest yield of Limonin and Hesperidin from Lime peel powder (*Citrus aurantifolia*). *Molecules, 27*(3), 820.

Pollini, L., Cossignani, L., Juan, C., & Mañes, J. (2021). Extraction of phenolic compounds from fresh apple pomace by different non-conventional techniques. *Molecules, 26*, 4272.

Pop, C., Suharoschi, R., & Pop, O. L. (2021). Dietary fiber and prebiotic compounds in fruits and vegetables food waste. *Sustainability, 13*, 7219.

Ranjha, M. M. A. N., Irfan, S., Lorenzo, J. M., Shafique, B., Kanwal, R., Pateiro, M., Arshad, R. N., Wang, L., Nayik, G. A., Roobab, U., & Aadil, R. M. (2021). Sonication, a potential technique for extraction of phytoconstituents: A systematic review. *PRO, 9*, 1406.

Ravindran, R., Hassan, S. S., Williams, G. A., & Jaiswal, A. K. (2018). A review on bioconversion of agro-industrial wastes to industrially important enzymes. *Bioengineering, 5*, 93.

Restrepo-Serna, D. L., Solarte-Toro, J. C., & Cardona-Alzate, C. A. (2022). A biorefinery approach for an integral valorisation of Avocado peel and seeds through supercritical fluids. *Waste and Biomass Valorization, 13*, 3973–3988.

Rodrigues, L. A., Matias, A. A., & Paiva, A. (2021). Recovery of antioxidant protein hydrolysates from shellfish waste streams using subcritical water extraction. *Food and Bioproducts Processing, 130*, 154–163.

Rodríguez García, S. L., & Raghavan, V. (2022). Green extraction techniques from fruit and vegetable waste to obtain bioactive compounds: A review. *Critical Reviews in Food Science and Nutrition, 62*(23), 6446–6466.

Saad, N., Louvet, F., Tarrade, S., Meudec, E., Grenier, K., Landolt, C., Ouk, T. S., & Bressollier, P. (2019). Enzyme-assisted extraction of bioactive compounds from Raspberry (*Rubus idaeus L.*) Pomace. *Journal of Food Science, 84*, 1371–1381.

Sagar, N. A., Pareek, S., Sharma, S., Yahia, E. M., & Lobo, M. G. (2018). Fruit and vegetable waste: Bioactive compounds, their extraction, and possible utilization. *Comprehensive Reviews in Food Science and Food Safety, 17*(3), 512–531.

Sahoo, A., Sarkar, S., Lal, B., Kumawat, P., Sharma, S., & De, K. (2021). Utilization of fruit and vegetable waste as an alternative feed resource for sustainable and eco-friendly sheep farming. *Waste Management, 128*, 232–242.

Saini, A., Panesar, P. S., & Bera, M. B. (2019). Valorization of fruits and vegetables waste through green extraction of bioactive compounds and their nanoemulsions-based delivery system. *Bioresources and Bioprocessing, 6*, 26.

Salimi, N. (2021). Circular economy in agri-food systems. In J. Rezaei (Ed.), *Strategic decision making for sustainable management of industrial networks. Greening of industry networks studies* (Vol. 8). Springer.

Samborska, K., Barańska, A., Kamińska, P., KamińskaDwórznicka, A., & Jedlińska, A. (2019). Reduction of waste production during juice processing with a simultaneous creation of the new type of added-value products. *Environmental Protection and Natural Resources, 30*(3), 1–6.

Sánchez-Borrego, F. J., Álvarez-Mateos, P., & García-Martín, J. F. (2021). Biodiesel and other value-added products from bio-oil obtained from agri-food waste. *PRO, 9*, 797.

Santos, S. S., Rodrigues, L. M., Costal, S. C., & Madrona, G. S. (2019). Antioxidant compounds from blackberry (*Rubus fruticosus*) pomace: Microencapsulation by spray-dryer and pH stability evaluation. *Food Packaging and Shelf Life, 20*, 100177.

Santos-Zea, L., Gutierrez-Uribe, J. A., & Benedito, J. (2019). Effect of ultrasound intensification on the supercritical fluid extraction of phytochemicals from Agave salmiana bagasse. *Journal of Supercritical Fluids, 144*, 98–107.

Sarmadi, B. H., & Ismail, A. (2010). Antioxidative peptides from food proteins: A review. *Peptides, 31*, 1949–1956.

Šeregelj, V., Pezo, L. L., Šovljanski, O., Lević, S., Nedović, V. A., Markov, S. L., Tomić, A., Čanadanović-Brunet, J., Vulić, J. J., Šaponjac, V. T., & Ćetković, G. (2020). New concept of fortified yogurt formulation with encapsulated carrot waste extract. *LWT-Food Science and Technology, 138*, 110732.

Shang, X., Zhang, M., Hu, J., Zhang, Y., Yang, L., & Hou, X. (2022a). Chemical compositions, extraction optimizations, and in vitro bioactivities of flavonoids from Perilla Leaves (*Perillae folium*) by microwave-assisted natural deep eutectic solvents. *Antioxidants (Basel), 12*(1), 104.

Shang, X., Zhang, Y., Zheng, Y., & Li, Y. (2022b). Temperature-responsive deep eutectic solvents as eco-friendly and recyclable media for microwave extraction of flavonoid compounds from waste onion (*Allium cepa L.*) skins. *Biomass Conversion and Biorefinery, 14*, 3729–3738.

Sharma, P., Gaur, V. K., Kim, S. H., & Pandey, A. (2020a). Microbial strategies for bio-transforming food waste into resources. *Bioresource Technology, 299*, 122580.

Sharma, V., Nargotra, P., Sharma, S., & Bajaj, B. K. (2020b). Efficient bioconversion of sugarcane tops biomass into biofuel-ethanol using an optimized Alkali-Ionic liquid pretreatment approach. *Biomass Conversion and Biorefinery, 10*, 1–14.

Sharma, A., Kuthiala, T., Thakur, K., Thatai, K. S., Singh, G., Kumar, P., & Arya, S. K. (2022). Kitchenwaste: Sustainable bioconversion to value-added product and economic challenges. *Biomass Conversion and Biorefinery, 12*, 1–22.

Singh, S., Gupta, P., Sharma, V., Koul, S., Kour, K., & Bajaj, B. K. (2014). Multifarious potential applications of keratinase of *Bacillus Subtilis K-5*. *Biocatalysis and Biotransformation, 32*, 333–342.

Singh, A., Singhania, R. R., Soam, S., Chen, C. W., Haldar, D., Varjani, S., Chang, J. S., Dong, C. D., & Patel, A. K. (2022). Production of bioethanol from food waste: Status and perspectives. *Bioresource Technology, 360*, 127651.

Skinner, R. C., Gigliotti, J. C., Ku, K. M., & Tou, J. C. (2018). A comprehensive analysis of the composition, health benefits, and safety of apple pomace. *Nutrition Reviews, 76*, 893–909.

Socas-Rodríguez, B., Álvarez-Rivera, G., Valdés, A., Ibáñez, E., & Cifuentes, A. (2021). Food by-products and food wastes: Are they safe enough for their valorization? *Trends in Food Science and Technology, 114*, 133–147.

Squillaci, G., la Cara, F., Roseiro, L. B., Marques, I. P., & Morana, A. (2021). Agro-industrial wastes as bioactive molecules source. *Chemical Engineering Transactions, 86*, 37–42.

Tan, W. K., Lee, S. Y., Lee, W. J., Hee, Y. Y., Abedin, N. D., Abas, F., & Chong, G. H. (2021). Supercritical carbon dioxide extraction of pomegranate peel-seed mixture: Yield and modelling. *Journal of Food Engineering, 301*.

Theocharidou, A., Psomas, E., Koliouskas, A., & Ritzoulis, C. (2023). Yogurt products fortified with microwave-extracted peach polyphenols. *Gels, 9*(4), 266.

Tiwari, S., Upadhyay, N., Singh, A. K., Meena, G. S., & Arora, S. (2019). Organic solvent-free extraction of carotenoids from carrot bio-waste and its physico-chemical properties. *Journal of Food Science and Technology, 56*, 4678–4687.

Torres-Valenzuela, L. S., Ballesteros-Gómez, A., & Rubio, S. (2019). Green solvents for the extraction of high added-value compounds from agri-food waste. *Food Engineering Reviews, 12*, 83–100.

Tran, T. M. K., Akanbi, T., Kirkman, T., Nguyen, M. H., & Vuong, Q. V. (2022). Maximising recovery of bioactive compounds from coffee pulp waste using microwave-assisted extraction. *European Journal of Engineering and Technology Research, 7*, 1–6.

Tsegaye, B., Jaiswal, S., & Jaiswal, A. (2021). Food waste biorefinery: Pathway towards circular bioeconomy. *Food, 10*, 1174.

Villacís-Chiriboga, J., Voorspoels, S., Uyttebroek, M., Ruales, J., Van Camp, J., Vera, E., & Elst, K. (2021). Supercritical CO_2 extraction of bioactive compounds from Mango (*Mangifera indica L.*) peel and pulp. *Foods, 10*(9), 2201.

Wen, L., Zhang, Z., Sun, D. W., Sivagnanam, S. P., & Tiwari, B. K. (2020). Combination of emerging technologies for the extraction of bioactive compounds. *Critical Reviews in Food Science and Nutrition, 60*, 1826–1841.

Xu, K., Wu, C., Kou, X., Fan, G., Li, T., Sun, W., & Suo, A. (2022). Enzyme-assisted extraction of apricot polysaccharides: Process optimization, structural characterization, rheological properties and hypolipidemic activity. *Journal of Food Measurement and Characterization, 16*, 2699–2709.

Yaashikaa, P. R., Senthil Kumar, P., & Varjani, S. (2022). Valorization of agro-industrial wastes for biorefinery process and circular bioeconomy: A critical review. *Bioresource Technology, 343*, 126126.

Zaccardelli, M., Roscigno, G., Pane, C., Celano, G., Di Matteo, M., Mainente, M., Vuotto, A., Mencherini, T., Esposito, T., Vitti, A., & Falco, E. D. (2021). Essential oils and quality composts sourced by recycling vegetable residues from the aromatic plant supply chain. *Industrial Crops and Products, 162*, 113255.

Zhang, X., Liu, C., Chen, Y., Zheng, G., & Chen, Y. (2022). Source separation, transportation, pretreatment, and valorization of municipal solid waste: A critical review. *Environment, Development and Sustainability, 24*(10), 11471–11513.

Chapter 6
Construction and Demolition Waste Management: Bottlenecks, Regulations, and Policy Framework

Shiromi Karunaratne and Arshiya Kariapper

1 Introduction

Construction and Demolition Waste (CDW) is materials generated during the construction, renovation, and demolition of an infrastructure (Yuan, 2017; Zhao et al., 2022; Reis et al., 2021; Lai et al., 2016). CDW contributes to more than 30% of the total waste produced in the world and causes environmental and economic challenges (Papargyropoulou et al., 2011). It was reported by Menegaki and (2018) that more than 35% of CDW are annually disposed of in landfills. The amount of CDW escalates simultaneously with population growth and rapid urbanization. The United States (US), China, and the European Union (EU) are the three biggest CDW generators in the world (Kabirifar et al., 2020). The construction industry is responsible for using a large portion of natural resources and is the highest emitter of greenhouse gases in the world (IPCC Report, 2021; Yeheyis, 2013).

The composition of CDW depends upon the geographical position depending upon demographics, regional planning, available technologies, and so on (Reis et al., 2021). CDW consists of a spectrum of materials spanning from biodegradable materials such as wood, paper, and textile to hazardous materials such as adhesives, wood treated with pesticides, and asbestos (Liu et al., 2020; Cook et al., 2022; What is CD waste, 2020; Esa et al., 2017). According to studies, the uses of CDW as a resource and the research studies centered around CDW were extremely promising to achieve a sustainable CDWM (Qiao et al., 2022). The major challenge associated

S. Karunaratne (✉)
Department of Civil Engineering, Sri Lanka Institute of Information Technology, Malabe, Sri Lanka
e-mail: shiromi.k@sliit.lk

A. Kariapper
Department of Civil Engineering, University of Moratuwa, Katubedda, Moratuwa, Sri Lanka

V. Kumar et al. (eds.), *Environmental Engineering and Waste Management*,
https://doi.org/10.1007/978-3-031-58441-1_6

with CDWM is related to the awareness of stakeholders such as owners, contractors, and architects (Yuan, 2017; Myneni & Rajan, 2021; Mallinson, 2020; Ghaffar et al., 2020), inefficient waste handling, and disposal practices due to inadequate infrastructure and facilities. Illegal dumping due to inadequate waste-related legislation and the scarcity of effective recycling centers dedicated to handling CDW are some key examples of this paramount deficit (Yuan, 2017; Seror & Portnov, 2018; Liu et al., 2020; Hoang et al., 2020; Aslam et al., 2020; Zhao et al., 2022). Financial and technological constraints faced by many developing and transition economies are mainly due to lack of incentives and improper waste recycling plants that struggle with the staggering disparities between the cost and profit margins due to the costly treatment process of CDW or the cost embedded in implementing advanced technologies (Yuan, 2017; Kariapper et al., 2015; Galán et al., 2019; Aslam et al., 2020; Li et al., 2020). Thus, it is evident that effective CDWM is of paramount importance considering environmental, economic, and social aspects. Effective CDWM practices reduce the demand for natural material leading to its conservation and diverting materials from landfills while reducing greenhouse gas emissions, paving the way for a circular economy (Marzouk & Azab, 2014; Mah et al., 2018). The importance of robust policies, regulatory frameworks, and statutory laws is considered the key driving mechanisms to achieve efficient CDWM (Aslam et al., 2020; Hao et al., 2020; Tissera et al., 2022; Ghaffar et al., 2020).

This chapter extensively discusses the bottlenecks and challenges faced by decision-makers in incorporating effective CDWM and analyses the countless regulations and policy frameworks practiced by countries proficient in good CDWM practices. The chapter begins with an introduction to CDW, and Sect. 2 shines light on the factors driving CDW generation and the environmental impact it has. Furthermore, it emphasizes the importance of stringent CDWM requirements, given their far-reaching ecological, social, and economic ramifications, which will be elucidated in greater depth in the proceeding sections. Section 3 is based on "Challenges and Bottlenecks in Sustainable Management of Construction and Demolition Waste (CDW)" and delves into several pivotal facets of challenges unique to CDWM. It reveals the global scope of issues such as lack of awareness and understanding among stakeholders, and inefficient waste handling and disposal practices due to inadequate infrastructure and facilities and financial and technological constraints.

Section 4 is based on "Policy Frameworks, Regulations, and Legal Framework for CDW Management," which delineates critical domains essential for sustainable CDWM. These objectives encompass regulatory structures, policy aspirations, stakeholder involvement, circular construction principles, global collaboration, the cultivation of recycling markets, and more. Effective CDWM is paramount to reduce environmental impacts, promote resource preservation, and align with sustainable construction practices. Thus, EU, Australia, and the US were selected to discuss international and regional policy regulations in Sect. 4.2, while Singapore, Japan, and South Korea were selected to discuss under countries in Asia as discussed in Sect. 4.5. These regions and countries were selected in this study since they bolster effective and efficient CDWM practices (Aslam et al., 2020; Ghaffar et al., 2020; Wu et al., 2020; Zhao et al., 2022). Valuable lessons can be learned by analyzing the statutory and regulatory frameworks of countries and regions practicing apt CDWM (Agamuthu, 2008; Aslam et al., 2020; Ghaffar et al., 2020). The importance of applying circular economy in construction industry and implementation of proper data acquisition to bolster effective decision-making are highlighted throughout Sect. 4 as key factors that instigate policymakers' role in making robust and effective policies, and laws and enforce penalties to those noncompliant (Aslam et al., 2020; Hao et al., 2020; Tissera et al., 2022; Ghaffar et al., 2020).

In Sect. 5, an extensive discussion is carried out on the success stories, best practices, and some selected case studies. The effective implementation of rules and regulations; enforcement of taxes and levies on landfill disposal; granting of incentives to promote recycling and treatment of CDW; initiating standards, certificates, and award systems to boost consumer confidence; and maintaining apt online marketplaces to facilitate buying and selling of CDW were identified as the driving forces to achieve a successful CDWM (Aslam et al., 2020; Ghaffar et al., 2020; Wu et al., 2020; Zhao et al., 2022). It should be emphasized that the integration of environmental, financial, and social elements is paramount to achieve sustainability and is imperative to succeed. Section 6 based on the study provides valuable recommendations to establish a robust CDWM framework.

1.1 Objective and Scope of the Chapter

- To conduct a thorough investigation on CDW, probing into the sources contributing to its generation, composition, and unique characteristics and examining innovative initiatives harnessing CDW as a valuable resource.
- To identify the challenges and bottlenecks innate to CDWM.
- To analyze the established policy frameworks, regulatory measures, and legal structures practiced in countries having effective CDWM systems.
- To conduct an in-depth investigation of best practices and case studies of CDWM.
- To offer recommendations based on the research findings to achieve an effective CDWM system.

2 Understanding Construction and Demolition Waste

2.1 Definitions

There are various definitions for CDW depending on the perspectives of policymakers of different regions in the world (Wu et al., 2020). In general, CDW is materials resulting from construction, renovation, and demolition of infrastructure. It mainly constitutes a wide range of materials, including concrete, asphalt pavement, clay bricks, plastic, wood, glass, and ceramics (Yuan, 2017; Wu et al., 2020).

It also contains hazardous elements such as solvents, asbestos, and coal tar (Nguyen et al., 2021).

2.2 Composition and Characteristics

CDW consists of different categories of materials which depend upon its sources, size, and location. There is a distinctive disparity between CDW and municipal waste. Its characteristics vary due to regional factors such as demographics, population growth, regional planning, legislation, construction materials, and available technologies (Reis et al., 2021). Understanding these characteristics is of paramount importance for effective management of CDW.

CDW typically consists of non-biodegradable materials such as concrete, metal, clay bricks, plastic, glass, ceramics, asphalt, stone, and soil (Zhao et al., 2022; Nguyen et al., 2021), which are more than 10 mm in size. Slowly biodegradable materials such as wood, plastics, paper, rubber, leather, and textiles as well as less hazardous materials such as adhesives, coatings, cleaning agents, preservatives, paints, sealants, tars, plaster boards, wood treated with pesticides, asphalt shingles, asbestos, glass fibers, Mercury, oils, lubricants, waste acid, waste alkali, and gypsum are also present in CDW and may pose severe environmental catastrophes and health hazard if not handled correctly (Liu et al., 2020; Cook et al., 2022; What is CD waste, 2020; Esa et al., 2017). Figure 6.1 portrays the characteristic material composition in CDW (Santos et al., 2017; Yuan, 2017; Kourmpanis et al., 2008).

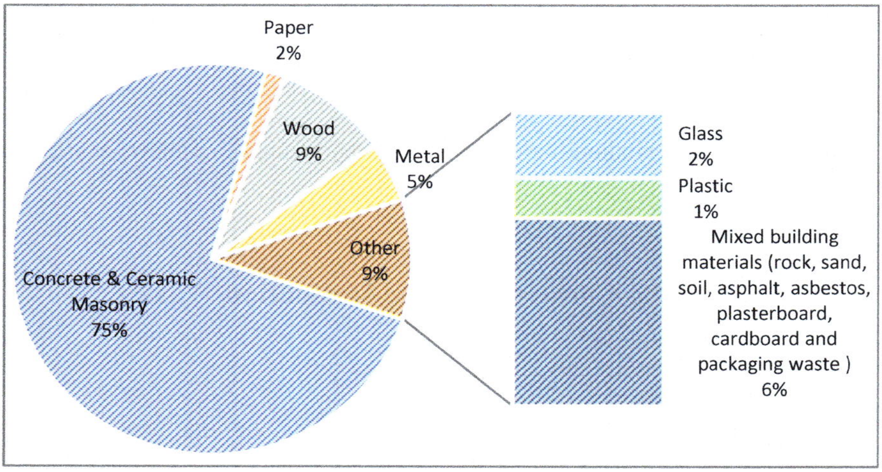

Fig. 6.1 General composition of construction and demolition waste (CDW). (Based on the data from Santos et al., 2017; Yuan, 2017; Kourmpanis et al., 2008; Islam et al., 2019; Alsheyab, 2022; Bergsdal et al., 2008)

2.3 *Causative Factors*

The major causative factors contributing to CDW are multifaceted. CDW emerges throughout the project lifecycles of any construction project (Saadi et al., 2016; Lai et al., 2016). CDW can be categorized into two sources of origin "human-generated origins through building life cycles" and "natural origins" (Reis et al., 2021), as represented in Fig. 6.2. Human-generated sources primarily comprise CDW generated from construction stage through operational stage to demolition stage of any infrastructure life cycle. The natural origins of CDW arise from environmental

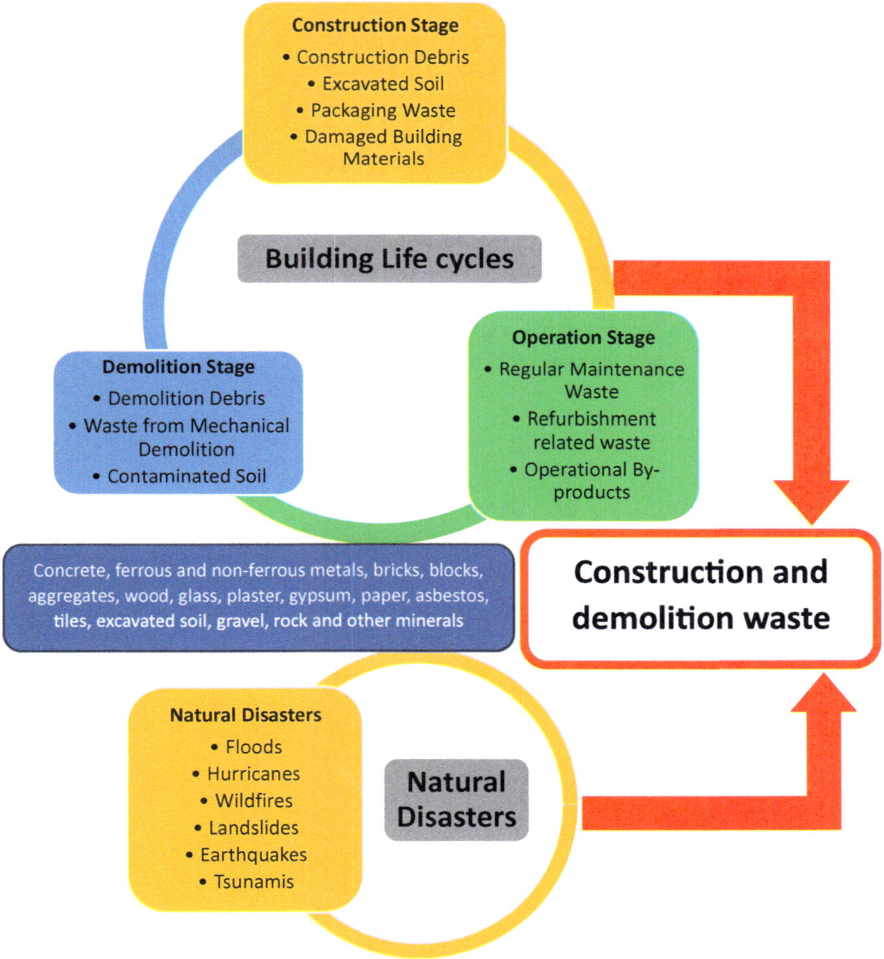

Fig. 6.2 Classification of construction and demolition waste (CDW) according to the source of origin and the life cycle of CDW generated in infrastructure

disasters. These natural disasters have the potential to generate vast and sudden quantities of CDW, presenting significant challenges for governments and posing major health hazards.

2.4 The Scale of the Problem Contribution by Countries

Around 90% of CDW is considered potentially recyclable, as indicated by Zhao et al. (2022) and Galán et al. (2019). However, this is not the case in many parts of the world. Figure 6.3 shows the CDW generation rate in some countries and their recycling rates. In the European Union (EU), CDW recovery rates in 2016 averaged around 89%, with considerable variation among member states, ranging from nearly 100% to under 5%, emphasizing the differences in CDWM practices (Hao et al., 2020). In southern European countries, CDW recycling rates are notably lower (Villoria et al., 2014). Differences in CDWM within the EU stem from factors such as landfill pricing, penalties for non-compliance, and the availability of cost-effective raw materials (Villoria et al., 2014). Countries like the Netherlands and Belgium, who portray advance CDWM measures, have achieved high recycling rates, demonstrating the substantial recycling potential of CDW (Villoria et al., 2014). In Australia, CDW recycling rates are influenced by regional factors, such as waste generation trends, cross-border waste influx, and the absence of a mandatory landfill levy in as witnessed in Tasmania (Zhao et al., 2022).

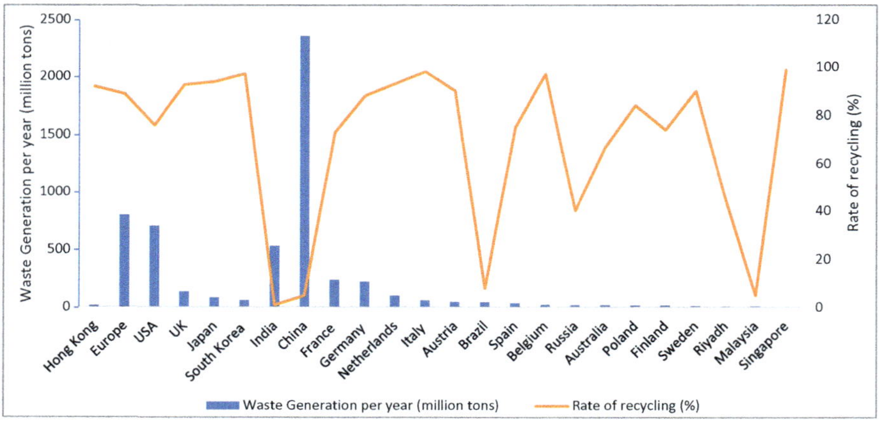

Fig. 6.3 Construction and demolition waste (CDW) generation and rate of recycling in various countries. (Based on the data from Jain, 2020; Wu et al., 2020; Wang et al., 2021; Zhao et al., 2022; Haider et al., 2022; Statistics Unit, Environmental Protection Department, 2022; US EPA, 2017; 28 incredible statistics about waste generation, 2021; UK statistics on waste, 2023; Liu et al., 2020; Recovery rate of construction and demolition waste in the EU-272018, n.d.; Hao et al., 2020; Nunes & Mahler, 2020; Zhang et al., 2020; Nguyen et al., 2021)

2.5 *Construction and Demolition Waste (CDW) as a Resource*

Considering CDW as a resource can contribute to cost savings in construction projects while minimizing the impact on the environment by reducing the use of natural resources. Proper CDWM contributes to a more circular and sustainable economy bringing multifaceted social benefits (Qiao et al., 2022). CDW may be subjected to upcycling or downcycling, depending on the purpose/value of the secondary product (Zhang et al., 2022). Reinforced concrete can be considered a suitable candidate for upcycling. Contaminated concrete waste is commonly repurposed as road base, backfill, or for engineering applications in landscaping, representing a prevalent downcycling approach.

Much research has been conducted to investigate the ability to replace natural aggregates using recycled aggregates generated from waste concrete or bricks to manufacture fresh concrete (Premasiri et al., 2013; Kien et al., 2013; Kariapper et al., 2015). Performance parameters, including replacement ratio, compressive strength, elastic modulus, and treatment methods, were assessed and found to have promising values (Premasiri et al., 2013; Kariapper et al., 2015). Furthermore, research is ongoing to investigate the utilization of glass and other CDW as aggregates in concrete production (Abdelli et al., 2020; Baikerikar et al., 2023). Table 6.1 shows the pathways of managing CDW as a resource and becomes apparent that CDW nurtures numerous untapped resource potentials.

3 Challenges and Bottlenecks in Sustainable Management of Construction and Demolition Waste (CDW)

Stakeholder awareness, fragmented responsibility, inefficient waste management practices, inadequate infrastructure, financial constraints, lack of localized case studies, uneven adherence to waste management laws, untapped potential of Building Information Modeling (BIM) in deconstruction, and limited research on circular construction applications are some of the challenges highlighted by Yuan (2017), Myneni and Rajan (2021), Mallinson (2020) and Ghaffar et al. (2020) that hinders an effective CDWM system.

3.1 *Lack of Awareness and Understanding Among Stakeholders*

Key internal stakeholders in the successful implementation of CDWM programs, including architects, engineers, engineering managers, quantity surveyors, and clients, play a pivotal role. Saadi et al. (2016) stress the need for these stakeholders to prioritize awareness and knowledge dissemination in establishing a robust CDWM

Table 6.1 Pathways of managing construction and demolition waste (CDW) as a resource

Strategy	Application	World/country statistics details	Reference/s
Reusing and repurposing	Salvaged and reused directly or repurposed for other applications	In the US materials and waste exchange markets facilitate the trade of used doors, windows, cabinets, plumbing fixtures, lumber, millwork, flooring, and fencing. These markets are overseen by state, local governments and/or private entities. They vary in scale depending on the materials involved and the extent of their transformation	US EPA (2016)
Recycling	Recycled and used as raw materials for new construction projects	The Netherlands, Japan, Singapore and South Korea are some countries that have introduced quality certificates for recycled aggregates. In Japan, the recycling rate of asphalt-concrete blocks and concrete blocks were 98.4% and 97.3% respectively in the year 2014	Wu et al. (2020), Hao et al. (2020)
		Recycling of metal waste (ferrous metals, aluminium, copper, lead, and zinc) is practised worldwide. The ability of metal to be directly used or melted down for reuse presents a sustainable opportunity for its multiple applications in construction work	Hamakareem (2020), Jordan (2021)
		Plaster and gypsum board recycling addresses resource scarcity, reduces costs, and enhances resource utilization. EU-27 member states have established a 70% target for gypsum recycling	Bumanis et al. (2022)
		Paper and paperboard are typically recycled into new paper products through purification processes as practiced in the USA, Singapore, and Japan. Various	Okeyinka et al. (2015), Tam and Tam (2006)

(continued)

Table 6.1 (continued)

Strategy	Application	World/country statistics details	Reference/s
		studies have investigated the use of waste paper in producing building materials, including fiber-reinforced cement composites, wall panels, bricks, building blocks, thin cement sheets, composite panels, low-density boards, and building insulation	
Waste-to-Energy Conversion	CDW that cannot be recycled or reused can be processed in waste-to-energy plants	In Singapore, a single landfill site exists, with incineration plants serving as the principal disposal solution due to limited land space	Wu et al. (2020)
		In Sweden, combustible waste is banned from being disposed of in landfills. Successful implementation of incineration of CDW has demonstrated improved CDWM practices in Sweden, the Netherlands, Japan, and Singapore	Wu et al. (2020), Hao et al. (2020)
		In Denmark and Belgium, the overall use of incineration for effective CDWM is primarily due to their notable dependence on wood materials in construction activities	Villoria and Osmani (2019)
		The bottom ash resulting from incineration is predominantly used as road materials	Tan et al. (2022)
		Fly ash resulting from incineration exhibits pozzolanic properties, making it a valuable building material that has the potential to enhance the compressive strength of concrete and geopolymer concrete. However, research is still underway	Tan et al. (2022), Poranek et al. (2021)
		In 2015, Timor-Leste's capital, Dili, initiated a waste-to-energy project	Hoang et al. (2020)

(continued)

Table 6.1 (continued)

Strategy	Application	World/country statistics details	Reference/s
		where a Plasma Furnace System was developed to tackle the increasing waste quantities, including CDW, deposited at the Tibar landfill	
Landfill	CDW that cannot be recycled, reused or cannot be processed in waste-to-energy plants need to be managed in sanitary landfills	In 2014, the percentage of non-recyclable CDW landfilled varied widely among countries. The Netherlands achieved efficient waste management, with less than 1% of CDW landfilled. In contrast, Belgium and Greece had high landfill rates, of 65% and 99% respectively	Villoria and Osmani (2019), Hao et al. (2020)
		Underground landfills, such as closed salt, manganese, and potassium mines are utilized to store stabilized fly ash which is a product of the incineration process. This is due to its high content of hazardous substances like heavy metals, furans, dioxins, and chlorides	Tan et al. (2022), Poranek et al. (2021)

system. According to Zhao et al. (2022) and Ghaffar et al. (2020), CDWM is also influenced by external stakeholders such as government bodies and the general public. These dynamics are significantly shaped by human elements such as awareness, attitudes, and behavioral norms on a global scale (Zhao et al., 2022). A critical global challenge lies in the limited awareness of circular construction principles among both internal and external stakeholders, generating more CDW (Saadi et al., 2016).

The perception of the client has a significant influence on waste management practices among designers, contractors, and subcontractors. Profit-driven client motivations undermine the importance of proper waste management practices. Furthermore, design alterations driven by client preferences often result in increased waste generation (Zhao et al., 2022).

Contractors may perceive waste management as a financial burden and tend to base decisions regarding CDW on landfill costs and penalties for illegal dumping, overlooking the environmental impact it has (Yuan, 2017; Agamuthu, 2008).

Further, lack of expertise and training among site operators stand as a pivotal factor influencing poor CDWM practices (Zhao et al., 2022; Vivian, 2011).

The construction sector faces CDWM challenges related to fragmented responsibility, facility management coordination, and failed afterlife approaches (Myneni & Rajan, 2021; Saadi et al., 2016; Yuan, 2017). Thus, a lack of understanding, unequal profit distribution, and a lack of incentives for afterlife design amongst stakeholders hinder the need for a circular economy (Myneni & Rajan, 2021).

The inadequate understanding of waste flows, essential for waste reduction, environmental strategies, and technical standards in CDWM emphasize the need for precise estimation of statistical data (Saadi et al., 2016; Mihai, 2019; Yuan, 2017). Official national reports often lack accurate CDW generation data due to the inefficiency of government officials. This inadequacy impedes effective CDWM decision-making in various countries, including China, Brazil, Bolivia, Mexico, and the EU (Yuan, 2017; Moschen et al., 2023; Ferronato et al., 2023). Precise CDW estimates from construction sites are crucial for planning container quantities, sizes, and spatial requirements for effective CDWM (Zhang et al., 2019; Villoria et al., 2014).

3.2 *Inefficient Waste Handling and Disposal Practices Due to Inadequate Infrastructure and Facilities*

Inefficient waste handling and disposal practices, coupled with inadequate infrastructure and facilities, pose substantial challenges to CDWM. This issue is a critical concern in recycling CDW, as it hinders the efficient handling of these materials. The lack of recycling centers, sorting facilities, or transportation networks committed to handling CDW are some key examples of this paramount deficit (Yuan, 2017; Seror & Portnov, 2018; Yuan, 2017; Liu et al., 2020; Hoang et al., 2020; Aslam et al., 2020; Zhao et al., 2022). In many countries, insufficient landfill capacities for CDW are a pressing issue, as noted by Yuan (2017). Developing nations, including Peru, Panama, and Malta, resort to improper disposal methods such as dumping CDW into the sea, public drainage systems, and mangroves (Colorado et al., 2022; Cilia, 2019; Disposal at Sea, 2019). In some instances, recycling facilities lack the capacity to handle substantial CDW generation, termed the "capacity scarcity scenario" (as in China), or face an excess of wasted building materials, known as the "capacity glut scenario" (as in Sweden) (Yuan, 2017; Liu et al., 2019). Inadequate tracking of waste-carrying vehicles also instigates private contractors to illegally dump CDW to evade landfill fees (Yuan, 2017; Seror & Portnov, 2018).

Australia manages CDW at the regional level, allowing interstate transport for recycling, landfilling, or export, particularly practiced in NSW, Queensland, and Victoria. This cross-border waste movement is influenced by local recycling infrastructure, differing landfill levies among states, and market dynamics. The USA also witnesses cross-jurisdiction waste movement, a practice not considered ideal for proper CDWM (Zhao et al., 2022; Wu et al., 2020).

3.3 Financial and Technological Constraints

The absence of robust mechanisms for financial incentives diminishes the motivation of project stakeholders to actively reduce waste during construction and demolition activities (Yuan, 2017). Despite the robust CDWM system in place in the UK, ambiguous financial justifications have been identified to impede decision-making and planning processes (Myneni & Rajan, 2021).

High waste disposal charges have effectively curbed CDW landfilling in EU countries. However, determining an appropriate fee, especially in less regulated waste disposal areas in developing countries or regions, is challenging and may lead to an increase in illegal dumping, as noted by Yuan (2017). Recycling plants in transition and developing economies often struggle financially because revenue from recycled sales does not cover operational costs (Yuan, 2017; Premasiri et al., 2013; Aslam et al., 2020).

As outlined by Yuan (2017), Galán et al. (2019), Maria et al. (2018) and Premasiri et al. (2013), some of the key challenges faced by the recycling plants can be identified as elevated production costs due to power and fuel consumption and investment costs associated with advanced sorting technologies which may not always lead to higher economic performance. Furthermore, dealing with highly mixed CDW requires complex separation techniques (Galán et al., 2019; Aslam et al., 2020). Intricate treatment procedures of CDW are needed to produce quality recycled products that can be used in construction activities, which results in higher unit costs compared to virgin raw materials which impedes their acceptance in the market (Yuan, 2017; Kariapper et al., 2015).

Even though civil engineering projects consists of valuable materials, end-of-life construction products often possess limited value, making their reuse economically unviable. This impedes the implementation of circular construction methods given that profit remains the primary motivator for developers and clients (Myneni & Rajan, 2021). However, adopting advanced recycling processes and selective demolition is hindered by high costs, lack of knowledge, and unclear regulations (Aslam et al., 2020; Li et al., 2020).

Recycled aggregates (RA) are a valuable construction raw material that can be produced from CDW. However, depending on the demolition methods used, it may contain impurities that affect concrete properties, emphasizing thorough purification processes to prevent issues like high water absorption due to high porosity and steel corrosion. The safety and feasibility of using high-quality CDW RA as a substitute for natural aggregates in concrete are still actively researched and not yet practiced in the industry in many parts of the world (Kariapper et al., 2015; Premasiri et al., 2013; Di Maria et al., 2018; Aslam et al., 2020; Zhang et al., 2023).

4 Policy, Regulations, Legal Frameworks, and Other Support Mechanisms for Construction and Demolition Waste Management (CDWM)

The absence of comprehensive legislative regulations that govern CDW and the assurance of efficient quality control measures are considered to be the main challenges in achieving sustainable CDWM. Thus, practices such as illegal dumping are particularly dominant in most developing countries such as India, Cyprus, the Czech Republic, and Malaysia (Agamuthu, 2008). Therefore, it is imperative to establish robust regulatory frameworks and provide adequate financial support to foster sustainable CDWM.

4.1 Objectives and Goals of Policy Development

CDWM involves reduction, recycling, and disposal of CDW in an efficient and sustainable manner. The policy objectives of CDWM are aimed at minimizing the environmental impact, promoting resource efficiency, and ensuring the health and safety of relevant stakeholders. It is of paramount importance to establish robust policies on CDWM and comparing social, economic, and managerial approaches of nations and regions with effective CDWM can inspire the development of innovative and relevant practices (Aslam et al., 2020).

Following key considerations should be considered in policy development to achieve effective CDWM in a country (Aslam et al., 2020; Hao et al., 2020; Tissera et al., 2022; Ghaffar et al., 2020):

1. Reduce waste generation through better design, planning, and construction practices.
2. Reuse and recycling of used materials to reduce the strain on virgin resources and enhance the circular economy.
3. Promoting the implementation of sustainable building practices in construction and renovation.
4. Accurate data collection and reporting to monitor recycling rates, to steer the enactment of efficient waste management regulations.
5. Having strict legislative and regulative enforcement to administer licensing, and penalties for non-compliance.
6. Promote innovation and adaptation of state-of-the-art technologies.

Emphasis should be placed on adopting a well-established urbanized circular construction and circular economy model with a complete management plan (Aslam et al., 2020; Ghaffar et al., 2020). Furthermore, when integrating a CDWM system into national law, various factors must be considered, such as the cost of virgin materials, landfill capacity, stakeholder involvement, public awareness, and disposal expenses (Villoria et al., 2014). These factors result in unique CDWM policies and regulations tailored to each country's social and political context.

4.2 Overview of International and Regional Policy Regulations

The 1992 United Nations Conference on Environment & Development (UNCED) Conference in Rio de Janeiro produced Agenda 21, for achieving sustainable development in the twenty-first century (UNCED, 1992). It endorsed novel strategies encircling education, resource preservation, and economic sustainability. Agenda 21 stressed waste reduction, reuse, and recycling to achieve the lowest influence on the environment. It promoted eco-friendly waste treatment technologies and recognized the construction sector's potential for environmental harm. The involvement of stakeholders in waste management decisions was further emphasized. Objectives and goals to achieve effective CDWM were promoted by this agenda (UNCED, 1992).

Countries such as the UK, the USA, Austria, and Germany exhibit strong and effective regulatory systems for CDW (Hao et al., 2020; Wu et al., 2020; Aslam et al., 2020; Villoria et al., 2014). Moreover, their ability to incorporate changes to address arising challenges and specific needs enhances the comprehensiveness of their policy frameworks (Zhao et al., 2022; Aslam et al., 2020; Villoria et al., 2014).

In 2008, the EU set new recycling targets for CDW, leading to the Waste Framework Directive (WFD) (2008/98/EC) and Landfill Directive (1999/31/EC) to promote a recycling-focused European society (Villoria et al., 2014). The directive mandates that at least 70% of nonhazardous CDW must be redirected for reuse, recycling, or recovery, including backfilling operations (Hao et al., 2020). By 2016, the majority of member states had successfully attained the EU's 70% CDW recovery rate goal (Waste Statistics, 2023; Hao et al., 2020).

In Australia, the federal government does not directly enact legislation governing CDWM. It is the responsibility of the authorities of the states and territories. Australia's DCCEEW in 2018 introduced a National Waste Policy, embracing the principles of a circular economy and having a vision extending up to 2030. The state and territory parliaments develop policies and waste management strategies tailored to their unique contexts. This collaborative framework serves as the legal reinforcement for CDWM practices in Australia (Zhao et al., 2022). For instance, most jurisdictions have implemented landfill levies and the majority have released waste management strategy documents that establish specific waste recovery or recycling targets except the Northern Territory (NT) (Zhao et al., 2022). In Australia, landfill levies are determined by various factors, including landfill location (metropolitan, regional, or rural), exemptions for specific materials, and levy zones (Zhao et al., 2022). According to Zhao et al. (2022), these strategy documents play a vital role in legislative frameworks even though they lack statutory power. However, it is important to acknowledge that the resistance to change is identified as the top barrier to CDWM in Australia (Zhao et al., 2022).

In the late 1960s and early 1970s, eight southern US states established the foundation for CDWM through the "Solid Waste Disposal Act of 1965," as noted by Aslam et al. (2020). To ensure compliance and effective CDWM, the US government mandates advanced frameworks and economic measures at the state level (Us Epa, 2013). In 1976, Congress passed the Resource Conservation and Recovery Act (RCRA), highlighting nonhazardous CDW issues, including construction, reduction, landfills, and safety, as outlined in sections 40 CFR 257 and 258.2 of the RCRA (Humber & Lingel, 1980). The Comprehensive Environmental Response Compensation and Liability Act (CERCLA) serves as the primary legislation governing CDW, placing responsibility on contractors (CERCLA, 1984; Aslam et al., 2020). The US has prioritized CDW research collaborating with its construction industry. Many states have initiated their own CDW regulations and surveillance systems, tailored to them (Aslam et al., 2020).

These countries and regions exhibit strong and effective regulatory systems for CDWM. Moreover, their ability to incorporate changes to address emerging challenges and specific needs enhances the comprehensiveness of their policy framework (Zhao et al., 2022; Aslam et al., 2020; Villoria et al., 2014).

Table 6.2, provides a summary of the existing legal statutes and regulations for various aspects of CDW-related activities within these nations.

4.3 National and Local Regulations and Policies in a Few Selected Countries in Asia

Within this section, the CDW regulations, policies, and management approaches of Singapore, Japan, and South Korea will be discussed as they have established efficient regulations and policies which had led to effective CDWM as a country. Other major countries in Asia such as India, China, and Malaysia have not been included as they have not reached very significant level of success at national level compared to Singapore, Japan, and South Korea.

Singapore's Building and Construction Authority (BCA) and the National Environment Agency (NEA) play central roles in CDWM, with a focus on sustainability. Japan stresses waste reduction at the design stage and mandates responsible waste handling by waste producers, backed by stringent legal frameworks. South Korea, on the other hand, has shifted its focus to waste minimization, guided by a dynamic "master plan" responsive to growing CDW needs. Each country's approach reflects its unique environmental priorities and legal structures, providing valuable insights into how nations address the challenges of CDWM.

Table 6.3 provides an overview of the existing legal statutes and regulations of various aspects of CDW-related activities within these nations.

Table 6.2 Regulatory frameworks in countries with well-established construction and demolition waste management (CDWM) legislations

Country	Laws and regulations	Obligations in legalization	Reference/s
Austria	Waste management act 2002 remediation of contaminated Sites. Austrian ordinance for tracking waste. List of waste ordinance. Separation of construction waste ordinance. Landfill ordinance. Hazardous Waste Ordinance End-of-Waste Act. Recycled Construction Materials Regulation.	Selective demolition On-Site CDW Sorting Mandate Segregated CDW collection Hazardous waste management plans	Hao et al. (2020), CDWM in Austria (2015), De Bock (2012), Soil protection law (2014)
Belgium	Belgium is divided into three main regions: Flemish, Brussels Capital, and Walloon. Each of these regions has its own set of regulations and legislation concerning waste management, including the recycling of CDW. Flemish Statute on environmental permits dated 28 June 1985. Brussels Environmental Permit Statute, dated 5 June 1997. Walloon Environmental Permit Statute, dated 11 March 1999. Federal Code on wellbeing at the workplace (Book VI, Title 3)		
Denmark	Environmental Protection Act No. 879 26/06/2010. Statutory Order No. 1309/2012. Statutory Order No. 1662/2010. Circular of 15 July 1985.		
Finland	Waste Act 646/2011. Land use and building Act 132/1999. Environmental protection Act 527/2014.		

(continued)

Table 6.2 (continued)

Country	Laws and regulations	Obligations in legalization	Reference/s
France	Law 2009–967. Decree n°2011–610. Law 2010–788. Decree n°2014–1501.		
Germany	Waste Disposal Act. The Circular Economy Act. Waste Classification Ordinance. Federal Soil Protection Act of 17 March 1998. Federal Soil Protection and Contaminated Sites Ordinance of 12 July 1999. Ordinance on the Management of Municipal Wastes. Landfill Directive.		
Luxembourg	Law of 21 March 2012 on management of waste. Grand-Ducal Regulation of 24 February 2003 on landfilling of waste.		
Netherlands	Environmental Protection Act The National Waste Plan The Decree on landfills and waste bans The Decree on notification of industrial and hazardous waste From Waste to Resource	Selective Demolition On-Site Sorting Mandate for CDW Segregated Collection of CDW Plans for Managing Hazardous CDW Elevated Fees for CDW Disposal at Landfills Established Markets for CDW Reuse/Recycling Enhanced Quality Control for CDW Promotion through Green Building Rating and Sustainable Procurement Systems	Wu et al. (2020), Hao et al. (2020)
Sweden	Ordinance on Environmental assessment SFS 2013: 251 Waste Ordinance SFS 2011:927 Building Code (SFS 2010: 900)	On-Site CDW Sorting Requirement Segregated Collection of CDW Plans for Managing Hazardous CDW Sustainable	Wu et al. (2020), Hao et al. (2020)

(continued)

Table 6.2 (continued)

Country	Laws and regulations	Obligations in legalization	Reference/s
	Swedish Ordinance on PCB (SFS 2007:19) The Regulation NFS 2004: 4	Procurement Practices Enforcement of Waste Landfilling Ordinance Elevated Charges for CDW Landfill Disposal Prohibition of Landfilling Combustible CDW Fractions Advanced Technologies for CDW Incineration Pre-Demolition Audits and Selective Demolition Implementation	
Australia	Protection of the Environment Operations (POEO) Act 1997 Waste Avoidance and Resource Recovery (WARR) Act 2001 Protection of the Environment Operations POEO (Waste) Regulation 2014 NSW Government Sustainability Policy 2008 Waste Management Policy (Used Packaging Materials) 2010 Waste Management Policy (Movement of Controlled Waste Between States and Territories) 2002 Waste Reduction and Recycling Act 2011 Recycling policy for buildings and civil infrastructure	Variations in CDW Regulations Across States In-Depth Insight into the Current CDW Management Status in Each State Well-Established Market for Reused Masonry Products	Wu et al. (2020)
United States	The-National Environmental Policy Act (NEPA): Act 1980–97 Laws of Pennsylvania: Arkansas Solid Waste Management Act 1971: Solid Waste Management Plan Act 1989: Resource Recovery & Management: The City's Diversion of Construction and	Official National Guidelines for CDW Management Varied CDW Regulations Among States Adoption of Source Reduction Measures Adoption of Deconstruction Practices Well-Developed Markets for CDW Trade	Aslam et al. (2020), Wu et al. (2020)

(continued)

Table 6.2 (continued)

Country	Laws and regulations	Obligations in legalization	Reference/s
	Demolition Debris Ordinance: Solid Waste, Infectious Waste Treatment and Construction and Demolition Debris (C&DD) Facility Licenses:		

4.4 Regulatory Frameworks for Construction and Demolition Waste Management (CDWM)

As per Wu et al. (2020), regulatory measures play a pivotal role in achieving sustainable CDWM practices. These regulations enforce mandatory requirements and compel the stakeholders to adopt effective CDWM techniques. Moreover, policies such as taxation on new materials, the establishment of end-of-waste criteria, and the setting of reuse/recycling targets serve as potent incentives for CDW recycling (Ghaffar et al., 2020).

The regulatory framework governing CDWM should encompass critical factors influencing recycling efforts. These factors encompass specific legislation, normative standards, environmental consciousness, raw material availability, sales of remanufactured products, and supply chain management (Ghaffar et al., 2020; Liu et al., 2020). Legislation should prioritize the reuse and recycling of materials by clients (Ghaffar et al., 2020). Furthermore, facilitating the sale of remanufactured products is integral to effective supply chain management in CDW recycling (Liu et al., 2020).

As recommended by Liu et al. (2020) and Wu et al. (2020), illegal dumping should be discouraged and proper waste disposal practices should be encouraged by the development of a landfill charging policy tailored to local socioeconomic conditions and consideration of higher landfilling fees for mixed CDW. These serve as an effective incentive for contractors to implement waste sorting at the source.

Several key initiatives, as proposed by Ghaffar et al. (2020) and Zhao et al. (2022), warrant consideration for the enhancement of CDWM:

1. Incorporating Circular Design Concepts: Regulatory frameworks should integrate circular design principles, including "design for deconstruction," and "design for reuse or recycle." This integration promotes resource efficiency and waste reduction, with durability, workability, and post-use aspects being key considerations during the design phase.
2. On-site Inspections and Audits: Mandatory on-site inspections and audits conducted prior to demolition should be enforced, as they are recognized measures for significantly reducing CDW.

Table 6.3 Regulatory Framework in Countries with Well-Established CDW Management Legislation

Country	Laws and regulations	Obligations in legalization	Reference/s
Singapore	Code of Practice of Demolition (SS 557), 2010 Environmental Public Health Act (EPHA) EPH (General Waste Collection) Regulation EPH (Toxic Industrial Waste) Regulation	Rigorous monitoring to deter illegal dumping Enforcement of a Comprehensive Waste Collection System Imposition of substantial waste disposal fees Effective waste segregation practices Adoption of Green Mark Initiatives Well-developed technologies and infrastructure for construction and demolition waste recycling Directory of recycling enterprises	Vishnoi (2023), Wu et al. (2020), Mee (2009)
Japan	Waste Management and Public Cleansing Law Construction Material Recycling Law Basic Act on Promotion of Recycling Oriented Society Proper Process on C&D Waste Recycled Aggregate for Concrete—Class M (JIS A 5022), Class H (JIS A 5021) and Class L (JIS A 5023) Construction Recycling Promotion Plan Governmental Funds for Waste Disposal	The company is required to submit its plan, which will then undergo examination and documentation by the relevant authorities before project construction or demolition can commence Stacking and landfilling CDW are strictly prohibited. Advanced construction and demolition waste recycling technologies and infrastructure Roster of recycling enterprises Public infrastructure projects are mandated to incorporate recycled materials	Mingxue et al. (2023), Wu et al. (2020)
South Korea	Business Waste Reduction Program Construction Waste Recycling Promotion Act Waste Control Act Act for the Promotion of Saving and Recycling of Resources Recycled Aggregate for Concrete Enforcement Regulations for the Promotion of Construction Waste Recycling Act	Obligatory incorporation of recycled aggregates Certification system ensuring the quality of recycled aggregates Waste exchange information management system	Mingxue et al. (2023), Wu et al. (2020)

4.5 *Role of Other Support Mechanisms for Sustainable Construction and Demolition Waste Management (CDWM)*

Implementation of financial subsidies and reward systems promote CDWM practices (Aslam et al., 2020; Ghaffar et al., 2020). Government subsidies, directed toward CDW recycling production and recycling units, are vital for fostering rational CDW recycling. Government loans and public procurement policies can assist companies in establishing CDW recycling plants (Hao et al., 2020). Given the initial capital requirements for CDW recycling facilities, private sector involvement also becomes essential (Hoang et al., 2020; Aslam et al., 2020).

Adoption of systematic data collection approaches, coupled with computer-aided technologies like WasteSpec and SmartStart, as observed in the USA and the UK (Aslam et al., 2020; Hoang et al., 2020), is supportive and increasingly critical on construction sites. Developing mature recycling markets is another economically sustainable approach to support CDWM. A mature market can be delineated into three stages, such as generation stage, trading stage, and recycling stage. Contractors are incentivized for on-site sorting due to opportunities for selling recyclable materials while effective information exchange is facilitated through online systems. A mature market for recycling stage will help the buyers of recycled products encounter more client inquiries. Disseminating information about potential CDW recyclers on official websites, as practiced in Japan, the US, South Korea, and Singapore, plays a significant role in promoting CDW recycling (Aslam et al., 2020).

According to Wu et al. (2020), the establishment of a government- or industry-authorized certification system for recycled products is also crucial. Such quality certification systems, as implemented in South Korea, Australia, Singapore, and the Netherlands, substantially reduce public concerns and bolster market demand for recycled products (Aslam et al., 2020; Wu et al., 2020). Furthermore, in Singapore, the Green Mark Scheme has been utilized to actively engage construction stakeholders in showcasing their eco-friendly practices (Wu et al., 2020).

5 Success Stories, Best Practices, and Case Studies

Sharing best practices will give a good insight into how the countries have successfully implemented CDWM. Before investigating case studies, a brief insight into the present status of the countries with successful CDWM schemes is presented.

In Germany, CDW disposal in landfill sites was banned by legislative enactments and proper treatment and reuse were encouraged (Agamuthu, 2008). In Sweden, there is a ban on combustible waste fractions being landfilled (Wu et al., 2020). Singapore has implemented a general waste collection system to recycle waste and categorize CDW. According to the National Environment Agency (NEA) in Singapore, 99% of construction debris was recycled in the year 2017 excluding steel re-bars, wood, and glass wastes (Wu et al., 2020). The Building and Construction

Authority (BCA) in Singapore has established special funds to encourage stakeholders to purchase demolition equipment that helps to recover concrete. Consequently, Green Mark, Singapore Green Building Certification Scheme was employed to foster the proactive engagement of construction stakeholders in showcasing their eco-friendly practices, with a special emphasis on waste management as a strategic advantage (Wu et al., 2020). In South Korea, it is mandatory to use recycled aggregates and the government has successfully implemented a quality standard and a quality certificate to boost the confidence of potential consumers (Wu et al., 2020). In the Netherlands, clients and contractors are not afraid to use recycled materials because of the quality labels being introduced and the use of recycled material can facilitate in achieving a green building certificate (Wu et al., 2020).

Successful CDWM initiatives enclose a range of strategies and practices aimed at minimizing waste generation and promoting recycling (Aslam et al., 2020; Ghaffar et al., 2020; Wu et al., 2020; Zhao et al., 2022). Successful CDWM initiatives require a holistic and integrated approach that considers environmental, economic, and social aspects to attain sustainable outcomes while minimizing the environmental footprint of construction and demolition activities (Aslam et al., 2020; Ghaffar et al., 2020; Wu et al., 2020; Zhao et al., 2022).

This section delves into the key aspects identified that contribute to the success of CDWM initiatives.

5.1 Success Stories of Countries with Effective CDW Policies

EU countries, including Austria, Belgium, Denmark, Finland, France, Germany, Luxembourg, the Netherlands, and Sweden, have well-established legal frameworks for sustainable CDWM (Hao et al., 2020). Poland, for instance, achieved a remarkable 92% CDW recovery rate, thanks to effective practices like landfill taxes (Villoria & Osmani, 2019; Hao et al., 2020). Stern control of landfilling, particularly for specific types of CDW, serves as a significant driver for enhanced management, as observed in Denmark, Germany, and the Netherlands (Villoria et al., 2014). According to a survey conducted by Zhao et al. (2022), the enforcement of landfill levies has proven highly effective in Australia. Similarly, findings from the Western Australia Department of Water and Environmental Regulation (Dave et al., 2021) indicate that increased landfill levies have led to reduced CDW disposal. New South Wales (NSW) and South Australia (SA) have comparatively high landfill levies for CDW, while Tasmania employs a voluntary levy approach in regional areas.

5.1.1 Germany

In Germany, there exists an initiative comprising six prominent market stakeholders, namely BBS, BDE, BGRB, BRB, HDB, and ZDB, committed to advancing the circular economy within the construction sector (CDWM in Germany, 2015). The country periodically publishes monitoring reports every 2 years. Germany has notably completed 14 successful projects in the regions of Rhineland Pfalz and BadenWürttemberg, utilizing recycled concrete in surface construction (CDWM in Germany, 2015). A noteworthy endeavor in Germany is the Gypsum to Gypsum project, which strives to revolutionize the European gypsum demolition waste market with the goal of achieving higher rates of gypsum waste recycling, thus contributing to the realization of a resource-efficient economy (CDWM in Germany, 2015). Furthermore, Cemex, a prominent partaker, has embarked on the development of concrete incorporating recycled aggregates for surface construction (CDWM in Germany, 2015). It is essential to highlight that Germany's stringent adherence to rules and regulations has effectively minimized major loopholes and illegal practices within the CDWM sector (CDWM in Germany, 2015; Agamuthu, 2008). The country has also witnessed numerous lawsuits pertaining to CDW management (CDWM in Germany, 2015). Furthermore, the "Deutsche Gesellschaft für Nachhaltiges Bauen (DGNB)" evaluates buildings and urban districts that display exceptional dedication to fulfilling sustainability goals (CDWM in Germany, 2015) where the level of adherence to the required criteria determines the accolade level attained by a project.

5.1.2 Spain

According to Villoria et al. (2014), Spain has established a CDW policy through the Royal Decree 105/2008 and the Second National Plan on CDW in 2008. Under this framework, the construction industry is legally obligated to recover certain types of CDW when quantities exceed a specified threshold, a practice also adopted by Austria. Additionally, Germany has successfully experimented with a national voluntary commitment within the construction industry to reduce CDW landfilling since 1996, yielding positive results (Villoria et al., 2014). The adoption of regulations and quality standards for materials from CDW is a prevailing measure implemented in Belgium, Germany, and Finland (Villoria et al., 2014).

5.1.3 The USA

The USA enforces individual developer accountability for CDWM, promoting a reward system for zero CDW emissions. Developers bear responsibility for CDW from production to disposal under state-specific laws. The "Design for Deconstruction" approach, practiced during building design and deconstruction in the USA, fosters long-term economic growth, reduces environmental impact, and enhances economic resources (Aslam et al., 2020; Design for Deconstruction, 2015). The Construction & Demolition Recycling Association (CDRA) plays a pivotal role in the CDW recycling sector, offering education, support, research, and advocacy (Doherty, 2023). It actively promotes and defends the environmentally responsible recycling of the vast quantities of recoverable Construction and Demolition (CD) materials generated annually in the US, encompassing materials like concrete, gypsum drywall, wood, shingles, and asphalt pavement, among others (Doherty, 2023). For those seeking CD recycling services, the CDRA's Directory of member companies is a valuable resource for locating CD recyclers in their respective areas (Doherty, 2023). Moreover, the CDRA's Annual Meeting, known as CD World, stands as the premier gathering for the CD industry. Attendees span a spectrum, from top-level executives of leading CD hauling and recycling enterprises to newcomers in the field and the vendors that cater to this market. CD World delivers informative educational sessions covering essential trends and topics, facilitates networking opportunities, and showcases the latest equipment and services available in the industry.

5.1.4 Singapore

Singapore's waste management, including CDW, is robust. CDW recyclables are separated and incinerated in waste-to-energy facilities (Wu et al., 2020; Hoang et al., 2020). Singapore imposes a substantial solid waste disposal fee of US$ 57 per ton to incentivize higher waste recycling rates (Wu et al., 2020). A network of licensed waste collectors, guided by tailored codes of practice, handles waste from commercial and industrial sources. Singapore's National Environment Agency (NEA) provides an online directory of waste-related organizations, accessible on its website, supporting effective CDWM (Hoang et al., 2020). Singapore incentivizes concrete recovery in demolition through funds (Wu et al., 2020). As per Vishnoi (2023), Singapore conducts pre-demolition consultations with the building's original plans and contractors to assess the structure and create a fitting demolition strategy. Material quantities are predetermined, and an estimation of salvageable materials is included in the demolition plan. Additionally, a catalog of approved recycling facilities for various waste types is provided before demolition commences (Vishnoi, 2023). The Building and Construction Authority (BCA) has launched Green Mark Schemes such as the "Green and Gracious Builder Scheme" to encourage green practices, including waste management (Wu et al., 2020; Anggadjaja, 2009). An accreditation system by the Waste Management and Recycling Association of Singapore (WMRAS) for recycled aggregate suppliers is also in place (Wu et al. 2020; Accreditation—WMRAS, 2023; Anggadjaja, 2009).

5.1.5 Japan

In Japan, the Aichi Prefecture Environment Award System is designed to recognize and celebrate outstanding practices in sustainability. These practices span innovative technologies, business ventures, activities, and educational initiatives. This award system recognizes a wider array of environmental contributions that go beyond the realm of construction and building design. The core objective is to inspire the adoption of new production methods and lifestyles, all in support of advancing a society that places a strong emphasis on recycling (Global Environment Centre Osaka, Japan, 2012). Wang and Qin (2018) noted that in the realm of construction waste recycling, the Japanese government has implemented an advanced system to facilitate the efficient recovery of construction waste. To stimulate enterprises to harness waste resources, the government offers financial support to those engaged in construction waste recycling endeavors (Japan Ministry of the Environment, 2016).

5.1.6 South Korea

In South Korea, an online marketplace is employed to manage CDW transparently and effectively. This system automatically records waste-related data, including collection, transportation, recycling, and landfilling (Wu et al., 2020; Mingxue et al., 2023). Authorized users can access this information for up to 3 years. The online platform offers features like waste bidding and GIS-based search engines, facilitating the convenient exchange of waste details, such as type, quality, quantity, and prices (Wu et al., 2020). Additionally, local subsidies may be granted by state governments to cover installation costs and guidance of waste recycling facilities, from the Ministry of Environment. Quality standards and a certification system for recycled aggregate products have also been implemented by the government to boost consumer confidence (Wu et al., 2020).

5.2 Case Studies

5.2.1 Redwood Community Action Agency Old Growth Redwood Recycling Project in California, USA (Redwood Community Action Agency Old Growth Redwood Recycling Project, 2001)

The Redwood Community Action Agency (RCAA) took the lead in launching the Old Growth Redwood Recycling (OGRR) project in response to a request for assistance from the California Department of Fish and Game (CDFG). The CDFG had a restoration project encompassing over 5000 acres of pasturelands in various county locations, aiming to transform them into wetland habitats. However, the presence of several large barns and structures posed a challenge as they did not align with the wetland habitat restoration plans.

Initially, CDFG's proposed solution was to burn these structures. Nevertheless, a collaborative effort with RCAA led to the inception of the Old Growth Redwood Recycling (OGRR) project. This initiative primarily aimed to meticulously deconstruct the barns, which were predominantly constructed using valuable old-growth redwood. The project entailed several key aspects:

Planning
OGRR aimed to deconstruct structures, salvage the valuable redwood lumber, and create a self-sustaining program.

Resources
RCAA utilized some of its unrestricted funds to jumpstart OGRR. Additional funding was secured through waste diversion grants, development grants, and support from various agencies and organizations.

Cost
OGRR's monthly operating costs were approximately $8000, with yearly costs totaling around $96,000. Anticipating the project to achieve financial self-sufficiency within its inaugural year of operation was impractical. A substantial infusion of funding was necessary to support the project during its initial startup phase and to enable it to eventually achieve financial sustainability.

Scale of Waste Management
OGRR deconstructed five buildings, diverting over 60,000 board feet of old growth redwood over 9 months. Approximately 39,000 board feet of recovered lumber were processed, with 13,000 board feet sold, primarily through wholesalers.

Technological Utilization
Deconstruction involved pulling nails, trimming ends, and grading material into categories. Processed material was sold, and wood waste was sent to a local co-generation plant for use as boiler fuel.

Methodological Approach
The project involved deconstructing buildings, salvaging old growth redwood, and processing salvaged lumber for sale. OGRR aimed to create jobs, develop worker skills through training programs, and become self-sustaining through deconstruction fees and wood sales. It also focused on reducing greenhouse gas emissions through source reduction.

Despite the financial challenges faced, OGRR served as a catalyst for future projects and a model for economic development and resource conservation.

Stakeholders Involvement
OGRR partnered with CalWorks, the Private Industry Council (PIC), California Department of Fish and Game, Humboldt County Environmental Services, other government agencies, and the private sector.

5.2.2 Looney Bins, Inc., a California-Based Enterprise, and Their Achievement in Recycling (Case Studies—Construction & Demolition Debris, 2016)

Challenges Prior to the Initiative
Looney Bins, Inc. grappled with the intricacies of efficiently managing and recycling CDW. The specific challenge centered around the appropriate disposal of diverse

materials, encompassing wood, cardboard, metal, plastic, and other recoverable items, often being disposed of in a manner that neglected recycling priorities.

Strategic Approach
In response to this challenge, Looney Bins meticulously crafted a strategic blueprint. This blueprint entailed forging partnerships with local Hollywood movie studios to meticulously deconstruct movie sets, consequently salvaging valuable materials. This initiative was able to divert much useful material away from landfills earning an income while donating the rest to the end-users who could reuse the same.

Resource Acquisition
Looney Bins Inc. received financial support from the California Integrated Waste Management Board's Recycling Market Development Zone (RMDZ) Program in 1999 to address the challenge they faced. With the funding, they were able to acquire necessary machinery such as wood grinders, other auxiliary machinery as well as working capital.

Scale of Waste Management
By 2003, expanding their operational footprint, Looney Bins Inc. was able to establish the Downtown Diversion site, a state-of-the-art facility capable of processing a wide spectrum of CDW. The facility was able to handle brick, asphalt, concrete, drywall, wood, cardboard, carpet, scrap metal, and more. Their target was to divert a minimum of 80% of incoming CDW away from landfills with a targeted annual quantity of 50,000 tons.

Technological Utilization
Looney Bins Inc.'s primary technological asset was the wood grinder, a pivotal tool that played an instrumental role in the meticulous processing of wood and drywall, transforming them into mulch.

Methodological Approach
The methodological blueprint pursued by Looney Bins Inc. revolved around the systematic identification of sources of construction and demolition waste, with a pronounced focus on materials originating from movie studio operations. This involved a methodical deconstruction of movie sets to extract valuable materials, subsequently destined for sale or donation across diverse markets.

Stakeholder Engagement
Key stakeholders in this transformative initiative encompassed Looney Bins Inc. itself, in collaboration with local Hollywood movie studios, the California Integrated Waste Management Board's RMDZ Program, entities that harnessed recovered materials, such as wood for reconstituted pallets, and overseas hospitals benefiting from reclaimed resources. These stakeholders played integral roles in the successful realization of the recycling program.

5.2.3 Environmental Service Management Group, Inc. Based in New Jersey (Buono, 2007)

Challenges Prior to the Initiative

Prior to implementing the initiative, the schools in New Jersey, specifically Adelphia & Ramtown Elementary Schools (the first public school in New Jersey to go green with LEED Rating System), O.C.C New Technology Center Building (the first higher education building project in New Jersey to go green with LEED Rating System), The Willow School (the first private schools in New Jersey to go green with LEED Rating System.), and BASF House (the first New Jersey home to adopt the LEED for Homes Rating System), were faced with challenges related to waste management, recycling, and environmentally sustainable construction practices. There was a need to enhance recycling and waste diversion efforts during construction, particularly in scenarios where materials like wood, metal, plastic, and paper were being discarded.

Strategic Approach

The schools and projects in question planned to adopt the U.S. Green Building Council's LEED Rating System. This system is renowned for promoting sustainable and environmentally friendly building practices. General contractors worked cooperatively to share waste and recycling services, fostering a collaborative approach to recycling and waste management. For each of these projects, a tailored recycling strategy was devised, including recycling of specific materials like wood, metal, plastic, and paper.

Resource Acquisition

The U.S. Green Building Council's LEED Rating System served as a critical resource, providing a framework for environmentally sustainable construction and waste management practices. The projects leveraged the cooperation of multiple prime contractors who shared waste and recycling services, optimizing resource utilization.

Cost

The recycling costs for each project were detailed, including the total expenditure and cost per ton for recycling various materials, as shown in Table 6.4.

Table 6.4 Recycling costs for each project

Project name	Total recycling cost ($)	Cost per ton ($)	Estimated landfill cost avoidance ($)	Direct cost savings ($)
Adelphia & Ramtown Elementary Schools	37,134	67	80,933	43,799
O.C.C New Technology Center Building	8747	51	22,934	14,178
The Willow School	19,857	36	65,384	45,527
BASF House	–	–	–	7500

Methodological Approach
The projects aimed at handling significant quantities of waste, with each specifying the types of materials recycled, such as wood, paper, metal, plastic, and more. The primary methodology revolved around adhering to the LEED Rating System's guidelines, which encompassed waste diversion goals and credits for exceeding landfill diversion standards.

Technologies Used
The projects employed varied recycling approaches, such as weekly collection of metal, wood, plastic, and paper in small quantities, demonstrating adaptability in technology utilization.

Stakeholder Engagement
Each project involved different stakeholders, including the schools or organizations undertaking the construction, contractors, and likely regulatory bodies promoting sustainable building practices.

6 Recommendations for Improving Construction and Demolition Waste Management (CDWM)

Effective CDWM poses a complex challenge in environmental sustainability, economic viability, and regulatory compliance. To enhance the efficiency and sustainability of CDWM, a holistic approach is imperative. Based on our research findings, we propose a set of recommendations designed to elevate CDWM practices. These recommendations enclose a diverse array of strategies, spanning from heightening awareness and infrastructure investment to promoting innovation and fortifying enforcement measures.

1. Implementation of source separation and segregation
   - During the predesign phase, construction plans should incorporate specific areas for waste separation. For example, designate spaces on the construction site for sorting materials into categories such as recyclables, hazardous waste, and non-recyclables.
   - In the construction phase, workers should follow established protocols for source separation, ensuring that waste generated on-site is divided appropriately.
   - In the demolition phase, separation should extend to the deconstruction process, enabling the efficient recovery of materials for reuse or recycling.
2. Creating a resilient recycling infrastructure
   - The redesign and construction phases involve planning and investing in recycling plants, sorting facilities, and materials recovery centers to facilitate the collection and processing of construction and demolition waste.

- The post-design and operation phases require the continued maintenance and upgrading of these facilities to adapt to evolving recycling needs and technologies.

3. Providing incentives or grants to promote effective CDWM

- During the predesign and post-design phases, governments and organizations can offer financial incentives to construction projects that adopt sustainable waste management practices. These incentives can include grants, subsidies, or tax benefits.
- Throughout the entire building life cycle, incentives can be provided for the use of recycled materials or the achievement of certain waste reduction targets, encouraging a circular economy approach.

4. Adopting cutting-edge technologies

- During the predesign and post-design phases, integrating modern technologies such as Building Information Modeling (BIM), Geographic Information Systems (GIS), and 3D printing can enhance the efficiency of CDWM processes and support the design of buildings with minimal waste generation.
- In the construction and operation phases, technologies like near-infrared sorting and GPS can be applied to streamline waste management and improve resource utilization.
- Over the entire building life cycle, continuous innovation and research can lead to the adoption of new technologies, reducing waste generation and enhancing recycling.

5. Implementation of regulatory framework that is updated regularly to have sustainable CDWM

- Throughout the entire building life cycle, the regulatory framework should ensure that waste management practices remain aligned with sustainability goals.
- Regular updates to waste management regulations are essential to adapt to changing technology, best practices, and environmental considerations.

6. Organizing public awareness and training initiatives for relevant stakeholders

- Awareness campaigns and training programs should target a diverse audience, including construction professionals, builders, and the general public, throughout the predesign, post-design, and construction phases.
- Continual awareness efforts during the operation and demolition phases promote responsible waste management practices, including reuse and recycling.

7. Waste auditing and tracking to help inform decision-making and identify areas for improvement in CDWM

- Conducting waste audits and tracking during the construction and operation phases provides data for evaluating the effectiveness of waste management practices.
- Findings from waste audits should inform decisions regarding waste reduction, recycling efficiency, and opportunities for improvement during the entire building life cycle.

8. Encouraging collaboration among stakeholders
   - Collaboration among stakeholders, including government agencies, construction companies, and recycling facilities, is essential during the entire building life cycle.
   - Effective coordination minimizes waste transport distances and enhances recycling rates.
9. Promoting the reuse and salvaging of materials from demolition projects
   - The demolition phase presents a unique opportunity to salvage and reuse materials, reducing the need for new resources.
   - Salvaged materials can find new life in other construction projects, extending their utility beyond the original building.
10. Supporting and acknowledging construction projects and companies that embrace sustainable practices via green building certification
    - Throughout the predesign, post-design, and construction phases, projects that achieve green building certifications can serve as examples and be recognized for their commitment to sustainability.
    - Such acknowledgements encourage others to follow suit in subsequent building projects.
11. Issuing quality certifications for recycled products
    - During the entire building life cycle, quality standards and certifications for recycled products promote trust in the use of recycled materials in construction.
    - Manufacturers and suppliers of recycled products should meet established quality criteria.
12. Conducting life cycle assessments to appraise the effectiveness of appropriate CDWM practices
    - Life cycle assessments can be performed during the predesign, post-design, and construction phases to evaluate the environmental impact of chosen materials and waste management practices.
    - The results inform decisions on materials selection and waste management.
13. Establishing advanced online markets for recycled CDW materials
    - Throughout the entire building life cycle, online platforms can be created to facilitate the exchange of recycled CDW materials.

- These platforms connect suppliers and consumers, promoting the reuse of materials and reducing waste generation.

These actions contribute to a holistic approach to CDWM throughout the building's life cycle, minimizing waste, conserving resources, and promoting environmental sustainability.

7 Conclusion

CDW encompasses a diverse mix of materials including hazardous elements resulting from construction, alteration, or demolition activities. Its composition varies due to regional factors. CDW generation is influenced by rapid urbanization. While approximately 90% of CDW is potentially recyclable, recycling rates vary widely among countries due to factors like landfill pricing, penalties, and raw material availability. Managing CDW as a resource offers environmental benefits and cost savings in construction projects. Strategies such as reusing, recycling, waste-to-energy conversion, and landfilling are employed in CDWM based on its characteristics and resource recovery potential.

CDWM faces multifaceted global challenges. Limited awareness of circular construction principles among stakeholders, profit-driven client preferences leading to increased waste, inadequate regulatory frameworks, fragmented responsibility in the construction sector, illegal dumping, and insufficient recycling infrastructure are key issues. Economic challenges include higher production costs for recycled products and a lack of financial incentives. Addressing these challenges requires collaboration among governments, the construction industry, the private sector and the public to promote awareness, improve regulations, invest in infrastructure, and incentivize sustainable CDWM.

Establishing comprehensive legislative regulations, clear policy objectives, and promoting compliance and incentives are crucial steps toward achieving sustainable CDWM globally. Each country can adapt these strategies to effectively address CDW challenges.

To advance CDWM practices, key recommendations include strengthening awareness, engaging stakeholders, investing in infrastructure, enhancing coordination, encouraging innovation, establishing monitoring systems, and strengthening enforcement. Lessons learned from countries with effective CDW policies provide valuable insights for shaping future CDWM practices globally. Addressing CDW challenges is essential for achieving sustainable outcomes while minimizing the environmental footprint.

References

28 incredible statistics about waste generation. (2021). *StoneCycling*. Available at: https://www.stonecycling.com/news/statistics-about-waste-generation/ (Accessed: September 1, 2023).

Abdelli, H. E., Mokrani, L., Kennouche, S., & de Aguiar, J. B. (2020). Utilization of waste glass in the improvement of concrete performance: A mini review. *Waste Management & Research, 38*(11), 1204–1213. https://doi.org/10.1177/0734242X20941090

Accreditation – WMRAS. (2023). *Org.sg*. Available at: https://www.wmras.org.sg/accreditation/ (Accessed: September 17, 2023).

Agamuthu, P. (2008). Challenges in sustainable management of construction and demolition waste. *Waste Management & Research, 26*(6), 491–492. https://doi.org/10.1177/0734242X08100096

Alsheyab, M. A. T. (2022). Recycling of construction and demolition waste and its impact on climate change and sustainable development. *International journal of Environmental Science and Technology, 19*(3), 2129–2138. https://doi.org/10.1007/s13762-021-03217-1

Anggadjaja, E. (2009). *Sustainable construction strategies: A Singapore perspective, Kit.edu*. Available at: https://www.iip.kit.edu/downloads/2009_Anggadjaja_Report_Singapore.pdf (Accessed: September 17, 2023).

Aslam, M. S., Huang, B., & Cui, L. (2020). Review of construction and demolition waste management in China and USA. *Journal of Environmental Management, 264*, 110445. https://doi.org/10.1016/j.jenvman.2020.110445

Baikerikar, A., Mudalgi, S., & Ram, V. V. (2023). Utilization of waste glass powder and waste glass sand in the production of Eco-Friendly concrete. *Construction and Building Materials, 377*, 131078. https://doi.org/10.1016/j.conbuildmat.2023.131078

Bergsdal, H., Bohne, R. A., & Brattebø, H. (2008). Projection of construction and demolition waste in Norway. *Journal of Industrial Ecology, 11*(3), 27–39. https://doi.org/10.1162/jiec.2007.1149

Bumanis, G., Zorica, J., Korjakins, A., & Bajare, D. (2022). Processing of gypsum construction and demolition waste and properties of secondary gypsum binder. *Recycling, 7*(3), 30. https://doi.org/10.3390/recycling7030030

Buono, M. D. (2007). *Case studies in cost effective construction waste management*. Available at: https://carpetrecovery.org/wp-content/uploads/2014/04/Buono_Recycling_Case_Studies.pdf (Accessed: October 14, 2023).

Case studies - Construction & demolition debris. (2016). *Epa.gov*. Available at: https://archive.epa.gov/region9/waste/archive/web/html/casestud.html (Accessed: October 14, 2023).

CDWM in Germany. (2015). *Europa.eu*. Available at: https://ec.europa.eu/environment/pdf/waste/studies/deliverables/CDW_Germany_Factsheet_Final.pdf (Accessed: September 23, 2023).

Cilia, R. (2019). *Dumping at sea of most construction waste is illegal, The Malta Independent*. Available at: https://www.independent.com.mt/articles/2019-11-17/local-news/Most-construction-waste-cannot-be-dumped-at-sea-6736216300 (Accessed: September 8, 2023).

Colorado, H. A., Muñoz, A., & Neves, M. S. (2022). Circular economy of construction and demolition waste: A case study of Colombia. *Sustainability, 14*(12), 7225. https://doi.org/10.3390/su14127225

Cook, E., Velis, C. A., & Black, L. (2022). Construction and demolition waste management: A systematic scoping review of risks to occupational and public health. *Frontiers in Sustainability, 3*, 924926. https://doi.org/10.3389/frsus.2022.924926

Dave, H., Mla, K., & Andrews, M. (2021). *Gov.au*. Available at: https://www.wa.gov.au/system/files/2021-10/DWER-annual-report-2020-21.pdf (Accessed: September 2, 2023).

De Bock, P. (2012). *Q&A on Environmental Law in Belgium*. Clifford Chance. Available at: https://www.cliffordchance.com/content/dam/cliffordchance/briefings/2012/12/qa-on-environmental-law-in-belgium.pdf (Accessed: October 13, 2023).

Delegations of Authority Under the Comprehensive Environmental Response, Compensation, and Liability Act (CERCLA). (1984). *United States Environmental Protection Agency*. Available at: https://nepis.epa.gov/ (Accessed: September 8, 2023).

Design for Deconstruction. (2015). *Epa.gov*. Available at: https://www.epa.gov/sites/default/files/2015-11/documents/designfordeconstrmanual.pdf (Accessed: September 8, 2023).

Di Maria, A., Eyckmans, J., & Van, A. K. (2018). Downcycling versus recycling of construction and demolition waste: Combining LCA and LCC to support sustainable policy making. *Waste Management, 75*, 3–21. https://doi.org/10.1016/j.wasman.2018.01.028

Disposal at sea. (2019). *ERA*. Available at: https://era.org.mt/topic/disposal-at-sea/ (Accessed: September 8, 2023).

Doherty, P. (2023). *The CDRA brings the construction and demolition recycling industry together, Cdrecycling.org*. Available at: https://www.cdrecycling.org/ (Accessed: September 23, 2023).

Esa, M. R., Halog, A., & Rigamonti, L. (2017). Developing strategies for managing construction and demolition wastes in Malaysia based on the concept of circular economy. *Journal of Material Cycles and Waste Management, 19*(3), 1144–1154. https://doi.org/10.1007/s10163-016-0516-x

Ferronato, N., Fuentes, S., Rocio, C., Guisbert, L., Edith, G., Conti, F., & Torretta, V. (2023). Construction and demolition waste recycling in developing cities: Management and cost analysis. *Environmental Science and Pollution Research International, 30*(9), 24377–24397. https://doi.org/10.1007/s11356-022-23502-x

Galán, B., Viguri, J. R., Cifrian, E., Dosal, E., & Andres, A. (2019). Influence of input streams on the construction and demolition waste (CDW) recycling performance of basic and advanced treatment plants. *Journal of Cleaner Production, 236*, 117523. https://doi.org/10.1016/j.jclepro.2019.06.354

Ghaffar, S. H., Burman, M., & Braimah, N. (2020). Pathways to circular construction: An integrated management of construction and demolition waste for resource recovery. *Journal of Cleaner Production, 244*, 118710. https://doi.org/10.1016/j.jclepro.2019.118710

Global Enviroment Centre Osaka, Japan. (2012). *Laws and Support Systems for Promoting Waste Recycling in Japan*. Available at: https://gec.jp/gec/en/Publications/LawSupportSystems.pdf (Accessed: September 20, 2023).

Haider, H., AlMarshod, S. Y., AlSaleem, S. S., Ali, A. A. M., Alinizzi, M., Alresheedi, M. T., & Shafiquzzaman, M. (2022). Life cycle assessment of construction and demolition waste management in Riyadh, Saudi Arabia. *International Journal of Environmental Research and Public Health, 19*(12), 7382. https://doi.org/10.3390/ijerph19127382

Hamakareem, M. I. (2020). *Construction wastes: Types, causes, and recycling strategies, The Constructor*. Available at: https://theconstructor.org/concrete/construction-waste-recycling/1088/ (Accessed: August 21, 2023).

Hao, J., Di Maria, F., Chen, Z., Yu, S., Ma, W., & Di Sarno, L. (2020). Comparative study of construction and demolition waste management in China and THE European Union. *Detritus, 13*, 114–121. https://doi.org/10.31025/2611-4135/2020.14029

Hoang, N. H., Ishigaki, T., Kubota, R., Yamada, M., & Kawamoto, K. (2020). A review of construction and demolition waste management in Southeast Asia. *Journal of Material Cycles and Waste Management, 22*, 315–325. https://doi.org/10.1007/s10163-019-00914-5

Humber, N., & Lingel, S. (1980). *The resource conservation and recovery act of 1976 how will it impact on resource recovery and conservation 1980, United States Environmental Protection Agency*. Available at: https://nepis.epa.gov/ (Accessed: September 8, 2023).

IPCC Report. Climate Change 2021: The Physical Science Basis, the Working Group I Contribution to the Sixth Assessment Report. (2021). Available online: https://www.ipcc.ch/report/sixth-assessment-report-working-group-i/ (Accessed: September 7, 2023).

Islam, R., Nazifa, T. H. A., Shanawaz, U. A. S. M., Salmiati, S., & Shahid, S. (2019). An empirical study of construction and demolition waste generation and implication of recycling. *Waste Management, 95*, 10–21. https://doi.org/10.1016/j.wasman.2019.05.049

Jain, M. S. (2020). Possible ways of re-utilization of the construction and demolition wastes in various construction sectors. In *Sustainable development and environment* (pp. 99–109). National Institute of Construction Management and Research.

Japan Ministry of the Environment. (2016). *Solid waste management and recycling technology of Japan - Toward a Sustainable Society -, Env.go.jp*. Available at: https://www.env.go.jp/recycle/circul/venous_industry/en/brochure.pdf (Accessed: September 17, 2023).

Jordan. (2021). *Construction and demolition waste: Disposal management problems and recycling solutions, Ftmmachinery.com*. Available at: https://www.ftmmachinery.com/blog/construction-and-demolition-waste-disposal-management-problems-and-recycling-solutions.html (Accessed: August 21, 2023).

Kabirifar, K., Mojtahedi, M., Wang, C., & Tam, V. W. Y. (2020). Construction and demolition waste management contributing factors coupled with reduce, reuse, and recycle strategies for effective waste management: A review. *Journal of Cleaner Production, 263*, 121265. https://doi.org/10.1016/j.jclepro.2020.121265

Kariapper, A., Karunaratne, S., & Nanayakkara, D. (2015). A study on how to improve the physical properties of recycled coarse aggregates with the use of rice husk ash- cement slurry. *Annual Session of IESL*. Institution of Engineers, Sri Lanka, pp. 1–8.

Kien, T. T., Thanh, L. T., & Lu, P. V. (2013). Recycling construction demolition waste in the world and in Vietnam. In *The International Conference on Sustainable Built Environment for Now and the Future*. Hanoi, pp. 247–256.

Kourmpanis, B., Papadopoulos, A., Moustakas, K., Stylianou, M., Haralambous, K. J., & Loizidou, M. (2008). Preliminary study for the management of construction and demolition waste. *Waste Management & Research, 26*(3), 267–275. https://doi.org/10.1177/0734242X07083344

Lai, Y. Y., Yeh, L. H., Chen, P. F., Sung, P. H., & Lee, Y. M. (2016). Management and recycling of construction waste in Taiwan. *Procedia Environmental Sciences, 35*, 723–730. https://doi.org/10.1016/j.proenv.2016.07.077

Li, C. Z., Zhao, Y., Xiao, B., Yu, B., Tam, V. W. Y., Chen, Z., & Ya, Y. (2020). Research trend of the application of information technologies in construction and demolition waste management. *Journal of Cleaner Production, 263*, 121458. https://doi.org/10.1016/j.jclepro.2020.121458

Liu, J., Nie, J., & Yuan, H. (2019). To expand or not to expand: A strategic analysis of the recycler's waste treatment capacity. *Computers & Industrial Engineering, 130*, 731–744. https://doi.org/10.1016/j.cie.2019.03.016

Liu, H., Long, H., & Li, X. (2020). Identification of critical factors in construction and demolition waste recycling by the grey-DEMATEL approach: A Chinese perspective. *Environmental Science and Pollution Research International, 27*, 8507–8525. https://doi.org/10.1007/s11356-019-07498-5

Mah, C. M., Fujiwara T., & Ho, C. S. (2018). Life cycle assessment and life cycle costing toward eco-efficiency concrete waste management in Malaysia. *Journal of Cleaner Production, 172*, 3415–3427.

Mallinson, K. (2020). *5 things you need to know about the Demolition Code of Practice, RVA Group*. Available at: https://www.rvagroup.org/2020/07/5-things-you-need-to-know-about-the-demolition-code-of-practice/ (Accessed: August 22, 2023).

Marzouk, M., & Azab, S. (2014). Environmental and economic impact assessment of construction and demolition waste disposal using system dynamics. *Resources, Conservation and Recycling, 82*, 41–49. https://doi.org/10.1016/j.resconrec.2013.10.015

Mee, Y. A. (2009). *Guidebook for best environmental practices*. Land Transport Authority. Available at: https://www.lta.gov.sg/content/dam/ltagov/industry_innovations/industry_matters/safety_health_environment/pdf/Construction_Waste_Guidebook.pdf (Accessed: September 17, 2023).

Menegaki, M., & Damigos, D. (2018). A review on current situation and challenges of construction and demolition waste management. *Current Opinion in Green and Sustainable Chemistry, 13*, 8–15. https://doi.org/10.1016/j.cogsc.2018.02.010

Mihai, F. C. (2019). Construction and demolition waste in Romania: The route from illegal dumping to building materials. *Sustainability, 11*(11), 3179. https://doi.org/10.3390/su11113179

Mingxue, M., Vivian, W. T., Khoa, N. L., Anthony, B., Wengui, L., & Xiangyu, W. G. (2023). Comparative analysis on international construction and demolition waste management policies and laws for policy makers in China. *Journal of Civil Engineering and Management, 29*(2), 107–130. https://doi.org/10.3846/jcem.2023.16581

Moschen, S. J., Kasper, T., & Huber-Humer, M. (2023). Critical review of the recovery rates of construction and demolition waste in the European Union - An analysis of influencing factors in selected EU countries. *Waste Management, 167*, 150–164. https://doi.org/10.1016/j.wasman.2023.05.020

Myneni, K. K., & Rajan, D. (2021). Impact of circular construction on demolition waste management in the Indian construction industry. *International Journal of Engineering Technologies and Management Research, 8*(1), 12–24. https://doi.org/10.29121/ijetmr.v8.i1.2021.846

Nguyen, H. G., Nguyen, D. T., Nghiem, H. T., Tran, V. C., Kato, A., Matsuno, A., Isobe, Y., Kawasaki, M., & Kawamoto, K. (2021). Current management condition and waste composition characteristics of construction and demolition waste landfills in Hanoi of Vietnam. *Sustainability, 13*(18), 10148. https://doi.org/10.3390/su131810148

Nunes, K., & Mahler, C. F. (2020). Comparison of construction and demolition waste management between Brazil, European Union and USA. *Waste Management & Research, 38*(4), 415–422. https://doi.org/10.1177/0734242X20902814

Okeyinka, O. M., Oloke, D. A., & Khatib, J. M. (2015). A review on recycled use of solid wastes in building materials. *World Academy of Science, Engineering and Technology International Journal of Civil and Environmental Engineering, 9*(12), 1578–1587.

Papargyropoulou, E., Preece, C. N., Padfield, R., & Abdullah, A. (2011). Sustainable construction waste management in Malaysia: a constructor's perspective. In *Proceedings of the MISBE 2011-international conference on management and innovation for a sustainable built environment.* Amsterdam, Netherlands. ISBN: 9789052693958.

Poranek, N., Łaźniewska, P. B., Czajkowski, A., & Pikoń, K. (2021). Possibilities of management of fly ash from municipal solid waste incineration plant in building industry in the circular economy. *IOP Conference Series. Materials Science and Engineering, 1203*(3), 032087. https://doi.org/10.1088/1757-899x/1203/3/032087

Premasiri, E. S. Y., Kariapper, A. B. Y., Abeysinghe, A., & Karunarathne, S. (2013). *Use of recycled aggregates for concrete production.* Department of Earth Resources Engineering, University of Moratuwa.

Qiao, L., Tang, Y., Li, Y., Liu, M., Yuan, X., Wang, Q., & Ma, Q. (2022). Life cycle assessment of three typical recycled products from construction and demolition waste. *Journal of Cleaner Production, 376*, 134139. https://doi.org/10.1016/j.jclepro.2022.134139

Recovery rate of construction and demolition waste in the EU-272018, by country. (n.d.). *Statista.* Available at: https://www.statista.com/statistics/1316268/recovery-rate-of-construction-and-demolition-waste-eu-by-country/ (Accessed: September 1, 2023).

Redwood Community Action Agency Old Growth Redwood Recycling Project. (2001). *Epa.gov.* Available at: https://archive.epa.gov/region9/waste/archive/web/pdf/2redwood.pdf (Accessed: October 14, 2023).

Reis, G. S. D., Quattrone, M., Ambrós, W. M., Grigore, C. B., & Hoffmann, S. C. (2021). Current applications of recycled aggregates from construction and demolition: A review. *Materials, 14*(7), 1700. https://doi.org/10.3390/ma14071700

Saadi, N., Ismail, Z., & Alias, Z. (2016). A review of construction waste management and initiatives in Malaysia. *Journal of Sustainability Science and Management, 11*(2), 101–114.

Santos, M. T., Lamego, P., & Frade, P. (2017). Management options for construction and demolition wastes from residential recuperation. *Waste and Biomass Valorization, 8*, 1679–1687. https://doi.org/10.1007/s12649-016-9675-1

Seror, N., & Portnov, B. A. (2018). Identifying areas under potential risk of illegal construction and demolition waste dumping using GIS tools. *Waste Management, 75*, 22–29. https://doi.org/10.1016/j.wasman.2018.01.027

Soil protection law. (2014). *Umweltbundesamt*. Available at: https://www.umweltbundesamt.de/en/topics/soil-land/soil-protection/soil-protection-law (Accessed: October 13, 2023).

Statistics Unit, Environmental Protection Department. (2022). *Monitoring of solid waste in Hong Kong waste statistics for 2021, Gov.hk*. Available at: https://www.wastereduction.gov.hk/sites/default/files/resources_centre/waste_statistics/msw2021_eng.pdf (Accessed: September 14, 2023).

Tam, V. W. Y., & Tam, C. M. (2006). A review on the viable technology for construction waste recycling. *Resources, Conservation and Recycling, 47*(3), 209–221. https://doi.org/10.1016/j.resconrec.2005.12.002

Tan, J., De Vlieger, J., Desomer, P., Cai, J., & Li, J. (2022). Co-disposal of construction and demolition waste (CDW) and municipal solid waste incineration fly ash (MSWI FA) through geopolymer technology. *Journal of Cleaner Production, 362*, 132502. https://doi.org/10.1016/j.jclepro.2022.132502

Tissera, W. D. A., Dahanayake, R., & Edirisinghe, V. (2022). The applicability of regulations for the disposal of construction and demolition waste in Sri Lanka. In *Proceedings of 10th World Construction Symposium 2022*. Building Economics and Management Research Unit (BEMRU), University of Moratuwa, pp. 821–832. https://doi.org/10.31705/wcs.2022.66

UK statistics on waste. (2023). *Gov.uk*. Available at: https://www.gov.uk/government/statistics/uk-waste-data/uk-statistics-on-waste (Accessed: September 5, 2023).

UNCED. (1992). *Sustainabledevelopment.un.org*. Available at: https://sustainabledevelopment.un.org/content/documents/Agenda21.pdf (Accessed: October 13, 2023).

US EPA, O. (2016). Best practices for reducing, reusing, and recycling construction and demolition materials. Available at: https://www.epa.gov/smm/best-practices-reducing-reusing-and-recycling-construction-and-demolition-materials (Accessed: August 22, 2023).

US EPA, O. (2017). Construction and Demolition debris: Material-specific data. Available at: https://www.epa.gov/facts-and-figures-about-materials-waste-and-recycling/construction-and-demolition-debris-material (Accessed: September 5, 2023).

Us Epa, O. P. (2013). Summary of the resource conservation and recovery act. Available at: https://www.epa.gov/laws-regulations/summary-resource-conservation-and-recovery-act (Accessed: November 2, 2023).

Villoria, S. P., & Osmani, M. (2019). A diagnosis of construction and demolition waste generation and recovery practice in the European Union. *Journal of Cleaner Production, 241*, 118400. https://doi.org/10.1016/j.jclepro.2019.118400

Villoria, S. P., del Río, M. M., Porras-Amores, C., & San-Antonio, G. A. (2014). Assessing the accumulation of construction waste generation during residential building construction works. *Resources, Conservation and Recycling, 93*, 67–74. https://doi.org/10.1016/j.resconrec.2014.10.004

Vishnoi, A. (2023). *C&D waste management in Singapore: A land-scarce challenge, Malba Project*. Available at: https://www.malbaproject.com/post/c-and-d-waste-management-singapore (Accessed: September 17, 2023).

Vivian, W. Y. T. (2011). Rate of reusable and recyclable waste in construction. *The Open Waste Management Journal, 4*, 28–32. https://doi.org/10.2174/1876400201104010028

Wang, Q. F., & Qin, S. (2018). Comparative analysis of construction waste recycling policies at home and abroad. In *Proceedings of the international workshop on environmental management, science and engineering*. SCITEPRESS - Science and Technology Publications, pp. 393–400.

Wang, B., Yan, L., Fu, Q., & Kasal, B. (2021). A comprehensive review on recycled aggregate and recycled aggregate concrete. *Resources, Conservation and Recycling, 171*, 105565. https://doi.org/10.1016/j.resconrec.2021.105565

Waste statistics. (2023). *Eurostat*. Available at: https://ec.europa.eu/eurostat/statistics-explained/index.php?title=Waste_statistics#Total_waste_generation (Accessed: September 15, 2023).

What is CD waste. (2020). *CT.gov - Connecticut's Official State Website*. Available at: https://portal.ct.gov/DEEP/Waste-Management-and-Disposal/Construction-and-Demolition-Waste/What-is-CD-Waste (Accessed: September 14, 2023).

Wu, Z., Yu, A. T. W., & Poon, C. S. (2020). Promoting effective construction and demolition waste management towards sustainable development: A case study of Hong Kong. *Sustainable Development, 28*(6), 1713–1724. https://doi.org/10.1002/sd.2119

Yeheyis, M., Hewage, K., & Alam, M. S. et al. (2013). An overview of construction and demolition waste management in Canada: a lifecycle analysis approach to sustainability. *Clean Technologies and Environmental Policy, 15*, 81–91. https://doi.org/10.1007/s10098-012-0481-6

Yuan, H. (2017). Barriers and countermeasures for managing construction and demolition waste: A case of Shenzhen in China. *Journal of Cleaner Production, 157*, 84–93. https://doi.org/10.1016/j.jclepro.2017.04.137

Zhang, N., Zheng, L., Duan, H., Yin, F., Li, J., & Niu, Y. (2019). Differences of methods to quantify construction and demolition waste for less-developed but fast-growing countries: China as a case study. *Environmental Science and Pollution Research International, 26*, 25513–25525. https://doi.org/10.1007/s11356-019-05841-4

Zhang, N., Duan, H., Miller, T. R., Tam, V. W. Y., Liu, G., & Zuo, J. (2020). Mitigation of carbon dioxide by accelerated sequestration in concrete debris. *Renewable and Sustainable Energy Reviews, 117*, 109495. https://doi.org/10.1016/j.rser.2019.109495

Zhang, C., et al. (2022). An overview of the waste hierarchy framework for analyzing the circularity in construction and demolition waste management in Europe. *Science of the Total Environment, 803*, 149892.

Zhang, N., Konyalıoğlu, A. K., Duan, H., Feng, H., & Li, H. (2023). The impact of innovative technologies in construction activities on concrete debris recycling in China: A system dynamics-based analysis. *Environment, Development and Sustainability*. https://doi.org/10.1007/s10668-023-03178-0

Zhao, X., Webber, R., Kalutara, P., Browne, W., & Pienaar, J. (2022). Construction and demolition waste management in Australia: A mini-review. *Waste Management & Research, 40*(1), 34–46. https://doi.org/10.1177/0734242X211029446

Chapter 7
Approaches Involved in the Treatment and Disposal of Petroleum Refinery Sludge

W. James Singha and Hemen Deka

1 Introduction

Petroleum or crude oil holds a paramount role as a primary energy source and a fundamental raw material for the synthesis of a wide array of petrochemical products (Jasim & Aljaberi, 2023). The extensive utilization of petroleum and its products has spurred the global production of crude oil witnessing a substantial increase to 4141 million tons in 2020, from the 2868 million tons in 1973 (Niu et al., 2022). The revolution in the petroleum industry significantly contributes to fostering economic growth. However, these remarkable expansion in the petroleum industry has brought about ecological challenges that cannot be overlooked. One major environmental concern is the generation of petroleum oily sludge (POS), the main residue or wastes resulting from diverse petrochemical production activities such as during refining, storage, transportation, and crude oil exploration (Hu et al., 2020). It is a semi-liquid/solid waste containing significant amounts of toxic pollutants such as heavy metals (HMs) and hydrocarbons including the polycyclic aromatic hydrocarbons (PAHs). Moreover, the presence of oil products and other noxious substances like benzene in POS poses significant risks to both the environment and human health (Hosseini et al., 2023; Hu et al., 2013). The production of POS is alarmingly increasing due to the growing demand for energy worldwide. As a consequence, various measures have been adopted in recent times to mitigate the production of POS, recover economically valuable oil, and subject it to requisite treatment prior to ultimate disposal in the environment.

The United States Environmental Protection Agency (US EPA) recognizes POS as a hazardous and intricate organic waste (Abouee Mehrizi et al., 2019). The need

W. J. Singha · H. Deka (✉)
Ecology and Environmental Remediation Laboratory, Department of Botany, Gauhati University, Guwahati, India
e-mail: hemendeka@gauhati.ac.in

V. Kumar et al. (eds.), *Environmental Engineering and Waste Management*,
https://doi.org/10.1007/978-3-031-58441-1_7

for sustainable waste management practices within the petroleum industry has never been more critical, aligning with global efforts to mitigate climate change and reduce the environmental impact of industrial processes. The environmental contamination arising from POS extends beyond refinery boundaries, permeating ecosystems, and impacting the quality of soil, water, and air (Shahbaz et al., 2023). Besides, the hazardous nature of POS extends beyond its ecological implications to pose a direct threat to human health (Borah & Deka, 2023; Asim et al., 2021). The exposure of toxic compounds presents in POS such as HMs and PAHS, whether through direct contact or inhalation, can result in a spectrum of health issues, ranging from respiratory problems to more severe long-term effects, including carcinogenic risks (Hu et al., 2020). Moreover, the improper disposal of POS into the environment exacerbates these risks exponentially. When POS is released without adequate treatment, its toxic components leach into soil and water systems, thereby contaminating ecosystems. The POS pollution can result in the accumulation of hazardous constituents in animals and plants, ultimately leading to food chain contamination. As a consequence, humans may be exposed to these harmful substances indirectly through the consumption of contaminated food and water (Liu et al., 2019). Besides, the persistence nature of the hazardous constituents of POS in the environment over an extended period further exacerbates this issue, creating a chronic and cumulative impact on ecosystems (Chen et al., 2023; Sinha et al., 2020). Aquatic life, soil quality, and air purity are all compromised when POS is carelessly disposed of without proper treatment measures. Consequently, the resultant environmental degradation has a cascading effect on biodiversity, ecosystem services, and, ultimately, on the health and well-being of communities in proximity to these contaminated sites (Hu et al., 2020). Thus, the need for environmental protection against harmful constituents of POS has become increasingly important given the continuous exploitation of petroleum resources and the consequent widespread environmental pollution. Consequently, the treatment and safe disposal of POS in the refinery industry have emerged as significant areas of research in recent years.

Current approaches of management of POS primarily focus on three key aspects: pollutant elimination, detoxification, and containment immobilization (Niu et al., 2022). The elimination strategy is designed to systematically and thoroughly break down the complex organic constituents present in POS by subjecting them to high-temperature treatment. This method is intended to facilitate the breakdown of the organic molecules into simpler forms, enhancing the efficiency of the elimination process and reducing the environmental impact of POS. Detoxification, on the other hand, involves the extraction of toxic elements from POS to mitigate its harmful effects, employing techniques such as biological or chemical degradation or physical separation (Khazaal & Ismail, 2021). The containment approach seeks to curb the spread of POS into the surrounding areas by either storing it in dedicated facilities or by solidifying it for immobilization using cementitious substances (Hu et al., 2020). Despite showcasing different levels of efficacy in minimizing the environmental repercussions of POS, the widespread adoption of these techniques is impeded by their associated high expenses (Snowden-Swan et al., 2022). Consequently, there is a growing interest in exploring sustainable and economically viable alternatives for

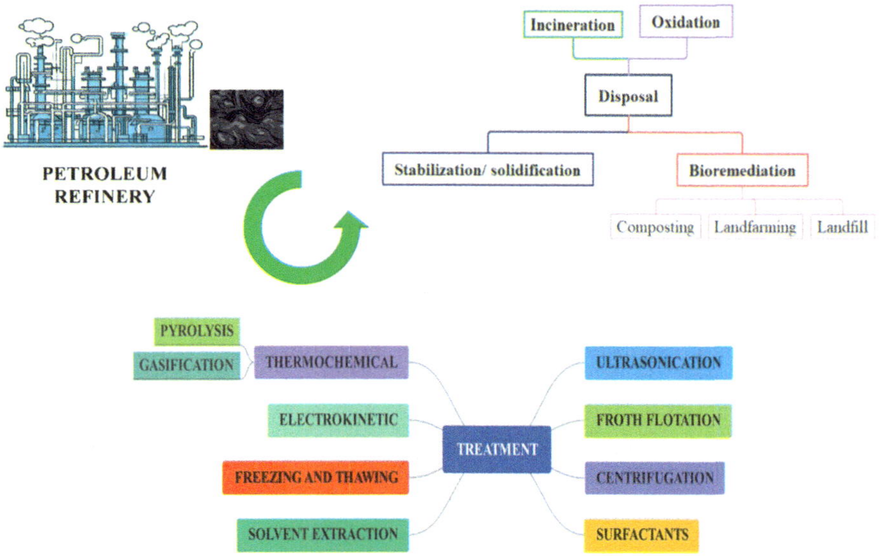

Fig. 7.1 Different treatments and disposal methods of petroleum oily sludge: schematic pathway from petroleum oily sludge generation to disposal

POS management. This includes exploring miscellaneous applications, synthesis of composite materials using POS as a waste material, and recovering carbonaceous adsorbents and fuels from POS (Jagaba et al., 2022; Sinha et al., 2020). These approaches not only address the environmental concerns associated with POS but also offer potential economic benefits through the repurposing of this waste material.

However, significant progress has been made in addressing the hazardous issue of POS through advancements in technology and the implementation of various treatment methods. Over time, various techniques have been developed for effectively treating POS by recovering residual oil and its safe disposal by reducing its pollutants profile. Besides, all these treatments and disposal approaches are gradually evolving from traditional methods like incineration and landfarming to advanced methods like ultrasonication, froth flotation, and many others. While several methods have been developed and tested, only a few of them showed satisfactory results as standalone treatments. These methods include biodegradation, chemical treatment, pyrolysis, ultrasonic treatment, solvent extraction, solidification/stabilization, incineration, and many others, as shown in Fig. 7.1 (Wang et al., 2021; Hu et al., 2013). The treatment of POS primarily focuses on the recovery of usable oil from the POS, while the disposal aspect aims to safely discard the remaining waste into the environment (Wang et al., 2021). Many traditional treatment technologies employed for treating and disposing of POS, such as incineration, landfilling, and biodegradation, have encountered several drawbacks such as high cost, instances of suboptimal performance, and its potential risk for pollution of environment

(Hu et al., 2020). Consequently, innovative treatment strategies have emerged, encompassing methods such as ultrasonic radiation, microwave radiation, electrokinetic, and many others (Murungi & Sulaimon, 2022). The different methods utilized in the treatment of POS are influenced by various factors including environmental regulations in the region, unique properties of the sludge, and the expenses involved in treatment. Moreover, thorough assessment of the physicochemical properties of POS is essential for advancing the economical technologies for the treatment of POS and understanding its potential detrimental impact on its surrounding biota (Niu et al., 2022). Conducting a composition analysis of the POS before commencing the treatment process is essential as distinctive origin of POS leads to variations in its characteristics, influencing the subsequent procedures. The composition of the sludge is usually evaluated using three key indicators: solid content, oil content (or organic matter content), and water content (Wang et al., 2018b), while the management of oil sediments involves three key aspects: reducing the volume, recovering valuable fuels, and properly disposing of the remaining waste (Hui et al., 2020). Hence, the characterization and toxicological evaluation of POS play a crucial role in acquiring the essential insights for the efficient management of POS originating from the petroleum industry.

At present, the main challenges and issue with regard to the treatment of POS is that it is a very complex compounds containing both noxious harmful product and useful resources (Chu et al., 2023). Consequently, the contemporary treatment techniques strategically focus on the dual objectives, that is, eliminating the harmful constituents while maximizing the recovery of the beneficial resources. Besides, the current state of many innovative treatment approaches remains predominantly confined to laboratory or experimental stages (Murungi & Sulaimon, 2022). Moreover, the main hindrance to the development of a single treatment methods for the POS is the variability in the characteristics of the POS necessitating further exploration and optimization for real-world implementation (Lin et al., 2023). The physicochemical properties and elemental composition of the POS reported in various studies are presented in Table 7.1. This diversity in POS characteristics has led to the simultaneous exploration of various treatment methods diverting attention from the pursuit of a unified treatment approach. Nevertheless, advanced analytical techniques and interdisciplinary collaborations are instrumental in deciphering the intricate dynamics POS, paving the way for effective, scalable, and environmentally sustainable solutions.

The POS is a hazardous material and thus associated with adverse environmental impacts. Hence, the petroleum sector is obligated by law to manage it properly. However, managing POS in a cost-effective manner is a significant challenge for the oil and/or petroleum industries. There has been a surge in research on POS with a focus on advancing fundamental understanding and fostering innovative technologies for its improved management. Besides, numerous studies have explored and reported the state of knowledge on POS, emphasizing treatment technologies designed to either reduce the environmental repercussions of POS or harness it for beneficial purposes such as fuel or material recovery. Thus, the management of POS remains a multifaceted problem, demanding comprehensive approaches for its

Table 7.1 Physicochemical properties of petroleum oily sludge

Parameters	Values	
pH	5.4–8.24	Ayotamuno et al. (2007), Aguelmous et al. (2018), Roy et al. (2018), Jerez et al. (2021), Xiao et al. (2019)
Ash content (%)	4.6–88.60	Zubaidy and Abouelnasr (2010), Gao et al. (2020), Wang et al. (2022), Al-Futaisi et al. (2007)
TKN(g/kg)	0.681–2.8	Jerez et al. (2021), Hentati et al. (2022)
Conductivity (μS / cm)	0–1.4	Roy et al. (2018)
Moisture content (%)	20–75.3	Al-Futaisi et al. (2007), Ayotamuno et al. (2007), Heidarzadeh et al. (2010), Mokhtar et al. (2018)
Organic matter (%)	24.7–77.9	Jasmine and Mukherjee (2015), Aguelmous et al. (2018)
K (mg/kg)	103–524.66	Zhang et al. (2013), Abouee Mehrizi et al. (2019), Jasmine and Mukherjee (2015)
Mg (mg/kg)	432–622	Zhang et al. (2013), Jasmine and Mukherjee (2015)
Ca (mg/kg)	30.5–3900	Zhang et al. (2013), Abouee Mehrizi et al. (2019), Jasmine and Mukherjee (2015), Roy et al. (2018)
Al (mg/kg)	64–1019	Hentati et al. (2022), Zhang et al. (2013), Abouee Mehrizi et al. (2019), Jasmine and Mukherjee (2015)
Na (mg/kg)	2.23–77.66	Zhang et al. (2013), Abouee Mehrizi et al. (2019), Roy et al. (2018)
Ni(mg/kg)	0.4–2700	Al-Futaisi et al. (2007), Niu et al. (2022), Heidarzadeh et al. (2010)
Cr(mg/kg)	3.9–480	Hu et al. (2013), Al-Futaisi et al. (2007), Wang et al. (2017), Roy et al. (2018)
Cu(mg/kg)	2.4—500	Al-Futaisi et al. (2007), Hu et al. (2013), Jasmine and Mukherjee (2015)
Zn(mg/kg)	34.7–6100	Al-Futaisi et al. (2007), Hu et al. (2013), Roy et al. (2018), Wang et al. (2017), Heidarzadeh et al. (2010)
Mn(mg/kg)	66	Jasmine and Mukherjee (2015)
Pb(mg/kg)	33.5–850	Hu et al. (2013), Al-Futaisi et al. (2007), Jasmine and Mukherjee (2015), Heidarzadeh et al. (2010)
Fe (mg/kg)	5324–60,200	Hu et al. (2013), Abouee Mehrizi et al. (2019), Jasmine and Mukherjee (2015)
Ba (mg/kg)	75–2136	Zhang et al. (2013), Jasmine and Mukherjee (2015)
As (mg/kg)	48	Jasmine and Mukherjee (2015)
Hg (mg/kg)	2.1–376.6	Niu et al. (2022), Al-Futaisi et al. (2007)
Cd (mg/kg)	7.6–100	Niu et al. (2022), Heidarzadeh et al. (2010)
Total petroleum hydrocarbon (g/kg)	125.9–900.3	Rondon-Afanador et al. (2023), Zhang et al. (2013), Roy et al. (2014), Roy et al. (2018), Jerez et al. (2021)

treatment and disposal to mitigate environmental risks and promote sustainable practices within the petroleum industry. Combining different treatment methods has emerged as a strategy to enhance the overall efficacy of POS treatment, addressing individual weaknesses and optimizing resource recovery. The exploration of these combinations adds a layer of complexity to the scientific inquiry, emphasizing the need for interdisciplinary collaboration and holistic approaches to tackle this formidable environmental challenge. Moreover, there is a notable gap in the literature concerning a thorough review that comprehensively addresses diverse treatment technologies elucidating their mechanisms and delineating their respective advantages and disadvantages. Thus, the present chapter aims to provide a brief overview on the different steps and methods involved in the treatment of the POS generated in the oil refinery industries along with its advantages and disadvantages. By examining the advantages and disadvantages of these methods and revisiting its mechanisms, the study aims to contribute to the scientific understanding of POS treatment technologies paving the way for more efficient and environmentally friendly practices in the petroleum industry. Moreover, the chapter seeks to offer insights that go beyond conventional understanding, providing a detailed view of the complexities involved in management of refinery POS.

2 Refinery Petroleum Oily Sludge (POS): Brief Overview

Petroleum oily sludge is the byproduct generated during the entire oil development process, starting from exploration to refining. It is characterized as a stable colloidal mixture comprising oil-in-water or water-in-oil emulsion, particulate matters, and other constituents (Guolin et al., 2011; Lin et al., 2018). The composition and properties of POS are intricately linked to the origin of crude oil and its physico-chemical attributes. For instance, sludge derived from oily wastewater displays a higher moisture content in contrast to sludge obtained from oilfield pit wastes (Niu et al., 2022). Besides, POS can be categorized into four types based on its formation mechanism: tank bottom sludge, drilling waste sludge, land-based POS, and refinery sludge (Teng et al., 2021). Nevertheless, the POS primarily originates from refinery wastewater treatment plants and oil storage tanks (Ubani et al., 2013).

POS contains HMs, polycyclic aromatics, and radioactive constituents that are considered to be highly carcinogenic and toxic (Gao et al., 2020). Besides, it contains compounds like volatile and semi-volatile organic carbons such as PAHs which are known to be highly genotoxic to living organisms (Prakash et al., 2015). Additionally, the total hydrocarbon concentration of the POS can vary significantly ranging from 5% to 86% depending upon its origin and composition namely cycloalkanes, branched alkanes, n-alkanes, polycyclic aromatic hydrocarbons, resins, asphaltenes, and heterocyclic compounds with oxygen, nitrogen, and sulfur. Furthermore, the concentration of HMs such as zinc (Zn), chromium (Cr), lead (Pb), cadmium (Cd), copper (Cu), and nickel (Ni) found in POS is significantly higher compared to the background value of soil (Jasmine & Mukherjee, 2015). Hence,

POS is a hazardous wastes harboring myriad of pollutants which can significantly harm the environment and its surrounded organism if proper treatments and disposal techniques are not followed.

3 Ecological and Human Health Implications of Petroleum Refinery Sludge

The repercussions of improper treatment and disposal of petroleum refinery POS extend far beyond mere environmental concerns as they pose a direct risk to ecosystems and human health. The classification of POS as a hazardous waste can be attributed to its composition of various noxious pollutants. Consequently, discharge of untreated POS has the potential to contaminate soil, water bodies, and air, disrupting the delicate balance of ecosystems (Fig. 7.2). Thus, improper treatment and disposal of petroleum refinery POS can have significant ecological and human health implications. The contamination in soil can result in hindering of plant growth, disturbance in soil biological activity, fertility, and its physicochemical properties (Niu et al., 2022). The PAHs and HMs found in the POS are recognized for their capacity to hinder crucial soil biological properties (microbial profile and enzyme activities) and alter physicochemical properties which serve as a key indicator to soil health and fertility. The alteration and inhibition of these factors can result in an overall reduction in soil quality and health (Singha et al., 2022; Borah & Deka, 2023; Chakravarty et al., 2022). This is an important concern as it

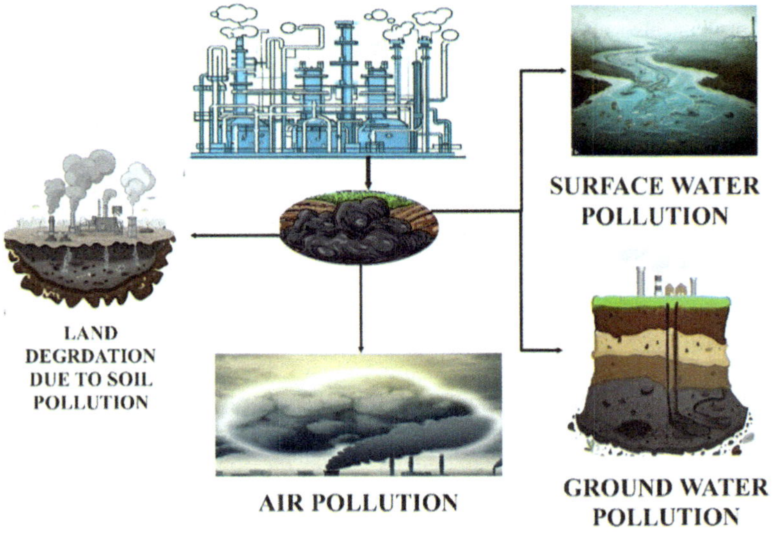

Fig. 7.2 Environmental implication of hazardous petroleum POS

renders the soil environment inhospitable for dependent organisms. For instance, the presence of HMs and PAHs in soil can significantly impede plant growth and development. These contaminants have adverse effects, including a reduction in seed germination, hindered root growth, diminished plant biomass, and the induction of morphological abnormalities and metabolic disorders in plants (Tiwari & Lata, 2018; Raklami et al., 2021; Molina & Segura, 2021). Excessive concentrations of HMs in soil not only lead to reduced yields but also hinder overall plant growth and development, inducing oxidative stress that damages cellular components and curtails plant growth. Furthermore, essential nutrients required in trace amounts, such as zinc, manganese, and copper, can become toxic to plants, resulting in symptoms like necrosis, cytotoxicity, chlorosis, alterations in the photosynthetic machinery, and a shortened lifespan (Arif et al., 2016). The presence of PAHs in soil is detrimental to plant growth and development. PAHs can inhibit photosynthesis, resulting in decreased plant growth and development (Molina & Segura, 2021).

Besides, this repercussion extends far beyond to its nearby water bodies posing a threat to aquatic organisms. Moreover, it can also result in groundwater pollution rendering it unsuitable for human consumption (Sonone et al., 2020). The introduction of pollutants into water bodies can significantly degrade water quality by increasing both the chemical oxygen demand and biological oxygen demand. This is a major concern as reduction of oxygen level in the water bodies can result in the depletion of aquatic life and an overall deterioration in the quality of the water bodies and their capacity to sustain the aquatic organisms. Furthermore, oily wastes can accumulate on the water surface, negatively impacting photosynthetic plants and organisms within aquatic ecosystems. This not only contributes to the degradation of the aquatic environment but also leads to the loss of species, potentially resulting in their eventual extinction (Abuhasel et al., 2021; Sattar et al., 2022). Over time, these pollutants can also enter the food chain which can result in biomagnification and deterioration in the organisms (Ali et al., 2019). Furthermore, incineration or open burning releases harmful gas pollutants which can result in air pollution and respiratory health issues to organisms near the surrounding areas (Hu et al., 2013). Additionally, the workers in the refinery industry who come into contact with POS face occupational health risks (Hejazi et al., 2003). The semi-volatile and volatile organic complexes present in POS including the genotoxic PAHs tend to have a collective negative effect on the central nervous system (CNS). Prolonged exposure to these compounds can result in fatigue, dizziness, headaches, memory loss, compromised respiratory system, and kidney disorders (Olasehinde & Olaniran, 2022; Rabani et al., 2020). Besides, the severity of these symptoms varies and depends upon the duration of exposure (Niu et al., 2022). The toxicity of the PAHs is also dependent upon its molecular weight. For instance, the high molecular weight PAHs (Dibenz[a,h]anthracene, Benzo[a]pyrene, Chrysene, and Pyrene) are recognized as carcinogens and mutagens, while the low molecular weight PAHs (Phenanthrene, fluorene, and acenaphthene naphthalene) usually display acute toxicity (Geier et al., 2018; Jasmine & Mukherjee, 2015). In addition, epoxide compounds produced during the metabolism of PAHs in the human body possess carcinogenic and mutagenic properties affecting the blood, skin, lungs, immune system, spleen, developing fetus, kidney, and liver (Ubani et al., 2013). Lastly, the persistent components of the POS such as HMs have a tendency to accumulate in the environment which can result in disruption of the natural processes including habitat contamination and degradation of the local biodiversity (Hu et al., 2013). In light of these issues mentioned above, proper management of POS has become an immediate necessity to ensure environmental protection and human well-being.

4 Treatment Methods: Brief Overview

Prior to disposal, POS must be treated so as to reduce its harmful effects to the environment. The treatment methods available for POS can be categorized into two groups: resource utilization treatments and harmless treatments. Besides, the treatment process employed for POS ideally consists of three distinct stages. The primary step involves reducing the quantity of POS produced by the refinery industry. Secondary steps include extraction of valuable oil from the available POS. Finally, any residual or unrecyclable POS should undergo harmless treatment methods to ensure environmental safety (Teng et al., 2021; Jafarinejad, 2017). Thus, the subsequent segments provide a concise summary of the diverse treatment approaches available for handling POS derived from refinery industries.

4.1 Centrifugation

The centrifugation process is extensively employed in treating POS due to its efficacy in separating oil from water and solid particles. It is a physical separation process that utilizes centrifugal force to segregate the components of a sludge mixture based on its density of the substance (Wang et al., 2018a; Lin et al., 2017a). However, prior to subjecting it to centrifugation process, the POS usually undergoes pre-treatment processes to decrease its viscosity which aids in improving the efficacy of the centrifugation process and save energy. This can be done by adding tensioactive chemicals, steam, organic solvents, demulsifying agents, or direct heating. This step is important because the higher the viscosity, the more difficult it is for particles to settle and separate (Hu et al., 2013).

During centrifugation process, the POS is placed in a centrifuge tube and rotated at high speed. The resulting centrifugal force causes the heavier particles to migrate toward the lower side of the tube, while the lighter particles remain at the surface levels (Hui et al., 2020). By the end of the treatment, the POS is ideally stratified into three distinct layers of oil, water, and solid residues. The oil is recovered from the top, followed by water, while the solid particles remain at the bottom of the tube (Zheng et al., 2017). The settlement distance of solid particles is influenced by the sludge viscosity, particle size, and centrifugation speed. Larger particles settle more quickly than smaller particles (Hui et al., 2020). However, the centrifugation speed is the final factor that affects settlement distance. The higher the speed, the greater is the centrifugal force leading to a quicker settling of particles at the bottom of the tube. The separated water and solid particles are then treated further to remove any residual oil or contaminants (Johnson & Affam, 2019). The water is typically treated with a bioremediation process, while the solid particles are incinerated. The oil is then recycled or sold. The efficiency of oil extraction during the centrifugation process can be influenced by diverse factors including temperature, speed, flocculent concentration, and duration of the treatment (Philemon & Benoît, 2013). Cambiella et al. (2006) reported that addition of some quantity of coagulant salt (0.01–0.5 M of $CaCl_2$) can increase the efficiency of oil water separation of POS by 92–96%. Li et al. (2023) employed a combination of centrifugation and froth flotation techniques to improve the separation of solids and oil materials in the treatment of POS. The study explored the impact of multiple variables such as impeller speed, the liquid to oil sludge ratio, duration, temperature, and flotation reagent with the objective of minimizing the residual oil content. The incorporation of flotation agents proved to be ineffective in reducing the oil content within the residues. However, the impeller speed had a significant effect in increasing the solid content in the oil phase resulting in lower concentration of oil. Moreover, Wang et al. (2018a) show that the movement pattern of solids, water, and oils in POS during centrifugation process was influenced by factors such as size distribution, surface topography, and lyophobic–lyophilic balance of the POS. Additionally, Zheng et al. (2017) combined multiple methods such as centrifugation, viscosity reduction, and demulsification to dewater and recover oil from high-viscosity POS. The results show that both the duration and speed of centrifugation play an important role in augmenting the dewatering and deoiling efficiency. Furthermore, the quality of the recovered oil and the efficiency of water removal were enhanced with increase in the quantity of the demulsifying solution. Overall, the combined process proves to be effective for the dewatering and oil recovery from high-viscosity POS. Thus, centrifugation is a well-established and environmentally friendly technique for effectively separating oil from the POS.

However, the complex nature of POS can make it challenging to achieve complete separation. Besides, generating a high centrifugal force to segregate the oil from the other components of the POS demands significant energy consumption. Furthermore, the centrifugation method faces limited implementation in refinery industries due to the requirement of high investment for the equipment involved in the process. Additionally, the supplementation of tensioactive chemicals and demulsifying agents during the pre-treatment stages also contributes to the high processing cost. In addition, it poses an environmental concern in the form of noise pollution (Islam, 2015; Johnson & Affam, 2019; Islam, 2015).

4.2 *Freezing and Thawing*

Freezing/thawing is a highly effective method for demulsification and simultaneous dewatering and oil recovery from POS (Niu et al., 2022). Demulsification is the process of separating water from the oil component in POS and is a crucial step in the process of oil recovery. In cold regions, freeze/thaw treatment has emerged as an effective demulsification method while treating POS (Hu et al., 2013). This can be attributed to the feasibility of natural freezing of the POS although certain factors such as temperature, freezing time, solid contents, presence of surfactants, salinity, and water content are also equally important for successful treatments (Johnson & Affam, 2019). This technique utilizes two mechanisms to achieve efficient demulsification. The first mechanism involves the freezing of the water phase before the oil phase. As the frozen water droplets expand in volume, they coalesce and disrupt the internal arrangement of the mixture. Subsequently, the decreasing temperature causes the gradual freezing of the oil. The second mechanism involves capturing of water droplets during the freezing process. The second mechanism revolves around capturing water droplets as the oil undergoes freezing, leading to the formation of a solid cage by exploiting the earlier freezing property of oil. Thus, the frozen water droplets exert pressure on the oil cage and burst as it increases its volume. This results in a mixture of water and oil, which are then segregated using the gravitational force and centrifugal force (Johnson & Affam, 2019). As the thawing process commences, the interfacial tension facilitates the coalescence of oil, leading to the separation of the oil–water mixture into distinct phases by the force of gravity, thereby allowing for the recovery of the oil (Hui et al., 2020). This technique ejects oil from the POS body as the ice front advanced during the freezing process. Furthermore, this physical conditioning method has shown promise in segregating the oily component from the sludge body, offering an alternative approach for handling POS in industries especially in cool regions where natural freezing occurs (Jean et al., 2001). A study conducted by Jean et al. (1999) demonstrated that freezing and thawing processes could effectively segregate over 50% of the oil content in POS. Additionally, the results also show that ultra-fast freezing did not contribute positively to oil separation. In a subsequent study, Jean et al. (2001) highlighted that the freeze/thaw treatment can efficiently remove the oily phase from POS by leveraging the physical forces associated with the motion of ice fronts. Moreover, the study reported that the freezing effects on POS were more pronounced at lower freezing speeds. Zhang et al. (2012) explored the efficacy of freeze/thaw and ultrasound combination in recovering oil from the POS. The result shows that both of these treatments could synergistically enhance each other efficiency. The ultrasound treatment played a key role in improving oil recovery by facilitating the enhanced desorption of oily components from solid particles, while freeze/thaw contributed to the separation of oil and water within the multiphase system. Likewise, Hu al. (2015) illustrated that integrating freeze/thaw process with the solvent extraction methods provides an effective and viable strategy for handling hazardous high-moisture oily waste. The results show an elevated concentration of

TPH in the recovered oil, showcasing the feasibility of this approach for management of POS. Furthermore, Chen et al. (2023) demonstrated that the freeze and thaw treatment methods prove effective for removing water from POS. The results show that the ratio of water removal from the water-in-oil (W/O) emulsion is influenced by factors such as thawing temperature and rate, freezing time and temperature, and initial water content. Besides, the study reported that the water removal ratio increases with higher initial water content, and milder thawing processes yield better dewatering results.

While freeze/thaw treatment shows promise for dewatering and oil recovery, it is important to consider factors like required freezing time and related costs when considering its industrial application. Compared to other methods like surfactants, mechanical centrifugation, and flotation, freezing/thawing may have limitations in terms of cost and economic feasibility (Hu et al., 2013).

4.3 Thermochemical Conversion

The thermochemical treatment involves the utilization of thermal processes to modify the compositions of POS such as its biological and physicochemical properties. This treatment approach aims to generate gas, liquid fuel, and solid fuel along with other supplementary products that are recyclable suitable conditions. The primary goal of thermochemical approach is to harness the energy content of the POS while minimizing its detrimental effects on the environment to meet increasingly strict regulations (Jiang et al., 2021). Thus, the subsequent sections provide a basic working mechanism of the two commonly employed thermochemical methods in refinery industries for handling POS.

4.3.1 Pyrolysis

Pyrolysis is a method where complex substances are heated to extremely high temperatures (500–1000 °C) within an oxygen deprived chamber to chemically transform or decompose them down into one or multiple valuable components, as shown in Fig. 7.3 (Mathew et al., 2018). It is an efficient thermal decomposition technique for organic substances serving the purpose of energy recovery and pollution mitigation prior to disposal (Gao et al., 2020). This method typically utilizes microwave radiation, thermal fluids, elevated temperature flue gases, and electric power as their energy sources. Comparatively, pyrolysis exhibits the ability to achieve resource and energy recovery with heightened efficiency and reduced *environmental* repercussions thereby showing greater potential for industrial utilization (Hu et al., 2020). Besides, the burning of substances under high temperatures and oxygen-free environment results in the production of syngas, liquid oil, and char, each offering specific benefits and practical uses. The generated syngas and liquid oil may be utilized as alternative raw materials or as fuels in various industries

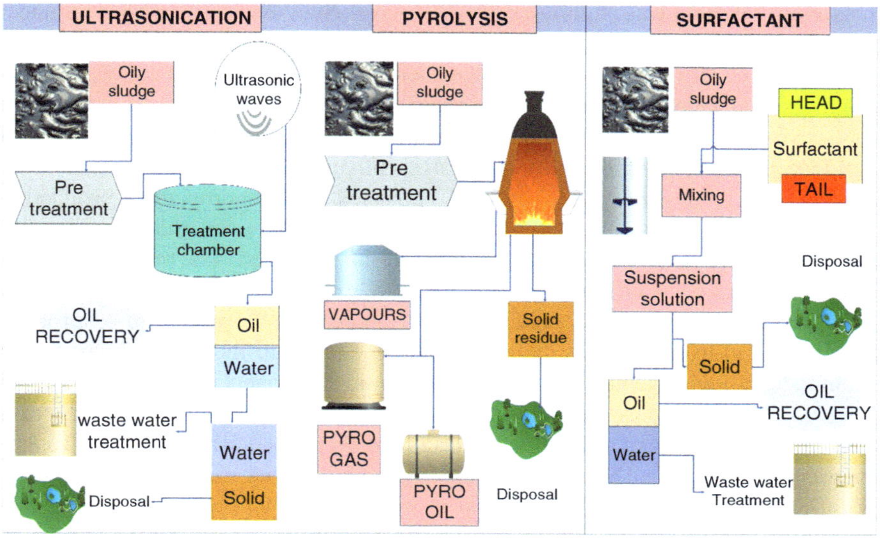

Fig. 7.3 Flowchart showing the treatment process for ultrasonication, pyrolysis, and surfactant methods

contributing to energy sustainability and reducing reliance on petrochemical sources (Murungi & Sulaimon, 2022).

The pyrolysis techniques can be categorized into four phases based on different processes occurring in different ranges of temperatures. Initially, water evaporation takes place within the range of 27–107 °C. Subsequently, light organic components vaporize within the thermal range of 107–317 °C. The third phase occurs within the temperature of 317–827 °C, during which disintegration of heavy to medium constituents takes place. Finally, the fourth stage encompasses the diminution and disintegration of other inorganic substances and coke at a temperature ranging from 827 to 927 °C (Hui et al., 2020). Besides, the efficiency of this technique is dependent on multiple factors including chemical additives, heating rate, temperature, and physicochemical characteristics of the POS (Hu et al., 2013). Furthermore, it can be broadly classified into three distinct types such as microwave pyrolysis, catalytic pyrolysis, and conventional pyrolysis. In addition to its resource recovery capabilities, pyrolysis offers effective pollution control. The char produced during the pyrolysis process can immobilize/fix the toxic HMs present in the POS thereby mitigating their harmful effects on the environment (Wang et al., 2021). Enhancing and optimizing the efficiency of pyrolysis techniques to control pollutant emissions and maximize oil yield represent a sustainable approach for realizing cost benefits associated with this method. Liu et al. (2021a) explored the performance of a microwave reactor during the pyrolysis of POS under both power control mode and temperature control mode. The study reported that at a temperature of 550 °C, the peak oil yield reached 72.55%, whereas under a power supply of 800 W, the maximum oil yield was slightly lower at 71.47%. Notably, when comparing power control mode with temperature control mode, the oil derived from the latter exhibited elevated levels of oxygenated compounds and heterocycles. Additionally, the study highlighted an increased retention of HMs in the POS char under temperature control mode. Similarly, Lin et al. (2018) examined the efficacy of co-pyrolysis of rice husk and POS in enhancing the quality of pyrolysis oil and reducing sulfur emissions in a fixed bed reactor. The result reported enhanced light hydrocarbon production, reduced heavy fractions, and oxygenated compounds in the oil. Besides, co-pyrolysis of the waste materials resulted in increased gas products (C1-C2 hydrocarbons, CO_2, CO, and H_2), inhibited H2S release, and decomposed thiophene in solid residues. The collaborative impact on improving the quality of oil products was attributed to the catalytic influence of ash and alkali metals originating from biomass, facilitating subsequent reactions of liquid components. Likewise, Gao et al. (2020) studied the impact of incorporating fly ash additives on the product distribution in the pyrolysis of POS. The inclusion of fly ash additives led to a 13.85% increase in oil yield and a 2.24% increase in gas yield. Moreover, the introduction of fly ash contributed to the improvement of pyrolysis product quality and facilitated the stabilization of metals. Additionally, Liu et al. (2021b) investigated the capacity of the catalyst (FeMg layer double hydroxide) in inhibiting the emission of nitrogen- and sulfur-containing pollutants during pyrolysis of POS. The study reported that the catalyst was effective in curbing the emission of nitrogen- and sulfur-containing pollutants. Besides, the study reported that the primary sources of S-containing pollutants were identified as disulfides-S and aliphatic, while the primary contributors to N-containing pollutants were identified as amide and heterocyclic-N species. The catalyst exhibited the capability to decrease emissions by facilitating the conversion of these pollutants into N_2 and H_2O. Consequently, a systematic investigation into pyrolysis techniques is crucial for unlocking their potential and achieving optimal yields from this hazardous waste.

The primary benefit of pyrolysis lies in its ability to minimize the volume of POS while simultaneously recovering energy and valuable resources. Additionally, the solid char produced during pyrolysis can be utilized as an adsorbent, soil conditioner, or even incorporated into construction materials, further enhancing the economic value and sustainability of the procedure (Hu et al., 2020; Murungi & Sulaimon, 2022). While pyrolysis has proven to be a viable treatment method, certain challenges remain. These include high equipment costs, complex operational requirements, product stability concerns, and a need for further understanding of reaction kinetics and mechanisms. Lastly, POS typically holds a substantial level of water content and the expenses associated with the pre-pyrolysis dehydration process are high (Li et al., 2020).

4.3.2 Gasification

Gasification process stands as a clean and sustainable thermochemical procedure that transforms carbon-derived materials like coal or coke into a highly combustible gas called syngas by employing air or steam as gasifying agents. The resulting syngas acts as a flexible energy source and fuel comprising hydrocarbon, hydrogen, carbon monoxide and dioxide, methane, and many others (Chu et al., 2021; Gonzalez et al., 2019). Notably, syngas contains a substantial proportion of hydrogen which not only serves as a clean fuel characterized by high energy density but also possesses potential as a superior chemical raw material (Santos & Alencar, 2020). Besides, the gasification of POS offers the dual advantage of producing clean energy while maximizing energy efficiency. Nevertheless, the chemical makeup and energy capacity within the resulting syngas are profoundly impacted by factors such as the input materials, operational parameters, configuration of the gasifier, and the specific combinations of gasification agents (Gonzalez et al., 2019; De Lasa et al., 2011).

This treatment involves introducing the sludge into a gasifier which is a high-temperature, high-pressure vessel, and under controlled chemical reactions. The dried sludge is introduced into a gasification reactor where it is subjected to high temperatures (700–1100 °C) in an oxygen-limited environment (partial oxidation). This thermal decomposition process converts the organic components of the sludge into synthesis gas or syngas (Jiang et al., 2021; Kamyab et al., 2022; Syed-Hassan et al., 2017). The generation of syngas involves a sequence of intricate reactions between gases and solids, as well as interactions between different gases. This method comprises multiple steps like drying at 70–200 °C, pyrolysis at 350–500 °C, oxidation at 1000–1100 °C, and reduction process. The POS are subjected to initial drying stage prior to gasification for removal of free and bound water (Jiang et al., 2021). During gasification, some heavier hydrocarbons present in the sludge can be converted into tarry residues. These tars are typically condensed and collected separately for further treatment or disposal. Likewise, any solid materials or ash generated during the gasification procedure are also removed (Zhang et al., 2010). Gasification treatment offers numerous benefits when it comes to dealing with POS. One advantage is its cleanliness, as it generates minimal amounts of ash and other pollutants during the process. Additionally, the produced syngas has versatile applications ranging from fuel for combustion and electricity generation to serving as a feedstock for chemical production. Furthermore, gasification treatment is an efficient method as it allows for the recovery of a significant amount of energy from the POS (Santiago et al., 2021). These beneficial aspects of gasification process and its efficiency have been reported in several pieces of literature. A study conducted by Li et al. (2022a) showed that the gas production from POS increased with longer residence times and higher reaction temperatures during supercritical water gasification. Additionally, the study highlighted the stabilization of HMs during the treatment process through the transformation from an oxidizable and ionic form to a residual fraction form. Likewise, Wang et al. (2023) examined the efficacy of a two-step method utilizing Raney-Ni catalyst for catalytic gasification and a subsequent desorption process to achieve efficient gasification of POS under mild conditions. The approach achieved an impressive result with a 93.87% carbon gasification efficiency and a 99.57% oil removal efficiency. Besides, the dual-step approach exhibited superior performance compared to the single-step method with increasing treatment concentration. Moreover, the study also elucidated the mechanism of the dual-step supercritical water gasification process for POS, highlighting high oil removal efficiency in the initial step using supercritical water in the desorption unit, followed by the effective gasification of high concentration oil facilitated by the Raney–Ni catalyst at a low temperature. Furthermore, Li et al. (2022b) investigated the stabilization, risk assessment, and distribution of HMs during the gasification of POS utilizing supercritical water. The findings revealed that at higher reaction temperatures and longer residence times, concentrations of Cu, Zn, and Cr were lower. Moreover, effective stabilization of Cr (98.6%), Zn (84.2%), and Cu (96.2%) was achieved at a temperature of 600 °C. Therefore, gasification techniques demonstrate a promising and environmentally friendly approach for the utilization of POS, offering potential for the stabilization of pollutants, including heavy metals.

However, there are certain challenges associated with gasification treatment of POS. One concern is the cost involved in setting up and operating gasification plants, which can be relatively high. Moreover, the corrosive nature of gasifiers necessitates regular maintenance to ensure their proper functioning. Lastly, the syngas produced during the process can be flammable and explosive, requiring careful handling and safety precautions (Arena, 2012). Overall, gasification treatment of POS presents a promising technology for managing this hazardous waste. It is a clean, versatile, and efficient process that can convert the sludge into a valuable resource. However, it is crucial to address the challenges associated with cost, maintenance, and safety to fully harness the benefits of gasification treatment in POS management.

4.4 Surfactants

Surfactants are extensively employed in the removal of organic pollutants from POS primarily due to their capacity to effectively treat large volumes of contaminants. Surfactants are those compounds that possess the ability to decrease the interfacial or surface tension between various liquid interfaces and liquid–solid interfaces, thereby facilitating the efficient removal of organic pollutants from the POS. They are amphibolic compounds consisting of heads (hydrophilic) and tails (hydrophobic) (Johnson & Affam, 2019). The process relies on the concept of chemical cleaning exploiting the water phase solution and the hydrophilic group of the surfactants to enhance the solubility of petroleum hydrocarbons in the POS as indicated in Fig. 7.3 (Hui et al., 2020; Murungi & Sulaimon, 2022). During the oil recovery process, the hydrophobic groups of surfactants accumulate at the juncture between oil and water resulting in reduction in surface tension and viscosity of the POS. This reduction in turn improves the movement capability of petroleum hydrocarbons between the water and oil stages facilitating efficient liquid–solid separation, water–oil demulsification, and recovery of oil (Seo et al., 2018; Hui et al., 2020). The alterations in surface tension and viscosity of the POS serve as key indicators of the method effectiveness. Based on the types of surfactants employed during the removal process, this method maybe classified as either biological or chemical methods. However, the former option is often favored due to its cost-effectiveness, biodegradability, lower likelihood of causing secondary pollution, and higher efficiency (Murungi & Sulaimon, 2022; Akinyemi, 2020). Corexit 9527, Afonic 1412-7, Triton X-100, sodium dodecyl sulfate, and Tween 80 are some known surfactants that are capable of enhancing of PHCs in water medium (Johnson & Affam, 2019; Hu et al., 2013). Additionally, bio-surfactants are produced by bacteria and yeast utilizing diverse ranges of substrates such as waste materials, alkanes, oils, and sugars. Some examples of biosurfactants are phospholipids, fatty acids, glycolipids, polymeric and particulate compounds, lipopeptides, and neutral lipids. The majority are classified as either neutral or anionic, while only few surfactants containing amine groups fall under the cationic category (Mulligan, 2009). Besides, this method is commonly employed as a preliminary treatment before subjecting the POS to other recovery techniques like flotation methods, electric methods, and mechanical centrifugation to maximize the crude oil recovery efficiency (Shao et al., 2014). A study conducted by AL-Doury (2019) reported that water separation, percent sludge removal, and oil recovery improve with the rise in surfactant temperature and dose. Besides Zhang et al. (2021) explored the viability of a combined treatment involving persulfate and surfactants for the treatment of POS. The principle behind the treatment was that surfactants initially elute organics from the POS and the activated persulfate further degrades the residual POS. The results demonstrated a significant enhancement in the degradation efficiency of POS with a removal rate reaching 94.6%. Besides, the oil content in the residual POS reduced to 0.57% meeting discharge standards. Additionally, the mechanistic analysis revealed that the surfactant enhanced oil solubility by reducing surface tension, while the sulfate and hydroxyl radicals produced by activated persulfate efficiently degraded intricate organic compounds into smaller molecules, ensuring the effective degradation of oil sludge. Likewise, Bao et al. (2021) explored the treatment potential of a mixture of complex biosurfactants, specifically anionic lipopeptides (LT) and non-anionic sophorolipids (SL) for separating oil from POS. The results demonstrated synergistic effects between LT and SL, reducing the critical micelle concentration (CMC) and enhancing crude oil solubilization. The ideal LT to SL mass ratio was found to be 8: 2. Additionally, the study investigated the impact of various washing conditions, including concentration, temperature, time, stirring speed, and liquid/sludge ratio on the oil removal rate. The optimal conditions were reported as 100 mg/L concentration, a temperature of 55 °C, a time of 3 h, a liquid/sludge ratio of 1:3, and a stirring speed of 300 rpm. Nevertheless, several factors can influence the outcome of this method, including the surfactant type, liquid–solid ratio, dosage, pH, cost, toxicity levels, and many others (Murungi & Sulaimon, 2022; Li et al., 2020; Hui et al., 2020).

The utilization of chemical surfactants presents distinct advantages in terms of rapid action and cost efficiency. However, there is a concern related to their potential ecological toxicity and their tendency to resist natural degradation. As a result, recent attention has shifted toward formulating biosurfactants that demonstrate enhanced environmental compatibility and superior biodegradability. (Johnson & Affam, 2019; Hu et al., 2013).

4.5 Ultrasonication

Ultrasonication stands as a highly efficient technique for effectively separating solids from liquids within suspensions of high concentration by inducing destabilization within the mixture of oil and water. This technology primarily relies on the acoustic cavitation effect and mechanical vibration of ultrasonic waves to alter the state and physical properties of POS (Hui et al., 2020; Murungi & Sulaimon, 2022). Figure 7.3 shows the process of ultrasonication methods employed during the treatment of POS. In this process, ultrasonic waves are employed to create cycles of rarefaction and compression within the treatment chamber. During the rarefaction cycle, molecules are pulled apart, leading to a negative pressure. Conversely, positive pressure is created as the molecules are pressed together during the compression stage (Johnson & Affam, 2019). The generation of negative pressure during this process creates microbubbles that grow until they become unstable and collapse with great force. This collapse results in generation of the shockwaves and produces high levels of temperature and pressure within a mere fraction of microseconds. This phenomenon is known as cavitation, and it results in the reduction of the viscosity and elevation of the emulsion system temperature. As a result, the mass transfer of the liquid phase is enhanced, leading to the destabilization of the water–oil mixture (Chung & Kamon, 2005; Hu et al., 2013; Pilli et al., 2011). Besides, the mechanical vibration and cavitation effect induced by the radiation can effectively remove the oil droplets adhered to the solid particle surfaces resulting in a reduction of crude oil content on the particle surfaces (Hui et al., 2020). Apart from cleaning the surface of solid particles, ultrasonic irradiation has the unique ability to penetrate into hard-to-reach regions within a multiphase system surpassing the limitations of other treatment methods. This unique mechanism known as ultrasonic leaching facilitates the easier entry of solvents or leaching reagents into the interior pores of solids thereby enhancing the mass transfer of contaminants through the solid matrix (Hu et al., 2013). The application of ultrasonic irradiation causes smaller droplets within an emulsion to move more rapidly compared to larger droplets. This heightened mobility results in an increased collision frequency, enabling the formation of droplet aggregates and coalescence. Consequently, this mechanism plays a pivotal role in promoting the effective separation of the water and oil phases (Islam, 2015). Besides, the agitation arising from the irradiation can result in the generation of different hydrogen radicals which are able to effectively dismantle the complex hydrocarbons into its simpler constituents (Murungi & Sulaimon, 2022). Moreover, Gao et al. (2015) investigated the impact of ultrasonic conditions (25, 50, and 100 kHz) on the extraction of various oil components from POS. The result shows that the ultrasonic washing exhibited optimal performance at 25 kHz in aspects of oil extraction efficiency and rate. Besides, the study reported that employing dual frequencies or standing waves did not influence the efficiency of the oil removal rate. However, a minimum ultrasonic intensity of 0.33 W/cm^2 was crucial for surpassing the threshold energy required for effective oil washing. Consequently, the optimal settings for removing oil from POS were identified as a water/sludge ratio of 2/1 (v/v), an intensity of 0.33 W/cm^2, and an ultrasonic frequency of 25 kHz. Likewise, Luo et al. (2020) demonstrated that the desorption of oil from solid particles enhances when lower ultrasound frequency was applied as it can result in the generation of bigger and more energetic cavitation bubbles. The study also indicated that increasing treatment time and acoustic intensity strengthens acoustic cavitation, improving oil recovery efficiency, although excessive acoustic intensity impedes cavitation bubble growth. Additionally, adjusting the pH and sludge-to-water ratio reduces the viscosity of slurry which in turn promotes the formation of $HSiO_3^-$, enhancing oil recovery efficiency. Furthermore, the mechanical effects induced by shock waves and microjets from acoustic cavitation were identified as breaking the H-bonds between solid particles and asphaltenes, thereby aiding in the recovery of oil from POS. Besides, multiple studies have reported enhancement in Ultrasonification treatment with supplementation of surfactants. Gao et al. (2018) found that ultrasonic washing assisted by surfactants (Span 80) could improve the efficiency of oil removal by 82–90%. However, single ultrasonic washing of POS could remove only 25.0% to 42.0% of oil content and double washing increased removal to 46.0% to 60.7%. Subsequent washing cycles did not yield further improvement in oil removal rate. Similarly, Lin et al. (2021) investigated the impact of three surfactant types, Na_2CO_3 (cosurfactant), Tween 80 (non-ionic), and AES (anionic) on the structure, surface composition, and oil components of solid particles in POS during ultrasonic treatment of POS. The study observed that each surfactant exhibited distinct oil removal abilities and selectivity for the four oil components (asphaltenes, resins, aromatics, and saturates). AES demonstrated the overall best performance, while Tween 80 showed a preference for lighter components, and the Na_2CO_3 favored heavier components, particularly resins. The AES, with excellent foaming properties, aided effective oil separation, while Tween 80 solubilized oil at the solid surface, assisting in oil removal. Both AES and Na_2CO_3 ionized anions in water, promoting the removal of resins to 64.6% and 38.9%, respectively. In contrast, Tween 80 exhibited lower removal efficiency for resins at only 13.2% as it cannot ionize in water. For the removal of asphaltenes, aromatics, and saturates, Tween 80 and AES outperformed the Na_2CO_3. The optimal treatment times for Na_2CO_3, Tween 80, and AES were determined to be 20 mins, 30 mins, and 40 mins, respectively. The study also noted that ultrasonic treatment altered the morphology and wettability of solid particles, making them smaller and rougher. Therefore, the choice of surfactants plays a pivotal role in the efficiency of ultrasonication treatment technologies. Furthermore, Hanxuan et al. (2023) investigated the impact of nanomaterials in the ultrasonic treatment of POS. The nano-β-CD@Fe_3O_4 magnetic material synthesized from β-CD and $FeCl_3$-$6H_2O$ and induced changes in oil composition under ultrasound exposure, leading to decreased viscosity and increased fluidity. The results showed that optimal treatment conditions were established at a Nano-β-CD@Fe_3O_4 concentration of 0.5%, a temperature of 60 °C, and an ultrasonic frequency of 60 Hz. This configuration achieved a remarkable 90.17% oil removal efficiency within 45 minutes. Moreover, oil removal efficiency remained high at 85.65% in the second run after recovering the nanomaterials from first treatment, achieved through magnetic separation of Nano-β-CD@Fe_3O_4, ensuring the reusability of the nanomaterials.

Nonetheless, the efficacy and successful implementation of this method are reliant on several factors such as additives, temperature, treatment time, sound frequency and its intensity, salt or other impurities, and many others (Hui et al., 2020; Johnson & Affam, 2019). The utilization of ultrasound technology offers a dual advantage: swift and efficient treatment of POS within short timeframes, coupled with the assurance of zero secondary pollution. However, this ultrasonic treatment technology is currently in the experimental and laboratory phase and is very costly for large-scale application (Johnson & Affam, 2019).

4.6 *Solvent Extraction*

Solvent extraction is a commonly employed technique for the treatment of POS. It involves the utilization of specific extracting agents to dissolve and separate organic components from water or soil matrices. Its function involves the removal of semi-volatile and non-volatile organic compounds from POS (Johnson & Affam, 2019; Hu et al., 2013). This method operates on the concept of "similar miscibility," where solvents showing similar properties like that of crude oil are carefully chosen for extracting petroleum substances from POS. This procedure can effectively recover oil and is in alignment with the principle of revering valuable crude oil (Li et al., 2020). The process begins by thoroughly mixing the sludge with the selected solvent, allowing for the removal of different impurities, water, and other solid particles. The resultant mixture is then subjected to distillation process where the extraction and recovery of the crude oil take place from the solvent mixture. The selection of an appropriate solvent is crucial in solvent extraction methods. Different solvents such as methyl ethyl ketone (MEK), toluene, sodium lignosulfonate, hexane, xylene, and many others have been tested for their efficacy in recovering oil from the POS. Moreover, alternative solvents such as supercritical fluids like supercritical ethanol and water have also been explored for improving the extraction process (Hui et al., 2020; Teng et al., 2021; Murungi & Sulaimon, 2022). Through comparative studies, researchers have identified the optimal conditions for achieving maximum oil recovery. Solvents like MEK have shown high recovery rates, but they may result in impurities in the recovered oil, such as sulfur and carbon residue. Therefore, purification processes may be required (Zubaidy & Abouelnasr, 2010). Additionally, the combination of different solvents like toluene and MEK has been investigated to optimize the recovery of hydrocarbons from POS (Nezhdbahador et al., 2018; Teng et al., 2021). The efficiency of solvent extraction treatment is influenced by various factors like the ratio between sludge and solvent, pH, extraction pressure and temperature, stirring strength, types of solvent used, stirring time, and many others (Niu et al., 2022; Teng et al., 2021; Hu et al., 2013). For example, higher temperature can expedite the extraction process but may cause the evaporation of PHCs and solvents, whereas lower temperatures reduce extraction costs but may compromise oil recovery efficiency. Besides, distillation is favored at lower pressures to facilitate solvent evaporation at a lower temperature. This not only conserves heating expenses but also prevents solvent degradation. Furthermore, augmenting the solvent quantity in the sludge mixture can yield a greater volume of oil with improved quality. This is achieved by diminishing the generation of high molecular weight hydrocarbons and ash content (Zubaidy & Abouelnasr, 2010; Hu et al., 2013). Hu et al. (2016) conducted an investigation comparing the efficacy between mechanical shaking extraction (MSE) and two ultrasonic-assisted solvent extraction processes (UAEB and UAEP) of recovering oil from POS, utilizing three extraction solvents: MEK, ethyl acetate (EA), and cyclohexane (CHX). Under optimal conditions, UAEP treatment achieved a rapid oil recovery of 68.8% within 20 seconds, while UAEB treatment demonstrated an oil recovery of 62.6% within 15 minutes, surpassing the 62.0% obtained by MSE after a 60-minute extraction. Thus, the results show that UAEB and UAEP were much more efficient, effective, and improved approach for POS recycling. Furthermore, many studies have explored the potential of ionic liquids (ILs) in enhancing the oil extraction process and recovery of oil from POS. IL exhibit distinctive characteristics, such as prolonged reusability, excellent solubility, robust thermal stability, and nonvolatility. Consequently, ILs find extensive utilization in separation and extractions processes leading to reductions in reaction durations and solvent consumptions (Zhu et al., 2020). Tian et al. (2019) examined the potential of incorporating ionic liquid (IL) into the cyclohexane solvent to enhance oil recovery from POS. The findings revealed that addition of IL achieved comparable or higher total petroleum hydrocarbon (TPH) recovery (96.92%) within a significantly lesser duration and at a lower solvent/sludge ratio, all while consuming less energy. Moreover, the recovered oil exhibited a higher F3 fraction level and a calorific value similar to crude oil, indicating its suitability for use as a petrochemical feedstock and energy source. Thus, the IL-enhanced solvent extraction method presents to be much more efficient and effective for recovering oil from the POS all the while consuming lesser amount of solvent. Similarly, Zhu et al. (2020) studied the efficacy of the hydrophobic ionic liquid [Bmmim][PF6] in improving the recovery of oil from POS. The findings indicated that the incorporation of [Bmmim][PF6] to the extraction solvent led to a 15% increase in oil recovery, a 38% increase in asphaltenes, and a 14% increase in resins. Furthermore, the results demonstrated a substantial reduction in solvent consumption, approximately 60% less in each instance. Hence, these outcomes suggested that the recovery of oil from POS utilizing the solvent extraction technique could be improved with the incorporation of the IL as IL not only improves its efficiency but also reduces the cost of treatment by reducing the amount of solvent required for the treatment process.

Thus, the solvent extraction technique is a simple and efficient method for treating POS generated from the refinery industries. The advantages of this method lie in its ease of operation, ability to segregate the sludge into solid impurities and valuable hydrocarbons, and high oil recovery efficiency resulting in a substantial reduction in waste volume for disposal. Nevertheless, there are certain challenges linked to this method. One of the major limitations is the cost of the extracting agent which can hinder the scalability of the process. Besides, the use of volatile and toxic solvents not only leads to solvent loss but also poses environmental risks necessitating the use of tightly sealed equipment. Additionally, the presence of residual extractant in the extracted sludge can contribute to secondary pollution requiring continuous treatment processes (Teng et al., 2021; Johnson & Affam, 2019; Murungi & Sulaimon, 2022; Hui et al., 2020).

4.7 Electrokinetic Method

Electrokinetic technology relies on the synergistic effects of electrophoresis, electromigration, and electroosmosis mechanism for effectively treating POS. The application of an electric field can result in the separation of solid and aqueous phases, with solid components concentrating at the anode, and hydrocarbons and water accumulating at the cathode. This process enables the effective separation of oil, water, and mud facilitating a three-phase treatment method (Wei et al., 2015). The approach utilizes a low-intensity direct current (DC) applied across a permeable medium with a pair of electrodes. As a consequence, electrons migrate from regions of lower concentration to higher concentration within the liquid phase of the permeable medium (Johnson & Affam, 2019). This process involves three steps mechanisms for effective separation. Firstly, under the influence of the electrical field, colloidal aggregates within the POS are disrupted, causing the solid and colloidal particles to migrate toward the anode via electrophoresis. Simultaneously, the liquid phase encompassing oil and water migrates toward the cathode through electro-osmosis. This separation is beneficial in breaking down the emulsifiers and barriers that stabilize the water droplets, allowing for enhanced coalescence. Secondly, the process of electrocoagulation takes place for the segregated solid particles near the anode area resulting in higher amounts of sediments and solid particles. Lastly, the separated liquid phase devoid of fine solids and colloidal particles forms an unstable secondary emulsion of oil-in-water. Through the charging and agglomeration of droplets near the cathode, this emulsion gradually coalesces forming two distinct phases of oil and water (Murungi & Sulaimon, 2022; Islam, 2015). Jahromi and Elektorowicz (2018) investigated the application of a novel controller system based on Percolation Theory to improve the demulsification of POS and optimize the efficiency of oil–water phase separation. The study revealed that integrating the controller system into a linear horizontal electrokinetic system significantly increased the oil-to-water volumetric ratio and reduced the energy consumption of the process. Additionally, it enhanced the separation of oil and water in POS and resulted in a twofold reduction in the cost of phase operation. Furthermore, the study demonstrated that water production was closely linked to exposure time and voltage gradient, with the attachment of the controller to the electrokinetic system yielding a much higher recovery rate of water. Nevertheless, the efficiency of the electrokinetic method is dependent on various factors, including resistance, pH, electrical potential, electrode spacing, voltage gradient, additives, and many others. Besides, the addition of reagents or surfactants can be employed to enhance the rate of contaminant removal near the electrodes. In contrast to alternative oil recovery techniques like pyrolysis and centrifugation, the electrokinetic process demands a lower amount of energy (Johnson & Affam, 2019; Hui et al., 2020).

However, the application of the electrokinetic method for recovering oil from POS has been limited to laboratory-scale experiments. Besides, the efficacy of recovering oil from POS with elevated levels of solid particles and high viscosity is relatively low (Hui et al., 2020). While its effectiveness has been demonstrated in laboratory settings, further research and evaluation are needed to determine its viability for large-scale applications.

4.8 Froth Flotation

Froth flotation technique relies on surface chemistry principles to effectively separate fine solid particles from a liquid suspension of POS. This method encompasses the capturing of particulate matter or oil droplets onto air bubbles dispersed within a water mixture. Subsequently, these oil–water mixture experiences buoyant forces causing them to ascend and ultimately amass within a froth layer (Fig. 7.4) (Hu et al., 2013). In this method, an adequate quantity of water is added to POS to create a liquid slurry. Additionally, extraction solvents or surfactants are frequently incorporated to enhance the efficiency of recovering oil from the POS (Johnson & Affam, 2019). Next, air is introduced into the slurry to generate bubbles. The air bubbles then migrate toward the surface and come into contact with the oil droplets present in the liquid slurry. The thin water film situated between the air and oil bubble

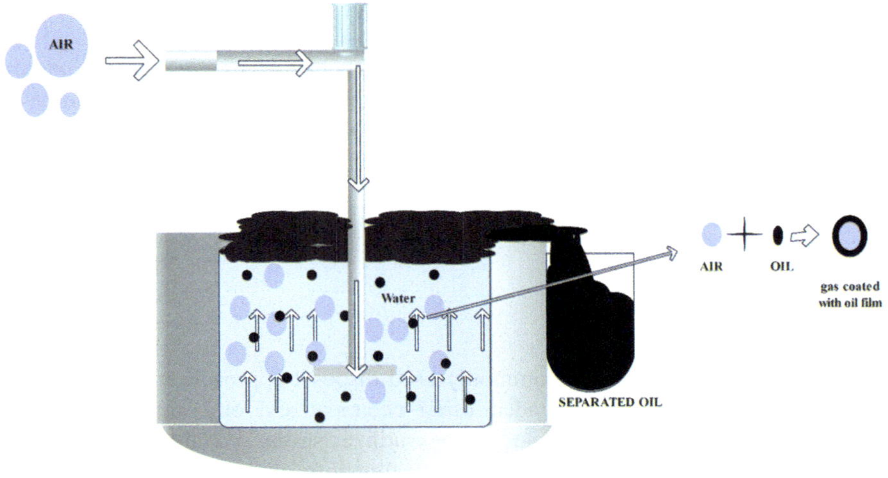

Fig. 7.4 Froth flotation treatment methods

gradually becomes thinner until it reaches a critical thickness, causing the film to break and allowing the oil to spread onto the air bubbles. Eventually, cluster of air bubbles with oil on its surface rises rapidly to the upper layer of the slurry mixture. The amassed oil on the surface can then be removed and stored for further refinement ultimately achieving the purpose of oil recovery principle (Hui et al., 2020). This method provides a straightforward and cost-efficient solution for POS treatment, but it is generally more successful when used for POS with lower viscosity. The efficiency of oil recovery in froth flotation technique is dependent on a range of factors such as flotation duration, surfactants, air bubble size, temperature, salinity, pH levels, and properties of the POS like density, solid content, and viscosity (Johnson & Affam, 2019). Li et al. (2021) examined the efficacy of froth flotation at ambient temperature for the separation of oil and solids in POS. The findings suggested that the segregation of the solid component and the oily components of the POS could be accomplished through conventional froth flotation technology under high shear force conditions, given appropriate optimization and the use of an inorganic salt. The resultant oil concentration in the oil-separated top (OST) and oil-separated bottom (OSS) residue was successfully reduced to 1.2% and 0.6%, respectively, indicating the potential of froth flotation as a future low-cost, large-scale treatment technology for POS.

Nevertheless, the froth flotation method requires a significant amount of water, which can lead to challenges in treating POS especially those with higher viscosity and lower moisture content. Besides, it is necessary to reduce sludge thickness and eliminate large solid particles before subjecting it to froth flotation treatment (Hu et al., 2013).

5 Disposal Methods

The disposal methods aim to safely and responsibly discard the treated POS into the environment while minimizing harm to the surroundings following the recovery of any usable oil. Thus, the following sections described the most commonly employed disposal techniques which include oxidation, incineration, stabilization, and biodegradation methods.

5.1 Oxidation

Oxidation is one of the most commonly used disposal approaches for POS where organic contaminants are degraded using chemical and other oxidizing agents (Johnson & Affam, 2019). This technology utilizes strong oxidants such as ozone, Fenton's reagent, and hydrogen peroxide to transform the macromolecular hydrocarbons found in POS into smaller and less toxic substances. The efficiency of this procedure relies on the interaction between the oxidizing agents and the contaminants (Ossai et al., 2020). The process of chemical oxidation includes the incorporation of reactive chemicals to oily wastes enabling the oxidation of organic complexes into H_2O and CO_2 or their conversion into other safe constituents like inorganic salts (Hu et al., 2013). Oxidation are achieved through the utilization of hypochlorite, ultrasonic irradiation, ozone, persulfate, permanganate, and Fenton's reagent which can generate abundant radicals like hydroxyl radicals (OH•) capable of rapidly reacting with a broad spectrum of organic and inorganic substances (Rivas, 2006).

Fenton oxidation is a process that uses a combination of ferrous iron and hydrogen peroxide to oxidize organic pollutants. Hydroxyl radicals are produced through the reaction between hydrogen peroxide and ferrous iron, which serves as a reducing agent. Hydroxyl radicals are highly reactive and can degrade a wide range of organic pollutants. Fenton oxidation, on the other hand, is characterized by its simplicity, low running costs, and environmental friendliness. However, it requires a significant amount of reagents and corrosion-resistant equipment (Sani et al., 2020). Photocatalytic oxidation (PO), wet air oxidation (WAO), and supercritical water oxidation (SCWO) are among the advanced oxidation techniques that have seen growing interest in recent years. SCWO method utilizes the supercritical water as a reaction medium along with different catalysts like air, oxygen, and hydrogen peroxide to efficiently treat POS by reducing both oil viscosity and concentration of the heavy component in the separated sludge (Guolin et al., 2011; Hu et al., 2020). Beyond the critical point of water pressure and temperature, SCWO initiates a free radical reaction that efficiently degrades the organic matter. It offers several advantages, including fast and thorough degradation, wide applicability, minimal external interference, and absence of secondary pollution (Guolin et al., 2011). It is a promising technology as it utilized supercritical water as an oxidant to degrade organic pollutants. Supercritical water is water that is heated to a pressure and temperature beyond its critical point. In these settings, the characteristics of the water are that of the intermediate between those of a gas and a liquid. This makes it an effective oxidant for a wide range of organic pollutants (Bermejo & Cocero, 2006). Using supercritical water as a medium, it facilitates the transformation of crude oil emulsifiers in the POS into valuable oil products, resulting in a homogeneous single-phase mixture and reducing recombination fractions, surfactants, and impurities. This process effectively overcomes the challenges associated with the material transmission between the stages.

However, the harsh reaction conditions and the need for corrosion-resistant equipment contribute to higher costs, and heavy metal removal remains a challenge. Some of the major factors that affect the oxidation efficacy are pressure, time, reaction temperature, ratio of oxidant to demand, alkali concentration, and solution pH (Vadillo et al., 2013). In conclusion, oxidation technologies provide effective solutions for degrading organic contaminants in petroleum sludge. While each method has its unique benefits and challenges, advancements in equipment size, reagent quantities, and cost reduction are necessary for large-scale implementation (Johnson & Affam, 2019; Murungi & Sulaimon, 2022).

5.2 Incineration

In this method, the combustibility of oil present in POS is utilized within a controlled environment along with secondary fuel and extra air to eradicate the detrimental constituents and accomplish the complete decomposition of the POS (Hu et al., 2020). This technique has the ability to minimize POS volume swiftly, exhibiting a high degree of its eco-friendliness and robust flexibility to diverse raw materials. The POS containing a significant amount of water is typically subjected to pre-treatment to reduce its water content. In addition, the incineration process frequently utilizes supplementary fuels like coal to enhance the heating power (Hu et al., 2020). The fluidized bed and rotary kiln incinerators are the prevailing choices in this approach for waste disposal.

Fluidized bed incinerators and rotary kiln incinerators are distinct waste disposal systems. In fluidized bed incinerators, the waste materials are placed atop a mineral grain layer and burned in a fluidized state, agitated by an upward airstream. This method is effective for pre-treated waste (Makarichi et al., 2018). On the other hand, rotary kiln incinerators use an inclined rotating cylinder to transport and uniformly burn waste, displaying adaptability to diverse waste types and high moisture content. Although rotary kilns boast high combustion efficiency, up to 90%, they have drawbacks such as potential air pollution due to unsorted waste burning and the need for feedstock drying to prevent hazardous pollutant emissions (Rajendran et al., 2021). Fluidized bed incinerators are often preferred for their superior pollution control, efficient handling of pre-treated waste, and reduced emission of hazardous pollutants, making them a more environmentally friendly choice, especially for municipal solid waste disposal (Hu et al., 2013). The different factors influencing the incineration process include the treatment duration, types of additional fuels, temperature, types of incinerators, and properties of the POS (Niu et al., 2022). It involves the complete combustion of POS in an incinerator operating at temperatures exceeding 700 °C (Aguelmous et al., 2019). However, this also varies and is dependent on the incinerators used during the treatment process. The ignition temperature of a fluidized bed incinerator can range from 730 to 760 °C, and the treatment duration can last for several days, while the temperature in rotary kiln incinerator ranges from 980 to 1200 °C with a treatment duration of approximately 30 minutes only (Hu et al., 2013). As a response to the variable efficacy observed in this treatment approach, there has been a growing interest in research focused on the co-combustion of POS with biomass. The aim is to optimize operating conditions and enhance treatment efficiency (Niu et al., 2022). For instance, Liu et al. (2018a, b) explored into the characteristics and kinetics of co-combusting POS with litchi peels through thermogravimetric analysis. The study analyzed the interactions between POS and litchi peels at different temperatures and found that they had both inhibitory and promotional effects on the combustion process depending on the temperature range. Similarly, Cao et al. (2022) investigated the impact of incorporating cornstalk (CS) during the incineration of POS. The simultaneous combustion of CS and POS exhibited a synergistic effect, which means that the incineration of the two materials together was more efficient than the incineration of each material separately. Besides, the thermogravimetry analysis further confirmed this synergistic effect by revealing that the combined burning of CS and POS has a higher combustion efficiency and lower activation energy than the combustion of each material separately. Moreover, the NO and SO_2 emissions of the co-combustion of POS and CS are lower than the emissions of the combustion of POS alone. Additionally, Deng et al. (2016) investigated the effect of incorporating plant biomass during the incineration of POS using the thermogravimetry method. The study concluded that adding biomass to POS can enhance the combustion quality and decrease the ash content and burnout temperature of the mixtures. These diverse research endeavors contribute to a better understanding of the co-combustion process and aim to improve its effectiveness in treating POS.

Incineration offers the advantages of yielding steam for generating power and stands as the swiftest and most operative means to attain the resource utilization, harmlessness, and reduction objectives associated with POS treatment (Murungi & Sulaimon, 2022). However, the elevated heavy metal content within POS necessitates caution during incineration as these metals can act as airborne pollutants or remain in unstable forms within the resultant bottom ash presenting significant ecological hazards (Hu et al., 2013). Moreover, incineration necessitates the utilization of additional fossil fuels to sustain the desired combustion temperature resulting in the production of hazardous ash residues and unwanted gaseous emissions (Hui et al., 2020; Gong et al., 2018).

5.3 *Stabilization and Solidification*

Stabilization and solidification (S/S) is a cost-effective and expeditious treatment method for converting hazardous POS into a less soluble and immobilized form. It operates on the fundamental principles of solidification, which involves containing hazardous waste within a solid cement-like structure and stabilization, which securely binds and immobilizes contaminants in a durable and insoluble configuration, thereby reducing their harmful effects (Hu et al., 2013). This method is particularly compatible for waste materials rich in heavy metals, as it stabilizes pollutants by immobilizing them (Malviya & Chaudhary, 2006). Through these synergistic mechanisms, the treatment method aptly confines the hazardous constituents ensuring their enduring stability and safeguarding against potential environmental ramifications (Murshid et al., 2018). These methods incorporate the curative agents to stabilize the structure of POS and transform the hazardous components into a less toxic form ultimately achieving the goal of safe treatment (Niu et al., 2022). The solidifying agents added into the POS results in a cascade of steady and permanent chemical and physical reactions leading to the development of solid materials. Among the solidifying agents, stale and fresh cement kiln dust are regarded as the most effective one for making fill materials with impressive compressive strength. The ideal weight mixture ratio of fresh kiln dust to sludge is determined as 1:1, whereas for stale kiln dust it is recommended at 2:1 (Guolin et al., 2011).

The S/S technique for managing hazardous waste employing cement as a treatment agent for hazardous waste involves a systematic three-stage approach: (1) addressing the chemical contaminants through chemical reactions between cement hydration products and the contaminants themselves; (2) taking in contaminants that are present on the surface of hydrated cement products and physically absorbing them; and (3) encapsulating the contaminated soil or waste to create a solidified matrix with reduced permeability. This comprehensive approach ensures effective immobilization and containment of hazardous contaminants, minimizing their environmental impact and promoting long-term stability (Johnson & Affam, 2019). These techniques not only expedite the process but also immobilize organic pollutants and high concentrations of heavy metals within the solidified sludge facilitating easier transportation and mitigating the environmental and human health risks (Hu et al., 2013). The successful stabilization of POS is distinguished by the achievements of desired attributes such as strength, stability, free-liquid content, leachability, biodegradability, and durability when buried. Besides, the crucial parameters for evaluating the efficacy of solidification include the hardness of the solidified matrix and leakage rate which are contingent upon the type and dosage of the curing agent as well as the oil content (Murungi & Sulaimon, 2022). A study conducted by Karamalidis and Voudrias (2007) demonstrated the effectiveness of the S/S method in reducing the leaching of PAHs, alkanes, and TPH from petroleum oil sludge (POS) through macroencapsulation. The investigation highlighted that the leaching performance and mechanical properties of S/S samples are influenced by the type and content of cement. The study concluded that employing a cement-based S/S technique is a viable approach for immobilizing hydrocarbons in oil refinery sludge, thereby mitigating environmental risks associated with this waste. Likewise, Murshid et al. (2018) explored the impact of hydration days on heavy metal leaching and the strength of POS using the S/S treatment. The comparison between the 7th and 28th hydration days revealed a decrease in heavy metal leaching with increased hydration days. Correspondingly, there was a proportional increase in the compressive strength of the S/S samples with prolonged hydration days. The compressive strength of solidified/stabilized samples showed an increase with prolonged hydration days. Optimal conditions for the S/S method were identified as a cement/POS ratio of 1:1, a curing time of 28 days, and a pH of 12. The study suggests that the S/S method employing a cement matrix holds promise for treating petroleum sludge due to its effectiveness in immobilizing pollutants and reducing leaching.

However, despite its advantages, this method is hindered by concerns regarding the potential leaching of secondary pollutants into the environment once the method is implemented (Johnson & Affam, 2019). This can result in secondary soil and groundwater pollution through the leaching of harmful substances.

5.4 *Bioremediation*

Bioremediation is a process that utilizes microorganisms to remove environmental contaminants (Kumar et al., 2018, 2022a, b). It is widely used to restore oil-polluted environments by facilitating the microbial degradation of hydrocarbon contaminants (Agrawal et al., 2021). Bioremediation treatments account for approximately 25% of all remediation methods targeting petroleum hydrocarbon pollutants (Lukić et al., 2017). The efficacy of bioremediation for POS is influenced by various factors like presence of appropriate microorganisms, treatment duration, and concentration and composition of TPHs in the sludge (Abouee Mehrizi et al., 2019). The following sections highlight the most extensively utilized bioremediation techniques such as composting, land farming, and landfill disposal.

5.4.1 Landfarming

Landfarming is a well-established and highly effective method for biological soil remediation offering both ex situ and in situ treatment options (Lukić et al., 2017). It includes the precise application of POS onto the reactive soil layer, allowing the soil microbiota to act as a natural agent for actively degrading contaminants (Lukić et al., 2017). This natural process utilizes the inherent biological, chemical, and physical processes of the POS contaminated soil to effectively neutralize or treat the organic contaminants. The organic pollutants are primarily degraded through biological activity in addition to the chemical and physical activities like evaporation and photodegradation (Marin et al., 2005). To enhance treatment efficiency, optimal conditions such as pH, moisture level, fertilization, and aeration must be maintained to promote microbial population and ultimately increase its rate of contaminants degradation in the soil and sludge mixture (Hu et al., 2013). Likewise, nutrients, moisture, and oxygen are supplemented to enhance in situ biological decomposition. Moreover, to enhance the degradation process, contaminated soil is often supplemented with soil amendments like nutrients and bulking agents, or waste substrates to boost the indigenous microbial population (Lukić et al., 2017; Gan et al., 2009).

Landfarming offers several advantages in terms of its ability to handle large volumes of POS, little energy consumption, high success potential, simplicity of operation, and cost of operation. However, it demands a significant land area and is a time intensive procedure, often taking from a couple of months to years or longer, particularly for challenging compounds (Khan et al., 2004). Climate conditions, such as temperature and rainfall, also significantly affect biodegradation efficiency, posing challenges for landfarming in cold regions (Hu et al., 2013). Besides, landfarming raises environmental concerns as it emits volatile organic compounds which contribute to air pollution and the risk of groundwater pollution from leachate containing heavy metals, phenols, and PHCs (Maila & Cloete, 2004). The potentially toxic metals present in POS are non-degradable and can result in contamination of water, soil, and air. Additionally, their accumulation in plant tissues can contribute to the contamination of the food chain (Niu et al., 2022). Lastly, landfarming has advantages but must adhere to environmental regulations to minimize contamination risks. Therefore, risk assessment of landfarming is necessary to understand the potential risks associated with the method and to develop appropriate remediation strategies. It is particularly important for heavy metal leaching, which can pose a risk to the environment and human health.

5.4.2 Landfill Disposal

Landfilling is the cheapest and widely employed solid waste disposal method where POS is mixed with soil and allowed to undergo natural attenuation (Mishra et al., 2017). This technique involves the strategic placement of sludge wastes within a carefully constructed landfill area, employing thick layers of impermeable clay and synthetic liners to effectively isolate the sludge from both air and water (Hu et al., 2020). Additionally, a well-designed leachate collection system is implemented above the bottom liner to prevent the migration of contaminants into groundwater sources (Mishra et al., 2017). Secure landfilling has gained prominence in developed countries like Germany, Canada, UK, and USA as it offers a robust solution to address environmental concerns associated with other methods like landfarming (Hu et al., 2013; Bhattacharyya & Shekdar, 2003). However, the operation costs of secure landfilling tend to be higher. Despite the higher costs, landfilling is favored due to its energy efficiency, making it a more cost-effective and reliable bioremediation process when compared to composting (Murungi & Sulaimon, 2022). To ensure the protection of groundwater during landfill disposal, the implementation of impermeable linings is essential to prevent any potential leakage. Regular monitoring through routine sampling and testing of wells is employed to detect any signs of leaks and maintain the integrity of the containment system (Mishra et al., 2017). Aguelmous et al. (2018) observed a proliferation of microorganisms during the landfilling of POS, contributing to the reinforcement of the biodegradation process. Furthermore, there was an increase in both mesophilic and thermophilic bacteria during the intermediate stage, followed by a reduction toward the end of the process. The removal efficiency of TPH (C6 to C22) and total polyphenols reached 73% and 71.6%, respectively, resulting in reduction of its potential toxicity. These changes align with the effectiveness of the biotransformation process, indicating that landfilling of POS reduces the presence of toxic organic compounds, resulting in a stable final product. Nevertheless, the incomplete stabilization of POS during landfilling presents significant pollution risks, with the potential to contaminate groundwater and create health-related issues (Moltó et al., 2013).

5.4.3 Biopile/Composting

Biopile is a method of waste management wherein waste materials are stacked in piles or rows, typically 2 to 4 meters in height to facilitate the degradation of the waste through the activity of native or externally introduced microorganisms. If organic matters are incorporated into POS during this process, it is known as composting (Hu et al., 2013). This technique has garnered a worldwide attention as a means to substitute the landfarming which otherwise requires a large land area. It is an economical and efficient approach for treating POS, involving the thorough mixing of sludge with bulky materials and the introduction of bacteriological agents to promote microbial degradation. Composting can be achieved throughout using diverse methods such as bioreactors, piles/windrows, or a combination of both. However, the bioreactor mode is considered the most favorable for composting as it enables the optimal management of gas emission, nutrients, humidity, temperature, and oxygen (Aguelmous et al., 2019). This process significantly reduces pollutant levels and yields environmentally friendly byproducts such as CO_2, H_2O, and fatty acids. The organic matter in the soil performs a crucial function during the composting process, while additional bulking agents like wood chips, rice husks, and manure are often employed to enhance microbial activity (Murungi & Sulaimon , 2022; Niu et al., 2022). The effectiveness of this process is improved through consistent watering, aeration, and tilling. Besides, the porosity, moisture level, nutrient content, C:N:P ratio, aeration, pH of the POS, and temperature of the composting system are considered to be essential factors for the success of the composting process. Thus, the recommended ranges for the composting process are C:N:P ratio of 100:10:1, oxygen concentration between 5% and 15%, pH level within 6–7.5, and a moisture level of approximately 60% (Lukić et al., 2017; Juwarkar et al., 2010; Sayara et al., 2010). Moreover, the microorganisms responsible for oil degradation may not function efficiently when essential nutrients like phosphorus and nitrogen are scarce (Koolivand et al., 2017; Kulikowska, 2016; Kasinski et al., 2016; Ma et al., 2016; Makan, 2015). Thus, by adjusting parameters such as nutrient levels, oxygen availability, watering, and the use of suitable microorganisms and bulking agents, the remediation process can be optimized, leading to improved biodegradation of oil sludge. Composting proves particularly effective for hydrocarbon-rich wastes, such as tank bottoms and ground pit sludges, enhancing permeability, water retention, and soil quality. Furthermore, it is important to construct biopiles on impermeable bases to prevent environmental leakage (Murungi & Sulaimon, 2022).

However, composting has drawbacks such as requiring ample space although comparatively lesser than other land treatments, being susceptible to natural conditions like temperature and rainfall, and the potential risk of soil and groundwater contamination. The success of composting treatment relies on the activities of microorganisms, the optimal mixture of organic fertilizer and POS, additives and nutrients, decomposition duration, and the selection and amount of microbial agents (Aguelmous et al., 2019). Overall, the composting technique represents an operative and environmentally friendly solution for the treatment of POS.

6 Conclusion and Future Directions

The management of POS poses significant challenges due to its complex composition and environmental impact on its surroundings. Consequently, identification of an efficient advanced treatment technology for POS is a point of interest in recent times. The efficiency of various treatments approaches commonly utilized for treating POS is shown in Table 7.2. Several methods have shown great potential in recovering oil; however, their high operational costs limit their applicability on a larger scale. For instance, physical methods such as centrifugation effectively segregate the distinct phases within POS facilitating the recovery of oil and water. Thermal method like incineration, gasification, and pyrolysis offers efficient destruction of organic compounds and resource recovery through energy generation. Moreover, certain methods like froth flotation and centrifugation are more suitable for the managing POS with elevated moisture levels while the thermochemical methods require pre-treatment to decrease the moisture level of the POS. Thus, choosing the most optimal treatment approach for POS depends on several factors including the composition and characteristics of the POS, disposal standard, treatment cost, and its capacity. Due to the diverse criteria involved in technology selection, it becomes challenging to determine how well the available methods work overall. Consequently, an integrated approach has been explored in the advancement of POS treatment technologies. Thus, a combination of multiple approaches not only treats the POS effectively but also offers a solution to the limitations that arises while using a single treatment method. As a result, there has been a recent focus on exploring combinations of different techniques to enhance efficiency and maximize the recovery of valuable resources. Moreover, several methods have been developed to dispose of the residue after proper treatment of the POS. Biological methods such as bioremediation, landfarming, and landfilling utilized microorganisms to degrade hydrocarbons and mitigate the environmental footprint of POS. However, the different disposal methods employed across various refinery industries have its own limitations. For instance, S/S treatment and incineration methods are expensive. Besides, incineration process poses a risk for air pollution. Additionally, the utilization of land farming and filling techniques may not yield optimal results and can potentially result in the leaching of harmful heavy metals. Moreover, these methods are time-consuming in nature and demand large land areas. Therefore, it is crucial to devise an innovative technique which is environmentally sustainable and economically feasible for the proper disposal of this environmentally detrimental product arising from crude oil utilization.

Although the beneficial utilization of the POS seems promising, an in-depth study is necessary to ascertain whether the economic value derived from recovered fuels or materials outweighs the supplementary costs incurred for the additional treatment processes. Moreover, it is imperative to identify effective solutions for the diverse challenges encountered by various treatment technologies. Addressing the unique issues faced by each technology is essential for fostering advancements in the field and ensuring the development of robust and adaptable solutions for the hazardous

Table 7.2 Comparison of treatment technologies for management of petroleum oily sludge

Methods/ Technology	Process summary	Outcome and value-added products	Properties	Efficiency (%)
Solvent extraction	Use of solvents to dissolve and separate components of POS	Efficient removal of hydrocarbons; recovered oil for reuse or refining	Clean, refined oil suitable for processing	30–88
Pyrolysis	Thermal decomposition of POS in the oxygen free environment	Decomposition of sludge into syngas, biochar, and oil fractions	Syngas for energy; biochar for soil as carbon-rich soil conditioner	50–90
Surfactants	Emulsion destabilization using surfactants	Enhanced emulsion destabilization; improved oil recovery and reduced environmental impact	Reduced viscosity of recovered oil	70–90
Centrifugation	Rotational separation of oil from POS based on particle density	High oil recovery rate and efficient separation of solids and liquids; reduced sludge volume	Dry solids; liquid suitable for treatment	40–70
Freezing & thawing	Cyclic freezing and thawing to break down POS	Physical disintegration of sludge; improved subsequent treatments	Disrupted sludge for easier processing	50–70
Ultrasonication	Use of high frequency ultrasonic waves to enhance oil recovery from POS	Breakdown of sludge particles; improved separation and oil recovery rate and reduced environmental impact	Smaller, more manageable sludge particles	45–70
Electrokinetic	Use of electric fields to enhance oil recovery from POS by mobilizing ions	Enhanced migration of charged particles; improved separation	Recovered oil has high purity and can be used as fuel or as a raw material for other products	50–70
Froth flotation	Use of air bubbles to separate hydrophobic particles (oil) from POS	Effective removal of oily components; reduced sludge volume	Concentrated oil-rich froth for recovery	55–70

Ma et al. (2014), Khan et al. (2019), Da Silva et al. (2019), Gao et al. (2018), Wang et al. (2018b), Kariminezhad and Elektorowicz (2018), Hou et al. (2018), Kim et al. (2019), Lin et al. (2017b), Cheng et al. (2017), Taleghani et al. (2019), Shen et al. (2016), Feng et al. (2017), Liu et al. (2018b), Hui et al. (2020), Zubaidy and Abouelnasr (2010), Sahebnazar et al. (2018), Zhao et al. (2020), Huang et al. (2014), Shao et al. (2014), Liang et al. (2014), Ramirez and Collins (2018)

POS. The following points highlights some of the problems and its possible solutions:

- The considerable obstacle of high operational costs poses a limitation for the adoption of novel technologies for management of refinery oily sludge or POS. Thus, future research and development should aim at reducing the cost associated with these effective technologies. This endeavor is essential to not only mitigate operational expenses but also to enhance the scalability and broader applicability of these innovative solutions on a larger scale. Addressing the economic aspect of new technologies is pivotal for their successful integration into the existing framework of POS management in the refinery industry.
- The complexity of POS hinders the widespread applicability of certain treatment technologies. Therefore, the development and exploration of adaptable treatment solutions are crucial to overcome these obstacles and ensure effective management of POS on a broader scale.
- Future studies should further explore synergies between various treatment methods to develop an integrated approach that overcomes limitations associated with standalone techniques for treating POS and at the same time explore its feasibility of treating POS with diverse characteristics.
- Development of environmentally sustainable disposal methods for the treated residue of refinery POS with a focus on mitigating air and soil pollution is necessary. This involves the creation of disposal techniques that not only adhere to environmental standards but also actively contribute to minimizing the ecological footprint associated with the management of POS.

References

Abouee Mehrizi, E., Kermani, M., Farzadkia, M., Esarfili, A., & Ghorbanian, M. (2019). Study of improvement of bioremediation performance for the degradation of petroleum hydrocarbons in oily sludge by a chemical pretreatment strategy. *Journal of Material Cycles and Waste Management, 21*, 1052–1062.

Abuhasel, K., Kchaou, M., Alquraish, M., Munusamy, Y., & Jeng, Y. T. (2021). Oily wastewater treatment: Overview of conventional and modern methods, challenges, and future opportunities. *Water, 13*(7), 980.

Agrawal, N., Kumar, V., & Shahi, S. K. (2021). Biodegradation and detoxification of phenanthrene in in-vitro and in-vivo conditions by a newly isolated ligninolytic fungus *Coriolopsis byrsina* strain APC5 and characterization of their metabolites for environmental safety. *Environmental Science and Pollution Research, 29*, 61767–61782.

Aguelmous, A., El Fels, L., Souabi, S., et al. (2018). Petroleum sludge bioremediation and its toxicity removal by landfill in gunder semi-arid conditions. *Ecotoxicology and Environmental Safety, 166*, 482–487. https://doi.org/10.1016/j.ecoenv.2018.09.106

Aguelmous, A., El Fels, L., Souabi, S., Zamama, M., & Hafidi, M. (2019). The fate of total petroleum hydrocarbons during oily sludge composting: A critical review. *Reviews in Environmental Science and Bio/Technology, 18*, 473–493.

Akinyemi, O. (2020). Recent development on biological treatment of oil sludge from petroleum industry. *American Journal of Sciences and Engineering Research, 3*(3).

AL-Doury, M. M. I. (2019). Treatment of oily sludge produced from Baiji oil refineries using surfactants. *Petroleum Science and Technology, 37*(6), 718–726.

Al-Futaisi, A., Jamrah, A., Yaghi, B., & Taha, R. (2007). Assessment of alternative management techniques of tank bottom petroleum sludge in Oman. *Journal of Hazardous Materials, 141*(3), 557–564.

Ali, H., Khan, E., & Ilahi, I. (2019). Environmental chemistry and ecotoxicology of hazardous heavy metals: Environmental persistence, toxicity, and bioaccumulation. *Journal of Chemistry, 2019*, 1.

Arena, U. (2012). Process and technological aspects of municipal solid waste gasification. A review. *Waste Management, 32*(4), 625–639.

Arif, N., Yadav, V., Singh, S., Singh, S., Ahmad, P., Mishra, R. K., et al. (2016). Influence of high and low levels of plant-beneficial heavy metal ions on plant growth and development. *Frontiers in Environmental Science, 4*, 69.

Asim, N., Badiei, M., Torkashvand, M., Mohammad, M., Alghoul, M. A., Gasaymeh, S. S., & Sopian, K. (2021). Wastes from the petroleum industries as sustainable resource materials in construction sectors: Opportunities, limitations, and directions. *Journal of Cleaner Production, 284*, 125459. https://doi.org/10.1016/j.jclepro.2020.125459

Ayotamuno, M. J., Okparanma, R. N., Nweneka, E. K., et al. (2007). Bio-remediation of a sludge containing hydrocarbons. *Applied Energy, 84*, 936–943. https://doi.org/10.1016/j.apenergy.2007.02.007

Bao, Q., Huang, L., Xiu, J., Yi, L., & Ma, Y. (2021). Study on the treatment of oily sludge in oil fields with lipopeptide/sophorolipid complex bio-surfactant. *Ecotoxicology and Environmental Safety, 212*, 111964.

Bermejo, M. D., & Cocero, M. J. (2006). Supercritical water oxidation: A technical review. *AICHE Journal, 52*(11), 3933–3951.

Bhattacharyya, J. K., & Shekdar, A. V. (2003). Treatment and disposal of refinery sludges: Indian scenario. *Waste Management & Research, 21*(3), 249–261.

Borah, G., & Deka, H. (2023). Crude oil associated heavy metals (HMs) contamination in agricultural land: Understanding risk factors and changes in soil biological properties. *Chemosphere, 310*, 136890.

Cambiella, A., Benito, J. M., Pazos, C., & Coca, J. (2006). Centrifugal separation efficiency in the treatment of waste emulsified oils. *Chemical Engineering Research and Design, 84*(1), 69–76.

Cao, Y., Liu, Y., Li, Z., Zong, P., Hou, J., Zhang, Q., & Gou, X. (2022). Synergistic effect, kinetics, and pollutant emission characteristics of co-combustion of polymer-containing oily sludge and cornstalk using TGA and fixed-bed reactor. *Renewable Energy, 185*, 748–758.

Chakravarty, P., Chowdhury, D., & Deka, H. (2022). Ecological risk assessment of priority PAHs pollutants in crude oil contaminated soil and its impacts on soil biological properties. *Journal of Hazardous Materials, 437*, 129325.

Chen, H., Wang, X., Liang, H., Chen, B., Liu, Y., Ma, Z., & Wang, Z. (2023). Characterization and treatment of oily sludge: A systematic review. *Environmental Pollution*, 123245.

Cheng, S., Wang, Y., Fumitake, T., Kouji, T., Li, A., & Kunio, Y. (2017). Effect of steam and oil sludge ash additive on the products of oil sludge pyrolysis. *Applied Energy, 185*, 146–157.

Chu, Z., Gong, Z., Wang, Z., Zhang, H., Liu, L., Wu, J., & Wang, J. (2021). Experimental study on gasification of oil sludge with steam and its char characteristic. *Journal of Hazardous Materials, 416*, 125713.

Chu, Z., Li, Y., Zhang, C., Fang, Y., & Zhao, J. (2023). A review on resource utilization of oil sludge based on pyrolysis and gasification. *Journal of Environmental Chemical Engineering*, 109692.

Chung, H. I., & Kamon, M. (2005). Ultrasonically enhanced electrokinetic remediation for removal of Pb and phenanthrene in contaminated soils. *Engineering Geology, 77*(3–4), 233–242.

da Silva, D. C., dos Santos Lucas, C. R., de Moraes Juviniano, H. B., de Alencar Moura, M. C. P., de Castro Dantas, T. N., & Neto, A. A. D. (2019). Analysis of the use of microemulsion systems to treat petroleum sludge from a water flotation unit. *Journal of Environmental Chemical Engineering, 7*(1), 102934.

De Lasa, H., Salaices, E., Mazumder, J., & Lucky, R. (2011). Catalytic steam gasification of biomass: Catalysts, thermodynamics and kinetics. *Chemical Reviews, 111*(9), 5404–5433.

Deng, S., Wang, X., Tan, H., Mikulčić, H., Yang, F., Li, Z., & Duić, N. (2016). Thermogravimetric study on the co-combustion characteristics of oily sludge with plant biomass. *Thermochimica Acta, 633*, 69–76.

dos Santos, R. G., & Alencar, A. C. (2020). Biomass-derived syngas production via gasification process and its catalytic conversion into fuels by Fischer Tropsch synthesis: A review. *International Journal of Hydrogen Energy, 45*(36), 18114–18132.

Feng, W., Yin, Y., de Lourdes Mendoza, M., Wang, L., Chen, X., Liu, Y., et al. (2017). Freeze-thaw method for oil recovery from waste cutting fluid without chemical additions. *Journal of Cleaner Production, 148*, 84–89.

Gan, S., Lau, E. V., & Ng, H. K. (2009). Remediation of soils contaminated with polycyclic aromatic hydrocarbons (PAHs). *Journal of Hazardous Materials, 172*(2–3), 532–549.

Gao, Y., Ding, R., Wu, S., Wu, Y., Zhang, Y., & Yang, M. (2015). Influence of ultrasonic waves on the removal of different oil components from oily sludge. *Environmental Technology, 36*(14), 1771–1775.

Gao, Y. X., Ding, R., Chen, X., Gong, Z. B., Zhang, Y., & Yang, M. (2018). Ultrasonic washing for oily sludge treatment in pilot scale. *Ultrasonics, 90*, 1–4.

Gao, N., Li, J., Quan, C., & Tan, H. (2020). Product property and environmental risk assessment of heavy metals during pyrolysis of oily sludge with fly ash additive. *Fuel, 266*, 117090.

Geier, M. C., Chlebowski, A. C., Truong, L., Massey Simonich, S. L., Anderson, K. A., & Tanguay, R. L. (2018). Comparative developmental toxicity of a comprehensive suite of polycyclic aromatic hydrocarbons. *Archives of Toxicology, 92*, 571–586.

Gong, Z., Wang, Z., & Wang, Z. (2018). Study on migration characteristics of heavy metals during oil sludge incineration. *Petroleum Science and Technology, 36*(6), 469–474.

Gonzalez, A. M., Lora, E. E. S., & Palacio, J. C. E. (2019). Syngas production from oil sludge gasification and its potential use in power generation systems: An energy and exergy analysis. *Energy, 169*, 1175–1190.

Guolin, J., Mingming, L., & Tingting, C. (2011). Prospects for development of oily sludge treatment. *Chemistry and Technology of Fuels and Oils, 47*(4), 312–326.

Hanxuan, S., Yan, Y., Weiru, Z., Bibiche, E. E. A. F., Qingwen, Z., & Jixiang, G. (2023). Synthesis of nano-β-CD@ Fe3O4 magnetic material and its application in ultrasonic treatment of oily sludge. *Ultrasonics Sonochemistry, 92*, 106256.

Heidarzadeh, N., Gitipour, S., & Abdoli, M. A. (2010). Characterization of oily sludge from a Tehran oil refinery. *Waste Management & Research, 28*(10), 921–927.

Hejazi, R. F., Husain, T., & Khan, F. I. (2003). Landfarming operation of oily sludge in arid region—Human health risk assessment. *Journal of Hazardous Materials, 99*(3), 287–302.

Hentati, D., Abed, R. M., Abotalib, N., El Nayal, A. M., Ashraf, I., & Ismail, W. (2022). Biotreatment of oily sludge by a bacterial consortium: Effect of bioprocess conditions on biodegradation efficiency and bacterial community structure. *Frontiers in Microbiology, 13*, 998076.

Hosseini, M. K., Liu, L., Hosseini, P. K., Lee, K., & Miao, J. (2023). Performance evaluation of a pilot-scale membrane filtration system for oily wastewater treatment: CFD modeling and scale-up design. *Journal of Water Process Engineering, 52*, 103570.

Hou, Y., Qi, S., You, H., Huang, Z., & Niu, Q. (2018). The study on pyrolysis of oil-based drilling cuttings by microwave and electric heating. *Journal of Environmental Management, 228*, 312–318.

Hu, G., Li, J., & Zeng, G. (2013). Recent development in the treatment of oily sludge from petroleum industry: A review. *Journal of Hazardous Materials, 261*, 470–490.

Hu, G., Li, J., & Hou, H. (2015). A combination of solvent extraction and freeze thaw for oil recovery from petroleum refinery wastewater treatment pond sludge. *Journal of Hazardous Materials, 283*, 832–840.

Hu, G., Li, J., Huang, S., & Li, Y. (2016). Oil recovery from petroleum sludge through ultrasonic assisted solvent extraction. *Journal of Environmental Science and Health, Part A, 51*(11), 921–929.

Hu, G., Feng, H., He, P., Li, J., Hewage, K., & Sadiq, R. (2020). Comparative life-cycle assessment of traditional and emerging oily sludge treatment approaches. *Journal of Cleaner Production, 251*, 119594.

Huang, Q., Han, X., Mao, F., Chi, Y., & Yan, J. (2014). A model for predicting solid particle behavior in petroleum sludge during centrifugation. *Fuel, 117*, 95–102.

Hui, K., Tang, J., Lu, H., Xi, B., Qu, C., & Li, J. (2020). Status and prospect of oil recovery from oily sludge: A review. *Arabian Journal of Chemistry, 13*(8), 6523–6543.

Islam, B. (2015). Petroleum sludge, its treatment and disposal: A review. *International Journal of Chemical Science, 13*(4), 1584–1602.

Jafarinejad, S. (2017). Solid-waste management in the petroleum industry. *Petroleum Waste Treatment and Pollution Control*, 269–345.

Jagaba, A. H., Kutty, S. R. M., Lawal, I. M., Aminu, N., Noor, A., Al-dhawi, B. N. S., et al. (2022). Diverse sustainable materials for the treatment of petroleum sludge and remediation of contaminated sites: A review. *Cleaner Waste Systems, 2*, 100010. https://doi.org/10.1016/j.clwas.2022.100010

Jahromi, A. F., & Elektorowicz, M. (2018). Electrokinetically assisted oil-water phase separation in oily sludge with implementing novel controller system. *Journal of Hazardous Materials, 358*, 434–440.

Jasim, M. A., & AlJaberi, F. Y. A. (2023). Treatment of oily wastewater by electrocoagulation technology: A general review (2018-2022). *Journal of Electrochemical Science and Engineering, 13*(2), 361–372.

Jasmine, J., & Mukherji, S. (2015). Characterization of oily sludge from a refinery and biodegradability assessment using various hydrocarbon degrading strains and reconstituted consortia. *Journal of Environmental Management, 149*, 118–125.

Jean, D. S., Lee, D. J., & Wu, J. C. S. (1999). Separation of oil from oily sludge by freezing and thawing. *Water Research, 33*(7), 1756–1759.

Jean, D. S., Chu, C. P., & Lee, D. J. (2001). Freeze/thaw treatment of oily sludge from petroleum refinery plant. *Separation Science and Technology, 36*(12), 2733–2746.

Jerez, S., Ventura, M., Molina, R., Pariente, M. I., Martínez, F., & Melero, J. A. (2021). Comprehensive characterization of an oily sludge from a petrol refinery: A step forward for its valorization within the circular economy strategy. *Journal of Environmental Management, 285*, 112124.

Jiang, G., Xu, D., Hao, B., Liu, L., Wang, S., & Wu, Z. (2021). Thermochemical methods for the treatment of municipal sludge. *Journal of Cleaner Production, 311*, 127811.

Johnson, O. A., & Affam, A. C. (2019). Petroleum sludge treatment and disposal: A review. *Environmental Engineering Research, 24*(2), 191–201.

Juwarkar, A. A., Singh, S. K., & Mudhoo, A. (2010). A comprehensive overview of elements in bioremediation. *Reviews in Environmental Science and Bio/Technology, 9*, 215–288.

Kamyab, H., Yuzir, A., Ashokkumar, V., Hosseini, S. E., Balasubramanian, B., & Kirpichnikova, I. (2022). Review of the application of gasification and combustion technology and waste-to-energy technologies in sewage sludge treatment. *Fuel, 316*, 123199.

Karamalidis, A. K., & Voudrias, E. A. (2007). Cement-based stabilization/solidification of oil refinery sludge: Leaching behavior of alkanes and PAHs. *Journal of Hazardous Materials, 148*(1–2), 122–135.

Kariminezhad, E., & Elektorowicz, M. (2018). Effect of various electrokinetic treatment regimes on solids surface properties and thermal behavior of oil sediments. *Journal of Hazardous Materials, 353*, 227–235.

Kasinski, S., Slota, M., Markowski, M., & Kaminska, A. (2016). Municipal waste stabilization in a reactor with an integrated active and passive aeration system. *Waste Management, 50*, 31–38.
Khan, F. I., Husain, T., & Hejazi, R. (2004). An overview and analysis of site remediation technologies. *Journal of Environmental Management, 71*(2), 95–122.
Khan, M. K., Cahyadi, H. S., Kim, S. M., & Kim, J. (2019). Efficient oil recovery from highly stable toxic oily sludge using supercritical water. *Fuel, 235*, 460–472.
Khazaal, R. M., & Ismail, Z. Z. (2021). Bioremediation and detoxification of real refinery oily sludge using mixed bacterial cells. *Petroleum Research, 6*(3), 303–308. https://doi.org/10.1016/j.ptlrs.2021.05.003
Kim, J. H., Oh, J. I., Baek, K., Park, Y. K., Zhang, M., Lee, J., & Kwon, E. E. (2019). Thermolysis of crude oil sludge using CO2 as reactive gas medium. *Energy Conversion and Management, 186*, 393–400.
Koolivand, A., Rajaei, M. S., Ghanadzadeh, M. J., Saeedi, R., Abtahi, H., & Godini, K. (2017). Bioremediation of storage tank bottom sludge by using a two-stage composting system: Effect of mixing ratio and nutrients addition. *Bioresource Technology, 235*, 240–249.
Kulikowska, D. (2016). Kinetics of organic matter removal and humification progress during sewage sludge composting. *Waste Management, 49*, 196–203.
Kumar, V., Shahi, S. K., & Singh, S. (2018). Bioremediation: An eco-sustainable approach for restoration of contaminated sites. In J. Singh, D. Sharma, G. Kumar, & N. Sharma (Eds.), *Microbial bioprospecting for sustainable development*. Springer. https://doi.org/10.1007/978-981-13-0053-0_6
Kumar, V., Agrawal, S., Bhat, S. A., Américo-Pinheiro, J. H. P., Shahi, S. K., & Kumar, S. (2022a). Environmental impact, health hazards, and plant-microbes synergism in remediation of emerging contaminants. *Cleaner Chemical Engineering, 2*, 100030.
Kumar, V., Agrawal, S., Shahi, S. K., Motghare, A., Singh, S., & Ramamurthy, P. C. (2022b). Bioremediation potential of newly isolated *Bacillus albus* strain VKDS9 for decolourization and detoxification of biomethanated distillery effluent and its metabolites characterization for environmental sustainability. *Environmental Technology & Innovation, 26*, 102260.
Li, X., Zhang, F., Guan, B., Sun, J., & Liao, G. (2020). Review on oily sludge treatment technology. In *IOP conference series: Earth and environmental science* (Vol. 467, No. 1, p. 012173). IOP Publishing.
Li, W., Lin, H., Yang, Y., Shang, Z., Li, Q., Ma, Y., et al. (2021). Enhanced separation of oil and solids in oily sludge by froth flotation at normal temperature. *PRO, 9*(12), 2163.
Li, L., Cao, W., Wang, G., Peng, P., Liu, S., Jin, H., et al. (2022a). Experimental and kinetic study of heavy metals transformation in supercritical water gasification of oily sludge. *Journal of Cleaner Production, 373*, 133898.
Li, L., Cao, W., Peng, P., Wang, G., Liu, S., Jin, H., et al. (2022b). Distribution, risk assessment and stabilization of heavy metals in supercritical water gasification of oily sludge. *Process Safety and Environmental Protection, 168*, 591–600.
Li, W., Ma, Y., Feng, X., Alheshibri, M., Bu, X., Ma, G., et al. (2023). Optimization of an oil recovery process from oily sludge using a combined technique of froth flotation and centrifugal treatment. *Journal of Cleaner Production, 400*, 136752.
Liang, J., Zhao, L., Du, N., Li, H., & Hou, W. (2014). Solid effect in solvent extraction treatment of pre-treated oily sludge. *Separation and Purification Technology, 130*, 28–33.
Lin, Z., Wang, W., & Huang, R. (2017a). Study of oily sludge treatment by centrifugation. *Desalination and Water Treatment, 68*, 99–106.
Lin, B., Wang, J., Huang, Q., Ali, M., & Chi, Y. (2017b). Aromatic recovery from distillate oil of oily sludge through catalytic pyrolysis over Zn modified HZSM-5 zeolites. *Journal of Analytical and Applied Pyrolysis, 128*, 291–303.
Lin, B., Huang, Q., & Chi, Y. (2018). Co-pyrolysis of oily sludge and rice husk for improving pyrolysis oil quality. *Fuel Processing Technology, 177*, 275–282.

Lin, Z., Xu, F., Wang, L., Hu, L., Zhu, L., Tan, J., et al. (2021). Characterization of oil component and solid particle of oily sludge treated by surfactant-assisted ultrasonication. *Chinese Journal of Chemical Engineering, 34*, 53–60.

Lin, F., Yu, H., Li, J., Zygourakis, K., Li, R., Cheng, Z., et al. (2023). Investigation on the interaction between oil compositions and soil minerals with the targets of resource recovery and harmless disposal of oily sludges by pyrolysis. *ACS ES&T Engineering, 3*(5), 734–744.

Liu, C., Liu, J., Sun, G., Xie, W., Kuo, J., Li, S., et al. (2018a). Thermogravimetric analysis of (co-) combustion of oily sludge and litchi peels: Combustion characterization, interactions and kinetics. *Thermochimica Acta, 667*, 207–218.

Liu, C., Zhang, Y., Sun, S., Huang, L., Yu, L., Liu, X., et al. (2018b). Oil recovery from tank bottom sludge using rhamnolipids. *Journal of Petroleum Science and Engineering, 170*, 14–20.

Liu, X., Yao, T., Lai, R., Xiu, J., Huang, L., Sun, S., et al. (2019). Recovery of crude oil from oily sludge in an oilfield by sophorolipid. *Petroleum Science and Technology, 37*(13), 1582–1588. https://doi.org/10.1080/10916466.2019.1594286

Liu, Y., Yu, H., Jiang, Z., Song, Y., Zhang, T., Siyal, A. A., et al. (2021a). Microwave pyrolysis of oily sludge under different control modes. *Journal of Hazardous Materials, 416*, 125887.

Liu, W. J., Shao, Z. G., & Xu, Y. (2021b). Emission characteristics of nitrogen and sulfur containing pollutants during the pyrolysis of oily sludge with and without catalysis. *Journal of Hazardous Materials, 401*, 123820.

Lukić, B., Panico, A., Huguenot, D., Fabbricino, M., van Hullebusch, E. D., & Esposito, G. (2017). A review on the efficiency of landfarming integrated with composting as a soil remediation treatment. *Environmental Technology Reviews, 6*(1), 94–116.

Luo, X., Gong, H., He, Z., Zhang, P., & He, L. (2020). Research on mechanism and characteristics of oil recovery from oily sludge in ultrasonic fields. *Journal of Hazardous Materials, 399*, 123137.

Ma, Z., Gao, N., Xie, L., & Li, A. (2014). Study of the fast pyrolysis of oilfield sludge with solid heat carrier in a rotary kiln for pyrolytic oil production. *Journal of Analytical and Applied Pyrolysis, 105*, 183–190.

Ma, J., Yang, Y., Dai, X., Chen, Y., Deng, H., Zhou, H., et al. (2016). Effects of adding bulking agent, inorganic nutrient and microbial inocula on biopile treatment for oil-field drilling waste. *Chemosphere, 150*, 17–23.

Maila, M. P., & Cloete, T. E. (2004). Bioremediation of petroleum hydrocarbons through landfarming: Are simplicity and cost-effectiveness the only advantages? *Reviews in Environmental Science and Bio/Technology, 3*, 349–360.

Makan, A. (2015). Windrow co-composting of natural casings waste with sheep manure and dead leaves. *Waste Management, 42*, 17–22.

Makarichi, L., Jutidamrongphan, W., & Techato, K. A. (2018). The evolution of waste-to-energy incineration: A review. *Renewable and Sustainable Energy Reviews, 91*, 812–821.

Malviya, R., & Chaudhary, R. (2006). Factors affecting hazardous waste solidification/stabilization: A review. *Journal of Hazardous Materials, 137*(1), 267–276.

Marin, J. A., Hernandez, T., & Garcia, C. (2005). Bioremediation of oil refinery sludge by landfarming in semiarid conditions: Influence on soil microbial activity. *Environmental Research, 98*(2), 185–195.

Mathew, A. K., Abraham, A., Mallapureddy, K. K., & Sukumaran, R. K. (2018). Lignocellulosic biorefinery wastes, or resources? In *Waste biorefinery* (pp. 267–297). Elsevier.

Mishra, H., Karmakar, S., Kumar, R., & Singh, J. (2017). A framework for assessing uncertainty associated with human health risks from MSW landfill leachate contamination. *Risk Analysis, 37*(7), 1237–1255.

Mokhtar, N. M., Ethaib, S., & Omar, R. (2018). Effects of microwave absorbers on the products of microwave pyrolysis of oily sludge. *Journal of Engineering Science and. Technology, 13*(10), 3313–3330.

Molina, L., & Segura, A. (2021). Biochemical and metabolic plant responses toward polycyclic aromatic hydrocarbons and heavy metals present in atmospheric pollution. *Plants, 10*(11), 2305.

Moltó, J., Barneto, A. G., Ariza, J., & Conesa, J. A. (2013). Gas production during the pyrolysis and gasification of biological and physico-chemical sludges from oil refinery. *Journal of Analytical and Applied Pyrolysis, 103*, 167–172.

Mulligan, C. N. (2009). Recent advances in the environmental applications of biosurfactants. *Current Opinion in Colloid & Interface Science, 14*(5), 372–378.

Murshid, N., Kamil, N. A. F. M., & Kadir, A. A. (2018, April). Treatment of petroleum sludge by using solidification/stabilization (S/S) method: Effect of hydration days to heavy metals leaching and strength. In *IOP conference series: Earth and environmental science* (Vol. 140, No. 1, p. 012054). IOP Publishing.

Murungi, P. I., & Sulaimon, A. A. (2022). Petroleum sludge treatment and disposal techniques: A review. *Environmental Science and Pollution Research, 29*(27), 40358–40372.

Nezhdbahadori, F., Abdoli, M. A., Baghdadi, M., & Ghazban, F. (2018). A comparative study on the efficiency of polar and non-polar solvents in oil sludge recovery using solvent extraction. *Environmental Monitoring and Assessment, 190*, 1–9.

Niu, A., Sun, X., & Lin, C. (2022). Trend in research on characterization, environmental impacts and treatment of oily sludge: A systematic review. *Molecules, 27*(22), 7795.

Olasehinde, T. A., & Olaniran, A. O. (2022). Neurotoxicity of polycyclic aromatic hydrocarbons: A systematic mapping and review of neuropathological mechanisms. *Toxics, 10*(8), 417.

Ossai, I. C., Ahmed, A., Hassan, A., & Hamid, F. S. (2020). Remediation of soil and water contaminated with petroleum hydrocarbon: A review. *Environmental Technology & Innovation, 17*, 100526.

Philemon, Z. B. O., & Benoît, N. M. (2013). Treatment of conditioned oily sludge from Cameroon petroleum refinery by centrifugation. *International Journal of Environmental Sciences, 3*(5), 1373–1382.

Pilli, S., Bhunia, P., Yan, S., LeBlanc, R. J., Tyagi, R. D., & Surampalli, R. Y. (2011). Ultrasonic pretreatment of sludge: A review. *Ultrasonics Sonochemistry, 18*(1), 1–18.

Prakash, V., Saxena, S., Sharma, A., Singh, S., & Singh, S. K. (2015). Treatment of oil sludge contamination by composting. *Journal of Bioremediation & Biodegradation, 6*(3), 1–6.

Rabani, M. S., Habib, A., & Gupta, M. K. (2020). Polycyclic aromatic hydrocarbons: Toxic effects and their bioremediation strategies. In *Bioremediation and biotechnology, Vol 4: Techniques for noxious substances remediation* (pp. 65–105). Springer Nature.

Rajendran, N., Gurunathan, B., Han, J., Krishna, S., Ananth, A., Venugopal, K., & Priyanka, R. S. (2021). Recent advances in valorization of organic municipal waste into energy using biorefinery approach, environment and economic analysis. *Bioresource Technology, 337*, 125498.

Raklami, A., Oubane, M., Meddich, A., Hafidi, M., Marschner, B., Heinze, S., & Oufdou, K. (2021). Phytotoxicity and genotoxicity as a new approach to assess heavy metals effect on Medicago sativa L.: Role of metallo-resistant rhizobacteria. *Environmental Technology & Innovation, 24*, 101833.

Ramirez, D., & Collins, C. D. (2018). Maximisation of oil recovery from an oil-water separator sludge: Influence of type, concentration, and application ratio of surfactants. *Waste Management, 82*, 100–110.

Rivas, F. J. (2006). Polycyclic aromatic hydrocarbons sorbed on soils: A short review of chemical oxidation-based treatments. *Journal of Hazardous Materials, 138*(2), 234–251.

Rondon-Afanador, C., Pinilla-Meza, G., Casallas-Cuervo, F. C., Diaz-Vanegas, C., Barreto-Gomez, D., Benavides, C., et al. (2023). Bioremediation of heavy oily sludge: A microcosms study. *Biodegradation, 34*(1), 1–20.

Roy, A., Pal, S., Kazy, S. K., Sarkar, P., Sar, P., & Ghoshal, A. K. (2014). Characterization of culturable bacterial communities in petroleum hydrocarbon contaminated sludge of oil refineries and oil exploration sites. *Journal of Environmental Research and Development, 8*(3), 451.

Roy, A., Sar, P., Sarkar, J., Dutta, A., Sarkar, P., Gupta, A., et al. (2018). Petroleum hydrocarbon rich oil refinery sludge of North-East India harbours anaerobic, fermentative, sulfate-reducing, syntrophic and methanogenic microbial populations. *BMC Microbiology, 18*(1), 1–22.

Sahebnazar, Z., Mowla, D., Karimi, G., & Yazdian, F. (2018). Zero-valent iron nanoparticles assisted purification of rhamnolipid for oil recovery improvement from oily sludge. *Journal of Environmental Chemical Engineering, 6*(1), 917–922.

Sani, S., Dashti, A. F., & Adnan, R. (2020). Applications of Fenton oxidation processes for decontamination of palm oil mill effluent: A review. *Arabian Journal of Chemistry, 13*(10), 7302–7323.

Santiago, Y. C., González, A. M., Venturini, O. J., & Maya, D. M. Y. (2021). Assessment of the energy recovery potential of oil sludge through gasification aiming electricity generation. *Energy, 215*, 119210.

Sattar, S., Hussain, R., Shah, S. M., Bibi, S., Ahmad, S. R., Shahzad, A., et al. (2022). Composition, impacts, and removal of liquid petroleum waste through bioremediation as an alternative clean-up technology: A review. *Heliyon, 8*(10), 11101. https://doi.org/10.1016/j.heliyon.2022.e11101

Sayara, T., Sarrà, M., & Sánchez, A. (2010). Effects of compost stability and contaminant concentration on the bioremediation of PAHs-contaminated soil through composting. *Journal of Hazardous Materials, 179*(1–3), 999–1006.

Seo, S., Mastiani, M., Mosavati, B., Peters, D. M., Mandin, P., & Kim, M. (2018). Performance evaluation of environmentally benign nonionic biosurfactant for enhanced oil recovery. *Fuel, 234*, 48–55.

Shahbaz, M., Rashid, N., Saleem, J., Mackey, H., McKay, G., & Al-Ansari, T. (2023). A review of waste management approaches to maximise sustainable value of waste from the oil and gas industry and potential for the State of Qatar. *Fuel, 332*, 126220. https://doi.org/10.1016/j.fuel.2022.126220

Shao, J., Lin, H., & Zhang, K. (2014). Swarm robots reinforcement learning convergence accuracy-based learning classifier systems with gradient descent (XCS-GD). *Neural Computing and Applications, 25*(2), 263–268.

Shen, Y., Chen, X., Wang, J., Ge, X., & Chen, M. (2016). Oil sludge recycling by ash-catalyzed pyrolysis-reforming processes. *Fuel, 182*, 871–878.

Singha, W. J., Borah, G., & Deka, H. (2022). Physico-chemical, biological and heavy metal status of spent oil-contaminated soils in the vicinity of garages in and around Guwahati city, Assam, India. *Current Science (00113891), 123*(10), 1246.

Sinha, B., Roy, S., & Kumar, K. (2020). Bioremediation of oily sludge: A case base analysis to sustainable supply chain. *Resources, Environment and Sustainability, 2*, 100008. https://doi.org/10.1016/j.resenv.2020.100008

Snowden-Swan, L. J., Li, S., Jiang, Y., Thorson, M. R., Schmidt, A. J., Seiple, T. E., et al. (2022). *Wet waste hydrothermal liquefaction and biocrude upgrading to hydrocarbon fuels: 2021 state of technology (No. PNNL-32731)*. Pacific Northwest National Lab. (PNNL), Richland, WA (United States). https://doi.org/10.2172/1863608

Sonone, S. S., Jadhav, S., Sankhla, M. S., & Kumar, R. (2020). Water contamination by heavy metals and their toxic effect on aquaculture and human health through food chain. *Letters in Applied NanoBioScience, 10*(2), 2148–2166.

Syed-Hassan, S. S. A., Wang, Y., Hu, S., Su, S., & Xiang, J. (2017). Thermochemical processing of sewage sludge to energy and fuel: Fundamentals, challenges and considerations. *Renewable and Sustainable Energy Reviews, 80*, 888–913.

Taleghani, S. T., Jahromi, A. F., & Elektorowicz, M. (2019). Electro-demulsification of water-in-oil suspensions enhanced with implementing various additives. *Chemosphere, 233*, 157–163.

Teng, Q., Zhang, D., & Yang, C. (2021). A review of the application of different treatment processes for oily sludge. *Environmental Science and Pollution Research, 28*, 121–132.

Tian, Y., McGill, W. B., Whitcombe, T. W., & Li, J. (2019). Ionic liquid-enhanced solvent extraction for oil recovery from oily sludge. *Energy & Fuels, 33*(4), 3429–3438.

Tiwari, S., & Lata, C. (2018). Heavy metal stress, signaling, and tolerance due to plant-associated microbes: An overview. *Frontiers in Plant Science, 9*, 452.

Ubani, O., Atagana, H. I., & Thantsha, M. S. (2013). Biological degradation of oil sludge: A review of the current state of development. *African Journal of Biotechnology, 12*(47), 6544–6567.

Vadillo, V., Sánchez-Oneto, J., Portela, J. R., & Martinez de la Ossa, E. J. (2013). Problems in supercritical water oxidation process and proposed solutions. *Industrial & Engineering Chemistry Research, 52*(23), 7617–7629.
Wang, J., Liu, T. L., Huang, Q. X., Ma, Z. Y., Chi, Y., & Yan, J. H. (2017). Production and characterization of high quality activated carbon from oily sludge. *Fuel Processing Technology, 162*, 13–19.
Wang, J., Han, X., Huang, Q., Ma, Z., Chi, Y., & Yan, J. (2018a). Characterization and migration of oil and solids in oily sludge during centrifugation. *Environmental Technology, 39*(10), 1350–1358.
Wang, Z., Gu, S., & Zhou, L. (2018b). Research on the static experiment of super heavy crude oil demulsification and dehydration using ultrasonic wave and audible sound wave at high temperatures. *Ultrasonics Sonochemistry, 40*, 1014–1020.
Wang, Z., Gong, Z., Wang, Z., Li, X., & Chu, Z. (2021). Application and development of pyrolysis technology in petroleum oily sludge treatment. *Environmental Engineering Research, 26*(1).
Wang, L., Xu, Y., Zhao, Z., Zhang, D., Lin, X., Ma, B., & Zhang, H. (2022). Analysis of pyrolysis characteristics of oily sludge in different regions and environmental risk assessment of heavy metals in pyrolysis residue. *ACS Omega, 7*(30), 26265–26274.
Wang, G., Li, J., Li, X., Kou, J., Ge, Z., Li, L., et al. (2023). Experimental study on supercritical water gasification of oily sludge using a continuous two-step method. *Journal of Hazardous Materials, 455*, 131619.
Wei, Y. L., Lv, L., Yang, Z. G., & Gao, Z. Q. (2015). Advances on recycling technique for oily sludge. *Oilfield Chemistry, 32*(1), 151–158.
Xiao, W., Yao, X., & Zhang, F. (2019). Recycling of oily sludge as a roadbed material utilizing phosphogypsum-based cementitious materials. *Advances in Civil Engineering, 2019*, 2019.
Zhang, L., Xu, C. C., & Champagne, P. (2010). Overview of recent advances in thermo-chemical conversion of biomass. *Energy Conversion and Management, 51*(5), 969–982.
Zhang, J., Li, J., Thring, R. W., Hu, X., & Song, X. (2012). Oil recovery from refinery oily sludge via ultrasound and freeze/thaw. *Journal of Hazardous Materials, 203*, 195–203.
Zhang, J., Li, J., Thring, R., & Liu, L. (2013). Application of ultrasound and Fenton's reaction process for the treatment of oily sludge. *Procedia Environmental Sciences, 18*, 686–693.
Zhang, Q., Jiang, Q., Bai, Y., Li, H., Xue, J., Gao, Y., & Cheng, D. (2021). Optimization and mechanism of oily sludge treatment by a novel combined surfactants with activated-persulfate method. *Science of the Total Environment, 800*, 149525.
Zhao, M., Wang, X., Liu, D., Li, Z., Guo, S., Zhu, W., et al. (2020). Insight into essential channel effect of pore structures and hydrogen bonds on the solvent extraction of oily sludge. *Journal of Hazardous Materials, 389*, 121826.
Zheng, X., Ying, Z., Cui, J., Wang, B., Chen, J., & Zhang, Q. (2017). Simultaneous dewatering and recovering oil from high-viscosity oily sludge through the combination process of demulsification, viscosity reduction, and centrifugation. *Energy & Fuels, 31*(12), 14401–14407.
Zhu, L., Lin, Z., Tan, J., Hu, L., & Zhang, T. (2020). Application of hydrophobic ionic liquid [Bmmim][PF6] in solvent extraction for oily sludge. *Chinese Journal of Chemical Engineering, 28*(9), 2294–2300.
Zubaidy, E. A., & Abouelnasr, D. M. (2010). Fuel recovery from waste oily sludge using solvent extraction. *Process Safety and Environmental Protection, 88*(5), 318–326.

Chapter 8
Waste Management in COVID-19 Pandemic: Impacts, Challenges, and Perspectives

Manoj Kumar Srinivasan, Nivedha Jayaseelan, Briska Jifrina Premnath, Harish Krishnan, Kamalesh Balakumar Venkatesan, Saravanan Alamelu, and Sathish-Kumar Kamaraj

1 Introduction

The emergence of Coronavirus disease 2019 (COVID-19, also known as the coronavirus pandemic) has triggered an unprecedented level of instability within global social and economic systems, resulting in widespread changes across various industries. These changes encompass initiatives like social distancing measures, restructuring of public health systems, and adjustments in the production and distribution of hygiene products. According to a study by Zambrano-Monserrate et al. (2020), COVID-19 has both direct and indirect effects on the ecosystem (Zambrano-Monserrate et al., 2020). As of September 20, 2021, Worldometer.info reported a staggering 229,350,549 confirmed COVID-19 cases and over 4,706,669 fatalities (Info, 2020). Developing a vaccine for the virus is a laborious and intricate process, typically taking 10–15 years, and it ideally involves collaboration between the public and commercial sectors. During the vaccine manufacturing process, priority is given to safety over efficacy, requiring rigorous clinical trials (Sarkodie & Owusu, 2020). These trials are conducted in various phases, starting with the preclinical phase. Apart from vaccine distribution, managing an effective supply chain demands significant time and effort. Numerous flaws and deficiencies in the global socioeconomic, healthcare, and environmental sectors have been made clear by the COVID-19 pandemic (Sarkodie & Owusu, 2020). Positively, the pandemic's

M. K. Srinivasan · N. Jayaseelan · B. J. Premnath · H. Krishnan · K. B. Venkatesan · S. Alamelu
Department of Biochemistry and Biotechnology, Faculty of Science, Annamalai University, Chidambaram, TN, India

S.-K. Kamaraj (✉)
Instituto Politécnico Nacional (IPN)- Centro de Investigación en Ciencia Aplicada y Tecnología Avanzada (CICATA), Unidad Altamira, Manzano, Industrial Altamira, Altamira, Tamps, Mexico

V. Kumar et al. (eds.), *Environmental Engineering and Waste Management*,
https://doi.org/10.1007/978-3-031-58441-1_8

nationwide lockdowns have resulted in beaches, cleaner rivers, and skies throughout the world (Zambrano-Monserrate et al., 2020; Gardiner, 2020; Urban & Nakada., 2021). The negative effects of rising solid waste output and decreased recycling efforts, though, are causing more people to express concern as they may have medium- to long-term effects (Zambrano-Monserrate et al., 2020; Dey et al., 2023).

Both sanitation workers and policymakers are distressed as a result of the pandemic's disruption of the dynamics of trash generation (Mallapur, 2020). In many developing nations, improper waste management procedures have made it possible for the coronavirus to spread through waste-handling procedures. According to allegations, poor collecting practices may cause the virus to infect municipal solid waste (MSW), increasing the chance of transmission. As a result, good waste treatment and disposal increasingly play a crucial role in effective disaster response (Mallapur, 2020).

This involves the collection, storage, sorting, treatment, transportation, and appropriate disposal of waste, along with critical factors such as staff training, disinfection protocols, and staff safety, all of which are integral components of efficient waste management (UNEP, 2020). Managing MSW appropriately during COVID-19 has exacerbated challenges in dealing with Health and Medical Waste (H&MW). Estimates indicate a substantial increase in healthcare waste (HCW), with approximately 3.4 kg per person per day being generated. COVID-19-related HCW production has reached rates of 2.5 kilograms per day in low-income nations, ranging from 2.0 to 2.2 kg per bed per day in Mexico, 2.23 kg per bed per day in Indonesia, and 2.85 kg per day in Thailand. Approximately 15% of waste generated in HCW facilities is considered hazardous, while the remaining 85% is not (Skrzyniarz et al., 2022).

Guidelines for the pandemic's management of infectious and non-infectious medical waste have been developed by the World Health Organization (WHO). More than 80% of non-infectious HCW must be managed as part of MSW (World Health Organization, 2020b). In addition, the International Solid Waste Association (ISWA) has identified three overarching Solid Waste Management (SWM) priorities: maintaining the efficiency of treatment, recycling, and disposal facilities; preventing public health risks brought on by improper management; preventing contamination and cross-infections; modifying recycling practices; and guaranteeing the safe treatment and disposal of H&MW (ISWA, 2020).

Over time, the operations of MSW management are likely to face increasing challenges, which is just one of the numerous highly detrimental outcomes of the COVID-19 outbreak (Kulkarni & Anantharama, 2020). The WHO recommends that waste treatment and disposal methods in healthcare facilities incorporate effective elimination of the coronavirus using standard biocidal chemicals and controlled heat treatment (Kampf et al., 2020). Variations in the current waste management system, local environmental conditions, viral resilience, and proper treatment of MSW outside healthcare facilities must all be carefully considered. Developing nations are particularly vulnerable to coronavirus transmission due to improper waste management, insufficient use of personal protective equipment, and other adverse conditions (Mol & Caldas, 2020).

Ineffective SWM, including informal waste collectors who collect, sort, and classify recyclable materials from the general population, further compounds the risk of health issues in certain communities (Cruvinel et al., 2019). The entire population is in danger from improper treatment of household waste contaminated with the coronavirus that is produced by patients getting at-home care, municipal personnel, and other formal workers. Therefore, this study examines MSW management practices in both developed and developing nations during pandemic conditions, with a focus on sound waste and risk management (Thomson, 2020).

There are significant knowledge gaps in understanding the long-term environmental impact of COVID-19-related waste generation, specifically regarding the efficacy of recycling single-use plastics, the potential environmental hazards posed by improper disposal of personal protective equipment (PPE), and the effectiveness of different waste management strategies in preventing environmental pollution and health risks. More research is needed to evaluate the sustainability and environmental implications of current waste management practices, especially in developing countries.

The chapter comprehensively covers the impact of the COVID-19 pandemic on waste generation, emphasizing the surge in single-use plastic waste, challenges in waste disposal, and the need for sustainable waste management strategies. It explores the implications on the global waste management cycle, the varying government responses to the waste crisis, and the critical role of effective healthcare waste management systems. Additionally, the chapter provides recommendations for future policies and practices to address the escalating issue of single-use plastics, advocating for a holistic approach involving behavioral changes, institutional shifts, and innovative technological solutions.

2 COVID-19 and Waste Generation

The emergence of the novel coronavirus Severe Acute Respiratory Syndrome Coronavirus 2 (SARS-CoV-2), responsible for the COVID-19 pandemic that originated in late 2019, triggered a swift global spread of the virus. It manifested in a range of symptoms, spanning from mild to severe, encompassing fever, coughing, and respiratory distress (Osuchowski et al., 2020). Its transmission occurred primarily through respiratory droplets, prompting governments worldwide to implement lockdowns, social distancing measures, and mask mandates as countermeasures. The rapid development and distribution of vaccines brought hope for control, yet challenges persisted in the form of new variants and vaccine hesitancy. This unprecedented global health crisis led to significant economic disruptions, strained healthcare systems, and resulted in millions of reported cases and fatalities worldwide (Seo, 2022).

The pandemic ushered in a paradigm shift in trash production, notably marked by a substantial surge in plastic waste, such as single-use food packaging and PPE like masks, gloves, respirators, and syringes. Factors contributing to reduced single-use

Table 8.1 The estimated quantity of masks to be used during a pandemic (Nzediegwu & Chang, 2020)

Continents	Total daily facemask (million)	Weight of total daily facemask used (tons)
Asia	3716.20	1486.48
Africa	922.22	368.89
Europe	884.71	353.88
North America	489.05	195.62
Oceania	45.43	18.17
South America	544.39	217.75

plastic consumption included declining fossil fuel costs and skepticism regarding the effectiveness of recycling (Klemeš et al., 2020, Silva et al., 2020). The high demand for PPE necessitated disposable masks, exacerbating waste generation (Bauchner et al., 2020). Additionally, restaurants and cafes shifted to single-use packaging materials to prevent viral transmission (Naughton, 2020, Single-use plastic in the pandemic 2020). Conversely, the surge in remote work, online shopping, and home-based food consumption led to increased household waste during lockdowns, while waste from large gatherings decreased (Nzeadibe & Ejike-Alieji, 2020; Sharma et al., 2020). The Politecnico di Torino estimated a significant rise in mask and glove demand during the restart of public activities in phase 2 (WWF, 2020).

Several environmental groups, such as Operation Mer Propre in France, Thames21 in London, and Oceans Asia in Hong Kong, raised concerns about the environmental pollution resulting from improper disposal of masks and PPE in water bodies (Thames, 2020). Masks typically use hydrophobic materials like polypropylene (PP) or more expensive options like polyurethane (PUR) or polyacrylonitrile (PAN) for protective layers (Earth.Org, 2020). The discarded plastic waste from hand sanitizers, surgical masks, tissue papers, and plastic-based masks exacerbated landfill pressures and river and ocean pollution. These pollutants disrupted natural habitats and posed risks to aquatic life (Saadat et al., 2020). Moreover, the virus could potentially contaminate water bodies when it adheres to discarded masks. The World-Wide Fund for Nature (WWF) Italy estimated that ten million masks could be littered in the environment within a month, with the potential for a disastrous scenario if only 1% were improperly disposed of, resulting in massive plastic pollution (Singh et al., 2022). Improper disposal of biomedical waste not only led to groundwater and soil pollution but also adversely impacted ecosystems (Sharma et al., 2020). The global mask usage rates were detailed in Table 8.1.

Notably, the risk of contamination from waste and wastewater was greater in impoverished countries due to inadequate waste management practices, in contrast to more advanced industrialized nations with safer and more environmentally friendly waste management systems. In underdeveloped regions, much waste is discarded in unsafe landfills, attracting unprotected scavengers, which could potentially facilitate disease transmission and hinder contact tracing efforts (Sharma et al., 2020; Nzediegwu & Chang, 2020; Solid Waste Management, 2019). Figure 8.1 represents the waste generation from quarantined home, temporary medical facility, health care facility, and household during pandemic.

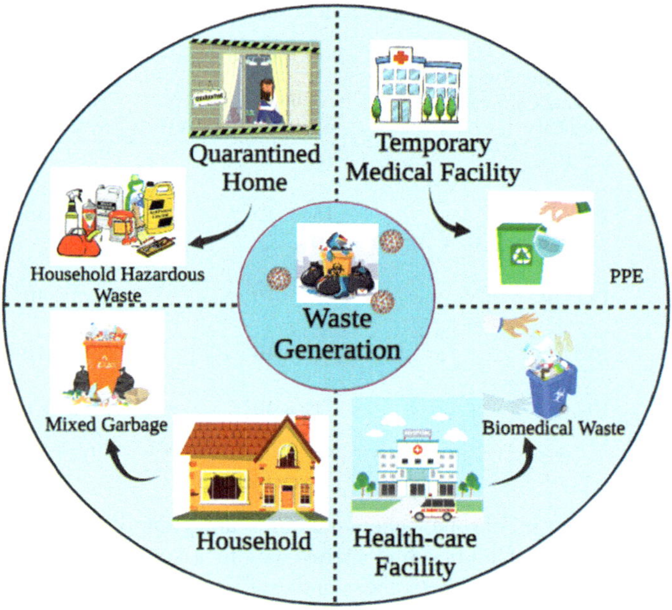

Fig. 8.1 Waste generation from different sectors

The WHO categorized "healthcare waste" into several categories, including infectious waste (containing contagious materials, bodily fluids, and patient blood), pathological waste (comprising body parts and human tissues), sharp metallic debris (such as needles and surgical blades), chemical waste (including mercury, sterilant, and disinfectants), cytotoxic waste (with genotoxic properties), and pharmaceutical waste (consisting of contaminated and expired medications). Biomedical waste, in general, posed risks to those handling it and was highly contagious (Health-care Waste, 2020). The pandemic led to an unprecedented increase in hospital admissions and COVID-19 testing, resulting in a substantial uptick in hospital and laboratory waste (Table 8.2). Managing this waste was crucial to prevent accumulation due to its potential for rapid transmission.

3 Waste Disposal Challenges

3.1 Impact on the Entire Waste Management Cycle

The COVID-19 pandemic has unquestionably disrupted the waste management cycle across the entire nation, primarily due to the absence of returns and ongoing challenges faced by recycling facilities. This crisis has particularly impacted

Table 8.2 Amount of medical waste expected to increase in Asian countries based on the experience of China (Adb, 2020)

City, Country	Population (world population review), (number)	Additional medical waste (kg per day)	Total waste production over 60 days (kg)
Hanoi, Vietnam	8×10^6	160×10^3	9.6×10^9
Kuala Lumpur, Malaysia	7.7×10^6	154×10^3	9.24×10^9
Manila, Philippines	14×10^6	280×10^3	16.8×10^9
Bangkok, Thailand	10.5×10^6	210×10^3	12.6×10^9
Jakarta, Indonesia	10.6×10^6	212×10^3	12.75×10^9

recycling businesses in Europe, where facilities are grappling with a significant crisis. These recycling facilities find themselves in a socioeconomic dilemma as they contend with the cost-effectiveness and environmental impact of using virgin plastic versus recycled plastic, leading to the inevitable incineration of recyclable materials, further harming the environment and reducing plastic recovery rates (Yousefi et al., 2021; Baran, 2021).

In Oregon, for instance, the Oregon Beverage Recycling Cooperative (OBRC) witnessed a staggering 45% decline in returns for recycling in April 2019 compared to the previous year, primarily due to Americans staying home during the pandemic. Despite the eventual reopening of redemption centers, return quantities continued to dwindle. Similarly, recycling facilities in California and Michigan faced similar challenges, with a decline in recyclable collection (Attaran, 2020). In Vienna, where approximately 500 visitors are received daily, recyclable collection has dipped by 10–15%, resulting in the utilization of only 4 out of 16 recycling units (Muñoz-Soto et al., 2023).

In the Netherlands, the supply of sorted cloth and plastic packaging materials cannot keep pace with the demand for recycled products. Processing and sorting businesses are discouraged from receiving donations, as only a few are willing to purchase materials at higher prices. A mere 40% of thrift stores are open due to reduced demand for recycled textiles, causing inventory overflow (Maharana et al., 2022).

Even before the pandemic, one of the Indonesia's primary challenges for both on-site and off-site activities was the operational capacity of existing HCW treatment and disposal facilities. With rural and remote areas now grappling with COVID-19 outbreaks, they lack access to proper HCW management facilities, while the rapid increase in infectious waste generation exacerbates the issue. Similar challenges are being faced in Sri Lanka and Kenya due to the swift surge in COVID-19 waste production and the inadequacy of efficient waste treatment systems (Skrzyniarz et al., 2022).

3.2 Government's Responses on Handling the Waste Generation

There was a noticeable increase in household garbage output during the COVID-19 outbreak. In the United States, domestic residential waste reached a peak, surpassing normal levels by approximately 20%, and in some regions, this increase exceeded 30%, as reported by SWANA in late April 2020. To efficiently handle the growing volumes of residential waste and recyclables, some local councils temporarily suspended curbside recycling programs (Penteado & de Castro, 2021).

Due to travel restrictions and widespread fear, the pandemic reduced waste deposit returns while increasing the amount of mixed waste at curbsides (Kahlert & Bening, 2020). Lockdowns caused a global decline in industrial production at the same time, greatly limiting the availability of high-value post-industrial waste materials (Kahlert & Bening, 2020; Mukherjee et al., 2023). This crisis affected the sector in many different ways (Kahlert & Bening, 2020).

Food waste increased significantly during the initial lockdown phase in the United States as a result of the closure of colleges, restaurants, and other large-scale consumers (Kim et al., 2020). The management of food waste at the household, institutional, and industrial levels is covered in several sources (Hantoko et al., 2021).

A large number of seasonal crops were wasted as a result of India's statewide lockdown, including summer fruits and vegetables, barley, paddy, and wheat (Harris et al., 2020). The Indian Government created economic incentive programs and made use of food delivery apps to address this issue by managing food efficiently and minimizing waste (Harris et al., 2020).

Urban waste management policies and practices, such as health and personnel protection measures, general waste sector rules, and waste treatment standards, were challenged by the COVID-19 pandemic (Sharma et al., 2023). In response, governments set up procedures to efficiently reduce and separate garbage without endangering the public's health and safety (Pinto et al., 2022).

Due to more individuals eating and drinking at home, Tokyo had an increase in domestic fuel waste of 3.10% (Jecker & Takahashi, 2021). With a focus on waste stream prioritization, temporary waste storage extension, waste segregation, and modifying municipal solid waste incinerators to handle COVID-19-contaminated trash, the British Government published instructions for waste collectors and local authorities (DEFRA, 2020).

The European Commission developed a waste management guidance document addressing the COVID-19 pandemic, emphasizing the importance of maintaining consistency with EU law in managing municipal solid waste, including recycling, collection, and separation. The document also stressed the preservation of biodegradable waste and residual waste processing frequency if facility reductions were necessary (European Commission, 2020).

For those involved in municipal solid waste management, safety guidelines were provided by the US Occupational Safety and Health Administration (OSHA) in response to the COVID-19 crisis, recognizing the critical nature of MSW management services (US OSHA, 2020). Developing nations such as India, Malaysia, and Vietnam published guidelines for handling waste from infected households and medical waste, but the importance of managing municipal waste during the pandemic should not be overlooked (Nghiem et al., 2020). Municipal waste management services play a crucial role in preventing the spread of infectious diseases, although they are often underemphasized in public health crisis communication responses.

The Central Pollution Control Board (CPCB) reports that between June and September 2020, India created almost 18,000 tons of biological waste due to COVID-19. Hazmat suits, syringes, headgear, plastic coverings, gloves, face masks, and other PPE used by both patients and healthcare professionals are included in this garbage. The amount of biomedical waste associated with COVID-19 has been increasing, with India producing roughly 3025.41 tons in June, 4253.46 tons in July, and 5238.45 tons in August. This amounted to almost 101 tons per day in June, and in June 2020, there were roughly 609 metric tons of biomedical waste produced every day. The number of COVID-19 instances and the rise in waste generation go hand in hand, according to The Energy and Resources Institute (TERI) (Ahuja, 2020).

In September 2020, Gujarat, Tamil Nadu, and Maharashtra generated approximately 543.78 tons, 622.89 tons, and 542.82 tons of COVID-19-related biomedical waste, respectively. To address the treatment and disposal of this waste nationwide, about 198 Common Biomedical Waste Treatment Facilities (CBWTFs) were employed, as reported by the CPCB (Ahuja, 2020).

4 Sustainable Waste Management Strategies

An effective healthcare waste management system in healthcare facilities includes the evaluation of waste management options, assessment of the waste stream and current environmental practices, implementation of institutional policies, and development of waste management plans and guidelines that outline staff roles and responsibilities. Sustainable medical waste management requires the development of a waste management organization, the allocation of financial and personnel resources, the implementation of a plan, monitoring, continual training, continuous improvement, and assessment. Effective organization and techniques play a vital role in healthcare waste management and creating a dedicated waste management team or committee is recommended. In low-income communities, an infection control committee with a member responsible for medical waste management should

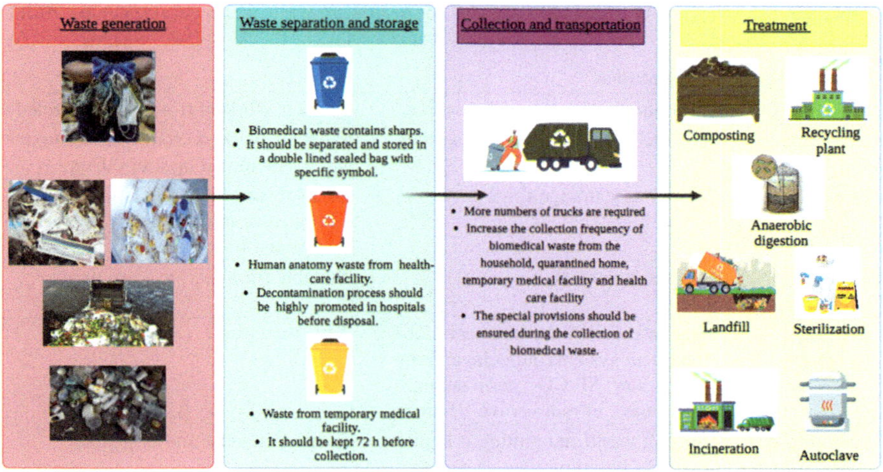

Fig. 8.2 Solid waste management trends and practices during COVID-19 pandemic. (Modified from Hantoko et al. (2021))

be established. Regular assessments and keeping staff informed about procedures and changes are crucial (Bharsakade et al., 2021; Thakur, 2021). According to Rodriguez-Pardo et al. (2020), efficient healthcare resource utilization can lead to reduced waste production (Rodríguez-Pardo et al., 2020). Figure 8.2 provides a sensible plan for managing medical solid waste during the COVID-19 pandemic.

Although different nations have distinct classifications for medical solid waste, the majority follow WHO guidelines. Segregation, which entails separating distinct types of waste at the source based on their classification, is a key component of effective healthcare waste management (Omar et al., 2012). By separating recyclables from non-hazardous garbage, proper segregation can dramatically reduce the overall waste volume (Mosquera et al., 2014; Windfeld & Brooks, 2015). Utilizing containers that are clearly identified with the type of trash and its weight allows for segregation. In order to discriminate between various waste kinds, infectious trash is often stored in plastic-lined cardboard boxes, plastic bags, or other leak-proof containers with particular performance requirements. In most cases, infectious trash is placed in yellow or red bags, whereas ordinary medical waste is placed in translucent or black bags. Containers for disposing of sharps are robust, leak-proof, break-resistant, and puncture-resistant. The right positioning and labeling of these containers are crucial for maximizing the efficiency of segregation. Placing infectious waste bins next to conventional garbage cans in places where both types of waste are produced encourages effective segregation. Healthcare professionals can be reminded to use the proper number of garbage cans and to display diagrams of efficient segregation on walls in areas with several containers.

The unique COVID-19 virus presents a number of difficult problems, including managing the rising amount of medical hazardous waste and the greater infection

Table 8.3 An overview of measured strategies for healthcare waste treatment (Das et al., 2021)

Taken measurements	Implications	Impacts
Safety	PPE and hand sanitizer for workers	Protects workers from being infected
Disinfectant	Disinfecting waste prior to sorting	Prevents workers from contracting the virus and spreading it to others
Storage	Temporary storage for 9 days	Helps to handle excess waste and prevent it from spreading the virus, which can persist for up to 9 days
Treating location	On-site treatment	Suitable for handling extra waste
Treating technique	Sterilwave (an ultra-compact mobile treatment system), autoclave, burn incinerator, SF-CO_2 sterilization, microwave or radio-wave treatment	Kills the virus
Handling	Sorting waste and putting it in different bags based on type of waste	Easier to handle and dispose of
Social distancing	1.5 m among workers and public during waste handling	Prevents infection and spread
Training	Using PPE and sanitization	Use of these will help protect from infection
Advertisement of guidelines	Social media and display in workplaces	People will become more aware of health measures

rate. During the COVID-19 pandemic, temporary healthcare waste treatment and transportation facilities have been constructed to efficiently manage this material and prevent transmission. Hospitals and other healthcare facilities may send their waste directly to treatment facilities for disposal, or they may first send it through temporary transit facilities. Temporary trash treatment and transportation facilities are crucial for effective hospital waste management due to the unpredictability of waste creation and the fluctuating infection rates (Yu et al., 2020). Alternative technologies can be used to manage medical waste, such as autoclaves and high-temperature burn incinerators (Ilyas et al., 2020; World Health Organization, 2020c). Utilizing SF-CO_2 sterilizing technology can lessen the danger of coming into contact with infectious medical waste (Hossain et al., 2011). The COVID-19 virus is efficiently destroyed on-site by Sterilwave, an ultra-compact technique, while processing medical waste, lowering its weight and enabling disposal as standard municipal garbage (Asim et al., 2021). In order to manage the added burden of medical waste during the pandemic, mobile treatment equipment can also be helpful.

When handling medical waste, PPE is necessary, including heavy-duty gloves, long-sleeved gowns, boots, goggles or masks, and face shields. It is crucial to properly wash your hands after disposing of garbage, and outdated PPE should be sealed in a bag for safe cleaning, either on or offshore. Typically, 10% lime slurry is present in the cleaning solution (World Health Organization, 2020a). The strategies for managing medical waste during the COVID-19 pandemic are listed in Table 8.3.

The proportion of recyclable materials can rise with effective waste management in the healthcare industry (Moreira & Günther, 2013). During the COVID-19 epidemic, autoclaving in particular can increase the amount of recyclable material and decrease landfill trash (Zand & Heir, 2020). Autoclaves use steam sterilization to disinfect contaminated waste efficiently, making it more suitable for recycling. Disinfectants like alcohol and sodium hypochlorite can also render the COVID-19 virus inactive, and storing medical waste for nine days in appropriate containment facilities can prevent the virus from spreading among those handling the waste. Additionally, medical waste can be used to generate electricity through pyrolysis and combustion, producing value-added products like incinerator ash made from medical waste. Organic waste can be converted into compost and bioenergy, contributing to the circular economy (Rowan & Laffey, 2021; Sobiecka et al., 2014; Abylkhani et al., 2021; Velvizhi et al., 2020). Considering these possibilities, repurposing medical waste for beneficial purposes and pandemic-related economic gains can provide significant value to the circular economy.

A "healthcare waste management strategy" is a term used to describe how generated garbage is managed for disposal within a facility. This plan often includes a number of important elements such as: (1) adherence to applicable laws, (2) defining staff roles, (3) classifying and characterizing healthcare waste, (4) specified handling procedures, and (5) training for workers involved. In response to the sizeable and infectious volume of trash produced during the COVID-19 pandemic, several governments have implemented various measures.

To effectively control and prevent infections, healthcare waste management programs must include new measures in response to the pandemic. Depending on their capabilities, resources, and commitment, some nations have embraced the best management practices (Zhou et al., 2020; Zand & Heir, 2021). There are many organizations that have offered guidance on how to handle medical waste in a way that satisfies sustainability and safety standards (Dharmaraj et al., 2022). It is recommended that the management of medical solid waste produced during the COVID-19 pandemic be improved in accordance with the recommendations made by a number of European Union (EU) member states, who have classified this material as contagious. It should be possible to store garbage temporarily in sufficient facilities in circumstances when there are problems with incineration or disposal capacity. Such trash needs to be kept in safe locations with only authorized employees having access to sealed containers. Applying disinfectants to interior and outdoor surfaces and strictly following the advised safety precautions are both necessary to reduce the spread of the virus (Cremonesi et al., 2020; Cesaro & Pirozzi, 2020).

During the COVID-19 pandemic in Hubei, China, healthcare solid waste generated in hospitals, including infected waste, was segregated and double-bagged by waste disposal personnel. Before being placed in temporary healthcare storage within the hospital premises, it was disinfected using a 0.5% chlorine solution. The specific disposal method for medical waste varied depending on the hospital and its waste management infrastructure. Some waste underwent sterilization through autoclave or irradiation before being disposed of in authorized landfills.

Certain hospitals conducted on-site incineration, while others transported waste to designated remote locations for incineration. Mobile incineration and autoclaving systems were also established to manage the surge in medical waste during the outbreak. Alternative disposal methods, such as utilization of cement kilns and industrial furnaces, were proposed. Limited secure storage facilities were used to temporarily store excess medical waste. Rigorous record-keeping was maintained, and designated vehicles were employed for healthcare solid waste transportation, with loading areas being secured, cleaned, and separated from drivers to minimize infection risk (Liang et al., 2021).

According to guidelines set forth by the Centers for Disease Control and Prevention (CDC), medical waste generated by COVID-19 patients in the USA is treated in the same manner as waste from other patients. Therefore, this material is handled as standard medical waste and doesn't require special management (Shooshtarian et al., 2022).

In response to the COVID-19 epidemic, the Philippines has enacted a specific law regarding the disposal of medical waste. Luzon Island has designated transporters and places for the treatment, storage, and disposal (TSD) of medical waste. In order to streamline the handling process, registered transporters must upload a number of documents online, including an official request letter, certificates of transporter and TSD registration, a transport management plan, transportation routes, schedules, and agreements between healthcare waste generators, transporters, and treatment staff. Each waste transport truck is conspicuously marked with the transporter's name and identification number, a placard that is visible from 15 meters away, the waste classification designation, the garbage number, and the placard. Registered carriers must also submit a compliance and completion report of the transportation, which must be certified by a TSD and healthcare department official (Quintero, 2021).

During the present pandemic, Jordan's healthcare waste management is directed by three major principles: minimizing unnecessary healthcare waste, separating routine waste from hazardous materials, and ensuring sufficient treatment to reduce risks to the public and healthcare personnel. PPE such as ultra-filtered masks (Nano), head covers, shoes, fluid-resistant long sleeves, protective glasses, leather gloves, and full-face shields are required for personnel handling COVID-19-contaminated medical waste. Local legislation, particularly the new cleaning and disinfection criteria for COVID-19-affected areas, are strictly enforced by the authorities. To stop the transmission of infection, trash from COVID-19 patients is immediately disposed of every day, and temporary and permanent storage places, containers, and medical waste bags are all sterilized (Kang et al., 2021).

During the epidemic, healthcare waste management remains a key global concern. Before medical waste can be safely handled, transported, and disposed of, it must be properly collected, sorted, and labeled. It is believed that effective healthcare waste management requires the utilization of personal protective equipment, disinfection, and training (Kudli et al., 2021). Sharps, pathological waste, and infectious waste from confirmed COVID-19 patients should be safely collected and stored in containers with the appropriate labeling. Healthcare waste collectors are expected to wear PPE, including long-sleeved shirts, boots, masks, thick gloves, face shields, and goggles. They should wash their hands after disposing of trash (Koniorczyk et al., 2022).

The examples above illustrate that different countries adopt varying strategies for healthcare waste management. As the COVID-19 pandemic has led to more individuals receiving medical care at home due to overwhelmed hospitals, healthcare waste generated in households often goes unmanaged. It is crucial to establish appropriate management strategies despite a country's economic conditions, with a focus on sustainable healthcare waste management, including recycling and reducing landfill waste.

5 Future Perspectives and Recommendations

Lockdowns and prohibitions may gradually loosen, and we would notice a greater reliance on plastic. Increased human mobility and an uptick in economic activity would encourage the use of personal protective equipment and single-use plastics, which would increase garbage generation. Governments around the world, especially in developing nations, are faced with a tremendous challenge as they try to control the growing problem of single-use plastic trash as a result of this spike. Without adequate planning and essential legislative intervention, we risk triggering a new public health emergency due to environmental litter.

Addressing this growing concern is imperative, but we must also recognize that the pressing climate crisis cannot be traded for the resolution of the COVID-19 situation, as per the Climate Action Tracker in 2020 (Vanapalli et al., 2021). To tackle the plastic problem and safeguard our environment, a comprehensive, global effort from both businesses and governments may be required, unless technological advancements offer better alternatives (Konov, 2020). Businesses and governments should allocate increased funding to promote awareness of the circular economy, recycling, and material reuse (Abidah et al., 2021), Implementing global regulatory incentives for recycling and reuse, alongside private and public investments in physical infrastructure (Konov, 2020).

Merely disseminating information through awareness campaigns is unlikely to change behavior, as seen in research. For example, campaigns highlighting the safety of sanitation workers and proper disposal of PPE may indirectly influence behavior by setting societal norms (Huang, 2016). A recent study shows that voice prompts can modify context and encourage pro-environmental actions, suggesting that reminders or prompts can effectively prompt desired actions, rather than convincing individuals to change their behavior (Ohtomo & Ohnuma, 2014). By reshaping norms and standards, we can reduce our reliance on single-use plastics (McKenzie-Mohr, 2011). Increased dialog about the issue and potential solutions can prompt companies to seek and provide alternatives, fostering collective action (McKenzie-Mohr, 2011).

Discussions about how to reduce plastic pollution frequently center on the potential switch from conventional plastic to bioplastic. However, studies show

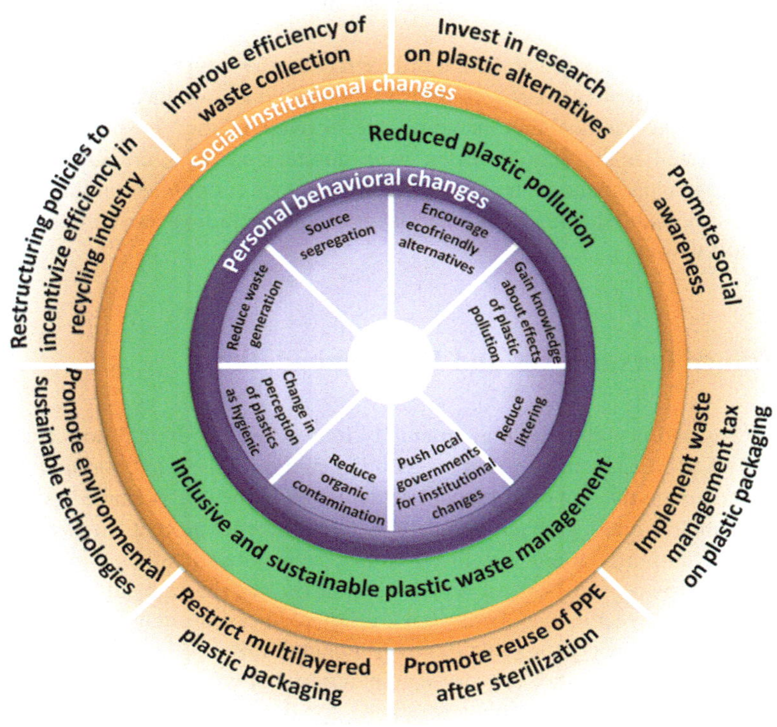

Fig. 8.3 A strategy aimed at mitigating plastic pollution and fostering a more inclusive and sustainable approach to managing plastic waste (Vanapalli et al., 2021)

that biodegradable and oxo-biodegradable plastics can survive and work for a long time in marine and soil settings. Even though biodegradable bags may disappear quickly in saltwater, they endure in soil environments, raising doubts about their ability to reduce marine litter significantly (Napper & Thompson, 2019). Moreover, there is evidence that bioplastics may degrade into micro and nanoplastics, further endangering ecosystems (Balestri et al., 2017, Shrut & Kutralam-Muniasamy, 2019, Sintim et al., 2020). Microplastic buildup in the environment poses a greater threat than intact litter, emphasizing the importance of proper disposal and integrated solutions (Willis et al., 2018). Pro-environmental efforts must involve all stakeholders, including legislators, scientists, industry, and the general public, to avoid shifting blame (Heidbreder et al., 2019).

To reshape the plastic waste management industry, society must initiate essential institutional and behavioral changes. A suitable institutional framework and policy-level directives will support this transition. Achieving sustainable and inclusive plastic waste management requires simultaneous personal behavioral changes and social institutional shifts, as depicted in Fig. 8.3.

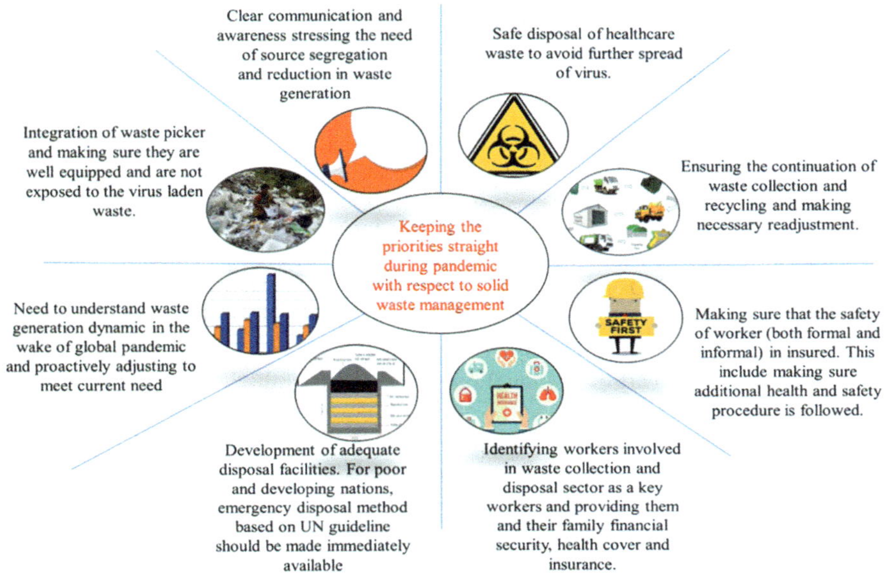

Fig. 8.4 Prioritizing waste management amidst the pandemic (Sharma et al., 2020)

The pandemic's waste management priorities are visually depicted in Fig. 8.4.

Here are some short- and long-term proposals for governments to prevent the rising usage and disposal of single-use plastics following the pandemic, based on thorough studies:

1. Create rules that address psychological and behavioral barriers, such as the general skepticism of products that have been reused or recycled and the promotion of the idea that single-use plastic is a problem rather than a solution.
2. Provide houses with specific-colored bags in which they can place biomedical waste and PPE kits (masks, gloves) in sealed bags. Install colored trash cans in public spaces to improve rubbish collection.
3. Promote innovation for reusable products that have been properly cleaned or disinfected by encouraging investments in the design of circular products suited for both hygiene and recycling.
4. Encourage research into environmentally friendly methods for recycling sophisticated plastic packaging, particularly those that address hygienic issues. Invest in cutting-edge technology like the chemical recycling of mixed plastic trash to produce high-quality resins and chemicals.
5. Before introducing bioplastics to the market, thoroughly verify their ability to be recycled and biodegraded, and set up the necessary infrastructure for efficient bioplastic waste disposal. Encourage the creation of reusable, eco-friendly PPE.
6. Include lessons on plastic pollution in environmental science courses to develop environmentally conscious students.

7. Promote local manufacturing and consumption in times of economic hardship by providing incentives like tax breaks for locally produced goods.
8. Restructure regulations to encourage recycling plant efficiency by identifying, awarding, and motivating other plants.
9. Support effective garbage collection and sorting business models by offering tax incentives and public praise for source separation.
10. Implement rules that will lessen the use of multi-layer packaging and promote homogeneous plastic packaging materials for simpler recycling, maybe by the imposition of a waste management levy on packaging with low recycling potential.
11. Promote entrepreneurial efforts in plastic waste management, encourage businesses to sponsor start-up projects in waste management and recognize their contributions as part of extended producer responsibility (EPR).

6 Conclusion

An urgent global appeal has been made due to the COVID-19 pandemic's spreading reach and negative effects on both public health and economic development. It demands that waste management, which includes dangerous materials, medical facility waste, and home garbage, be acknowledged as an essential public function. This understanding is essential for minimizing the pandemic's potential negative consequences on environmental sustainability and general health. The spread of COVID-19 through waste management has been fought against using precise rules. They support the use of high-temperature incineration facilities with temperatures approaching 1000 °C for the disposal of residual trash, which includes products like tissues, handkerchiefs, and various organic and packaging materials. This methodical procedure guarantees the virus's total and secure eradication. These actions are in line with the UNEP's dedication to encouraging environmentally friendly trash disposal methods. Due to the COVID-19 pandemic and the careless usage of single-use plastics, there are now greater environmental concerns around the world. Therefore, it is essential to concentrate future research initiatives on creating biodegradable and ecologically friendly personal protective equipment, such as face masks, gloves, and protective overalls. This tactical change intends to advance the cause of environmentally friendly manufacturing and consumption, ultimately lowering the environmental impact of pandemic response activities.

References

Abidah, H. N., Ismah, H. A., Irmayanti, S., Nurika, G., & Wikurendra, E. A. (2021). The effectivity of solid medical waste management in pandemic era. *Journal of Public Health for Tropical and Coastal Region, 4*(3), 98–107. https://doi.org/10.14710/jphtcr.v4i3.10618

Abylkhani, B., Guney, M., Aiymbetov, B., Yagofarova, A., Sarbassov, Y., Zorpas, A. A., Venetis, C., & Inglezakis, V. (2021). Detailed municipal solid waste composition analysis for Nur-Sultan City, Kazakhstan with implications for sustainable waste management in Central Asia. *Environmental Science and Pollution Research, 28*, 24406–24418. https://doi.org/10.1007/s11356-020-08431-x

Adb. (2020). *Managing infectious medical waste during the COVID-19 pandemic*. https://www.adb.org/publications/managing-medical-waste-covid19

Ahuja, A. (2020). *India generated over 18,000 tonnes of COVID-19 related bio-medical waste in 4 Months, experts call to reduce, reuse and segregate*. https://swachhindia.ndtv.com/india-generated-over-18000-tonnes-of-covid-19-related-bio-medical-waste-in-4-months-experts-call-to-reduce-reuse-and-segregate-52901/

Asim, N., Badiei, M., & Sopian, K. (2021). Review of the valorization options for the proper disposal of face masks during the COVID-19 pandemic. *Environmental Technology & Innovation, 23*, 101797. https://doi.org/10.1016/j.eti.2021.101797

Attaran, M. (2020). 3D printing role in filling the critical gap in the medical supply chain during COVID-19 pandemic. *American Journal of Industrial and Business Management, 10*(05), 988. https://doi.org/10.4236/ajibm.2020.105066

Balestri, E., Menicagli, V., Vallerini, F., & Lardicci, C. (2017). Biodegradable plastic bags on the seafloor: A future threat for seagrass meadows? *Science of the Total Environment, 605*, 755–763. https://doi.org/10.1016/j.scitotenv.2017.06.249

Baran, B. (2021). The COVID-19 pandemic and the problem of plastic waste in the EU. *European Research Studies Journal, 24*(Special Issue 3), 894–904. https://doi.org/10.35808/ersj/2531

Bauchner, H., Fontanarosa, P. B., & Livingston, E. H. (2020). Conserving supply of personal protective equipment—a call for ideas. *JAMA, 323*(19), 1911–1911. https://doi.org/10.1001/jama.2020.4770

Bharsakade, R. S., Acharya, P., Ganapathy, L., & Tiwari, M. K. (2021). A lean approach to healthcare management using multi criteria decision making. *Opsearch, 58*, 610–635. https://doi.org/10.1007/s12597-020-00490-5

Cesaro, A., & Pirozzi, F. (2020). About the effects of Covid-19 on solid waste management. *TeMA-Journal of Land Use, Mobility and Environment*, Special Issue – Covid-19 vs City-20, 59–66. https://doi.org/10.6092/1970-9870/6904

Cremonesi, P., Sartini, M., Spagnolo, A. M., Adriano, G., Zsirai, E., Patrone, C., Cevasco, I., & Cristina, M. L. (2020). Transformation of a ferry ship into a ship hospital for COVID-19 patients. *International Journal of Environmental Research and Public Health, 17*(23), 8976. https://doi.org/10.3390/ijerph17238976

Cruvinel, V. R. N., Marques, C. P., Cardoso, V., Novaes, M. R. C. G., Araújo, W. N., Angulo-Tuesta, A., Escalda, P. M. F., Galato, D., Brito, P., & Da Silva, E. N. (2019). Health conditions and occupational risks in a novel group: Waste pickers in the largest open garbage dump in Latin America. *BMC Public Health, 19*(1), 1–15. https://doi.org/10.1186/s12889-019-6879-x

Das, A. K., Islam, M. N., Billah, M. M., & Sarker, A. (2021). COVID-19 and municipal solid waste (MSW) management: A review. *Environmental Science and Pollution Research, 28*, 28993–29008. https://doi.org/10.1007/s11356-021-13914-6

DEFRA-Department for Environment Food & Rural Affairs, Government of UK. (2020). *UK statistics on waste*. Dos Santos, R.E., Dos Sant. https://www.gov.uk/government/statistics/uk-waste-data/uk-statistics-on-waste

Dey, S., Anand, U., Kumar, V., Kumar, S., Ghoraj, M., Ghosh, A., Kant, N., Suresh, S., Bhattacharya, S., Bontempi, E., Bhat, S. A., & Dey, A. (2023). Microbial strategies for degradation of microplastic generated from COVID-19 healthcare waste. *Environmental Research, 216*, 114438. https://doi.org/10.1016/j.envres.2022.114438

Dharmaraj, S., Ashokkumar, V., Chew, K. W., Chia, S. R., Show, P. L., & Ngamcharussrivichai, C. (2022). Novel strategy in biohydrogen energy production from COVID-19 plastic waste: A critical review. *International Journal of Hydrogen Energy, 47*(100), 42051–42074. https://doi.org/10.1016/j.ijhydene.2021.08.236

Earth.Org-Another side effect of COVID-19: The surge in plastic pollution | Earth.Org – past | present | future (2020). https://earth.org/covid-19-surge-in-plastic-pollution/,%202020

European Commission. (2020). *Waste management in the context of the coronavirus crisis*. https://projects2014-2020.interregeurope.eu/smartwaste/news/news-article/8320/waste-management-in-the-context-of-covid-19-crisis/

Gardiner, B. (2020). Pollution made COVID-19 worse. Now, lockdowns are clearing the air. *National Geographic, 8*.

Hantoko, D., Li, X., Pariatamby, A., Yoshikawa, K., Horttanainen, M., & Yan, M. (2021). Challenges and practices on waste management and disposal during COVID-19 pandemic. *Journal of Environmental Management, 286*, 112140. https://doi.org/10.1016/j.jenvman.2021.112140

Harris, J., Depenbusch, L., Pal, A. A., Nair, R. M., & Ramasamy, S. (2020). Food system disruption: Initial livelihood and dietary effects of COVID-19 on vegetable producers in India. *Food Security, 12*(4), 841–851. https://doi.org/10.1007/s12571-020-01064-5

Health-care Waste, WHO. (2020). https://www.who.int/news-room/fact-sheets/%20detail/health-care-waste

Heidbreder, L. M., Bablok, I., Drews, S., & Menzel, C. (2019). Tackling the plastic problem: A review on perceptions, behaviors, and interventions. *Science of the Total Environment, 668*, 1077–1093. https://doi.org/10.1016/j.scitotenv.2019.02.437

Hossain, M. S., Santhanam, A., Norulaini, N. N., & Omar, A. M. (2011). Clinical solid waste management practices and its impact on human health and environment–A review. *Waste Management, 31*(4), 754–766. https://doi.org/10.1016/j.wasman.2010.11.008

Huang, H. (2016). Media use, environmental beliefs, self-efficacy, and pro-environmental behavior. *Journal of Business Research, 69*(6), 2206–2212. https://doi.org/10.1016/j.jbusres.2015.12.031

Ilyas, S., Srivastava, R. R., & Kim, H. (2020). Disinfection technology and strategies for COVID-19 hospital and bio-medical waste management. *Science of the Total Environment, 749*, 141652. https://doi.org/10.1016/j.scitotenv.2020.141652

Info W. (2020). *Coronavirus Update*. Accessed on 28 August 2020. https://www.worldometers.info/coronavirus/

ISWA (International Solid Waste Association). (2020). *COVID-19 response international knowledge sharing on waste management*. https://www.iswa.org/iswa/COVID-19/

Jecker, N. S., & Takahashi, S. (2021). Shaming and stigmatizing healthcare workers in Japan during the COVID-19 pandemic. *Public Health Ethics, 14*(1), 72–78. https://doi.org/10.1093/phe/phab003

Kahlert, S., & Bening, C. R. (2020). Plastics recycling after the global pandemic: Resurgence or regression? *Resources, Conservation, and Recycling, 160*, 104948. https://doi.org/10.1016/j.resconrec.2020.104948

Kampf, G., Todt, D., Pfaender, S., & Steinmann, E. (2020). Persistence of coronaviruses on inanimate surfaces and their inactivation with biocidal agents. *Journal of hospital infection, 104*(3), 246–251. https://doi.org/10.1016/j.jhin.2020.01.022

Kang, A., Ren, L., Hua, C., Song, H., Dong, M., Fang, Z., & Zhu, M. (2021). Environmental management strategy in response to COVID-19 in China: Based on text mining of government open information. *Science of the Total Environment, 769*, 145158. https://doi.org/10.1016/j.scitotenv.2021.145158

Kim, K., Kim, S., & Park, C. Y. (2020). *Food Security in Asia and the Pacific amid the COVID-19 Pandemic (No. 139)*. Asian Development Bank. https://doi.org/10.22617/BRF200176-2

Klemeš, J. J., Van Fan, Y., Tan, R. R., & Jiang, P. (2020). Minimising the present and future plastic waste, energy and environmental footprints related to COVID-19. *Renewable and Sustainable Energy Reviews, 127*, 109883. https://doi.org/10.1016/j.rser.2020.109883

Koniorczyk, M., Bednarska, D., Masek, A., & Cichosz, S. (2022). Performance of concrete containing recycled masks used for personal protection during coronavirus pandemic. *Construction and Building Materials, 324*, 126712. https://doi.org/10.1016/j.conbuildmat.2022.126712

Konov, D. (2020) *COVID-19 is forcing us to rethink our plastic problem*. World Econ. Forum. https://weforum.org/agenda/2020/05/covid-19-is-forcing-us-to-rethink-our-plastic-problem/

Kudli, L. P., Jaishankar, S. S., & Iyer, R. D. (2021). Biomedical waste management during the COVID-19 Pandemic–Indian Scenario. *Journal of Indian Association for Environmental Management (JIAEM), 41*(1), 41–53. http://op.niscpr.res.in/index.php/JIAEM/article/view/45615

Kulkarni, B. N., & Anantharama, V. (2020). Repercussions of COVID-19 pandemic on municipal solid waste management: Challenges and opportunities. *Science of the Total Environment, 743*, 140693. https://doi.org/10.1016/j.scitotenv.2020.140693

Liang, Y., Song, Q., Wu, N., Li, J., Zhong, Y., & Zeng, W. (2021). Repercussions of COVID-19 pandemic on solid waste generation and management strategies. *Frontiers of Environmental Science & Engineering, 15*, 1–18. https://doi.org/10.1007/s11783-021-1407-5

Maharana, N., Patnaik, L. P., Mishra, B. B., Chaudhury, S. K., & Mohanty, J. (2022). A review of challenges and approaches to effective medical solid waste management during the COVID-19 pandemic in India. *International Journal of Circular Economy and Waste Management (IJCEWM), 2*(1), 1–17. https://doi.org/10.4018/IJCEWM.309986

Mallapur, C. (2020). *Sanitation workers at risk from discarded medical waste related to COVID-19*. IndiaSpend. https://www.indiaspend.com/sanitation-workers-at-risk-from-discarded-medical-waste-related-to-covid-19-681414. Accessed on 26 April 2020.

McKenzie-Mohr, D. (2011). *Fostering sustainable behavior: An introduction to community-based social marketing*. New society publishers.

Mol, M. P. G., & Caldas, S. (2020). Can the human coronavirus epidemic also spread through solid waste? *Waste Management & Research, 38*(5), 485–486. https://doi.org/10.1177/0734242X20918312

Moreira, A. M. M., & Günther, W. M. R. (2013). Assessment of medical waste management at a primary health-care center in São Paulo, Brazil. *Waste Management, 33*(1), 162–167. https://doi.org/10.1016/j.wasman.2012.09.018

Mosquera, M., Andrés-Prado, M. J., Rodríguez-Caravaca, G., Latasa, P., & Mosquera, M. E. (2014). Evaluation of an education and training intervention to reduce health care waste in a tertiary hospital in Spain. *American Journal of Infection Control, 42*(8), 894–897. https://doi.org/10.1016/j.ajic.2014.04.013

Mukherjee, S., Nagariya, R., Baral, M. M., Patel, B. S., Chittipaka, V., Rao, K. S., & Rao, U. A. (2023). Blockchain-based circular economy for achieving environmental sustainability in the Indian electronic MSMEs. *Management of Environmental Quality: An International Journal, 34*(4), 997–1017. https://doi.org/10.1108/MEQ-03-2022-0045

Muñoz-Soto, V. D., Dávila-Gálvez, S., Pérez, A. E., Rojo-González, L., Valenzuela-Levi, N., & Vásquez, Ó. C. (2023). COVID-19, waste production and municipal recycling programs: Insights from Chile to the global south. *Science of the Total Environment, 899*, 165388. https://doi.org/10.1016/j.scitotenv.2023.165388

Napper, I. E., & Thompson, R. C. (2019). Environmental deterioration of biodegradable, oxo-biodegradable, compostable, and conventional plastic carrier bags in the sea, soil, and open-air over a 3-year period. *Environmental Science & Technology, 53*(9), 4775–4783. https://doi.org/10.1021/acs.est.8b06984

Naughton, C. C. (2020). Will the COVID-19 pandemic change waste generation and composition?: The need for more real-time waste management data and systems thinking. *Resources, Conservation, and Recycling, 162*, 105050. https://doi.org/10.1016/j.resconrec.2020.105050

Nghiem, L. D., Morgan, B., Donner, E., & Short, M. D. (2020). The COVID-19 pandemic: Considerations for the waste and wastewater services sector. *Case Studies in Chemical and Environmental Engineering, 1*, 100006. https://doi.org/10.1016/j.cscee.2020.100006

Nzeadibe, T. C., & Ejike-Alieji, A. U. (2020). Solid waste management during Covid-19 pandemic: Policy gaps and prospects for inclusive waste governance in Nigeria. *Local Environment, 25*(7), 527–535. https://doi.org/10.1080/13549839.2020.1782357

Nzediegwu, C., & Chang, S. X. (2020). Improper solid waste management increases potential for COVID-19 spread in developing countries. *Resources, Conservation, and Recycling, 161*, 104947. https://doi.org/10.1016/j.resconrec.2020.104947

Ohtomo, S., & Ohnuma, S. (2014). Psychological interventional approach for reduce resource consumption: Reducing plastic bag usage at supermarkets. *Resources, Conservation and Recycling, 84*, 57–65. https://doi.org/10.1016/j.resconrec.2013.12.014

Omar, D., Nazli, S. N., Subramaniam, A., & Karuppannan, L. (2012). Clinical waste management in district hospitals of Tumpat, Batu Pahat and Taiping. *Procedia-Social and Behavioral Sciences, 68*, 134–145. https://doi.org/10.1016/j.sbspro.2012.12.213

Osuchowski, M. F., Aletti, F., Cavaillon, J. M., Flohe, S. B., Giamarellos-Bourboulis, E. J., Huber-Lang, M., Relja, B., Skirecki, T., Szabo, A., & Maegele, M. (2020). *SARS-CoV-2/COVID-19: Evolving reality, global response, knowledge gaps, and opportunities*. Shock (Augusta, Ga.). https://doi.org/10.1097/SHK.0000000000001565

Penteado, C. S. G., & de Castro, M. A. S. (2021). Covid-19 effects on municipal solid waste management: What can effectively be done in the Brazilian scenario? *Resources, Conservation and Recycling, 164*, 105152. https://doi.org/10.1016/j.resconrec.2020.105152

Pinto, A. D., Jalloul, H., Nickdoost, N., Sanusi, F., Choi, J., & Abichou, T. (2022). Challenges and adaptive measures for US municipal solid waste management systems during the COVID-19 pandemic. *Sustainability, 14*(8), 4834. https://doi.org/10.3390/su14084834

Quintero, A. F. (2021). COVID-19 PPE waste management: A looming ecological crisis in The Philippines? *Eubios Journal of Asian & International Bioethics, 31*(7), 369–372. https://web.s.ebscohost.com/abstract?direct=true&profile=ehost&scope=site&authtype=crawler&jrnl=11732571&AN=154350500&h=vOpPX5FyJG8u04hqR4nfRE67lexNHuBqt2XFJDQD29mIEKWqlM87jnLPN3B1NbuWxKDrnjGsqX4DyHG7dDjjew%3d%3d&crl=c&resultNs=AdminWebAuth&resultLocal=ErrCrlNotAuth&crlhashurl=login.aspx%3fdirect%3dtrue%26profile%3dehost%26scope%3dsite%26authtype%3dcrawler%26jrnl%3d11732571%26AN%3d154350500

Rodríguez-Pardo, J., Fuentes, B., de Leciñana, M. A., Campollo, J., Castaño, P. C., Ruiz, J. C., Herrero, J. E., Leal, R. G., Núñez, A. G., Cerezo, J. G., & Martínez, A. M. (2020). Acute stroke care during the COVID-19 pandemic. ictus Madrid program recommendations. *Neurología (English Edition), 35*(4), 258–263. https://doi.org/10.1016/j.nrleng.2020.04.007

Rowan, N. J., & Laffey, J. G. (2021). Unlocking the surge in demand for personal and protective equipment (PPE) and improvised face coverings arising from coronavirus disease (COVID-19) pandemic–implications for efficacy, re-use and sustainable waste management. *Science of the Total Environment, 752*, 142259. https://doi.org/10.1016/j.scitotenv.2020.142259

Saadat, S., Rawtani, D., & Hussain, C. M. (2020). Environmental perspective of COVID-19. *Science of the Total Environment, 728*, 138870. https://doi.org/10.1016/j.scitotenv.2020.138870

Sarkodie, S. A., & Owusu, P. A. (2020). Investigating the cases of novel coronavirus disease (COVID-19) in China using dynamic statistical techniques. *Heliyon, 6*(4). https://doi.org/10.1016/j.heliyon.2020.e03747

Seo, S. N. (2022). Pandemic analysis II: Governmental actions during the pandemic—Lockdown or no lockdown? In *The economics of pandemics: Exploring globally shared experiences* (pp. 115–153). Springer International Publishing. https://doi.org/10.1007/978-3-030-91021-1_4

Sharma, H. B., Vanapalli, K. R., Cheela, V. S., Ranjan, V. P., Jaglan, A. K., Dubey, B., Goel, S., & Bhattacharya, J. (2020). Challenges, opportunities, and innovations for effective solid waste management during and post COVID-19 pandemic. *Resources, Conservation and Recycling, 162*, 105052. https://doi.org/10.1016/j.resconrec.2020.105052

Sharma, R., Chakraborty, P., & Barman, S. R. (2023). Decontamination strategies and Technologies for tackling COVID-19 hospitals and related biomedical waste. In *Modern approaches in waste bioremediation: Environmental microbiology* (pp. 57–99). Springer International Publishing. https://doi.org/10.1007/978-3-031-24086-7_4

Shooshtarian, S., Caldera, S., Maqsood, T., & Ryley, T. (2022). Evaluating the COVID-19 impacts on the construction and demolition waste management and resource recovery industry: Experience from the Australian built environment sector. *Clean Technologies and Environmental Policy, 24*(10), 3199–3212. https://doi.org/10.1007/s10098-022-02412-z

Shruti, V. C., & Kutralam-Muniasamy, G. (2019). Bioplastics: Missing link in the era of Microplastics. *Science of the Total Environment, 697*, 134139. https://doi.org/10.1016/j.scitotenv.2019.134139

Silva, A. L. P., Prata, J. C., Walker, T. R., Campos, D., Duarte, A. C., Soares, A. M., Barcelò, D., & Rocha-Santos, T. (2020). Rethinking and optimising plastic waste management under COVID-19 pandemic: Policy solutions based on redesign and reduction of single-use plastics and personal protective equipment. *Science of the Total Environment, 742*, 140565. https://doi.org/10.1016/j.scitotenv.2020.140565

Singh, E., Kumar, A., Mishra, R., & Kumar, S. (2022). Solid waste management during COVID-19 pandemic: Recovery techniques and responses. *Chemosphere, 288*, 132451. https://doi.org/10.1016/j.chemosphere.2021.132451

Sintim, H. Y., Bary, A. I., Hayes, D. G., Wadsworth, L. C., Anunciado, M. B., English, M. E., Bandopadhyay, S., Schaeffer, S. M., DeBruyn, J. M., Miles, C. A., & Reganold, J. P. (2020). In situ degradation of biodegradable plastic mulch films in compost and agricultural soils. *Science of the Total Environment, 727*, 138668. https://doi.org/10.1016/j.scitotenv.2020.138668

Skrzyniarz, M., Sajdak, M., Zajemska, M., Iwaszko, J., Biniek-Poskart, A., Skibiński, A., Morel, S., & Niegodajew, P. (2022). Plastic waste management towards energy recovery during the COVID-19 pandemic: The example of protective face mask pyrolysis. *Energies, 15*(7), 2629. https://doi.org/10.3390/en15072629

Sobiecka, E., Obraniak, A., & Antizar-Ladislao, B. (2014). Influence of mixture ratio and pH to solidification/stabilization process of hospital solid waste incineration ash in Portland cement. *Chemosphere, 111*, 18–23. https://doi.org/10.1016/j.chemosphere.2014.03.057

Solid Waste Management, The World Bank. (2019). https://www.worldbank.org/en/topic/urbandevelopment/brief/solid-waste-management

Thakur, V. (2021). Framework for PESTEL dimensions of sustainable healthcare waste management: Learnings from COVID-19 outbreak. *Journal of Cleaner Production, 287*, 125562. https://doi.org/10.1016/j.jclepro.2020.125562

Thames-Coronavirus plastic pollution: What we can do about it -Thames21, (2020). https://www.thames21.org.uk/2020/06/coronavirus-plastic-pollu%20tion-what-we-can-do-about-it/

Thomson, B. (2020). The COVID-19 pandemic: A global natural experiment. *Circulation, 142*(1), 14–16. https://doi.org/10.1161/CIRCULATIONAHA.120.047538

UNEP, B. (2020). Waste management an essential public service in the fight to beat COVID-19. In *The United Nations Environment Programme (UNEP) and The Basel convention*. https://www.unep.org/news-and-stories/press-release/waste-management-essential-public-service-fight-beat-covid-19

Urban, R. C., & Nakada, L. Y. K. (2021). COVID-19 pandemic: Solid waste and environmental impacts in Brazil. *Science of the Total Environment, 755*, 142471. https://doi.org/10.1016/j.scitotenv.2020.142471

US OSHA. (2020). *COVID-19. The occupational safety and health administration*. Accessed 17 May 20. https://www.osha.gov/SLTC/COVID-19/controlprevention

Vanapalli, K. R., Sharma, H. B., Ranjan, V. P., Samal, B., Bhattacharya, J., Dubey, B. K., & Goel, S. (2021). Challenges and strategies for effective plastic waste management during and post COVID-19 pandemic. *Science of the Total Environment, 750*, 141514. https://doi.org/10.1016/j.scitotenv.2020.141514

Velvizhi, G., Shanthakumar, S., Das, B., Pugazhendhi, A., Priya, T. S., Ashok, B., Nanthagopal, K., Vignesh, R., & Karthick, C. (2020). Biodegradable and non-biodegradable fraction of municipal solid waste for multifaceted applications through a closed loop integrated refinery platform: Paving a path towards circular economy. *Science of the Total Environment, 731*, 138049. https://doi.org/10.1016/j.scitotenv.2020.138049

Willis, K., Maureaud, C., Wilcox, C., & Hardesty, B. D. (2018). How successful are waste abatement campaigns and government policies at reducing plastic waste into the marine environment? *Marine Policy, 96*, 243–249. https://doi.org/10.1016/j.marpol.2017.11.037

Windfeld, E. S., & Brooks, M. S. L. (2015). Medical waste management–A review. *Journal of Environmental Management, 163*, 98–108. https://doi.org/10.1016/j.jenvman.2015.08.013

World Health Organization. (2020a). *Water, sanitation, hygiene and waste management for COVID-19: Technical brief, 03 March 2020 (No. WHO/2019-NcOV/IPC_WASH/2020.1)*. World Health Organization. https://www.who.int/emergencies/diseases/novel-coronavirus-2019/technical-guidance-publications

World Health Organization. (2020b). *Water, sanitation, hygiene, and waste management for the COVID-19 virus: Interim guidance, 23 April 2020 (No. WHO/2019-nCoV/IPC_WASH/2020.3)*. World Health Organization. https://www.who.int/publications/i/item/WHO-2019-nCoV-IPC-WASH-2020.4

World Health Organization. (2020c). *Water, sanitation, hygiene, and waste management for the COVID-19 virus: Interim guidance, 19 March 2020 (No. WHO/2019-nCoV/IPC_WASH/2020.2)*. World Health Organization. https://iris.who.int/handle/10665/331499?locale-attribute=pt&show=full

WWF-Discarding of masks and gloves requires responsibility WWF Italy. (2020). https://www.wwf.it/pandanews/ambiente/nello-smaltimento-di-mascherine-e-guanti-serve-responsabilita/

Yousefi, M., Oskoei, V., Jonidi Jafari, A., Farzadkia, M., Hasham Firooz, M., Abdollahinejad, B., & Torkashvand, J. (2021). Municipal solid waste management during COVID-19 pandemic: Effects and repercussions. *Environmental Science and Pollution Research, 28*(25), 32200–32209. https://doi.org/10.1007/s11356-021-14214-9

Yu, H., Sun, X., Solvang, W. D., & Zhao, X. (2020). Reverse logistics network design for effective management of medical waste in epidemic outbreaks: Insights from the coronavirus disease 2019 (COVID-19) outbreak in Wuhan (China). *International Journal of Environmental Research and Public Health, 17*(5), 1770. https://doi.org/10.3390/ijerph17051770

Zambrano-Monserrate, M. A., Ruano, M. A., & Sanchez-Alcalde, L. (2020). Indirect effects of COVID-19 on the environment. *Science of the Total Environment, 728*, 138813. https://doi.org/10.1016/j.scitotenv.2020.138813

Zand, A. D., & Heir, A. V. (2020). Emerging challenges in urban waste management in Tehran, Iran during the COVID-19 pandemic. *Resources, Conservation, and Recycling, 162*, 105051. https://doi.org/10.1016/j.resconrec.2020.105051

Zand, A. D., & Heir, A. V. (2021). Emanating challenges in urban and healthcare waste management in Isfahan, Iran after the outbreak of COVID-19. *Environmental Technology, 42*(2), 329–336. https://doi.org/10.1080/09593330.2020.1866082

Zhou, F., Yu, T., Du, R., Fan, G., Liu, Y., Liu, Z., Xiang, J., Wang, Y., Song, B., Gu, X., & Guan, L. (2020). Clinical course and risk factors for mortality of adult inpatients with COVID-19 in Wuhan, China: A retrospective cohort study. *The Lancet, 395*(10229), 1054–1062. https://doi.org/10.1016/S0140-6736(20)30566-3

Chapter 9
Circular Cement Decarbonisation: Towards a Net-Zero Built Environment

Ana Laura De la Colina Martínez and David Joaquín Delgado Hernández

1 Introduction

Human actions have clearly caused global warming due to the large amount of GHG emissions, namely methane (CH_4), carbon dioxide (CO_2) and nitrous oxide (N_2O). Regarding global GHG emissions over the last years, it is recognised that they continue to increase despite the Paris Agreement (PA). Today, GHG emissions are higher than ever before. At the global level, it is estimated that the temperature will reach an increase of more than 3 °C in this century, that is, well beyond the intention of the PA to limit global warming so that it does not rise beyond 1.5 °C (2020 United Nations Environment Programme, 2020; Christensen & Olhoff, 2019). The increase in global GHG emissions is due to unsustainable energy and land use, resulting from economic models and lifestyles that determine consumption and production patterns in countries and regions (Calvin et al., 2023). It is estimated that half of global GHG emissions is due to the extraction and manufacture of materials. GHG emissions occur mainly due to the energy used in machinery, facilities, transportation and chemical reactions (Wang et al., 2022).

In the case of cement production, its GHG emissions are mainly due to the chemical reaction of calcination, which also demands high amounts of combustible fossils. The cement industry is the second largest industrial emitter, accounting for around 8% of global CO_2 emissions and is the third major industrial energy consumer. For instance, in 2019, the cement industry emitted around 2400 million

A. L. De la Colina Martínez (✉)
Doctora en Ciencia de Materiales por la Facultad de Química, Universidad Autónoma del Estado de México, Paseo Colón esquina Paseo Tollocan S/N, Toluca, Estado de México, Mexico

D. J. Delgado Hernández
Facultad de Ingeniería, Universidad Autónoma del Estado de México, Avenida Universidad S/N, Cerro de Coatepec, Ciudad Universitaria, Toluca, Estado de México, Mexico

V. Kumar et al. (eds.), *Environmental Engineering and Waste Management*,
https://doi.org/10.1007/978-3-031-58441-1_9

tons of CO_2 ($MtCO_2$) or 2.4 gigatonnes of CO_2 ($GtCO_2$), of which 60% (1.5 $GtCO_2$) will come from the chemical calcination process (process emissions), 25% (0.60 $GtCO_2$) from fossil fuels (direct production) and 15% (0.36 $GtCO_2$) from electricity (indirect production) (Hasanbeigi, 2021; International Energy Agency, 2023). Emissions Database for Global Atmospheric Research (EDGAR) considers that in 2019 emissions from global cement manufacturing grew by 5.1%, while the Global Carbon Project (GCP) estimates 3.3% (2020 United Nations Environment Programme, 2020). Cement is projected to continue to grow and would be around 20% higher than today by 2050 if no efforts are made to optimise its use in buildings and construction (Watson et al., 2023).

Globally, the best chance of avoiding catastrophic climate change is to limit the temperature increase to 1.5 °C above pre-industrial intensity with zero CO_2 emissions by 2050 (Watson et al., 2023). Therefore, the pathways used by the Science Based Targets initiative (SBTi) guides and encourages companies to reduce their GHG emissions in the short term by around 2030. In the longer term, they must reduce GHG emissions by 90% by 2050 at the latest. Furthermore, to achieve net zero emissions, residual emissions (10%) will remain and will need to be removed from the atmosphere and permanently stored (Watson et al., 2023). Besides, circular strategies can reduce GHG emissions, and they must be applied to the products whose life cycle generates large quantities of GHG (the built environment, mobility and food systems), such as the cement industry (Wang et al., 2022).

Could clinker dominance decline? Will it be replaced by new materials? If the cement industry does not consider SBTi and circular strategies, it will be highly responsible for failing to prevent catastrophic climate change. Cement companies, need to take radical measures to reduce the large amount of GHG emissions it generates year after year (Linde, 2023). Even though the rate at which the sector can decarbonise is limited by its process (geogenic) of calcining limestone in clinker production (Science Based Target, 2022).

1.1 The Challenges of SBTi's Net Zero Standard

According to IPPC (2022) and Hausfather (2018), about 2345 $GtCO_2$ have accumulated in the atmosphere since pre-industrial times. If this trend continues (see Fig. 9.1), it is estimated that a temperature increases of 3 °C will be reached shortly after 2110, 2 °C before 2060 and 1.5 °C by 2035. The estimated "carbon budget" for this last temperature increase (1.5 °C) is 500 $GtCO_2$ (with a base year of 2020). This 500 $GtCO_2$ will be rapidly depleted on current global trends (of almost 40 $GtCO_2$ emitted annually). Then the temperature increase cannot be kept below 1.5 °C by 2050.

In line with the SBTi Net Zero Standard, companies must meet some very significant challenges. By 2030, they must halve their GHG emissions. By 2050, they must reduce their emissions by 90%. In addition, the residual emissions (10%) that remain from year to year will have to be neutralised (stored) to achieve constant

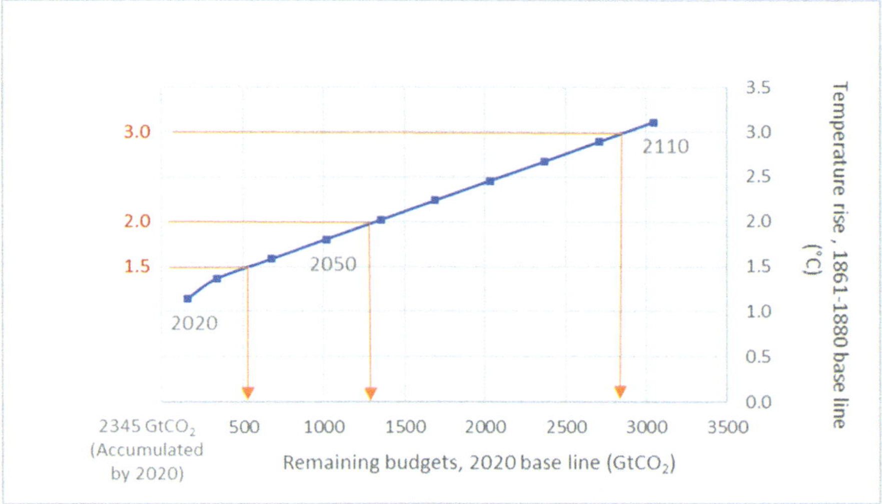

Fig. 9.1 Carbon budgets estimating temperature increases of 1.5, 2 and 3 °C. Pointing to 1.5 °C with a carbon budget of 500 $GtCO_2$

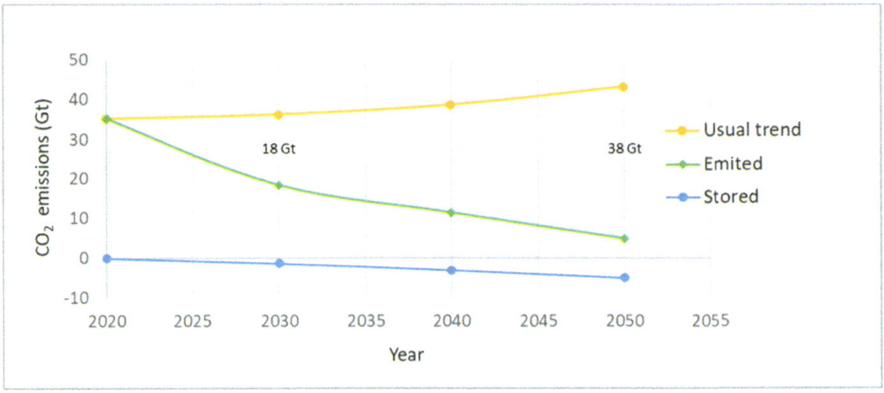

Fig. 9.2 The rate at which global CO_2 emissions must be avoided and stored to keep global temperatures below 1.5 °C

net zero emissions. Furthermore, circularity must be applied to all products, especially those whose life cycle generates large quantities of GHG (mobility, food systems and the built environment). In this regard, according to (Blaufelder et al., 2021; Statista, 2023a) to keep the global temperature below 1.5 °C, it is essential to avoid and store CO_2 emissions, at the rate suggested in Fig. 9.2. Globally, for example, 18 $GtCO_2$ should be avoided and 1.3 $GtCO_2$ stored by 2030. Similarly, 38 $GtCO_2$ should be avoided and 5 $GtCO_2$ stored by 2050.

Table 9.1 CO_2 emission budgets from 2020 to 2050 for the energy, transport, industry and buildings sectors

Sector	CO_2 emissions in 2019 ($GtCO_2$)	CO_2 budgets from 2020 to 2050 ($GtCO_2$)
Energy supply	15.3	115–146
Electricity and heat	13.8	102–133
Transport	8.3	100–129
Road	6.1	73–91
Aviation	1.0	15–19
Maritime	0.9	12–16
Industry	8.9	134–153
Cement	2.5	35–41
Iron and steel	2.5	20–40
Chemicals	1.3	13–26
Buildings	3	30–41
Residential	2	20–30
Service	1	10–11
Total	35.5	450–480

Today, in the context of cement industry, the cement production accounts for almost 8% of global anthropogenic CO_2 emissions (in 2002, it generated 5% of man-made CO_2 (World Business Council for Sustainable Development, 2005)). About 50% of these emissions come from the production of clinker, 40% from the burning of fuels to heat the cement kiln and the remaining 10% from transport and electricity use.

Table 9.1 shows the remaining global budget of about 500 $GtCO_2$ to limit global warming to 1.5 °C, and the cement industry budget of 35–41 $GtCO_2$ (Science Based Target, 2021). The SBTi estimated the intensity pathway for the global cement industry from 2020 to 2050 using the budget 35–41 $GtCO_2$ (see Fig. 9.3).

Figure 9.3 shows the global cement industry targets, expressed in tons of CO_2 per ton of cement (t CO_2/t cement). However, if the targets suggested by this trajectory are not met, a new scratch will be needed, which could become increasingly steep to meet the 1.5 °C target (Science Based Target, 2022).

The Cement Sustainability Initiative (CSI) which brings together 24 major cement producers with operations in more than 100 countries has identified four levers (energy efficiency, alternative fuels, reducing the clicker-to-cement and carbon storage) that can reduce carbon emissions in the cement industry (Carter, 2023).

1.2 A Lost Decade—2010–2020

It is fair to say that since the early twentieth century, there have been major advances in the technology of composite materials and chemical admixtures for the Portland

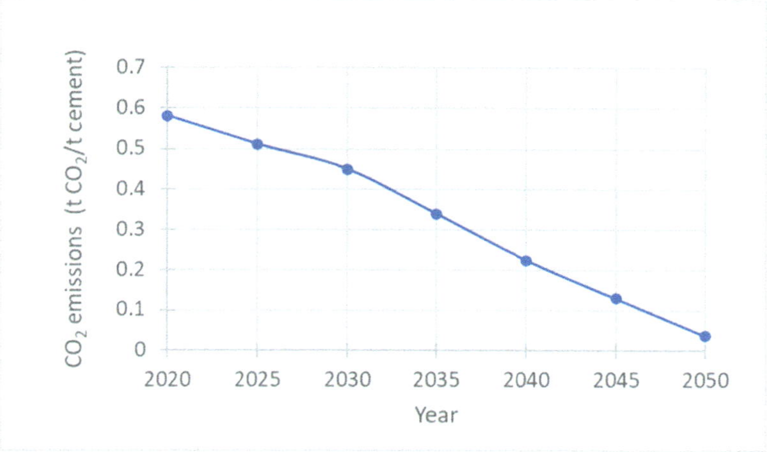

Fig. 9.3 Global cement industry targets based on a budget of 35–41 $GtCO_2$ to limit global warming to 1.5 °C

cement-based construction industry. The resulting functionalities, together with the economics of raw materials, have made concrete the material that occupies the number one place in the world. However, during the twentieth century, both the composition and production of clinker have remained surprisingly unchanged, despite the huge environmental impact of Portland cement production and consumption (Biernacki et al., 2017).

It is known that in Britain, Joseph Aspdin patented an artificial cement in 1824, which became known as "Portland cement". However, it could not be produced on a large scale, and it was not until 1844 that it was decided to calcine the mixture until it was partially vitrified. This significantly improved the Portland manufacturing process and for the first time produced clinker, the basis of today's synthetic cement. The cementitious superiority of the new Portland cement over natural cements was soon evident, and large-scale production would soon make the synthetic cement industry one of the most important globally (Centro de Estudios y Experimentación de Obras Públicas, 2022).

The addition of metals such as iron to concrete improved tensile strength and this increased the range of applications. William Wilkinson, when obtaining the first patent of this process, in 1854, initiated, without knowing it, the change of the face of the planet. Francois Hennebique then patented a new building system in 1892, which, over the next 10 years, facilitated the construction of more than 40,000 structures (Science Museum, 2021). Since the early twentieth century, cement has become one of the strategic products for the economic growth of any country. The twentieth century can be considered as the era of concrete (Science Museum, 2021).

From 2010 to 2020, the main greenhouse gas (GHG) emitters were China, the United States, the 27 members of the European Union + the United Kingdom and

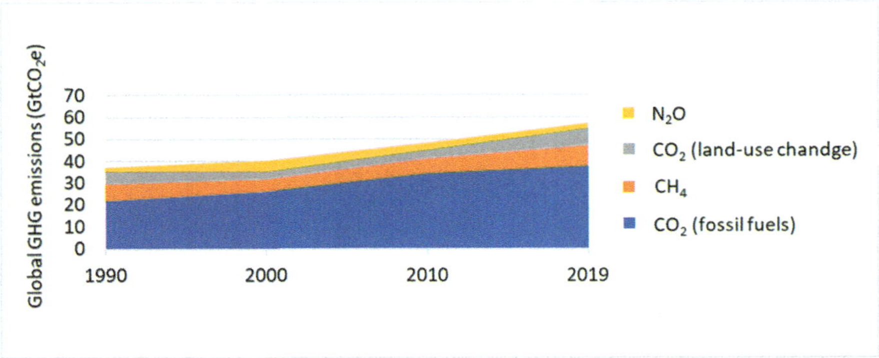

Fig. 9.4 Global GHG emissions show a continuous increase since 1990

India accounted for 55% of total GHG emissions. If the contribution of the Russian Federation, Japan and the international transport sector is added to this group, these seven emitters account for 65% of total emissions. It is also estimated that G20 members account for 78% of total emissions (2020 United Nations Environment Programme, 2020).

Global GHG emissions and their origin can be seen in Fig. 9.4, showing a continuous increase since 1990. It is observed that CO_2 emissions of fossil origin are the most preponderant, compared to emissions of CH_4, N_2O, fluorinated gases and emissions produced by land use change (CO_2, CH_4 and N_2O) (National Energy Technology Laboratory, 2023). That is, despite being a decade of augmenting social and political attention on the issue of global warming, and having achieved the PA, GHG emissions are higher than ever. The long-awaited decrease in GHG is not noticed. So it can be said that the period from 2010 to 2019 can be considered as a lost decade (Christensen & Olhoff, 2019).

On the other hand, the global cement industry emitted around 2.4 $GtCO_2$ in 2019, of which 60% (1.5 $GtCO_2$) came from the chemical calcination process (process emissions), 25% (0.60 $GtCO_2$) from fossil fuel use (direct emissions) and 15% (0.36 $GtCO_2$) from electricity use (indirect emissions) (Hasanbeigi, 2021; International Energy Agency, 2023). See Fig. 9.5. EDGAR estimates that emissions from cement manufacturing in 2019 grew by 5.1%, while the GCP estimates by 3.3%.

The Emissions Gap Report (EGR) 2020 finds that the planet is trending to rise by more than 3 °C during this century, well away from the objectives of the PA (which is pursuing 1.5 °C) (2020 United Nations Environment Programme, 2020).

Figure 9.6 shows global emissions from using fossil fuels and, for contrast, emissions from the chemical reaction from cement production; emissions from burning fossil fuels (for heating chemical reaction) are considered as emissions of the fuel. Total emissions from these sources for 2019 were more than 35,000 million tons of CO_2 (Fakta o klimatu, 2023).

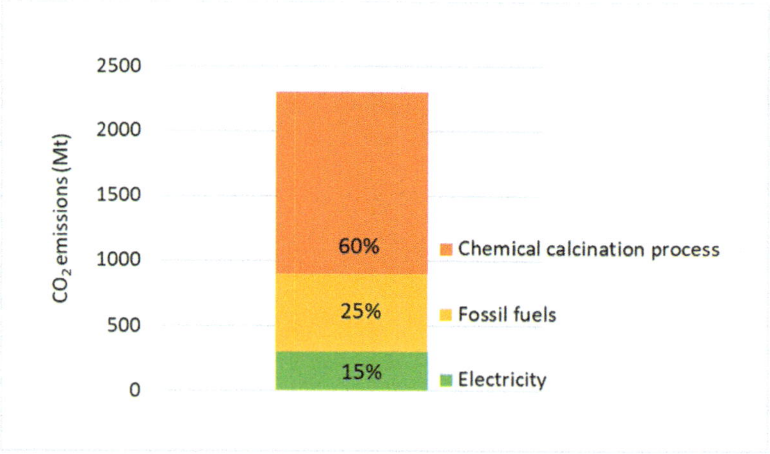

Fig. 9.5 The global cement industry emitted around 2.4 $GtCO_2$ in 2019 (60% came from calcination, 25% from fossil fuel use and 15% from electricity use)

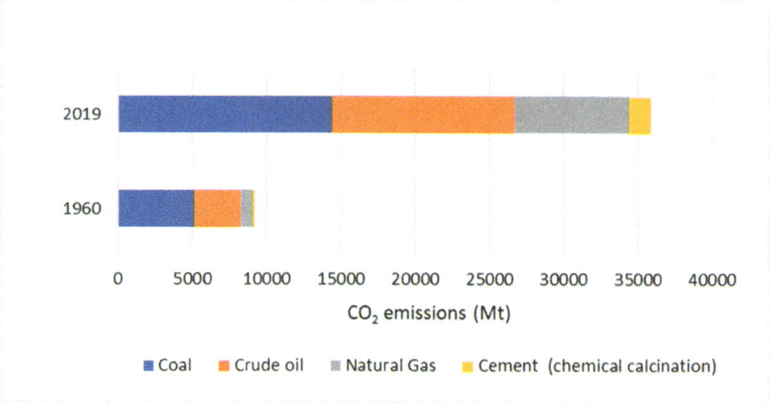

Fig. 9.6 Comparison of global CO_2 emissions from global cement production in 1960 and 2019 (chemical calcination, burning coal, crude oil, and natural gas)

Cement is generated by warming a combination of limestone and clay to 1450 °C. About 60% of total CO_2 emissions come from the chemical reaction ($CaCO_3 \rightarrow CaO + CO_2$). The other 40% of emissions are due to fuels and electricity. Total emissions represent about 2.8 billion tons of CO_2, (8%) of global CO_2 releases (Fakta o klimatu, 2023).

The following results from the U.S. cement industry, referring to 2015, show that the California cement industry was far from being a low-CO_2 emissions company. These factories were, not long ago, the largest consumers of coal and petroleum coke

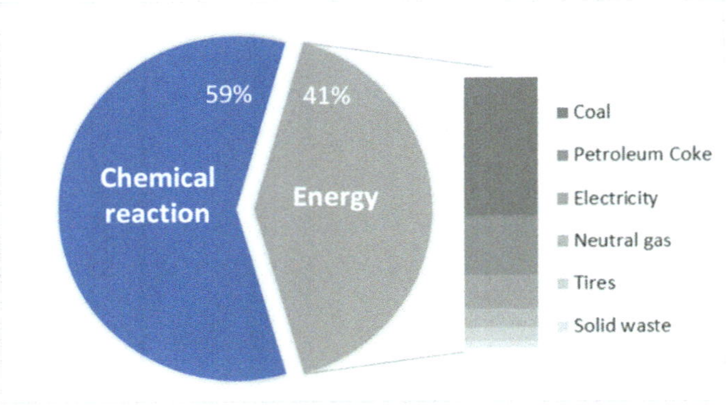

Fig. 9.7 Percentage of CO_2 emissions in 2015 from the California cement industry, which was the largest consumer of coal and petroleum coke in California

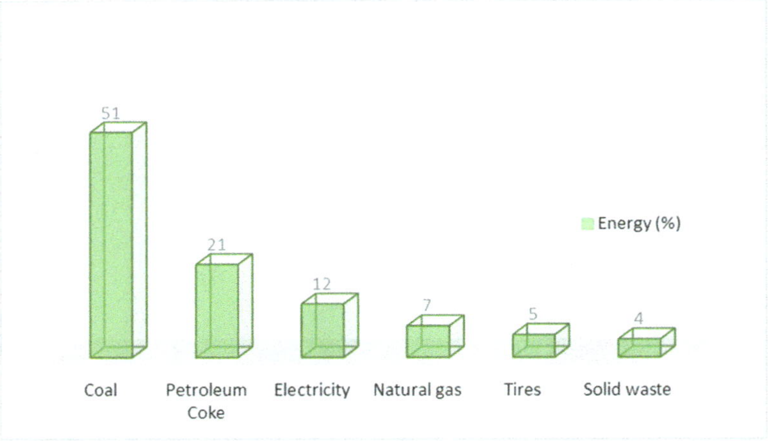

Fig. 9.8 The largest amount of energy used in California's cement factories (74%) came from highly polluting fuels such as coal and petroleum coke

in California; in fact, they have higher CO_2 emissions per ton of cement (Hasanbeigi, 2019) (see Fig. 9.7).

Figure 9.8 shows that more than 70% of the energy used in California's cement factories came from coal and petroleum coke (highly polluting fuels). Whilst natural gas has been used in low proportions (7%) (Hasanbeigi, 2019).

As can be seen in Fig. 9.9, the main challenge in terms of CO_2 emissions from cement production is a profound transformation of the clinker production process, because even if the cement industry could use clean energy to replace fuels and electricity, it could only eliminate 35–40% of the total net CO_2 emissions.

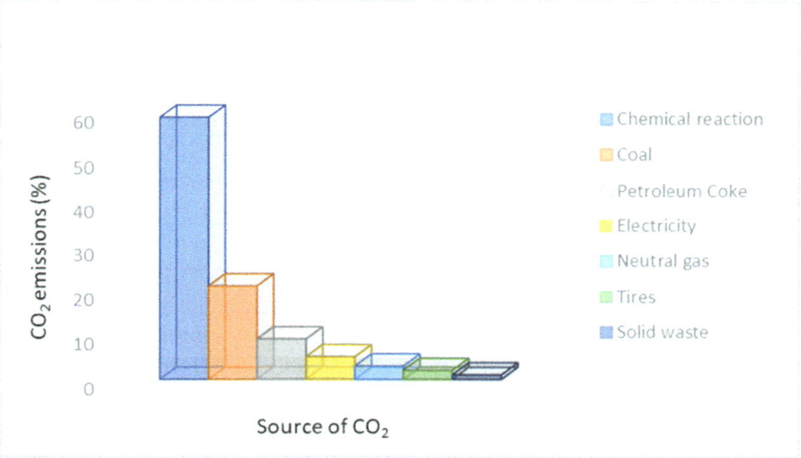

Fig. 9.9 The main challenge in terms of CO_2 emissions from cement production is a profound transformation of the clinker production process

Regarding fuels used in the cement industry, it should be noted that in 2019, globally, almost 8000 million tons of coal were extracted. Among the countries that benefited most from this resource are China (46%), India (10%) and the United States (8%). As for the natural gas extracted, in this same year, the amount is summarised in 4000 billion m^3, that is, almost 3200 million tons (Fakta o klimatu, 2023).

The 2020 Emissions Gap Report shows that neither the policies implemented over the past decade nor the climate action plans (Nationally Determined Contributions, or NDCs) at the United Nations Climate Change Conference 21 (COP21) in 2015 are sufficient to keep global warming below the temperature parameters set out in the PA (2020 United Nations Environment Programme, 2020).

Yet it is remarkable that more and more countries are determined to reach net-zero emissions around mid-century. It is also true that, in 2019, global GHG emissions increased again and stood at an all-time high of 59.1 Gt of CO_2 equivalent ($GtCO_2e$). Under this perspective, the litmus test will have to adopt, by 2030, much more ambitious NDCs. Even with the full implementation of the NDCs proposed in the PA, we head for a 3.2 °C warmer world at the end of this century (2020 United Nations Environment Programme, 2020).

1.3 The Challenge of Avoiding CO_2 Emissions in the Near Term—By 2030

The Emissions Gap Report highlighted the fact that global emissions need to fall by 45–50% by 2030 to ensure temperatures do not rise above 1.5 °C by 2100. Short-term target, set in 2019 by SBTi's Net Zero Standard, aims to halve GHG emissions by 2030 to keep the temperature rise below 1.5 °C (Watson et al., 2023).

In this regard, in 2021, during the Climate Change Conference 26 (COP26), countries agreed to accelerate their actions to reduce their emissions by 2030. And even though many countries have NDCs in line with 1.5 °C, they are still not enough (UNEP, 2023). In 2022, United Nations Environment Programme (UNEP) presented a proposal in search of climate stability, based on six sectors: Energy; Industry; Buildings and Cities, Agriculture and Food; Forests and land use; and Transportation (UNEP, 2023). Besides, some countries—led by India—at the Climate Change Conference 27 (COP27) wanted to reach an agreement to phase out all fossil fuels, but this proposal did not bear fruit. In addition, the Intergovernmental Panel on Climate Change warned of catastrophic consequences that can only be avoided if sharp and urgent GHG cuts are carried out (The guardian, 2022).

The cement, steel and chemicals sectors are decarbonising through circularity and efficient use of materials and energy. Certain low-carbon hydrogen technologies, produced by electrolysis, biomass, and biogas, are replacing the use of fossil fuels in the industrial production of large-scale heat (World Business Council for Sustainable Development, 2021). Some countries and regions have taken the following policy initiatives in the last 2 years to decarbonise cement production. China launched the Carbon Peak Implementation Plan for the Building Materials Industry (in 2022) to achieve a reduction in its CO_2 emissions by 2030 through clean and green energy (Morgado et al., 2023). The six countries (France, the Netherlands, Germany, the United Kingdom, Sweden, and the United States) have been recognised by the World Economic Forum (WEF) and the Global Cement and Concrete Association (GCCA) as leaders in adopting low-carbon concrete and construction regulations and implementing supportive public procurement policies (Morgado et al., 2023). On the other hand, by 2030, a significant number of cement kilns will end their useful life. This is expected to facilitate investment in infrastructure and innovative technologies that will lead the cement industry to a rapid net-zero transition by 2050. At the policy level, this transition is expected to be supported by policies (rules, regulations and strategic legislation) that boost the circular economy, new clean technologies, the use of green cement, CO_2 infrastructure and carbon pricing (Pissaro, 2023).

Despite the existence of the Sustainable Development Goals and the Paris Agreement, unprecedented global risks continue to emerge because change is not happening at the speed or scale required. A major social and environmental tipping point is approaching. There is a unique window of opportunity for action over the next decade, but it is closing fast (World Business Council for Sustainable Development, 2021). Today, in 2023, the focus is on how quickly nations can reduce GHG emissions to avert the climate crisis and its devastating consequences for the world. Then the world must act, and act now (UNEP, 2023).

Finally, according to World Business Council for Sustainable Development (2021), among the actions based on the circular economy that need to be taken between 2020 and 2030 to contribute to reducing the GHG emissions of the cement sector are the following

- *Products:* Ensure that product life cycles are extended as far as possible, encouraging maintenance and renewal. Integrate circularity and the use of "next lives" of the product, from design to end of life. Invest in innovations that replace non-renewable and fossil-based materials with biological materials that are sustainable, circular and carbon-neutral. Consider the natural, social, and human capital costs of products. Promote the collection and processing of materials, particularly in developing countries.
- *Energy:* Align with net zero carbon targets. Use net zero carbon energy in all operations. Invest in nature-based solutions to remove emissions from the atmosphere. Shift fossil fuel use to low-carbon energy, using cutting-edge technologies such as carbon capture and storage.
- *Transport:* Adopt an ambitious mobility policy that promotes electric energy. Expand the use of low-carbon fuels for heavy-duty and long-distance transport.

1.4 The Challenge of Avoiding CO_2 Emissions in the Long Term—By 2050

From the beginning of the Industrial Revolution (eighteenth century), until 2022, atmospheric CO_2 went from 280 to 417 ppm (in 2022). This concentration is projected to increase to 450 ppm by 2050 if no action is taken (Hack et al., 2022). Besides, cement demand is projected to continue rising and would be some 20% higher than today by 2050 (Science Based Target, 2022).

The long-term goal, set by SBTi's Net Zero Standard in 2019, is for companies to reduce GHG emissions by 90% by 2050, with the aim of having only 10% of GHG emissions remaining by 2050 (Watson et al., 2023). SBTi help cement companies to implement the level of climate ambition required to meet the 1.5 °C goal of the PA.

The global cement industry's targets, based on a budget of 35–41 $GtCO_2$ to limit global warming to 1.5 °C, would require the implementation of several measures and approaches, the combination of which will be the key to achieving a significant reduction in the sector's GHG emissions.

- *Use of low-carbon fuels:* Replace conventional fossil fuels with alternatives such as natural gas, biogas, renewable energies, and waste that can be used as fuels.
- *Energy efficiency:* Reduce energy consumption in cement manufacturing processes, improving energy efficiency.

- *Use of alternative raw materials:* Replace clinker with materials such as fly ash, slag, pozzolan, metakaolin or other industrial by-products.
- *Carbon capture and storage*: Separate the CO_2 emitted by the cement industry and transport it to a safe storage site, to isolate it from the atmosphere for a long term.
- *Circular economy*: Promote the reuse and recycling of construction and demolition waste, as well as other materials, to reduce the demand for new raw materials and waste generation.
- *Innovation in new materials*: Develop new types of cement with a lower carbon footprint, such as cements that have low CO_2 emissions, geopolymeric cements and cements based on new production processes.
- *Design and construction optimisation*: Encourage sustainable design and construction practices that reduce the amount of cement used and maximise its efficiency, without compromising the quality and durability of structures.
- *Sectoral collaboration and commitment*: Foster collaboration between the cement industry, governments, environmental organisations and other relevant actors to establish common goals, exchange best practices and promote the adoption of more sustainable technologies and processes.

To address the global climate challenge, the cement industry, through the GCCA, is publishing its "2050 concrete in the low-carbon built environment roadmap", which sets out a clear long-term path for concrete that adopts a circular economy approach. This roadmap, which will be a key reference document for the sector, includes measures which

1. Promote investment in state-of-the-art technology for new and retrofit plants.
2. Facilitate increased use of waste (alternative fuels) and by-products (raw materials).
3. Support the research and development of breakthrough technologies as well as the acceleration and scaling-up of proven efficient low-carbon technologies, with a particular focus on CCUS and new and alternative binders. Policies should help mitigate the risk through investment mechanisms.
4. Promote cooperation between government and industry to develop CO_2 transport and storage infrastructure and to implement circular economy strategies, promoting waste avoidance, collection and sorting, pre-treatment, recovery, recycling, and co-processing.
5. Drive the demand for sustainable building materials by helping to stimulate market demand for innovative products by construction contractors and customers.
6. Support life-cycle assessment-based methodologies, tools and databases to enable a whole-life based approach to procurement. Appropriate sustainability assessment methods using life cycle analysis are to be preferred in public and private tendering.
7. Recognise at national level the uptake of CO_2 by existing concrete in the built environment.

8. Energy performance of buildings calculation methods should be sophisticated enough to take account of thermal mass.
9. Electricity systems should facilitate demand response, that is interaction between the grid and households, where the consumer enjoys a share of the cost savings.
10. Enable revision of building codes and regulations to facilitate the adoption of innovations without jeopardising safety and durability and recognising the increased need for resilience in the built environment.
11. Establish the means of recognising that the resilience of the built environment can contribute to favourable social and economic benefits for society.
12. Establish the means of recognising that concrete can contribute to favourable emission benefits in other sectors of the economy.
13. Access to recycled concrete for utilisation for recarbonation.

1.5 The Challenge of Neutralising Residual CO_2 Emissions—Beyond 2050

Although most firms will decrease emissions by at least 90%, some remaining emissions can persist (10%). These emissions must be "neutralised" to reach net-zero emissions (Watson et al., 2023). Neutralisation of residual emissions refers to the actions that enterprises take to capture CO_2 from the atmosphere to suppress the impact of emissions that remain constant.

1.5.1 Carbon Capture and Storage (CCS)

CCS is presented as a suitable technology for CO_2 neutralisation. It consists of capturing the CO_2 emitted by industrial facilities, transporting it to a storage site and injecting it into a geological formation located in the seabed for permanent storage (Diario Oficial de la Unión Europea, 2009). Since 2016, some standards and technical reports have been available by the ISO technical committee for different areas of CCS technologies. CCS standards give support to GHG emissions reduction technology (Nasehi, 2020).

Large, commercial-scale projects of CO_2 storage are currently happening across the world (National Energy Technology Laboratory, 2023). For example, in Norway, since 1996, the Sleipner field has been used for the capture and storage of CO_2 (Equinor, 2019).

The international research project in Weyburn Saskatchewan, Canada (2000–2012), developed technological solutions for the design, implementation and monitoring of projects whose objective is to geologically store CO_2 (Her Majesty the Queen in Right of Canada, 2013).

Denmark opens, in the former Nini West oil field, the first CO_2 cemetery at 1800 metres under the North Sea. The "Greensand" project, led by the companies Ineos and Wintershall Dea, will allow to store up to eight million tons of CO_2 "per year" from 2030 (Cepeda Minaya, 2023). The CO_2 that will be stored will be captured in Belgium and Germany and transported by boat to the platform (France24, 2023). The storage potential of the Danish subsurface is 22 Gigatonnes, about 700 years of Danish CO_2 emissions at current levels, according to the Geological Survey of Denmark and Greenland (Cepeda Minaya, 2023).

The North Sea is home to many pipelines that have been left empty after decades of oil and gas production. Near Greensand, the French giant Total Energies will explore the possibility of burying five million tons of CO_2 per year more than two kilometres under the seabed until 2030. A pioneer of the CCS, neighbouring Norway will also receive tonnes of liquefied CO_2 from Europe in the coming years. There are currently more than 200 projects in development around the world (France24, 2023).

Large-scale efforts are also currently underway in China, Australia and Europe. To date, more than 200 CO_2 capture and/or storage operations (including in-development and completed) have been carried out worldwide (National Energy Technology Laboratory, 2023).

According to (France24, 2023) the technique involves risks, warns the research centre, citing the risk of leaks with catastrophic consequences. "CCS should not be used to maintain the current level of CO_2 production, but it is necessary to limit CO_2 in the atmosphere". Besides, according to Zeng et al. (2022), corrosion is the main related problem when injecting CO_2, given the characteristics of this gas as a highly corrosive agent.

In the European Union, Directive 2009/31/EC on the geological storage of CO_2 provides a regulatory framework. It applies to capacities of 100 kilotons or more per year. One of the key provisions of the Directive is that it sets minimum requirements for the selection of storage sites, ensuring that there is no significant risk of harm to health or the environment. In addition, the Directive regulates all cases of geological storage of CO_2. This includes the storage part of any carbon dioxide removal activities that store pure gaseous/supercritical CO_2, that is bioenergy with carbon capture and storage (BECCS) and direct air capture with carbon storage (DACCS) (Carbon Gap, 2023).

The risk analysis in CO_2 geological storage projects carried out by Rodriguez-Dono and Espí Rodríguez (2011) shows that there is no extreme or very high risk that the site is not suitable for storage. Nevertheless, if there is a high risk due to a failure in the injection capacity (reduced injectivity) which would mainly affect the costs of the project (0.5–1 Million USD). All this could imply a low suitability for the geological storage of CO_2, suggesting that control mechanisms and preventive actions should be introduced to reduce this risk. It is also expected that, if CO_2 leaks occur, they could be 7 every 100 years, as shown in Table 9.2.

According to (France24, 2023), the level of CO_2 stored is very small compared to the large amount of CO_2 emissions produced. Although CCS was long classified as a complicated and expensive technology, today the Intergovernmental Panel on Climate Change (IPCC) and the International Energy Agency consider it a necessary technology.

Table 9.2 If CO_2 leaks occur, 7 events are predicted every 100 years, as specified in the table

Type of CO_2 leak	Failure	Overfilling	Dissolution	Overpressure A	Overpressure B	Defects	Abandonment
Place of affectation	In the installation pipes	Overfilling during operation	Dissolution in fluid flow	Injectivity lower than calculated	Injection flow rate higher than due	Corrosion in the well casing	Abandoned wells
Sea surface	1						
Geological warehouse		1	1	1	1	1	1

On the other hand, research is currently being carried out on different types of underground formations for the geological storage of CO_2, such as saline formations, unmineable coal seams, basalt formations and organic-rich shales (National Energy Technology Laboratory, 2023). For example, Aquistore is the largest industry, globally, that studies the storage of CO_2 in deep salt formations. Aquistore began injection in 2015 (The Petroleum Technology Research Centre, 2023).

Within the framework of a circular economy, it could be said that the capture of CO_2 through its mineralisation offers interesting alternatives for the cement industry, since it facilitates the formation of industrial minerals such as calcite and dolomite which have numerous commercial uses (Sanz et al., 2022).

1.5.2 Carbon Capture Utilisation and Storage (CCUS)

Today, the CCS is comprised in the CCUS. In CCUS, captured CO_2 can be injected into deep geological formations for storage, or it can be utilised in situ or compressed and transported by pipeline, ship, rail or truck for use in a variety of applications (IEA, 2021a).

Although, in the last decade (2010–2020) there have been cancellations of high-impact CCUS projects and government support programmes that did not yield results. CCUS by the end of 2021 had an annual capacity surpassing 40 $MtCO_2$. This needs to increase to 1600 $GtCO_2$ by 2030 to be coherent with a line to net zero by 2050. 2021 has generated unprecedented momentum behind CCUS. More than 100 new CCUS plants have been announced and the global project pipeline for CO_2 capture capacity is on track to quadruple. Growing interest in low-carbon hydrogen production has resulted in nearly 50 facilities under development to capture CO_2 from hydrogen-related processes. The investment environment for CCUS has improved significantly, for example, since early 2020, governments and industry have committed more than $25 billion specifically to CCUS projects and programmes. CCUS projects under development are led by the United States with 40.5%, followed by Europe with 35.7%. In the same context, fuel supply (excluding clean hydrogen) leads the applications with 40%, followed by clean hydrogen with 24% (IEA, 2021b).

Clean hydrogen produced from renewable or nuclear energy, or from carbon captured, can help decarbonise chemicals, iron and steel, and long-distance transport. Clean hydrogen can also support the electrical system by allowing energy to be stored for days, weeks or months. However, faster action is needed to create demand for low-emission hydrogen and to stimulate investment that can accelerate the scale-up of production and drive down the cost of technologies for clean hydrogen production (with carbon capture) and use, such as electrolysers and fuel cells (IEA, 2023).

1.6 Avoiding and Storing CO_2 Emissions—Beyond the Value Chain

It refers to investments related to projects that are focused on avoiding and storing GHG emissions. These investments go beyond the company's value chain. In this context, companies can invest in mitigation actions (Watson et al., 2023), for example, through the carbon credit, in the voluntary carbon market (A Primer on the Voluntary Carbon Markets, 2022). Every carbon credit is endorsed by third parties and represents one metric ton of CO_2 (or CO_2e) that has been avoided or stored (CarbonCure, 2022) or will be avoided or stored (Favasuli, 2021). There are many companies, which after generating a carbon credit, can sell it to companies that support low-carbon initiatives or that cannot avoid or store their own emissions. Basically, a carbon credit is a bargaining chip. In today's voluntary carbon markets, the price of a carbon credit can range from a few cents to $300 (for technology-based phase-out projects like CCUS). Today it is seeing more organisations tap into the voluntary carbon markets (A Primer on the Voluntary Carbon Markets, 2022). It is estimated that the market for carbon credits could be worth more than $50 billion by 2030 (Favasuli and Vandana, 2021).

There are two categories of carbon credit in the voluntary carbon markets. The avoidance category and the storage category (CarbonCure, 2022).

1. *The avoidance category* includes projects that avoid CO_2 emissions such as renewable energy, energy efficient buildings, fuel efficiency and stoves. It also includes projects that avoid deforestation, such as the United Nations-backed framework known as REDD+, which aims to curb climate change by halting the destruction of forests through reducing emissions from deforestation and forest degradation. Also in this category are projects to avoid agricultural emissions (through proper nutrition).
2. *The storage category* includes projects that capture and store CO_2. They can focus on nature, for example, to remove and sequester carbon through trees (reforestation, afforestation, and wetland management). They can also be based on forestry or agriculture or focused on direct air capture or carbon capture and storage (Favasuli and Vandana, 2021). And even CO_2 mineralised concrete falls into this category (CarbonCure, 2022).

According with (Favasuli and Vandana, 2021), the main participants in carbon markets are:

3. *Project Developers*: This group develops projects to eliminate or avoid CO_2 emissions. Facilitating the creation of CCs. Projects can range from intrusive scale to something much smaller.
4. *Standards*: It generally refers to non-governmental organisations (NGOs) that certify projects when they meet the objectives set and the agreed volume of emissions.

5. *Retail traders and brokers*: To connect supply to demand, as with other commodity markets, traders buy many credits directly from the supplier, which are grouped into portfolios that can contain hundreds or thousands of tonnes of CO_2 equivalent. These packages are sold to brokers or end buyers, usually with some commission.
6. *End buyers*: They are companies, or individual consumers, seeking to offset some or all their GHG emissions. Initially, the first buyers were Apple, Google, airlines, oil and gas companies. Over time, other industrial sectors, including finance, have joined the market. Even through the PA, it has been easier for countries to buy these credits (A Primer on the Voluntary Carbon Markets, 2022).

In the case of cement, the carbon mineralisation solution, such as CarbonCure, uses a proven methodology to add an exact amount of recycled CO_2 to concrete. When the CO_2 reacts with the cement present in the concrete mixture, it becomes a new mineral, so the CO_2 is permanently stored, and in this way, it is removed from the atmosphere. This methodology also improves the compressive strength of concrete, which allows to reduce both cement content and carbon emissions. In this way, mineralised concrete reduces carbon emissions present in the atmosphere. Following the quantification of the captured CO_2, a carbon credit is created and can then be sold in CarbonCure's global customer network to more than 100 dollars by one metric ton of CO_2e (\$100/mtCO_2e).

2 Positive Impacts That the Cement Industry Will Bring

2.1 Built Environment

In the world, urban areas have about 4400 million inhabitants, which represents more than 50% of the world's population. Globally, by 2045, the urban population will increase to six billion people (1.5 times). If this trend continues, the urban population will double by 2050 and nearly 7 out of 10 people will live in cities. In addition, nearly one billion poor people around the world live in informal urban settlements. One in four people live in areas at high risk of flooding. Exposure is particularly high in the densely populated and rapidly urbanising river basins and coastal areas of developing countries, which account for 89% of the world's flood exposure. Building cities that "work"—that are green, resilient and inclusive—requires intense policy coordination and investment decisions (Worldbank, 2023).

Once a city is built, the physical structure and patterns of land use are often impossible to modify for generations, resulting in unsustainable expansion (Worldbank, 2023). The total volume of cement production worldwide amounted to an estimated 4.1 billion tons in 2022 (Statista, 2023b), see Fig. 9.10, which was elaborated with the data of (International Energy Agency, 2023). Worldwide, 30 Gt of concrete is used each year (Nature, 2021).

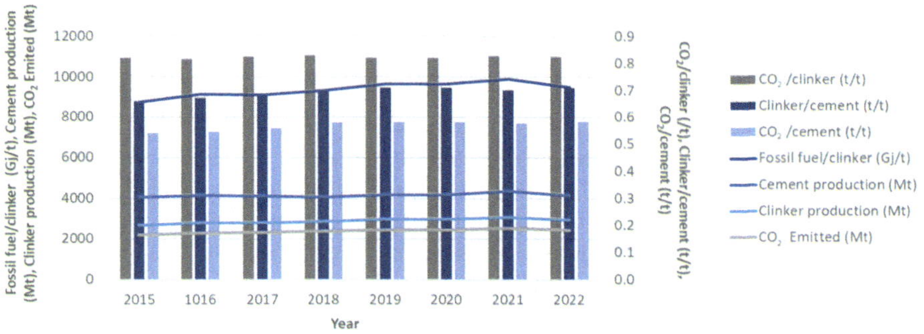

Fig. 9.10 CO_2 emited due to global cement production between 2015 and 2022

Globally, cities cover about three percent of the Earth's surface (15.3 million km^2) but consume 60–80% of energy and emit 75% of carbon (United Nations, 2020). By 2030, urban growth could add up to 1.2 million km^2 of new built area to the world (Worldbank, 2023).

Urban areas are the main generators of GHG. It is estimated that more than 70% of these global emissions are generated in cities. The emissions of these gases are mainly due to the burning of fossil fuels used for heating, cooling, industrial processes, and transport of goods and people (Dasgupta et al., 2022).

On a global scale, the increase in urban building land is a significant contributor to carbon dioxide emissions across different geographical distributions and city sizes. Both the construction and operation of cities have led to changes in both land use and the deep atmosphere (Peng et al., 2022). In general, the materials used in buildings, roads and other infrastructure change the nature of the surface in cities, altering the flow of energy, water and air. As a result, cities are generally warmer than rural areas (EPA, 2023).

As for the hardened surfaces typical of cities, these include asphalt roads, cement concrete, brick pavements and building roofs. Due to their properties, these surfaces will alter the properties of the surface soil, the distribution of solar radiation, the ambient temperature and the water cycle, which in turn will influence public health (Barnes et al., 2002). For example, impervious surfaces block water permeation and reduce the accessibility of water to be stored both in the soil and subsoil. In addition, water evaporation (an important link of the hydrological cycle) on impervious surfaces is intermittent, as it depends on the available water, so it is closely related to the number of rainy days (Zhou et al., 2021). On the other hand, evaporation from permeable surfaces is continuous, so rainwater can seep in, and rainwater, once infiltrated, can pass between the permeable layer, and evaporate into the air. There is concern that the proportion of permeable surfaces in urban areas is negligible compared to impervious surfaces (Zhou et al., 2021).

In this sense, the cement industry, through permeable concrete production, could decrease the environmental impact caused by impervious surfaces present in every city in the world.

2.2 Nature

The rapid overpopulation of the Earth is notoriously influencing global environmental transformations, derived from atmospheric warming induced by both GHG, deforestation and the detriment of biodiversity (Peng et al., 2022). According to FAO, nearly 80% of global deforestation is caused by agriculture (Bodo et al., 2021).

The massive deforestation of tropical forests directly impacts the area subject to deforestation and the compromised ecosystems. Promoting the loss of plant and animal species, soil fertility, water quality, etc. Similarly, the environmental consequences of deforestation affect territories due to the degradation of watersheds and the loss of biodiversity, so deforestation ends up affecting all countries and inhabitants of the planet, thus becoming a global dilemma (Ministerio de Educación, 2023).

Forests cover almost 1/3 of the land globally, which is equivalent to 40.6 million km^2 (or 31% of the entirety land area). In other words, there are about 5400 m^2 of forest for every person on the planet. More than half (54%) of the world's forests are in just five territories: Russia (20%), Brazil (12%), Canada (9%), the United States (8%) and China (5%). The rest of the world has 18.7 (46%) million km^2 (FAO, 2020). See Fig. 9.11.

A total of 93% of the world's forest area is made up of naturally regenerated forests and 7% is planted. Forests around the world preferentially inhabit the tropical region. Of the most important climatic regions, the tropical zone has the maximum percentage of forest at 45% (FAO, 2020), see Fig. 9.12.

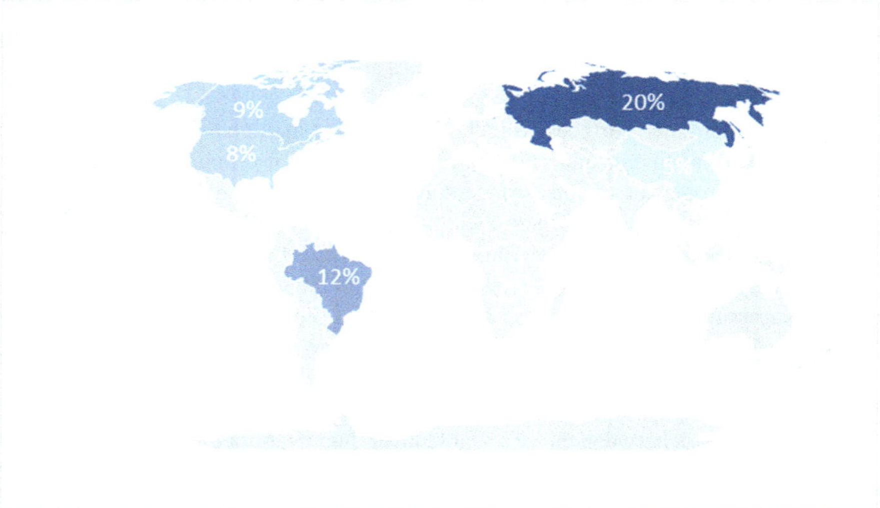

Fig. 9.11 More than half of the world's forests are in just five territories: Russia, Brazil, Canada, the United States and China

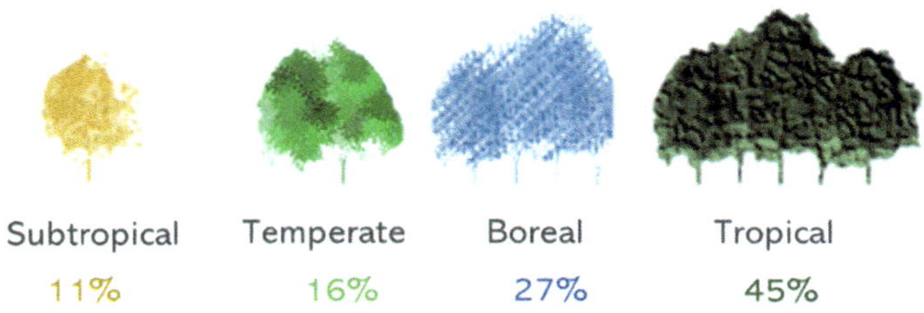

Fig. 9.12 Global percentage of forests by climatic domain. The tropical zone has the maximum percentage of forest, at 45%

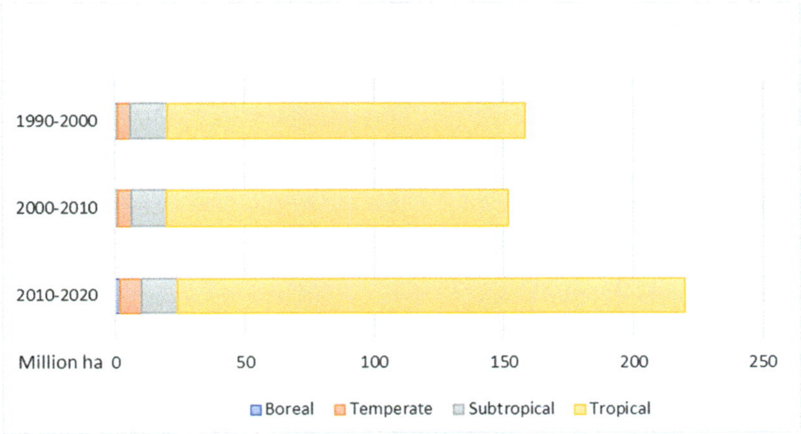

Fig. 9.13 Deforestation rate, by climatic area, between 1990 and 2020 (in million ha)

An estimated 420 million hectares of forest have been lost worldwide to deforestation since 1990. Deforestation has been increasing in the last three decades, as shown in the Fig. 9.13, which shows the deforestation level, by climatic area, for three periods covering 1990–2020 (FAO, 2020).

Mitigating climate change by expanding forests is an interesting alternative due to their high capture potential of CO_2 (planted forests and woodlots were found to have the highest CO_2 removal rates, ranging from 4.5 to 40.7 CO_2 tons/ha per year during the first 20 years of growth) and the possible cost advantages over other mitigation alternatives (the economic value of the forest is five times the costs of its conservation) (Bernal et al., 2018).

Forests help keep global temperatures from rising. They account for 92% of the world's terrestrial biomass and store around 400 $GtCO_2$. Deforestation prevents the absorption of up to 5 $GtCO_2$ per year or about 10% of global GHG emissions. Carbon credits provide income to forest management communities, which also helps to prevent illegal logging and land use change for crops and livestock. More than

620 individual REDD+ projects and programmes were implemented in 2022. Although the majority of these are funded by international organisations such as United Nations-REDD and the World Bank, they are also present in the voluntary carbon market. For example, more than 400 million REDD+ carbon credits were issued in 2022, representing a quarter of the total credits issued in the REDD+ market. Most REDD+ projects are in South America. Others are in sub-Saharan Africa, Asia-Pacific and Central America (Jennifer, 2023).

In this regard, the cement industry could invest in such programmes as the REDD + described above.

2.3 *Biodiversity*

Human activity has affected more than 70% of all ice-free land. Currently, one billion species are at risk of extinction. In most terrestrial habitats, native species have declined considerably in the last century. The main cause of the extinction of species is due to overpopulation that demands food, raw materials and cities to live in (United Nations, 2023a). Besides, soil pollution, in all its forms, has a lasting impact on soil resilience and biodiversity, as well as productivity. About 95% of the food comes from fertile soil which is a finite resource (United Nations, 2023b). If this trend continues, the enormous threat it poses to future generations is clear.

Businesses, as part of society, have an inescapable responsibility, and they need to stop this unprecedented loss of flora and fauna. Cement occupies the number one place among the materials used in the world. The extraction of raw materials for cement processing as well as its high levels of GHG emissions generates significant environmental impacts, so the cement industry carries great risks for biodiversity and ecosystems.

Until now, the conservation, restoration and sustainable management of habitats and ecosystems are recognised as the most effective way to protect biodiversity. One example is the Great Green Wall, which is an initiative that seeks to restore biodiversity and rehabilitate Africa's degraded landscapes, by restoring 100 million hectares of land. More than 100 countries have committed to reducing and reversing degradation on more than 4.5 million km^2 of land by 2030, of which 2.5 million km^2 is agricultural land (United Nations, 2023a).

The cement and concrete industry have committed to carbon reduction for 20 years, since the formation of the Cement Sustainability Initiative, (a series of tools to address sustainability and the care of biodiversity, in each of the stages of the extraction of raw materials. These tools include the environmental and social impact assessment and quarry rehabilitation, focused on sustainable development and biodiversity (International Union for Conservation of Nature and Natural Resources, 2014)) and the adoption of the CO_2 Monitoring Protocol in 2001. Following its formation in 2018, the GCCA has taken responsibility for this work and adopted the Sustainability Charter which has the following key guidelines (GCCA, 2020):

1. Monitoring and reporting of CO_2 and other emissions.
2. Co-processing of fuels and materials.
3. Biodiversity and quarry rehabilitation.
4. Health and safety.
5. Water management.

Given the widely recognised links between biodiversity and climate change, the cement industry reaffirms its commitment to improving biodiversity in its operations. The new Cement, Concrete and Aggregates Biodiversity Policy outlines four key actions to ensure ecosystems can thrive (GCCA, 2020):

- Formulate and implement effective and progressive quarry rehabilitation and biodiversity management plans.
- Track, monitor, report and provide assurance through key performance indicators.
- Highlight the strong sustainability attributes of concrete, such as its high versatility, which enables material efficiency in line with nature-based solution initiatives.
- Work in partnership to scale up efforts, building strong links with civil society organisations.

2.4 Society

The cement industries, by providing employment, also carry certain social benefits that complement the worker's salary. That is, these are services or products whose cost falls directly on the company. In this perspective, for example, in the United States, there are 74 cement manufacturing businesses as of 2023. There are 11,457 people employed in this industry. The average business in this industry now employs more workers than it did 5 years ago. However, the number of these businesses has declined 10.7% per year on average, over the 5 years between 2018 and 2023 (Industry Statistics—United States, 2023).

Both dust and respirable cement particles are the two causes that threaten the health of workers in the cement industry to a greater extent. Some studies show that the concentrations of lead, chromium and cadmium in these particles had low risk rates for all plant workplaces except the kiln and material grinding site, where workers were exposed to higher cadmium and lead concentrations. Therefore, strategies are recommended to reduce exposure to heavy metals in these sectors and thus reduce the risk (Pouransari et al., 2020). Adverse health effects for the population living near cement plants were investigated. Almost all studies showed respiratory illnesses and symptoms associated with exposure to the cement plant (Raffetti et al., 2019).

On the other hand, cement companies can lead to a variety of social outcomes related to sustainable activities, biodiversity management, educational and research activities, as well as the creation of local groups, by adopting an Integrated

Biodiversity Management System (IBMS) (Luo et al., 2021). A Rehabilitation Plan can also promote the social dimensions. For example, it can offer opportunities focused on the employment of local workers, in the areas of rehabilitation, enrichment and restoration of ecosystems (efficiency in the use of water and aesthetic management of the environment). Besides, the plan may also require different land use, such as forest management, regulated fishing and ecological recreation (International Labour Organization, 2020).

3 Outlook for the Cement Industry

According to (GCCA, 2020), to address the global climate challenge, the cement industry is adopting a circular economy approach and, through the GCCA, is publishing its vision of what cement will look like in the future.

3.1 CO_2 *from Clinker Production*

High rates of alternative/biomass fuels (or even possibly green energy in electrified kilns) will replace fossil fuels still in use today. (In general, there is no technical barrier to achieving rates of 100% replacement of conventional fuels in cement production, the barriers being related to local conditions). CO_2 from limestone calcination will be captured to carbonate fresh or recycled concrete.

3.2 *Less Clinker in Cement*

Today's conventional additives, including slags, fly ash, natural and artificial pozzolans, silica fume, limestone fillers, calcined clays, rice husk ash and many others, are added to reduce the amount of clinker in the cement. The clinker of the future will be more reactive, allowing less clinker to be used in cement. Cement will be produced with an optimised particle size distribution of its constituents. The addition of supplementary cementitious materials such as calcined clay, carbonated recycled concrete fines and selected limestone fillers will increase. Cement will also be blended to optimise its performance.

3.3 *Less Cement in Concrete*

Concrete constituents will include new alternative binder concepts. Production will be digitally controlled through data analytics and artificial intelligence. Less water will be utilised because of the optimisation of the paste. Value differentiation will no longer be based solely on strength performance.

3.4 Mineral Carbonation

During its life cycle, concrete reacts with CO_2 from the air, resulting in its carbonation or carbon uptake. This process occurs at different rates, e.g. it is rapid in thin and porous applications (masonry exposed to air can be completely carbonated in a few years), but slower in reinforced concrete and thicker elements. As expected, the carbonation of concrete increases as the surface area exposed to air increases. For example, in reinforced concrete structures that are demolished and exposed to air. Finally, there are several binders that emit lower amounts of CO_2 during production (the raw material is burned at lower temperatures) and others that use a significant amount of CO_2 during curing. However, the use of these new binders is currently limited to niche markets (GCCA, 2024).

3.5 Recycling

Concrete recycling should be encouraged through stricter regulation and together with more technologically advanced demolition. This will avoid sending concrete to landfill and maximise CO_2 absorption.

3.6 CO_2 Capture

CO_2 capture is still expensive today, but the technology is improving. Some CO_2 capture technologies have been developed to date: calcium looping, oxyfuel, chemical absorption direct capture, mineral adsorption and CO_2 utilisation technologies. The binder of the future will therefore capture its own emitted CO_2 during production, use, recycling and upcycling. In other words, it will enable the complete closure of the CO_2 loop.

4 Conclusions

According to the 2015 PA, it is vital for the survival of life on the planet that CO_2 emissions are reduced by 50% by 2030 and by 90% by 2050. Industries such as cement, whose products are obtained through chemical reactions that produce CO_2 as one of their by-products, and which also require constant and large quantities of resources, require profound and substantial changes. Not only the use of renewable energies to replace fossil and non-renewable fuels (car oils and tyres). They will also have to change their raw materials, processes, technologies and, of course, their products. It should be noted that 60% of the excessive amount of CO_2 emitted into the atmosphere by the cement industry is due to the calcination of its raw materials and 25% to the intensive use of fossil fuels. Thus, 85% of its GHG emissions are at risk. This makes its decarbonisation process a major challenge. This is because the calcination of limestone at 1450 °C is an insurmountable aspect due to the nature of the chemical reaction resulting from the nature of the raw materials used.

The concern arises from the fact that, between 2015 and 2022, not only will CO_2 emissions from this industry increase by 10% (from 2198 to 2428 Mt), but also the ratio of clinker to cement will increase by 8% (from 0.66 to 0.71), and energy consumption from fossil fuels will increase by 10% (from 8759 to 9506 Gj). This could suggest that the cement industry is not committed to the environment, biodiversity or climate change after the PA, despite the initiatives it claims to have taken.

In conclusion, the cement industry will not reduce the production of clinker in the coming decades but will seek to store the CO_2 it emits through calcination in the form of carbonates (mineralisation), so that cement in the form of concrete will reach a point where, through constant use and over-recycling, it will absorb (store) all the CO_2 it has emitted in this way, thus closing the emissions cycle. It will also make extensive use of renewable fuels to replace the fossil fuels that will have been depleted by then. However, for the sake of the survival of the entire planet, it would be a priority to eliminate the calcination of limestone without delay.

References

2020 United Nations Environment Programme. (2020). *Emissions Gap Report 2020.*

A Primer on the Voluntary Carbon Markets. (2022).

Barnes, K. B., Iii, J. M. M., Roberge, M. C., & Roberge, M. (2002). *Impervious surfaces and the quality of natural and built environments.* Department of Geography and Environmental Planning Towson.

Bernal, B., Murray, L. T., & Pearson, T. R. H. (2018). Global carbon dioxide removal rates from forest landscape restoration activities. *Carbon Balance and Management, 13*(1), 22. https://doi.org/10.1186/s13021-018-0110-8

Biernacki, J. J., Bullard, J. W., Sant, G., Brown, K., Glasser, F. P., Jones, S., Ley, T., et al. (2017). Cements in the 21 St century: Challenges, perspectives, and opportunities. *Journal of the American Ceramic Society, 100*(7), 2746–2773. https://doi.org/10.1111/jace.14948

Blaufelder, C., Levy, C., Mannion, P., & Pinner, D. (2021). A blueprint for scaling voluntary carbon markets to meet the climate challenge.

Bodo, T., Gimah, B. G., & Seomoni, K. J. (2021). Deforestation and habitat loss: Human causes, consequences and possible solutions. *Journal of Geographical Research, 4*(2), 22–30. https://doi.org/10.30564/jgr.v4i2.3059

Calvin, K., Dasgupta, D., Krinner, G., Mukherji, A., Thorne, P. W., Trisos, C., Romero, J., et al. (2023). IPCC, 2023: Climate Change 2023: Synthesis Report. Contribution of Working Groups I, II and III to the Sixth Assessment Report of the Intergovernmental Panel on Climate Change. In Core Writing Team, H. Lee, & J. Romero (Eds.). IPCC, Geneva, Switzerland. Edited

by P. Arias, M. Bustamante, I. Elgizouli, G. Flato, M. Howden, C. Méndez-Vallejo, J. J. Pereira, et al. https://doi.org/10.59327/IPCC/AR6-9789291691647
Carbon Gap. (2023). *CCS Directive*. In a Nutshell.
CarbonCure. (2022). Carbon credits: The new revenue opportunity for concrete producers.
Carter, D. (2023). How CO_2 emissions can be reduced in cement industry? Proper guidance for you.
Centro de Estudios y Experimentación de Obras Públicas. (2022). Development of Natural Cements and Invention of Artificial Portland Cement.
Cepeda Minaya, D. (2023). Así es el primer cementerio de carbono marino europeo contra el cambio climático.
Christensen, J., & Olhoff, A. (2019). *Lessons from a Decade of Emissions Gap Assessments*. United Nations Environment Programme, Nairobi.
Dasgupta, S., Lall, S., & Wheeler, D. (2022). Cutting global carbon emissions: Where do cities stand?
Diario Oficial de la Unión Europea. (2009). *Directiva 2009/31/CE del Parlamento Europeo y del Consejo de 23 de abril de 2009 relativa al almacenamiento geológico de dióxido de carbono.*
EPA. (2023). Learn About Heat Islands.
Equinor. (2019). Sleipner Partnership Releases CO_2 Storage Data.
Fakta o klimatu. (2023). Global CO_2 Emissions from Fossil Fuels & Cement Production.
FAO. (2020). *Global Forest Resources Assessment 2020.*
Favasuli, S. (2021). Recognizing limitations and potential of REDD+ credits may offer way out of a destructive debate.
Favasuli, S., & Vandana, S. (2021). Voluntary carbon markets: How they work, how they're priced and who's involved.
France24. (2023). Dinamarca abre en el Mar del Norte el primer cementerio de CO_2 importado.
GCCA. (2020). GCCA Climate Ambition Statemen.
GCCA. (2024). Carbon Uptake.
Hack, J., Maeda, N., & Meier, D. M. (2022). Review on CO_2 capture using amine-functionalized materials. *ACS Omega*. https://doi.org/10.1021/acsomega.2c03385
Hasanbeigi, A. (2019). California's cement industry: Failing the climate challenge.
Hasanbeigi, A. (2021). Global cement industry's GHG emissions.
Hausfather, Z. (2018). Analysis: Why the IPCC 1.5C report expanded the carbon budget. *Carbon Brief.*
Her Majesty the Queen in Right of Canada. (2013). International energy agency greenhouse gas Weyburn-Midale CO_2 monitoring and storage project.
IEA. (2021a). About CCUS.
IEA. (2021b). Carbon capture in 2021: Off and running or another false start?
IEA. (2023). Hydrogen.
Industry Statistics -United States. (2023). Cement manufacturing in the US.
International Energy Agency. (2023). Cement.
International Labour Organization. (2020). Promoting employment and decent work in a changing landscape.
International Union for Conservation of Nature and Natural Resources. (2014). Biodiversity management in the cement and aggregates sector.
IPPC. (2022). *Climate Change 2022.*
Jennifer, L. (2023). What is REDD+? Development, issues, and solutions.
Linde. (2023). Linde Engineering.
Luo, M., Zhang, S., Wei, H., Zhou, P., Zhou, Y., Chen, Y., Li, Z., & Zhang, J. (2021). Ecological restoration of the Dabaoshan mine based on IUCN biodiversity management guidelines. *Earth Science Frontiers, 28*(4), 90–94. https://doi.org/10.13745/j.esf.sf.2020.10.9
Ministerio de Educación. (2023). Las múltiples ecalas de los problemas ambientales.
Morgado, A., Hugues, P., & Vass, T. (2023). Cement.
Nasehi, M. (2020). ISO standards for carbon capture utilization and storage.
National Energy Technology Laboratory. (2023). Carbon Storage FAQS.

Nature. (2021). Concrete needs to lose its colossal carbon footprint. *Nature, 597*(30 September 2021), 593–594.
Peng, J., Zheng, Y., & Liu, C. (2022). The impact of urban construction land use change on carbon emissions: Evidence from the China land market in 2000–2019. *Land, 11*(9), 1440. https://doi.org/10.3390/land11091440
Pissaro, F. (2023). Concrete policies to underpin the cement transition.
Pouransari, M., Mohammadyan, M., & Mohammadpour, R. (2020). Risk assessment of occupational exposure to heavy metals (lead, chromium, cadmium) in cement respirable particles in a cement factory. *Iran Occupational Health, 17*(1), 103–115.
Raffetti, E., Treccani, M., & Donato, F. (2019). Cement plant emissions and health effects in the general population: A systematic review. *Chemosphere, 218*(March 2019), 211–222. https://doi.org/10.1016/j.chemosphere.2018.11.088
Rodriguez-Dono, A., & Espí Rodríguez, J. A. (2011). *Análisis de riesgos en proyectos de amacenamiento geológico de CO2: selección de emplazamientos adecuados, costes y análisis de los resgos asociados a este tipo de proyectos.*
Sanz, J., Tomasa, O., Jimenez-Franco, A., & Sidki-Rius, N. (2022). Elements and mineral resources. In *Springer textbooks in earth sciences, geography and environment.* Springer International Publishing. https://doi.org/10.1007/978-3-030-85889-6
Science Based Target. (2021). Pathways to Net-Zero SBTi Technical Summary.
Science Based Target. (2022). Cement science based target setting guidance.
Science Museum. (2021). Building the modern world: Concrete and our environment.
Statista. (2023a). Forecast of carbon dioxide emissions worldwide from 2018 to 2050.
Statista. (2023b). Cement production worldwide from 1995 to 2022.
The guardian. (2022). What are the key outcomes of Cop27 climate summit? *The Guardian.*
The Petroleum Technology Research Centre. (2023). Aquistore project reaches +500,000 tonnes of CO_2 stored.
UNEP. (2023). How do countries measure greenhouse gas emissions?
United Nations. (2020). World cities day 2020: Better city, better life.
United Nations. (2023a). Biodiversity.
United Nations. (2023b). World Environment Day: Tackling plastic pollution from the ground up.
Wang, K., Costanza-van denBelt, M., Heath, G., Walzberg, J., Curtis, T., Barrie, J., Schröder, P., Lazer, L., & Altamirano, J. C. (2022). Circular economy as a climate strategy: Current knowledge and calls-to-action.
Watson, E., Chang, A., Carrillo Pineda, A., Anderson, C., Cummis, C., & Stevenson, M. (2023). *SBTi Corporate Net-Zero Standard.*
World Business Council for Sustainable Development. (2005). The Cement Sustainability Initiative Progress Report June 2005.
World Business Council for Sustainable Development. (2021). Vision 2050.
Worldbank. (2023). Desarrollo Urbano.
Zeng, D., Dong, B., Zhang, S., Yi, Y., Huang, Z., Tian, G., Yu, H., & Sun, Y. (2022). Annular corrosion risk analysis of gas injection in CO_2 flooding and development of oil-based annulus protection fluid. *Journal of Petroleum Science and Engineering, 208*(Part (January 2022)), 109526. https://doi.org/10.1016/j.petrol.2021.109526
Zhou, J., Liu, J., Chu, Q., Wang, H., Shao, W., Luo, Z., & Zhang, Y. (2021). Mechanisms and empirical modeling of evaporation from hardened surfaces in urban areas. *International Journal of Environmental Research and Public Health, 18*(4), 1790. https://doi.org/10.3390/ijerph18041790

Chapter 10
Novel Approaches for Greener Synthesis of Extremozymes Using Agro/Food Waste

Freny Shah, Bablesh Ranawat, Vishwa Patel, Shailesh Kumar Patidar, Rohan Mani Thomas, and Sandhya Mishra

1 Introduction

Extremozymes are enzymes produced by extremophilic microbes (e.g., fungi, bacteria, and algae) surviving in extreme environmental ranges of temperatures, pH, pressure, and high salt or organic/solvent concentrations (Elleuche et al., 2014). Recently, several efforts have started to shift petroleum matrices into sustainable and renewable feedstock-based industries for a greener production process, ultimately resulting in high-value products by eliminating the chances of environmental pollution (Shah et al., 2021). However, the manufacture of industrially important chemicals more specifically enzymes, at lower prices is a difficult task as enzymes are more frequently used in aqueous solutions and it is observed that natural substrates do not always provide the chosen and useful product (Woodley, 2019). To address this challenge, researchers are working for an efficient solution. For instance, protein engineering, genetic engineering, and several chemical modification approaches can be applied for improving enzyme stability and activity. Modification in the size of the active site groove along with the alteration in amino acid sequences may provide enhanced effectiveness to fine-tune the activity of the enzyme not only towards the non-native substrate but also helps to elevate the stability of the enzyme under harsh or unnatural environmental conditions (Strohmeier et al., 2011; Liszka

F. Shah · V. Patel · R. M. Thomas · S. Mishra (✉)
Applied Phycology and Biotechnology Division, CSIR-Central Salt and Marine Chemicals Research Institute, G.B. Marg, Bhavnagar, Gujarat, India
e-mail: smishra@csmcri.res.in

B. Ranawat
Department of Earth Sciences, Indian Institute of Technology, Mumbai, Maharashtra, India

S. K. Patidar
Department of Environmental Science, Central University of Rajasthan, Bandar Sindri, Kishangarh, Rajasthan, India

V. Kumar et al. (eds.), *Environmental Engineering and Waste Management*,
https://doi.org/10.1007/978-3-031-58441-1_10

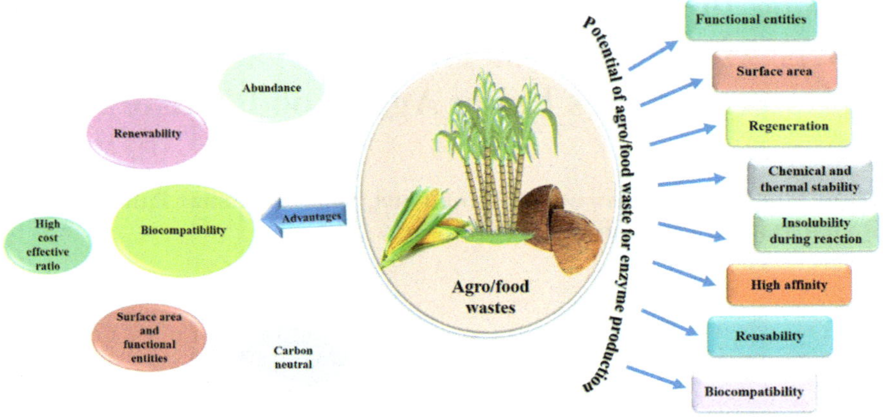

Fig. 10.1 Significance with potential of agro/food waste for extremozymes production

et al., 2012; Reetz, 2016). To meet the global demand, we must create a trend of using sustainable, renewable and widely available natural resources such as agro/ food waste, which results in a zero-waste strategy (Arevalo-Gallegos et al., 2017; Dwivedi et al., 2022; Dhanker et al., 2022). Agronomical/food wastes are widely available and are also bio-renewable, which are promising resources for their suitability as substrates for the production of enzymes by reducing the production cost and making the entire process economical and environmentally green. Figure 10.1 shows the significance of agro/food waste in the production of enzymes.

Agronomical/food waste can be defined as the leftover or unused residues acquired from farming and other food and poultry-based industries along with other raw materials such as vegetables, crops, fruits, and milk products (Saharia et al., 2024). The waste generated from these agricultural commodities is very useful substrate. However, their economic importance is comparatively less for the production of high-value products, but these can also serve the purpose of waste utilization. Further, these wastes can be categorized into non-lignocellulosic (e.g., eggshells and their membrane) and lignocellulosic (e.g., spent grains, coconut fiber, corn stover, sugarcane bagasse, wheat bran, and coffee ground waste). As mentioned above, agro/food waste generated from the agricultural industries creates pollution but, if reused, their compounds can serve as a great benefit. Agricultural and food wastes are comprised majorly of the cellulosic backbone and on the basis of their sources, they may consist of the high carbon content along with the other multifunctional groups. The properties of agro/food waste have been evaluated, which led them for their application in several areas such as biodiesel generation, effluent management, and for the production of enzymes and phytochemicals (Yahya et al., 2015). Therefore, utilization of agro/food-based wastes as an inexpensive source of natural substrates is increasing nowadays for the production of enzymes from extremophilic micro/macroorganisms (Shah et al., 2021).

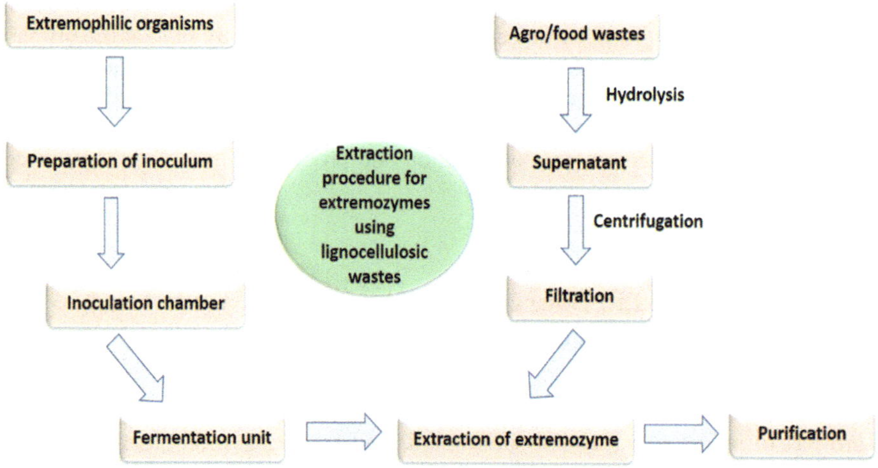

Fig. 10.2 Schematic representation for production of extremozymes from agro/food waste

In recent years, the processing and management of agro/food waste for their applicability in several fields such as the manufacture of enzymes has gained the interest of researchers (Mostafa et al., 2018). The overall process for the extraction of extremozymes using lignocellulosic waste materials has been explained through a schematic diagram (Fig. 10.2).

2 Problem and Solution

The demand for enzymes in different industries is continuously increasing, and the use of a pure form of carbon and nitrogen source for enzyme production is expensive. Thus, organic waste as a low-cost substrate is used to produce enzymes, mitigate the market demand, and reduce the production cost. Besides, this action helps in decreasing organic waste and finally pollution (Shrestha et al., 2022). It has been seen that in common practice, lignocellulosic/food /agro waste is either discharged to the environment in its native form without undergoing proper treatment or simply burnt off resulting in the environmental load and landfills. To solve this major obstacle, lignocellulosic wastes should be used as potential raw materials for production of high-value products. This attempt might require better technology along with application of agro-waste management approaches. The current scenario suggests that the estimation of food/agro-based waste generation is increasing day by day. As per the data, global food waste generation per annum is approximately 30% for cereals, 20% for oil seeds, and 40–50% for rooted crops. Other research data proved a food waste generation of about 95–115 kg/year per capita by European and North American consumers. Consumers from South Asia and sub-Saharan Africa discarded around 6–11 kg/year (Corrado & Sala, 2018). It has been observed that in a developing country, food waste can be seen to be significantly high at the primary stages of the food value chain, which is caused by financial and technical problems in harvesting techniques, storage, and facilities for cooling. Some research has been actively done on the use of lignocellulosic waste for extraction of phytochemicals, production of biodiesel, enzyme preparation and production as well as wastewater treatment.

3 Composition and Source of the Agro/Food Waste Used for the Production of Extremozymes

Agro/food waste is categorized as lignocellulosic wastes produced massively by agro-based industries, forestry, household garbage, and food-based industries (Fig. 10.3).

The disposal of these types of wastes is a challenge as it causes environmental pollution. Generally, lignocellulosic wastes are disposed of by burning them, which leads to the generation of methane and ammonia ultimately causing the greenhouse gas emissions globally resulting in pollution (Anwar et al., 2014). Presently, lignocellulosic/food/agro-based biomasses have gained the research impetus due to their greater availability and bio-renewability (Bilal et al., 2017; Shah et al., 2019, 2021). A great number of agro/food waste can be used for the transformation of high-value products such as fine chemicals, biofuels, and enzymatic support and also could be used for the production of enzymes.

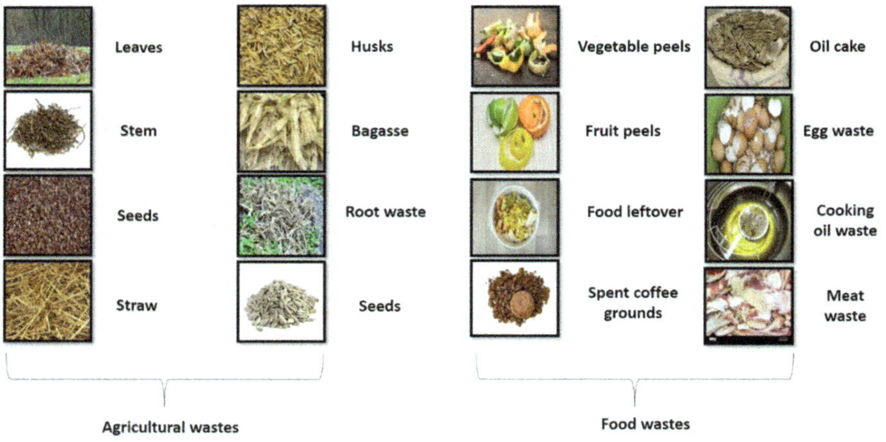

Fig. 10.3 Variety of agro/food waste used as a substrate for the production of extremozymes

4 Utilization of Agro/Food Waste to Produce Value-Added Compounds

Waste generated by food or agricultural industries is one of the major environmental issues being faced by us. These wastes can be generated at various stages during the production and processing of food and even at the consumption stage. Improper waste disposal may cause significant sanitary and environmental issues in a lot of countries (Garcia-Garcia et al., 2017). Also, a few parts of several food products are inedible to humankind for example, banana peel or spent coffee grounds and they eventually become part of the waste. Therefore, we need to bring some sustainable approach to attain the proper and feasible options for the management of such waste.

Growing of microorganism utilizing agricultural wastes can be one of the great solutions toward the waste management and further exploiting those organisms for the production of several high-value products may be a great step towards waste management.

The application of lignocellulosic wastes as a growth constituent for microorganisms constitutes a great alternative for the production of enzymes with robust properties also. It has been established that the enzymes extracted from the extremophilic microorganisms are more active and stable when compared to the enzymes derived from animal and plant sources and hence, they have wide biotechnological applications. Besides, they have several unique characteristics such as wide biochemical diversity, gene manipulation sensitivity, and most importantly large-scale production in a shorter duration making the extremophilic microorganisms as an alternate and natural source for enzyme production. Hence, the catalyst industries could use novel microorganisms for several enzyme productions to meet the global need for enzymes having robust characteristics. Some examples of extremozymes production using lignocellulosic wastes are mentioned below.

4.1 *Amylase*

For the production of α-amylase from *Gibberella fujikuroi* which has some characteristics that can withstand harsh environmental or industrial conditions, many agro-wastes such as wheat bran, wheat straw, rice straw, jowar straw, and gram husk have been tested (Mulimani & Ramalingam, 2000). It has been reported that α-amylase from thermophilic *Bacillus subtilis* showed 7.3-fold higher enzyme production when wheat bran used as substrate (Baysal et al., 2003). In another study, α-amylase production from *Aspergillus oryzae* showed 20% enhanced yield when spent coffee grains were used as the carbon source at lower pH (pH 5) (Francis et al., 2003). High titers of thermostable α-amylase were obtained by Soni et al. (2003) using *Bacillus* sp. and wheat bran as the sole carbon source. Another study done by Bhatt et al. (2020) suggested the maximum amylase production when moong husk (4%) and soybean cake (2%) were used as a substrate. Ramachandran et al. (2004) obtained thermostable α-amylase from fungi *A. oryzae* supplementing coconut oil cake as a substrate. Mamo and Gessesse (1999) reported the production of two α-amylases (AmyI and AmyII) by thermophilic *Bacillus* strain WN11 using raw potato and corn starch as a carbon source. Production of amylase using thermotolerant *Bacillus cereus* was reported by Vijayaraghavan et al. (2015) when the production medium was supplemented with cow dung as a substrate. Mihajlovski et al. (2016) reported improved production of β-amylase through *Paenibacillus chitinolyticus* CKS1 under extreme optimized condition using beet pulp and molasses as the substrate. Several cost-effective carbon substrates such as potato peel, rice husk, and wheat bran were used for the production of α-amylase by thermophilic *B. subtilis* (Shukla & Kar, 2006). Maximum yield of extracellular amylase by *Baillus siamensis* F2 was observed when the production medium was supplemented with different agro/food waste such as sweet potato peel, rice husk, and sugarcane bagasse. The extracted amylase was further utilized for the production of bioethanol (Rathod et al., 2023). Ayinde et al. (2023) reported the maximum amylase production using potato peel as the substrate via solid-state fermentation. Agro/food waste has attracted significant attention over the globe for a long, as such waste is generally dumped or burned that ultimately causes a serious threat to humans as well as the environment. To overcome this serious issue, Naik et al. (2023) have demonstrated the use of agricultural waste for the cost-effective and eco-friendly production of amylase.

4.2 Cellulase

The production of cellulase that withstand the robust environmental conditions from extremophilic microbes has caught the eye nowadays due to their high stability and efficiency in catalyzing several industrial processes. In a study, *Virgibacillus salaries* strain BM-02, a halophilic bacteria was found to utilize several non-pre-treated agro-wastes like corn stover, cellulose, wheat bran, and sugarcane bagasse for the maximum yield of cellulase (Yousef & Mawad, 2023). Cellulose-hydrolyzing organisms significantly use carbohydrates as their energy source and are unable to utilize lipids or proteins as an energy substrate for their growth (Lynd et al., 2002). Cellulase is obtained through green process of fermentation of agro-waste using cellulolytic microbes, for example, *Trichoderma reesei, Aspergillus niger,* and *Penicillium funiculosum.* Cellulolytic microbes have the competence of biosynthesizing cellulase enzymes act on cellulose and converting cellulose (polymer) into glucose (monomer). Cellulase enzymes may be extracted from the above microbes and purified for the industrial application. This process becomes greener when extraction is also carried out by green chemicals apart from the raw bio-substrates (Ezejiofor et al., 2014).

It has been reported that the high titre production of cellulase can be effectively carried out by *Penicillium* sp. when the production media was supplemented with rice straw, and they found that extracted cellulase was optimally stable and active even at 65 °C at acidic pH (4.0–5.0) (Picart et al., 2007). Annamalai et al. (2014) observed cellulase which was showing extremely high enzyme activity around 4040.45 U/mL when *Bacillus carboniphilus* CAS 3 was supplemented with 6.27 g/L rice bran and 2.52 g/L of yeast extract at pH 9, and temperature 50 °C. Liu et al. (2011) grown *Aspergillus fumigates* Z5 under solid-state fermentation (SSF) for thermostable cellulase production to degrade agricultural wastes for ethanol production, which yielded 0.112 g bioethanol/g of dry substrate, signifying its potential for the production of bioethanol. Annamalai et al. (2013) reported an extracellular haloalkali-tolerant cellulase from *Bacillus halondurans* CAS 1 when using lignocellulosic waste as the carbon source. The enzyme was stable and active at 60 °C and pH 9.0 (Annamalai et al., 2013). Another study by Olajuyigbe and Ogunyewo (2016) suggested the maximum cellulase production, that is 285.7 U/mL was recorded at 96 h of cultivation period when inoculated with 2.5% lignocellulosic waste at pH 6.0 and temperature 80 °C. Zhang et al. (2023) reported the maximum cellulase production (13.503 U/ml) in the presence of rice straw by *Raoultella terrigena*. Maximum yield of cellulase (0.21 μmol/ ml) was observed by *Paenibacillus polymyxa* strain H210517 when the production medium was supplemented with *Zea maize* as a carbon source (Hassan et al., 2022).

4.3 *Tannase*

Tannase can be produced and biosynthesized using microbes acting on raw biomaterials such as tannin-rich plant materials like tea leaves, coffee beans, and grape pomace. Popular tannase-producing microorganisms such as *Aspergillus niger, Aspergillus oryzae, and Penicillium chrysogenum* are capable to produce industrial scale tannase enzymes that can break down tannins into gallic acid and glucose in the bioreactors. The process involves the use of raw biomaterials as a substrate for the growth of these microorganisms under optimal environmental conditions (Lekshmi et al., 2021).

Tannase is a member of hydrolytic enzyme family. The major application of tannase is in the food and pharma industries. Tannase plays an important role in the stabilization of malt polyphenols. It can also be used in the tannery effluent treatment, in fruit juice, and beer clarification by preventing the maderisation in wine and fruit juices (Aguilar & Sanchez, 2001). Palm kernel cake and tamarind seed powder were utilized by Sabu et al. (2005) for the production of tannase from thermophilic *Aspergillus niger*. de Lima et al. (2014) used agricultural waste for the production of tannase from *Penicillium montanense* URM6286. It recorded the maximum activity of tannase around 41.64 IU/ml at pH 9.0 and temperature 50 °C after 72 h of incubation. It was also observed that the tannase produced from *Penicillium montanense* URM6286 was stable over a wide temperature and pH range. Several agro/food wastes have been studied extensively for the production of tannase from the *Aspergillus terreus* such as coffee husk, green gram husk, red gram husk, coconut powder, cotton seed waste, tamarind powder, wheat bran, rice bran, cashew apple bagasse, and corn powder. Among all the agro/food waste, maximum tannase production was observed when the production medium was supplemented with wheat bran at acidic pH 3.5 after 72 h of incubation time (Malgireddy & Nimma, 2015). *Penicillium verrucosum* produces maximum tannase after 96 h of incubation/fermentation period when coffee pulp was used as a substrate (Bhoite & Murthy, 2015). Mohan et al. (2014a, b) used RSM for the maximum tannase production by thermophilic fungi *Aspergillus foetidus* utilizing red gram husk as the substrate at pH 5.5 and temperature 35.5 after 97 h of fermentation period. Shah et al. (2023) reviewed the use of several lignocellulosic waste material for optimum tannase production through solid state fermentation technique. They used pomegranate peel, banana peel waste, or potato peel waste for tannase production.

4.4 Xylanase

Xylanases are majorly used for juice and wine clarification, for the quality improvement of bakery products, and also in bio-bleaching (Bhat, 2000). Many agricultural waste biomass had been evaluated by Gawande and Kamat (1999) like rice straw, wheat bran, sugarcane bagasse, and soya bean hulls for the production of xylanse by *Aspergillus terreus* and *Aspergillus niger*. Highest yield of xylanase was recorded when the production media was supplemented with wheat bran. This result was outstanding and opened the door to replacing costly substrate with the cheaper ones that too with high product titres. Previously xylan was used for the production of xylanase but cost-wise xylan is costlier than wheat bran which led to the high production cost of the enzyme. Moreover, both the strains of *Aspergillus* used in the study produced a higher quantity of xylanase with a lower level of cellulase ultimately making the purification very easier. It was observed by Milagres et al. (2004) that a high amount of thermostable xylanase around 1597 U/g was produced by *Thermoascus aurantiacus* after 10 days of fermentation period. Sonia et al. (2005) recorded maximum xylanase production (11,855 U/g) using thermophilic fungi *Thermomyces lanuginosus* while utilizing sorghum straw as a substrate. Several agricultural wastes were evaluated for the production of xylanse and it was found that rice bran was showing maximum xylanase production using thermophilic *Bacillus* sp. (Virupakshi et al., 2005). Azin et al. (2007) reported the maximum xylanase production having xylanase activity of 479.7 U/g with thermo-acidophilic *Trichoderma longibrachiatum*. They used a combination of wheat bran and wheat straw (7:3) for the xylanase production. Soliman et al. (2012) reported the maximum cellulase production 42.5 U/g from *Trichoderma viride* at pH 5.5, temperature 35 °C, and after 2 days of fermentation period using barely bran as the substrate.

Izidoro and Knob (2014) utilized several lignocellulosic waste materials such as brewer's spent grain, soya bean peel, passion fruit peel, orange peel, wheat bran, corn straw, sugarcane straw, pineapple peel, and orange peel for the xylanase production by thermophilic fungus *Aspergillus niger*. The study suggested that maximum activity of xylanase was recorded at acidic pH 5.0 and at 30 °C after 5 days of incubation period when brewer's spent grain was utilized as a substrate. Guler and Özçelik (2023) reported the maximum yield (23.47 U/ml) of xylanse by novel *Bacillus* sp. that is, *Bacillus halotolerans* DSM 8802 through submerged fermentation technique when the production medium was supplemented with sourdough and wheat bran. Rice straw waste was optimized for the production of xylanase using *Bacillus altitudinis* RS3025 by Ketsakhon et al. (2023).

4.5 Protease

Agri-food waste arises at any point in the food distribution chain during the harvesting and processing of agri-material to derive value-added products. Such wastes decompose readily, generate odors, and sometimes act as a potential source to cause infections. Conventional disposal methods of food waste ultimately result in many problems such as environmental pollution, due to the liberation of toxic chemical compounds like dioxin and furans. Extracted grains, cereal residues, vegetable waste, and fruit seed, skin, rind, and pomace are the principal agricultural wastes and contain good sources of valuable bioactive compounds, such as carotenoids, polyphenols, dietary fibers, vitamins, enzymes, proteins, sugars, minerals, and oils. Because of their high nutritional composition, these residues are not depicted as "wastes" but are considered as raw materials for the fermentative production of other commercial products (Siragam & Guntuku, 2023). Protease has very wide applications in different industrial matrices like in food, detergent, pharmaceuticals, leather, and silk industries. Protease is also used to recover silver from used X-ray films (Jisha et al., 2013). Tunga et al. (2003) recorded that wheat bran was the best substrate for the production of protease having alkaline serine properties when *Aspergillus pasticus* was used for enzyme production. The extracted protease was found to be active and stable in several detergents. Prakasham et al. (2006) evaluated several lignocellulosic wastes such as red gram, wheat bran, chickpea, green gram, for the higher protease production by *Bacillus* sp. They found out that the maximum production of alkaline protease was with the green gram husk. Protease produced by *Bacillus* sp. was studied by Sathishkumar et al. (2015) using different agricultural wastes. The extracted protease was found to be active and stable at 50 °C and pH 9.0. It was also solvent and surfactant stable.

5 Use of Agro/Food Waste as a Carbon Source

Different forms of agro-waste are being produced worldwide in huge amounts which pollute the environment (Mehmood et al., 2021). Carbon sources such as cellulose, hemicellulose, and lignin are known to be good and effective absorbent for a varied range of pollutants, consisted by agro-industrial waste that is easily available (Aravani et al., 2022). Agro-industrial wastes produce such easily available potentials which are economical and proven effectively useful for the purposes like being a substrate for enzyme production (Nair et al., 2022). The agricultural and food wastes such as sugarcane bagasse, nutshell, coir pith, peanuts, coconut shell, and fruit peels are rich in organic matter and nutrients which can provide a suitable carbon source for the growth of certain enzyme-producing bacteria.

Different studies have shown that these agricultural waste materials have been of great effect and use for the production of some extremozymes such as protease, lipase, and cellulase.

Domestic, garden, and agro-industrial wastes have lignocellulose and starch as the major content. About half of the plant matter is composed of lignocelluloses which is the most abundant renewable organic matter in soil (Tufail et al., 2021). Lignocellulose has cellulose (35–50%), hemicellulose (20–35%), and lignin (15–25%) that are strongly connected together (Baruah et al., 2018), which serves as a suitable nutrient source for the production of cellulase.

Oil cakes (oil meals) are by-product of oil extraction from seeds, which is the most widely utilized substrate among all the agricultural waste residues for lipase production. They comprise residual nutrients that can serve as both carbon and nitrogen sources and have been reported to be good substrates for microbial enzyme production (Ramachandran et al., 2007).

Food wastes such as wheat bran, lentil husk, different oil cakes, rice barn, jackfruit seed powder, deoiled jatropha seed cakes, coffee pulps, and corncobs provide suitable carbon and nitrogen sources for protease production (Rao and Rathod, 2019).

6 Advantages of Using Agro/Food Waste Over Traditional Carbon Sources

For the manufacture of enzymes, agro/food waste has shown to be more effective than conventional carbon sources. The mass production of food-grade extremozymes is in industry demand. Depending on the source and accessibility of the waste material, the environment offers a variety of agro/food waste (Donner et al., 2020). Every day, the agricultural and food processing sectors produce enormous amounts of agro-wastes, including crop residues and agro-industrial waste (Aravani et al., 2022).

These agricultural by-products, which are cheap, renewable sources of carbon and energy, are widely available as agro/food waste materials. Typically, they are made of lignocellulosic material. Worldwide, agricultural waste is valued for its economic and cultural significance. A great source of carbon compared to traditional carbon resources is agro/food waste because it is cheap, sustainable, and environmentally benign. Every day, agricultural activity produces tonnes of waste biomass, of which only 10% is utilized as raw materials in the industry (Hasan & Ammenberg, 2019). According to reports, leftover vegetable biomass can play a significant role in the synthesis of biomolecules from extremophilic bacteria, including enzymes and biopolymers. An affordable alternative to synthetic growth media for the generation of microbial enzymes and compounds has been the utilization of food waste material.

7 Mechanism of Extremozyme Synthesis Using Agro/Food Waste

The agro/food waste such as bagasse, bran, fruit peels, and oil peels are produced in very large amounts these days (Ahmad & Zaidi, 2020). The untreated agro/food waste accumulation causes pollution and health hazards. These agro/food waste materials have been proven as a rich source of proteins, carbohydrates, lipids, and other essential minerals for the production of some food-grade extremozymes (Gaonkar & Furtado, 2021). There are several microbes known which can degrade proteins, lipids, and carbohydrate from animal and plant waste material residues to build up their cellular components to excrete some important enzymes. One of the major drawbacks at large-scale extremozyme production is a need for pure carbon and nitrogen sources which follows much expense and thus accounts for almost 50% of the total cost of enzyme production (Singh & Porwal, 2022); hence, low cost-effective material has to be used as a nutrient source for enzyme production.

The very first step for the synthesis of extremozymes is the selection and screening of suitable extremophiles. Strain improvement is also a crucial step to enhance the extremozyme production with the agro/food waste nutrient medium. Selection of the most efficient fermentation technique for the chosen strain of extremophilic bacterial strain (Shah et al., 2021). Genome editing, genetic engineering, and directed evolution are the frontier techniques to ameliorate the desired traits of the tolerant strains for producing the exoenzymes efficiently (Gong et al., 2022; Madhavan et al., 2022; Taha et al., 2022; Mesbah, 2022; Ye et al., 2023). Furthermore, enzyme for industrial production from microorganisms is mostly conducted via submerged fermentation, subjected to various physicochemical parameters. Optimization of physico-chemical environment for production of extremozymes and poly-extremozymes at industrial scale is also one of the crucial steps. Valorization of the agro/food waste is also as important parameter to synthesize the extremozyme with multiple valuable biochemicals.

8 Comparison with Other Methods for Extremozyme Production

8.1 Comparison with Traditional Synthesis Methods

An in-depth study has been focused on the utilization of cellulolytic enzymes as a competitive technique for the efficient deconstruction of widely accessible cellulosic material. A key tactic for the efficient utilization of readily available cellulosic substrates, in particular for bio-refineries, could be the search for stable and robust inherent cellulolytic enzymes from varied and unique habitats that are of microbial origin (Thapa et al., 2020). Advanced approaches must be developed to comprehend how enzymes and substrates interact. Cellulase-lignin non-productive adsorption is influenced by hydrophobicity, hydrogen bonds, and electrostatic interactions. The non-productive adsorption can also be influenced by the lignin content, distribution, and structural features of the substrate (Yuan et al., 2021). It is now possible to understand the whole protein content of the cellulosome and its expression levels to identify the regulatory mechanism, thanks to genomics and proteomics!! An in-situ surface measurement method that is sensitive to changes in surface mass and viscoelasticity is the quartz crystal microbalance with dissipation (QCM-D). It tracks minute changes in mass and viscoelasticity on a thin film surface (Wang et al., 2022). The technology of confocal laser scanning microscopy (CLSM) allows researchers to see how cellulase and cellulose connect to one another (Nagl et al., 2022). A high-resolution imaging method, atomic force microscopy (AFM) can directly measure the cellulase–lignin contact force and provide surface pictures. AFM has the unique capacity to detect forces between particular materials and molecules, thanks to the manipulation or design of AFM probes with particular chemical or biochemical capabilities (Muraille et al., 2017; Chandrashekar et al., 2022). A cutting-edge technique that permits real-time monitoring and investigation of interactions between proteins and biomolecules is surface plasma resonance (SPR). To get quantifiable and repeatable kinetic parameters of enzyme hydrolysis, this method only needs a small sample (Steglich et al., 2022). The improvement of the cutting-edge method aids in the kinetic study of the enzymatic degradation of lignocellulosic biomass. Along with thermogravimetry, attenuated total reflectance-Fourier transform infrared (FTIR-ATR), solid-state nuclear magnetic resonance (NMR), molecular dynamics simulations (MD), and other techniques, these analyses are also frequently used. Additionally, some models are used to clarify how an enzyme interacts with its substrate. For instance, PACER (Plant Cell Wall Model for the Analysis of Non-Catalytic and Enzymatic Reactions) can be used to examine how an enzyme moves through lignocellulosic composites and how it functions. The kinetic behavior of lignocellulosic biomass hydrolysis is explained by semi-mechanistic models. These kinetic investigations are primarily concerned with the interaction between the substrate and cellulase (Monschein et al., 2022; Monclaro et al., 2022).

8.2 Comparison with Other Green Synthesis Methods

A viable alternative that can address both economic and environmental concerns is the use of agro/food waste as a carbon source when compared to other green synthesis techniques like the use of renewable energy sources or non-toxic solvents. One of the developing techniques is without a doubt the creation of sustainable enzymes to increase the bio-conversion efficiency of cellulosic biomass into different value-added products since this gives higher specificity with eco-friendly and green products and less substrate loss. Green synthesis can be seen as an emerging technology, which can be sustainable over the long term if met the requirements based on the three pillars of people, planet, and profit. To put it another way, we shouldn't consume our natural resources at a rate that causes them to run out, and we shouldn't produce residues at a rate that outpaces how quickly the ecosystem can absorb them. Unquestionably, one of the burgeoning techniques is the creation of sustainable enzymes to increase the bio-conversion efficiency of cellulosic biomass into various value-added products. This approach gives improved specificity with eco-friendly and green products and less substrate loss. An in-depth study has been focused on the utilization of cellulolytic enzymes as a competitive technique for the affordable deconstruction of readily accessible cellulosic material. A key tactic for the successful utilization of readily available cellulosic substrates, in particular for bio-refineries, could be the search for stable and strong intrinsic cellulolytic enzymes evidently of microbial origin from various and unique habitats (Bugg et al., 2020).

9 Advantages and Disadvantages of the Approach

The development of environmentally friendly and sustainable methods for the conversion of waste biomass into biofuels, common chemicals, and new bio-based products like bioplastics is one of the major challenges facing chemistry and biology today. This task is motivated by the urgent need to mitigate climate change. Firstly, the major motive is reducing waste by effectively using green and inexpensive raw materials. Secondly, to avoid poisonous and/or dangerous compounds, particularly solvents are of great concern to health, safety, and the environment. Third, switching to renewable biomass instead of fossil fuels like crude oil, coal, or natural gas will reduce the burden on non-renewable energy sources. In contrast to end-of-pipe waste clean-up, green chemistry focuses primarily on pollution prevention and cleaner production through replacing the raw materials. Sustainable development includes the so-called three pillars of sustainability, that is social, environmental, and economic factors. There are many appealing catalytic techniques for the sustainable manufacture of platform commodity chemicals through the valorization of waste biomass, involving both biological and chemical catalysis. Disadvantages include the need for optimization and process control as well as potential variability in the composition and quality of agro/food waste (Sheldon, 2014). The market for cellulolytic enzymes may undergo considerable change as a result of the potential for enhanced enzyme properties with greater physical stability and increased efficiency. For discovering novel cellulolytic enzymes with enhanced functionality, cutting-edge technologies are crucial. A thorough understanding of the potential source of cellulolytic enzymes and their effective production in the future could be strengthened by the use of a wide range of molecular and genetic tools and techniques, including but not limited to genetic engineering, strain enhancement by mutagenesis, structure-guided recombination approach in corporation with synthetic expression system, statistical optimization, solid-state fermentation, and consolidated bio-processing conceptions. To transform the most abundant bio-resource into value-added green goods useful for various aspects and industries, there are vast avenues still to be explored and innovative solutions just waiting to be developed.

10 Potential Applications and Future Research

For application in procedures like grease esterification, hydrolysis, transesterification, inter-esterification, and organic biosynthesis, thermostable lipases are also necessary for industrial and biotechnological operations. These thermostable lipases have a wide range of potential uses in the chemical, pulp/paper, food, detergent, cosmetics, fragrance, and pharmaceutical industries (Hasan et al., 2006). Another class of industrially significant biocatalysts that has drawn a lot of attention is deep-sea thermostable polymer-degrading enzymes including agarases, amylases, and cellulases. Agarases have been successfully used to create agar-oligosaccharides, which have a variety of biological and physiological properties that are advantageous to human health and hence have prospective applications in the food and nutraceutical industries. Agarases catalyze the hydrolysis of agar (Gao et al., 2019; Fu & Kim, 2010). One of the most intriguing classes of enzymes for industrial processes is the thermostable amylolytic group because they are crucial for the hydrolysis of starch at high temperatures, accelerating operations, and lowering the danger of contamination (Prakash & Jaiswal, 2010). Industrial lingo-cellulolytic processes frequently employ thermophilic enzymes, with thermostable cellulases finding employment in pulp and paper, food, animal feed, textile, and other industries (Dalmaso et al., 2015).

Low-temperature-adapted enzymes have several advantages for commercial applications and have been applied in a variety of fields, including food processing, molecular biology, and precision chemical synthesis (Gomes & Steiner, 2004). The food and feed sectors may benefit from cold-active enzymes since it's important to prevent spoilage because it could modify the nutritional content and flavor of the original thermosensitive substrates and products. Because enzymes must be used in a series of reactions and must be deactivated after serving their purpose, cold-adapted enzymes are also helpful in molecular biology. Proteases, lipases, amylases, and cellulases are examples of cold-active hydrolases that are of interest to the pulp and paper, biofuel, and detergent sectors because they can lower energy use and production costs (Santiago et al., 2016).

Additionally, halophiles create enzymes that are both active and stable in the presence of salts. These enzymes use several adaption mechanisms and display extremely high stability in the presence of both low water activity and high salt concentrations. According to structural studies, the surfaces of the molecules are where non-halophilic and halophilic proteins differ most from one another. The food, detergent, textile, bioremediation, and biosynthetic sectors benefit greatly from the use of halophilic enzymes because of their activity and stability not just at high salt concentrations but also in the presence of organic solvents (Yin et al., 2015; Schreck & Grunden, 2014). Halophilic amylases are suitable for industrial applications because of their high activity and stabilities in difficult conditions, especially for the treatment of wastewater containing high salt concentrations and starch residues. Esterases are one type of lipolytic halophilic enzyme that has a very high potential for use in the synthesis of biodiesel, polyunsaturated fatty acids, and food. Some of these had high salinity activity and polar solvent functionality, making them appropriate for usage in the chemical, pharmaceutical, and biofuel sectors (Ferrer et al., 2005).

Piezophilic enzymes have received very little study thus far, and to further our understanding, many more computational analyses and experiments on various enzymes from various piezophiles are needed (Ichiye, 2018). Piezophilic proteins, on the other hand, have demonstrated significant efficacy in a variety of industrial processes, with applications for food production in particular, where high pressures are used for the processing and sterilization of food components (Balny & Hayashi, 1996). Piezophilic alpha-amylase has been demonstrated to make tetrasaccharide from malto-oligosaccharide and trisaccharide instead of maltobiose at high pressures with little energy, which is a beneficial reaction in the food processing sector. Enzymes that take part in these metabolic pathways under high pressure may be extremely useful in industry. Piezophilic endonucleases might be useful in the biotechnology sector as well (Rosenbaum et al., 2012).

11 Areas for Future Research and Development

These unique biocatalytic processes will benefit from extremozymes, making them more efficient, precise, accurate, and environmentally benign. Over the past two decades, there has been significant growth in the number of studies on extremophiles and associated extremozymes, due to their unique characteristics and tremendous potential in commercial applications. Extremophiles have unique dietary needs and harsh growing conditions, which make the isolation and maintenance of these organisms tough. This limited utilization of extremozymes is mostly a result of these factors. Fortunately, next-generation sequencing technologies now make it simple to get extremophilic genomes and metagenomes, providing an ever-growing resource for the discovery of new extremozymes. Additionally, despite the existence of enzyme engineering tools, the enzyme optimization process continues to be a barrier to the creation of novel industrial processes inspired by extremozymes.

Extremophiles will therefore be used more frequently in the industry as a result of the simultaneous development of protein engineering technologies that will aid in the further modification and refinement of biocatalytic characteristics. Additionally, significant developments in our understanding of the genetics, physiology, metabolism, and enzymology of extremophiles are anticipated, which will help us better comprehend the uses of their biocatalysts. Thus, extremozymes will be able to significantly influence a variety of sectors in the future as modern molecular tools and sampling methods progress.

12 Perspective and Annotations

The usage and disposal of agricultural and food waste materials is currently a burning issue worldwide to address the issue of environmental sustainability. Moreover, exploitation of this kind of waste may lead to some tremendous value-added applications. The reutilization of agro-wastes has opened a wide horizon in the field of white biotechnology. These agricultural or food wastes can replace the costly chemicals that were previously used as a substrate for the production of enzymes. The major characteristics of these waste materials are wide availability, greater biocompatibility, and lower cost which ultimately achieve the dual goal by producing the high titer of enzymes with robust characteristics and make the entire process environmentally green and feasible. It has been established that industrial processes are performed in harsh conditions such as high temperature, pH, and salinity. Hence, the production of such enzymes that withstand extreme industrial conditions is necessary. These desire the use of natural lignocellulosic waste material as a substrate which can serve as an alternative to already used chemical substances. Additionally, this may provide a useful strategy for enhanced enzyme production from the extremophilic organisms that ultimately produce the extremozymes. This study may provide the basic information on the production of extremozymes using agro/food-based wastes in the future while expecting significant improvement on the current study and better developmental technologies with higher turnover numbers and requisite yields.

13 Conclusion

In summary, we can say that the application of lignocellulosic waste materials for the production of extremozymes may contribute to an economically environmentally greener process. The produced extremozymes can be beneficial not only for the industries but they may open the doors for further genetical or metagenomic studies on the extremozymes. The nanoscale developments in the technology may contribute to the production of extremozymes having more robust characteristics. Consecutively, the bioengineered enzymatic cues using agro−/food-based wastes could motivate people to understand the economic advantage of these waste materials and increase the global interest in agro/food derived wastes as a great substitute for the currently present chemicals for efficient extremozyme production.

Acknowledgements FS and SM would like to acknowledge the CSIR Emeritus Scientist Scheme for the financial support during the preparation of this book chapter. The Director, CSIR-CSMCRI and present DC, Applied Phycology and Biotechnology Division are deeply acknowledged for their encouragement. BR would like to thank IIT-Bombay for providing financial support through I-PDF. VP would like to thank CSMCRI for giving an opportunity to work as dissertation student. SKP would like to acknowledge the Central University of Rajasthan for financial support.

Declaration of Interest The authors would like to declare no conflict of interest.

Declaration of Generative AI and AI Assisted Technologies in the Writing Process During the preparation of this work the author(s) did not used any AI or Ai assisted technology.

References

Aguilar, C. N., & Gutiérrez-Sánchez, G. (2001). Sources, properties, applications and potential uses of tannin acyl hydrolase. *Food Science and Technology International, 7*(5), 373–382.

Ahmad, F., & Zaidi, S. (2020). Potential use of agro/food wastes as biosorbents in the removal of heavy metals. *Emerging Contaminants*.

Annamalai, N., Rajeswari, M. V., Elayaraja, S., & Balasubramanian, T. (2013). Thermostable, haloalkaline cellulase from *Bacillus halodurans* CAS 1 by conversion of lignocellulosic wastes. *Carbohydrate Polymers, 94*(1), 409–415.

Annamalai, N., Rajeswari, M. V., & Balasubramanian, T. (2014). Enzymatic saccharification of pretreated rice straw by cellulase produced from *Bacillus carboniphilus* CAS 3 utilizing lignocellulosic wastes through statistical optimization. *Biomass and Bioenergy, 68*, 151–160.

Anwar, Z., Gulfraz, M., & Irshad, M. (2014). Agro-industrial lignocellulosic biomass a key to unlock the future bio-energy: A brief review. *Journal of Radiation Research and Applied Sciences, 7*(2), 163–173.

Aravani, V. P., Sun, H., Yang, Z., Liu, G., Wang, W., Anagnostopoulos, G., & Papadakis, V. G. (2022). Agricultural and livestock sector's residues in Greece & China: Comparative qualitative and quantitative characterization for assessing their potential for biogas production. *Renewable and Sustainable Energy Reviews, 154*, 111821.

Arevalo-Gallegos, A., Ahmad, Z., Asgher, M., Parra-Saldivar, R., & Iqbal, H. M. (2017). Lignocellulose: A sustainable material to produce value-added products with a zero waste approach—A review. *International Journal of Biological Macromolecules, 99*, 308–318.

Ayinde, E. O., Adejumo, T. O., & Adeoyo, O. R. (2023). Production, optimization and characterization of amylase using solid state fermentation method. *Frontiers in Environmental Microbiology, 9*(1), 9.

Azin, M., Moravej, R., & Zareh, D. (2007). Production of xylanase by *Trichoderma longibrachiatum* on a mixture of wheat bran and wheat straw: Optimization of culture condition by Taguchi method. *Enzyme and Microbial Technology, 40*(4), 801–805.

Balny, C., & Hayashi, R. (Eds.). (1996). *High pressure bioscience and biotechnology*. Elsevier.

Baruah, J., Nath, B. K., Sharma, R., Kumar, S., Deka, R. C., Baruah, D. C., & Kalita, E. (2018). Recent trends in the pretreatment of lignocellulosic biomass for value-added products. *Frontiers in Energy Research, 6*, 141.

Baysal, Z., Uyar, F., & Aytekin, Ç. (2003). Solid state fermentation for production of α-amylase by a thermotolerant *Bacillus subtilis* from hot-spring water. *Process Biochemistry, 38*(12), 1665–1668.

Bhat, M. (2000). Cellulases and related enzymes in biotechnology. *Biotechnology Advances, 18*(5), 355–383.

Bhatt, K., Lal, S., Srinivasan, R., & Joshi, B. (2020). Bioconversion of agriculture wastes to produce α-amylase from bacillus velezensis KB 2216: Purification and characterization. *Biocatalysis and Agricultural Biotechnology, 28*, 101703.

Bhoite, R. N., & Murthy, P. S. (2015). Biodegradation of coffee pulp tannin by *Penicillium verrucosum* for production of tannase, statistical optimization and its application. *Food and Bioproducts Processing, 94*, 727–735.

Bilal, M., Asgher, M., Iqbal, H. M., Hu, H., & Zhang, X. (2017). Biotransformation of lignocellulosic materials into value-added products—A review. *International Journal of Biological Macromolecules, 98*, 447–458.

Bugg, T. D., Williamson, J. J., & Rashid, G. M. (2020). Bacterial enzymes for lignin depolymerisation: New biocatalysts for generation of renewable chemicals from biomass. *Current Opinion in Chemical Biology, 55*, 26–33.

Chandrashekar, A., Belardinelli, P., Bessa, M. A., Staufer, U., & Alijani, F. (2022). Quantifying nanoscale forces using machine learning in dynamic atomic force microscopy. *Nanoscale Advances, 4*(9), 2134–2143.

Corrado, S., & Sala, S. (2018). Food waste accounting along global and European food supply chains: State of the art and outlook. *Waste Management, 79*, 120–131.

Dalmaso, G. Z. L., Ferreira, D., & Vermelho, A. B. (2015). Marine extremophiles: A source of hydrolases for biotechnological applications. *Marine Drugs, 13*(4), 1925–1965.

Dhanker, R., Rawat, S., Chandna, V., Deepa, Kumar, R., Das, S., Sharma, A., & Kumar, V. (2022). Recovery of silver nanoparticles and management of food wastes: Obstacles and opportunities. *Environmental Advances, 9*, 100303.

Donner, M., Gohier, R., & de Vries, H. (2020). A new circular business model typology for creating value from agro-waste. *Science of the Total Environment, 716*, 137065.

Dwivedi, S., Tanveer, A., Yadav, S., Anand, G., & Yadav, D. (2022). Agro-wastes for cost effective production of industrially important microbial enzymes: An overview. *Microbial Biotechnology: Role in Ecological Sustainability and Research*, 435–460.

Elleuche, S., Schröder, C., Sahm, K., & Antranikian, G. (2014). Extremozymes—Biocatalysts with unique properties from extremophilic microorganisms. *Current Opinion in Biotechnology, 29*, 116–123.

Ezejiofor, T. I. N., Enebaku, U. E., & Ogueke, C. (2014). Waste to wealth-value recovery from agro-food processing wastes using biotechnology: A review. *British Biotechnology Journal, 4*(4), 418.

Ferrer, M., Golyshina, O. V., Chernikova, T. N., Khachane, A. N., Dos Santos, V. A. M., Yakimov, M. M., & Golyshin, P. N. (2005). Microbial enzymes mined from the Urania deep-sea hypersaline anoxic basin. *Chemistry & Biology, 12*(8), 895–904.

Francis, F., Sabu, A., Nampoothiri, K. M., & Ramachandran, S. (2003). Use of response surface methodology for optimizing process parameters for the production of α-amylase by *Aspergillus oryzae*. *Biochemical Engineering Journal, 15*, 107–115.

Fu, X. T., & Kim, S. M. (2010). Agarase: Review of major sources, categories, purification method, enzyme characteristics and applications. *Marine Drugs, 8*(1), 200–218.

Gao, B., Li, L., Wu, H., Zhu, D., Jin, M., Qu, W., & Zeng, R. (2019). A novel strategy for efficient agaro-oligosaccharide production based on the enzymatic degradation of crude agarose in *Flammeovirga pacifica* WPAGA1. *Frontiers in Microbiology, 10*, 1231.

Gaonkar, S. K., & Furtado, I. J. (2021). Valorization of low-cost agro-wastes residues for the maximum production of protease and lipase haloextremozymes by *Haloferax lucentensis* GUBF-2 MG076078. *Process Biochemistry, 101*, 72–88.

Garcia-Garcia, G., Woolley, E., Rahimifard, S., Colwill, J., White, R., & Needham, L. (2017). A methodology for sustainable management of food waste. *Waste and Biomass Valorization, 8*(6), 2209–2227.

Gawande, P. V., & Kamat, M. Y. (1999). Production of *Aspergillus* xylanase by lignocellulosic waste fermentation and its application. *Journal of Applied Microbiology, 87*, 511–519.

Gomes, J., & Steiner, W. (2004). The biocatalytic potential of extremophiles and extremozymes. *Food Technology and Biotechnology, 42*(4), 223–225.

Gong, C., Cao, L., Fang, D., Zhang, J., Awasthi, M. K., & Xue, D. (2022). Genetic manipulation strategies for ethanol production from bioconversion of lignocellulose waste. *Bioresource Technology, 352*, 127105.

Güler, F., & Özçelik, F. (2023). Screening of xylanase producing *Bacillus* species and optimization of xylanase process parameters in submerged fermentation. *Biocatalysis and Agricultural Biotechnology, 51*, 102801.

Hasan, A. M., & Ammenberg, J. (2019). Biogas potential from municipal and agricultural residual biomass for power generation in Hazaribagh, Bangladesh–A strategy to improve the energy system. *Renewable Energy Focus, 29*, 14–23.

Hasan, F., Shah, A. A., & Hameed, A. (2006). Industrial applications of microbial lipases. *Enzyme and Microbial Technology, 39*(2), 235–251.

Hassan, O., Khalil, M. S., & Elnawawy, A. S. (2022). *Biodegradation of agricultural wastes by cellulase producing bacteria.*

Ichiye, T. (2018, December). Enzymes from piezophiles. In *Seminars in cell & developmental biology* (Vol. 84, pp. 138–146). Academic Press.

Izidoro, S. C., & Knob, A. (2014). Production of xylanases by an *Aspergillus niger* strain in wastes grain. *Acta Scientiarum: Biological Sciences, 36*(3), 313–319.

Jisha, V. N., Smitha, R. B., Pradeep, S., Sreedevi, S., Unni, K. N., & Sajith, S. (2013). Versatility of microbial proteases. *Advances in Enzyme Research, 1*(3), 39–51.

Ketsakhon, P., Thammasittirong, A., & Thammasittirong, S. N. R. (2023). Adding value to rice straw waste for high-level xylanase production using a new isolate of *Bacillus altitudinis* RS3025. *Folia Microbiologica, 68*(1), 87–99.

Lekshmi, R., Nisha, S. A., Vasan, P. T., & Kaleeswaran, B. (2021). A comprehensive review on tannase: Microbes associated production of tannase exploiting tannin rich agro-industrial wastes with special reference to its potential environmental and industrial applications. *Environmental Research, 201*, 111625.

Lima, J. S. D., Cruz, R., Fonseca, J. C., Medeiros, E. V. D., Maciel, M. D. H. C., Moreira, K. A., & Motta, C. M. D. S. (2014). Production, characterization of tannase from *Penicillium montanense* URM 6286 under SSF using agroindustrial wastes, and application in the clarification of grape juice (*Vitis vinifera* L.). *The Scientific World Journal, 2014*.

Liszka, M. J., Clark, M. E., Schneider, E., & Clark, D. S. (2012). Nature versus nurture: Developing enzymes that function under extreme conditions. *Annual Review of Chemical and Biomolecular Engineering, 3*, 77–102.

Liu, D., Zhang, R., Yang, X., Wu, H., Xu, D., & Tang, Z. (2011). Thermostable cellulase production of *Aspergillus fumigates* Z5 under solid-state fermentation and its application in degradation of agricultural wastes. *International Biodeterioration & Biodegradation, 65*(5), 717–725.

Lynd, L. R., Weimer, P. J., Van Zyl, W. H., & Pretorius, I. S. (2002). Microbial cellulose utilization: Fundamentals and biotechnology. *Microbiology and Molecular Biology Reviews, 66*(3), 506–577.

Madhavan, A., Arun, K. B., Sindhu, R., Nair, B. G., Pandey, A., Awasthi, M. K., & Binod, P. (2022). Design and genome engineering of microbial cell factories for efficient conversion of lignocellulose to fuel. *Bioresource Technology*, 128555.

Malgireddy, N. R., & Nimma, L. N. R. (2015). Optimal conditions for production of tannase from newly isolated *Aspergillus terreus* under solid state fermentation. *European Journal of Biotechnology and Bioscience, 3*(2), 56–64.

Mamo, G., & Gessesse, A. (1999). Purification and characterization of two raw-starch digesting thermostable α-amylases from a thermophilic *Bacillus*. *Enzyme and Microbial Technology, 25*, 433–438.

Mehmood, T., Nadeem, F., Qamar, S. A., Bilal, M., & Iqbal, H. M. (2021). Bioconversion of agro-industrial waste into value-added compounds. *Sustainable Bioconversion of Waste to Value Added Products*, 349–368.

Mesbah, N. M. (2022). Industrial biotechnology based on enzymes from extreme environments. *Frontiers in Bioengineering and Biotechnology, 10*, 870083.

Mihajlovski, K. R., Radovanovic, N. R., Veljovic, D. N., Siler-Marinkovic, S. S., & Dimitrijevic-Brankovic, S. I. (2016). Improved β-amylase production on molasses and sugar beet pulp by a novel strain *Paenibacillus chitinolyticus* CKS1. *Industrial Crops and Products, 80*, 115–122.

Milagres, A. M. F., Santos, E., Piovan, T., & Roberto, I. C. (2004). Production of xylanase by *Thermoascus aurantiacus* from sugar cane bagasse in an aerated growth fermentor. *Process Biochemistry, 39*, 1387–1391.

Mohan, S. K., Viruthagiri, T., & Arunkumar, C. (2014a). Application of response surface methodology for the production of tannase using *Aspergillus foetidus* MTCC (3557). *International Journal of Pharmaceutical Development and Technology, 4*(3), 151–156.

Mohan, S. K., Viruthagiri, T., & Arunkumar, C. (2014b). Statistical optimization of process parameters for the production of tannase by *Aspergillus flavus* under submerged fermentation. 3. *Biotechnology, 4*(2), 159–166.

Monclaro, A. V., Silva, C. D. O. G., Gomes, H. A. R., de Souza Moreira, L. R., & Ferreira Filho, E. X. (2022). The enzyme interactome concept in filamentous fungi linked to biomass valorization. *Bioresource Technology, 344*, 126200.

Monschein, M., Jurak, E., Paasela, T., Koitto, T., Lambauer, V., Pavicic, M., & Master, E. R. (2022). PACER: A novel 3D plant cell wall model for the analysis of non-catalytic and enzymatic responses. *Biotechnology for Biofuels and Bioproducts, 15*(1), 1–11.

Mostafa, N. A., Farag, A. A., Abo-dief, H. M., & Tayeb, A. M. (2018). Production of biodegradable plastic from agricultural wastes. *Arabian Journal of Chemistry, 11*(4), 546–553.

Mulimani, V. H., & Ramalingam, P. G. N. (2000). Amylase production by solid state fermentation: A new practical approach to biotechnology courses. *Biochemical Education, 28*, 161–163.

Muraille, L., Aguié-Béghin, V., Chabbert, B., & Molinari, M. (2017). Bioinspired lignocellulosic films to understand the mechanical properties of lignified plant cell walls at nanoscale. *Scientific Reports, 7*(1), 44065.

Nagl, M., Haske-Cornelius, O., Skopek, L., Bausch, F., Pellis, A., Bauer, W., & Guebitz, G. M. (2022). Mechanistic investigation of the effect of endoglucanases related to pulp refining. *Cellulose, 29*(4), 2579–2598.

Naik, B., Kumar, V., Rizwanuddin, S., Chauhan, M., Gupta, A. K., Rustagi, S., et al. (2023). Agro-industrial waste: A cost-effective and eco-friendly substrate to produce amylase. *Food Production, Processing and Nutrition, 5*(1), 30.

Nair, L. G., Agrawal, K., & Verma, P. (2022). An overview of sustainable approaches for bioenergy production from agro-industrial wastes. *Energy Nexus, 6*, 100086.

Olajuyigbe, F. M., & Ogunyewo, O. A. (2016). Enhanced production and physicochemical properties of thermostable crude cellulase from *Sporothrix carnis* grown on corn cob. *Biocatalysis and Agricultural Biotechnology, 7*, 110–117.

Picart, P., Diaz, P., & Pastor, F. I. J. (2007). Cellulases from two *Penicillium* sp. strains isolated from subtropical forest soil: Production and characterization. *Letters in Applied Microbiology, 45*, 108–113.

Prakash, O., & Jaiswal, N. (2010). α-Amylase: An ideal representative of thermostable enzymes. *Applied Biochemistry and Biotechnology, 160*, 2401–2414.

Prakasham, R. S., Rao, C. S., & Sarma, P. N. (2006). Green gram husk—An inexpensive substrate for alkaline protease production by *Bacillus* sp. in solid-state fermentation. *Bioresource Technology, 97*, 1449–1454.

Ramachandran, S., Patel, A. K., Nampoothiri, K. M., & Francis, F. (2004). Coconut oil cake—A potential raw material for the production of α-amylase. *Bioresource Technology, 93*, 169–174.

Ramachandran, S., Singh, S. K., Larroche, C., Soccol, C. R., & Pandey, A. (2007). Oil cakes and their biotechnological applications–A review. *Bioresource Technology, 98*(10), 2000–2009.

Rao, P., & Rathod, V. (2019). Valorization of food and agricultural waste: A step towards greener future. *The Chemical Record, 19*(9), 1858–1871.

Rathod, B. G., Pandala, S., & Poosarla, V. G. (2023). A novel halo-acid-alkali-tolerant and surfactant stable amylase secreted from halophile *Bacillus siamensis* F2 and its application in waste valorization by bioethanol production and food industry. *Applied Biochemistry and Biotechnology*, 1–21.

Reetz, M. T. (2016). What are the limitations of enzymes in synthetic organic chemistry? *The Chemical Record, 16*(6), 2449–2459.

Rosenbaum, E., Gabel, F., Durá, M. A., Finet, S., Cléry-Barraud, C., Masson, P., & Franzetti, B. (2012). Effects of hydrostatic pressure on the quaternary structure and enzymatic activity of a large peptidase complex from *Pyrococcus horikoshii*. *Archives of Biochemistry and Biophysics, 517*(2), 104–110.

Sabu, A., Pandey, A., Daud, M. J., & Szakacs, G. (2005). Tamarind seed powder and palm kernel cake: Two novel agro residues for the production of tannase under solid-state fermentation by *Aspergillus niger* ATCC 16620. *Bioresource Technology, 96*, 1223–1228.

Saharia, M., Dey, G., & Kumar, V. (2024). Vermiremediation of plant agro waste to recover residual nutrients and improve crop productivity. In *Earthworm technology in organic waste management: Recent trends and advances* (pp. 79–113). Elsevier.

Santiago, M., Ramírez-Sarmiento, C. A., Zamora, R. A., & Parra, L. P. (2016). Discovery, molecular mechanisms, and industrial applications of cold-active enzymes. *Frontiers in Microbiology, 7*, 1408.

Sathishkumar, R., Ananthan, G., & Arun, J. (2015). Production, purification and characterization of alkaline protease by ascidian associated Bacillus subtilis GACAS8 using agricultural wastes. *Biocatalysis and Agricultural Biotechnology, 4*, 214–220.

Schreck, S. D., & Grunden, A. M. (2014). Biotechnological applications of halophilic lipases and thioesterases. *Applied Microbiology and Biotechnology, 98*, 1011–1021.

Shah, F., Ranawat, B., & Mishra, S. (2019). An approach toward cellulase production, bioconversion, and utilization. In *Advanced bioprocessing for alternative fuels, biobased chemicals, and bioproducts* (pp. 207–223). Woodhead Publishing.

Shah, F., Ranawat, B., Dubey, S., & Mishra, S. (2021). Optimization of fermentation conditions for higher cellulase production using marine *Bacillus licheniformis* KY962963: An epiphyte of *Chlorococcum* sp. *Biocatalysis and Agricultural Biotechnology, 35*, 102047.

Shah, N. A. A., Mansor, A., & Manikam, R. V. S. (2023). Systematic review on characterization of Tannase from agricultural by-products. *Industria: Jurnal Teknologi dan Manajemen Agroindustri, 12*(1).

Sheldon, R. A. (2014). Green and sustainable manufacture of chemicals from biomass: State of the art. *Green Chemistry, 16*(3), 950–963.

Shrestha, S., Chio, C., Khatiwada, J. R., Kognou, A. L. M., & Qin, W. (2022). Formulation of the agro-waste mixture for multi-enzyme (pectinase, xylanase, and cellulase) production by mixture design method exploiting *Streptomyces* sp. *Bioresource Technology Reports, 19*, 101142.

Shukla, J., & Kar, R. (2006). Potato peel as a solid state substrate for thermostable α-amylase production by thermophilic *Bacillus* isolates. *Journal of Microbiology and Biotechnology, 22*, 417–422.

Singh, R., & Porwal, P. (2022). Innovative technologies for enzyme production from extremophilic microbes. In *Microbial Extremozymes* (pp. 31–38). Academic Press.

Siragam, S. L., & Guntuku, G. (2023). Proteases production using agri-food wastes through microbial applications. In *Microbial bioprocessing of agri-food wastes* (pp. 209–258). CRC Press.

Soliman, H. M., Sherief, A. A., & El-Tanash, A. B. (2012). Production of xylanase by *Aspergillus niger* and *Thrichoderman viride* using some agriculture residues. *International Journal of Agricultural Research, 7*(1), 46–57.

Soni, S. K., Kaur, A., & Gupta, J. K. (2003). A solid state fermentation based bacterial α-amylase and fungal glucoamylase system and its suitability for the hydrolysis of wheat starch. *Process Biochemistry, 39*, 185–192.

Sonia, K. G., Chadha, B. S., & Saini, H. S. (2005). Sorghum straw for xylanase hyperproduction by *Thermomyces lanuginosus* (D2W3) under solid state fermentation. *Bioresource Technology, 96*, 1561–1569.

Steglich, P., Lecci, G., & Mai, A. (2022). Surface plasmon resonance (SPR) spectroscopy and photonic integrated circuit (PIC) biosensors: A comparative review. *Sensors, 22*(8), 2901.

Strohmeier, G. A., Pichler, H., May, O., & Gruber-Khadjawi, M. (2011). Application of designed enzymes in organic synthesis. *Chemical Reviews, 111*(7), 4141–4164.

Taha, E. A., Lee, J., & Hotta, A. (2022). Delivery of CRISPR-Cas tools for in vivo genome editing therapy: Trends and challenges. *Journal of Controlled Release, 342*, 345–361.

Thapa, S., Mishra, J., Arora, N., Mishra, P., Li, H., O'Hair, J., & Zhou, S. (2020). Microbial cellulolytic enzymes: Diversity and biotechnology with reference to lignocellulosic biomass degradation. *Reviews in Environmental Science and Bio/Technology, 19*, 621–648.

Tufail, T., Saeed, F., Afzaal, M., Ain, H. B. U., Gilani, S. A., Hussain, M., & Anjum, F. M. (2021). Wheat straw: A natural remedy against different maladies. *Food Science & Nutrition, 9*(4), 2335–2344.

Tunga, R., Shrivastava, B., & Banerjee, R. (2003). Purification and characterization of a protease from solid state cultures of *Aspergillus parasiticus*. *Process Biochemistry, 38*, 1553–1558.

Vijayaraghavan, P., Kalaiyarasi, M., & Vincent, S. G. P. (2015). Cow dung is an ideal fermentation medium for amylase production in solid-state fermentation by *Bacillus cereus*. *Journal of Genetic Engineering and Biotechnology, 13*, 111–117.

Virupakshi, S., Gireesh Babu, K., Gaikwad, S. R., & Naik, G. R. (2005). Production of a xylanolytic enzyme by a thermoalkaliphilic *Bacillus* sp. JB-99 in solid state fermentation. *Process Biochemistry, 40*, 431–435.

Wang, P., Wang, Q., Liu, T., Guo, J., Jin, Y., Xiao, H., & Song, J. (2022). Exploring the promoting mechanisms of bovine serum albumin, lignosulfonate, and polyethylene glycol for lignocellulose saccharification from perspective of molecular interactions with cellulase. *Arabian Journal of Chemistry, 15*(7), 103910.

Woodley, J. M. (2019). Accelerating the implementation of biocatalysis in industry. *Applied Microbiology and Biotechnology, 103*(12), 4733–4739.

Yahya, M. A., Al-Qodah, Z., & Ngah, C. Z. (2015). Agricultural bio-waste materials as potential sustainable precursors used for activated carbon production: A review. *Renewable and Sustainable Energy Reviews, 46*, 218–235.

Ye, J. W., Lin, Y. N., Yi, X. Q., Yu, Z. X., Liu, X., & Chen, G. Q. (2023). Synthetic biology of extremophiles: A new wave of biomanufacturing. *Trends in Biotechnology*.

Yin, J., Chen, J. C., Wu, Q., & Chen, G. Q. (2015). Halophiles, coming stars for industrial biotechnology. *Biotechnology Advances, 33*(7), 1433–1442.

Yousef, N. M., & Mawad, A. M. (2023). Characterization of thermo/halo stable cellulase produced from halophilic *Virgibacillus salarius* BM-02 using non-pretreated biomass. *World Journal of Microbiology and Biotechnology, 39*(1), 22.

Yuan, Y., Jiang, B., Chen, H., Wu, W., Wu, S., Jin, Y., & Xiao, H. (2021). Recent advances in understanding the effects of lignin structural characteristics on enzymatic hydrolysis. *Biotechnology for Biofuels, 14*, 1–20.

Zhang, S., Wang, Z., Shen, J., Chen, X., & Zhang, J. (2023). Isolation of an acidophilic cellulolytic bacterial strain and its cellulase production characteristics. *Agriculture, 13*(7), 1290.

Chapter 11
Microplastic Waste Management in Textile Industry Wastewater

Surya Singh, Sankar Chakma, and Vishal Diwan

1 Introduction

The word *textile* is originated from a Latin word '*textilis*' which means woven. Textile manufacturing is one of the oldest activities of humans, which has progressed tremendously after the onset of industrial era. The textile industry is also one of the largest employment-generating sectors worldwide, which is expected to grow at a compound annual growth rate of 4% for the financial years from 2021 to 2030 (Grand view research, 2022a). Moreover, owing to the increasing population and fast fashion trends, the textile industry is supposed to grow further in future. To fulfill the increasing demand, now-a-days textile industry uses not only natural fibers, such as cotton, flax, jute, and bamboo, but also synthetic fibers such as rayon, nylon, acrylic, and polyester. According to an estimation, the market size of synthetic fiber is estimated to grow at the compounding annual growth rate of 5.1% for the financial year from 2022 to 2030 (Grand view research, 2022b).

Textile manufacturing involves various processes, starting from yarn manufacturing (spinning); fabric manufacturing (weaving/knitting); dyeing, printing, and finishing; up to garment manufacturing (Fig. 11.1). All these processes utilize excessive amount of water and thus produce huge quantity of wastewater as well. Moreover, during the manufacturing process, textile material is washed several times which results in the generation of microplastics, in the form of 'microfibers.' Thus, wastewater/effluent generated from the industry comprises of excessive amount of

S. Singh (✉) · V. Diwan
Division of Environmental Monitoring and Exposure Assessment (Water & Soil), ICMR – National Institute for Research in Environmental Health, Bhopal, India
e-mail: suryasingh.nireh@icmr.gov.in

S. Chakma
Department of Chemical Engineering, Indian Institute of Science Education and Research Bhopal, Bhopal, India

V. Kumar et al. (eds.), *Environmental Engineering and Waste Management*,
https://doi.org/10.1007/978-3-031-58441-1_11

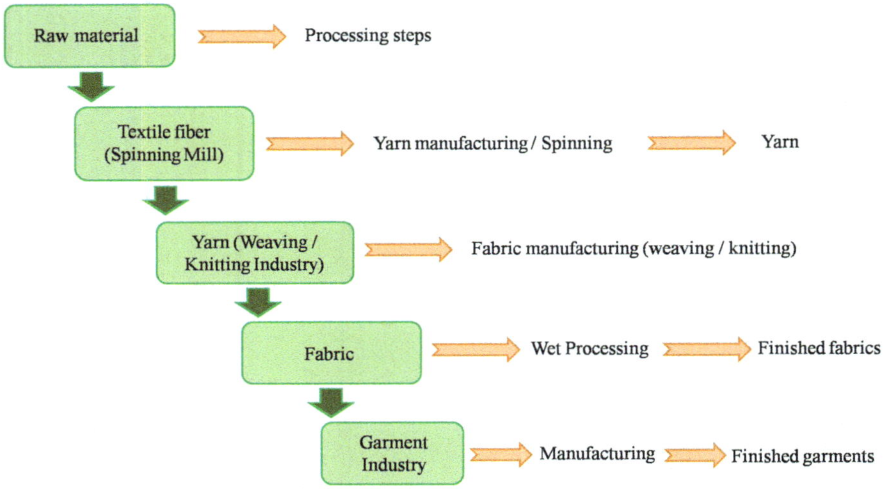

Fig. 11.1 Textile industry processes

(natural/synthetic) microfibers as well, which constitutes a significant share of the microplastic waste.

Microplastics are the emerging contaminants which are defined as small plastic particles in the size range between 1.0 μm and 5.0 mm (Frias & Nash, 2019). These plastic particles may be of any shape, texture, color, and/or composition (Singh et al., 2022). Fibers are also one of the types of microplastics, named as microfibers. As textile industries are one of the biggest producers of fibers, these industries share a major portion in the overall microplastic/microfiber pollution (Singh & Biswas, 2023). Additionally, as the fashion trends are evolving continuously in recent times, textile industries are trying to reach the masses in the least possible time, that too, at an affordable cost. This is often made possible by using cheaper raw materials and poor quality of textile manufacturing. To keep the cost low, synthetic polymeric fibers such as acrylic, polypropylene, and polyester are utilized in fashionable textiles. These synthetic fibers contribute approximately 63% of the total production of fiber in the textile industries (Young, 2019). These fibers and the associated processes become the source for microplastic fibers during the wearing, laundering, and disposal. Although microfibers are generated from both the natural and synthetic materials used for textile, the fibers of natural origin are easily biodegradable in the environment. Therefore, fibers originating from synthetic textile materials are the major cause of concern. It has been estimated that washing process of synthetic textiles results in approximately 35% of the total microplastics that are released annually into the oceans, though this estimate varies across the studies (Boucher & Friot, 2017). In terms of mass, literature shows that approximately 200,000–500,000 tonnes of microplastic particles in marine environment are contributed by the textile industries (EEA, 2022). Not only oceans but microplastics have also been reported from all types of environmental matrices such as rivers, lakes, glaciers (Ambrosini et al., 2019; Bergmann et al., 2019), ponds (Aves et al., 2022), fresh water (Zhang et al., 2021), groundwater (Singh & Bhagwat, 2022), soil (Singh et al., 2023, 2024), atmosphere, and food items. Due to such an ubiquitous presence, microplastics are inhaled and ingested resulting in their penetration inside the human body as well. The presence of microplastics has been reported in the human saliva (Abbasi & Turner, 2021), lungs (Jenner et al., 2022), blood (Leslie et al., 2022), colon (Ibrahim et al., 2020), stool (Schwabl et al., 2019), urine (Pironti et al., 2023), placenta (Ragusa et al., 2021), and breast milk (Ragusa et al., 2022) indicating their entry inside the human body. Nevertheless, microplastics act as efficient vectors of various contaminants, such as metals, chemicals, pigments, pharmaceuticals, and microbes, which further pose negative implications onto human health (Singh et al., 2022; Upadhyay et al., 2022). With the increasing growth of textile industries, fraction of synthetic textile fibers is also expected to increase, thereby contamination in the water resources as well. Moreover, the release of microfibers also depends upon various process parameters and textile/fiber parameters. Therefore, to understand the mechanisms of eliminating microfibers from the textile industry effluent, familiarity with various processes and textile parameters is essential.

Globally, research in the area of microplastic contamination in textile wastewater is limited. Moreover, there is a lack of awareness as well as sufficient data to project the increasing threat due to textile microfibers. Hence, this chapter discusses the fundamentals of textile industry and its various processes which are responsible for microfiber generation. Several factors that affect the microfiber generation in the textile industry processes are also discussed. At the end, possible removal mechanisms of the microfibers from textile wastewater are elaborated. It is expected that this discussion would inculcate interest as well as awareness among the readers to work in the direction of textile industry microfibers.

2 Overview of Textile Industry Processes

Fiber generation in textile industries is an inevitable process. Fiber generation is the outcome of detachment of damaged, loose, and/or fiber fragmentation from the surface of textiles. This process is also called as '*microfiber shedding*.' This shedding may take place owing to any kind of mechanical stress on the textile material or the abrasion of the textile's surface. Moreover, various kinds of chemical treatments during the processing period of textiles as well as during the laundering process also render damage to fibers. Thus, fibers are generated during every stage of the textile life cycle, viz. production, consumption, and disposal.

2.1 Manufacturing Stage

In the manufacturing stage, textile materials are processed to improve its quality. This involves the treatment of textile materials with various chemicals and colors collectively including sizing, desizing, bleaching, dyeing, printing, and finishing. Thus, this process helps to smoothen the fiber and to increase the wettability. Dyeing process helps to provide color to the fabric making it wearable. In this process, both mechanical and chemical forces are applied to the fabric resulting in the breakage of fibers. Upon further exposure to heat and cold treatments, fibers get detached from the textiles and become part of the wastewater effluent. The presence of microfibers has been detected in the wastewater generated after the dyeing and printing processes (Zhou et al., 2020).

2.2 Consumption Stage

The important sources of microfiber generation during the 'use' or 'consumption' phase are the wearing and washing processes. Upon wearing the finished garments, these are subjected to abrasion, stress, stretching, and other modes of tension which causes damage to the textile fibers. Upon continuous use, fibers become weak and ultimately break down from the garments. Moreover, the amount of microfiber generated also depends upon the habits of wearer/user, socioeconomic status, pollution, traffic, season (summers/winters), etc. (Kaya et al., 2018). Similarly, during the washing process, certain mechanical stress is imposed onto the fabrics besides the chemical stress, which leads to microfiber generation. A study was done to determine the microfiber generation during the washing procedure and it was estimated that a total of 13,62,770 fibers were released out of 6 kg of garments washed (Napper & Thompson, 2016). However, it is important to note here that fiber generation also depends upon the type of fabric, type of chemical/detergent applied, temperature during the washing process, and intensity of mechanical force applied (Periyasamy & Tehrani-Bagha, 2022).

2.3 Disposal or Dumping Stage

At the last phase of the textile life cycle, used textile materials/garments are dumped as solid waste which ultimately gets placed in the landfill sites. Moreover, fast-changing fashion trends have further led to the excessive purchasing and use of garments for short duration only. It has been reported that in the last 15 years, the usage duration of clothes has reduced by approximately 36% (Ellen Foundation, 2017). This over-consumerism is responsible for the disposal of garments at a fast rate, thus resulting in increased microfiber production. Upon reaching the landfills, the textiles are also exposed to various other sorts of materials, biotic and abiotic agencies, which further enhance the textile deterioration leading to microfiber generation. These textile materials not only contaminate the surface water and groundwater resources but also air and soil.

3 Wastewater Generation in Textile Industries

Textile industry is considered one of those industries that contribute the most pollution in the world. Textile industry uses a huge amount of energy as well as water, approximately 93 billion cubic meters in a year. The extent of water utilization in the textile industry can be understood by considering the example that about 2700 l of water is used for the production of just one t-shirt (European Parliament, 2020). Moreover, this industry is responsible for causing 20% of global water pollution (Hudd, 2022). The textile industry comprises of a range of raw materials (natural/synthetic), processes, and machinery to design the final product which is commercialized in the market. The wastewater generated from this industry is mainly derived from textile materials' wet processing. The wet processing involves the application of chemicals, pigments/dyes, sizing/de-sizing, bleaching, printing, finishing, laundering, etc. All these processes utilize high volume of water which is ultimately discharged as effluent (wastewater). Moreover, all these processes are responsible for shedding a significant quantity of microfibers which become part of the effluent. It has been estimated that 0.5 million tonnes of microfibers are generated every year from the washing processes of textiles, which accounts for 35% of the primary microplastics released into the environment (European Parliament, 2020). However, this amount may vary depending upon various factors such as fabric properties, yarn properties, method of laundering and, therefore, the amount of microfibers released may be increased as well.

In Indian scenario, textile industries serve as one of the important sectors responsible for water consumption as well as pollution (Pattnaik et al., 2018). These industries may consume the freshwater as much as 3000 m^3/day (Sarayu & Sandhya, 2012). According to a study, water consumption in the Indian textile industries is more than double as compared to the global water consumption in such industries (Restiani & Khandelwal, 2016) which indicates that the amount of wastewater generation shall also be high accordingly. Notably, the wet processing unit of the textile industry (Fig. 11.1) has the highest water utilization as well as the highest potential for causing pollution (Restiani & Khandelwal, 2016). These units generally result in effluents having high pH, suspended/dissolved solids, color, BOD and COD, chromium, sulfide, etc. Available discharge standards for these parameters offer a significant reduction in curtailing the toxicity of textile effluent (CPCB, 2016). However, as on date, the effluent characteristics of textile wastewater do not consider microfibers (CPCB, 2016), which indicates that awareness as well as research is pertinent in this area.

4 Factors Affecting Microfiber Shedding in Textile Wastewater

4.1 Textile Parameters Affecting Microfiber Shedding

There are various properties of textile/fabric materials which vary among one another. Depending upon the origin of raw material, its manufacturing, physical and chemical treatments, and finishing procedures, etc., the shedding of microfiber is altered (Fig. 11.2). There are varieties of fibers, such as natural, synthetic, and/or regenerated. Microfiber shedding of different types of fibers such as cotton, rayon, and polyester has been studied. It was found that all types of materials shed the microfibers; however, more shedding was reported in cotton and rayon compared to polyester fibers (Zambrano et al., 2019). As fiber shedding is associated with fuzz formation, it might be speculated that the varied fuzz formation nature of different types of fibers may result in different shedding behavior. Moreover, the hydrophilic nature of fibers (e.g., cotton) may further enhance the shedding (Cesa et al., 2020). Another fiber property that affects microfiber generation is the length of the fiber. Shorter length of fibers (also called as staple fibers) is generally more prone to dissociate from the yarn as compared to longer filaments (Spencer, 2001). On the contrary, longer filaments avoid shedding until there are any external actions applied. The higher shedding in case of cotton fibers compared to rayon can also be explained on the basis of fiber length. Cotton fibers are mostly staple in nature and thus more prone to produce microfibers (Zambrano et al., 2019; Yang et al., 2019). Nevertheless, it is also important to mention that the higher shedding of natural fibers, such as cotton, is not an issue of concern as it gets easily biodegraded in

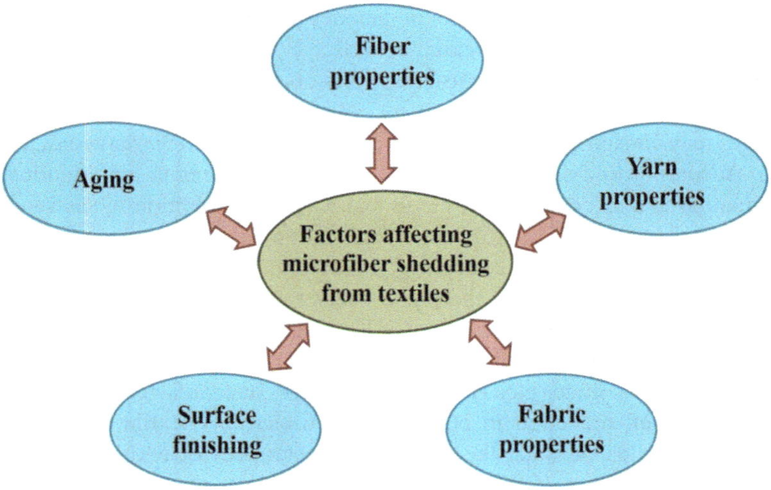

Fig. 11.2 Textile parameters affecting microfiber shedding

environment. On the other hand, synthetic fibers pose risk to the environment as well as human health.

Apart from fiber properties, the properties of yarn also affect the microfiber generation from the textile materials. Yarn is the result of interlocking of fibers during the spinning process and thus, the method of spinning affects the quality of yarn. There are various spinning methods, such as air-jet spinning, rotor spinning, ring spinning, and compact spinning. Cai et al. studied and reported that shedding of microfibers was more in the case of rotor yarns compared to ring-spun and air-jet yarns. This is so because in the last two processes, the short fibers are removed (Cai et al., 2020). Moreover, the tensile strength of the yarn also plays a significant role. Lower tensile strength of yarn results into the easy breakage of fibers during use/consumption/disposal and vice-versa. It is also noticed that protrusion of small fibers in the yarns may result in increased generation of microfibers. Though hairiness in the yarn helps increase the softness and warmness, it increases the pollution load from the textile industry.

Once the yarn is produced and processed, it further results in the production of fabric. Microfiber shedding has been reported at this stage as well depending upon the fabric parameters and properties. Tightness and structure of fabric, abrasion, and pilling formation have been reported to affect the microfiber generation. Yarns are converted into the fabric using inter-weaving (weaving), inter-twining, and inter-looping (knitting) (Spencer, 2001); among which knitted fabrics are prone to shed more microfibers owing to their loose construction (Yang et al., 2019). Pilling is one more important aspect that affects the microfiber shedding. In pilling, the protruding fibers are entangled and form fiber pills or balls. Piling is found to be more common in the synthetic fibers compared to natural fibers (Napper & Thompson, 2016), thus making synthetic fibers more prone to shed microfibers. Another property of fabric that is important for microfiber shedding is abrasion resistance. According to a study, the abrasion resistance of textile material may depend upon the fineness of the fiber, type and count of yarn, and fabric construction (Fatma et al., 2003). The higher the abrasion resistance, lesser would be the chances for microfiber development.

Surface finishing and aging are two more parameters that may affect the shedding behavior of microfibers from textiles. Surface finishing to textiles is often given either through mechanical means or chemical means. Mechanical finishing includes singeing, calendaring, raising, and brushing, while chemical treatments involve the use of various chemicals to smoothen and strengthen the fabric. Cai et al. demonstrated that fabrics having mechanical finishing treatment result in more microfiber generation, compared to those receiving no treatment. This happens because mechanical treatment results in the loosening of fibers/yarn which enhances the shedding (Cai et al., 2020). But the chemical finishing results in binding of the fiber's protrusions within the surfaces, thus lowering the amount of shedding (De Falco et al., 2019). Aging of the textile material relates to the time difference between the textile formation and its current use. As textile use/consumption puts various types of stresses on to the material, it is expected that aging may cause substantial changes in the microfiber shedding properties. It has been reported that garments put through mechanical aging result in more fiber shedding; viz. approximately 25% higher than the unaged garments (Hartline et al., 2016).

4.2 Laundering Parameters Affecting Microfiber Shedding

Laundering parameters are also crucial in determining the microfiber shedding in the wastewater generated from textile industries. Once the garments are finished, they are washed several times before bringing it into the market. During washing, parameters such as the concentration of detergents, temperature of water used in washing, number of washes, softener agents, and quality of water do affect the amount of microfiber generated.

4.2.1 Effect of Detergent

Detergents are the salts of long-chain carboxylic acids which are water soluble and help in removing dust, dirt, stains, soil, etc. from the garments. These contain surfactants that help reduce surface tension of the liquid (water), which further enhances the wettability. Due to the increased wettability and mechanical forces applied during washing (either manual or through machines), the dirt particles get broken down and come out of the textile materials. Therefore, the type and amount of detergent applied to the textiles affect the microfiber generation. In an interesting study, researchers tested the effect of detergents on the polyester, acrylic, and polyester cotton blend. The study was carried out using a bio- and non-bio detergent as well as without detergent. It was found that the textile materials exposed to the non-bio detergents resulted in the highest amount of microfiber shedding, followed by the bio-detergent, and no detergent (Napper & Thompson, 2016; Almroth et al., 2018). However, another study reported contradictory results that detergents have minimal effects on microfiber generation; rather, the drying process poses more stress on the fibers causing their shedding (Pirc et al., 2016).

4.2.2 Effect of Washing Temperature and Washing Cycles

Temperature during the washing plays an important role in determining the amount of microfiber shed from the textile materials because high temperature causes the reactions between the laundry ingredients and textile structure to increase. These increased reactions result in loosening of the fibers, thus causing more shedding. Researchers have reported that an increase in washing temperature from 40 to 75 °C results in increased shedding of microfiber (De Falco et al., 2019). This is because the surface hydrolysis of textile material increases due to higher temperature. Further, the effect of higher washing temperature on the microfiber shedding was also studied in the absence of detergent to exclude the effect of detergent used. The chosen textile materials were polyester, acetate fiber, and amide; while the temperature was varied from 30 up to 60 °C. It was deduced that in case of polyester, the increase in temperature from 30 to 40 °C did not significantly affect the microfiber shedding. However, shedding was increased significantly when the temperature was elevated up to 60 °C. Nevertheless, the amount of shedding from acetate fiber and amide was more than the polyester fibers. This investigation, therefore, suggests that the washing carried out at mild temperature shall result in less microfibers and vice-versa (Yang et al., 2019).

Apart from washing temperature, washing cycle also has the potential to affect microfiber shedding as it further imposes mechanical stress onto the fibers. Napper and Thompson studied and reported the microfiber shedding with different washing cycles for the first time. Acrylics, polyester, and cotton-polyester blend were tested for continuous four times washing. Results indicated that initial washings lead to more microfibers generation for all types of textile materials. For example, in this particular case, the first wash resulted in 2.79 mg of fibers, while 1.63 mg of fibers were produced in the last wash (Napper & Thompson, 2016). In general, it was found that in the first five washes, approximately one-tenth of the microfibers are shedded (Almroth et al., 2018).

4.2.3 Effect of Usage of Softener

Softeners are generally used in the washing process to minimize wrinkles, reduce static-charge build-up, and provide a soft touch to the fabric. As the use of detergent removes oils/waxes from the fibers upon repeated washing, use of softener helps to achieve lubrication and thus makes the fabric soft and smooth. Therefore, it is imperative to assess the impact of softener on fabric. However, there are not many studies on the effect of softeners on the microfiber generation, as the application of softener is optional and depends upon the discretion of the user. De Falco et al. studied the influence of fabric softener on the laundering of polyester fibers and found that use of softener reduces the amount of microfiber shed significantly, compared to normal laundry without softener. The amount of microfiber shed with the use of softener was found to be 4,000,000 while in absence of softener, 6,000,000 microfibers were released, thus showing approximately 33% increase. The reason for the reduction in microfiber generation was attributed to the decrease in friction between the textile fibers upon the use of softener (De Falco et al., 2018). Moreover, there are some studies that report that softeners do not play a significant role in reducing microfiber shedding (Napper & Thompson, 2016). A study conducted by Lant et al. also supports that the use of softeners does not reduce microfiber generation. In this study, different commercial fabrics were evaluated through applying European as well as North American styles of testing. However, in none of the cases, a significant decrease in the microfiber quantity was observed upon use of softener (Lant et al., 2020). Therefore, more studies are required in this area to arrive onto something conclusive.

4.2.4 Effect of Washing Method Used

Apart from the use of softener, one more parameter that has been found to affect microfiber generation is the method of washing. Two very common types of washing methods involve the use of front-loading washing machines and top-loading washing machines. Therefore, researchers have conducted the study using these two types of washing machines with new and aged garments. For both these garment types, front-load washing machines resulted in 220 mg of microfibers on average, while top-loading washing machines resulted in 1906 mg (Hartline et al., 2016). This difference in the amount of microfibers generation was found to be due to the different mechanisms of agitation provided in the washing machine. It was speculated that the central agitation mechanism in the top-loading washing machines causes more abrasion compared to the front-loading washing machine, thus resulting in higher microfiber generation (Hartline et al., 2016). However, more studies are required to conclude the specific cause of higher microfiber generation in the top-loading washing machines. Another study on the method of washing found that higher acceleration during the washing process leads to more microfibers for all types of textile materials (Zambrano et al., 2019). Further, it has also been found in some studies that laboratory models of washing machines shed more microfibers, while other studies reported more microfibers generation in household washing machines (Napper & Thompson, 2016; Almroth et al., 2018; De Falco et al., 2018, 2019). Hence, there is a need for further research in this area by adopting standardized conditions, so that results among various studies may be compared.

4.2.5 Effect of Water Hardness and Amount of Water Used

Apart from the properties of fibers and washing parameters, the quality and quantity of water also affect microfiber shedding. Though there are limited studies in this area, it has been reported that hard water results in more microfiber shedding and vice-versa. Higher abrasion in the hard water was elucidated as the possible reason for more microfiber shedding (De Falco et al., 2018). In a preliminary study, the application of different amounts of water volumes with different agitation speeds was also tested. It was found that a combination of higher water volume and lower agitation resulted in the highest microfiber generation (Yang et al., 2019; Kelly et al., 2019). The cause of this result was the high hydrodynamic pressure created by the high amount of water, which further inflicts pressure on the fibers resulting in detachment. However, research in this direction is limited and hence there is a need for further studies to arrive at something conclusive.

5 Elimination Techniques for Microfibers

Wastewater treatment plants are generally considered as the ultimate repositories of textile microfibers/microplastics. However, as discussed in the previous sections, microfiber shedding from textiles does occur during the complete life cycle of the material. Therefore, in order to eliminate the microfibers from environment, suitable measures need to be undertaken at each of the steps, viz. production, consumption, and disposal phase. There can be three mechanisms through which the pollution due to textile microfibers in the environment can be minimized: (a) proactive approach—practicing the philosophy of less and sustainable consumption, (b) preventive approach—reducing the amount of microfibers generated at various steps of textile processing through adopting best practices, and (c) corrective approach—removal/elimination of microfibers in the wastewater treatment plants (and thereafter in sludge) before discharging the effluent into surface water sources (Fig. 11.3).

5.1 Proactive Approach

This approach involves the theory of sustainable consumption, which follows the idea of only needful consumption, rather than over-consumption. In order to follow this, textile materials/garments need to be used and reused as long as possible before discarding them. This would reduce the amount of microfiber's release into the landfills and thereafter into soil and groundwater (Singh & Bhagwat, 2022). After the sufficient use, materials need to be recycled as much as possible. There are techniques through which new fibers can be produced from the used textile materials. For example, cellulose waste can be converted into new fibers after dissolving in a suitable solvent followed by wet spinning (Sixta et al., 2015). Similarly, polyester materials can also be recycled into new fiber. This kind of approach will

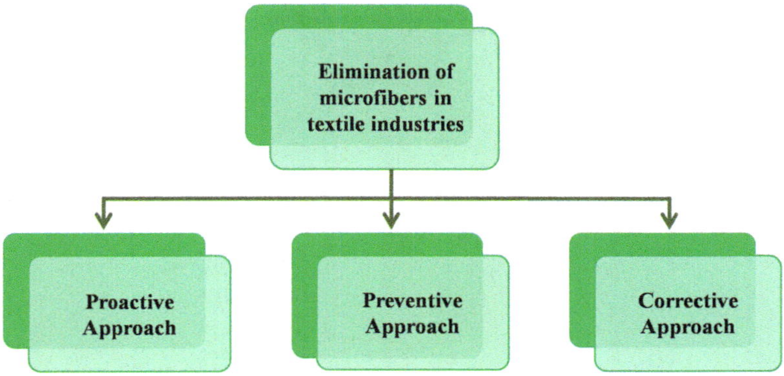

Fig. 11.3 Elimination techniques for microfibers in textile industry

lead to encourage sustainability and help in minimizing the release of microfiber into the environment. Moreover, promotion of bio-based fibers obtained from renewable sources is also required over synthetic fibers (Periyasamy & Tehrani-Bagha, 2022). Bio-based fibers tend to easily degrade in the environment, thereby solve the issue of microfiber persistence. Some of the examples of bio-based fibers are cotton, viscose rayon, and poly-lactic acid.

5.2 *Preventive Approach*

Preventive approach deals with the minimization of microfibers release during textile processing and use. Various factors have been outlined in the previous sections, which affect the microfiber generation at various stages of textile processing. Therefore, by adopting the best possible practices, the amount of microfibers may be reduced up to a significant extent (Fig. 11.4). This involves practices such as washing of clothes using mild liquid detergent at mild temperature

Fig. 11.4 Best possible practices to minimize the amount of microfibers generated in textile industry

with less mechanical agitation. Although practices such as use of softener and use of front-loading washing machines depend upon personal choice, they might further reduce the amount of microfibers shed. Short laundering time, use of soft water, and less volume of water further help to minimize the microfiber shedding.

5.3 Corrective Approach

After minimizing the amount of microfibers generated to the maximum possible extent, the remaining amount becomes part of the wastewater which is finally directed to the effluent treatment plants/common effluent treatment plants. Though these treatment plants are not designed to remove microfibers, but some of the portion gets removed while undergoing different stages of the effluent treatment (Singh et al., 2021). In the primary treatment process, grit removal, flotation, and sedimentation help to remove microfibers, especially which are large in dimension. Approximately 90% of the microfibers are removed at this stage. Further, various aerobic treatment methods are used under the secondary treatment process, in which microbial mass is employed to treat the water. Many of the microfibers get entangled in these microbial masses, and thereby get removed along with the sludge. If advanced treatment options are available, such as coagulation and filtration, membrane bioreactors, ultra-filtration, and/or nano-filtration, some more microfibers can be removed (Table 11.1). A detailed description of various conventional and advanced treatment mechanisms applied in effluent/wastewater treatment plants (ETPs/WWTPs) for the removal of microplastics/microfibers is presented in Table 11.1, along with the efficiency of each process. Thus, depending upon the stages available in the effluent treatment plants and processes applied, approximately 76–98% of microfibers are removed (Belzagui & Gutierrez-Bouzan, 2022). However, it is important to note here that effluent treatment plants process millions of liters of water per day; thus, despite having high removal efficiency, these plants release significant amount of microfibers in the freshwater resources and contaminate them. Further, globally there are insufficient operational effluent treatment plants resulting in the treatment of only a part of the total textile effluent generated, while rest of the effluent goes untreated and directly mixes with the surface water sources. Therefore, even after having high removal efficiency, microfibers contaminate the environmental matrices. Another important aspect here is that the amount of microfibers removed from the treatment plants is disposed off as sludge. This sludge is invariably applied onto the soil, from where the microfibers may leach into the deeper soil layers as well as groundwater (Singh & Bhagwat, 2022). Therefore, suitable interventions are required to eliminate the microfibers from textile wastewater as well as sludge.

Table 11.1 Removal mechanisms of microplastics (MPs) in effluent/wastewater treatment plants

Process description	Major mechanism	Lowest size of microplastic particle removed/ Finest mesh	Efficiency (%)	Advantages	Challenges	Reference
Wastewater treatment plant processes	Skimming, settling of the entrapped microplastics	300 μm	99.9	Conventional process, no additional cost	Not possible to remove MPs of size <300 μm	Magnusson and Noren, (2014)
Wastewater treatment plant processes	Primary, secondary, and tertiary	100 μm	99.9	Conventional process, no additional cost	Not possible to remove MPs of size <100 μm	Carr et al. (2016)
Wastewater treatment plant processes	Secondary treatment	20 μm	95.6	MBR process exhibited greater overall efficiency	Complete retention is not possible	Michielssen et al. (2016)
	Tertiary treatment		97.2			
	Membrane bioreactor		99.4			
Wastewater treatment plant processes	Membrane bioreactor (MBR)	250 μm	99.3	MBR process helped to retain more microplastics compared to conventional activated sludge process	Not possible to remove MPs of size <250 μm	Lares et al. (2018)
Al and Fe salt	Coagulation	<0.5 mm	45.34 ± 3.93	Simple process, doesn't require additional set-up	Low efficiency	Ma et al. (2019)
Adsorption using biochar	Morphologically controlled mechanism (stuck, trapped, entangled)	10 μm	>95%	Low cost and efficient	Process is slow and results in obstruction of the pores with time; costly; regeneration is tough	Wang et al. (2020a)
Filtration with granular activated carbon (combined with coagulation and sedimentation)	Physical properties (size and shape)	1–5 μm	56.8–60.9%	Efficient to remove plastic particles of nano-size range	Process is slow and results in obstruction of the pores with time; costly; regeneration is tough	Wang et al. (2020b)

Pulse clarification with filtration	Entrapment in sludge blanket formed due to coagulation floats	<100 μm	85%	Removal efficiency is comparable to the other treatment plants having advanced processes	Complete retention is not possible	Sarkar et al. (2021)
Bioinspired molecules	Mechanical capture mechanism driven by the hydrophobic and van der Waals interactions	–	–	Flexible; possibility to remove different types of plastic particles in wastewater stream	Method yet to be established for practical purposes	Herbort and Schuhen (2017)
Photocatalytic micromotors	Phoretic interaction and shovelling/pushing interactions	–	–	Self-propelled devices works efficiently independent of the fuel	Selectivity of micromotors for microplastics is crucial	Wang et al. (2019)
Metal organic framework (MIL-100 Fe) nanoparticles incorporated into polysulfone matrix	Membrane separation	40 μm	–	Removal of MPs (polyethylene and polyvinyl chloride) from textile wastewater	Tested for only textile wastewater	Gnanasekaran et al. (2021)
Photocatalytic degradation	Photocatalysis using *n*-TiO_2 semiconductor	MPs extracted from commercial exfoliating scrub	Max. mass loss of 6.4%	Environment-friendly technology	High cost, requirement of specific set-up	Ariza-Tarazona et al. (2019)
Algal degradation	Electrostatic charge onto the microplastic particles and algal surfaces	20 μm	94.5%	No chemical, electrical, and mechanical operations	Efficiency will vary owing to physiological and topographical differences on the seaweed surface	Sundbæk et al. (2018)
Fungal degradation	Biodegradation	–	–	Biological process, no toxic side-products	Study investigated only low-density polyethylene. Method yet to be established for other polymers	Kunlere et al. (2019)
Bacterial degradation	Biodegradation	–	50–61%			

(continued)

Table 11.1 (continued)

Process description	Major mechanism	Lowest size of microplastic particle removed/ Finest mesh	Efficiency (%)	Advantages	Challenges	Reference
				Biological process, no toxic side-products	Method yet to be established for wide variety of polymers	Rajandas et al. (2012)
Combinatorial method	Electrocoagulation, electroflotation, and membrane filtration	MPs collected from WWTPs filtered through mesh of diameter 26 μm	100%	Short retention time, high efficiency	Application of pressure creates additional cost, membrane rupture results in high replacement frequency	Akarsu et al. (2021)

Adapted from Singh et al. (2021)

6 Challenges and Future Perspectives

Textile industries are ever-growing considering the increasing population growth and fast-changing fashion trends. Further, transition in consumers' dressing habits from multiple-uses to few uses has led to the scenario that almost 5% of the total volume in landfills is comprised of textile wastes only (Utebay et al., 2020). Data of US Environmental Protection Agency (EPA) shows that textile waste management involves four types of options, among which landfilling is the most sought option (Fig. 11.5) (EPA, 2022). These dumped textile waste materials upon exposure to various abiotic and biotic factors result in the generation of microfibers/ microplastics. Therefore, the challenge is not only about the microfiber removal from water; rather, challenges include excessive generation of textile material and thereby microfibers as well. Apart from the generation of microfibers, excessive water utilization and pollution, chemicals usage, and energy utilization in textile industries are other challenges to deal with. Nevertheless, textile materials lying in landfills also generate greenhouse gases (such as CH_4) and leach out harmful chemicals/dyes/pigments, etc., apart from microfibers, upon decomposition, in the soil and groundwater, which further enhances ecological and human health risks.

In this context, recycling of textile materials plays an important role because it not only increases the life of base material (EPA, 2022) but also lessens the negative impacts on environment in terms of low CO_2 emissions, reduced water consumption, etc. Moreover, recycling might also result in saving excessive use of chemicals, energy, and other harmful processes that are mandatory in producing a new fiber. Therefore, circular economy approach needs to be targeted in order to reduce the

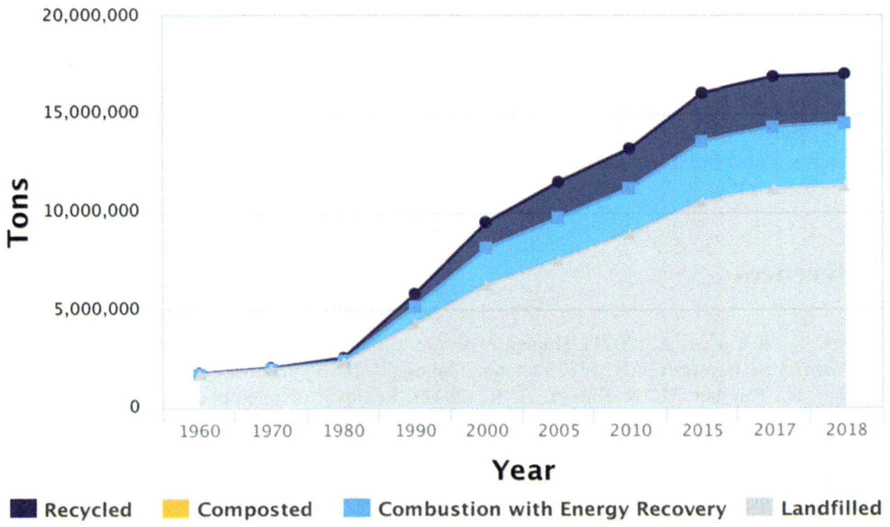

Fig. 11.5 Textile waste management: 1960–2018. (Source: US EPA, 2022)

negative environmental impacts of textile industries. Further, awareness as well as research in the direction of microfiber generation, detection/quantification methods, treatment methods, etc., needs to be strengthened. Therefore, following points must be considered by researchers working in the areas of textile waste management:

- Development of standard sampling techniques for microfibers/nanofibers from the environmental samples (air/water/soil).
- Development and validation of standard methodology for identification and analysis of microfibers/nanofibers produced from different stages of textile production/consumption/disposal.
- Development of effective wastewater treatment technologies to capture the microfibers/nanofibers.
- Awareness and dissemination activities to promote sustainable utilization of textile materials.
- Promotion of circular economy approach in specific context to textile industries.

7 Conclusion

Textile industries are recognized as one of the important sources of microfibers/ microplastics, which contaminate the water resources either through direct discharge or through the effluent treatment plants. Considering the fast-growing fashion trends, corresponding growth in the textile sector, and the extent of contamination from a single textile unit, it is imperative to consider this issue. It can be said that, though there is lack of any specific treatment technique for removal of entire microfibers from effluent, a significant portion still gets removed through existing techniques in the effluent treatment plants. However, these techniques will prove to be useful only when all the effluent generated gets the chance to pass through the treatment facilities. Further, there are various proactive and preventive measures that need to be adopted to minimize the microfibers generation at the source itself. In this regard, awareness and behavioral change are the key issues to be considered.

References

Abbasi, S., & Turner, A. (2021). Human exposure to microplastics: A study in Iran. *Journal of Hazardous Materials, 403*, 123799. https://doi.org/10.1016/j.jhazmat.2020.123799

Akarsu, C., Kumbur, H., & Kideys, A. E. (2021). Removal of microplastics from wastewater through electrocoagulation—Electroflotation and membrane filtration processes. *Water Science & Technology, 84*, 1648–1662. https://doi.org/10.2166/wst.2021.356

Almroth, B. M. C., Astrom, L., Roslund, S., Petersson, H., Johansson, M., & Persson, N. K. (2018). Quantifying shedding of synthetic fibers from textiles; a source of microplastics released into the environment. *Environmental Science and Pollution Research, 25*, 1191–1199. https://doi.org/10.1007/s11356-017-0528-7

Ambrosini, R., Azzoni, R. S., Pittino, F., Diolaiuti, G., Franzetti, A., & Parolini, M. (2019). First evidence of microplastic contamination in the supraglacial debris of an alpine glacier. *Environmental Pollution, 253*, 297–301. https://doi.org/10.1016/j.envpol.2019.07.005

Ariza-Tarazona, M. C., Villarreal-Chiu, J. F., Barbieri, V., Siligardi, C., & Cedillo-Gonzalez, E. I. (2019). New strategy for microplastic degradation: Green photocatalysis using a protein-based porous N-TiO_2 semiconductor. *Ceramics International, 45*, 9618–9624. https://doi.org/10.1016/j.ceramint.2018.10.208

Aves, A. R., Revell, L. E., Gaw, S., Ruffell, H., Schuddeboom, A., Wotherspoon, N. E., LaRue, M., & Mc Donald, A. J. (2022). First evidence of microplastics in Antarctic snow. *The Cryosphere, 16*, 2127–2145. https://doi.org/10.5194/tc-16-2127-2022

Belzagui, F., & Gutierrez-Bouzan, C. (2022). Review on alternatives for the reduction of textile microfibers emission to water. *Journal of Environmental Management, 317*, 115347. https://doi.org/10.1016/j.jenvman.2022.115347

Bergmann, M., Mutzel, S., Primpke, S., Tekman, M. B., Trachsel, J., & Gerdts, G. (2019). White and wonderful? Microplastics prevail in snow from the Alps to the Arctic. *Science Advances, 5*, eaax1157. https://doi.org/10.1126/sciadv.aax1157

Boucher, J., & Friot, D. (2017). *Primary microplastics in the oceans: A global evaluation of sources*. IUCN Gland, Switzerland. Available online at https://portals.iucn.org/library/sites/library/files/documents/2017-002-En.pdf. Accessed on 13 Nov 2022.

Cai, Y., Mitrano, D. M., Heuberger, M., Hufenus, R., & Nowack, B. (2020). The origin of microplastic fiber in polyester textiles: The textile production process matters. *Journal of Cleaner Production, 267*, 121970. https://doi.org/10.1016/j.jclepro.2020.121970

Carr, S. A., Liu, J., & Tesoro, A. G. (2016). Transport and fate of microplastic particles in wastewater treatment plants. *Water Research, 91*, 174–182. https://doi.org/10.1016/j.watres.2016.01.002

Cesa, F. S., Turra, A., Checon, H. H., Leonardi, B., & Baruque-Ramos, J. (2020). Laundering and textile parameters influence fibers release in household washings. *Environmental Pollution, 257*, 113553. https://doi.org/10.1016/j.envpol.2019.113553

CPCB. (2016). *Standards for discharge of effluents from textile industry*. Ministry of Environment, Forest and Climate Change. Available online at https://cpcb.nic.in/displaypdf.php?id=SW5kdXN0cnktU3BlY2lmaWMtU3RhbmRhcmRzL0VmZmx1ZW50LzQxMi5wZGY. Accessed on 13 Oct 2023.

De Falco, F., Gullo, M. P., Gentile, G., Pace, E. D., Cocca, M., Gelabert, L., Brouta-Agnesa, M., Rovira, A., Escudero, R., Villalba, R., Mossotti, R., Montarsolo, A., Gavignano, S., Tonin, C., & Avella, M. (2018). Evaluation of microplastic release caused by textile washing processes of synthetic fabrics. *Environmental Pollution, 236*, 916–925. https://doi.org/10.1016/j.envpol.2017.10.057

De Falco, F., Cocca, M., Guarino, V., Gentile, G., Ambrogi, V., Ambrosio, L., & Avella, M. (2019). Novel finishing treatments of polyamide fabrics by electrofluidodynamic process to reduce microplastic release during washings. *Polymer Degradation & Stability, 165*, 110–116. https://doi.org/10.1016/j.polymdegradstab.2019.05.001

EEA. (2022). *Microplastics from textiles: Towards a circular economy for textiles in Europe*. European Environment Agency. Available online at https://www.eea.europa.eu/publications/microplastics-from-textiles-towards-a. Accessed on 27 Nov 2022.

Ellen MacArthur Foundation. (2017). *A new textiles economy: redesigning fashion's future*. Available online at https://ellenmacarthurfoundation.org/a-new-textiles-economy. Accessed on 15 Nov 2022.

EPA. (2022). *Textiles: Material—specific data*. Available online at https://www.epa.gov/facts-and-figures-about-materials-waste-and-recycling/textiles-material-specific-data. Accessed on 15 Oct 2023.

European Parliament. (2020). *The impact of textile production and waste on the environment*. Available online at https://www.europarl.europa.eu/news/en/headlines/society/2020120

8STO93327/the-impact-of-textile-production-and-waste-on-the-environment-infographic. Accessed on 23 Nov 2022.

Fatma, K., Onder, E., & Ozipek, B. (2003). Influence of varying structural parameters on abrasion characteristics of 50/50 wool/polyester blended fabrics. *Textile Research Journal, 73*, 980–984. https://doi.org/10.1177/004051750307301108

Frias, J. P. G. L., & Nash, R. (2019). Microplastics: Finding a consensus on the definition. *Marine Pollution Bulletin, 138*, 145–147. https://doi.org/10.1016/j.marpolbul.2018.11.022

Gnanasekaran, G., Arthanareeswaran, G., & Mok, Y. S. (2021). A high-flux metal-organic framework membrane (PSF/MIL-100(Fe)) for the removal of microplastics adsorbing dye contaminants from textile wastewater. *Separation and Purification Technology, 277*, 119655. https://doi.org/10.1016/j.seppur.2021.119655

Grand view research. (2022a). *Textile market size, share & trends analysis report 2022–2030*. Report ID: GVR-1-68038-736-0. Available online at https://www.grandviewresearch.com/industry-analysis/textile-market. Accessed on 12 Nov 2022.

Grand view research. (2022b), *Synthetic fiber market size, share & trends analysis report 2022–2030*. Report ID: GVR-2-68038-029-3, Available online at https://www.grandviewresearch.com/industry-analysis/synthetic-fibers-market. Accessed on 13 Nov 2022.

Hartline, N. L., Bruce, N. J., Karba, S. N., Ruff, E. O., Sonar, S. U., & Holden, P. A. (2016). Microfiber masses recovered from conventional machine washing of new and aged garments. *Environmental Science and Technology, 50*, 11532–11538. https://doi.org/10.1021/acs.est.6b03045

Herbort, A. F., & Schuhen, K. (2017). A concept for the removal of microplastics from the marine environment with innovative host-guest relationships. *Environmental Science and Pollution Research, 24*, 11061–11065. https://doi.org/10.1007/s11356-016-7216-x

Hudd, A. (2022), *Dyeing for fashion: why the clothes industry is causing 20% of water pollution*. Available online at https://www.euronews.com/green/2022/02/26/dyeing-for-fashion-why-the-fashion-industry-is-causing-20-of-water-pollution. Accessed on 11 Nov 2022.

Ibrahim, Y. S., Anuar, S. T., Azmi, A. A., Khalik, W. M. A. W. M., Lehata, S., Hamzah, S. R., Ismail, D., Ma, Z. F., Dzulkarnaen, A., Zakaria, Z., Mustaffa, N., Sharif, S. E. T., & Lee, Y. Y. (2020). Detection of microplastics in human colectomy specimens. *JGH Open, 5*, 116–121. https://doi.org/10.1002/jgh3.12457

Jenner, L. C., Rotchell, J. M., Bennett, R. T., Cowen, M., Tentzeris, V., & Sadofsky, L. R. (2022). Detection of microplastics in human lung tissue using μFTIR spectroscopy. *Science of the Total Environment, 831*, 154907. https://doi.org/10.1016/j.scitotenv.2022.154907

Kaya, A. T., Yurtsever, M., & Bayraktar, S. C. (2018). Ubiquitous exposure to microfiber pollution in the air. *The European Physical Journal Plus, 133*, 488. https://doi.org/10.1140/epjp/i2018-12372-7

Kelly, M. R., Lant, N. J., Kurr, M., & Burgess, J. G. (2019). Importance of water-volume on the release of microplastic fibers from laundry. *Environmental Science & Technology, 53*, 11735–11744. https://doi.org/10.1021/acs.est.9b03022

Kunlere, I. O., Fagade, O. E., & Nwadike, B. I. (2019). Biodegradation of low density polyethylene (LDPE) by certain indigenous bacteria and fungi. *International Journal of Environmental Studies, 76*, 428–440. https://doi.org/10.1080/00207233.2019.1579586

Lant, N. J., Hayward, A. S., Peththawadu, M. M. D., Sheridan, K. J., & Dean, J. R. (2020). Microfiber release from real soiled consumer laundry and the impact of fabric care products and washing conditions. *PLoS One, 15*, e0233332. https://doi.org/10.1371/journal.pone.0233332

Lares, M., Ncibi, M. C., Sillanpaa, M., & Sillanpaa, M. (2018). Occurrence, identification, and removal of microplastic particles and fibers in conventional activated sludge process and advanced MBR technology. *Water Research, 133*, 236–246. https://doi.org/10.1016/j.watres.2018.01.049

Leslie, H. A., van Velzen, M. J. M., Brandsma, S. H., Vethaak, A. D., Garcia-Vallejo, J. J., & Lamoree, M. H. (2022). Discovery and quantification of plastic particle pollution in human blood. *Environment International, 163*, 107199. https://doi.org/10.1016/j.envint.2022.107199

Ma, B., Xue, W., Hu, C., Liu, H., Qu, J., & Li, L. (2019). Characteristics of microplastic removal via coagulation and ultrafiltration during drinking water treatment. *Chemical Engineering Journal, 359*, 159–167. https://doi.org/10.1016/j.cej.2018.11.155

Magnusson, K., & Noren, F. (2014). *Screening of microplastic particles in and downstream a wastewater treatment plant* (pp. 1–22). IVL Swedish Environmental Research Institute, C55.

Michielssen, M. R., Michielssen, E. R., Ni, J., & Duhaime, M. B. (2016). Fate of microplastics and other small anthropogenic litter (SAL) in wastewater treatment plants depends on unit processes employed. *Environmental Science: Water Research & Technology, 2*, 1064–1073. https://doi.org/10.1039/C6EW00207B

Napper, I. E., & Thompson, R. C. (2016). Release of synthetic microplastic plastic fibers from domestic washing machines: Effects of fabric type and washing conditions. *Marine Pollution Bulletin, 112*, 39–45. https://doi.org/10.1016/j.marpolbul.2016.09.025

Pattnaik, P., Dangayach, G. S., & Bhardwaj, A. K. (2018). A review on the sustainability of textile industries wastewater with and without treatment methodologies. *Reviews on Environmental Health, 33*, 163–203. https://doi.org/10.1515/reveh-2018-0013

Periyasamy, A. P., & Tehrani-Bagha, A. (2022). A review on microplastic emission from textile materials and its reduction techniques. *Polymer Degradation and Stability, 199*, 109901. https://doi.org/10.1016/j.polymdegradstab.2022.109901

Pirc, U., Vidmar, M., Mozer, A., & Krzan, A. (2016). Emissions of microplastic fibers from microfiber fleece during domestic washing. *Environmental Science and Pollution Research, 23*, 22206–22211. https://doi.org/10.1007/s11356-016-7703-0

Pironti, C., Notarstefano, V., Ricciardi, M., Motta, O., Giorgini, E., & Montano, L. (2023). First evidence of microplastics in human urine, a preliminary study of intake in the human body. *Toxics, 11*, 40. https://doi.org/10.3390/toxics11010040

Ragusa, A., Svelato, A., Santacroce, C., Catalano, P., Notarstefano, V., Carnevali, O., Papa, F., Rongioletti, M. C. A., Baiocco, F., Draghi, S., D'Amore, E., Rinaldo, D., Matta, M., & Giorgini, E. (2021). Plasticenta: First evidence of microplastics in human placenta. *Environmental International, 146*, 106274. https://doi.org/10.1016/j.envint.2020.106274

Ragusa, A., Notarstefano, V., Svelato, A., Belloni, A., Gioacchini, G., Blondeel, C., Zucchelli, E., De Luca, C., D'Avino, S., Gulotta, A., Carnevali, O., & Giorgini, E. (2022). Raman microspectroscopy detection and characterization of microplastics in human breast milk. *Polymers, 14*, 2700. https://doi.org/10.3390/polym14132700

Rajandas, H., Parimannan, S., Sathasivam, K., Ravichandran, M., & Yin, L. S. (2012). A novel FTIR-ATR spectroscopy based technique for the estimation of low-density polyethylene biodegradation. *Polymer Testing, 31*, 1094–1099. https://doi.org/10.1016/j.polymertesting.2012.07.015

Restiani, P., Khandelwal, A. (2016), Water governance mapping report: Textile industry water use in India. *Sustainability Outlook*. Available online at https://siwi.org/wp-content/uploads/2017/06/Water-Governance-Mapping-Report-INDIA.pdf. Accessed on 13 Oct 2023.

Sarayu, K., & Sandhya, S. (2012). Current technologies for biological treatment of textile wastewater—A review. *Applied Biochemistry and Biotechnology, 167*, 645–661. https://doi.org/10.1007/s12010-012-9716-6

Sarkar, D. J., Sarkar, S. D., Das, B. K., Praharaj, J. K., Mahajan, D. K., Purokait, B., Mohanty, T. R., Mohanty, D., Gogoi, P., Kumar, V. S., Behera, B. K., Manna, R. K., & Samanta, S. (2021). Microplastics removal efficiency of drinking water treatment plant with pulse clarifier. *Journal of Hazardous Materials, 413*, 125347. https://doi.org/10.1016/j.jhazmat.2021.125347

Schwabl, P., Koppel, S., Konigshofer, P., Bucsics, T., Trauner, M., Reiberger, T., & Liebmann, B. (2019). Detection of various microplastics in human stool. *Annals of Internal Medicine, 171*, 453–458. https://doi.org/10.7326/M19-0618

Singh, S., & Bhagwat, A. (2022). Microplastics: A potential threat to groundwater resources. *Groundwater for Sustainable Development, 19*, 100852. https://doi.org/10.1016/j.gsd.2022.100852

Singh, S., & Biswas, M. K. (2023). Management strategies for single-use plastics: Lessons to learn from Indian approach of minimizing microplastic waste, *Environmental Science Advances, 2*, 1680–1695. https://doi.org/10.1039/D3VA00222E

Singh, S., Kalyanasundaram, M., & Diwan, V. (2021). Removal of microplastics from wastewater: Available techniques and way forward. *Water Science & Technology, 84*, 3689–3704. https://doi.org/10.2166/wst.2021.472

Singh, S., Trushna, T., Kalyanasundaram, M., Tamhankar, A. J., & Diwan, V. (2022). Microplastics in drinking water: A macro issue. *Water Supply, 22*, 5650–5674. https://doi.org/10.2166/ws.2022.189

Singh, S., Chakma, S., Alawa, B., Kalyanasundaram, M., & Diwan, V. (2023). Identification, characterization, and implications of microplastics in soil: A case study of Bhopal, Central India. *Journal of Hazardous Materials Advances, 9*, 100225. https://doi.org/10.1016/j.hazadv.2022.100225

Singh, S., Chakma, S., Alawa, B., Kalyanasundaram, M., & Diwan, V. (2024). Assessment of microplastic pollution in agricultural soil of Bhopal, Central India. *Journal of Material Cycles and Waste Management, 26*, 708–722. https://doi.org/10.1007/s10163-023-01805-6

Sixta, H., Michud, A., Hauru, L., Asaadi, S., Ma, Y., King, A. W. T., Kilpelaninen, I., & Hummel, M. (2015). Ioncell – F: A high-strength regenerated cellulose fiber. *Nordic Pulp & Paper Research Journal, 30*, 43–57. https://doi.org/10.3183/npprj-2015-30-01-p043-057

Spencer, D. J. (2001). *Knitting technology—A comprehensive handbook and practical guide* (3rd ed.). Woodhead Publishing Ltd..

Sundbæk, K. B., Koch, I. D. W., Villaro, C. G., Rasmussen, N. S., Holdt, S. L., & Hartmann, N. B. (2018). Sorption of fluorescent polystyrene microplastic particles to edible seaweed *Fucus vesiculosus*. *Journal of Applied Phycology, 30*, 2923–2927. https://doi.org/10.1007/s10811-018-1472-8

Upadhyay, R., Singh, S., & Kaur, G. (2022). Sorption of pharmaceuticals over microplastics' surfaces: Interaction mechanisms and governing factors. *Environmental Monitoring and Assessment, 194*, 803. https://doi.org/10.1007/s10661-022-10475-0

Utebay, B., Celik, P., & Cay, A. (2020). Textile wastes: Status and perspectives. In A. Korlu (Ed.), *Waste in textile and leather sectors*. IntechOpen. https://doi.org/10.5772/intechopen.92234

Wang, L., Kaeppler, A., Fischer, D., & Simmchen, J. (2019). Photocatalytic TiO_2 micromotors for removal of microplastics and suspended matter. *ACS Applied Materials & Interfaces, 11*, 32937–32944. https://doi.org/10.1021/acsami.9b06128

Wang, Z., Lin, T., & Chen, W. (2020a). Occurrence and removal of microplastics in an advanced drinking water treatment plant (ADWTP). *Science of the Total Environment, 700*, 134520. https://doi.org/10.1016/j.scitotenv.2019.134520

Wang, Z., Sedighi, M., & Lea-Langton, A. (2020b). Filtration of microplastic spheres by biochar: Removal efficiency and immobilisation mechanisms. *Water Research, 184*, 116165. https://doi.org/10.1016/j.watres.2020.116165

Yang, L., Qiao, F., Lei, K., Huiqin, L., Kang, Y., Cui, S., & An, L. (2019). Microfiber release from different fabrics during washing. *Environmental Pollution, 249*, 136–143. https://doi.org/10.1016/j.envpol.2019.03.011

Young, S. (2019). *The fabrics with the worst environmental impact revealed, from polyester to fur*. Available online at https://www.independent.co.uk/climate-change/sustainable-living/fast-fashion-sustainable-worst-fabrics-b1855935.html. Accessed on 13 Nov 2022.

Zambrano, M. C., Pawlak, J. J., Daystar, J., Ankeny, M., Cheng, J. J., & Venditti, R. A. (2019). Microfibers generated from the laundering of cotton, rayon, and polyester based fabrics and their aquatic biodegradation. *Marine Pollution Bulletin, 142*, 394–407. https://doi.org/10.1016/j.marpolbul.2019.02.062

Zhang, Y., Gao, T., Kang, S., Allen, S., Luo, X., & Allen, D. (2021). Microplastics in glaciers of the Tibetan Plateau: Evidence for the long-range transport of microplastics. *Science of the Total Environment, 758*, 143634. https://doi.org/10.1016/j.scitotenv.2020.143634

Zhou, H., Zhou, L., & Ma, K. (2020). Microfiber from textile dyeing and printing wastewater of a typical industrial park in China: Occurrence, removal and release. *Science of the Total Environment, 739*, 140329. https://doi.org/10.1016/j.scitotenv.2020.140329

Chapter 12
Agri-Food Waste Management and Treatment Approaches for Environmental Sustainability

Pinki Koiri and Suchismita Das

1 Introduction

Agri-food systems are undergoing rapid transformation to withstand the food insecurity issues faced all over the world. Intensive farm concepts are formulated and practiced to get a better grip on the roaring food demand (Shinde et al., 2022). All the residues produced within the agricultural farm to its gate are known as agri wastes, and the wastes further generated from these agricultural products right from transporting from the agricultural gate till the customer's plate are called food wastes. All these wastes are one of the major agendas to be addressed to achieve sustainability transitions. By 2025, 44% of the total global waste would be produced by agrifood waste sector (Romero-Perdomo & González-Curbelo, 2023). Significant quantities of different types of livestock manures, crop residues, post-harvest wastes, agri-industrial wastes, kitchen/domestic wastes, food service waste, and food retail wastes compose the wholesome of agri-food wastes. In 2020, farm-to-gate wastes had world emission of nearly about 7.5 billion tonnes CO_2 equivalent (World Food and Agriculture Statistical Yearbook, 2022). According to the Food and Agricultural Organisation (FAO 2022), food loss and waste account for 8–10% of global greenhouse gases (GHGs) thus leading to an unsustainable environment and climate change (UNEP Food Waste Index Report, 2021). To meet emission norms and global climate goals, it is recommended to bring down the agrifood waste and carbon to reduced levels immediately to mitigate global warming and climate change risk (Yuniarti et al., 2023). According to the World Meteorological Organisation (WMO), global mean temperatures have risen by 1.15 °C (United in Science Report, 2023). Warm weather all-round the year and the El Nino phenomenon are persistent from the year 2020 to 2023. Decreases in ice cover or cryosphere, rise in

P. Koiri · S. Das (✉)
Department of Life Science and Bioinformatics, Assam University, Silchar, Assam, India

V. Kumar et al. (eds.), *Environmental Engineering and Waste Management*,
https://doi.org/10.1007/978-3-031-58441-1_12

sea water levels, and extreme weather are some of the radical changes occurring due to increased GHGs emissions.

To cope with the extreme losses, sustainable utilization of these agrifood wastes has to be focused on current research where deep emphasis has to be given to the concept of reduce, reusing, and recycle (Dhanker et al., 2022; Saharia et al., 2024). Agri-food wastes compiled in stocks and piles are to be handled efficiently by using either traditional or modern technologies. For example, recent work on cow dung has been done to transform animal waste into biochar (Lei et al., 2022). It can also be co-valorised for the generation of value-added products such as biogas and digestate (Erraji et al., 2023) and biooil (Mariyam et al., 2022). Further, novel innovative technologies, for example, enzyme-assisted extraction, ultrasound-assisted extraction, solid–liquid extraction, submerged extraction, microwave assisted extraction, and bioreactors were also applied to fruit wastes (peels and seeds) (Ali et al., 2022) and oilseed wastes (Morya et al., 2022) for the extraction of valuable compounds such as enzymes, phenolic compounds, flavonoids, and bioactive compounds. Apart from these, traditional approaches like composting/vermicomposting (Waqas et al., 2023) and mushroom cultivation (Ejigu et al., 2022) were used from past times. Therefore, the main objective of this study is to discuss and promote the ways of treating and managing agri-food wastes. Along with it, the current review highlights the basic advantages and disadvantages of the various technical processes. This chapter specifically intends to provide a brief outline of the varieties of agri-food wastes, inclusive of their sources. It also comes up with the environmental hazards and impacts raised by these wastes. The review is novel as very few prior studies have focussed on comprehensive state-of-art technologies regarding the non-wastage and conversion of agrifood wastes to form value-added products. The latest reports starting from 2020 to 2023 are focussed and reviewed to provide current information and estimates regarding the agri-food wastes. This chapter is important for the scientific community to understand the problem of agrifood waste dumping and disposal and to mitigate the issue using reliable technologies and management strategies.

2 Agricultural Wastes: Definition, Types, and Sources

It is defined as waste generated by farm practices such as livestock rearing and cropping in farm yards and fields. It also comprises the residues generated during the entire cropping period surpassing the harvest by-products as well as the trash produced during the processing and packaging of agri-products in industries. Notably, developing nations mainly encounter disability in managing these kinds of wastes generated, both due to the shortage of cropping period and lack of training and knowledge. As a consequence, the residues are burnt or dumped locally by the farmers.

There are mainly three major types of agricultural wastes generated—livestock manures, post-harvest waste, and agri-processed industrial wastes.

2.1 Livestock Manures

Animal manures are theoretically defined as either semi-solid or liquid farmyard waste products and are considered as outstanding sources of nutrients such as nitrogen, phosphorus, and potassium (NPK) for agricultural fields and crops. Manure comprises a heterogenous mixture of animal feces, urine, raw fodder (grass, hay, straw, etc.), bedding materials such as rice husk and wood dust, waste feed drink (rice and wheat bran mixture), and shedwash waters. The domesticated animals such as cows, bulls, sheep, pigs, horses, poultry and rabbits are constant sources of manure production. The biomass of the manure is derived from the food eaten. The useless cellulose and fibres after digestion are eliminated from the animal's body in the form of manure. However, the generation of huge steeps of manures is mainly persistent in rural areas.

In India, Chhattisgarh, Karnataka, Gujarat, and Tamil Nadu are the major organic manure-producing states. According to the National Centre for Organic and Natural Farming, 42,940,832.474 million tonnes of organic manure was produced in India during the year 2020–2021 (Annual Report 2020–2021 National Centre of Organic Farming). However, the year 2019–2020 resulted in the generation of 60,594,104 million tonnes of organic manure, which was recorded as the highest in the entire decade span (https://ncof.dacnet.nic.in/StatusOrganicFarming). This upsurge in organic productivity lies mainly in the increment or augmentation of livestock sector. It is reported by Animal Husbandry India, that the livestock sector is noted to continuously grow at the compound annual growth rate of 7.9% from the year 2014–15 to the year 2020–2021.

2.2 Post Harvest Agricultural Wastes

Post-harvest waste comprises all the leftovers that are generated during agrarian and retailing activities of agri-based products. Among the various agri-residues, crop stubbles, stalks of pulses, husk stover of crops such as rice, wheat, maize, barley, seedpods, stovers, leaf litter, fibres, banana leaves, and stalk, weeds, etc. These products are mainly composed of cellulose, hemicellulose, and lignin. This lignin is the sole reason for the sturdy structure of the biomass of wastes. However, the principal sources of these agri-based wastes are large agricultural farms, processing mills, agricultural industries, wood industries, etc. The notable state contributors to crop residue generation in India are Uttar Pradesh, Punjab, Maharashtra, Bihar, Haryana, West Bengal, etc. The pre-processing, transport, and storage of crop residues will continue to raise the concern of agricultural waste management annually. Commonly, harvesting, threshing, drying, and cleaning are the chief activities subject to increasing the load of agricultural waste every year. These are generally used as fodder, as domestic fuel in rural and underprivileged societies as roofing materials for sheds and houses, as raw material in pulp, cardboard industries, etc.

(Abdurrahman et al., 2020). The leftover quantity is eventually set to fire to quickly get rid of past season waste and to prepare the land for the next planting. With the advancement of agricultural approaches, in addition to the new cropping patterns or by mixed harvesters, has not only promised higher yield but also resulted in the production of millions of tons of residues seasonally.

2.3 *Agri-Industrial Wastes*

This category refers to the residual plant-based materials that are produced during the entire processing of agricultural products. The by-products here mainly include cereal bran (rice, wheat), molasses and sugarcane bagasse from the sugar industry, coffee peel, tea wastes (discarded leaves, buds, and stems of tea), nut shells, paper and pulp residues, textile fibres of jute, coir, cakes from different types of oilseeds, etc. The wheat and rice bran find their massive use as fodder for ruminants. However, sugarcane bagasse and oil cakes are the major processed residues among all. India being one of the largest producers of sugar, generates about 75–90 million tonnes of sugarcane bagasse per annum (Quereshi et al., 2022). In the case of oil cakes, countries namely China, the European Union, Canada, and Ukraine are the dominant producers of oilseeds and oilcake (OECD FAO Agricultural Outlook 2023–2032). Oil cakes are produced in enormous quantities during the crushing of lignocellulosic substrate especially from seeds such as canola, olive, mustard, sesame, coconut, rapeseed, sunflower, cotton seed, soybean, groundnut, peanut, falx, kernel, castor, karanja, and neem. The cakes derived from these seeds are divided into two types—edible and inedible cakes. The edible cakes are rich in protein content, bioactive compounds, and other nutrients and are commonly used for feeding cows, camels, goats, and fishes. Whereas inedible cakes that are highly toxic but rich in macronutrients (N, P and K) are used as natural fertilizers in fields.

3 Food Wastes: Definition, Types, and Sources

Food wastes are defined as unused food items that were potentially eaten stuffs especially leftovers, spoiled food or fruit and vegetable peels, kitchen waste generated and discarded as garbage. Food waste can be generated at any level of the food supply chain—cooking, processing, distribution, and sale. According to the United Nations, food loss and food waste undermine the sustainability of our food system. The food market stability, food security, and sustainable development are in question with the increase in continuous increase in food waste generation value. According to the UNEP Food Waste Index Report, 17% of food sold worldwide in 2019 was wasted and dumped.

There are mainly three major sectors in which food wastes are generated—domestic or household waste, food waste during food service, and food retail.

3.1 Domestic/Household Wastes

These types of waste refer to the rich organic residues produced by domestic consumers of food products daily. It comprises of vegetable wastes (peels, rotten vegies, stems), fruit peels, fish wastes (scales, rotten fish, bones), egg shells, chicken feathers, cooked and uncooked food, used tea bags, spoilt food, fermented foods, food grains, etc. The main sources of kitchen waste are houses and hostels. With the alarming increase in population, the production rate of household waste is also increasing (Meng et al., 2022). Household food waste is generated approximately 68,780,163 tonnes/year (Sinha & Tripathi, 2021). It leaves back undesirables imprint on the environment like foul smell, increased attack from flies and rodents, and increased release of GHGs. The major contributors are potato peels, onion peels, pods of peas and pulses, peels of pulses, foliage of cauliflower and cabbage, non-edible portions of citrus fruits (orange and lemon), peels of banana, mango, watermelon, coconut shells, egg shells, etc. In rural areas, feeding these vegetable and fruit wastes to cattle is a very common practice, but in urban areas, all the wastes end up being dumped in municipality areas.

3.2 Food Service Waste

Wastes generated during the preparation, serving, selling, and consumption of food at food service stalls, cafeterias, eateries, hotels, and restaurants form the second category of food waste. It is produced as a consequence of overcooking by chefs in the kitchen and also by leaving food in plates for customers or clients. According to a study, about 17.5% of food prepared by the food services wind up being in the trash (Lins et al., 2021). With the increase in population, the chances of consuming home-cooked food by metropolitan city inhabitants are decreasing. Before the coronavirus disease of 2019 (COVID-19) pandemic, food service facilities were in tremendous profit with easy mobility and give-and-take situations. The Economic Research Service of the United States Department of Agriculture (USDA) revealed that the food service had supplied $1.06 trillion worth of food in 2019. This can indirectly indicate the amount the food waste generated in this sector.

3.3 Food Retail

Wastes of food stuffs that occur in retail or wholesale business are included in this category. Large grocery and food sections of shopping malls are the raisers of this waste. Along with these, importers, retailers, and wholesalers are also involved. The major driving factors of waste generation in the food retail sector are the lack of deep freezers, packaging defects, overstocking due to misunderstanding the present food demand, and failure to follow minimum food safety standards. It occurs commonly in metros, cities, and in small cities also. India ranked 71 among 152 other countries in the United Nations Conference (Trade and Development's Business-to-Consumer (B2C) E-Commerce Index, 2020). According to Statista, 63% of the total retail market is contributed by the food and grocery sector. Therefore, the generation of large amounts of food product wastage is quite obvious with damaged or expired food products, destroyed packaging, and microbial contamination.

4 Environmental Impacts

The changes due to improper dumping of agri-food wastes into huge heaps and ill management are enormously triggering harmful impacts to cause a heated environment. Human beings, animals, fish, and environment (land, water, and air) are the victims of this global issue. Figure 12.1 shows the environmental impacts caused due to improper treatment and management of agri-food wastes, thus ultimately hampering the components of the biosphere.

4.1 Greenhouse Gas Emission

Any gas that absorbs and traps sun's radiant energy in the atmosphere without releasing it back to space is known as a GHG. They are responsible for causing the greenhouse effect as less energy is released from earth's atmosphere in comparison to absorption, resulting in raised global surface temperature. In the United Nations Framework Convention on Climate, six GHGs mentioned are carbon

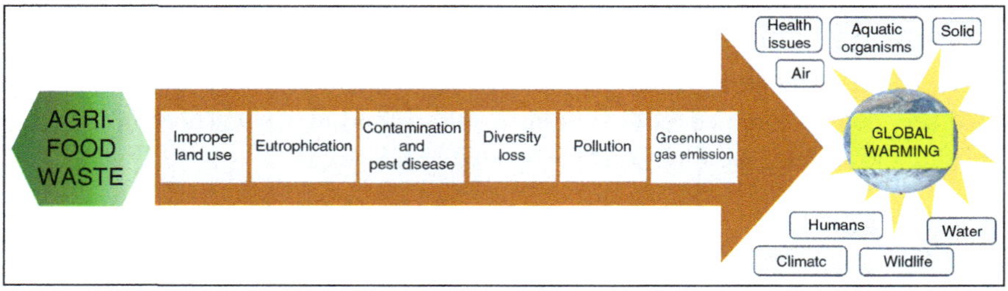

Fig. 12.1 The environmental impacts of agri-food wastes causing global warming and climate change. Major effects like greenhouse gas emission, pollution (air, water, and soil), eutrophication, improper land use, biodiversity loss, pest attack, and microbial contamination in the environment are imposing threats to the Earth in the form of global warming and climate change. As an outcome, the inhabitants that is wildlife (plants and animals), aquatic animals, humans, and microbes along with their abiotic component (soil, air, water) are suffering substantially in terms of health issues and habitat destruction

dioxide, methane, nitrous oxide, hydrofluorocarbons, perfluorocarbons, and sulphur hexafluoride. The former three gases are released from agricultural activities, fossil fuel combustion, industries, waste dumping, power generation, etc. However, of all the sectors, we will be discussing the agri-food waste sector. According to the FAO, food wastes generate 8–10% of the total GHGs. On the other hand, agriculture and land use change contribute to a quarter of total GHGs (Laborde et al., 2021). The formation and release of carbon dioxide, methane, and nitrous oxide gases originated from improper disposal and burning of agri-food wastes. According to the Centers for Disease Control and Prevention (CDC), the massive impact posed by GHGs and global warming on the planet is endless health effects with respiratory and cardiovascular disruptions and diseases, premature death due to heat waves, increased infection caused by pathogens to host utilizing contaminated food and water. Hence, there is a prompt need for paying more attention to the seriousness of the mitigation of climate change. Proper disposal and valorisation of food and agri-waste products can reduce emissions to a great extent.

4.2 *Eutrophication*

Eutrophication can be defined as the process of gradual increase of organic nutrients especially nitrogen and phosphorus along with other limiting factors in an aquatic system. It results in increased aquatic plants and algal growth. It can be seen to occur naturally in lakes, streams, ponds, and fisheries. The productivity of the entire ecosystem increases initially during eutrophication as more nutrients are available for utilization. Manures, poultry wastes, and other organic residues from farm and domestic discharge, industrial waste, fertilizers, detergents, animal wastes, and sediments enter into water reservoirs (Pawar et al., 2020). This is considered an acute aquatic hazard. It occurs naturally due to the gradual accumulation of wastes into water bodies. However, the process can be provoked to progress further or accelerated by human activities like allowing the flow of wastewater into nearby water sources. The effects of the eutrophication process in a water resource are lower water quality, turbidity, sunlight, reduced dissolved oxygen, harmful toxins, and gases released, which are harmful to both fish and humans. The factors that contribute to enhanced eutrophication are slow water current, pH of 7.4–8.5, increased nitrogen and phosphorus content, and temperature around 23 °C as high temperature stimulates microbial activity at the sediment bed (Diatta et al., 2020).

4.3 *Biodiversity Loss*

Biodiversity refers to all the diversification of life forms that are present on the Earth. In other words, the summation of all species belonging to all the genera living in that area whether discovered or not indicates biodiversity of that particular place.

Regions that are highly rich in species or life forms are known as biodiversity hotspots. India is one of the nations which is fully loaded with rich biodiversity. Fortunately, four hotspots are located in the Indian subcontinent. Soil biodiversity is mainly affected by bad management of crop stubbles and wastes. However, when the crop residues are blazed, they cause a great loss in soil biodiversity loss especially in agricultural fields. The helpful microorganisms that are present in the top soil die due to elevated temperature during the burning process. The Blazing fire for a considerable long period is reported to decrease the population size of beneficial bacteria from fields. Fungi are also hampered but they recover faster.

4.4 Pest and Infectious Disease

Dumping organic wastes in open areas offers a breeding ground for flies, food, and waterborne pathogens. The decomposition and fermentation provide adequate conditions for the origin and colonization of pathogens. Large heaps of waste including animal excreta, plant-based residues, and wastewater are acute sources of pathogens. Pathogens can be defined as microorganisms of bacterial, viral, fungal, protozoan, or parasitic origin and are responsible for causing disease in hosts either animal, human, or plant. They are borne from contaminated sites of organic waste disposal and stagnant wastewater flown from dumpsites. Some pathogen-caused infectious diseases are plague, typhoid, yellow fever, malaria, dengue, encephalitis, dysentery, and diarrhoea (Centers for Disease Control and Prevention, CDC). Hepatitis- A, B, and C virus, poliovirus, and F+ coliphages are also found to exist in these dumped areas (Addy et al., 2022). The pathogens get transmitted by agents and end up infecting hosts through food and water. Chronic diseases like respiratory, dermatological, and gastrointestinal problems are also prevalent in waste-related workers due to their regular improper dumping and incineration activities without safety measures (Addy et al., 2022).

4.5 Pollution

The condition in which the surrounding air, water, land, or environment gets polluted or contaminated with various toxic substances beyond the dissipation limit of nature is called pollution. The major pollutants are noxious gases, smog, dust, aerosols, microplastics, heavy metals, chemicals, pesticide runoffs, industrial effluents, wastewater from farms, etc. However, among all the sectors, agriculture and food waste are the only two moderate sectors pitching in pollution in India. Interestingly, six of the major air pollutants recognised by the United States Environmental Protection Act (US EPA) are carbon monoxide, nitrogen oxides (NO and NO_2), sulphur dioxide, ozone, particulate matter, and lead.

According to the latest forecast of the Food and Agricultural Organisation for world cereal production in 2022, 2777 million tonnes of cereals were produced. This upward trend cools down the matter of food insecurity but the stubble associated with such a huge production rate could be boundless. From the year 2000 to 2014, about one-third of the total quantity of crop residues in South Asia was burnt in the agricultural field there itself (Lin & Begho, 2022). The northern states especially the Indo–Gangetic plain persist on dual cropping systems (rice, wheat) all around the year. Therefore, combustion of the stubbles remains the only solution, to prepare the farmland for planting the next batch of crops. In 2016, the trend of burning crop residues in India was recorded the highest in the entire decade, after which subsequent downdrift was observed till 2019. Suddenly, an unexpected rise in burning paddy residues was reported in 2020 (Ravindra et al., 2022). All these inhuman activities increase air pollution levels, consequently creating health issues. Open-air burning, which is a frequent practice in agriculture, is known to produce Polycyclic Aromatic Hydrocarbons (PAHs) and particulate matter $PM_{2.5}$. Their exposure causes respiratory diseases such as asthma, cough, and breathing issues. The national capital territory Delhi and north-western states namely Punjab, Haryana, and Uttar Pradesh are associated with major burnings of agricultural residues.

Moreover, both surface water and ground water increase the chances of contamination by indiscriminate or improper disposal of agri-residues (Bala et al., 2023). Leachates of organic residue decomposition contain toxic levels of heavy metals (Essien et al., 2022). Moreover, the quality of top soil also gets degraded by improper landfill and leachate.

4.6 *Improper Land Use*

Land use encompasses all the human actions related to the use of land for obtaining resources and for carrying out daily activities. In this context, proper usage of land describes good farming practices that gradually change the type of land towards better fertility and soil type. Maintenance is the key to attaining a better quality of top layer of soil. Nevertheless, the pressure on land is increasing with the development and industrialization, especially from the food and agricultural sectors. Soil fertility is hampered by burning crop residues. The pH of the soil inclines towards lower pH indicating a high concentration of H^+ ion. Soil acidification occurs during the transformation and cycling of carbon, nitrogen, and sulphur fertilizer reactions (Uwiragiye et al., 2023). It gives on to decreased microbial activity and reduced population, less organic matter decomposition, slow soil nutrient cycling, and less plant nutrient absorption, ultimately hampering the productivity.

5 Waste Treatment Strategies

Treatment of waste is referred to a process in which the volume and toxicity of any type of waste can be reduced. No matter how much wastes get reused, recycled, and managed, ultimately treatment becomes mandatory for the last-stage wastes in one way or the other. The main objective for treating wastes of any origin is to reduce the level of liquid and gaseous pollutants that would be released into the environment. Figure 12.2 shows the various possible treatment strategies of agri-food wastes including thermal, chemical, and biological techniques.

Thermal treatment methods require a lot of energy resources to reduce the content of waste. Large units are required to set up which is time-consuming. Similarly, during the chemical treatment, the wastes are treated with the mechanism of altering the toxic nature of pollutants or the by-products with the help of chemical reactions. It requires knowledge of chemistry as well as it is also avoided because maximum of the chemicals is restricted for use and are not available to common people. The biological system of treatment of wastes and pollutants is affordable, site-specific approach, low cost, environmentally sound, and less spacious. In bioaugmentation, bio stimulation and mycoremediation, the microbes feed on the waste materials. Mixed cultures, pure cultures, fungal strains, and genetically modified organisms are

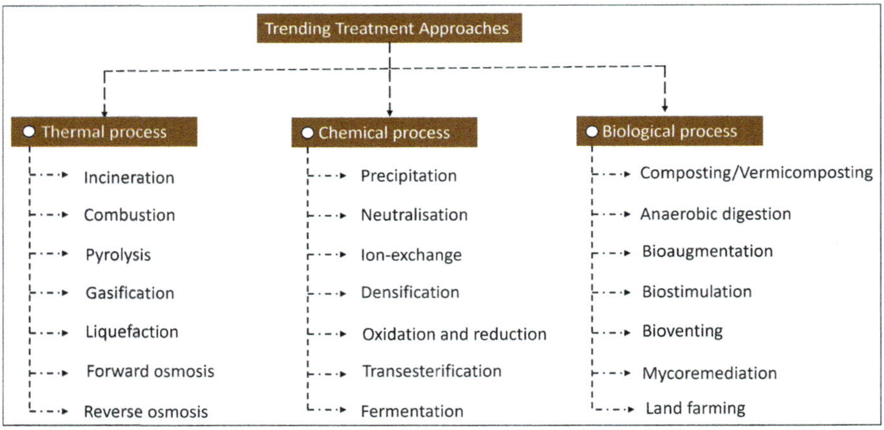

Fig. 12.2 A flowchart depicting the three main categories for treatment of agri-food wastes that can be employed to prevent pathogen colonization and disease transmission from organic wastes. Thermal processes such as incineration, combustion, pyrolysis, gasification, liquefaction, and forward and reverse osmosis are being traditionally used. Chemical treatments such as precipitation, neutralisation, ion-exchange, densification, oxidation, reduction, transesterification, and fermentation require dedicated treatment setups with chemical experts. The biological treatment process such as composting, vermicomposting, anaerobic digestion, bioaugmentation, biostimulation, bioventing, mycoremediation, and landfarming can be done in contaminated sites with the help of microorganisms

the agents in the biological treatment of wastes. Therefore, in this way, the biological system can be accustomed to maintain environmental sustainability.

6 Management Approaches

The management process of any type of waste involves carrying out the complete collection, assortment, transportation, handling, treatment, recycling, innovation and disposal steps in a sequence initiating from the point of generation of wastes till the stage of permanent and final disposal. The wastes that cannot be further utilised in any sense are required to be properly treated and disposed of. There are certain outcomes related to proper and improper waste management highlighted in Fig. 12.3. Anyway, the management strategy tries to provide green solutions to unhealthy and toxic pollutants. However, the management process has to be cost-effective besides being socially acceptable and practicable. The extreme of economic, social, and environmental cannot be taken. The merged areas of these factors can give a sustainable environment and circular bioeconomy.

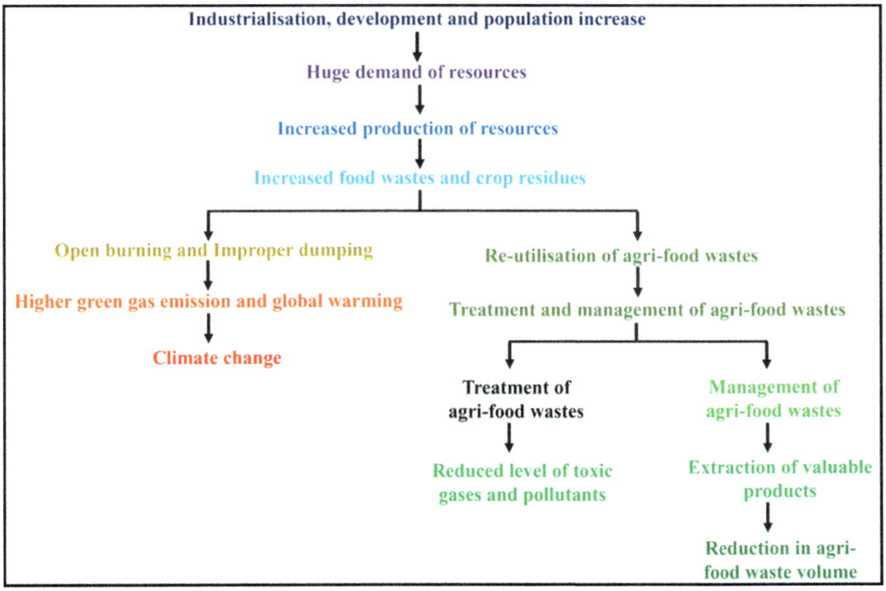

Fig. 12.3 A flowchart showing the consequences of improper versus proper management and treatment of agri-food wastes. High population density continuously pumps in food resources generating huge steeps of agri-food wastes. Improper management by burning and open disposal emit greenhouse gas and toxic substances resulting in global warming and climate change. On the contrary, proper management by reutilising the organic wastes to extract value added products is a highly appreciated initiative that produces less toxicity and reduces the volume of organic wastage

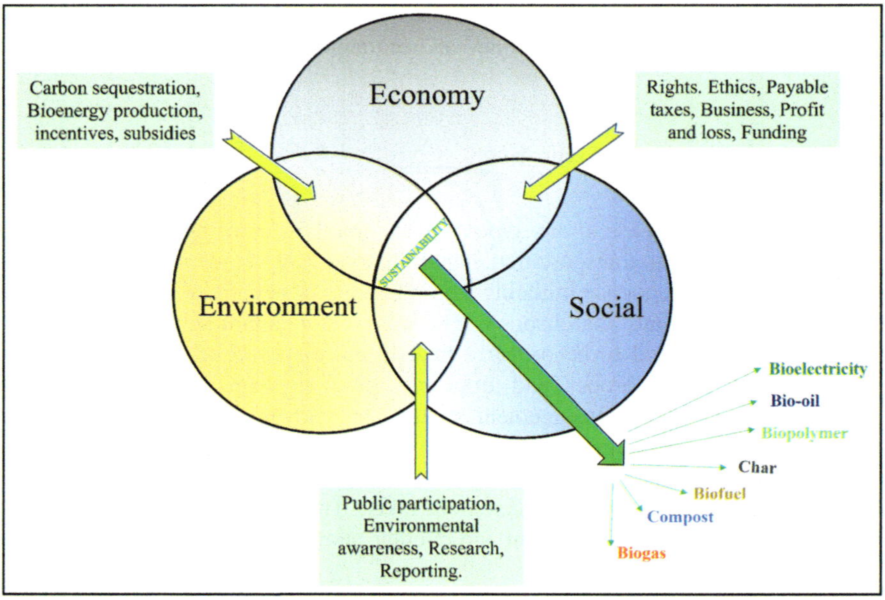

Fig. 12.4 Diagram showing the three components of sustainable management: social, economic, and environmental sectors and how their cooperation can produce sustainable products. The socio-economic factors comprising rights, taxes, ethics, business, trade, marketing, funding, interest, and culture; socio-environmental factors like public participation, societal cooperation, environmental awareness, research, reporting, and climatic concerns environmental economic factors such as carbon sequestration, bioenergy transformation, incentives, subsidies, environmental policies, and resource allocation can get associated together with proper planning and management to achieve sustainability and circular bioeconomy

The Fig. 12.4 shows the components of sustainable management of wastes along with the various contributing factors of socio-economic, socio-environmental, and economic-environmental origin. Upcycling, recovery, and valorisation are eminent components of an ideal management programme. Integrated management of waste is also a noteworthy and fashionable dimension in the present world. Along with basic technical management and technological advancement, research, reporting, policy and law enactment, educational programmes, public awareness, and cooperation also play a significant role. In the case of agri-food wastes, the lignocellulosic and organic wastes can be handled in several ways. Some of the sustainable methods of managing agri-food waste are discussed briefly in this chapter. Table 12.1 shows some of the management processes of agri-wastes to derive value-added products.

Table 12.1 Various trending management strategies of agri-food wastes

Value added product	Specification/conditions	Agri-food waste source	Uses/function	References
Biochar	Pyrolysis at 300, 400, 500 and 600 °C in N_2 environment at a heating rate of 4 °C/ min for 2 h	Kitchen wastes	For Cd (II) removal	Xu et al. (2021)
	Fe^{3+}impregnation and pyrolysis at 200, 400 and 600 °C	Cattle manure	Adsorption of Congo red and Methylene blue	Zhu et al. (2020)
	Slow pyrolysis 550 °C at a ramp rate of 2 °C min^{-1}	Agricultural wastes	For the removal of Cd (II) and Cu (II) from mine water	Bandara et al. (2020)
Substrate from mushroom cultivation	*Pleurotus ostreatus* (Jacq.) P. Kumm, P969 strain was inoculated at a rate of two tea spoons of inoculant per kg of substrate.	Cassava, maize, common bean, banana, soybean flour and cow dung	Edible mushroom production and also an income source	Kazige et al. (2022)
	20 g of spawn was added in 2 kg of substrate to grow *Lentinula edodes* (Shiitake mushroom)	Rice straw and sugarcane bagasse	Delignification of substrates and heavy metals uptake from dairy plant and sugar mill wastewaters	Kumar et al. (2022)
	Ganoderma lucidum was grown by spawning agricultural substrates at a ratio of 4%	Wood chips, wheat bran, soybean meal, date palm leaf, olive oil extraction waste	Production of medicinal mushroom	Kurd-Anjaraki et al. (2022)
Bioactive compounds	Mixture of ethanol: water (96:4 v/v) with a solid: liquid ratio of 1:100 (w/v) for 24 h at 40 °C under orbital stirring of 250 rpm	Tomato (peel), potato (peel), cabbage, eggplant (peel and stem), strawberry (leaves), pear (peel), kiwi (peel), garlic (white peels), orange (peel), tangerine (peel), beet (leaves, peel and stems)	Phenolic compounds like caffeic acid, chlorogenic acid, p-coumaric acid and protocatechuic acid were most frequently identified	Martínez-Inda et al. (2023)

(continued)

Table 12.1 (continued)

Value added product	Specification/conditions	Agri-food waste source	Uses/function	References
	Extraction of omega −3 fatty acids	Fish leftovers and their processing by-products	Eicosapentaenoic acid (EPA) and docosahexaenoic acid (DHA) are used in nutraceutical application	Alfio et al. (2021)
	Ultrasound-assisted extraction (UAE) and Enzyme-assisted ultrasound extraction (EAUE)	Mango peels	Phenolic compounds	Sharif et al. (2023)
Bio-based polymers	Blends prepared using CMC (waste derived), gelatin, agar and varied concentrations of glycerol (1.5%, 2% and 2.5%)	Sugarcane bagasse	Biodegradable carboxymethyl cellulose (CMC) based material used in sustainable packaging	Yaradoddi et al. (2020)
	Chicken feathers were disinfected and dissolved in 0.1 M sodium hydroxide solution to get precipitated keratin. Later, this keratin powder, starch and glycerol mixed in different proportions resulting bio-plastic films.	Chicken feather waste	Fabrication of keratin-starch bio-composite film for food packaging	Oluba et al. (2021)
	Extracted with methanol, and dissolved (5,10,15,20% v/v) with poly vinyl alcohol, starch and sorbitol	Pineapple peel	Active biodegradable packaging material	Kumar et al. (2021)
Extraction of enzymes	α-amylase enzyme was produced from rice bran using *Bacillus tequilensis* TB5	Rice bran	α-amylase is used in starch, brewing liquor and beer industry	Paul et al. (2020)
	Fungal biosynthesis of amylase and proteases using *Rhizopus oligosporus*	Food waste such as rice, beans, noodles, meat, chicken, pork, fish, vegetables, salad, fruits, wheat bran and sugarcane bagasse.	These enzymes are low cost and help in the digestion of carbohydrates and proteins	Escaramboni et al. (2022)
	Production of liquid cellulolytic enzymes using genetically modified	Coffee husk	Cellulases are used in converting ligno-	Coral et al. (2022)

(continued)

Table 12.1 (continued)

Value added product	Specification/conditions	Agri-food waste source	Uses/function	References
	Trichoderma reesei fungus		cellulosic bio-masses into biofuel	
Bioelectricity	Zinc cathode and copper anode was used in the microbial fuel cell (MFC) and the microorganisms identified were *Pichia guilliermondii,Candida boidinii* and *Brevibacilluslaterosporus*	Fruit wastes (blackberry, dragon fruit and noni)	Electric current of (25.78 ± 0.89) mA and maximum average open circuit voltage of (1.17 ± 0.12) V was reported from dragon fruit MFC.	Rojas-Flores et al. (2020)
	Zinc cathode and copper anode was used in the single chamber microbial fuel cell (MFC) and the microorganisms identified were s *Pseudomonaeruginosa, Acinetobacter bereziniae, Stenotrophomonas maltophilia*, and *Yarrowialipolytica*	Onion wastes	Maximum peaks of 4.459 ± 0.0608 mA and 0.991 ± 0.02 V of current and voltage were generated	Segundo et al. (2022)
	Benthic microbial fuel cell with one cathode and three anodes was prepared along with yam waste and metal supplemented pond water	Yam (*Dioscorea alata) waste*	Energy efficiency obtained was 0.0744 mW/m^2 along with remediation of Pb^{2+} (90.14%), Cd^{2+} (88.00%) and Cr^{3+} (90.34%)	Fadzli et al. (2021)

6.1 Reduce, Reuse, and Recycle

3Rs concept of reduce, reuse, and recycle is subjected to use minimal quantity, to utilize a product again and again, and also to repetitively upcycle a used material with the help of new and innovative ideas. In this way, wastes are transformed into new, attractive items or into respective products of interest. Recycling can provide immediate relief from waste, but practising or establishing a culture of creating less waste is more effective for achieving long-term consequences (Bigdeloo et al., 2021). Prevention of waste generation is mostly required at the individual level.

Almost, all agricultural residues are lignocellulosic, so they are comparatively harder and more stable for short storage. The food waste cannot be preserved or stored for a long time. They can be usable until the start of the decomposition and fermentation process. So, rather than holding the organic matter for a certain period, it can be publicly urged to reduce the waste generation quantity as far as possible. It can be done by only buying limited items from the food market and groceries and consuming the bought food very soon before it gets destroyed. Kitchen wastes can be diverted towards producing organic fertilizer or feeding cattle. Leftover unused food of good eatable quality, that is produced from restaurants, hotels, food services, and houses can be distributed daily in slum areas or in children's homes. Stray animals such as cows and dogs as well as poultry and fish can also be fed the leftover food by approaching interested non-governmental organisation (NGO) workers. Moreover, spoilt cooked food can be fed to pigs. Agriculture and animal wastes can be used to prepare animal feed to tackle the continuous generation of agri-food wastes. Wastes from the retail sector can be also managed. The products that have an expiry date in a year can be targeted to be sold out with the help of a discount strategy or with buy one get one free. Unitedly, the central governing authority, local governments, and private recyclers should also actively take part in the 3R waste management system.

6.2 Animal Feed

It is a very viable option to provide food to hungry humans and animals. The production of feed is done by formulating nutrients from agri-food wastes, which ultimately results in increasing the growth rates and productivity of animals. In this management strategy, the agri-food wastes are mixed with adequate nutritional doses of vitamins, minerals, amino acids, and fatty acids per kilogram weight. Better productivity is achieved in the form of fortified milk, eggs, meat, and fur. Raw, partially processed or fully processed agri-food waste products such as milling co-products, legume peels, oilseed cakes, animal proteins, brewers' grain, and spare food grains are the common feedstuffs chosen for livestock. The basic non-vegetarian food wastes are collected from slaughter houses of meat animals and gradually processed along with other base products like wheat and maize millings to obtain dietary supplement. Heat pre-treatment is required to reduce the chances of contamination and diseases from animal-based feedstocks. Some food-borne infections prevalent in animal waste-derived feeds are bovine spongiform encephalopathy (BSE) and dioxin contamination. For these reasons, antibiotics are also added as a precaution in subtherapeutic doses in animal feeds, but it acts as a disadvantage owing to drug resistance of the disease-causing pathogens (Mutua et al., 2020).

According to the International Market Analysis Research and Consulting Group (IMARC Group), the Indian animal feed market size will increase from 956.7 billion at present to 1578 billion by 2028. However, if proper management of agri-food wastes can be targeted it can result in higher economic growth along with achieving a circular bioeconomy. The majority of the animal feed is mainly prepared for poultry followed by cattle and aquafeed. Notably, Andhra Pradesh, Karnataka, and Tamil Nadu are the leading states of India in the production of animal feed.

6.3 Composting

Composting refers to the aerobic breakdown or degradation of organic debris such as kitchen wastes besides dry leaves, manure, and garden waste, at an ambient temperature and moisture level to form a black, stable, carbon-rich fertilizer that can be ultimately applied to growing crops and plants. The initial oxidation of organic matter is followed by the mineralisation of slowly degradable molecules and the humification of lignocellulosic compounds. The addition of prepared compost not only improves the texture and quality of soil also minimizes the chances of eutrophication in nearly freshwater sources. The main degrading agents are fungi (molds and yeast), bacteria (mesophilic, thermophilic, and actinomycetota), earthworms, millipedes, snails, slugs, and beetles. There are various types of composting depending on the place of decomposition and the microorganisms involved. The open-air composting approach involves collecting and accumulating agri-food wastes in a particular place and allowing the mixture to get broken down with the help of aerobic microbes in due course of time. In direct composting, a burrow or pit is prepared in the ground and all the organic wastes are chopped, dumped into it, and covered. Less foul smell and flies are subjected to that location. However, there is a separate technique that involves degrading the organic mixture in a bucket/tumbler/drum instead of pit. The green and brown organic waste is inoculated with worms inside the container. But, due to rise of temperature in the container the worms eventually die. Another method known as effective microorganism (EMO) composting involves the degradation and fermentation of degradable wastes with the help of effective microorganisms such as photosynthetic bacteria, lactobacillus, streptomyces, actinomycetes, and yeast. Yet another famous and efficient method of composting is vermicomposting. This technique involves the decomposition and degradation with the help of worms. These worms feed on the matter and produce their castings, humus, poop, or feces.

By composting agri-food waste, the burden of organic waste dumping and emission of greenhouse gases can be reduced. Also, the usage of chemical fertiliser can be decreased by simultaneously increasing the production and utilisation of commercial-level organic compost from agri-food waste. As the benefit–cost ratio of compost is higher than 1 (Mandpe et al., 2020), so for utilising this huge amount of organic waste, more composting plants are required to be constructed to sort, segregate and decompose green and brown garbage.

6.4 Anaerobic Digestion

The bio methanation process of utilisation of agri-food waste and wastewater to produce biochemical and other valuable by-products such as biogas and biofuel (bioethanol, bioethanol). Biowaste is processed by microbes and/or externally added cultures. Biomass sources that can be used for anaerobic digestion are crop residues, livestock manures, food waste, forestry residues, and industrial and municipal wastes including solid waste, sewage, and sludge.

Hydrolytic fermentation bacteria initially hydrolyse the organic residues with the help of hydrolytic enzymes. Then the hydrolysed residues are broken down to form volatile fatty acids, hydrogen, carbon dioxide, and ammonia. The process is known as acidogenesis. Then the acetogens produce large acetic acid compounds. At the last stage, the production of methane, hydrogen, carbon dioxide, hydrogen sulphide, siloxanes and water from acetate occurs (Gong et al., 2021).

Anaerobic co-digestion can also be performed by mixing food wastes with cellulosic or plant-based substrate (agri-waste) or with waste-activated sludge (Pan et al., 2021). The treatment and management of the organic portion of municipal solid waste can be correlated to the environmental impact (Habagil et al., 2020). The consumption of energy is increasing day by day. In May, 20.0422 million tonnes of crude oil were imported to India (Ministry of Commerce and Industry, India). These huge number of imports causes economic loss along with increased environmental pollution problems. Therefore, by recycling organic fractions of waste for generating bioenergy to avoid dependence on fossil fuels. Using the energy to dispose of waste should be side parted and instead the waste to energy generation can be done. The digestate produced at the end of the anaerobic digestion process can be judiciously used for organic fertiliser (Bandgar et al., 2021).

6.5 Production of Bioactive Compounds

The agri-industrial and food wastes are packed with many bioactive compounds. These bioactive compounds are trace constituents mainly found in plant-based residues and food products thus responsible for promoting health and other metabolic processes. It is biologically active in the sense that it can cause signalling or response in a cell or organism (Gomes-Araújo et al., 2021). Some plant-based bioactive compounds are phenolic compounds (p-coumaric acid), flavonoids (quercetin), alkaloids (catechin), volatiles, essential minerals, tannins, vitamins (A and E), carotenoids, phytosterols, β-glucan, lycopene, etc. and animal-based bioactive peptides namely amino acids and omega-3 polyunsaturated fatty acids (PUFA) from animal wastes (Ben-Othman et al., 2020). Single cell protein can also be prepared with the help of *Saccharomyces cerevisiae inoculated* from fruit wastes. Many of these compounds have various health benefits and therapeutic uses. They have antioxidant, anticancer, antidiabetic, anti-inflammatory, cardiovascular and neuroprotectant. So, the utilisation of agri-food wastes is necessary for the betterment of the environment and medicinal field.

The bioactive compounds are required to be extracted from agri-food wastes. There are many extraction techniques of bioactive components enlisted as—solid–liquid extraction, subcritical water extraction, membrane technology, solid phase micro-extraction, pressured liquid extraction, surfactant medicated techniques, enzyme-assisted extraction (EAE), ultrasound-assisted extraction (UAE), supercritical fluid extraction (SFE), and deep eutetic solvent (DES) extraction (Gomes-Araújo et al., 2021).

6.6 Substrates for Growing Edible Fungi (Mushrooms)

Mushrooms are a group of fleshy macroscopic fungi with a distinctive fruiting body. They live on dead matter and degrade them with the help of specific enzymes to get their food. Notably some mushrooms such as button mushrooms, oyster mushrooms, and shitake mushrooms have a rich source of quality proteins, essential amino acids, vitamins, and minerals. Due to high nutritive value, they are cultivated enormously in some countries like China, the USA, Spain, France, Europe, South Korea, Germany, Japan, and Italy. Mushrooms are easy to grow on agri-food residues with little effort. In this technique, non-consumable agri-food waste without getting dumped and burnt can be sustainably converted to edible food products. Agricultural wastes such as crop stubbles and agri-industrial processed wastes have lignin, cellulose, and hemicellulose in composition. The enzymes digest these lignocellulose materials in order to avail food for the mushroom and flourish the growth. Rice straw, wheat straw, maize straw, rice husk, rice bran, sugarcane bagasse & leaves, cotton waste, corn stover, maize leaves, bamboo waste, grass waste, tea waste, fruit peels like banana, pineapple, mango peel, etc. are used as potential substrates for growing oyster mushrooms. For growing mushroom on agri-food residues, the substrates are required to be sterilised and inoculated with mycelium and packed in a sealed bag. After 10–15 days, numerous primordia will be formed through small holes which further develop into mature fruiting bodies. These can be freshly plucked and consumed after cooking. Moreover, self-economy can also be boosted by selling the mushrooms.

6.7 Production of Biochar

Heavy metals are basic elements that have a high density compared to water and an atomic number greater than 20. They are known to produce free radicals that cause lipid peroxidation, and protein and nucleic acid modifications by initiating free radical-mediated chain reaction. So, the heavy metals when consumed very often and at high doses are responsible for detrimental health issues. The crops and fishes grown on heavy-metal contaminated soil and water, in turn, uptake the heavy metals into them. When these crop plants are consumed by herbivores or other primary consumers, the heavy metals enter the food chain. However, the initial accumulation of heavy metals in the environment (soil, water, crops, primary consumers) can be due to natural as well as human activities. Volcanic eruption and weathering of metallic rocks are the principal natural sources of heavy metals, whereas, undertreated sewage, sludge, industrial productions, extraction, mining, pesticide

Fig. 12.5 A diagram showing the right usage of agri-food wastes to prepare biochar and remediation of heavy metals. The agricultural and food wastes generated after processing and cooking can be chopped into smaller pieces, sun-dried or oven dried, crushed into powder, pyrolyzed at high temperatures, and the biochar obtained can be applied to agricultural lands specifically in heavy metal, organic, and inorganic contaminated sites to prevent accumulation of pollutants in the crops or plants

application, chemical fertilizers, textiles run-offs, etc. are the anthropogenic sources. Many heavy metals namely cadmium, chromium, copper, lead, inorganic mercury, nickel, selenium, and zinc are known to have toxic effects when they accumulate at high levels in the environment according to UNEPA.

To combat this problem, biochar can be appointed for bioremediation of a heavy metal contaminated area. In this context, biochar is defined as a carbonaceous thermal decomposable product generated from agricultural of food waste biomass. This technique can be used to immobilize the heavy metals from a specific contaminated site and also prevent crop and plants to uptake the heavy metals, organic as well as inorganic pollutants. The biochar works on the adsorption mechanism based on pore filling, cation exchange, co-precipitation, functional groups complexation, precipitation, hydrogen bonding, hydrophobic effect, etc. Any type of crop or food residues having lignocellulosic content can be utilised for producing high-quality biochar. Risk husk, sugarcane bagasse, corn cob, rice stubble, wheat straw, vegetable and fruit waste (peels), tea leaves waste, fish scales, pulse straw, etc. have been reported to give biochar products mediating successful reduction and elimination of contaminants. The biomass contents are pyrolyzed in the absence of oxygen at varying temperatures (300–900 °C). Figure 12.5 shows the overall preparation of biochar from agri-food wastes and its application in the soil to limit the heavy metal uptake by crop plants. The physio-chemical properties that improvised when the biomass gets combusted at high temperatures, thereby adsorbing heavy metal ions within biochar particles and ultimately reducing the impact crop plants, water, soil, etc.

6.8 *Biorefinery*

This sustainable process processes the agri-food wastes to convert them into green products such as bioenergy and bio-based products. Due to the lack of knowledge, awareness, and training about valorisation of agri-food wastes, the valuable wastes are getting uselessly dumped and destroyed on a large scale. In this type of framework, waste-to-resource and sustainable energy are ensured. However, irregular supply as well as non-availability of raw residues can affect the production process and system as well. In case of cheap and abundant availability of organic wastes, then the process becomes efficient for the respective producers also. The sequential procedure of the biorefinery method includes the pre-treatment and processing of the agri-food biomass, followed by extraction and further developing ideas to improve the quality and continuously generate energy (Velvizhi et al., 2022). Some of the main products of biorefinery are bio-fuels (bio-oil, bio jet oil, green diesel), hydrocarbons, and syngas products (bio methanol, bioethanol, fischer-tropsch liquids). There are four methods of attaining the valorised products—thermal, chemical, catalytic, and physical.

Energy security continuously seeks the development of an alternative reliable source of energy before the depletion of non-renewable resources. One of the various governmental policies using ethanol blended petrol (EBP) is brought into action from April 2023. The International Energy Agency (IEA) mentions that the global demand bio-fuels will grow by 41 million litres (28%) by 2026. So, to mitigate the demand, import exchange savings along with the frequent inflation and deflation of fuel prices can be solved by developing the biorefinery field.

6.9 *Generating Bio-Electricity*

The electricity demand was maximum at 223,254.00 MW in June, 2023 concerning the current decade (Census and Economic Information Center, 2023). In 2022–2023, the power requirement was 15,11,847 MU; however, the availability was 15,04,264 MU. Notably, most of the electricity power (49.1%) is generated by combusting coal in power plants (Ministry of Power, Government of India). Burning coal not only depletes the non-renewable resources but also postures hamper to the sphere. The issues regarding power generation and supply can be met if bioelectricity is produced from agri-food wastes to achieve a greener future (Power Sector at a Glance All India, 2023).

Bioelectricity is referred as electric current that is derived from biomass of organic and lignocellulosic origin. The agri-food wastes are blended and fermented anaerobically to produce fermentation liquids. These liquids can be used as fuels for running a catalytic fuel cell. Food waste and spent animal bedding were used for electricity generation (Dahunsi, 2020). When the volatile fatty acids get accumulated and the production of enormous organic acid occurs, an elongated period is required for the conversion of food wastes to finally form methane. The microbes are utilised as catalyst to generate electricity in agri-food mixed microbial fuel cell. However, various types of pre-treatments of the wastes such as physical, thermal, chemical, biological, and also thermochemical are done to increase the efficiency of electricity generation (Pandit et al., 2021). Cooked food that are rich in carbohydrates can be hydrolysed and oxidised so that electron transfer can occur from glucose molecule to the anode.

6.10 Enzymatic Hydrolysis

It is an outstanding technique that involves the production of a large number of industrially important enzymes. The enzymes produced are the by-products formed during microbial growth: amylase, cellulase, tannase, xylanase, protease, and laccase. α-amylase and β-amylase are best derived from wheat bran substrate and molasses & sugar beet pulp, respectively. Cellulase can be derived from microbial growth substituted with the pulp of rice straw, bagasse, and banana wastes. Tannase can be derived from Indian gooseberry leaves pulp as substrate. It helps in the production of glucose and gallic acid from tannic acid. Xylanase can be derived from wheat bran and brewers spent grain. It is used in the clarification of juice and wine and also as a component in animal feed. Protease can be derived from culture substituted with wheat bran, green gram husk, coffee pulp wastes, and corn cobs. It can be used in pharmaceuticals, detergent, and silk production. Laccase can be derived using sugarcane bagasse as substrate for microbial growth. This enzyme has applications in bio pulping, bio bleaching, and in industrial wastewater treatment.

6.11 Production of Bio-Based Polymer

A polymer composed of chemical compounds originating from renewable organic matter such as plant wastes or food wastes or microorganisms. Some well-known feedstocks are sugarcane, corn, jackfruit, wood chips, food waste, banana peels, potato peels, sea weed, lemon peels, sawdust, maize, etc. Microorganism source is yeast. This polymer can be used as an alternative to plastics for the purpose of food product packaging, disposable bowls, straws, crockery, cutlery, etc. The degradation ability of these polymers is very high. The degradation is of a three-step process—

biodeterioration (change in physio-chemical properties), biofragmentation (breakdown of highly polymerised chain into smaller units) and assimilation (the small units formed are utilised by the microorganisms for their nutritional purpose, and the end products results namely carbon dioxide, water, and microbial biomass) (Acquavia et al., 2021). The agri-food wastes consist of starch, cellulose, aliphatic compounds, fatty acids, and proteins. The compounds are extracted from wastes of vegetables, fruits, animals, and cooked food, and then further processed to attain a highly polymerised bio-based product. The bio-based polymers can be categorised depending on the originating compound such as starch-based, cellulose-based, protein-based, aliphatic-based, lipid-based, chitin-based, or genetically modified bio-based polymers. Starch-based polymer is derived from corn starch for example green dot plastics. It is used in food service packaging and kitchen bags. Cellulose-based polymer is derived from cellulose for example cellophane. It is used in food packaging. Protein-based polymer is derived from protein sources such as gluten, casein, and animal protein for example gelatin. It is used in protein-based films, coatings, and packages. Aliphatic compounds constitute polymers such as polylactic acid (PLA), polyglycolic acid (PGA), polyhydroxy butyrate (PHB), and poly-ξ-caprolactone. All these polymers have medicinal and pharmacological applications. Some other examples of bio-based polymers are bio-based polyurethanes, bio-based carbonates, bio-based polyethylene (bio-PE), bio-based propylene (bio-PET), bio-based nylons, etc.

6.12 *Proper Incineration*

This is the least preferable technique that focuses on reducing the volume of non-degradable as well as degradable wastes by their absolute combustion in between 700 and1400°C in specialised incinerating plants. The waste-to-resource conversion efficiency is less than 60%. The unmanageable solid and dry agri-food wastes can be targeted for incineration. The components of an incinerator plant are—a feed system, a combustion furnace or kiln, post combustion chamber (heat recovery system), a boiler, turbine, and a stack. Figure 12.6 shows the components of the incineration plant and the subsequent generation of electricity from the steam obtained by incineration. To avoid air pollution, devices are installed in the incinerators and emission chimneys(stack), so that the gases are released to be clear and non-toxic. Despite this precautionary measure, toxic gases like dioxins and furans get emitted. The major end products of these exothermic processes are heated, ash (10–25% fly ash and 75–90% bottom ash), carbon dioxide, and water vapour or steam. The steam can be used for subsequent power generation and the ashes can be utilised in constructing bridges, roads, and motorways (Qazi & Abushammala, 2020).

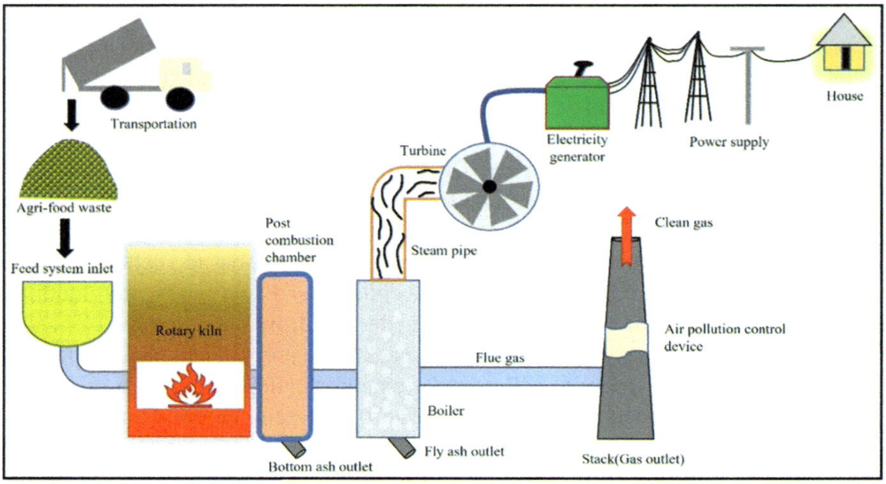

Fig. 12.6 Diagram showing the structure of an incineration plant and successive electricity generation from the steam. The unwanted agri-food wastes are first collected from fields, farms, processing plants, markets and municipal areas, and then transported to nearby incineration treatment plants. These waste materials are dumped into the inlet system from where they get combusted in a rotary kiln at high temperatures between 800 and 1400 °C. The burnt ash along with the gases produced is passed to the post-combustion chamber to cool down. The bottom ash is flushed out after appropriate treatment. Then the residue is transferred to a boiler chamber where the mixture is boiled to produce steam. The fly ash is also eliminated at this stage. The steam is sent to the turbine, which in turn rotates to produce electricity. The flue gas formed during the whole process is trapped and not allowed to be released directly before treatment. They are flown to the gas outlet chimney, where these are passed through the air pollution control device and after successful detoxification, they get released into the air

6.13 *Landfill*

Until recently, the agri-food residues were dumped openly for landfilling purpose. However, due to increased awareness and increased efforts at present, the chances of landfilling are decreasing day by day. A landfilling dumping site in an area where all the wastes are collected from the surrounding area and stocked in heaps in open areas. The manure storage lagoons and organic waste dumpsites are the main drivers of methane and carbon dioxide emissions into the atmosphere. According to the US EPE 2021, manure and municipal solid wastes lead to 9% and 14% of methane emissions. To achieve a greener and healthy environment, improper landfilling should not be done. The dumpsites should be managed and covered to reduce emissions.

7 Challenges in Managing and Treating Agri-Food Wastes

Generation of sustainable products is not enough for a circular bioeconomy as these commodities also need to be effectively marketed and globalized. Currently, global green technology and sustainability market size reached US$16.2 billion in 2022, which is forecasted to reach US$49.1 billion by 2018 (The International Market Analysis Research and Consulting Group IMARC). The reduced market size is due to the less adoption and acquaintance of sustainable and green products by customers. The Government of India approved the Waste to Energy Programme under the Umbrella Scheme of the National Bioenergy Programme from the financial year 2021 to 2026. The main objective of this program is to support the setup for the biogas, bioCNG, electricity, and syngas production from wastes. Despite the collaboration of government and citizens to mitigate environmental problems, there are certain limitations that obstruct the full-swing management and treatment process. These challenges are prevalent in India as our system lacks some major components of the entire waste management and treatment chain. Some key points that reflect the difficulties to the waste workers are:-

(i) Non-availability of agri-food wastes all around the year. The wastes are either generated in low quantities at non-uniform time and place, or sometimes seasonally in enormous quantities only at certain agricultural dominant locations.
(ii) Some agri-food wastes are being valorised extensively whereas some other wastes remain underutilised.
(iii) The rapid-decaying nature and short durability of agri-food wastes is posing a very serious problem. The time period for handling and transportation has to be very rapid and time-limited.
(iv) The mixed composition of agri-food wastes especially those collected from marketplaces and agri-processing areas is very complex to separate. The mixture also creates technical issues during processing.
(v) The management labourers/workers are neither skilfully trained nor equipped well all across the country.
(vi) The gas/liquid filter or treatment devices installed in the plants get out of order after a certain duration of time. Replacement of these devices is not done regularly. This leads to the release of toxic gases and contaminated liquids in air and water resulting in environmental pollution.
(vii) Maintenance of these setup installations is not regularly done.
(viii) During the valorisation process, the quantity of important compounds gets reduced. The cost–benefit ratio is high indicating economic loss.
(ix) The safety of the enzymes and other nutraceutical compounds is not up to the mark after every cycle of valorisation. Therefore, it is required to be tested every time to avoid mycotoxin contamination upon consumption of products.
(x) New waste to energy plant setup for treatment and management requires huge capital investments.

(xi) Lack of post installation inspection, remote monitoring, and product generation data reporting platform limits the purpose of setting up the energy recovery plants. These data ought to be shared to any reputed national or international platform.

(xii) Lack of unity to participate in filling biomethanation plants by manures in rural communities.

8 Support Systems

Collaborative partnership between the public, waste workers, municipalities, and government is foremost necessary for the effective and reliable treatment and management of agri-food wastes as well as for organic wastewater. Infrastructural support that is plants for agri-food waste treatment should be set up in the agriculturally based states. Nowadays, biogas plants are supplied by block development offices in some localities but its use to the maximum is not being done by users. Improved transportation carrier vehicles can help the waste workers in their management plan. Exposure such as training, public awareness, and educational programs for the waste workers and the public including organisations, societies, and educational institutions can be planned by the district administration section. Both rural and urban areas deserve equal attention for campaigns for treatment and management. Short-term as well as long-term incentives/subsidies to the farmers and domestic dwellers can be announced by governing body to increase their cooperation and partnership during house-to-house collection, waste assortment, and delivery of organic wastes. Information technology can also be used as a medium to disperse information related to the importance of treatment and management of agri-food wastes. In addition, social media and mass media like television and popular newspapers should also be targeted to ascertain regular campaigning on the concept and related impacts of agri-food waste mismanagement.

9 Preventive Measures

- At the initial stage, artificial intelligence should be approached in order to have a forecast report of food markets well before time. Then this report can be shared to the farmer community to ensure the right crop selection. So, this method will definitely help in the production of food in adequate quantity and hence less food wastage and a more sustainable environment.
- More valorisation units should be set up in the agriculturally based regions. Thereafter, all the crop residues should be actively brought to these units to transform waste into value-added products.

- Farmers can form a residue managing committee, which will look after the proper channelisation of agricultural wastes and evacuate the fields before the next cropping season.
- In domestic homes, food should be cooked in such a way that no leftover is produced after any meal. In case of less quantity of leftover, it should be distributed to other hungry people or fed to cattle and stray dogs. The spoilt food can be fed to pigs.
- The degradable and non-degradable wastes should not be mixed to facilitate easy collection, assortment, and management of food and kitchen wastes.
- Moreover, agricultural farm manures, beddings, and waste water drainage should not be connected to a nearby pond. Manures should be actively dumped into a biogas plant in a local area.
- Farmers should be trained in self-management of agri-food wastes.
- Kitchen waste should never be discarded away; instead, it should be invested into composting.
- Domestic waste should never be discarded into nearby water bodies to prevent eutrophication.
- More waste collectors/workers/pickers should be recruited for agri-food waste-related management processes.
- The workers should always use safety kits or measures to prevent chronic or fatal health issues.

10 Summary

In India, agri-food wastes are neither managed nor treated properly. This is obvious because if these wastes were managed for energy or resource recovery, then the huge import of crude oil from other countries would not have been so tremendous. Therefore, through this chapter, there is an urge to everyone to selflessly follow the reduce, reuse, and recycle concept of waste management at individual level. In addition, huge heaps of manure, crop residues, and waste foods from agricultural farms can be utilised efficiently. Small changes done by every person can definitely bring a bigger change. Normal people, farmers, waste workers, non-governmental organisations, municipalities, and street sweepers along with local, state, and central government should fight unitedly against the alarming issue of improper agri-food waste management and thus ultimately aim to reduce greenhouse gas emissions by at least 10–15% at national level by 2030.

References

Abdurrahman, M. I., Chaki, S., & Saini, G. (2020). Stubble burning: Effects on health & environment, regulations and management practices. *Environmental Advances, 2*, 100011. https://doi.org/10.1016/j.envadv.2020.100011

Acquavia, M. A., Pascale, R., Martelli, G., Bondoni, M., & Bianco, G. (2021). Natural polymeric materials: A solution to plastic pollution from the agro-food sector. *Polymers, 13*(1), 1–39. https://doi.org/10.3390/polym13010158

Addy, R., Kalamdhad, A., & Goud, V. V. (2022). Insight on the prevalence of pathogens present in the municipal solid waste of sanitary landfills, dumpsites, and leachate. In *Fate of biological contaminants during recycling of organic wastes* (pp. 279–295). Elsevier. https://doi.org/10.1016/B978-0-323-95998-8.00006-6

Alfio, V. G., Manzo, C., & Micillo, R. (2021). From fishwaste to value: An overview of the sustainable recovery of omega-3 for food supplements. *Molecules, 26*(4), 1002. https://doi.org/10.3390/molecules26041002

Ali, A., Riaz, S., Sameen, A., Naumovski, N., Iqbal, M. W., Rehman, A., et al. (2022). The disposition of bioactive compounds from fruit waste, their extraction, and analysis using novel technologies: A review. *Processes, 10*(10), 2014. https://doi.org/10.3390/pr10102014

Annual Report 2020-2021 National Centre of Organic Farming.

Bala, S., Garg, D., Sridhar, K., Inbaraj, B. S., Singh, R., Kamma, S., Tripathi, M., & Sharma, M. (2023). Transformation of agro-waste into value-added bioproducts and bioactive compounds: Micro/Nano formulations and application in the agri-food-pharma sector. *Bioengineering, 10*(2), 152. https://doi.org/10.3390/bioengineering10020152

Bandara, T., Xu, J., Potter, I. D., Franks, A., Chathurika, J. B. A. J., & Tang, C. (2020). Mechanisms for the removal of Cd (II) and Cu (II) from aqueous solution and mine water by biochars derived from agricultural wastes. *Chemosphere, 254*, 126745. https://doi.org/10.1016/j.chemosphere.2020.126745

Bandgar, P. S., Jain, S., & Panwar, N. L. (2021). Current status, opportunities and challenges in anaerobic digestion in Indian context: An overview. *Bioresource Technology Reports, 16*, 100830. https://doi.org/10.1016/j.biteb.2021.100830

Ben-Othman, S., Jõudu, I., & Bhat, R. (2020). Bioactives from agri-food wastes: Present insights and future challenges. *Molecules, 25*(3), 510. https://doi.org/10.3390/molecules25030510

Bigdeloo, M., Teymourian, T., Kowsari, E., Ramakrishna, S., & Ehsani, A. (2021). Sustainability and circular economy of food wastes: Waste reduction strategies, higher recycling methods, and improved valorization. *Materials Circular Economy, 3*(3), 1–9. https://doi.org/10.1007/s42824-021-00017-3

Coral, D. A., Leonardo, V., Óscar, F. C., & James, J. S. (2022). Process design and techno - economic assessment of cellulolytic enzymes production from coffee husk through process simulation. *Biomass Conversion and Biorefinery, 0123456789*, 1–21. https://doi.org/10.1007/s13399-022-03130-8

Dahunsi, S. O. (2020). Electricity generation from food wastes and spent animal beddings with nutrients recirculation in catalytic fuel cell. *Scientific Reports, 10*(1), 10735. https://doi.org/10.1038/s41598-020-67356-0

Dhanker, R., Rawat, S., Chandna, V., Deepa, Das, S., Sharma, A., Kumar, V., & Kumar, R. (2022). Recovery of silver nanoparticles and management of food wastes: Obstacles and opportunities. *Environmental Advances, 9*, 100303. https://doi.org/10.1016/j.envadv.2022.100303

Diatta, J., Waraczewska, Z., Grzebisz, W., Niewiadomska, A., & Tatuśko-Krygier, N. (2020). Eutrophication induction via N/P and P/N ratios under controlled conditions—Effects of temperature and water sources. *Water, Air, and Soil Pollution, 231*(4), 1–18. https://doi.org/10.1007/s11270-020-04480-7

Ejigu, N., Sitotaw, B., Girmay, S., & Assaye, H. (2022). Evaluation of oyster mushroom (*Pleurotus ostreatus*) production using water hyacinth (*Eichhornia crassipes*) biomass supplemented with agricultural wastes. *Biomass Conversion and Biorefinery, 2022*. https://doi.org/10.1155/2022/9289043

Erraji, H., Asehraou, A., Tallou, A., & Rokni, Y. (2023). Assessment of biogas production and fertilizer properties of digestate from cow dung using household biogas digester. *Biomass Conversion and Biorefinery, 0123456789*, 1–7. https://doi.org/10.1007/s13399-023-03818-5

Escaramboni, B., Castelli, B., Mateus, G., Abe, M., & Abel, D. (2022). Food waste as a feedstock for fungal biosynthesis of amylases and proteases. *Waste and Biomass Valorization, 13*(1), 213–226. https://doi.org/10.1007/s12649-021-01511-0

Essien, J. P., Ikpe, D. I., Inam, E. D., Okon, A. O., Ebong, G. A., & Benson, N. U. (2022). Occurrence and spatial distribution of heavy metals in landfill leachates and impacted freshwater ecosystem: An environmental and human health threat. *PLoS One, 17*(2), e0263279. https://doi.org/10.1371/journal.pone.0263279

Fadzli, F. S., Rashid, M., Yaqoob, A. A., Nasir, M., & Ibrahim, M. (2021). Electricity generation and heavy metal remediation by utilizing yam (*Dioscorea alata*) waste in benthic microbial fuel cells (BMFCs). *Biochemical Engineering Journal, 172*(April), 108067. https://doi.org/10.1016/j.bej.2021.108067

Gomes-Araújo, R., Martínez-Vázquez, D. G., Charles-Rodríguez, A. V., Rangel-Ortega, S., & Robledo-Olivo, A. (2021). Bioactive compounds from agricultural residues, their obtaining techniques, and the antimicrobial effect as postharvest additives. *International Journal of Food Science, 2021*, 1–13. https://doi.org/10.1155/2021/9936722

Gong, C., Singh, A., Singh, P., & Singh, A. (2021). Anaerobic digestion of agri-food wastes for generating biofuels. *Indian Journal of Microbiology, 61*(4), 427–440. https://doi.org/10.1007/s12088-021-00977-9

Habagil, M., Keucken, A., & Sárvári Horváth, I. (2020). Biogas production from food residues—The role of trace metals and co-digestion with primary sludge. *Environments, 7*(6), 42. https://doi.org/10.3390/environments7060042

https://ncof.dacnet.nic.in/StatusOrganicFarming

International Market Analysis Research and Consulting Group.

Kazige, O. K., Chuma, G. B., Lusambya, A. S., Mondo, J. M., Balezi, A. Z., Mapatano, S., & Mushagalusa, G. N. (2022). Valorizing staple crop residues through mushroom production to improve food security in eastern Democratic Republic of Congo. *Journal of Agriculture and Food Research, 8*, 100285. https://doi.org/10.1016/j.jafr.2022.100285

Kumar, P., Eid, E. M., Al-Huqail, A. A., Širić, I., Adelodun, B., Abou Fayssal, S., Valadez-Blanco, R., Goala, M., Ajibade, F. O., Choi, K. S., & Kumar, V. (2022). Kinetic studies on delignification and heavy metals uptake by Shiitake (Lentinula edodes) Mushroom cultivated on agro-industrial wastes. *Horticulturae, 8*(4), 316. https://doi.org/10.3390/horticulturae8040316

Kumar, P., Tanwar, R., Gupta, V., Upadhyay, A., & Kumar, A. (2021). Pineapple peel extract incorporated poly (vinyl alcohol) -corn starch film for active food packaging: Preparation, characterization and antioxidant activity. *International Journal of Biological Macromolecules, 187*(July), 223–231.

Kurd-Anjaraki, S., Ramezan, D., Ramezani, S., Samzadeh-Kermani, A., Pirnia, M., & Shahi, B. Y. (2022). Potential of waste reduction of agro-biomasses through Reishi medicinal mushroom (*Ganoderma lucidum*) production using different substrates and techniques. *Acta Ecologica Sinica, 42*(1), 90–101. https://doi.org/10.1016/j.chnaes.2021.04.010

Laborde, D., Mamun, A., Martin, W., Piñeiro, V., & Vos, R. (2021). Agricultural subsidies and global greenhouse gas emissions. *Nature Communications, 12*(1), 2601. https://doi.org/10.1038/s41467-021-22703-1

Lei, Y., Guo, X., Jiang, M., Sun, W., He, H., Chen, Y., Thummavichai, K., Ola, O., Zhu, Y., & Wang, N. (2022). Co-ZIF reinforced cow manure biochar (CMB) as an effective peroxymonosulfate activator for degradation of carbamazepine. *Applied Catalysis B: Environmental, 319*, 121932. https://doi.org/10.1016/j.apcatb.2022.121932

Lin, M., & Begho, T. (2022). Crop residue burning in South Asia: A review of the scale, effect, and solutions with a focus on reducing reactive nitrogen losses. *Journal of Environmental Management, 314*, 115104. https://doi.org/10.1016/j.jenvman.2022.115104

Lins, M., Puppin Zandonadi, R., Raposo, A., & Ginani, V. C. (2021). Food waste on foodservice: An overview through the perspective of sustainable dimensions. *Food, 10*(6), 1175. https://doi.org/10.3390/foods10061175

Mandpe, A., Kumari, S., & Kumar, S. (2020). Composting: A sustainable route for processing of biodegradable waste in India. In *Organic waste composting through nexus thinking: Practices, policies, and trends* (pp. 39–60). https://doi.org/10.1007/978-3-030-36283-6_3

Mariyam, S., Alherbawi, M., Rashid, N., Al-Ansari, T., & McKay, G. (2022). Bio-oil production from multi-waste biomass co-pyrolysis using analytical Py–GC/MS. *Energies, 15*(19), 7409. https://doi.org/10.3390/en15197409

Martínez-Inda, B., Esparza, I., Moler, J. A., Jiménez-Moreno, N., & Ancín-Azpilicueta, C. (2023). Valorization of agri-food waste through the extraction of bioactive molecules. Prediction of their sunscreen action. *Environmental Management, 325*, 116460. https://doi.org/10.1016/j.jenvman.2022.116460

Meng, Q., Liu, H., Zhang, H., Xu, S., Lichtfouse, E., & Yun, Y. (2022). Anaerobic digestion and recycling of kitchen waste: A review. *Environmental Chemistry Letters, 20*(3), 1745–1762. https://doi.org/10.1007/s10311-022-01408-x

Morya, S., Menaa, F., Jiménez-López, C., Lourenço-Lopes, C., BinMowyna, M. N., & Alqahtani, A. (2022). Nutraceutical and pharmaceutical behavior of bioactive compounds of miracle oilseeds: An overview. *Food, 11*(13), 1824. https://doi.org/10.3390/foods11131824

Mutua, F., Sharma, G., Grace, D., Bandyopadhyay, S., Shome, B., & Lindahl, J. (2020). A review of animal health and drug use practices in India, and their possible link to antimicrobial resistance. *Antimicrobial Resistance and Infection Control, 9*(1), 1–13. https://doi.org/10.1186/s13756-020-00760-3

Oluba, O. M., Obi, C. F., Akpor, O. B., Ojeaburu, S. I., Ogunrotimi, F. D., Adediran, A. A., & Oki, M. (2021). Fabrication and characterization of keratin starch biocomposite film from chicken feather waste and ginger starch. *Scientific Reports, 11*(1), 8768. https://doi.org/10.1038/s41598-021-88002-3

Pan, S. Y., Tsai, C. Y., Liu, C. W., Wang, S. W., Kim, H., & Fan, C. (2021). Anaerobic co-digestion of agricultural wastes toward circular bioeconomy. *Iscience, 24*(7). https://doi.org/10.1016/j.isci

Pandit, S., Savla, N., Sonawane, J. M., Sani, A. M. D., Gupta, P. K., Mathuriya, A. S., Rai, A. K., Jadhav, D. A., Jung, S. P., & Prasad, R. (2021). Agricultural waste and wastewater as feedstock for bioelectricity generation using microbial fuel cells: Recent advances. *Fermentation, 7*(3), 169. https://doi.org/10.3390/fermentation7030169

Paul, J. S., Beliya, E., Tiwari, S., Patel, K., Gupta, N., & Jadhav, S. K. (2020). Production of biocatalyst α-amylase from agro-waste 'rice bran'by using Bacillus tequilensis TB5 and standardizing its production process. *Biocatalysis and Agricultural Biotechnology, 26*, 101648. https://doi.org/10.1016/j.bcab.2020.101648

Pawar, R. S., Zambare, N. S., Wagh, V. M., & Panaskar, D. B. (2020). A review of eutrophication. *Studies in Indian Place Names (UGC Care Journal), 40*(88), 1–3. ISSN: 2394-3114.

Power Sector at a Glance ALL INDIA. (2023).

Qazi, W. A., & Abushammala, M. F. M. (2020). Multi-criteria decision analysis of waste-to-energy technologies. In *Waste-to-Energy* (pp. 265–316). Academic Press. https://doi.org/10.1016/B978-0-12-816394-8.00010-0

Quereshi, S., Naiya, T. K., Mandal, A., & Dutta, S. (2022). Residual sugarcane bagasse conversion in India: Current status, technologies, and policies. *Biomass Conversion and Biorefinery, 12*(9), 3687–3709. https://doi.org/10.1007/s13399-020-00871-2

Ravindra, K., Singh, T., & Mor, S. (2022). COVID-19 pandemic and sudden rise in crop residue burning in India: Issues and prospects for sustainable crop residue management. *Environmental Science and Pollution Research, 29*(2), 3155–3161. https://doi.org/10.1007/s11356-021-17550-y

Rojas-Flores, S., de La Cruz Noriega, M., Benites, S. M., Aguirre Gonzales, G., Salvador Salinas, A., & Silva Palacios, F. (2020). Generation of bioelectricity from fruit waste. *Energy Reports, 6*, 37–42. https://doi.org/10.1016/j.egyr.2020.10.025

Romero-Perdomo, F., & González-Curbelo, M. Á. (2023). Integrating multi-criteria techniques in life-cycle tools for the circular bioeconomy transition of agri-food waste biomass: A systematic review. *Sustainability, 15*(6), 5026. https://doi.org/10.3390/su15065026

Segundo, R. F., de La Cruz-Noriega, M., Otiniano, N. M., Benites, S. M., Esparza, M., & Nazario-Naveda, R. (2022). Use of onion waste as fuel for the generation of bioelectricity. *Molecules, 27*(3), 625. https://doi.org/10.3390/molecules27030625

Saharia, M., Dey, G., & Kumar, V. (2024). Vermiremediation of plant agro waste to recover residual nutrients and improve crop productivity. In *Earthworm technology in organic waste management: Recent trends and advances* (pp. 79–113). Elsevier.

Sharif, T., Bhatti, H. N., Bull, I. D., & Bilal, M. (2023). Recovery of high-value bioactive phytochemicals from agro-waste of mango (*Mangifera indica* L.) using enzyme-assisted ultrasound pretreated extraction. *Biomass Conversion and Biorefinery, 13*(8), 6591–6599. https://doi.org/10.1007/s13399-021-01589-5

Shinde, R., Shahi, D. K., Mahapatra, P., Singh, C. S., Naik, S. K., Thombare, N., & Singh, A. K. (2022). Management of crop residues with special reference to the on-farm utilization methods: A review. *Industrial Crops and Products, 181*, 114772. https://doi.org/10.1016/j.indcrop.2022.114772

Sinha, S., & Tripathi, P. (2021). Trends and challenges in valorisation of food waste in developing economies: A case study of India. *Case Studies in Chemical and Environmental Engineering, 4*, 100162. https://doi.org/10.1016/j.cscee.2021.100162

The Organisation for Economic Cooperation and Development FAO Agricultural Outlook 2023–2032-FAO Agricultural Outlook 2023–2032.

Trade and Development's Business-to-Consumer (B2C) E-Commerce Index. (2020).

United in Science Report. (2023).

United Nations Environment Programme UNEP Food Waste Index Report. (2021).

Uwiragiye, Y., Ngaba, M. J. Y., Yang, M., Elrys, A. S., Chen, Z., & Zhou, J. (2023). Spatially explicit soil acidification under optimized fertilizer use in Sub-Saharan Africa. *Agronomy, 13*(3), 632. https://doi.org/10.3390/agronomy13030632

Velvizhi, G., Balakumar, K., Shetti, N. P., Ahmad, E., Kishore Pant, K., & Aminabhavi, T. M. (2022). Integrated biorefinery processes for conversion of lignocellulosic biomass to value added materials: Paving a path towards circular economy. *Bioresource Technology, 343*, 126151. https://doi.org/10.1016/j.biortech.2021.126151

Waqas, M., Hashim, S., Humphries, U. W., Ahmad, S., Noor, R., Shoaib, M., Naseem, A., Hlaing, P. T., & Lin, H. A. (2023). Composting processes for agricultural waste management: A comprehensive review. *PRO, 11*(3), 731. https://doi.org/10.3390/pr11030731

World Food and Agriculture Statisical Yearbook. (2022).

Xu, C., Tan, X., Zhao, J., Cao, J., Ren, M., Xiao, Y., & Lin, A. (2021). Optimization of biochar production based on environmental risk and remediation performance: Take kitchen waste for example. *Journal of Hazardous Materials, 416*, 125785. https://doi.org/10.1016/j.jhazmat.2021.125785

Yaradoddi, J. S., Banapurmath, N. R., Ganachari, S. V., Soudagar, M. E. M., Mubarak, N. M., Hallad, S., Hugar, S., & Fayaz, H. (2020). Biodegradable carboxymethyl cellulose based material for sustainable packaging application. *Scientific Reports, 10*(1), 21960. https://doi.org/10.1038/s41598-020-78912-z

Yuniarti, R., Masudin, I., Rusdiansyah, A., & Handayani, D. I. (2023). Model of multiperiod production-distribution for closed-loop supply chain considering carbon emission and traceability for agri-food products. *International Journal of Industrial Engineering and Operations Management, 5*(3), 240–263. https://doi.org/10.1108/IJIEOM-10-2022-0045

Zhu, Y., Yi, B., Hu, H., Zong, Z., Chen, M., & Yuan, Q. (2020). The relationship of structure and organic matter adsorption characteristics by magnetic cattle manure biochar prepared at different pyrolysis temperatures. *Journal of Environmental Chemical Engineering, 8*(5), 104112. https://doi.org/10.1016/j.jece.2020.104112

Part III
Environmental Technologies

Chapter 13
Landfill Biomining of Legacy Waste Dumpsites in India: Process, Challenges and Future Perspective

Arghya Ghosh and Suresh A. Kartha

1 Introduction

Most developing countries have traditionally relied on open dumping as a prevalent municipal solid waste (MSW) disposal strategy. It thrives due to insufficient political will to alter current waste disposal practices and a lack of relevant technical, financial and human resources. In 2016, metropolitan areas of India produced 62 million tonnes of MSW, of which only 15–20% was recycled, and the rest was disposed of in landfills and dumpsites (Cheela et al., 2021). This has led to a situation where the older dumpsites in metropolitan India become mountains of legacy waste. It has been estimated that legacy waste dumpsites take up more than 10,000 hectares of metropolitan space in urban India (CPHEEO, 2016). Previously, dumpsite locations were randomly selected near the outskirts of the city without any consideration of the local hydrogeological and environmental features (Natalia et al., 2021). However, due to urban sprawl, these waste disposal sites have now become an integral part of the city, posing a threat not only to the environment and public health but also to the visual appeal of the cities (Cheela et al., 2023). In addition, the requirement of space for the daily influx of fresh MSW and processing of accumulated waste at these legacy waste dumpsites puts a burden on the carrying capacity of the metropolitan areas. It leads to issues related to the sustainability of the urban land (Saravanan & Ramesh, 2023). To free up much-needed space in overcrowded urban areas, where land is expensive, the problem of legacy waste dumpsites must be resolved decisively. In this context, the principle of landfill mining (LFM) can be used as a tactical solution to meet the demands for extra land for future MSW management and to rehabilitate the population inside the boundaries of the municipality (Chandana et al., 2021).

A. Ghosh (✉) · S. A. Kartha
Department of Civil Engineering, Indian Institute of Technology Guwahati, North Guwahati, Assam, India

V. Kumar et al. (eds.), *Environmental Engineering and Waste Management*,
https://doi.org/10.1007/978-3-031-58441-1_13

LFM has been proclaimed as a strategy to dismantle the landfilled mass trapped within the old landfills or dumpsites through the use of unit operations such as excavating, screening, separating and processing to recover secondary raw materials, which contribute to the principles of circular economy and also help to achieve sustainable development goals (Singh & Chandel, 2020). Apart from resource recovery, other motives of LFM are avoidance of post-closure costs, redevelopment of landfills or dumpsites, and economic and social benefits through pollution mitigation (Burlakovs et al., 2021). Many LFM projects have been carried out and practiced worldwide for over 50 years to reduce the environmental burden caused by the existing landfill site and to re-introduce buried resources into the material cycle. European and Central Asian countries have conducted more LFM projects than the USA, South Asia, East Asia and the Pacific (Jain et al., 2022). According to previous studies, the approaches of land reclamation for urban development, recycling of landfilled materials, energy recovery and creation of new landfill spaces were found to be the dominant factors determining the feasibility of LFM projects (Ghosh & Kartha, 2020; Esguerra et al., 2021).

Considering the importance and proven benefits of LFM, the Government of India has recently launched the nation's most extensive clean-up program, Swachh Bharat Mission 2.0 (SBM 2.0), to financially sponsor and encourage the urban local bodies (ULBs) for the implementation of LFM projects as a remedy to the problems arising from legacy waste dumpsites. In India, LFM is also known as landfill biomining (LFBM), which entails excavating, stabilizing, and screening previously dumped waste into several recoverable components (Mohan & Joseph, 2020). Using the potential LFBM, old dumpsites can be renovated with proper sanitation facilities to reduce environmental risks (Ghosh & Kartha, 2023a). However, the number of LFBM projects is unexpectedly low compared to the number of existing legacy waste dumpsites in urban India, mainly due to the difficulties in proving the feasibility of the many candidate cases. Like many other cutting-edge waste management systems, the economic viability of LFBM also appears to be a barrier to its wider adoption. Lack of detailed information prevents the development of evidence-based LFBM regulations, which in turn makes it difficult to better understand the material stocks in legacy waste dumpsites and their reuse potentials for the entire country. Unique operational, infrastructural and utilization challenges further advocate this complexity. Due to the lack of suitable guidelines for categorizing and reusing the excavated waste, the viability of LFBM as a long-term plan for managing legacy waste dumpsites is still up for question.

Previous studies on LFBM concentrated on one or a small number of particular cases, unable to provide a comprehensive picture of the potential of LFBM on a regional or national scale. Therefore, this chapter aims to introduce readers to the current scenario of legacy waste dumpsite management in India, the workflow process for successful LFBM operation, and also point out multiple challenges that need to be encountered during the implementation of an LFBM project. Moreover, this study also represents the future directions of LFBM by assessing the ground reality of several LFBM projects. In this context, a systematic literature review was performed using the multidisciplinary scientific databases by selecting appropriate keywords such as 'open dumpsites AND India', 'dumpsite reclamation AND India', 'landfill mining AND India', 'biomining AND India', etc. Furthermore, the government guidelines and municipality reports on solid waste management, and LFBM were also evaluated to extract significant information and experiences obtained during legacy waste dumpsites management. However, there is a shortage of available information on this topic.

2 Problems and Issues Related to Legacy Waste Dumpsites in Urban India

The un-scientific operations, size and quantity of the legacy waste at the abandoned dumpsites in India have become ticking time bombs with various long-term environmental, public health and social challenges (CSE, 2020). In the absence of liner, daily cover and gas collection systems, these dumpsites pose serious environmental and health hazards to the neighbourhood by leachate seeping into the groundwater, vector-borne diseases, foul odour, air pollution, etc (Ghosh & Kartha, 2022; Dixit et al., 2023). Waste rotting in the high-rises of airless heaps generates methane, a greenhouse gas with 28–36 times more global warming potential than carbon dioxide (Choudhary et al., 2020). Besides exacerbating global climate change, the highly flammable methane gas trapped at the dumpsites often leads to fire outbreaks and supports the burning of combustible waste such as plastics, textiles and oily rags (Chavan et al., 2022). As a result, most of the large dumpsites in India are smoldering from intermittent hidden fires and generate harmful smoke. In addition to the environmental and public health losses, these existing dumpsites also cause economic and social losses due to their unesthetic appearance and stigma around dumpsites. This issue is further complicated due to the expansion of cities in terms of area, and these dumpsites, originally located on the outskirts of the cities, are now almost in the heart of cities. Moreover, the presence of these dumpsites encourages the ULBs to further dump their MSW at these already exhausted sites. All these issues create immense public pressure on the ULBs, and several Public Interest Litigations (PILs) have been filed at various courts (CSE, 2020).

3 Evolution of Policies for Dumpsites Management in India

In this part of the article, we look at the evolution of dumpsite management policies in India. With a shifting emphasis on sanitation and health for the public as well as environmental safety, the dumpsites management policies in India have occurred in three stages, as shown in Fig. 13.1. The Indian government formulated the new 'Solid Waste Management Rules (SWM), 2016', based on the knowledge acquired over the past 14 years after the MSW regulations were first announced in 2000

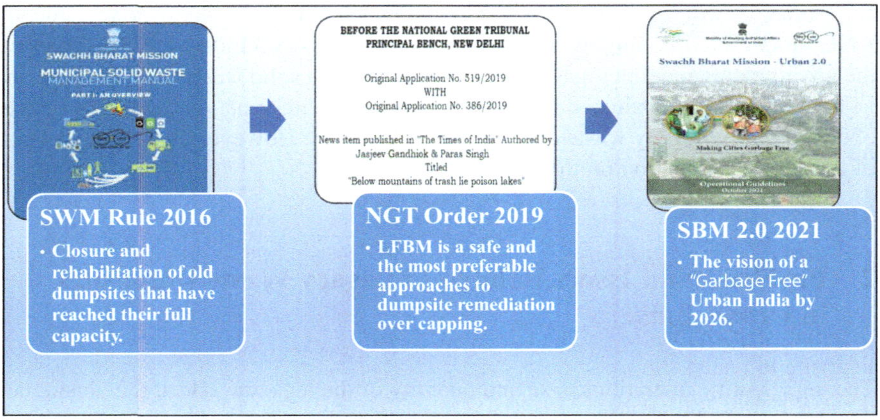

Fig. 13.1 Evolution of dumpsite management policies in India

(CPHEEO, 2016). Several guidelines have been established to handle old dumpsites under the SWM Rules of 2016. As per clause J of schedule I, old dumpsites that have exceeded capacity or will not receive further waste should be considered for closure and rehabilitation either by LFBM or should be scientifically covered by landfill capping standards to stop future environmental harm (CPHEEO, 2016). Since it is impossible to install a bottom liner below an existing dumpsite, from an environmental point of view, the site may still remain a source of groundwater contamination even after capping. Therefore, capping may not be the most relevant choice for managing legacy waste dumpsites in an environmentally sustainable manner. In this context, the hon'ble National Green Tribunal (NGT) of India ordered (Order No: OA 519/2019 dated 17/7/2019) ULBs to rehabilitate the legacy waste dumpsites using the LFBM technique as a favoured option over capping (NGT, 2019). The order also emphasized that there may be very few circumstances where LFBM is not feasible. Following the order of the hon'ble NGT, the Central Pollution Control Board (CPCB) and the Central Public Health and Environmental Engineering Organization (CPHEEO) issued standard operating procedures for the implementation of LFBM of legacy waste dumpsites (CPCB, 2019; CPHEEO, 2020). In order to achieve 'garbage-free cities', the government of India has recently allocated a massive financial investment of Rs 1,41,600 crore (about 171.2 billion USD) towards its largest flagship clean-up programme SBM 2.0 (2021). The mission primarily focuses on employing LFBM to clean up all legacy waste dumpsites in the country. According to the SBM 2.0 operational standards, cities with residents under one million have to clear legacy waste sites by March 31, 2023, and cities with residents of one million or more must clean up their dumpsites by March 31, 2024 (SBM 2.0, 2021). It is expected that, with help from SBM 2.0's funding, LFBM of legacy waste dumpsites will explode across the country in the coming few years.

4 Process and Technology of Landfill Biomining

The LFBM process at the dumpsite resembles a conveyor system with a series of processing devices. Successful execution of any intended LFBM project consists of the following sequence of steps, as depicted in Fig. 13.2.

4.1 Planning and Site Preparation

The first step of the LFBM process involves a precursor study of the existing dumpsite to analyse the technical parameters and feasibility of conducting such an operation. The work starts with a contour survey (Total Station Survey) or drone mapping of the entire project site and is then followed by random trial pit excavation to estimate the quantity, composition and stabilization status of legacy waste to be bio-remediated. An initial baseline environmental survey is also essential for the assessment of different site-specific parameters such as geological features, soil type, surface hydrology, rainfall, proximity of surface and groundwater resources, wind direction, etc. Based on these preliminary surveys, an execution plan of the LFBM process for a particular dumpsite is prepared, and required resources are mobilized to build up necessary infrastructures such as a temporary shed and weighbridge,

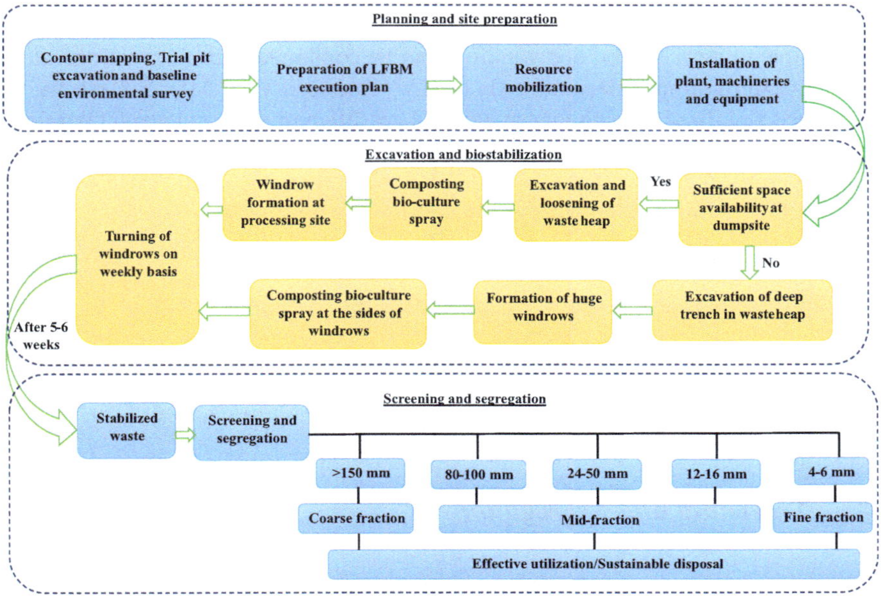

Fig. 13.2 General workflow process of LFBM. (*Source:* Compiled by authors from CPCB (2019), CPHEEO (2020))

screening and segregation facility, storage place for the segregated wastes, stormwater and leachate management system, etc.

4.2 Excavation and Bio-stabilization

The biological stability of the legacy waste is essential to avoid the unacceptable generation of nuisances during its handling and processing. It can be presumed that the biodegradable content of legacy waste has achieved satisfactory stabilization in decades after dumping (Somani et al., 2020). However, prevalent anaerobic conditions within the dumpsites prohibit the complete degradation of organic matter even after a 15- to 25-year period of dumping (Bisht et al., 2022).

Bio-stabilization of legacy waste is generally conducted using the conventional aerated windrow technique. Depending on the availability of space within the project site, windrows are formed either by placing the excavated waste in new space or cutting a trench in the existing heap of waste. If sufficient space is available, a heap of legacy waste is first excavated, and then composting bio-cultures (diluted leachate solution or dilute solution of 5% fresh cow dung or specially formulated bio-culture) are sprayed so that all the entrapped methane is removed from the heap, and the decomposing microbes get dispersed into the loosened waste. After that, the loosened waste is moved to form several long parallel windrows of about 2 m in height to get a maximum surface area to volume. Repeated turning of these wind rows every week is necessary to ensure that the innermost waste also gets exposed as much of it as possible to air. Usually, one windrow can be combined with the next one at the second or third turning, depending on their volume reduction. After 5–6 weeks, the waste gets stabilized enough for screening to get different sizes and weight fractions. If space does not permit, deep vertical trenches (about 2–2.5 m deep) are dug out from the top of the legacy waste heap at 1.5- to 2-m intervals to form giant windrows for aerating the waste. Composting bio-cultures are then sprayed on all the sides of the trenches to get microbes to reach newly exposed surfaces for a faster decomposition process. These huge windrows are turned weekly until a significant reduction in waste heap height is observed, indicating the stabilized status of the waste. Bio-stabilization of legacy waste can reduce its volume by 35–40%, and the material will be dry enough for further processing. Sometimes, seed germination tests are also conducted to ensure the stabilization status of the waste.

4.3 Screening and Segregation

The stabilized waste is screened and segregated by a sequence of trommel or vibrating screens of different size openings laid out in a functional conveyor system. Trommel screens are more widely used than vibrating screens for effective separation of the materials that may be physically attached. Generally, 150 mm,

80 to 100 mm, 24 to 50 mm, 12 to 16 mm and 4 to 6 mm size sieves are used for fractionating the waste into different categories such as coarse fraction, mid-fraction and fine fraction. Apart from different-size screens, air density separators or air classifiers are also utilized to separate lightweight fractions from heavier materials. The segregated waste is then packed and transported from the dumpsite for effective utilization and sustainable disposal according to CPCB guidelines (CPCB, 2019).

5 Current Status of Legacy Waste Dumpsites Management in India

According to the online public dashboard data of SBM 2.0, there are around 2241 legacy waste dumpsites (those having legacy waste ≥1000 Tonnes) in the country distributed among different states and union territories (UTs), as shown in Fig. 13.3. Due to differences in population and geography, there are comparatively more dumpsites in the various states than in the UTs. Tamil Nadu has the highest number of legacy waste dumpsites, while the union territories—Ladakh and Andaman and Nicobar Islands have the lowest number of legacy waste dumpsites. A total of 11 states have more than 100 legacy waste dumpsites. Further, the data on the total amount of legacy waste, amount of legacy waste remediated, and amount of legacy waste to be remediated in different states and UTs are presented in Table 13.1. Maharashtra has the highest amount of legacy waste, while the union territory Andaman and Nicobar Islands has the lowest amount of legacy waste. In terms of

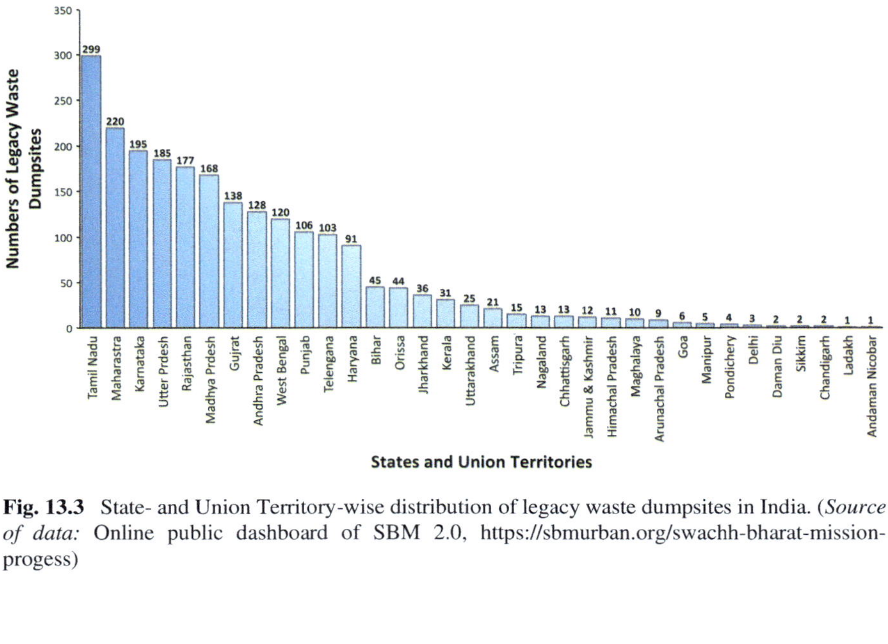

Fig. 13.3 State- and Union Territory-wise distribution of legacy waste dumpsites in India. (*Source of data:* Online public dashboard of SBM 2.0, https://sbmurban.org/swachh-bharat-mission-progess)

Table 13.1 State- and Union Territory-wise legacy waste remediation status in India

States and Union Territories	Total Amount of Legacy Waste (Lakh Tonne)	Legacy Waste Remediated (Lakh Tonne)	Legacy Waste to be Remediated (Lakh Tonne)
Andaman Nicobar	0.07	0.01	0.06
Andhra Pradesh	85.9	11.18	74.72
Arunachal Pradesh	0.36	0.17	0.19
Assam	25.49	0	25.49
Bihar	26.13	4.15	21.98
Chandigarh	12.77	5.24	7.53
Chhattisgarh	7.18	4.21	2.97
Daman Diu	1.42	0	1.42
Delhi	203	76.98	126.02
Goa	18.09	1.37	16.72
Gujrat	235.98	105.74	130.24
Haryana	107.35	48.29	59.06
Himachal Pradesh	2.68	0.66	2.02
Jharkhand	31.16	0.35	30.81
Jammu & Kashmir	18.86	2.34	16.52
Karnataka	182.64	0.02	182.62
Kerala	12.6	1.15	11.45
Nagaland	8.62	0	4.82
Ladakh	1.32	0	1.32
Madhya Pradesh	61.86	8.74	53.12
Maharashtra	531.96	149.83	382.13
Manipur	1.6	0	1.6
Meghalaya	4.82	0	4.82
Orissa	32.96	2.71	30.25
Punjab	74.3	21.57	52.73
Pondicherry	12.09	0.6	11.49
Rajasthan	76.06	2.33	73.73
Sikkim	2.86	0	2.86
Tamil Nadu	233	74.56	158.44
Telangana	153.29	121.4	31.89
Tripura	4.44	0.2	4.24
Uttarakhand	18.43	0	18.43
Utter Pradesh	104.44	55.75	48.69
West Bengal	146.32	7.88	138.44
Total	2440	707	1733

Source of data: Online public dashboard of SBM 2.0, https://sbmurban.org/swachh-bharat-mission-progess

the percentage of legacy waste remediated, the Telangana state has the highest percentage of remediated legacy waste, while in the states of Assam, Nagaland, Manipur, Meghalaya, Uttarakhand and union territories Daman Diu and Ladakh, the remediation activities are at the beginning stage. From Table 13.1, it can also be inferred that about 29% ($\approx$ 707 Lakh Tonne) of the total amount of legacy waste ($\approx$ 2440 Lakh Tonne) is currently remediated through LFBM. According to the most recent data provided to the CPCB, LFBM activity is underway at over 500 legacy waste dumpsites, and LFBM has finished over 550 dumpsites (Thakkar et al., 2023).

6 Problems and Challenges Encountered in LFBM of Legacy Waste Dumpsites

Despite its many advantages, major constraints exist to LFBM and dumpsite reclamation. Real-world implementation of these strategies, however, might be difficult for the stakeholders as each dumpsite presents its distinctive challenges. There are a lot of unknowns and unforeseen events that could happen during the excavation and processing of legacy waste. The major challenges can be categorized as operational, infrastructural and utilization, as shown in Table 13.2.

6.1 Operational

Major operational challenges faced during LFBM operation include the continuous inflow of fresh MSW in the legacy waste dumpsite, inadequate guidelines for sampling and characterization of legacy waste, management of space for

Table 13.2 Challenges in legacy waste dumpsites management through LFBM

Challenges	Description
Operational	Continuous inflow of fresh solid waste in the legacy waste dumpsite Non-availability of space as well as power and water connections in the existing dump yard for setting up of the plant Delay in processing due to diverse climatic conditions throughout the year
Infrastructural	Lack of experienced companies and contractors for execution of LFBM work Non-availability of customized machineries for removal and segregation of legacy waste Lack of skilled labour in LFBM work
Utilization	Very low or no market value for the segregated coarse and fine fractions Non-availability of industries accepting recyclables Segregated combustible fractions are heavily polluted with inert substances Fine fractions are generally contaminated with heavy metals Absence of proper guidelines for possible utilization of the reclaimed land after LFBM

stabilization and processing of legacy waste within the dumpsite, and lack of process functionality due to weather-related issues occurred during monsoon season.

Clearing legacy waste dumpsites is ineffectual if the entrance of fresh waste to the existing abandoned dumpsite is not stopped or managed properly. Usually, there is a lack of substitute strategy for managing fresh MSW. Therefore, the process of LFBM becomes a Sisyphean work since local authorities keep adding fresh MSW to legacy waste dumps while they undergo a cleaned-up operation. Lack of information on the composition and characteristics of legacy waste often makes it challenging for ULBs to establish proper waste disposal strategies and cost estimation. Space management is another great operational challenge for all LFBM projects since aeration, stabilization, and screening need to be performed within the constraints of an already overflowing dumpsite. The environment, terrain and climatic circumstances of mountainous regions also make it difficult to plan LFBM of legacy waste dumpsites in these areas. A prolonged rainy season in the Northeast's hilly regions hampers the effective functionality of LFBM.

6.2 Infrastructural

The CPCB recommendations on 'Disposal of Legacy Waste 2019' offer specific guidelines on the LFBM process and the methods through which the excavated waste can be processed properly. This needs modern equipment and a comprehensive strategy to maintain the segregated waste quality. However, there is a lack of experienced contractors with sophisticated machinery and modernized technology for the effective execution of LFBM work. Unskilled or inadequately trained labourers who do not comprehend the requirements or even the fundamentals of LFBM operation present another management difficulty. Since most municipal tenders are required to exhibit prior years of expertise in solid waste management and an adequate economic background, new start-up companies in LFBM encounter hurdles to compete in municipal tenders.

6.3 Utilization

LFBM requires careful planning for the destination of screened fractions, i.e. combustible, non-combustible and fine fraction (FF), to reduce the mass of mixed waste to useful components. There is no money to be made from any of these fractions. Most of the time, metropolitan authorities struggle to come up with financially sound, effective utilization choices for the screened fractions. The segregated combustible fraction (SCF) is often heavily polluted with inert substances and has an elevated moisture level, which makes them less acceptable for incineration plants (Ghosh & Kartha, 2023a). Even though there is possibility for the application of SCF as an alternative fuel, it may necessitate additional facilities.

As a result, in the neighbourhood of the dumpsite, there could not be any interested buyers for SCF. For instance, many cement manufacturing units lack the necessary pre-processing facility to co-process SCF that has been reclaimed by LFBM. Apart from SCF, the vast majority of excavated waste from legacy waste dumpsites in India contains FF (Rathod & Muthukkumaran, 2022). Therefore, the effectiveness of LFBM depends heavily on its valorization potential. However, the abundance of heavy metals complicates the management and utilization of FF (Ghosh & Kartha, 2023b). Most LFBM projects involve disposing of FF in low-lying locations as earth-fill material. However, this strategy for bulk utilization of FF has the potential to contaminate subsoil and groundwater due to the leaching of heavy metals from FF (Somani et al., 2020). The SWM Rules, 2016, make no reference to the possible utilization of the land that has been recovered after LFBM (CPHEEO, 2016). CPCB regulations state that the cleared dumpsites cannot be used for human habitation for a minimum period of 15 years after they have been cleaned up (CPCB, 2019). As a result, ULBs are left with many uncertainties regarding the final use of the recovered land after LFBM.

7 Future Perspective

The LFBM technique is not an entirely new concept; in fact, it has been in existence in India during 2002–2003 (Mohan & Joseph, 2020). However, this technique was not extensively followed across the country until direction in the form of planned policy was indicated in the SWM Rules of 2016. In the coming years, LFBM of legacy waste dumpsites will take off across the country with funding from SBM 2.0. LFBM has substantial potential for reclaiming land and resources from abandoned dumpsites. The primary advantages of reclamation are the reduction of pollution risk from poorly managed dumpsites, soil reclamation, redevelopment of the existing dumpsites, and the creation of an environment-friendly business model through the production of value-added products from the excavated legacy waste. An ecosystem of technologically driven approaches for recovering land and dumped waste as a resource has been developed as a result of legacy waste management. It is anticipated that in the future, more Indian municipalities will use the LFBM approach to deal with the overflowing legacy waste dumpsites. However, it is essential to remember that unless there is any legal mechanism supporting and promoting the sustainability of LFBM enterprises, there will not be sufficient confidence in their long-term viability.

Despite being mandated by SBM 2.0 to clear the decades-old dumpsites in cities, a comprehensive plan for future use of the recovered land after LFBM is lacking. Therefore, an extensive policy based on scientific and research-based learning is crucial for the reuse of recovered land by ULBs. Such a strategy must also include required social, environmental and monitoring measures, similar to the Environmental Impact Assessment (EIA). LFBM initiatives also have the potential to produce income from recoverable fractions as well as employment opportunities. However, it is necessary to determine the available market for the recoverable portion. The government can mandate the purchase of recovered materials from LFBM. There is an obvious and pressing need for cooperation between the government, industries and academia to work together to create a circular economy-based economic model for legacy waste dumpsite management. At this moment, LFBM initiatives in the country are in the infant stages. It is crucial to develop complete guidelines and regulations for the ULBs considering the enormity of LFBM operations that could be conducted in India over the coming 2 to 3 years. Some proposed solutions to address the current challenges of LFBM for its long-term viability are mentioned hereafter.

- A city-wide MSW management plan based on the recovery of resources and circular economy principles is essential for the effective clean-up of legacy waste dumpsites.
- Ensuring that recovered fractions are utilized to their fullest potential through proper treatment.
- Environment-specific test standards and procedures must be developed and strictly adhered to for the effective use of recovered portions.
- For government contracts, it is essential to mandate the usage of recyclable materials and goods.
- Sanitary landfills need to be constructed to dispose of non-recyclables from LFBM safely.
- Improved knowledge and evaluation skills on LFBM should be a top priority for ULBs, government officials and pollution control agencies.

8 Conclusion

The high population density, lack of sufficient area for establishing new waste disposal facilities and lack of suitable infrastructure for effectively minimizing waste at the source make legacy waste dumpsite management a pressing issue in developing countries like India. The continual accumulation of legacy waste at dumping grounds has a hazardous influence on the surrounding ecosystem. Therefore, it is crucial to start clearing legacy waste dumpsites today and ensure fresh MSW is managed appropriately. LFBM has emerged as an efficient technical tool for dealing with the mountains of legacy waste sustainably, with the added bonus of recovering valuable components from the waste and reclaiming valuable space. With the help of SBM 2.0's funding, LFBM of legacy waste dumpsites will explode across the country in the coming years. However, multiple challenges during real-life implementations of LFBM operation were also experienced and reported in this book chapter. There needs to be more understanding of LFBM and legacy waste dumpsite management between the government, industries and academia. Despite the intended goal of SBM 2.0 of clearing abandoned dumpsites from cities, a coherent policy direction for the utilization of reclaimed land and recovered fractions after LFBM needs to be improved. It is essential to create a city-wide MSW management plan based on the concepts of recovery of resources and circular economy principles for effective clean-up of legacy waste dumpsites through LFBM.

References

Bisht, A., Kamboj, N., & Kamboj, V. (2022). Groundwater quality and potential health risk assessment in the vicinity of solid waste dumping sites of quaternary shallow water aquifers of Ganga Basin. *Water, Air, & Soil Pollution, 233*(12), 485.

Burlakovs, J., Kriipsalu, M., Vincevica-Gaile, Z., Grinfelde, I., & Grinberga, L. (2021). Material recovery and revitalisation of landfills: Multitasking approach striving to 'beyond the zero waste'. *SGEM International Multidisciplinary Scientific GeoConference, 21*, 285–292.

Central Pollution Control Board (CPCB). (2019). *Guidelines for disposal of legacy waste (Old Municipal Solid Waste)*. Central Pollution Control Board, Ministry of Environment, Forest and Climate Change. Available at: https://cpcb.nic.in/uploads/LegacyWasteBiomining_guidelines_29.04.2019.pdf (accessed on 24 May 2023).

Central Public Health and Environmental Engineering Organization (CPHEEO). (2016). *Municipal solid waste management manual*. Available: http://cpheeo.gov.in/cms/manual-on-municipal-solid-waste-444 management-2016.php, (accessed on 20 May 2023).

Central Public Health and Environmental Engineering Organization (CPHEEO). (2020). *Advisory on Landfill Reclamation*. Central Public Health and Environmental Engineering Organization, Ministry of Housing and Urban Affairs. Available at: https://www.niua.org/csc/assets/pdf/key-documents/phase-2/Waste/SBM-Advisory-on-Landfill-Reclamation.pdf (accessed on 24 May 2023).

Centre for Science and Environment (CSE). (2020). *Clean it right: Dumpsite management in India*. Centre for Science and Environment. Available at: https://www.cseindia.org/content/downloadreports/10487 (accessed 24 May 2023).

Chandana, N., Goli, V. S. N. S., Mohammad, A., & Singh, D. N. (2021). Characterization and utilization of landfill-mined-soil-like-fractions (LFMSF) for sustainable development: A critical appraisal. *Waste and Biomass Valorization, 12*, 641–662.

Chavan, D., Arya, S., & Kumar, S. (2022). Open dumping of organic waste: Associated fire, environmental pollution and health hazards. In *Advanced Organic Waste Management* (pp. 15–31). Elsevier.

Cheela, V. R. S., John, M., & Dubey, B. (2021). Quantitative determination of energy potential of refuse derived fuel from the waste recovered from Indian landfill. *Sustainable Environment Research, 31*, 1–9.

Cheela, V. R. S., John, M., Biswas, W., & Dubey, B. (2023). Environmental impact evaluation of landfill mining of legacy waste with on-site sorting using life cycle assessment. *Environmental Science and Pollution Research, 30*(11), 30033–30047.

Choudhary, A., Kumar, A., & Kumar, S. (2020). National municipal solid waste energy and global warming potential inventory: India. *Journal of Hazardous, Toxic, and Radioactive Waste, 24*(4), 06020002.

Dixit, A., Singh, D., & Shukla, S. K. (2023). Assessment of human health risk due to leachate contaminated soil at solid waste dumpsite, Kanpur (India). *International Journal of Environmental Science and Technology*, 1–16.

Esguerra, J. L., Laner, D., Svensson, N., & Krook, J. (2021). Landfill mining in Europe: Assessing the economic potential of value creation from generated combustibles and fines residue. *Waste Management, 126*, 221–230.

Ghosh, A., & Kartha, S. A. (2020). Assessment of feasibility and viability of landfill mining of open dumpsites in India. In *Fifth Symposium on Circular Economy and Urban Mining*.

Ghosh, A., & Kartha, S. A. (2022). Simulation of heavy metals migration in groundwater from a non-engineered landfill to an eco-sensitive wetland. In *AGU Fall Meeting Abstracts* (Vol. 2022, pp. H13A-07).

Ghosh, A., & Kartha, S. A. (2023a). Composition and characteristics of excavated materials from a legacy waste dumpsite: Potential of landfill biomining. *Environmental Research and Technology, 6*(2), 108–117.

Ghosh, A., & Kartha, S. A. (2023b). Assessment of total content and leaching potential of heavy metals from Landfill-mined-Fine-Fractions under various re-use scenarios. In *International Conference on Solid Wastes 2023: Waste Management in Circular Economy and Climate Resilience (ICSWHK2023)* (pp. 578–581).

Jain, M., Kumar, A., & Kumar, A. (2022). Landfill mining: A review on material recovery and its utilization challenges. *Process Safety and Environmental Protection, 169*, 948–958.

Mohan, S., & Joseph, C. P. (2020). Biomining: An innovative and practical solution for reclamation of open dumpsite. In *Recent developments in waste management: Select proceedings of recycle 2018* (pp. 167–178). Springer Singapore.

Natalia, S., Krutova, A., Mozzhegorova, Y., & Polygalov, S. (2021). The possibility of secondary resource recovery during waste disposal site reclamation. *Detritus, 14*, 108.

National Green Tribunal (NGT). (2019). Order dated 17th July 2019 on OA 519/2019 and 386/2019 regarding "Below Mountains of Trash Lie Poison Lakes". https://greentribunal.gov.in/gen_pdf_test.php?filepath=L25ndF9kb2N1bWVudHMvRWZpbGluZ19kb2N1bWVudHMvbmd0ZG9jL2Nhc2Vkb2MvMDcwMTEwMjAwMjE3MjAxOS8wNC8wMS8yNS8wNF8yNV8wMDFfMTU2OTQ4MTA0MjQyOC5wZGY (accessed on 24 May 2023).

Rathod, D., & Muthukkumaran, K. (2022). Sustainable reuse potential of landfill mining waste retrieved from urban mining sites in South India. *Journal of Material Cycles and Waste Management, 24*(6), 2582–2597.

Saravanan, G., & Ramesh, S. T. (2023). Statistical analysis of seasonal variation in the characteristics of soil like material and refuse derived fuel recovered from landfill mining. *Stochastic Environmental Research and Risk Assessment*, 1–20.

Singh, A., & Chandel, M. K. (2020). Effect of ageing on waste characteristics excavated from an Indian dumpsite and its potential valorisation. *Process Safety and Environmental Protection, 134*, 24–35.

Somani, M., Datta, M., Ramana, G. V., & Sreekrishnan, T. R. (2020). Contaminants in soil-like material recovered by landfill mining from five old dumps in India. *Process Safety and Environmental Protection, 137*, 82–92.

Swachh Bharat Mission 2.0 (SBM 2.0). (2021). *Making Cities Garbage Free.* Available at: https://sbmurban.org/storage/app/media/pdf/swachh-bharat-2.pdf (accessed on 24 May 2023).

Thakkar, A., Semwal, N., & Gargava, P. (2023). Biomining of legacy waste dump sites in India—Overview. *International Journal of Ecology and Environmental Sciences, 49*, 23–30.

Chapter 14
Harnessing Microbial Recycling of Organic Wastes in a Circular Waste Management System for Greenhouse Gas Reduction

S. C. Viñarta, M. V. Angelicola, B. Garolera, and P. M. Fernández

1 Introduction

1.1 The Management of Organic Wastes and Climate Change

During recent decades, the generated amounts of solid wastes have exponentially increased almost all over the world. This increase is attributed and correlated primarily with population growth. However, modern lifestyles and standards due to economic development and the associated increase in urbanization have greatly led to accelerating waste generation (Chen et al., 2016; Maina et al., 2017; Soobhany, 2019). According to estimates from the United Nations, by the year 2050, the world's population could reach 9.7 billion (Randive et al., 2021). However, this growth will not occur without consequences. In the current context, the

S. C. Viñarta
Planta Piloto de Procesos Industriales Microbiológicos (PROIMI-CONICET), Tucumán, Argentina

Facultad de Ciencias Exactas y Naturales (FACEN), Universidad Nacional de Catamarca (UNCA), Catamarca, Argentina

M. V. Angelicola · B. Garolera
Planta Piloto de Procesos Industriales Microbiológicos (PROIMI-CONICET), Tucumán, Argentina

P. M. Fernández (✉)
Planta Piloto de Procesos Industriales Microbiológicos (PROIMI-CONICET), Tucumán, Argentina

Facultad de Ciencias Exactas y Naturales (FACEN), Universidad Nacional de Catamarca (UNCA), Catamarca, Argentina

Facultad de Bioquímica, Química y Farmacia, Universidad Nacional de Tucumán (UNT), Tucumán, Argentina
e-mail: pmfernandez@conicet.gov.ar

V. Kumar et al. (eds.), *Environmental Engineering and Waste Management*,
https://doi.org/10.1007/978-3-031-58441-1_14

demand for food will result in a significant increase in agricultural production, which could have a significant impact on the environment, reducing the availability of water resources, promoting the expansion of agricultural field boundaries, increasing the use of fertilizers and pesticides, and generating considerable amounts of effluents, wastes, and byproducts.

The most abundant types of organic wastes worldwide typically include food waste, garden, and agricultural residues, paper and cardboard waste, as well as wood waste (Dhanker et al., 2022; Kumar & Verma, 2023; Saharia et al., 2024). These organic materials constitute a significant portion of the wastes generated globally. Moreover, in many regions, sewage sludge is also a substantial source of organic wastes. Proper management of these organic wastes is essential for mitigating environmental pollution and harnessing their potential in terms of resources and energy (Kumar & Thakur, 2020; Olujobi et al., 2022; Kumar et al., 2021a). In particular, vinasse and sugarcane sludge produced by the sugar and alcohol industry, olive mill wastewater, and wastes generated by intensive animal farming for consumption are just some examples that already pose problems regarding their final disposal (Ahmed et al., 2019; Kumar et al., 2021b, c; Kumar et al., 2022a, b). In addition to the above, organic wastes has the potential to contribute to the greenhouse effect and climate change due to its natural decomposition in the absence of oxygen. When these wastes accumulate in landfills or improper disposal systems, the lack of oxygen promotes the release into the atmosphere of methane (CH_4), carbon dioxide (CO_2), nitrous oxide (N_2O), and fluorinated gases. These gases trap heat from the sun, leading to the greenhouse effect, which in turn contributes to global warming and climate change (Rajak, 2021).

This situation raises questions that we must address in the coming years: How can we balance growing production demand with conserving natural resources and environmental health? Is it possible to change our view of industrial, agricultural, and livestock byproducts, effluents, and wastes? Instead of considering them an undesirable and hazardous burden, could we envision them as potential resources? The answers to these questions give rise to concepts such as "biorefinery" and "circular economy"(Mandpe et al., 2023), which reshape our approach to these waste products and generate solutions that not only create new sources of income for agro-industries but also help alleviate the pressure they exert on the environment.

It is evident that an ideal waste treatment technology should enable the management of substantial wastes quantities while yielding high-value-added products. However, in the present context, a compromise solution between these two factors is necessary. There are methods that allow for the utilization of substantial waste quantities with minimal prior treatment, almost immediately upon their generation. Nevertheless, these methods often yield relatively modest economic returns, such as the fertigation of productive soils with vinasse or the incineration of bagasse in boilers for energy production (Mainardis et al., 2022; Miemczyk et al., 2022). Conversely, other technologies involve a moderate level of waste treatment before their reuse. Examples of these technologies include composting, anaerobic digestion of waste for biogas production, the production of activated carbon (or biochar) from solid waste in the olive industry, or the concentration of vinasse for subsequent combustion in boilers. Although these technologies have the potential to generate products with a higher added value compared to the aforementioned methods, they may not be as efficient in handling extensive waste volumes (Ahmed et al., 2019).

The most explored biotechnological applications in recent decades can be categorized into a third category, which enables the treatment of small quantities of waste while simultaneously generating high-value-added products. For instance, vinasse has proven to be an economical growth medium for the cultivation of microorganisms with various biotechnological applications, such as the cultivation of bioinoculants, the production of lipids and enzymes, or the cultivation of biomass for the formulation of balanced feeds for fish and farm animals (Thanigaivel et al., 2022).

Wastes management is essential for effectively addressing waste recovery and disposal processes. Organic wastes, being non-reusable, necessitates recovery. A significant challenge in this recovery process lies in reducing waste volume, minimizing the negative aspects associated with waste disposal across various sectors, and curbing land usage for waste disposal (Khalil et al., 2019). In alignment with this approach, considering that the organic fraction constitutes a substantial portion of the generated waste, the biodegradable component could be recycled and utilized as a valuable source of nutrients instead of being lost through improper disposal or inadequate treatment.

This problem requires multifaceted solutions, for that reason, innovative waste management practices can play a crucial role in reducing methane emissions, promoting resource efficiency, and mitigating the overall impact of human activities on the climate. These solutions align with broader sustainability goals and contribute to a more sustainable and resilient future.

1.2 Energy Recovery

The concepts of valorization and recycling are becoming increasingly essential on a global scale. In the context of fruit and vegetable processing, the by-products, wastes, and effluents often contain substantial quantities of proteins, sugars, and lipids, as well as unique aromatic and aliphatic compounds. Consequently, these materials can be viewed as cost-effective and readily available resources for the production of enhanced-value chemicals and biomaterials (Federici et al., 2009; Murthy & Naidu, 2012). In particular, "energy recovery"refers to the process of harnessing energy from various types of waste materials, particularly those containing organic or combustible components. Its primary objectives are to enhance the circularity of raw materials, extend their utility, and complete both the economic and ecological cycles of resource flows (Malinauskaite et al., 2017). This approach has gained widespread acceptance as a solution to pressing global challenges, including climate change, environmental degradation, and resource depletion. The European Union advocates for a shift from "waste management" to "sustainable material management". The objective is to integrate circular economy principles,

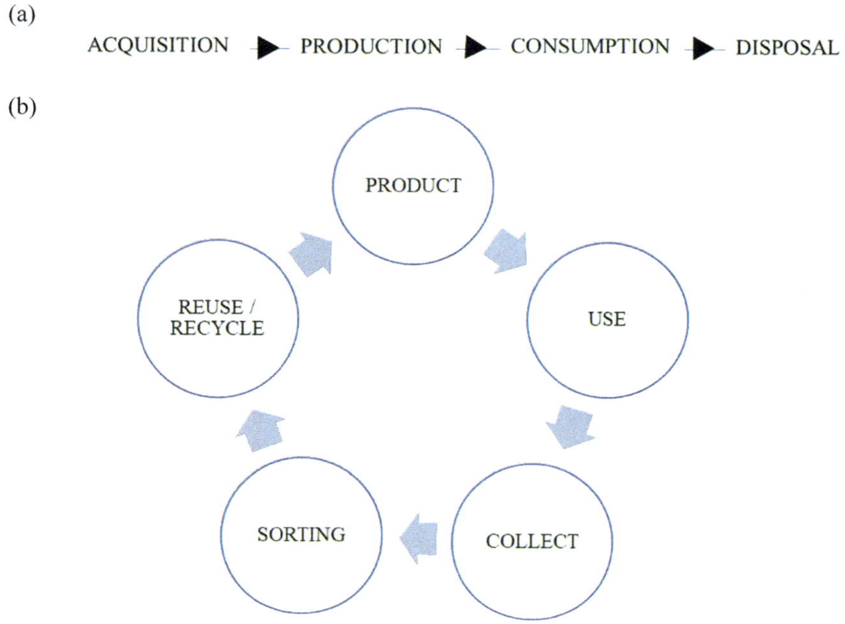

Fig. 14.1 Linear economy (**a**) *versus* closed-loop economic system (**b**), commonly known as a circular economy (CE)

expand the adoption of renewable energy, enhance energy efficiency, and reduce reliance on imported resources, thereby securing long-term economic prospects and competitiveness (Malinauskaite et al., 2017).

Tackling such intricate issues necessitates a holistic approach, departing from the prevailing linear economic model of acquisition, production, consumption, and disposal (Fig. 14.1a). The linear economy model leads to the inefficient use of scarce resources, increases harmful emissions, and yields increasing amounts of waste from the entire value chain including the extraction of resources, manufacturing, transportation, and consumption (Kiørboe et al., 2015). Instead, there is a shift toward embracing a closed-loop economic system, commonly known as a circular economy (CE), which has gained significant traction (Fig. 14.1b). This economic approach seeks not only to reduce raw material consumption and waste generation but also to enhance the recycling of utilized resources, extending their lifespan and thus completing both economic and ecological cycles. The CE represents a viable model with the objective of emulating natural cycles by closing the product life cycle. The transition to CE model requires a methodological, more flexible, decentralized approach, particularly for management in emerging economies (Paes et al., 2019; Traven, 2019).

Additionally, it promotes collaborative projects, the production of bio-based chemicals and energy, and the generation of job opportunities and investment possibilities. This, in turn, the development of a new business model and value chain centered around organic wastes, thereby supporting companies dedicated to sustainable solutions.

Various techniques can be employed for energy recovery from wastes. Among these, physical processing methods like separation, shredding, and drying of biodegradable wastes, known as refuse-derived fuel, can be utilized (Mateus et al., 2023). Additionally, biological processes that involve the utilization of microorganisms to produce biogas via anaerobic digestion and nutrient-rich compost through composting can also be implemented.

As an example of agricultural waste valorization, the utilization of agricultural residues, such as olive mill wastewater (OMWW), to produce high-quality biochemicals and recover valuable metabolites through biotechnological processes, represents a key challenge for the olive oil industry. In this regard, Ahmed and colleagues' review (2019) summarizes the various methodologies employed in utilizing OMWW to generate high-value compounds of significant importance to the biofuel, pharmaceutical, cosmetic, chemical, food, and agricultural industries. The authors suggest that the valorization of this substantial waste stream, particularly through a biorefinery platform, could substantially enhance the environmental sustainability aspects of the entire industry while simultaneously contributing to its economic viability.

For that, the implementation of a waste management system founded on microbial recycling within a circular economy framework yields multifaceted benefits. Not only does it confer environmental advantages by curtailing waste and resource conservation but it also engenders substantial economic benefits. This encompasses the generation of employment opportunities, reduction in expenditure through cost savings, and the creation of revenue streams. These combined factors significantly bolster the trajectory toward a more sustainable and resilient development paradigm (Mandpe et al., 2023).

2 Organic Wastes: Characteristics and Types

Organic wastes can be essentially described as any biodegradable material or unused by-product originating from plant or animal sources. The primary categories of organic wastes encompass household food wastes, agricultural residues, industrial by-products, and human and animal wastes. The types, characteristics, and applications of organic wastes are summarized in Table 14.1.

The degradation of organic wastes is contingent upon its composition and moisture content, typically occurring over time due to the activity of various organisms. The composition of organic wastes is diverse and dependent on the nature of the original materials (Yazid et al., 2017).

Table 14.1 Types, characteristics, and applications of organic wastes (including industrial by-products)

Organic wastes	Organic wastes types	Applications	References
Food wastes	Includes kitchen waste, products not consumed in the retail trade and food industry: oil, water, spoiled food, and food waste, market waste; vegetable and fruit peels, leaves, stalks, and pods; beverage industry (pulps and juices for vegetables and fruits, whey and others)	Glucoamylase and cellulase enzyme production; Aromatic esters production; Ethanol by fermentation; Succinic acid production for laundry detergents, plastics, and drug production; Renewable chemicals and biofuels production; Compost, value-added compounds (phytochemicals, antioxidants, color pigments, and nutrients); Polysaccharides and encapsulating materials for emulsions, bioplastics, biofuels, various chemicals, and additives.	Chandukishore et al. (2023) Su et al. (2022) Poe et al. (2020) Chandukishore et al. (2023) Poe et al. (2020) Pérez-Marroquín et al. (2023) Hameed et al. (2023)
Agricultural residues	Include leaf litter, seed pods, stalks, stems, straws, husks, weeds of soybean, sunflower, tobacco, peanut, maize, rice, wheat, sorghum, and other crops; sugarcane bagasse	Production of biocoal, biochar, biobricks, biofuels (biohydrogen, biomethane, bioethanol, biobutanol, biodiesel), organic acids and bioelectricity; Cellulolytic, hemicellulolytic, and ligninolytic enzymes production; Edible animal feedstuff; Composite materials for application in building elements, furniture, fishnet or as activated carbons; Soil and plant treatment compost	Koul et al. (2022) Szymczak et al. (2021) Colletti et al. (2020) Abu Yazid et al. (2017) Mohajerani et al. (2019) Shahrashoub and Bakhtiari (2021) De Corato (2020)
Human and animal manure	Slaughterhouse and livestock manure (pigs, goats, sheep, cattle, horses, ducks, chickens, ducks, quails, rabbits, pigeons and others); rumen, blood content, and carcasses of animals; human urine and faeces	Agricultural soil fertilization; Biogas and digestate production; Methane production; Biochar and activated carbon production; Biofertilization with mature compost; Biofuels sources	Obileke et al. (2021) Ngo et al. (2022) Khalil et al. (2019) Metson et al. (2020)
Sewage sludge	Organic matter rich in C, P and N, dioxins, pesticides, nanomaterials, microplastics and "emerging" pollutants (pharmaceuticals, hormones,	Biogas production by sludge fermentation; Liquid and solid end products (e.g., biodiesel and biochar)	Cecconet and Capodaglio (2022) Marques et al. (2023)

(continued)

Table 14.1 (continued)

Organic wastes	Organic wastes types	Applications	References
	endocrine disruptors, perfluoroalkylated and polyfluoroalkylated substances (PFAS), linear alkylbenzenesulphonate (LAS), alkylphenols (AP), phthalates (PAE), polycyclic aromatic hydrocarbons (PAH) and polychlorinated biphenyls (PCB)); industrial wastewater	Fertigation of agricultural soils	Viñarta et al. (2022)
Industrial by-products	Molasses, peels (orange, potato, cassava), pulps (orange, apple, mango, guava, pomegranate, pineapple, papaya, tomato etc.), oil-seed cakes (palm kernel cake, groundnut, soybean, mustard, coconut), By-products from the fuel and chemical Glycerol	Polymeric nanomaterials manufacture Use as biomineralization additives	Diacono et al. (2019) Picot-Allain and Emmambux (2023) Zhang et al. (2023) Angelicola et al. (2023) Sineli et al. (2022)

3 Microorganisms Involved in Organic Wastes Recycling

Microorganisms assume a central role in the recycling of organic materials. This includes a wide diversity of bacteria, fungi, and other microorganisms that break down complex organic compounds into simpler forms that can be readily used by plants and various other organisms (Gougoulias et al., 2014). Microorganisms are equipped with enzymatic machinery that can effectively break down stubborn organic compounds into high-value resources. Their exceptional adaptability and biodiversity enable them to thrive in diverse environments, rendering them indispensable in the recycling of organic wastes. As a result, dynamic microbial interactions serve to reduce the environmental footprint associated with waste generation and also present a sustainable outlook for waste management within a circular system (Verstraete et al., 2022). The key microorganisms involved in this process include:

Bacteria Aerobic bacteria require oxygen and are responsible for the initial stages of decomposition, while anaerobic bacteria work in the absence of oxygen, as in the digestion of organic matter in landfills or during anaerobic digestion. Bacteria play a pivotal role in the composting process, either naturally inhabiting organic wastes or being deliberately introduced to enhance the process. Although a diversity of bacterial genera is involved in this process, the most commonly encountered ones belong to the phyla *Proteobacteria*, *Firmicutes*, *Actinobacteria*, and *Bacteroidetes* (Vásquez Castro and Millones Chanamé, 2021).

Fungi Fungi, including molds and yeasts, contribute to the decomposition of organic wastes. They are particularly important in breaking down materials like wood and cellulose. Filamentous fungi offer several advantages in the production of hydrolytic and ligninolytic enzymes, among others (Cho et al., 2009; Wang et al., 2019). They can thrive in a wide range of temperatures, osmotic conditions, and pH levels. Additionally, many fungal species have the ability to secrete a high concentration of enzymes into the culture medium and can grow in both liquid and solid media. Their adaptability to industrial substrates, including agricultural residues, makes them an ideal choice for harnessing these carbon and nitrogen-rich resources.

Among the fungal genera known for their laccase production are *Pleurotus (P. ostreatus, P. pulmonarius)* and *Trametes* (*T. versicolor, T. hirsuta*). Fungi of the *Aspergillus* genus are widely recognized for their ability to produce various enzymes, including cellulases, pectinases, xylanases, proteases, and amylases (Sindhu et al. 2019; Melnichuk et al., 2020). In the food industry, the yeast *Kluyveromyces marxianus* has been successfully employed (Fonseca et al., 2008). This yeast can produce various lytic enzymes, including inulinase and beta-galactosidase. In another context, yeast species of the *Kluyveromyces* genus have shown value in recycling waste from the paper industry, which contains cellulose. These yeasts can metabolize cellulose and produce ethanol through solid substrate fermentation, comparable to *Saccharomyces cerevisiae* in terms of thermotolerance. In the textile sector, it has been demonstrated that *K. marxianus* can carry out decolorization by absorbing dyes into its biomass, although the process is not fully understood (Fonseca et al., 2008).

Actinobacteria These microorganisms are responsible for the decomposition of tougher organic materials, such as lignin and cellulose. They contribute to the decomposition of materials like dead plant tissue.

Within this phylum, the genus *Streptomyces* stands out for its versatile morphology and physiology. Several strains of *Streptomyces* have demonstrated their ability to produce oxytetracycline and other antibiotics using solid agricultural waste as substrates, such as pineapple peels, wheat bran, apple pomace, rice bran, tapioca powder, and orange peels (Asagbra et al., 2005; Ezejiofor et al., 2012).

Archaea Some archaea species are involved in the anaerobic breakdown of organic matter, particularly in marine environments or in the energy production from agricultural wastes (Ziganshin et al., 2011; Kabaivanova et al., 2022; Wegener et al., 2022).

Algae Algae can contribute to the decomposition of organic materials in aquatic environments, such as in the breakdown of dead aquatic plants and organic debris (Chew et al., 2018; Ling et al., 2019).

Table 14.2 Microbial taxa in organic wastes recycling processes

Organic wastes/ industrial by-products	Microorganisms	Process	Reference
Animal manure, food waste, and agricole waste.	Aerobic, anaerobic, and methanogenic bacteria	Anaerobic digestión, fermentation, and composting	Nakasaki et al. (2019); Duan et al. (2020); Qin et al. (2021); Immonen et al. (2020); Mellyanawaty et al. (2021)
Agricole waste and animal manure	Actinomycetes	Fermentation, composting, cultivation	Li et al. (2019); Schalchli et al. (2021); Pereliaeva et al. (2022)
Agricultural waste and food waste	Fungi	Fermentation	Wang et al. (2019); Melnichuk et al. (2020); Karim et al. (2020); Ozdal and Kurbanoglu (2019)
Food, agricultural waste, and animal manure	Microalgae	Composting, cultivation	Chew et al. (2018); Ling et al. (2019)
Diverse organic waste	Bacteria and Archea	Decomposition of solid waste, Anaerobic digestion	Yang et al. (2020); Jo et al. (2019)
Agricultural waste	Bacteria, Actinobacteria, and Fungi	Composting	Hernández-Lara et al. (2022)
Agricultural waste	Bacteria, Archea, and Fungi	Anaerobic digestion	Fan et al. (2022)
Agricultural waste	Bacteria, protozoa, and fungi	Composting	Ahmad and Sur (2023)

Protozoa Protozoa are single-celled organisms that feed on bacteria and other microorganisms in the decomposition process. They help control bacterial populations and nutrient cycling (Ahmad & Sur, 2023).

In Table 14.2, the taxa of microorganisms involved in various organic wastes recycling processes are summarized. In a circular economy model, these microbial processes are essential for closing nutrient and resource loops, minimizing waste generation, and maximizing the utilization of organic materials in various applications.

4 Case of Study: Organic Wastes as Substrate of Oleaginous Microorganisms

Organic wastes management represents a growing challenge in the search for sustainable environmental practices. Oleaginous microorganisms have emerged as a promising alternative for environmentally friendly solutions. These microorganisms possess a unique metabolic capability to accumulate lipids (more than 20%, w/w of their dry biomass) from organic substrates (Cho & Park, 2018). These microbial lipids (also called "single cell oils," SCOs) can be used as renewable biofuels, among other applications (Viñarta et al., 2022). This has generated considerable interest in utilizing oleaginous microorganisms to recycle organic wastes. Especially, oleaginous yeasts (Viñarta et al., 2016; Viñarta et al., 2020; Maza et al., 2020; Sineli et al., 2022; Angelicola et al., 2023) and certain bacterial strains (Wells et al., 2015), have demonstrated an outstanding ability to store reserve neutral lipids using organic residues as substrates (Cho & Park, 2018).

The use of oleaginous microorganisms in organic wastes recycling offers multiple environmental benefits including wastes and greenhouse gas emissions reduction (Viñarta et al., 2022). Moreover, this approach aligns with the principles of the circular economy by converting organic wastes into valuable resources.

This section describes the utilization of oleaginous microorganisms as a sustainable and environmental strategy for organic waste recycling. It reviews the types of organic wastes suitable for microbial recycling, environmental advantages, economic viability, and the challenges and opportunities associated with this innovative approach.

4.1 Organic Wastes as Alternative Carbon Sources for Microbial Oils

The improper disposal and accumulation of organic wastes not only pose environmental hazards but also contribute significantly to greenhouse gas emissions.

The exploration of organic wastes as alternative carbon sources for the production of microbial oils has gained significant attention in recent years. Particularly, oleaginous yeasts and bacteria, were extensively investigated for these purposes (Cho & Park, 2018).

A variety of organic wastes were employed as feedstocks for oleaginous microorganisms. Food waste, in particular, has been a focus due to its high organic content and abundance, making it an environmentally sustainable resource for SCOs production. Additionally, agricultural residues and lignocellulosic biomass have been explored as suitable substrates for microbial oil generation (Table 14.3), emphasizing the versatility and potential of these microorganisms in promoting sustainable waste-to-oil conversion processes (Gupta & Parkhey, 2015).

Diverse types of organic wastes, including agricultural residues, food waste, and wastewater, among others can be processed by oleaginous yeasts (Table 14.3). It highlights the efficiency of oleaginous yeasts in converting organic wastes into valuable lipids (Chatterjee & Mohan, 2018; Di Fidio et al., 2019). However, the characteristics and nature of the organic wastes used as substrate can influence the survival of the oleaginous microorganisms and/or the lipid accumulation (Table 14.3) such as type and concentration of substrate, carbon-to-nitrogen ratio,

Table 14.3 Characteristics, composition, and main limitations of the organic wastes used as carbon sources for SCOs production by oleaginous yeasts

Organic waste	Composition and characteristics	Limitations
Food waste	Food processing waste expired or spoiled food and restaurant residues contain high amounts of organic matter, including carbohydrates, lipids, and proteins, making them suitable for microbial utilization and lipid production.	Variable composition and impurities, which can affect biomass and lipid production. Additionally, the presence of contaminants, such as preservatives and additives, can hinder microbial growth and product quality.
Agricultural residues	Agricultural by-products such as straw, husks, stalks, and bagasse are often used as substrates for oleaginous yeasts. These wastes are rich in carbohydrates and lignocellulosic materials, providing a sustainable and readily available source for microbial growth and lipid accumulation.	Expensive enzymatic pre-treatments are required for sugar liberation from agricultural residues. Degradation products released during enzymatic pre-treatment can inhibit the growth of oleaginous yeasts. Some lignocellulosic enzymes employed for sugar liberation are inefficient.
Lignocellulosic biomass	Lignocellulosic materials derived from forestry residues, sawdust, or agricultural waste are complex materials that can be utilized as substrates for oleaginous yeasts.	These complex substrates require additional pre-treatment to release fermentable sugars for yeast metabolism and subsequent lipid accumulation. Enzymatic pre-treatments are expensive and non-efficient in some cases Compounds of degradation released during enzymatic treatment can act as inhibitors, significantly impacting oleaginous microorganisms.
Wastewater and sewage sludge:	Organic waste from wastewater treatment plants, including sewage sludge and various industrial effluents, can be utilized as substrates for oleaginous yeasts. These wastes contain complex organic compounds, providing an abundant and cost-effective source for microbial growth and lipid production.	Using wastewater and sewage sludge as feedstock poses limitations due to inconsistent nutrient content and the presence of toxic compounds. Effective pre-treatment and careful monitoring are necessary to mitigate these challenges and optimize lipid production.
Oil and fat residues	Residual oil and fat wastes from cooking oil production, rendering facilities, or food processing industries can be employed as substrates for oleaginous yeasts. These wastes often contain significant amounts of lipids, facilitating enhanced lipid accumulation by yeasts.	While oil and fat residues are rich in lipids, their utilization may be restricted by impurities and contaminants. Contaminants like heavy metals and organic pollutants can negatively impact both microbial growth and lipid quality, necessitating rigorous purification steps and microbial strain selection for optimal outcomes.

(continued)

Table 14.3 (continued)

Organic waste	Composition and characteristics	Limitations
Crude glycerol	Crude glycerol is generated as a by-product during biodiesel production, mainly consists of glycerol, methanol, and salts.	The presence of impurities in crude glycerol may hinder cell growth and affect lipid accumulation.
Volatile fatty acids, VFAs	VFAs can be obtained by biological synthesis through anaerobic fermentation of organic waste and have been recently proposed as a low-cost and environmentally friendly alternative to produce SCO.	High concentrations of VFAs can inhibit and severely affect oleaginous yeasts during fermentation.
Algal biomass	Algae-based biomass, including microalgae or macroalgae, can serve as excellent substrates for oleaginous yeasts. Algal biomass contains substantial quantities of carbohydrates, lipids, and proteins, providing an ideal nutrient source for lipid accumulation in yeasts.	Complex composition, including cell wall components like cellulose and hemicellulose. These require specialized enzymes for efficient saccharification, increasing process complexity and cost.

and cultivation conditions (Viñarta et al., 2016; Viñarta et al., 2020). The study of these aspects is essential for each bioprocess.

Volatile fatty acids, VFAs produced during anaerobic microbial fermentation of organic wastes have been also proposed as substrates of oleaginous yeasts to produce SCOs (Llamas et al., 2020). Crude glycerol, produced as the main by-product during biodiesel industrial manufacturing, is considered an important raw material for microbial oils production by oleaginous yeasts (Angelicola et al., 2023; Sineli et al., 2022) and promotes a model of circular economy. However, high concentrations of VFAs or the impurities present in crude glycerol can inhibit cell growth and influence lipid synthesis of oleaginous microorganisms (Llamas et al., 2020; Gao et al., 2016).

It is important to note, however, that research in this field continues to advance, and investigations are ongoing into alternative sources of organic wastes. Examples include lignocellulosic materials from agricultural and forestry residues, further expanding the possibilities of microbial recycling and promoting environmental sustainability.

The most studied oleaginous microorganisms for the recycling of organic wastes are typically oleaginous yeasts, with a particular focus on genera such as *Yarrowia, Rhodotorula, Lipomyces,* and *Cryptococcus. Yarrowia lipolytica*, a well-studied oleaginous yeast, has demonstrated remarkable potential in converting a wide range of organic wastes, including food waste, agricultural residues, and industrial by-products, into microbial oils (Abdel-Mawgoud & Stephanopoulos, 2020; Ramesh et al., 2020). Similarly, *Rhodotorula glutinis*, *Lipomyces starkeyi,* and *Cryptococcus curvatus* (recently renamed *Trichosporon oleaginosus*) have displayed efficient lipid accumulation from various organic substrates, contributing to their extensive examination as microbial oil producers (Viñarta et al., 2022; Sineli et al., 2022; Angelicola et al., 2023; Chatterjee & Mohan, 2018).

Among the bacterial species, *Rhodococcus* have emerged as prominent candidates for this purpose. *Rhodococcus opacus*, a notable oleaginous bacterium, has garnered attention for its ability to thrive on diverse organic wastes streams, including lignocellulosic materials, plant biomass, and glycerol-rich substrates (Wells et al., 2015; Chatterjee et al., 2020).

4.2 Microbial Oils Applications

Microbial oils are found to be an interesting alternative source for the oleochemical industry. SCOs display high similarities with plant oils in terms of fatty acid (FA) composition (Di Fidio et al., 2020). The FA profile of microbial oils can vary according to the microorganism and carbon source, making them highly versatile for a variety of industrial applications including biofuel (Viñarta et al., 2020; Khoo et al., 2020), pharmaceutical and food supplements (Bharathiraja et al., 2017; Viñarta et al., 2016), or wax esters (Papadaki et al., 2017), among other bioproducts (Tomás-Pejó et al., 2021).

Single-cell oils offer a wide range of applications, from chemical uses in the production of biodegradable materials and chemicals to nutritional applications in functional foods, dietary supplements, and infant formulas. Their versatility and sustainability make them valuable resources in addressing both industrial and nutritional needs while contributing to a more environmentally friendly future.

We describe the applications of SCOs in all fields. However, since carbon source is used to influence strongly the subsequent industrial application, organic wastes cannot be used as carbon source for SCOs produced for medical and nutritional applications (Tomás-Pejó et al., 2021). For that reason, potential applications of microbial oils from organic wastes have been associated mainly with biofuels and oleochemical products. The applications of SCOs for chemical and nutritional uses are summarized in Table 14.4.

4.3 Economic Viability and Environmental Benefits of Single Cell Oils (SCOs)

Significant progress has been made in the development of processes involving oleaginous yeasts, particularly in the past decade. Extraordinarily, high lipid productivities have been achieved without genetic modifications to yeasts. The emergence of techno-economic analyses (TEA) and life-cycle assessments (LCA) has further bolstered the credibility of the yeast lipid concept. Nevertheless, the primary challenge in the commercialization of yeast lipids as a sustainable alternative to

Table 14.4 Main Chemical and nutritional applications of SCOs

Chemical applications of SCOs	Nutritional applications of SCOs
SCOs has found numerous applications in the chemical industry due to their unique lipid composition and properties. These applications include:	SCOs also has significant nutritional value and are used in various dietary and nutritional products:
Biodegradable Polymers: SCOs can serve as feedstocks for the production of biodegradable polymers and plastics. Their bio-based nature and versatility make them an environmentally friendly alternative to traditional petroleum-based plastics.	*Functional Foods*: SCOs are incorporated into functional foods to enhance their nutritional content. They are a source of essential fatty acids, including omega-3 and omega-6, which have numerous health benefits.
Specialty Chemicals: The fatty acids derived from SCOs can be used as precursors for various specialty chemicals, including detergents, lubricants, and surfactants. These chemicals have a wide range of industrial applications.	*Dietary Supplements*: SCOs are used to produce dietary supplements, such as capsules or liquid supplements, providing consumers with a convenient way to access beneficial fatty acids, vitamins, and other nutrients.
Oleochemicals: SCOs is a source of oleochemicals, which are natural compounds derived from fats and oils. Oleochemicals are used in the production of cosmetics, soaps, and other personal care products	*Infant Formulas*: SCOs are included in infant formulas to mimic the fatty acid composition of breast milk, which is crucial for infant development. They provide essential nutrients for healthy growth.
Green Solvents: Some SCOs can be converted into green solvents that serve as eco-friendly alternatives to traditional volatile organic compounds (VOCs). These solvents find applications in various industrial processes.	*Animal Feed*: SCOs are used in animal feed formulations to enhance the nutritional quality of feed. This can lead to improved animal health and the production of nutrient-enriched animal products.
	Pharmaceuticals: SCOs are being investigated for their potential pharmaceutical applications, such as in the development of lipid-based drug delivery systems and treatments for various health conditions.

vegetable oils like palm oil or biodiesel derivatives is the cost of lipid production. Therefore, the successful commercialization of a yeast lipid depends on continued advancement in process and product design, as well as effective marketing and market intervention (Abeln & Chuck, 2021).

Intensive research efforts have been dedicated to the efficient utilization of organic wastes as a carbon source for microbial lipids. This approach is poised to significantly enhance production process profitability and stimulate the growth of a bio-based economy (Tomás-Pejó et al., 2021).

Employing oleaginous yeasts for sustainable organic wastes management presents a spectrum of challenges and opportunities. Substrate cost, microbial productivity, and downstream processing have been identified as the primary challenges in the commercial production of microbial oils (Parsons et al., 2020).

Some important obstacles as the feedstock price and availability, and the downstream processes, such as harvesting, cell disruption, and lipid extraction to be overcome for rendering SCOs production economically viable (Yellapu et al., 2018).

Figure 14.2 shows a general and simplified scheme of the stages involved in SCOs production bioprocesses by oleaginous yeasts from organic wastes. Some methods of cell disruption and SCOs extraction are also summarized.

In the context of organic wastes recycling, the utilization of oleaginous yeasts exploits their metabolic aptitude to metabolize and degrade the organic compounds found in these waste materials. This metabolic activity allows the yeasts to accumulate lipids, which can subsequently be extracted and employed in the production of biodiesel.

The current microbial processes are oriented at maximizing microbial oil production from cost-effective substrates such as lignocellulosic sugars, volatile fatty acids, and glycerol. Furthermore, downstream procedures like harvesting and lipid extraction play a critical role in determining the cost-effectiveness of the overall process (Tomás-Pejó et al., 2021).

The economic feasibility of SCOs production from organic wastes is currently constrained by limitations in lipid productivity and the necessity to optimize culture conditions to enhance lipid accumulation. However, the implementation of integrated strategies, coupled with innovative approaches like genetic engineering, strain selection, or co-production of metabolites, can contribute to the maximization of lipid production and further enhance the sustainability of the process (Viñarta et al., 2022).

The utilization of oleaginous yeasts for organic wastes recycling brings about significant environmental benefits. These include the reduction of greenhouse gas emissions by diverting organic wastes from landfills and harnessing the produced lipids (SCOs) as a renewable energy source. Furthermore, it emphasizes the potential to mitigate soil and water pollution by efficiently converting organic wastes into valuable lipids.

Regardless of its economic viability, the use of oleaginous yeasts for organic wastes recycling can be regarded as a sustainable environmental alternative. This approach offers numerous advantages from a sustainability perspective. Firstly, the employment of oleaginous yeasts facilitates the conversion of organic wastes into valuable products, reducing reliance on fossil fuels and contributing to efforts aimed at mitigating climate change. Additionally, recycling organic wastes using oleaginous yeasts aids in reducing waste accumulation in landfills, subsequently leading to decreased greenhouse gas emissions and preventing soil and water contamination.

It is worth emphasizing that the utilization of oleaginous yeasts in organic wastes recycling is still in the research and development phases. Further optimization of processes and the identification of the best yeast strains for maximizing lipid production are essential.

The utilization of organic wastes as alternative carbon sources for oleaginous microorganisms not only contributes to waste reduction and the reduction of greenhouse gas emissions but also fosters the concept of a circular economy. This approach provides a sustainable and environmentally responsible method for microbial oil production.

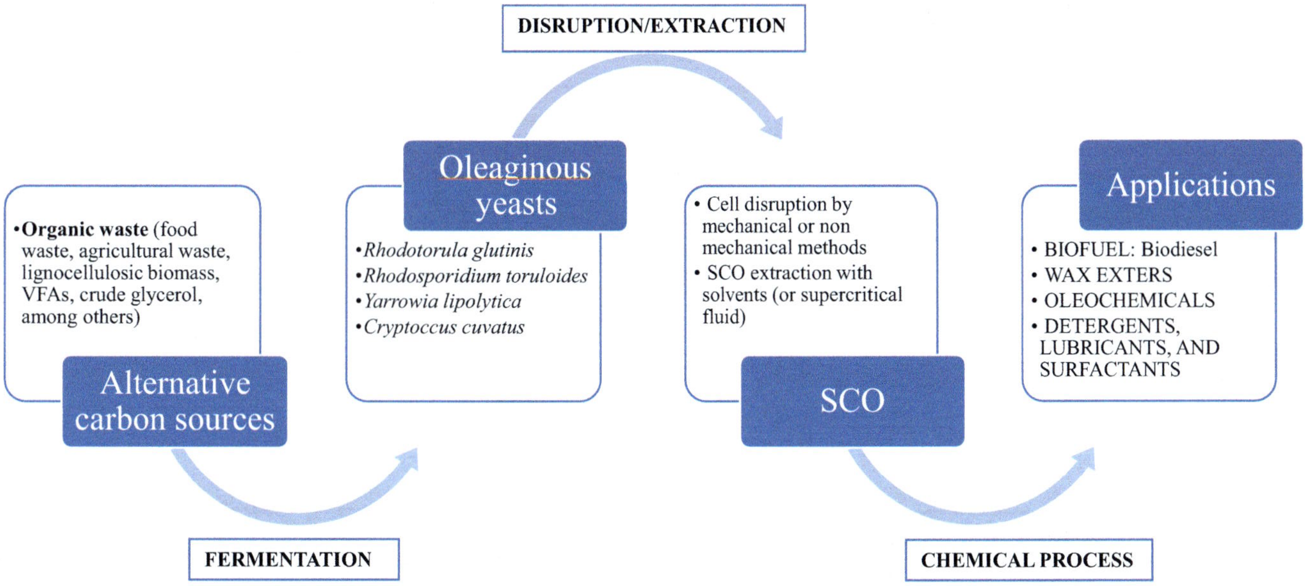

Fig. 14.2 General and simplified scheme of stages involved in SCOs production bioprocesses by oleaginous yeasts from organic wastes. Organic wastes more used as carbon sources and the more representative oleaginous yeasts are summarized and listed. Some methods of cell disruption and SCOs extraction are also summarized. Only applications of SCOs potentially produced from organic wastes are detailed

5 Challenges, Opportunities, and Future Perspectives

In the of organic wastes recycling, circular economy, and greenhouse gas reduction, various challenges and opportunities stand before us. Technological advancements are required to develop and refine efficient and sustainable waste management processes. On the other hand, there is an urgent need to implement policies and campaigns to organize solid waste recycling, involving rag pickers and improving waste collection efficiency for positive social impacts. Legislative, regulatory support, and economically competitive processes are crucial to ensure that recycling and circular economy principles are implemented effectively. Additionally, the development and maintenance of infrastructure for waste collection, transportation, and processing present logistical challenges.

It is also important to mention that there are substantial opportunities. Organic wastes can be transformed into renewable resources, including biogas, organic fertilizers, and bio-based materials. Embracing a circular economy approach can help close the nutrient loop and minimize resource waste. These initiatives have the potential to significantly reduce greenhouse gas emissions, particularly through methane capture in landfills and biogas production. The transition to a circular economy is also an economic opportunity, fostering innovation and the development of more effective waste management technologies. Moreover, this transition creates job opportunities in waste management and resource recovery sectors. To boost the implementation of the CE concept, it is necessary to develop a suitable platform for proper interaction and exchange of ideas at the national and international levels.

Looking into the future, we anticipate increased adoption of circular economy principles and organic wastes recycling as environmental consciousness continues to grow. Investment in research and development will drive the improvement of recycling and treatment technologies. While academic staff can conceptualize ideas, academic institutions often struggle to execute them due to financial constraints and a lack of practical experience. Collaborations with businesses can expedite the implementation of the Circular Economy model. Stricter environmental regulations are likely to drive the adoption of more sustainable practices. Innovative technologies, such as biotechnology and bioenergy, are expected to open new avenues in organic wastes management. Global collaboration will become increasingly important in waste management and greenhouse gas reduction efforts.

Addressing these challenges, seizing opportunities, and considering future perspectives will play a pivotal role in achieving sustainable organic wastes management, advancing the circular economy, and reducing greenhouse gas emissions.

6 Conclusion

The actual challenge is to maintain a delicate equilibrium between the demands of industrial production and the preservation of our finite natural resources and the environment. The reuse of agricultural waste unequivocally represents a sector poised for significant development in the near future. While there are already available alternatives, the most promising technologies are still in the experimental phases. The available technologies offer a diverse landscape, ranging from relatively simple and cost-effective methods to innovative biotechnological applications. Their implementation will undoubtedly require a substantial effort from the scientific and technological sectors and a significant investment from the private sector. However, the economic and environmental benefits are considerable.

To address this challenge, adopting visionary strategies like the biorefinery and circular economy is crucial. This innovative approach not only enhances economic prospects for the agro-industry but also fosters greater environmental sustainability and social responsibility. Such collaboration and wastes management focus are essential for a more balanced and sustainable future in the agro-industrial sector and society as a whole.

The use of microorganisms in organic wastes recycling aligns with the principles of the circular economy and offers several environmental benefits, including wastes reduction and the mitigation of greenhouse gas emissions. Furthermore, repurposing organic wastes may provide significant assistance in addressing concurrent disposal issues, such as environmental contamination and sewer system obstructions.

References

Abdel-Mawgoud, A. M., & Stephanopoulos, G. (2020). Improving CRISPR/Cas9-mediated genome editing efficiency in *Yarrowia lipolytica* using direct tRNA-sgRNA fusions. *Metabolic Engineering, 62*, 106–115. https://doi.org/10.1016/j.ymben.2020.07.008

Abeln, F., & Chuck, C. J. (2021). The history, state of the art and future prospects for oleaginous yeast research. *Microbial Cell Factory, 20*, 221. https://doi.org/10.1186/s12934-021-01712-1

Abu Yazid, N., Barrena, R., Komilis, D., & Sánchez, A. (2017). Solid-state fermentation as a novel paradigm for organic waste valorization: A review. *Sustainability, 9*(2), 224. https://doi.org/10.3390/su9020224

Ahmad, A., & Sur, S. (2023). Biodegradable solid waste management by microorganism: Challenge and potential for composting. *International Journal of Recycling Organic Waste in Agriculture, 12*(4), 735–745. https://doi.org/10.30486/ijrowa.2023.1958655.1465

Ahmed, P. M., Fernández, P. M., Castellanos, L. I., & Pajot, H. F. (2019). Exploitation alternatives of olive mill wastewater: Production of value-added compounds useful for industry and agriculture. *Biofuel Research Journal, 6*(2), 980–994. https://doi.org/10.18331/BRJ2019.6.2.4

Angelicola, M. V., Fernández, P. M., Van Nieuwenhove, C. P., Aybar, M. J., Figueroa, L. I. C., & Viñarta, S. C. (2023). Bioconversion of commercial and crude glycerol to single-cell oils by the Antarctic yeast *Rhodotorula glutinis* R4 as a biodiesel feedstock. *Biocatalysis and Agricultural Biotechnology, 47*, 102544. https://doi.org/10.1016/j.bcab.2022.102544

Asagbra, A. E., Sanni, A. I., & Oyewole, O. B. (2005). Solid-state fermentation production of tetracycline by *Streptomyces* strains using some agricultural wastes as substrate. *World Journal of Microbiology and Biotechnology, 21*, 107–114.

Bharathiraja, B., Sridharan, S., Sowmya, V., Yuvaraj, D., & Praveenkumar, R. (2017). Microbial oil – A plausible alternate resource for food and fuel application. *Bioresource Technology, 233*, 423–432. https://doi.org/10.1016/j.biortech.2017.03.006

Cecconet, D., & Capodaglio, A. G. (2022). Sewage sludge biorefinery for circular economy. *Sustainability, 14*(22), 14841.

Chandukishore, T., Das, S., Narasimhulu, K., & Prabhu, A. A. (2023). Biovalorisation of mixed food waste through newly isolated thermo tolerant fungal cell factories: A step toward transforming waste to wealth. *Sustainable Chemistry and Pharmacy, 35*, 101230.

Chatterjee, S., & Mohan, S. V. (2018). Microbial lipid production by *Cryptococcus curvatus* from vegetable waste hydrolysate. *Bioresource Technology, 254*, 284–289. https://doi.org/10.1016/j.biortech.2018.01.079

Chatterjee, A., DeLorenzo, D. M., Carr, R., & Moon, T. S. (2020). Bioconversion of renewable feedstocks by *Rhodococcus opacus*. *Current Opinion in Biotechnology, 64*, 10–16. https://doi.org/10.1016/j.copbio.2019.08.013

Chen, P., Xie, Q., Addy, M., Zhou, W., Liu, Y., Wang, Y., Chen, Y., Li, K., & Ruan, R. (2016). Utilization of municipal solid and liquid wastes for bioenergy and bioproducts production. *Bioresource Technology, 215*, 163–172.

Chew, K. W., Chia, S. R., Show, P. L., Ling, T. C., Arya, S. S., & Chang, J. (2018). Food waste compost as an organic nutrient source for the cultivation of *Chlorella vulgaris*. *Bioresource Technology, 267*, 356–362. https://doi.org/10.1016/j.biortech.2018.07.069

Cho N. S., Wilkolazka A. J., Staszczak M., Cho H. Y., & Ohga S. (2009). The role of laccase from white rot fungi to stress conditions. *Journal of the Faculty of Agriculture, Kyushu University, 54*(1), 81–83. https://doi.org/10.5109/14041

Cho, H. U., & Park, J. M. (2018). Biodiesel production by various oleaginous microorganisms from organic wastes. *Bioresource Technology, 256*, 502–508. https://doi.org/10.1016/j.biortech.2018.02.010

Colletti, A., Attrovio, A., Boffa, L., Mantegna, S., & Cravotto, G. (2020). Valorisation of by-products from soybean (*Glycine max* (L.) Merr.) processing. *Molecules, 25*(9), 2129.

De Corato, U. (2020). Agricultural waste recycling in horticultural intensive farming systems by on-farm composting and compost-based tea application improves soil quality and plant health: A review under the perspective of a circular economy. *Science of the Total Environment, 738*, 139840.

Dhanker, R., Rawat, S., Chandna, V., Deepa, Das, S., Sharma, A., Kumar, V., & Kumar, R. (2022). Recovery of silver nanoparticles and management of food wastes: Obstacles and opportunities. *Environmental Advances, 9*, 100303. https://doi.org/10.1016/j.envadv.2022.100303

Di Fidio, N., Dragoni, F., Antonetti, C., De Bari, I., Galletti, A. M. R., & Ragaglini, G. (2020). From paper mill waste to single cell oil: Enzymatic hydrolysis to sugars and their fermentation into microbial oil by the yeast Lipomyces starkeyi. *Bioresource technology, 315*, 123790. https://doi.org/10.1016/j.biortech.2020.123790

Di Fidio, N., Liuzzi, F., Mastrolitti, S., Albergo, R., & Bari, I. D. (2019). Single cell oil production from undetoxified *Arundo donax* L. hydrolysate by *Cutaneotrichosporon curvatus*. *Journal of Microbiology and Biotechnology, 29*(2), 256–267. https://doi.org/10.4014/jmb.1808.08015

Diacono, M., Persiani, A., Testani, E., Montemurro, F., & Ciaccia, C. (2019). Recycling agricultural wastes and by-products in organic farming: Biofertilizer production, yield performance and carbon footprint analysis. *Sustainability, 11*(14), 3824.

Duan, Y., Awasthi, S. K., Liu, T., Pandey, A., Zhang, Z., Kumar, S., & Awasthi, M. K. (2020). Succession of keratin-degrading bacteria and associated health risks during pig manure composting. *Journal of Cleaner Production, 258*, 120624. https://doi.org/10.1016/j.jclepro.2020.120624

Ezejiofor, T. I. N., Duru, C. I., Asagbra, A. E., Ezejiofor, A. N., Orisakwe, O. E., Afonne, J. O., & Obi, E. (2012). Waste to wealth: Production of oxytetracycline using *Streptomyces* species from household kitchen wastes of agricultural produce. *African Journal of Biotechnology, 11*(43), 10115–10124.

Fan, Q., Fan, X., Fu, P., Li, Y., Zhao, Y., & Hua, D. (2022). Anaerobic digestion of wood vinegar wastewater using domesticated sludge: Focusing on the relationship between organic degradation and microbial communities (archaea, bacteria, and fungi). *Bioresource Technology, 347*, 126384. https://doi.org/10.1016/j.biortech.2021.126384

Federici, F., Fava, F., Kalogerakis, N., & Mantzavinos, D. (2009). Valorisation of agro-industrial by-products, effluents and waste: Concept, opportunities and the case of olive mill wastewaters. *Journal of Chemical Technology & Biotechnology: International Research in Process, Environmental & Clean Technology, 84*(6), 895–900.

Fonseca, G. G., Heinzle, E., Wittmann, C., & Gombert, A. K. (2008). The yeast *Kluyveromyces marxianus* and its biotechnological potential. *Applied Microbiology and Biotechnology, 79*, 339–354.

Gao, Z., Ma, Y., Wang, Q., Zhang, M., Wang, J., & Liu, Y. (2016). Effect of crude glycerol impurities on lipid preparation by *Rhodosporidium toruloides* yeast 32489. *Bioresource Technology, 218*, 373–379.

Gougoulias, C., Clark, J. M., & Shaw, L. J. (2014). The role of soil microbes in the global carbon cycle: Tracking the below-ground microbial processing of plant-derived carbon for manipulating carbon dynamics in agricultural systems. *Journal of the Science of Food and Agriculture, 94*(12), 2362–2371. https://doi.org/10.1002/jsfa.6577

Gupta, P., & Parkhey, P. (2015). A two-step process for efficient enzymatic saccharification of rice straw. *Bioresource Technology, 173*, 207–215.

Hameed, A., Anwar, M. J., Perveen, S., Amir, M., Naeem, I., Imran, M., Hussain, M., Ahmad, I., Afza, M. I., Inayat, S., & Awuchi, C. G. (2023). Functional, industrial and therapeutic applications of dairy waste materials. *International Journal of Food Properties, 26*(1), 1470–1496.

Hernández-Lara, A., Ros, M., Cuartero, J., Bustamante, M. Á., Moral, R., Andreu-Rodríguez, F. J., Fernández, J. A., Egea-Gilabert, C., & Pascual, J. A. (2022). Bacterial and fungal community dynamics during different stages of agro-industrial waste composting and its relationship with compost suppressiveness. *Science of the Total Environment, 805*, 150330. https://doi.org/10.1016/j.scitotenv.2021.150330

Immonen, M., Maina, N. H., Wang, Y., Coda, R., & Katina, K. (2020). Waste bread recycling as a baking ingredient by tailored lactic acid fermentation. *International Journal of Food Microbiology, 327*, 108652. https://doi.org/10.1016/j.ijfoodmicro.2020.108652

Jo, Y., Hwang, K., & Lee, C. (2019). Enhancing anaerobic digestion of vegetable waste and cellulose by bioaugmentation with rumen culture. *Membrane Water Treatment, 10*(3), 213–221. https://doi.org/10.12989/mwt.2019.10.3.213

Kabaivanova, L., Hubenov, V., Dimitrova, L., Simeonov, I., Wang, H., & Petrova, P. (2022). Archaeal and bacterial content in a two-stage anaerobic system for efficient energy production from agricultural wastes. *Molecules, 27*(5), 1512.

Karim, A., Gerliani, N., & Aïder, M. (2020). *Kluyveromyces marxianus*: An emerging yeast cell factory for applications in food and biotechnology. *International Journal of Food Microbiology, 333*, 108818. https://doi.org/10.1016/j.ijfoodmicro.2020.108818

Khalil, M., Berawi, M. A., Heryanto, R., & Rizalie, A. (2019). Waste to energy technology: The potential of sustainable biogas production from animal waste in Indonesia. *Renewable and Sustainable Energy Reviews, 105*, 323–331.

Khoo, K. S., Chew, K. W., Yew, G. Y., Leong, W. H., Chai, Y. H., Show, P. L., & Chen, W. H. (2020). Recent advances in downstream processing of microalgae lipid recovery for biofuel production. *Bioresource Technology, 304*, 122996.

Kiørboe, N., Sramkova, H., & Krarup, M. (2015). *Moving towards a circular economy: Successful Nordic business models*. Nordic Council of Ministers.

Koul, B., Yakoob, M., & Shah, M. P. (2022). Agricultural waste management strategies for environmental sustainability. *Environmental Research, 206*, 112285.

Kumar, V., & Thakur, I. S. (2020). Extraction of lipids and production of biodiesel from secondary tannery sludge by in situ transesterification. *Bioresource Technology Reports, 11*, 100446. https://doi.org/10.1016/j.biteb.2020.100446

Kumar, V., & Verma, P. (2023). A critical review on environmental risk and toxic hazards of refractory pollutants discharged in chlorolignin waste of pulp and paper mills and their remediation approaches for environmental safety. *Environmental Research, 236*, 116728. https://doi.org/10.1016/j.envres.2023.116728

Kumar, V., Srivastava, S., & Thakur, I. S. (2021a). Enhanced recovery of polyhydroxyalkanoates from secondary wastewater sludge of sewage treatment plant: Analysis and process parameters optimization. *Bioresource Technology Reports, 15*, 100783. https://doi.org/10.1016/j.biteb.2021.100783

Kumar, V., Ferreira, L. F. R., Sonkar, M., & Singh, J. (2021b). Phytoextraction of heavy metals and ultrastructural changes of *Ricinus communis* L. grown on complex organometallic sludge discharged from alcohol distillery. *Environmental Technology & Innovation, 22*, 101382. https://doi.org/10.1016/j.eti.2021.101382

Kumar, V., Shahi, S. K., Ferreira, L. F. R., Bilal, M., Biswas, J. K., & Bulgariu, L. (2021c). Detection and characterization of refractory organic and inorganic pollutants discharged in biomethanated distillery effluent and their phytotoxicity, cytotoxicity, and genotoxicity assessment using *Phaseolus aureus* L. and *Allium cepa* L. *Environmental Research, 201*, 111551. https://doi.org/10.1016/j.envres.2021.111551

Kumar, V., Ameen, F., Islam, M. A., Agrawal, S., Motghare, A., Dey, A., Shah, M. P., Américo-Pinheiro, J. H. P., Singh, S., & Ramamurthy, P. C. (2022a). Evaluation of cytotoxicity and genotoxicity effects of refractory pollutants of untreated and biomethanated distillery effluent using *Allium cepa*. *Environmental Pollution, 300*, 118975. https://doi.org/10.1016/j.envpol.2022.118975

Kumar, V., Agrawal, S., Shahi, S. K., Motghare, A., Singh, S., & Ramamurthy, P. C. (2022b). Bioremediation potential of newly isolated *Bacillus albus* strain VKDS9 for decolourization and detoxification of biomethanated distillery effluent and its metabolites characterization for environmental sustainability. *Environmental Technology & Innovation, 26*, 102260. https://doi.org/10.1016/j.eti.2021.102260

Li, X., Shi, X. S., Lu, M. Y., Zhao, Y. Z., Li, X., Peng, H., & Guo, R. B. (2019). Succession of the bacterial community and functional characteristics during continuous thermophilic composting of dairy manure amended with recycled ceramsite. *Bioresource Technology, 294*, 122044. https://doi.org/10.1016/j.biortech.2019.122044

Ling, Y., Sun, L., Wang, S., Ki Lin, C. S., Sun, Z., & Zhou, Z. (2019). Cultivation of oleaginous microalga *Scenedesmus obliquus* coupled with wastewater treatment for enhanced biomass and lipid production. *Biochemical Engineering Journal, 148*, 162–169. https://doi.org/10.1016/j.bej.2019.05.012

Llamas, M., Dourou, M., González-Fernáandez, C., Aggelis, G., & Tomás-Pejó, E. (2020). Screening of oleaginous yeasts for lipid production using volatile fatty acids as substrate. *Biomass and Bioenergy, 138*, 105553.

Maina, S., Kachrimanidou, V., & Koutinas, A. (2017). A roadmap towards a circular and sustainable bioeconomy through waste valorization. *Current Opinion in Green and Sustainable Chemistry, 8*, 18–23.

Mainardis, M., Cecconet, D., Moretti, A., Callegari, A., Goi, D., Freguia, S., & Capodaglio, A. G. (2022). Wastewater fertigation in agriculture: Issues and opportunities for improved water management and circular economy. *Environmental Pollution, 296*, 118755.

Malinauskaite, J., Jouhara, H., Czajczyńska, D., Stanchev, P., Katsou, E., Rostkowski, P., Thorne, R. J., Colón, J., Ponsá, S., Al-Mansour, F., Anguilano, L., Krzyżyńska, R., López, I. C., Vlasopoulos, A., & Spencer, N. (2017). Municipal solid waste management and waste-to-energy in the context of a circular economy and energy recycling in Europe. *Energy, 141*, 2013–2044.

Mandpe, A., Paliya, S., Gedam, V. V., Patel, S., Tyagi, L., & Kumar, S. (2023). Circular economy approach for sustainable solid waste management: A developing economy perspective. *Waste Management & Research, 41*(3), 499–511. https://doi.org/10.1177/0734242X221126718

Marques, M. V. A., Lopes, B. C., Silvério, T. H. R., von Sperling, M., & Neves, T. D. A. (2023). Persistence of pathogens and bacterial community dynamics in tropical soil after application of raw sewage. *Scientific Reports, 13*(1), 13435.

Mateus, M. M., Cecílio, D., Fernandes, M. C., & Neiva Correia, M. J. (2023). Refuse derived fuels as an immediate strategy for the energy transition, circular economy, and sustainability. *Business Strategy and the Environment, 32*(6), 3915–3926.

Maza, D. D., Viñarta, S. C., Su, Y., Guillamón, J. M., & Aybar, M. J. (2020). Growth and lipid production of *Rhodotorula glutinis* R4, in comparison to other oleaginous yeasts. *Journal of Biotechnology, 310*, 21–31.

Mellyanawaty, M., Nakakoji, S., Tatara, M., Marbelia, L., Prijambada, I. D., Budhijanto, W., & Ueno, Y. (2021). Enrichment of thermophilic methanogenic microflora from mesophilic waste activated sludge for anaerobic digestion of garbage slurry. *Journal of Bioscience and Bioengineering, 132*(6), 630–639. https://doi.org/10.1016/j.jbiosc.2021.09.005

Melnichuk, N., Braia, M. J., Anselmi, P. A., Meini, M. R., & Romanini, D. (2020). Valorization of two agroindustrial wastes to produce alpha-amylase enzyme from *Aspergillus oryzae* by solid-state fermentation. *Waste Management, 106*, 155–161. https://doi.org/10.1016/j.wasman.2020.03.025

Metson, G. S., Feiz, R., Quttineh, N. H., & Tonderski, K. (2020). Optimizing transport to maximize nutrient recycling and green energy recovery. *Resources, Conservation & Recycling: X, 9*, 100049.

Miemczyk, J., Carbone, V., & Howard, M. (2022). Learning to implement the circular economy in the agri-food sector: A multilevel perspective. In *Circular economy supply chains: From chains to systems* (pp. 283–301). Emerald Publishing Limited.

Mohajerani, A., Hui, S. Q., Mirzababaei, M., Arulrajah, A., Horpibulsuk, S., Abdul Kadir, A., Rahman, M. T., & Maghool, F. (2019). Amazing types, properties, and applications of fibres in construction materials. *Materials, 12*(16), 2513.

Murthy, P. S., & Naidu, M. M. (2012). Sustainable management of coffee industry by-products and value addition—A review. *Resources, Conservation and Recycling, 66*, 45–58.

Nakasaki, K., Hirai, H., Mimoto, H., Quyen, T. N. M., Koyama, M., & Takeda, K. (2019). Succession of microbial community during vigorous organic matter degradation in the primary fermentation stage of food waste composting. *Science of the Total Environment, 671*, 1237–1244. https://doi.org/10.1016/j.scitotenv.2019.03.341

Ngo, T., Shahsavari, E., Shah, K., Surapaneni, A., & Ball, A. S. (2022). Improving bioenergy production in anaerobic digestion systems utilising chicken manure via pyrolysed biochar additives: A review. *Fuel, 316*, 123374.

Obileke, K., Mamphweli, S., Meyer, E. L., Makaka, G., & Nwokolo, N. (2021). Development of a mathematical model and validation for methane production using cow dung as substrate in the underground biogas digester. *PRO, 9*(4), 643.

Olujobi, O. J., Ufua, D. E., Olokundun, M., & Olujobi, O. M. (2022). Conversion of organic wastes to electricity in Nigeria: Legal perspective on the challenges and prospects. *International Journal of Environmental Science and Technology, 1–12*(19), 939–950. https://doi.org/10.1007/s13762-020-03059-3

Ozdal, M., & Kurbanoglu, E. B. (2019). Citric acid production by *Aspergillus Niger* from Agro-industrial by-products: Molasses and chicken feather peptone. *Waste and Biomass Valorization, 10*, 631–640. https://doi.org/10.1007/s12649-018-0240-y

Paes, M. X., de Medeiros, G. A., Mancini, S. D., de Miranda Ribeiro, F., & de Oliveira, J. A. P. (2019). Transition to circular economy in Brazil: A look at the municipal solid waste management in the state of São Paulo. *Management Decision, 59*(8), 1827–1840.

Papadaki, A., Mallouchos, A., Efthymiou, M. N., Gardeli, C., Kopsahelis, N., Aguieiras, E. C. G., Freire, D. M. G., Papanikolaou, S., & Koutinas, A. A. (2017). Production of wax esters via microbial oil synthesis from food industry waste and by-product streams. *Bioresource Technology, 245*, 274–282.

Parsons, S., Allen, M. J., & Chuck, C. J. (2020). Coproducts of algae and yeast-derived single cell oils: A critical review of their role in improving biorefinery sustainability. *Bioresource Technology, 303*, 122862.

Pereliaeva, E. V., Dmitrieva, M. E., Morgunova, M. M., Belyshenko, A. Y., Imidoeva, N. A., Ostyak, A. S., & Axenov-Gribanov, D. V. (2022). The use of Baikal psychrophilic actinobacteria for synthesis of biologically active natural products from sawdust waste. *Fermentation, 8*(5), 213. https://doi.org/10.3390/fermentation8050213

Pérez-Marroquín, X. A., Estrada-Fernández, A. G., García-Ceja, A., Aguirre-Álvarez, G., & León-López, A. (2023). Agro-food waste as an ingredient in functional beverage processing: Sources, functionality, market and regulation. *Foods, 12*(8), 1583.

Picot-Allain, M. C. N., & Emmambux, M. N. (2023). Isolation, characterization, and application of nanocellulose from agro-industrial by-products: A review. *Food Reviews International, 39*(2), 941–969.

Poe, N. E., Yu, D., Jin, Q., Ponder, M. A., Stewart, A. C., Ogejo, J. A., Wang, H., & Huang, H. (2020). Compositional variability of food wastes and its effects on acetone-butanol-ethanol fermentation. *Waste Management, 107*, 150–158.

Qin, S., Wainaina, S., Liu, H., Soufiani, A. M., Pandey, A., Zhang, Z., Awasthi, M. K., & Taherzadeh, M. J. (2021). Microbial dynamics during anaerobic digestion of sewage sludge combined with food waste at high organic loading rates in immersed membrane bioreactors. *Fuel, 303*, 121276. https://doi.org/10.1016/j.fuel.2021.121276

Rajak, J. (2021). A preliminary review on impact of climate change and our environment with reference to global warming. *International Journal of Environmental Science, 10*, 11–14.

Ramesh, A., Ong, T., Garcia, J. A., Adams, J., & Wheeldon, I. (2020). Guide RNA engineering enables dual purpose CRISPR-Cpf1 for simultaneous gene editing and gene regulation in *Yarrowia lipolytica*. *ACS Synthetic Biology, 9*, 967–971.

Randive, K., Raut, T., & Jawadand, S. (2021). An overview of the global fertilizer trends and India's position in 2020. *Mineral Economics, 34*, 371–384. https://doi.org/10.1007/s13563-020-00246-z

Saharia, M., Dey, G., & Kumar, V. (2024). Vermiremediation of plant agro waste to recover residual nutrients and improve crop productivity. In *Earthworm technology in organic waste management: Recent trends and advances* (pp. 79–113). Elsevier.

Schalchli, H., Hormazábal, E., Astudillo, Á., Briceño, G., Rubilar, O., & Diez, M. C. (2021). Bioconversion of potato solid waste into antifungals and biopigments using *Streptomyces* spp. *PLoS One, 16*(5), e0252113. https://doi.org/10.1371/journal.pone.0252113

Shahrashoub, M., & Bakhtiari, S. (2021). The efficiency of activated carbon/magnetite nanoparticles composites in copper removal: Industrial waste recovery, green synthesis, characterization, and adsorption-desorption studies. *Microporous and Mesoporous Materials, 311*, 110692.

Sindhu, R., Gnansounou, E., Rebello, S., Binod, P., Varjani, S., Thakur, I. S., Nair, R. B., & Pandey, A. (2019). Conversion of food and kitchen waste to value-added products. *Journal of environmental management, 241*, 619–630. https://doi.org/10.1016/j.jenvman.2019.02.053

Sineli, P. E., Maza, D. D., Aybar, M. J., Figueroa, L. I. C., & Viñarta, S. C. (2022). Bioconversion of sugarcane molasses and waste glycerol on single cell oils for biodiesel by the red yeast *Rhodotorula glutinis* R4 from Antarctica. *Energy Conversion and Management: X, 16*, 100331. https://doi.org/10.1016/j.ecmx.2022.100331

Soobhany, N. (2019). Insight into the recovery of nutrients from organic solid waste through biochemical conversion processes for fertilizer production: A review. *Journal of Cleaner Production, 241*, 118413.

Su, G., Ong, H. C., Fattah, I. R., Ok, Y. S., Jang, J. H., & Wang, C. T. (2022). State-of-the-art of the pyrolysis and co-pyrolysis of food waste: Progress and challenges. *Science of the Total Environment, 809*, 151170.

Szymczak, T., Cybulska, J., Podleśny, M., & Frąc, M. (2021). Various perspectives on microbial lipase production using agri-food waste and renewable products. *Agriculture, 11*(6), 540.

Thanigaivel, S., Priya, A. K., Kumar, P. S., Shiong, K. K., Hoang, T. K., Rajendran, S., & Soto-Moscoso, M. (2022). Exploration of effective biorefinery approach to obtain the commercial value-added products from algae. *Sustainable Energy Technologies and Assessments, 53*, 102450.

Tomás-Pejó, E., Morales-Palomo, S., & González-Fernández, M. C. (2021). Microbial lipids from organic wastes: Outlook and challenges. *Bio/Technology, 323*, 124612. https://doi.org/10.1016/j.biortech

Traven, L. (2019). Circular economy and the waste management hierarchy: Friends or foes of sustainable economic growth? A critical appraisal illustrated by the case of the Republic of Croatia. *Waste Management & Research, 37*, 1–2.

Vásquez, E., & Chanamé, C. E. M. (2021). Una revisin sobre la diversidad microbiana y su rol en el compostaje aerobio. *Aporte Santiaguino, 14*(2), 253–275.

Verstraete, W., Yanuka-Golub, K., Driesen, N., & De Vrieze, J. (2022). Engineering microbial technologies for environmental sustainability: Choices to make. *Microbiology and Biotechnology, 15*(1), 215–227. https://doi.org/10.1111/1751-7915.13986

Viñarta, S. C., Angelicola, M. V., Barros, J. M., Fernández, P. M., Mac Cormack, W., Aybar, M. J., & Figueroa, L. I. C. (2016). Oleaginous yeasts from Antarctica: Screening and preliminary approach on lipid accumulation. *Journal of Basic Microbiology, 56*, 1360–1368. https://doi.org/10.1002/jobm.201600099

Viñarta, S. C., Angelicola, M. V., Van Nieuwenhove, C. P., Aybar, M. J., & Figueroa, L. I. C. (2020). Fatty acids profiles and estimation of the biodiesel quality parameters from *Rhodotorula* spp. from Antarctica. *Biotechnology Letters, 42*, 757–772. https://doi.org/10.1007/s10529-020-02796-2

Viñarta, S. C., Maza, D. D., Fernández, P. M., Aybar, M. J., & Castellanos de Figueroa, L. I. C. (2022). Integrated production of biodiesel and industrial wastewater treatment by culturing oleaginous microorganisms (Chapter 4, Pag. 81–100). In V. Kumar & M. Kumar (Eds.), *Integrated environmental technologies for wastewater treatment and sustainable development*. ELSEVIER. ISBN: 9780323911801.

Wang, F., Xu, L., Zhao, L., Ding, Z., Ma, H., & Terry, N. (2019). Fungal laccase production from lignocellulosic agricultural wastes by solid-state fermentation: A review. *Microorganisms, 7*(12), 665. https://doi.org/10.3390/microorganisms7120665

Wegener, G., Laso-Pérez, R., Orphan, V. J., & Boetius, A. (2022). Anaerobic degradation of alkanes by marine archaea. *Annual Review of Microbiology, 76*, 553–577.

Wells, T., Wei, Z., & Ragauskas, A. (2015). Bioconversion of lignocellulosic pretreatment effluent via oleaginous *Rhodococcus opacus* DSM 1069. *Biomass and Bioenergy, 72*, 200–205.

Yang, S., Li, L., Peng, X., & Song, L. (2020). Leachate microbiome profile reveals bacteria, archaea and eukaryote dynamics and methanogenic function during solid waste decomposition. *Bioresource Technology, 320*, 124359. https://doi.org/10.1016/j.biortech.2020.124359

Yazid, A. N., Barrena, R., Komilis, D., & Sánchez, A. (2017). Solid-state fermentation as a novel paradigm for organic waste valorization: A review. *Sustainability, 9*(2), 224.

Yellapu, S. K., Bharti, Kaur, R., Kumar, L. R., Tiwari, B., Zhang, X., & Tyagi, R. D. (2018). Recent developments of downstream processing for microbial lipids and conversion to biodiesel. *Bioresource Technology, 256*, 515–528.

Zhang, Y., Hu, X., Wang, Y., & Jiang, N. (2023). A critical review of biomineralization in environmental geotechnics: Applications, trends, and perspectives. *Biogeotechnics, 1*(1), 100003.

Ziganshin, A. M., Schmidt, T., Scholwin, F., Il'inskaya, O. N., Harms, H., & Kleinsteuber, S. (2011). Bacteria and archaea involved in anaerobic digestion of distillers grains with solubles. *Applied Microbiology and Biotechnology, 89*, 2039–2052.

Chapter 15
Profiling of Microbial Community and Their Role in Solid Waste Treatment

Shubhra Singh and Douglas J. H. Shyu

1 Introduction to Solid Waste Treatment

1.1 Overview of Solid Waste Treatment

Solid waste treatment is a combined process that involves the collection, transportation, processing and disposal of solid waste. It begins with waste generation, which occurs in residential, commercial, institutional and industrial settings (Chander et al., 2023). Further, waste collection is carried out using bins and collection vehicles to gather waste from various sources. After collection, the waste is transported to treatment facilities or disposal sites, and before sending it to the treatment facilities, the segregation of waste is an important step where different types of waste are separated for appropriate handling and treatment (Wang et al., 2017). Various treatment methods, such as recycling, composting, incineration or landfilling, are carried out as a part of solid and wet waste management. The recycling method is the most recommended for different types of plastics or other materials that help convert them into reusable products, while composting involves the natural decomposition of organic waste, such as food waste from restaurants, hotels, etc. On the other hand, incineration involves the controlled combustion of waste to reduce its volume and generate energy which is gaining much attention due to the circular economy

S. Singh
Department of Tropical Agriculture and International Cooperation, National Pingtung University of Science and Technology, Pingtung, Taiwan

Department of Biological Science and Technology, National Pingtung University of Science and Technology, Pingtung, Taiwan

D. J. H. Shyu (✉)
Department of Biological Science and Technology, National Pingtung University of Science and Technology, Pingtung, Taiwan
e-mail: dshyu@mail.npust.edu.tw

V. Kumar et al. (eds.), *Environmental Engineering and Waste Management*,
https://doi.org/10.1007/978-3-031-58441-1_15

approach, while landfilling involves burying waste in designated areas with proper environmental safeguards (Hamer, 2003). Solid waste, such as municipal waste, uses landfilling as the indispensable type of disposal method in most parts of the world. However, developed nations and even developing nations have started practicing recycling as a preferred method for disposing of waste along with incineration to generate energy (Matsakas et al., 2017; Tang et al., 2023). Effective waste management also necessitates the implementation of policies and regulations by governments and regulatory bodies to ensure proper waste handling and disposal, in which the segregation of waste at the primary level plays an important role. Through these measures, solid waste management aims to protect public health, preserve the environment and promote sustainable development (Rastogi et al., 2020).

1.2 Importance of Microbial Community in Solid Waste Treatment

Solid waste consists of different types of organic and inorganic elements that cause air, soil and water pollution along with the depletion of environmental resources by contamination. As a solution, microbial communities are able to decompose, transform and stabilize organic matter in solid waste into more useful energy forms (Chander et al., 2023; Haruta et al., 2002; Herrmann & Shann, 1997). Here are the key reasons why microbial communities are important in solid waste treatment:

1.2.1 Decomposition and Organic Matter Breakdown

Microorganisms, such as bacteria and fungi, are adept at breaking down complex organic compounds present in solid waste. They secrete enzymes that help degrade organic matter into simpler substances, facilitating the decomposition process. Through their metabolic activities, microbes convert complex organic compounds into carbon dioxide, water and humus, which are more stable and less harmful (Vivas et al., 2009).

1.2.2 Composting

Microbes are essential for the composting process, which involves the controlled decomposition of organic waste materials. During composting, a diverse microbial community actively decomposes the organic waste, producing heat and facilitating the breakdown of complex compounds into stable and nutrient-rich compost. Certain groups of bacteria and fungi, particularly thermophilic organisms, are responsible for the high-temperature phase of composting, which helps eliminate pathogens and weed seeds (Xi et al., 2015).

1.2.3 Methane Production in Anaerobic Digestion

In anaerobic digestion, microorganisms break down organic waste in the absence of oxygen, producing methane-rich biogas. Methanogenic archaea, a specific group of microorganisms, convert complex organic compounds into methane and carbon dioxide through a series of biochemical reactions. This process not only helps in waste stabilization but also generates renewable energy that can be used for electricity or heat production (Hassard et al., 2015; Wang et al., 2017).

1.2.4 Nutrient Cycling

Microbial communities play a vital role in nutrient cycling within solid waste treatment systems. They facilitate the release, transformation and recycling of essential nutrients, such as nitrogen, phosphorus and potassium, present in organic waste. Microbes break down complex organic compounds into forms that can be readily taken up by plants or returned to the soil, promoting the nutrient balance and fertility of agricultural and horticultural applications (Kumar et al., 2017).

1.2.5 Bioremediation of Contaminated Waste

Microbes are capable of degrading and detoxifying hazardous substances present in certain types of solid waste. Some bacteria and fungi possess the ability to break down pollutants, such as petroleum hydrocarbons, heavy metals and various organic chemicals, through metabolic processes. Bioremediation techniques utilize microbial communities to treat contaminated soils, sediments and industrial waste, leading to the reduction of harmful pollutants and the restoration of environmental quality (Shah, 2021).

1.2.6 Control of Odour and Pathogens

Proper management of solid waste includes controlling odour generation and minimizing the presence of pathogens. Microbial activity helps in reducing unpleasant odours by breaking down odour-causing compounds. Additionally, certain microbial species can outcompete and inhibit the growth of pathogenic microorganisms, enhancing the safety of waste treatment processes and minimizing the risk of disease transmission (Ayilara et al., 2020; Yu, 2014).

Understanding and manipulating microbial communities in solid waste treatment systems can optimize waste degradation processes, enhance treatment efficiency and mitigate environmental impacts (de Jonge et al., 2020). Strategies such as bioaugmentation (Zeneli et al., 2019) (adding selected microbial strains) and biocontrol (Tóthné Bogdányi et al., 2021) (suppressing unwanted microorganisms) can be employed to enhance the performance of waste treatment systems. Additionally, advancements in molecular biology and microbial ecology research enable a deeper understanding of microbial communities, leading to the development of innovative and sustainable waste treatment technologies (Farooqi et al., 2021).

1.3 Purpose of Microbial Community Profiling

Microbial community profiling refers to the analysis and characterization of microbial populations present in a particular environment or sample. It involves identifying and quantifying the types and relative abundances of microorganisms within a microbial community (Rastogi et al., 2020). Studies on degradation dynamics for leachate treatment systems have indicated that microbial profiling is an important tool for the comprehensive analysis of chemical parameters and changes in the microbial community structure (dos Santos et al., 2022).

There are numerous purposes of microbial community profiling, such as the profiling allows researchers to assess the diversity of microorganisms present in a given environment. It provides insights into the richness (number of different species) and evenness (relative abundance of each species) of microbial populations (Lamb et al., 2011). Understanding microbial biodiversity is important for ecosystem functioning, as different microbial species perform various ecological functions, such as nutrient cycling, organic matter degradation and symbiotic interactions. Additionally, microbial community profiling can serve as a tool for monitoring and assessing the health and quality of various environments, including soil, water bodies, air and built environments (Bulgarelli et al., 2015). By analysing changes in microbial community composition and structure over time, researchers can detect disturbances, pollution or shifts in environmental conditions. Microbial indicators can provide early warning signs of environmental degradation or contamination. By analysing the composition and diversity of microbial communities associated with different habitats, researchers can investigate the relationships between microbes and host health, disease susceptibility and overall well-being. Microbiome research has implications for understanding the role of microorganisms in human and animal physiology, immune function and various diseases (Farooqi et al., 2021).

A very important aspect of community profiling contributes to the discovery and development of novel microbial resources for biotechnological applications. By identifying and characterizing microorganisms with specific functional traits or capabilities, such as the production of enzymes, bioactive compounds, or bioremediation potential, researchers can harness their potential for industrial processes, bioenergy production and environmental remediation (dos Santos et al., 2022). Microbial community profiling is useful for optimizing various biotechnological processes, such as wastewater treatment, composting and fermentation (Krishnan et al., 2023). By monitoring the microbial community composition and dynamics during these processes, researchers can identify key microbial players, assess their functional potential and make informed decisions to enhance process efficiency and stability. It also contributes to the field of microbial ecology and evolutionary biology (Gnaneswar Gude, 2015). It helps unravel ecological interactions, such as symbiotic relationships, competition and predation, among different microbial taxa. Comparative analyses of microbial communities from different habitats or geographic locations shed light on microbial adaptation, evolution and the role of environmental factors in shaping community structure. Microbial community profiling techniques, such as DNA sequencing, metagenomics and meta transcriptomics, have revolutionized our understanding of microbial diversity and function. These approaches provide valuable insights into the complex microbial world and its significance in various ecological, health-related and biotechnological contexts.

2 Techniques for Microbial Community Profiling

2.1 DNA Sequencing

DNA sequencing (as shown in Fig. 15.1) has revolutionized the field of microbial community profiling by providing a comprehensive and high-resolution view of the microbial diversity and composition within complex ecosystems. Microbial communities play critical roles in various environments, including soil, oceans, human body sites and industrial processes. DNA sequencing techniques allow researchers to analyse the genetic material (DNA) present in these microbial communities, enabling the identification and characterization of individual microbial species and their functional potential. There are different DNA sequencing methods employed for microbial community profiling, with each offering unique advantages and capabilities. Next-Generation Sequencing (NGS) platforms have become the cornerstone of microbial community analysis due to their high-throughput nature, cost-effectiveness and ability to generate vast amounts of sequence data. NGS technologies, such as Illumina sequencing, enable the simultaneous sequencing of millions of DNA fragments in a single run, facilitating comprehensive analysis of microbial communities (Mahato et al., 2017; Prakash et al., 2021).

Targeted gene sequencing is a widely used approach for microbial community profiling. The 16S ribosomal RNA (rRNA) gene is commonly targeted for bacteria and archaea, while the Internal Transcribed Spacer (ITS) region is often used for fungi. These conserved regions of the genome contain variable areas that can be used to differentiate between different microbial taxa. Amplicon sequencing involves amplifying the target region using polymerase chain reaction (PCR) and then subjecting the amplified DNA to NGS. The resulting sequence data can be analysed to determine the taxonomic composition and diversity of the microbial community (Liao et al., 2019). Shotgun metagenomic sequencing is another powerful approach for microbial community profiling. It involves sequencing all the DNA present in a sample, providing a comprehensive view of both the microbial and host genomes. This method allows for the analysis of functional potential and metabolic pathways

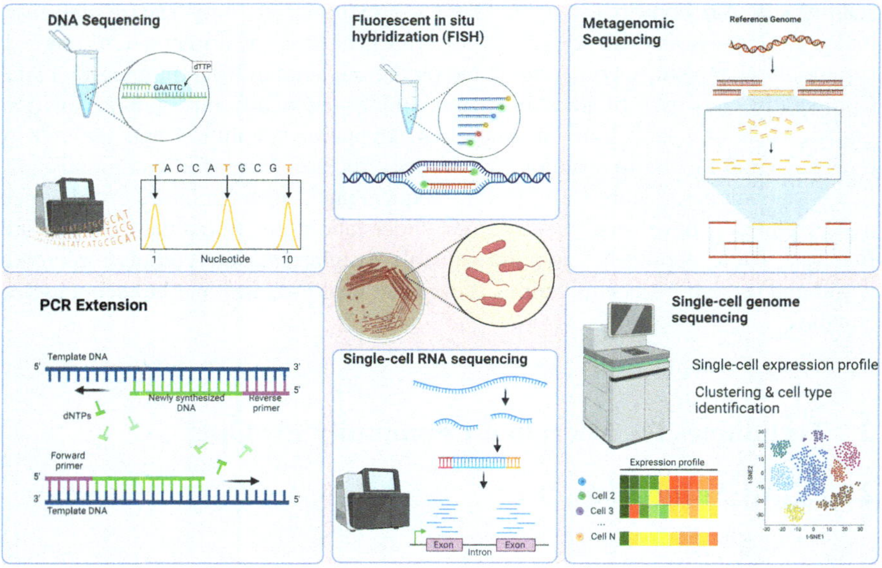

Fig. 15.1 Advanced Genomic Techniques: This figure illustrates various molecular biology methods including DNA sequencing, PCR extension, fluorescent in situ hybridization (FISH), single-cell RNA sequencing, metagenomic sequencing and single-cell genome sequencing. These genomic techniques enable researchers to sequence DNA, amplify specific DNA sequences, visualize gene locations, analyse individual cell gene expression, study microbial communities and examine single-cell genomes, respectively, advancing our understanding of genetics and biology. (Created with BioRender.com)

within the microbial community. However, shotgun metagenomic sequencing requires higher sequencing depth and computational resources compared to targeted gene sequencing (Loman & Pallen, 2015; Sidhu et al., 2017).

Bioinformatics analysis plays a crucial role in DNA sequencing-based microbial community profiling. Preprocessing steps involve quality control, trimming of adapter sequences and removal of low-quality reads. Taxonomic classification is performed by comparing the obtained sequences to reference databases and assigning each sequence to a specific taxonomic group. Operational Taxonomic Unit (OTU) clustering is commonly used to group similar sequences into taxonomic units, providing a measure of diversity within the microbial community. Functional annotation and pathway analysis can be performed by comparing the obtained sequences to databases of known functional genes and metabolic pathways (Shah et al., 2013).

2.2 PCR Amplification

PCR (Polymerase Chain Reaction) amplification (as shown in Fig. 15.1) is a fundamental technique used in microbial community profiling to selectively amplify specific regions of the microbial DNA, enabling subsequent analysis and characterization of the microbial community. PCR amplification is particularly employed in targeted gene sequencing approaches, such as 16S rRNA gene amplicon sequencing for bacteria and archaea and Internal Transcribed Spacer (ITS) region amplicon sequencing for fungi (Bhatia et al., 2015). The PCR amplification process involves a series of temperature-dependent steps that allow for the selective amplification of the target DNA region. The following steps are typically involved in PCR amplification for microbial community profiling:

(a) Denaturation: The PCR reaction starts by heating the DNA template containing the microbial DNA to a high temperature (~94–98 °C), causing the double-stranded DNA to denature into two single strands.
(b) Annealing: The reaction temperature is then lowered to allow for the binding of primers specific to the target DNA region. The primers are short DNA sequences that flank the target region and act as the starting points for DNA synthesis.
(c) Extension: The reaction temperature is increased to an optimal range (~68–72 °C) for DNA polymerase activity. The DNA polymerase enzyme synthesizes new DNA strands complementary to the single-stranded DNA template using the primers as a guide. This process extends the DNA sequence in the 5′ to 3′ direction.
(d) Cycling: Steps 2 and 3 are repeated for a specific number of cycles, typically ranging from 20 to 40 cycles. Each cycle doubles the amount of DNA, resulting in exponential amplification of the target region.

By carefully designing the primers, researchers can selectively amplify specific regions of the microbial DNA, such as hypervariable regions of the 16S rRNA gene. These regions contain sequence variations that allow for taxonomic classification and identification of microbial taxa. PCR amplification is a crucial step in microbial community profiling, as it allows for the selective amplification of target DNA regions from complex microbial communities. It enables the generation of sufficient quantities of DNA for downstream sequencing and subsequent analysis, providing valuable insights into the taxonomic composition and diversity of microbial communities in various environments.

2.3 Fluorescent In Situ Hybridization (FISH)

Fluorescent in situ hybridization (FISH) (as shown in Fig. 15.1) is a powerful technique used in microbial community profiling to visualize and identify specific microbial groups directly in their natural environment. Unlike other profiling methods that involve DNA extraction and sequencing, FISH allows for the detection and localization of target microbial cells based on their genetic composition using fluorescently labelled nucleotide probes (Calli et al., 2006). The process of FISH for microbial community profiling involves the following steps:

(a) Sample fixation: The microbial community sample is typically fixed using a chemical fixative, such as paraformaldehyde, to preserve the microbial cells in their original state.
(b) Probe design: Specific oligonucleotide probes are designed to target the genetic sequences of interest in the microbial community. These probes are typically 16–30 nucleotides in length and are labelled with a fluorescent dye, allowing for visualization under a fluorescence microscope.
(c) Hybridization: The fixed microbial sample is treated with fluorescently labelled probes. The probes hybridize to their complementary target sequences within the microbial cells, forming stable probe-target complexes.
(d) Washing: After hybridization, the sample is washed to remove unbound or non-specifically bound probes, reducing background noise.
(e) Visualization: The sample is observed under a fluorescence microscope equipped with appropriate filters for the excitation and emission wavelengths of the fluorescent dye. The labelled microbial cells appear as distinct fluorescent signals, allowing for their visualization and identification.

FISH remains a valuable tool for studying microbial communities, providing insights into their spatial distribution, interactions and functional roles in various ecosystems. It can be used in conjunction with other profiling techniques, such as DNA sequencing, to obtain a comprehensive understanding of microbial community structure and function.

2.4 Metagenomic Sequencing

For microbial community profiling, metagenomic sequencing (as shown in Fig. 15.1) provides a comprehensive and unbiased view of the genetic content within a microbial community. Unlike targeted gene sequencing approaches, such as 16S rRNA gene amplicon sequencing, metagenomic sequencing enables the analysis of all the DNA present in a sample, including the genomes of both the target microorganisms and their associated microbiota (Gnaneswar Gude, 2015; Mahato et al., 2017). For instance, the functional potential can only be linked by using high-quality metagenome-assembled genomes (MAG) for long and short-read sequencing in microbial species during recycling, water pollution and resource recovery in the wastewater industry. MAG can exponentially increase the rate of analysis by covering the community-based relative abundance to describe different relationships between lineages and their functional groups (Singleton et al., 2021). The process of metagenomic sequencing for microbial community profiling involves the following steps:

(a) DNA extraction: High-quality DNA is extracted from the microbial community sample using appropriate extraction methods. The extraction protocol should efficiently lyse microbial cells while preserving the integrity of the DNA.
(b) Library preparation: The extracted DNA is fragmented into smaller pieces and converted into sequencing libraries. This step involves end repair, adapter ligation and library amplification. Special attention is given to maintaining a representative and unbiased representation of the microbial genomes during library preparation.
(c) Sequencing: The prepared library is then subjected to high-throughput sequencing using next-generation sequencing (NGS) platforms. Commonly used NGS platforms include Illumina sequencing, which generates short reads, and PacBio or Nanopore sequencing, which provides longer reads. The choice of sequencing platform depends on factors such as the desired sequencing depth, genome assembly requirements and budget constraints.
(d) Quality control and data preprocessing: After sequencing, raw sequencing data are subjected to quality control measures, including adapter trimming and removal of low-quality reads. Preprocessing steps also include removing host DNA contamination and filtering out artifacts or non-microbial sequences.
(e) Metagenome Assembly and binning: The obtained long reads are typically assembled into longer contiguous sequences called contigs. Assembly aims to reconstruct the microbial genomes present in the sample. Binning, which involves grouping contigs based on their genomic characteristics, is performed to assign contigs to individual microbial taxa or species.
(f) Taxonomic and functional annotation: The assembled contigs or binned genomes are then subjected to taxonomic classification using reference databases, such as NCBI's non-redundant (NR) database, or specialized databases like the Integrated Microbial Genomes (IMG) database. Functional annotation involves predicting and assigning putative functions to genes and identifying functional pathways within the microbial community.
(g) Data analysis: The annotated metagenomic data is analysed using various bioinformatics tools and pipelines. This includes calculating microbial diversity indices, identifying differentially abundant taxa or genes, and assessing functional potentials and metabolic pathways within the microbial community.

Recently, a new approach of hybrid systems has been introduced where they use an iterative haplotype-resolved hierarchical clustering-based hybrid system (HCBNA) due to the high- complexity of error-prone long reads for the construction of genome. Using the HCBNA approach, researchers can nearly reconstruct the entire genome from complex ecosystems providing new insights into the complete metagenomics approach (Liu et al., 2021).

2.5 Single-Cell Genome Sequencing

This technology is cutting-edge and used in microbial community profiling to study the genomes of individual microbial cells within a complex community (as shown in Fig. 15.1). Unlike metagenomic sequencing, which provides a collective view of the genetic material present in a sample, single-cell genome sequencing enables the analysis of individual microbial genomes, allowing for a deeper understanding of microbial diversity, functional potential and ecological interactions at the single-cell level. However, recent advancement has presented a method to obtain viable bacteria-targeted single-cell genome sequencing called PMA-SAG-gel to identify whole genome sequences of surviving uncultured bacteria from microbial communities. According to the PMA-SAG gel approach, single amplified genomes (SAGs) existing in dead bacteria and selective sequencing of viable bacteria in the model samples (*Escherichia coli* and *Bacillus subtilis*) can recover the profile of the entire population under specific conditions (Hosokawa et al., 2022). The process of single-cell genome sequencing commonly starts with single-cell isolation, where individual microbial cells are physically isolated using techniques such as fluorescence-activated cell sorting (FACS), microfluidics or manual micromanipulation. This step ensures that each cell's genomic content can be analysed independently. Followed up by the whole genome amplification (WGA) of isolated single cells to amplify their genomic DNA. WGA methods, such as multiple displacement amplification (MDA) or multiple annealing and looping-based amplification cycles (MALBAC), are used to amplify the limited amount of DNA from a single cell. The next step is library preparation, in which the amplified DNA from each single cell is then prepared into sequencing libraries. This step involves fragmentation of the DNA, adapter ligation and PCR amplification to generate libraries suitable for sequencing. Then, the prepared single-cell libraries are subjected to high-throughput sequencing using NGS platforms, such as Illumina or PacBio. The choice of sequencing platform depends on factors such as desired sequencing depth, genome coverage and the need for long-read sequencing. Further, the obtained single-cell sequencing data undergoes bioinformatics analysis to assemble and annotate individual genomes. Since the amplified DNA from a single cell can contain artifacts and chimeric sequences, specialized bioinformatics tools and pipelines are used to address these challenges and reconstruct accurate genomes. In the end, the assembled single-cell genomes are compared to reference databases to assign taxonomic identities to each genome and predict their functional potential. This can involve using tools such as BLAST, HMMER, or specialized databases for taxonomic and functional annotation.

2.6 Single-Cell RNA Sequencing (scRNA-seq)

scRNA-seq (as shown in Fig. 15.1) is an emerging technique that holds great promise for microbial community profiling. While scRNA-seq is widely used in studying eukaryotic cells, its application to microbial communities is more recent but rapidly advancing. It enables the analysis of gene expression profiles at the single-cell level, providing valuable insights into the functional activities, metabolic states and cellular heterogeneity within microbial communities (Blattman et al., 2020).

The process of sequencing begins with single-cell isolation, where individual microbial cells are isolated using techniques such as FACS, microfluidics or manual picking. The goal is to capture a diverse range of cells representing different species or strains within the microbial community. Further, the isolated single cells are lysed, and RNA is extracted to capture the transcriptome of each individual cell. Special care is taken to minimize RNA degradation and maintain the integrity of the RNA molecules. The extracted RNA undergoes reverse transcription to convert the RNA into complementary DNA (cDNA). Amplification methods, such as PCR or in vitro transcription, are then employed to amplify the cDNA molecules for downstream analysis. The amplified cDNA is processed into sequencing libraries. This step involves fragmentation of the cDNA, adapter ligation and library amplification. Specialized protocols are often used to capture the full-length transcripts or specific RNA regions of interest. The prepared scRNA-seq libraries are subjected to high-throughput sequencing using NGS platforms. The choice of sequencing platform depends on factors such as desired sequencing depth, read length and the need for capturing splicing events or full-length transcripts. Moreover, the obtained scRNA-seq data undergoes bioinformatics analysis to characterize the gene expression profiles of individual cells. This involves aligning the sequencing reads to a reference genome or transcriptome, quantifying gene expression levels and performing downstream analysis such as clustering, differential gene expression and functional enrichment analysis. By analysing the gene expression profiles, individual microbial cells can be classified into different cell types or functional groups based on their transcriptional signatures. This allows for the identification of cell types, gene regulatory networks, metabolic pathways and functional activities within the microbial community (Haque et al., 2017; Hwang et al., 2018).

scRNA-seq holds immense potential for microbial community profiling, enabling the study of gene expression dynamics, cell functions and community-level interactions at a single-cell resolution. Continued advancements in scRNA-seq technologies and bioinformatics analysis methods will further enhance our understanding of microbial communities and their ecological roles.

2.7 Advantages and Limitations of Each Technique

DNA sequencing for microbial community profiling has numerous applications across various fields. Environmental microbiology allows the study of microbial diversity and ecosystem functioning in natural habitats, helping understand ecological processes and responses to environmental changes. In human microbiome research, it enables the characterization of the microbial communities residing in different body sites and their associations with health and disease. In agriculture and soil science, DNA sequencing helps assess soil microbial diversity, nutrient cycling and plant–microbe interactions. Furthermore, microbial community profiling using DNA sequencing is valuable in industrial processes, such as bioremediation, wastewater treatment and the optimization of biotechnological applications. Looking ahead, the future of DNA sequencing for microbial community profiling holds exciting prospects. Emerging technologies, such as long-read sequencing and single-cell sequencing, offer increased resolution and the ability to analyse individual microbial genomes within a community. Integration of multi-omics approaches, combining DNA sequencing with other 'omics' technologies like meta transcript, however, requires high technical based knowledge for data processing and interpretation (Mahato et al., 2017).

PCR amplification for microbial community profiling requires careful consideration of several factors to ensure accurate and representative results. One of the important factors is knowledge of primers that should be carefully designed to target the desired DNA region while minimizing non-specific amplification. They should be specific to the target taxa of interest and have minimal cross-reactivity with non-target sequences. In addition to that, optimal reaction conditions, including the annealing temperature and the concentration of primers, DNA template and DNA polymerase, should be determined through optimization experiments to ensure efficient and specific amplification. Multiple PCR replicates are typically performed to account for technical variability and ensure reproducibility of results. The length of the amplicon should be considered to ensure compatibility with downstream sequencing platforms and analysis pipelines.

FISH provides several advantages, such as specific detection and visualization of microbial groups or species of interest within a complex microbial community. The probes can be designed to target specific taxonomic groups or functional genes, providing detailed information about their spatial distribution. It has high-resolution imaging of microbial cells, providing insights into their morphology, spatial organization and interactions within the community. It also provides relatively quick results compared to DNA sequencing-based approaches, as it does not require DNA extraction, PCR amplification or sequencing. Additionally, it can be used for quantitative analysis by combining it with image analysis software to estimate the abundance and distribution of specific microbial groups within the community. However, there are a few limitations that must be considered while running FISH, such as the design of specific probes is crucial for target specificity. Cross-reactivity or non-specific binding of probes to non-target organisms can lead to false-positive signals. FISH is limited to the detection of microbial groups for which specific probes have been developed, and the coverage of probes may not encompass the entire microbial diversity present in a community. FISH may have limited penetration into dense or complex sample matrices, making it challenging to visualize microbial cells deep within biofilms or complex environmental samples (Huber et al., 2018).

Metagenomic sequencing allows for the analysis of the entire genetic content of the microbial community, providing insights into both the taxonomic composition and functional potential of the community. It also provides a high-resolution view of microbial community structure, enabling the identification of individual species or strains within the community that enables the prediction and characterization of functional genes and pathways within the microbial community, shedding light on the metabolic capabilities and potential interactions among community members. Metagenomic sequencing can also reveal previously unknown or unculturable microorganisms, expanding our understanding of microbial diversity from different sources. However, computational requirements in metagenomic data analysis require a substantial amount of storage, memory and processing power due to the large size of sequencing datasets. For instance, the complexity of microbial communities and the presence of low-abundance or rare taxa can pose challenges in accurately reconstructing their genomes and identifying their functions. Another challenge in this type of study includes fragmented genomes due to the presence of closely related strains or genetic variations within the microbial community, along with the cost, as this type of sequencing method is relatively expensive compared to targeted gene sequencing approaches, especially for large-scale studies or samples with high microbial diversity (Jansson & Hofmockel, 2018; Sharma & Singh, 2022).

Single-cell genome sequencing is a powerful technique that provides analysis of cellular heterogeneity within complex organisms along with the detection of a unique set of cells that has low abundance in the population. In addition to this, SCS enables the identification and characterization of genomic variations at the single-cell level. Overall, it can unravel the unknown genetic event. However, there is a certain limitation, as the technique can be technically challenging and requires specialized equipment and expertise. Moreover, it can be expensive and limited genome coverage due to the low amount of DNA obtained in a single cell. On the other hand, scRNA-seq allows for the analysis of gene expression profiles at the individual cell level, providing insights into cellular heterogeneity and functional diversity within microbial communities. It can also capture the gene expression profiles of rare or unculturable microorganisms that may play important roles in the community but are underrepresented in bulk sequencing approaches. By examining gene expression patterns, scRNA-seq can provide information about the active metabolic pathways, functional activities and responses to environmental conditions within individual cells. However, the amplification steps in scRNA-seq can introduce technical noise and biases, affecting the accuracy and reproducibility of gene expression quantification. As well as, analysing scRNA-seq data requires sophisticated bioinformatics tools and computational resources to handle the large volume of data generated and perform robust data processing, normalization and analysis. The transcriptome coverage in scRNA-seq may not be as comprehensive as bulk RNA-seq due to potential dropout events, where lowly expressed genes are missed or underestimated. Interpreting scRNA-seq data from microbial communities can be challenging due to the lack of well-annotated reference genomes or transcriptomes for all the organisms present, making it difficult to assign functional annotations accurately (Hwang et al., 2018).

3 Microbial Community in Composting

3.1 Role of the Microbial Community in Composting

The microbial community plays a fundamental role in the composting process, which is an important aspect of solid waste management. Composting is a controlled decomposition process that transforms organic waste materials into a nutrient-rich soil amendment called compost (Hamer, 2003). It is responsible for the breakdown and transformation of organic matter during composting and mostly consists of bacteria, fungi, actinomycetes and other microorganisms that initiate the decomposition of organic materials present in the compost pile. These microorganisms secrete enzymes that break down complex organic compounds, such as carbohydrates, proteins and lipids, into simpler substances. Bacteria are primarily responsible for the initial stages of decomposition, followed by fungi and actinomycetes. During composting, microbial activity generates heat as a by-product of organic matter decomposition. Thermophilic microorganisms, particularly heat-tolerant bacteria, and fungi, thrive in the high-temperature conditions of the compost pile. The heat produced helps create and sustain the thermophilic phase of composting, typically ranging from 40–70° Celsius (104–158° Fahrenheit). The high temperature is critical for the elimination of pathogens, weed seeds and potential plant pests present in the organic waste. The microbial community in composting plays a crucial role in nutrient cycling. As organic matter breaks down, microorganisms release nutrients such as nitrogen (N), phosphorus (P) and potassium (K) from the organic waste into forms that plants can readily absorb. Microbes mineralize organic compounds, converting them into inorganic forms that are bioavailable to plants. They also participate in the fixation of atmospheric nitrogen, making it accessible for plant growth. Along with the degradation of organic matter, certain microbial species actively break down complex polymers, such as lignin and cellulose, present in woody materials. These microorganisms, primarily fungi, possess enzymes capable of breaking down these resistant structures. Their activity leads to the breakdown of lignocellulosic materials, making them more accessible for further decomposition and nutrient release. The microbial community in composting also contributes to the reduction of odour during the process. As organic matter decomposes, unpleasant odourous compounds, such as volatile fatty acids and amines, can be generated. However, the metabolic activities of specific microbial species help in the transformation of these odourous compounds into more stable and less odourous substances, thereby reducing the overall odour of the compost pile. The microbial community is involved in the maturation and stabilization of compost. As the composting process progresses, the microbial activity decreases and the pile reaches a state of stability. The microbial community transitions from thermophilic to mesophilic microorganisms, and their activity shifts towards the final breakdown of remaining organic matter and the humification of compost. This process results in the formation of humus-like substances, which enhance the soil structure, water-holding capacity and nutrient retention capacity of the compost (Takaku et al., 2006; Vivas et al., 2009; Bhatia et al., 2015).

In order to achieve high-quality compost, understanding and managing the microbial community is crucial for optimizing the composting process. Factors such as moisture, oxygen levels, temperature, carbon-to-nitrogen ratio and proper turning of the compost pile influence the composition and dynamics of the microbial community. By creating favourable conditions for microbial growth and activity, composting can efficiently convert organic waste into a valuable soil amendment while minimizing the environmental impacts associated with waste disposal (Raviv, 2005).

3.2 Microbial Community Profiling in Composting

Profiling of microbial community involves the entire microbiome that plays an important role in understanding the frequently changing dynamics and functional potential of the population during composting, which is an essential process in solid waste management. It also allows the identification and characterization of the microbial taxa involved in the composting process. Most frequent process to characterize the microbial community is through DNA sequencing techniques, such as amplicon sequencing targeting the 16S rRNA gene for bacteria and archaea or ITS (Internal Transcribed Spacer) region for fungi. Using these techniques, researchers can assess the composition and relative abundance of different microbial groups throughout the composting stages and facilitate the classification of the dominant taxa contributing to organic matter decomposition (Hamer, 2003).

Composting is a dynamic process characterized by microbial successional changes, and profiling assists in monitoring and understanding the shifts in microbial populations over time, as mentioned before. It reveals how different microbial groups, such as initial colonizers, thermophiles, mesophiles and mature compost communities, succeed one another and contribute to the degradation of organic matter. This knowledge aids in optimizing composting conditions and predicting compost maturity. Deep understanding through metagenomics or meta-transcriptomic sequencing can provide insights into the functional potential of microbial populations by combining with functional annotation tools. Researchers are able to identify the presence of specific genes and pathways involved in organic matter degradation, nutrient cycling and odour reduction (Yu, 2014). This can help to comprehend the metabolic capabilities of microbial communities and optimize the composting conditions to enhance desired functional and biological activities (Aguilar-Paredes et al., 2023).

Monitoring the microbial community dynamics through profiling allows for the assessment of composting process efficiency and stability. Changes in microbial diversity, composition and functional potential can indicate the progress and quality of composting. For example, a decrease in pathogenic or potentially harmful microbial populations over time indicates successful pathogen reduction (Yu, 2014). By understanding the microbial groups and their functional roles, composting conditions, such as moisture, aeration and carbon-to-nitrogen ratio, can be adjusted to favour desired microbial activities and enhance organic matter degradation. Additionally, knowledge of the microbial community composition can guide the selection of bioaugmentation strategies, where specific microbial strains or consortia are added to enhance composting efficiency or address specific challenges, such as the elimination of pollutants from refinery solid waste that consists of total petroleum hydrocarbons (TPH) and polycyclic aromatic hydrocarbons (PAHs) (Zeneli et al., 2019). Certain microbial indicators, such as the presence of specific bacterial or fungal groups, can be used as indicators of compost stability, maturity and odour reduction. Profiling data, combined with physicochemical analyses, can provide a comprehensive evaluation of compost quality and ensure that it meets regulatory and industry standards for safe and effective use (Supaphol et al., 2011).

4 Microbial Community in Anaerobic Digestion

4.1 Role of the Microbial Community in Anaerobic Digestion

Anaerobic digestion is a key process in solid waste management for the conversion of organic waste into biogas and nutrient-rich digestate. The process relies on the metabolic activities of diverse groups of microorganisms to break down complex organic compounds (Yu, 2014; Matsakas et al., 2017). The initial step starts with anaerobic digestion includes hydrolysis, where complex organic matter, such as carbohydrates, proteins and lipids, are broken down into simpler compounds by hydrolytic microorganisms such as bacteria and archaea, produce extracellular enzymes that break down complex organic polymers into soluble monomers and small molecules, such as sugars, amino acids and fatty acids. This process makes the organic matter more accessible to subsequent microbial groups, which are divided into several stages acidogenesis, and the hydrolysis products are further metabolized by acidogenic microorganisms. Acidogens, predominantly bacteria, ferment the soluble monomers into volatile fatty acids (VFAs), alcohols and other organic acids. This step is crucial as it releases additional carbon sources and intermediate compounds that serve as substrates for the subsequent microbial groups in the anaerobic digestion process. The next step is acetogenesis which acts as a bridge between acidogenesis and methanogenesis, providing necessary precursors for methanogenic microorganisms. Acetogens, specialized bacteria, utilize the fermentation products produced during acidogenesis to convert them into acetate, hydrogen and carbon dioxide. Methanogenic microorganisms, predominantly archaea, carry out the final step of anaerobic digestion, where they convert acetate, hydrogen and carbon dioxide into methane (CH_4), the main component of biogas. Methanogens are strict anaerobes and are highly sensitive to environmental conditions. They are classified into acetoclastic methanogens, which use acetate as a substrate, and hydrogenotrophic methanogens, which utilize hydrogen and carbon dioxide. Anaerobic digestion also involves complex interactions between different microbial groups through syntrophic relationships. Syntrophic microorganisms work together in a mutually beneficial manner, where one group produces substrates for another group's growth through metabolic cooperation. For example, certain bacteria degrade complex organic compounds into hydrogen, carbon dioxide and volatile fatty acids, which are then utilized by hydrogenotrophic methanogens for methane production. The presence of a diverse and well-balanced microbial community ensures the efficient breakdown of organic waste and the production of high-quality biogas. A stable microbial community with a range of functional groups helps to maintain process performance, resist perturbations and recover from disturbances, ensuring a consistent and reliable anaerobic digestion process (Calli et al., 2006; Ventorino et al., 2018; de Jonge et al., 2020).

4.2 Dominant Microbial Species in Anaerobic Digestion

Anaerobic digestion is a complex process, as explained earlier, that involves diverse microbial communities that work together to convert organic waste into biogas and digestate. The dominant microbial species in anaerobic digestion can vary depending on the specific conditions, substrate composition and operational parameters. However, certain microbial groups and species are commonly found in anaerobic digesters. Methanogenic archaea are key players in anaerobic digestion, responsible for the production of methane (CH_4). Some dominant methanogenic species include *Methanobacterium* spp., *Methanosarcina* spp., *Methanococcus* spp., *Methanosaeta* spp., *Methanoculleus* spp. These methanogens utilize various substrates, such as acetate, hydrogen and carbon dioxide, to produce methane as a metabolic by-product (Wang et al., 2020). Consequently, acidogenic bacteria play a crucial role in the hydrolysis and fermentation stages of anaerobic digestion. They convert complex organic compounds into simpler molecules, such as volatile fatty acids (VFAs) and alcohols. Dominant acidogenic bacteria species may include *Clostridium* spp., *Bacteroides* spp., *Sytrophomonas* spp., *Syntrophobacter* spp. These bacteria contribute to the breakdown of complex organic matter and the production of intermediate products that are utilized by other microbial groups (Liu et al., 2016; Detman et al., 2021).

Acetogenic bacteria are responsible for the conversion of VFAs and alcohols into acetate, hydrogen and carbon dioxide. Some dominant acetogenic bacteria species include *Acetobacterium* spp., *Clostridium* spp., *Syntrophaceticus* spp. Acetogens play a crucial role in the acetogenesis stage, providing substrates for methanogens and facilitating the production of methane. During the acetogenic step, the products are acidogenesis are oxidized to substrate for methanogenesis; therefore, acetogenesis and methanogenesis are highly correlated due to syntrophic associations (Detman et al., 2021). Hydrogenotrophic bacteria utilize hydrogen and carbon dioxide as substrates for methane production. These bacteria form a syntrophic relationship with methanogens, providing them with hydrogen. Some dominant hydrogenotrophic bacteria species include *Syntrophomonas* spp., *Syntrophobacter* spp., *Desulfovibrio* spp. These bacteria contribute to the final step of methane production and help maintain a balanced microbial community in anaerobic digesters (Wang et al., 2015). However, it is important to note that the microbial composition in anaerobic digestion can be influenced by various factors, such as the feedstock used, digester design, temperature, pH and operating conditions. Additionally, microbial communities can vary between different stages of anaerobic digestion, such as hydrolysis, acidogenesis, acetogenesis and methanogenesis. The presence and relative abundance of specific microbial species depend on these factors, and optimization strategies can be employed to promote the growth of desired microbial groups and enhance process efficiency.

4.3 Microbial Community Profiling in Anaerobic Digestion

Microbial community profiling in anaerobic digestion is a valuable approach for studying and understanding the composition, dynamics and functional potential of microbial populations involved in the process (Table 15.1 provides the summary of the data). It provides insights into the microbial diversity, abundance and interactions within anaerobic digesters. The process of anaerobic digestion involves the degradation of complex organic matter by a consortium of microorganisms in the absence of oxygen. These microorganisms work in a coordinated manner to convert organic substrates into biogas, primarily composed of methane and carbon dioxide. Understanding the microbial community structure and function is crucial for optimizing anaerobic digestion performance, enhancing biogas production and identifying potential limitations or failures in the process. Microbial community profiling in anaerobic digestion typically involves the use of high-throughput DNA sequencing techniques, such as amplicon sequencing or metagenomic sequencing. These methods allow for the analysis of specific marker genes (e.g. 16S rRNA gene for bacteria or archaea) or the entire genetic content of the microbial community, respectively. By analysing the DNA sequences obtained from anaerobic digester samples, researchers can identify and quantify different microbial taxa present in the community. This information helps in determining the diversity and richness of the microbial community and tracking changes in community composition over time or under different operating conditions (Supaphol et al., 2011). Through metagenomic sequencing, researchers can predict the presence of functional genes involved in specific metabolic pathways, such as hydrolysis, acidogenesis, acetogenesis and

Table 15.1 Microbial community in aerobic and anaerobic digestion

Microbial community	Aerobic digestion	Anaerobic digestion
Dominant organisms	Aerobic bacteria (e.g. *Bacillus*, *Pseudomonas*, *Actinobacteria*)	Methanogenic Archaea (e.g. *Methanobacterium*, *Methanosarcina*, *Methanosaeta*)
	Fungi (e.g. *Aspergillus*, *Penicillium*)	Acidogenic bacteria (e.g. Clostridium, Bacteroides)
	Actinomycetes (e.g. *Streptomyces*)	Acetogenic bacteria (e.g. *Syntrophomonas*, *Acetobacterium*)
		Hydrolytic bacteria (e.g. *Clostridium*, *Bacteroides*)
		Homoacetogenic bacteria (e.g. *Acetobacterium*, *Acetomicrobium*)
Oxygen requirements	Requires oxygen for microbial activity	No oxygen is required; oxygen inhibits the process
Decomposition rate	Relatively fast due to efficient metabolism	Slower due to complex microbial interactions and stages
Temperature range	Mesophilic (20–40 °C)	Mesophilic (30–40 °C) and Thermophilic (50–60 °C)
Main products	Carbon dioxide, water, heat and stabilized organic matter	Biogas (mainly methane and carbon dioxide) and stabilized organic matter
Process characteristics	Highly efficient in reducing organic matter	Production of biogas can provide an energy source
	Odour production due to aerobic decomposition	Reduced odour due to lack of oxygen
	Pathogen reduction through high temperatures	Pathogen reduction through anaerobic conditions
Nutrient availability	Requires continuous supply of oxygen and nutrients	It can sustain with periodic feedings and nutrient recycling
pH level	Neutral to slightly alkaline (pH 6.5–8.5)	Neutral to slightly acidic (pH 6.5–7.5)
Examples of applications	Composting of organic waste	Biogas production from agricultural residues, food waste and wastewater sludge
	Treatment of organic-rich industrial wastewater	Anaerobic digestion of municipal solid waste
	Biological treatment of organic pollutants in soil	Anaerobic digestion of sewage sludge

methanogenesis. This knowledge aids in understanding the metabolic capabilities of different microbial groups and their contributions to overall process performance. Moreover, microbial community profiling allows for the identification of key microbial players that drive anaerobic digestion processes. Certain microorganisms have specialized functions and play essential roles in the degradation of specific substrates or the syntrophic interactions required for the conversion of complex organic matter into biogas. Understanding these key microbial players can help in designing targeted strategies for process optimization or troubleshooting. In addition to this, the profiling also provides insights into the functional potential of the community, such as advanced meta-omics techniques that include the identification of bio-markers to determine the activity and interactions among organisms during anaerobic digestion (Hashemi et al., 2021). For example, an a study, the most common archaea *Methanosarsina* sp. follows the acetoclastic pathway in the process of co-digestion of municipal solid waste and sewage sludge to enhance the production of biogas (Ghosh et al., 2019). In summary, microbial community profiling in anaerobic digestion offers valuable insights into the ecological interactions, metabolic potentials and functional roles of different microorganisms, enabling better process control and optimization for efficient and sustainable biogas production.

4.4 Improving Anaerobic Digestion Through Microbial Community Profiling

A deeper understanding of the metabolic pathways and interactions among different microbial groups enables the identification of key microbial players and their roles in organic matter degradation. This knowledge can transform process design and management strategies, ultimately leading to more sustainable and effective anaerobic digestion in solid waste management. The anaerobic process is a complex reduction process and involves multiple biochemical reactions to convert biomass into energy. Improving the conversions of complex organic substances into simple stabilized compounds is possible only by the group of microorganisms that follow various pathways to adapt to unfavourable conditions in the anaerobic digester, such as competition between sulphate-reducing bacteria and methane-forming bacteria for the same substrate. A group of researchers has found that *Methanosarcina* as a key microorganism that utilizes both acetoclastic and hydrogenotrophic pathways for methane production (Ziemiński and Frąc, 2012). It is important to analyse the composition, dynamics and functional potential of microbial populations and operational parameters can be optimized to create a favourable environment for desired microbial groups, which can be accomplished by DNA-based methods followed by cloning and fingerprinting. This optimization can lead to enhanced microbial activity, increased biogas production and improved process efficiency. Additionally, the identification of functional redundancy and potential bottlenecks within the microbial community allows for targeted interventions such as bioaugmentation or optimization of substrate composition. Also, early detection of microbial disturbances allows for timely corrective measures to maintain process stability (Yu, 2014).

5 Microbial Community in Aerobic Digestion

5.1 Role of the Microbial Community in Aerobic Digestion

The community of microbes plays a crucial role in aerobic digestion, which is a process used in solid waste management to break down organic waste in the presence of oxygen. Aerobic digestion relies on the metabolic activities of diverse microbial populations to decompose organic matter, such as bacteria and fungi utilize oxygen to break down complex organic compounds into simpler forms, including carbon dioxide, water and microbial biomass. These microorganisms produce extracellular enzymes that facilitate the degradation of organic materials, making them more accessible for further breakdown and nutrient release (Raviv, 2005). The microbial community in aerobic digestion plays a crucial role in nutrient recycling. As organic matter is decomposed, essential nutrients, such as nitrogen, phosphorus and sulphur, are released into the environment in forms that can be taken up by plants and other organisms. Microbes contribute to nutrient mineralization, transforming organic forms of nutrients into inorganic forms that are readily available for uptake by plants or for reuse within the microbial community. Aerobic digestion helps stabilize organic waste by reducing its volume and transforming it into a more stable form and breaking down complex organic compounds and transforming them into simpler, more stable forms, thereby reducing the organic waste's potential to emit odours and generate leachate. The microbial community's activities result in the conversion of labile organic matter into humus-like substances that are resistant to further decomposition. Microbes in aerobic digestion also contribute to biofiltration and odour control. As air is circulated through the aerobic digestion system, the microbial community can act as a biofilter, utilizing organic compounds and volatile substances present in the waste as a carbon source. This microbial biofiltration helps remove odourous compounds and pollutants, improving the air quality in and around the waste management facility. Monitoring the microbial community in aerobic digestion provides valuable information for process monitoring and optimization. Changes in microbial community composition, diversity or activity can serve as indicators of process health and performance. By assessing the microbial community through techniques like DNA sequencing or microbial biomarker analysis, operators can identify potential issues or imbalances and adjust operational parameters to optimize the aerobic digestion process (Yu, 2014). Therefore, understanding the roles of the microbial community in aerobic digestion allows for better management and optimization of solid waste treatment. By promoting a diverse and active microbial community, aerobic digestion processes can be more efficient, environmentally friendly, and economically viable in breaking down organic waste while minimizing potential negative impacts.

5.2 Dominant Microbial Species in Aerobic Digestion

In aerobic digestion during solid waste management, the dominant microbial species can vary depending on the specific conditions, waste composition and operational parameters. However, certain microbial groups and species are commonly found in aerobic digestion systems. Numerous bacterial species play significant roles in aerobic digestion, contributing to organic matter decomposition and nutrient cycling. The most dominant bacterial species include *Pseudomonas* spp., *Bacillus* spp., *Actinobacteria* spp., *Proteobacteria* spp., *Firmicutes* spp. These bacteria are known for their ability to degrade a wide range of organic compounds and produce extracellular enzymes that facilitate the breakdown of complex organic matter. Fungal populations also play a vital role in aerobic digestion, especially in the decomposition of lignocellulosic materials. Dominant fungal species commonly found in aerobic digestion systems include *Aspergillus* spp., *Penicillium* spp., *Trichoderma* spp., *Phanerochaete* spp., *Chaetomium* spp. These fungi are adept at breaking down lignin, cellulose and hemicellulose, contributing to the degradation of plant-based organic materials. Actinomycetes are filamentous bacteria that exhibit characteristics of both bacteria and fungi. They are important contributors to organic matter decomposition and are often abundant in aerobic digestion systems. Dominant actinomycete species include *Streptomyces* spp., *Micromonospora* spp., *Nocardia* spp., *Actinomyces* spp. *Actinomycetes* are known for their ability to degrade complex organic compounds and produce enzymes that aid in the breakdown of recalcitrant substances (Bhatia et al., 2015; Yang & Song, 2019).

5.3 Microbial Community Profiling in Aerobic Digestion

During aerobic digestion, it is important to analyse the composition, diversity and functional potential of microbial populations present in the system. This profiling technique allows the identification and quantification of different microbial taxa present in the aerobic digestion system. The knowledge of profiling provides a comprehensive view of the microbial diversity, including bacteria, fungi and other microorganisms, by characterizing the microbial community composition along with the ecological roles and potential contributions of different microbial groups. By analysing the genetic material, such as DNA or RNA, microbial functional genes and metabolic pathways involved in organic matter decomposition, nutrient cycling and other critical processes can be identified, allowing for the detection of shifts or imbalances in the community structure. Deviations from a stable microbial community can serve as early warning signs for process upsets, nutrient deficiencies, or inhibitory conditions. Monitoring the microbial community through profiling helps operators take timely corrective measures to maintain process stability. By understanding the composition and functional potential of the microbial community, process conditions such as temperature, pH, oxygen levels and substrate composition can be adjusted to promote the growth and activity of beneficial microbial groups. This optimization strategy enhances organic waste degradation, nutrient cycling and overall process efficiency. Therefore, profiling in aerobic digestion is a powerful tool for understanding the microbial ecology and functional potential of the microbial community.

5.4 Improving Aerobic Digestion Through Microbial Community Profiling

Understanding community profiling provides an introduction of unique methods to improve the digestion of sludge during solid waste treatment such as a two-stage sewage digestion process consisting of a thermophilic aerobic digester combined with mesophilic anaerobic digestion for the efficient sludge reduction and methane production has been efficiently used by researchers (Jang et al., 2013). Such changes can assist in adjusting factors such as temperature, pH, oxygen levels and substrate composition based on the microbial community's requirements, and process conditions can be optimized to enhance the activity and growth of beneficial microbial groups. This optimization strategy improves organic matter degradation, nutrient cycling and overall process efficiency. Profiling also helps in identifying key microbial species or groups that are crucial for effective aerobic digestion by monitoring the early detection of imbalances, inhibitory conditions or potential issues in aerobic digestion. Similar to the anaerobic process, changes in the microbial community composition in aerobic digestion can serve as indicators of process stability or disturbances. By identifying the specific microbial groups affected, appropriate corrective measures can be taken to restore process stability and prevent performance decline. It also serves as a valuable tool for real-time processes. By regularly analysing the microbial community, operators can assess the effectiveness of process interventions, evaluate the impact of operational changes, and ensure the desired microbial dynamics are maintained. This proactive approach allows for timely adjustments and interventions to maintain process efficiency and stability (Yu, 2014).

6 Applications of Microbial Community Profiling in Solid Waste Treatment

6.1 Improving Waste Treatment Processes

There are several ways to improve the solid waste treatment process, aiming for more effective and sustainable waste management, which is important to adapt in developing countries since the population is directly proportional to waste generation (Table 15.2 provides a summary of data). The foremost activity needed is implementing source segregation at the point of waste generation is a crucial step. Encouraging individuals and businesses to separate their waste into different categories, such as recyclables, organic waste and non-recyclables, allows for more efficient and targeted treatment methods. Promoting recycling and recovery of materials from the waste stream helps reduce the amount of waste sent to landfill. Implementing robust recycling programs and supporting the development of markets for recycled materials encourages the reuse of valuable resources and minimizes environmental impacts. As mentioned, composting is an effective method for treating organic waste. Encouraging the composting of food scraps, yard waste and other organic materials helps divert waste from landfills and produces nutrient-rich compost that can be used as a soil amendment in landscaping and agriculture. Anaerobic digestion is a process that converts organic waste into biogas and digestate. This process not only produces renewable energy but also reduces the volume of waste and eliminates harmful methane emissions from landfills. Investing in anaerobic digestion infrastructure and promoting its implementation can significantly improve solid waste treatment. Waste-to-energy technologies, such as incineration and gasification, can be utilized to generate electricity or heat from non-recyclable waste. These methods help reduce the volume of waste, minimize the need for landfill space, and provide a renewable energy source. Implementing advanced sorting technologies, such as optical sorting and automated systems, can improve the efficiency and accuracy of waste sorting and separation. Additionally, exploring and adopting innovative treatment technologies, such as plasma gasification or pyrolysis, can enhance the recovery of energy and resources from waste. Increasing public education and awareness about proper waste management practices, including recycling, composting and waste reduction, is vital. Engaging communities through educational campaigns and providing clear instructions on waste sorting and disposal can lead to higher participation rates and more effective waste treatment. Governments can play a crucial role in improving solid waste treatment through the implementation of effective regulations and policies. Continual research and development in waste treatment technologies, waste characterization and waste management strategies are essential. Investing in research and innovation can lead to the discovery of more efficient, cost-effective and environmentally friendly approaches to solid waste treatment (Hamer, 2003; Dhokhikah & Trihadiningrum, 2012).

6.2 Reducing the Waste Volume and Pathogen Levels

Reducing waste volume and pathogens levels are crucial steps in minimizing the environmental impacts and protecting public health. One important approach is to implement waste minimization and source reduction strategies, which focus on reducing the overall amount of waste generated. This can be achieved through initiatives such as promoting reusable products, encouraging responsible packaging

Table 15.2 Improving microbial community profiling at the industrial level during solid waste treatment

Application	Description	Benefits
Process optimization	Understanding the dominant microbial species and their activities, operators can adjust environmental conditions to enhance waste decomposition and reduce treatment time	Increased waste degradation rates Improved waste stabilization Energy and cost savings Enhanced process control and reliability
Troubleshooting	Monitoring changes can assist in troubleshooting solid waste treatment systems. Shifts in the community composition might indicate changes in the process	Early detection of process problems Minimization of treatment disruptions Improved system reliability and stability
Assessing treatment efficacy	Assessing the microbial communities in treated and untreated waste samples can determine the extent of organic matter degradation and the efficiency of pathogen removal	Quantification of treatment performance Data-driven decision-making Identification of potential treatment improvements
Predicting process outputs	Understanding the microbial community can aid in predicting the products of solid waste treatment	Enhanced process planning and optimization Estimation of energy recovery potential Anticipating system outputs and resource allocation
Monitoring pathogen reduction	Monitoring changes in pathogen-specific microbial populations helps assess the reduction of harmful microorganisms during treatment	Validation of treatment safety and effectiveness Compliance with regulatory standards Public health assurance
Studying microbial interactions	Microbial community profiling allows researchers to study the interactions between different microbial species during waste treatment	Optimization of microbial consortia Development of tailored treatment strategies Identification of potential synergistic microbial interactions
Assessment of environmental impact	Analysing the microbial community in the final compost or treated waste helps evaluate the environmental impact of the waste treatment process	Compliance with environmental regulations Safe and sustainable waste reuse or disposal Minimization of environmental risks and liabilities
Developing tailored treatment	Different waste materials may require unique microbial consortia to achieve optimal decomposition and resource recovery	Customization of treatment processes Maximization of waste resource recovery

(continued)

Table 15.2 (continued)

Application	Description	Benefits
		Efficient utilization of waste materials
Bioaugmentation strategies	Introducing these specific microbes, known as bioaugmentation, can enhance waste degradation and improve treatment performance	Augmentation of desirable microbial activities Improved treatment efficiency and effectiveness Remediation of recalcitrant waste components
Long-term monitoring	Tracking microbial changes over time provides valuable data for process stability, identifying trends and adjusting operational parameters for consistent performance	Evaluation of treatment system performance over time Trend analysis and identification of long-term patterns Decision support for system maintenance and upgrades

practices and implementing efficient manufacturing processes. By generating less waste at the source, the volume of waste requiring treatment is reduced. Recycling and material recovery programs are also effective in reducing waste volume and pathogen levels. These programs aim to divert recyclable materials from the waste stream and facilitate their transformation into new products. By implementing comprehensive recycling systems, including effective collection and sorting methods, valuable materials can be recovered and reused, minimizing the need for their disposal and reducing waste volume. Furthermore, the proper management of organic waste through composting plays a crucial role in waste volume reduction and pathogen control. Composting involves the decomposition of organic materials, such as food scraps and yard trimmings, under controlled conditions. This process not only reduces waste volume but also helps to eliminate or suppress pathogenic organisms through the heat generated during composting. In addition to waste volume reduction, it is essential to address pathogen levels during solid waste management. Proper waste handling practices, such as ensuring adequate storage, transportation and disposal facilities, help minimize the risk of pathogen contamination. Implementing stringent hygiene protocols, including personal protective equipment for waste handlers, can further mitigate the spread of pathogens (Yu, 2014; Ayilara et al., 2020).

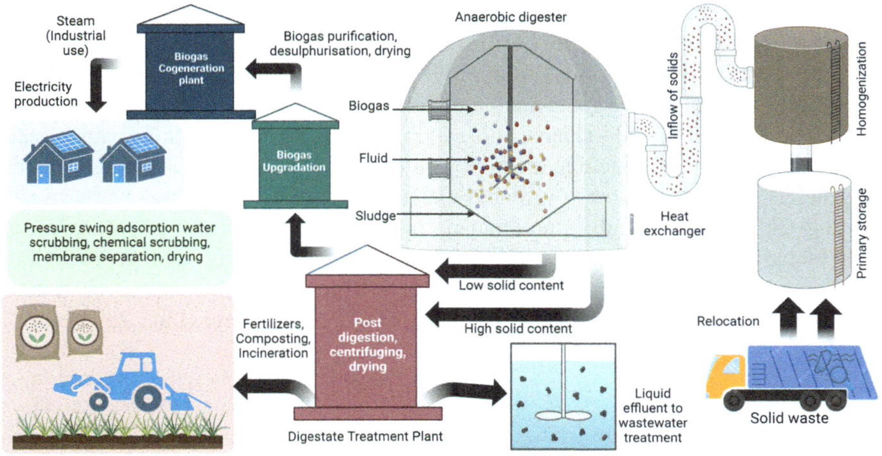

Fig. 15.2 The fate of solid waste: Tracing the journey from source to sink. The process starts with storage of solid waste followed up with homogenization and separated through an anaerobic digester which segregates the waste for utilization in agriculture purposes, wastewater treatment or industrial purposes. (Created with Biorender.com)

6.3 Producing Biogas and Other Value-Added Products or By-Products

Solid waste management processes can be designed to produce biogas (as shown in Fig. 15.2) and other value-added products or by-products, contributing to the sustainable utilization of waste resources. Anaerobic digestion is a commonly employed method in solid waste management to produce biogas. Organic waste, such as food waste, agricultural residues and sewage sludge, can be processed in anaerobic digesters (Shabir et al., 2023). During the digestion process, microorganisms break down the organic matter, producing biogas as a by-product. Biogas is primarily composed of methane and carbon dioxide, and it can be utilized as a renewable energy source for electricity generation, heating, or even as vehicle fuel. Anaerobic digestion also produces digestate, which is a nutrient-rich by-product. Digestate can be further treated and processed into compost or used as a soil amendment. Compost is a valuable product that can enhance soil fertility, improve moisture retention and promote plant growth. It serves as a sustainable alternative to chemical fertilizers, closing the nutrient loop and reducing the need for synthetic inputs (Tisserant et al., 2017). Advanced anaerobic digestion techniques, such as upgrading biogas to biomethane, offer additional opportunities for value creation. Biomethane is a purified form of biogas with a higher methane content. It can be injected into the natural gas grid, used as a vehicle fuel, or converted into heat and power. Biomethane production from solid waste management contributes to renewable energy generation and reduces greenhouse gas emissions (Hassard et al., 2015).

Biochar is a carbon-rich material produced through the pyrolysis or gasification of biomass, including certain types of solid waste. Biochar has high carbon stability and can be used as a soil amendment to improve soil fertility, water retention, and carbon sequestration. It provides a sustainable option for waste valorization and contributes to soil health and climate change mitigation (Gan et al., 2011). Solid waste management processes can include technologies for resource recovery, such as recovering metals, plastics and other valuable materials from the waste stream. Advanced sorting techniques, including automated sorting systems and optical sensors, can help separate recyclable materials for subsequent recycling and reprocessing. Resource recovery reduces the reliance on virgin materials, conserves natural resources and supports the circular economy (Tóthné Bogdányi et al., 2021).

7 Conclusion

7.1 Importance of Microbial Community Profiling in Solid Waste Treatment

In conclusion, microbial community profiling plays a crucial role in solid waste treatment processes. By analysing the composition, diversity and functional potential of microbial populations, it provides valuable insights into the dynamics and activities of microorganisms involved in waste degradation, nutrient cycling and overall process performance. Microbial community profiling helps in understanding the roles of different microbial groups, identifying key microbial species and assessing their functional capabilities. This knowledge enables process optimization, targeted interventions and troubleshooting to maintain process stability and improve waste treatment efficiency. Additionally, microbial community profiling facilitates the identification of potential issues or inhibitory factors affecting the microbial community, allowing for timely interventions and adjustments in operational parameters. It guides the development of tailored strategies such as bioaugmentation, which involves introducing specific microbial cultures to enhance desired metabolic activities. Furthermore, microbial community profiling serves as a powerful tool for process monitoring, control and decision-making. It enables real-time assessment of microbial dynamics, evaluates the effectiveness of interventions, and guides adjustments to operational parameters for optimal waste treatment. Overall, microbial community profiling in solid waste treatment provides critical insights that contribute to the development of sustainable and efficient waste management practices. By harnessing the power of microorganisms, we can enhance waste degradation, minimize environmental impacts, recover valuable resources and move towards a more sustainable and circular approach to waste management.

7.2 Future Research Directions and Prospects

Future research in solid waste management holds great potential for addressing current challenges and advancing sustainable practices. Continued research on waste-to-energy technologies, such as advanced incineration, gasification and pyrolysis, can lead to improved energy recovery and reduced environmental impacts. Developing more efficient and cost-effective methods for converting non-recyclable waste into energy will contribute to the transition to a low-carbon economy. Research focusing on circular economy principles can identify innovative strategies for waste reduction, resource recovery and closed-loop systems. This includes exploring novel recycling technologies, optimizing material flows and promoting product design for recyclability to minimize waste generation. Advancements in sorting and separation technologies can enhance the efficiency and accuracy of waste processing. Research efforts can focus on developing automated systems, artificial intelligence applications and sensor-based technologies for precise and high-throughput waste sorting, enabling improved recycling rates and resource recovery. Investigating alternative materials and packaging solutions with enhanced biodegradability and reduced environmental impacts can contribute to waste reduction. Research can explore biopolymers, bio-based materials and sustainable packaging designs to promote a shift towards more eco-friendly products. Further exploration of microbial and enzymatic treatment processes can optimize organic waste degradation and enhance biogas production. Research can focus on developing tailored microbial consortia, enzyme formulations and process conditions to maximize waste decomposition efficiency and reduce treatment times.

Research on effective communication strategies and behaviour change interventions can encourage responsible waste management practices at the individual and community levels. Understanding social and psychological factors that influence waste generation, recycling behaviour and acceptance of new waste management practices can inform targeted awareness campaigns and policy interventions. Conducting comprehensive life cycle assessments (LCA) can help evaluate the environmental impacts of different waste management strategies. Research can focus on refining LCA methodologies, incorporating new data on waste treatment technologies and assessing emerging environmental concerns, such as microplastic pollution and nanomaterials. Investigating the effectiveness of waste management policies, regulations and economic incentives can guide evidence-based decision-making. Research can examine policy frameworks, economic instruments and governance structures that promote sustainable waste management practices and encourage circular economy principles. Integration of smart technologies, Internet of Things (IoT) devices and data analytics can optimize waste management processes. Research can explore the application of digital tools for real-time monitoring, data-driven decision-making and predictive modelling to enhance operational efficiency and waste management performance. By pursuing these research directions, we can address the challenges of solid waste management, advance sustainable practices and contribute to a more circular and resource-efficient future. Collaborative efforts between researchers, industry stakeholders, policymakers and communities will be essential to drive innovation and implement effective solutions.

References

Aguilar-Paredes, A., Valdés, G., Araneda, N., Valdebenito, E., Hansen, F., & Nuti, M. (2023). Microbial community in the composting process and its positive impact on the soil biota in sustainable agriculture. *Agronomy, 13*(2). https://doi.org/10.3390/agronomy13020542

Ayilara, M. S., Olanrewaju, O. S., Babalola, O. O., & Odeyemi, O. (2020). Waste management through composting: Challenges and potentials. *Sustainability (Switzerland), 12*(11), 1–23. https://doi.org/10.3390/su12114456

Bhatia, A., Rajpal, A., Madan, S., & Kazmi, A. A. (2015). Techniques to analyze microbial diversity during composting-A mini review. *Indian Journal of Biotechnology, 14*(1), 19–25.

Blattman, S., Jiang, W., Oikonomou, P., & Tavazoie, S. (2020). Prokaryotic single-cell RNA sequencing by in situ combinatorial indexing. *Nature Microbiology, 176*(3), 139–148. https://doi.org/10.1053/j.gastro.2016.08.014.CagY

Bulgarelli, D., Garrido-Oter, R., Münch, P. C., Weiman, A., Dröge, J., Pan, Y., McHardy, A. C., & Schulze-Lefert, P. (2015). Structure and function of the bacterial root microbiota in wild and domesticated barley. *Cell Host and Microbe, 17*(3), 392–403. https://doi.org/10.1016/j.chom.2015.01.011

Calli, B., Durmaz, S., & Mertoglu, B. (2006). Identification of prevalent microbial communities in a municipal solid waste landfill. *Water Science and Technology, 53*(8), 139–147. https://doi.org/10.2166/wst.2006.244

Chander, A. M., Singh, N. K., & Venkateswaran, K. (2023). Microbial technologies in waste management, energy generation and climate change: Implications on earth and space. *Journal of the Indian Institute of Science, 103*(3), 833–838. https://doi.org/10.1007/s41745-023-00388-3

Detman, A., Bucha, M., Treu, L., Chojnacka, A., Pleśniak, Ł., Salamon, A., Łupikasza, E., Gromadka, R., Gawor, J., Gromadka, A., Drzewicki, W., Jakubiak, M., Janiga, M., Matyasik, I., Błaszczyk, M. K., Jędrysek, M. O., Campanaro, S., & Sikora, A. (2021). Evaluation of acidogenesis products' effect on biogas production performed with metagenomics and isotopic approaches. *Biotechnology for Biofuels, 14*(1), 1–25. https://doi.org/10.1186/s13068-021-01968-0

Dhokhikah, Y., & Trihadiningrum, Y. (2012). Solid waste management in Asian developing countries: Challenges and opportunities. *Journal of Applied Environmental and Biological Sciences, 2*, 329–335. https://www.textroad.com/pdf/JAEBS/J.%20Appl.%20Environ.%20Biol.%20Sci.,%202(7)329-335,%202012.pdf (Accessed on 28 July 2023)

Farooqi, Z. U. R., Kareem, A., Rafi, F., & Ali, S. (2021, January). Solid waste, treatment technologies, and environmental sustainability: Solid wastes and their sustainable management practices. In *Handbook of research on waste diversion and minimization technologies for the industrial sector* (pp. 35–57). IGI Global, Pennsylvania. https://doi.org/10.4018/978-1-7998-4921-6.ch003

Gan, Y., Liang, C., Hamel, C., Cutforth, H., & Wang, H. (2011). Strategies for reducing the carbon footprint of field crops for semiarid areas. A review. *Agronomy for Sustainable Development, 31*(4), 643–656. https://doi.org/10.1007/s13593-011-0011-7

Ghosh, P., Kumar, M., Kapoor, R., Kumar, S. S., Singh, L., Vijay, V., Vijay, V. K., Kumar, V., & Thakur, I. S. (2019). Enhanced biogas production from municipal solid waste via co-digestion with sewage sludge and metabolic pathway snalysis. *Bioresource Technology, 1*, 122275. https://doi.org/10.1016/j.biortech.2019.122275

Gnaneswar Gude, V. (2015). A new perspective on microbiome and resource management in wastewater systems. *Journal of Biotechnology & Biomaterials, 05*(02). https://doi.org/10.4172/2155-952x.1000184

Hamer, G. (2003). Solid waste treatment and disposal: Effects on public health and environmental safety. *Biotechnology Advances, 22*(1–2), 71–79. https://doi.org/10.1016/j.biotechadv.2003.08.007

Haque, A., Engel, J., Teichmann, S. A., & Lönnberg, T. (2017). A practical guide to single-cell RNA-sequencing for biomedical research and clinical applications. *Genome Medicine, 9*(1), 1–12. https://doi.org/10.1186/s13073-017-0467-4

Haruta, S., Kondo, M., Nakamura, K., Aiba, H., Ueno, S., Ishii, M., & Igarashi, Y. (2002). Microbial community changes during organic solid waste treatment analyzed by double gradient-denaturing gradient gel electrophoresis and fluorescence in situ hybridization. *Applied Microbiology and Biotechnology, 60*(1–2), 224–231. https://doi.org/10.1007/s00253-002-1074-9

Hashemi, S., Hashemi, S. E., Lien, K. M., & Lamb, J. J. (2021). Molecular microbial community analysis as an analysis tool for optimal biogas production. *Microorganisms, 9*(6), 1162. https://doi.org/10.3390/microorganisms9061162

Hassard, F., Biddle, J., Cartmell, E., Jefferson, B., Tyrrel, S., & Stephenson, T. (2015). Rotating biological contactors for wastewater treatment – A review. *Process Safety and Environmental Protection, 94*(C), 285–306. https://doi.org/10.1016/j.psep.2014.07.003

Herrmann, R. F., & Shann, J. F. (1997). Microbial community changes during the composting of municipal solid waste. *Microbial Ecology, 33*(1), 78–85. https://doi.org/10.1007/s002489900010

Hosokawa, M., Endoh, T., Kamata, K., Arikawa, K., Nishikawa, Y., Kogawa, M., Saeki, T., Yoda, T., & Takeyama, H. (2022). Strain-level profiling of viable microbial community by selective single-cell genome sequencing. *Scientific Reports, 12*(1), 1–12. https://doi.org/10.1038/s41598-022-08401-y

Huber, D., Voith von Voithenberg, L., & Kaigala, G. V. (2018). Fluorescence in situ hybridization (FISH): History, limitations and what to expect from micro-scale FISH? *Micro and Nano Engineering, 1*, 15–24. https://doi.org/10.1016/j.mne.2018.10.006

Hwang, B., Lee, J. H., & Bang, D. (2018). Single-cell RNA sequencing technologies and bioinformatics pipelines. *Experimental and Molecular Medicine, 50*(8), 1–14. https://doi.org/10.1038/s12276-018-0071-8

Jang, H. M,, Park, S. K., Ha, J. H., & Park, J. M. (2013). Microbial community structure in a thermophilic aerobic digester used as a sludge pretreatment process for the mesophilic anaerobic digestion and the enhancement of methane production. *Bioresource Technology, 145*, 80–89. https://doi.org/10.1016/j.biortech.2013.01.094

Jansson, J. K., & Hofmockel, K. S. (2018). The soil microbiome – From metagenomics to metaphenomics. *Current Opinion in Microbiology, 43*, 162–168. https://doi.org/10.1016/j.mib.2018.01.013

de Jonge, N., Davidsson, Å., la Cour Jansen, J., & Nielsen, J. L. (2020). Characterisation of microbial communities for improved management of anaerobic digestion of food waste. *Waste Management, 117*, 124–135. https://doi.org/10.1016/j.wasman.2020.07.047

Krishnan, R. Y., Manikandan, S., Subbaiya, R., Karmegam, N., Kim, W., & Govarthanan, M. (2023, October 2022). Recent approaches and advanced wastewater treatment technologies for mitigating emerging microplastics contamination—A critical review. *Science of the Total Environment, 858*, 159681. https://doi.org/10.1016/j.scitotenv.2022.159681

Kumar, V., Kumar, M., Sharma, S., & Prasad, R. (2017). *Probiotics and plant health* (pp. 1–600). https://doi.org/10.1007/978-981-10-3473-2

Lamb, E. G., Kennedy, N., & Siciliano, S. D. (2011). Effects of plant species richness and evenness on soil microbial community diversity and function. *Plant and Soil, 338*(1), 483–495. https://doi.org/10.1007/s11104-010-0560-6

Liao, H., Yu, K., Duan, Y., Ning, Z., Li, B., He, L., & Liu, C. (2019). Profiling microbial communities in a watershed undergoing intensive anthropogenic activities. *Science of the Total Environment, 647*, 1137–1147. https://doi.org/10.1016/j.scitotenv.2018.08.103

Liu, C., Li, H., Zhang, Y., Si, D., & Chen, Q. (2016). Evolution of microbial community along with increasing solid concentration during high-solids anaerobic digestion of sewage sludge. *Bioresource Technology, 216*, 87–94. https://doi.org/10.1016/j.biortech.2016.05.048

Liu, L., Wang, Y., Yang, Y., Wang, D., Cheng, S. H., Zheng, C., & Zhang, T. (2021). Charting the complexity of the activated sludge microbiome through a hybrid sequencing strategy. *Microbiome, 9*(1), 1–15. https://doi.org/10.1186/s40168-021-01155-1

Loman, N. J., & Pallen, M. J. (2015). Twenty years of bacterial genome sequencing. *Nature Reviews Microbiology, 13*(12), 787–794. https://doi.org/10.1038/nrmicro3565

Mahato, N. K., Gupta, V., Singh, P., Kumari, R., Verma, H., Tripathi, C., Rani, P., Sharma, A., Singhvi, N., Sood, U., Hira, P., Kohli, P., Nayyar, N., Puri, A., Bajaj, A., Kumar, R., Negi, V., Talwar, C., Khurana, H., et al. (2017). Microbial taxonomy in the era of OMICS: Application of DNA sequences, computational tools and techniques. *Antonie van Leeuwenhoek, International Journal of General and Molecular Microbiology, 110*(10), 1357–1371. https://doi.org/10.1007/s10482-017-0928-1

Matsakas, L., Gao, Q., Jansson, S., Rova, U., & Christakopoulos, P. (2017). Green conversion of municipal solid wastes into fuels and chemicals. *Electronic Journal of Biotechnology, 26*, 69–83. https://doi.org/10.1016/j.ejbt.2017.01.004

Prakash, A. A., Rajasekar, A., Sarankumar, R. K., AlSalhi, M. S., Devanesan, S., Aljaafreh, M. J., Govarthanan, M., & Sayed, S. R. M. (2021, January). Metagenomic analysis of microbial community and its role in bioelectrokinetic remediation of tannery contaminated soil. *Journal of Hazardous Materials, 412*, 125133. https://doi.org/10.1016/j.jhazmat.2021.125133

Rastogi, M., Nandal, M., & Khosla, B. (2020). Microbes as vital additives for solid waste composting. *Heliyon, 6*(2), e03343. https://doi.org/10.1016/j.heliyon.2020.e03343

Raviv, M. (2005, March). Production of high-quality composts for horticultural purposes: A mini review. *HortTechnology, 15*, 52–57.

dos Santos, V. H. J. M., Engelmann, P. d. M., Marconatto, L., Borge, L. G. d. A., Palhano, P. d. L., Augustin, A. H., Rodrigues, L. F., Ketzer, J. M. M., & Giongo, A. (2022, February). Exploratory analysis of the microbial community profile of the municipal solid waste leachate treatment system: A case study. *Waste Management, 141*, 125–135. https://doi.org/10.1016/j.wasman.2022.01.014

Shabir, I., Dash, K. K., Dar, A. H., Pandey, V. K., Fayaz, U., Srivastava, S., & R, N. (2023, January). Carbon footprints evaluation for sustainable food processing system development: A comprehensive review. *Future Foods, 7*, 100215. https://doi.org/10.1016/j.fufo.2023.100215

Shah, M. P. (2021). Removal of emerging contaminants through microbial processes. Springer Singapore. https://doi.org/10.1007/978-981-15-5901-3

Shah, V., Zakrzewski, M., Wibberg, D., Eikmeyer, F., Schlüter, A., & Madamwar, D. (2013). Taxonomic profiling and metagenome analysis of a microbial community from a habitat contaminated with industrial discharges. *Microbial Ecology, 66*(3), 533–550. https://doi.org/10.1007/s00248-013-0253-9

Sharma, P., & Singh, S. P. (2022). Identification and profiling of microbial community from industrial sludge. *Archives of Microbiology, 204*(4), 1–9. https://doi.org/10.1007/s00203-022-02831-y

Sidhu, C., Vikram, S., & Pinnaka, A. K. (2017, July). Unraveling the microbial interactions and metabolic potentials in pre- and post-treated sludge from a wastewater treatment plant using metagenomic studies. *Frontiers in Microbiology, 8*, 1–10. https://doi.org/10.3389/fmicb.2017.01382

Singleton, C. M., Petriglieri, F., Kristensen, J. M., Kirkegaard, R. H., Michaelsen, T. Y., Andersen, M. H., Kondrotaite, Z., Karst, S. M., Dueholm, M. S., Nielsen, P. H., & Albertsen, M. (2021). Connecting structure to function with the recovery of over 1000 high-quality metagenome-

assembled genomes from activated sludge using long-read sequencing. *Nature Communications, 12*(1), 1–13. https://doi.org/10.1038/s41467-021-22203-2
Supaphol, S., Jenkins, S. N., Intomo, P., Waite, I. S., & O'Donnell, A. G. (2011). Microbial community dynamics in mesophilic anaerobic co-digestion of mixed waste. *Bioresource Technology, 102*(5), 4021–4027. https://doi.org/10.1016/j.biortech.2010.11.124
Takaku, H., Kodaira, S., Kimoto, A., Nashimoto, M., & Takagi, M. (2006). Microbial communities in the garbage composting with rice hull as an amendment revealed by culture-dependent and -independent approaches. *Journal of Bioscience and Bioengineering, 101*(1), 42–50. https://doi.org/10.1263/jbb.101.42
Tang, L., O'Dwyer, J., Kimyon, Ö., & Manefield, M. J. (2023). Microbial community composition of food waste before anaerobic digestion. *Scientific Reports, 13*(1), 1–12. https://doi.org/10.1038/s41598-023-39991-w
Tisserant, A., Pauliuk, S., Merciai, S., Schmidt, J., Fry, J., Wood, R., & Tukker, A. (2017). Solid waste and the circular economy: A global analysis of waste treatment and waste footprints. *Journal of Industrial Ecology, 21*(3), 628–640. https://doi.org/10.1111/jiec.12562
Tóthné Bogdányi, F., Boziné Pullai, K., Doshi, P., Erdős, E., Gilián, L. D., Lajos, K., Leonetti, P., Nagy, P. I., Pantaleo, V., Petrikovszki, R., Sera, B., Seres, A., Simon, B., & Tóth, F. (2021). Composted municipal green waste infused with biocontrol agents to control plant parasitic nematodes-a review. *Microorganisms, 9*(10), 1–39. https://doi.org/10.3390/microorganisms9102130
Ventorino, V., Romano, I., Pagliano, G., Robertiello, A., & Pepe, O. (2018). Pre-treatment and inoculum affect the microbial community structure and enhance the biogas reactor performance in a pilot-scale biodigestion of municipal solid waste. *Waste Management, 73*, 69–77. https://doi.org/10.1016/j.wasman.2017.12.005
Vivas, A., Moreno, B., Garcia-Rodriguez, S., & Benitez, E. (2009). Assessing the impact of composting and vermicomposting on bacterial community size and structure, and microbial functional diversity of an olive-mill waste. *Bioresource Technology, 100*(3), 1319–1326. https://doi.org/10.1016/j.biortech.2008.08.014
Wang, H., Fotidis, I. A., & Angelidaki, I. (2015, July). Ammonia effect on hydrogenotrophic methanogens and syntrophic acetate-oxidizing bacteria. *FEMS Microbiology Ecology, 91*, 1–8. https://doi.org/10.1093/femsec/fiv130
Wang, X., Cao, A., Zhao, G., Zhou, C., & Xu, R. (2017). Microbial community structure and diversity in a municipal solid waste landfill. *Waste Management, 66*, 79–87. https://doi.org/10.1016/j.wasman.2017.04.023
Wang, P., Qiao, Z., Li, X., Su, Y., & Xie, B. (2020). Functional characteristic of microbial communities in large-scale biotreatment systems of food waste. *Science of the Total Environment, 746*, 141086. https://doi.org/10.1016/j.scitotenv.2020.141086
Xi, B., He, X., Dang, Q., Yang, T., Li, M., Wang, X., Li, D., & Tang, J. (2015). Effect of multi-stage inoculation on the bacterial and fungal community structure during organic municipal solid wastes composting. *Bioresource Technology, 196*, 399–405. https://doi.org/10.1016/j.biortech.2015.07.069
Yang, S., & Song, L. (2019). Succession of bacterial community structure and metabolic function during solid waste decomposition. *Bioresource Technology, 291*(266), 121865. https://doi.org/10.1016/j.biortech.2019.121865
Yu, D. (2014). Microbial community profiling of biodegradable municipal solid waste treatments – Aerobic composting and anaerobic digestion. Dissertation, University of Helsinki Finland.
Zeneli, A., Kastanaki, E., Simantiraki, F., & Gidarakos, E. (2019). Monitoring the biodegradation of TPH and PAHs in refinery solid waste by biostimulation and bioaugmentation. *Journal of Environmental Chemical Engineering, 7*(3), 103054. https://doi.org/10.1016/j.jece.2019.103054
Ziemiński, & Frąc, M. (2012) Methane fermentation process as anaerobic digestion of biomass: Transformations, stages and microorganisms, *African Journal of Biotechnology, 11*, 4127–4139, https://doi.org/10.5897/AJBX11.054

Chapter 16
Composting and Vermitechnology in Organic Waste Management

Shno Karimi, Taqi Raza, and Mouna Mechri

1 Introduction

During the twentieth century, agriculture witnessed the remarkable effects of the Green Revolution, which led to a significant increase in the use of industrially manufactured fertilizers such as urea and ammonium nitrate. The utilization of these fertilizers has significantly enhanced crop yields and agricultural productivity by increasing the nutrient levels in soils. However, this approach ended up fertilizing the crop instead of enriching the soil (Mupambwa et al., 2022). Regrettably, prioritizing high crop yield and neglecting soil health has led to soil degradation and increased nutrient requirements to sustain yields each season. There is an increasing recognition that soil is not only a physical resource but also a biologically active one that requires appropriate nutrient sources to sustain consistent productivity. As a result, there is a growing understanding that nutrients should be supplied alongside organic materials to nourish essential soil microorganisms, which play a crucial role in driving all biological processes within the soil. In modern agricultural cropping systems, there has been an increased emphasis on organic soil fertility management as a critical factor in ensuring sustainable agricultural production (Mupambwa & Mnkeni, 2018). On the contrary, bio-waste is considered the most significant contributor to soil health, especially in low and middle-income countries (Kaza

S. Karimi (✉)
Department of Soil Science, Faculty of Agriculture, Isfahan University of Technology, Isfahan, Iran
e-mail: karimi.s@ag.iut.ac.ir

T. Raza
Department of Biosystems Engineering & Soil Science, University of Tennessee, Knoxville, TN, USA

M. Mechri
National Institute of Field Crops, Boussalem, Tunisia

V. Kumar et al. (eds.), *Environmental Engineering and Waste Management*,
https://doi.org/10.1007/978-3-031-58441-1_16

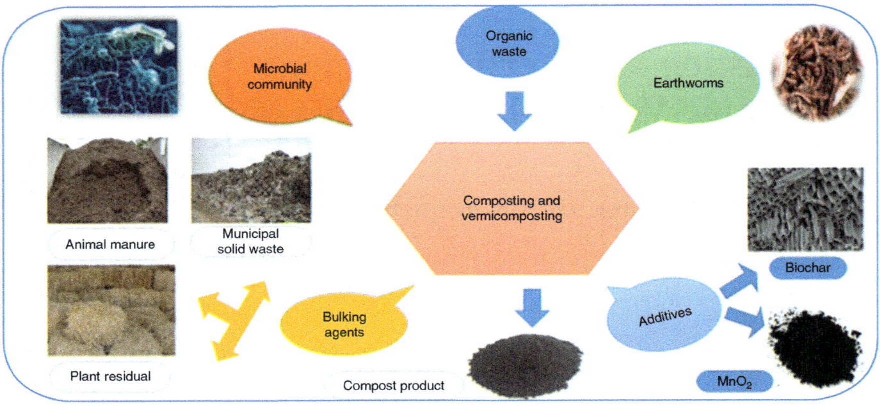

Fig. 16.1 The general process of composting and vermicomposting involves several technical steps

et al., 2018; Syed et al., 2022). However, the current waste management methods have not solved the challenges associated with existing waste collection and disposal issues (Bartolozzi et al., 2018). Mostly handling of waste is done by collecting and disposing but fails to address the ecosystem challenges. Reusing waste for human interests, especially in agriculture, requires innovative approaches to effectively manage bio-waste and extract value from it (Sindhu et al., 2019).

The utilization of organic waste often presents challenges due to issues related to its volume and unpleasant odor. Nevertheless, these challenges can be effectively addressed by converting organic waste into organic fertilizer to produce organic fertilizer from organic waste, specific requirements need to be met, including the composition of organic matter or substrate, a decomposer, and environmental factors that facilitate the conversion of waste into organic fertilizers such as compost and vermicompost (Saranraj et al., 2018). Composting is a widely used waste management strategy for recycling nutrients found in organic household or agricultural waste. Typically, compost is primarily made from agricultural by-products that come from animal and plant sources (Fig. 16.1). These materials contain vital nutrients (macro and microelements), along with numerous active minerals and organic compounds such as phytohormones, amino acids, polyamines, carbohydrates, and humic substances (Vukovic et., 2021; Ahmed et al., 2022).

Nevertheless, thermophilic composting (such as windrow composting) and vermicomposting (using earthworms) have different nutrient and enzyme compositions. Vermicomposting is considered to provide greater fertility advantages but is generally costlier compared to thermophilic composting methods (Bellitürk et al., 2023). The use of organic matter from these methods enhances soil fertility and promotes optimal plant growth and production (Ur Rehman et al., 2023; Manzoor et al., 2022). Syarifinnur et al. (2020) conducted that the use of compost increases total soil nitrogen contents by 54%. Additionally, the application of vermicompost improves total phosphorus by 196%, available phosphorus by 466.26%, total potassium by 110%, organic carbon by 42%, and pH by 12.50%. The utilization of organic fertilizers serves as an alternative to decrease reliance on inorganic fertilizers and provides a solution to mitigate environmental pollution. To improve composting practices, researchers have investigated different additives, particularly those rich in carbon. One widely recognized additive is biochar, which has proven to be effective in composting applications due to its efficacy (Gong et al., 2021; Yang et al., 2019). In a specific experiment conducted with cow manure compost as the test substrate, it was observed that the introduction of biochar and Nano carbon enhanced microbial activities (Qadir et al., 2024). Furthermore, the inclusion of straw facilitated the growth of earthworm larvae and the reproduction of cocoons. Nevertheless, it was discovered that the addition of biochar and Nano carbon resulted in a reduction of carbon (C) and nitrogen (N) loss, surpassing the effects of straw alone. Subsequent investigations were carried out through six control experiments, which concluded that the individual addition of Nano-carbon and straw was the most advantageous in achieving the highest level of vermicomposting development (Cao et al., 2021).

1.1 Organic Waste

Due to economic progress and societal advancement, the generation and accumulation of organic solid waste have become a major concern for the public (Zhao et al., 2020). Organic waste resulting from solid waste contributes significantly to environmental pollution. The traditional methods of organic solid waste disposal, such as incineration and landfill, have been found to cause significant environmental pollution. Treating organic waste through oxygen-driven biological processes produces high-quality fertilizers, and helps to protect the environment (Dominguez & Edwards, 2011). Aerobic composting and vermicomposting are widely recognized and accepted methods for the reuse of various waste materials (Dominguez & Edwards, 2011). Both aerobic processes effectively convert the input materials into valuable amendments or fertilizers that are rich in organic matter and nutrients. These end products can be used to enhance soil health and fertility in various agricultural and gardening applications (Lazcano & Domínguez, 2011). Unlike traditional composting, vermicompost is produced at intermediate temperatures through the combined action of earthworms and microorganisms. The presence of earthworms in the vermicomposting process enhances the breakdown of organic materials and promotes the development of a nutrient-rich end product (Dominguez & Edwards, 2011). Composting reduces environmental harm and creates valuable products from waste (Duan et al., 2020). Recently, compost and vermicompost applications have been widely utilized to enhance soil fertility (Table 16.1, Sinha et al., 2010). Compost or vermicompost made from diverse organic materials serves as a significant agricultural input, positively influencing various soil properties (Zandvakili et al., 2019). Organic materials play a crucial role in enhancing plant

Table 16.1 Comparison of nutrient composition of vermicompost and conventional manure

Nutrient	Vermicompost	Conventional manure
Total N (%)	1.6 ± 0.2	0.5 ± 0.2
Total C (%)	28 ± 3.3	36 ± 3.3
C:N ratio	1:18 ± 0.5	1:72 ± 0.5
Total P (%)	0.7 ± 0.1	0.2 ± 0.1
K (%)	0.8 ± 0.2	0.5 ± 0.2
Ca (%)	0.9 ± 0.3	0.5 ± 0.3
Mg (%)	0.2 ± 0.01	0.1 ± 0.01
Fe (ppm)	175.0 ± 10.2	146.5 ± 10.2
Cu (ppm)	5.0 ± 1.1	2.8 ± 1.1
Zn (ppm)	24.5 ± 0.2	14.5 ± 0.2
Mn (ppm)	96.6 ± 7.5	69.0 ± 7.5
pH (H_2O)	6.79 ± 1.4	5.43 ± 1.4
EC ($dSm^{-1)}$	1.2 ± 0.3	0.7 ± 0.3

(1) EC, electrical conductivity; (2) data are means ± standard error of means for three replicates

growth by positively impacting the physical, chemical, and biological properties of the soil. These materials contribute nutrients to the soil, which, after mineralization, are converted into mineral forms that can be readily absorbed by plants. This nutrient availability promotes healthy plant development and overall growth. Compost and vermicomposts, although not sufficient in nutrient content, significantly improve plant growth, yield, and quality due to their high organic matter content (Joshiand & Vig, 2010). the increase in soil organic matter is indeed the most significant contribution of these materials. During the decomposition process of organic matter, various decomposition products are released, which enhance the solubility of slowly released soluble nutrients in the soil. This increased solubility allows plants to access these nutrients more efficiently. Additionally, the organic matter acts as a protective barrier, safeguarding the nutrients against diverse soil conditions, such as leaching or nutrient fixation (Papathanasiou et al., 2012). Higher root growth, increase root contact and more surface area between roots and soil facilitate more water and minerals uptake for plants (Atiyeh et al., 2000). The nutrients in vermicompost, mainly N, P, and K, are considered directly available for plant growth (Edwards, 1995).

1.2 Composting

Compost and vermicomposts may not meet all plant nutrient requirements, but their organic matter significantly improves mineral nutrition, yield, and quality (Sanchez-Monedero et al., 2018). Composting typically involves three phases: initial activation, thermophilic phase, and maturation phase. The initial activation phase, lasting around 13 days, involves the mineralization of organic compounds by microbial communities. This process produces CO_2, NH_3, organic acids, and heat (Bernal et al., 2009; Raza et al., 2023). In the initial phase, the temperature of the composting pile rises, and during the thermophilic phase, it reaches its peak. The ideal temperature range for composting is between 40 and 65 °C (De Bertoldi et al., 1983), and temperature above 55 °C is important to kill pathogens. In the composting process, thermophilic microorganisms break down fats, cellulose, and lignin (Bernal et al., 2009). During the mesophilic phase, the temperature decreases as microbial activity reduces. Microorganisms transform organic matter into humic substance (HS) precursors like polyphenols, carboxyl acids, amino acids, reducing sugars, and polysaccharides. These precursors undergo polymerization through various pathways to form HS. HS, a complex compound in composting, greatly improves soil fertility due to its abundance of mature organic matter (Stevenson, 1994).

1.3 Vermitechnology

Vermicomposting, a cost-effective method in vermiculture biotechnology, transforms organic waste into nutrient-rich fertilizers that can be used as soil conditioners. Earthworms are essential in breaking down the substrate, aerating the material, increasing surface area for microbes, and modifying microbial activities to facilitate the decomposition of organic waste (Lavelle et al., 2006). The temperature during vermicomposting varies between 25 to 37 °C, depending on the earthworm species. Extreme temperatures, whether too low or too high, have a negative impact on earthworm activities and development. Four species commonly used in vermicomposting facilities include two tropical species, namely *Eudrilus Eugenia* (Hayawin et al., 2014) and *Perionyx excavations* (Suthar & Singh, 2008)] and two temperate ones; *Eisenia andrei* and *Eisenia fetida* (Alavi et al., 2017).

Vermicompost significantly improves soil quality, enhances microbial activity, and recycles the nutrient to produce high-quality crops (Nurhidayati et al., 2018). It serves as a biofertilizer as a substitute for chemical fertilizers and obtains better prices for organically produced crops (Ramnarain et al., 2019). Earthworms are critically beneficial for farming and are indicators of soil quality due to their role in soil health. They naturally produce organic material without and reduce the need for artificial inputs. Earthworms play a vital role in vermicomposting by aerating, mixing, grinding, fragmenting, enzymatically digesting, and facilitating microbial decomposition of the substrate within their intestines (Sharma et al., 2005). The presence and performance of earthworms are important indicators used to assess the success of the vermicomposting process (Suthar, 2006).

Vermicompost is rich in microorganisms, and this exceptional quality makes it a superior fertilizer. These properties are linked to nutrient cycling in the soil or growth substrates, and the provision of bioactive substances that enhance plant growth (Table 16.2).

Table 16.2 Chemical analysis of vermicompost and raw materials

Treatments	pH	EC (dS/m	OM (%)	C/N	N (%)	P_2O_5 (%)	K_2O (%)	Ca (%)	Mg (%)	Fe (%)	S (%)
Raw material	7.70	4.00	75.00	18.00	0.50	0.30	0.30	2	0.20	0.20	0.10
Vermicompost	6.50	3.00	48.00	12.00	1.40	1.0	1.00	4	0.60	0.55	0.45

2 Environmental Management

Considering the increase in global demand for organic materials and essential macronutrients such as nitrogen, phosphorus, potassium, and sulfur, organic wastes are considered valuable resources (Kulikowska & Bernat, 2021). Composting as an environmentally friendly technology, is being widely adopted to handle different types of organic waste and mitigate environmental pollution (Onwosi et al., 2017). There are several management techniques to remove contaminants, parasites, and pathogens from organic waste, including incineration, lime fixation, composting, and vermicomposting. However, some techniques involve a combustion process, which may produce dioxin, a highly toxic air pollutant. Even if these organic wastes still contain inorganic pollutants after treatment (Chen et al., 2012). According to the estimates by FAO (2011), globally one-third of total produced food for human consumption (approximately 1.6 billion tons) is wasted during processing and consumption annually.

Currently, recycling and composting help reduce the landfill burden. A Mauritius study found that 60% of total waste was kitchen and yard food waste (Mohee, 2002). Given the considerable organic waste generation, composting technology has become an appealing large-scale approach for managing a wide variety of organic wastes. Nonetheless, the composting process might not consistently result in high-quality products that offer considerable soil benefits concerning nutrient content and land improvement aspects. In the past few years, the idea of using earthworms to decompose organic wastes has garnered attention and interest from researchers. Considering that over 60% of the waste load at landfills comprises organic materials, this approach has gained significance (Mohee et al., 2010), Vermicomposting offers an effective solution for managing and processing waste, thereby alleviating the pressure on landfills. The resulting vermicompost can be utilized in gardens, commercial farms, and various agricultural industries, decreasing the reliance on chemical products (Dominguez & Edwards, 2011).

3 Composting Methods

Composting serves as a beneficial pre-treatment approach to mitigate the adverse impacts on soil conditions resulting from organic waste disposal. It aids in creating sanitized and stabilized products, which can be employed as organic fertilizers or soil improvement agents (Gabhane et al., 2012). The selection of a composting method is influenced by factors such as local circumstances, workforce, time constraints, financial resources, available land, and feedstock. A crucial objective in this process is to establish a conducive environment for microorganisms, primarily controlled by temperature and oxygen levels. Proper ventilation can be employed to regulate these conditions. Consequently, the efficiency and accessibility of ventilation systems emerge as crucial factors and design priorities. In active aeration systems, fans force air through the compost pile's free spaces, while passive aeration systems rely on convective heat exchange generated by temperature changes within the pile, resulting in natural air circulation. Proper ventilation not only conserves energy but also enhances compost quality. Nonetheless, an excessively high ventilation rate can negatively impact temperature maintenance, potentially leading to nutrient loss and decreased fertilizer efficiency. Conversely, inadequate ventilation may encourage localized anaerobic conditions, resulting in the production of substantial amounts of greenhouse gases (Awasthi et al., 2020).

The widely recognized commercial composting processes frequently employed in agriculture include:

(a) Thermophilic composting: This process involves maintaining high temperatures (40–70 °C) to accelerate the decomposition of organic materials and eliminate pathogens.
(b) Windrow composting: In this method, organic materials are piled in long, narrow rows (windrows), and turned periodically to ensure proper aeration and temperature control.
(c) Vermicomposting: This process utilizes earthworms to break down organic materials into a nutrient-rich end product called vermicompost.
(d) In-vessel composting: Here, organic materials are composted within an enclosed, controlled environment, such as a bin or container, allowing for better temperature and moisture management.

Composting comprises several stages that facilitate the recycling of nutrients from diverse waste sources, including domestic, agricultural, technological, and animal waste. Each waste type contributes distinct characteristics to enhance agricultural and horticultural production systems (Bellitürk et al., 2023). The composting process consists of four primary stages that contribute to the transformation of organic materials into a nutrient-rich product: (Borker et al., 2021).

Activation/Mesophilic stage: During this initial phase, the compost pile's temperature and microbial activity gradually increase. Mesophilic microorganisms, which thrive at moderate temperatures (30–45 °C), become dominant in breaking down the organic materials (Wurff et al., 2016). Thermophilic stage: As the compost pile's temperature rises to 45–70 °C, thermophilic microorganisms (thriving at high temperatures) become more active. This stage is crucial for pathogen elimination and the initial breakdown of complex organic compounds (Bhatia et al., 2012). Cooling/Second Mesophilic stage: After the thermophilic phase, the compost pile's temperature starts to decrease. The mesophilic microorganisms return, continuing the decomposition process as the compost cools down (Khan et al., 2022; Wurff et al., 2016). Final curing and maturation phase: This is the longest stage, where the compost matures and stabilizes. During this period, microbial activity slows down, and the compost's nutrients become more accessible for plant uptake. The compost reaches its optimal state for use in agriculture and horticulture when it is fully cured and matured (Wurff et al., 2016). In Windrow Composting, composting materials are mixed with a bulking agent and formed into long, parallel rows called windrows. These windrows typically range from 3 to 6 feet in height and 6 to 14 feet in width at the base. The dimensions depend on the equipment used for turning and mixing the windrows. This composting method takes place at open sites, allowing for ease of access and management.

Periodic turning and mixing of the windrows are essential for enhancing air circulation and reducing moisture levels within the pile. This promotes a more efficient and effective decomposition process, ultimately leading to a higher quality compost product. Vessel composting, in contrast to windrow composting, occurs within a container, allowing for more controlled environmental conditions. This method enables precise management of factors such as oxygen levels, airflow, and temperature, which can significantly impact the composting process. Vermicomposting is a specific type of composting that involves the use of earthworms to break down organic materials after the primary composting process has been completed. In this stage, the earthworms help further decompose the compost, improving its nutrient content and structure. After vermicomposting, the earthworms are separated from the finished compost through manual picking or other methods. This traditional practice ensures that the final product is enriched with earthworm castings, which are highly valued for their nutrient-dense and aerated properties, benefiting plant growth and soil health. The novelist technology in question refers to a rolling separator designed specifically for separating earthworms from compost. This machine boasts an efficiency rate of 83.79%, making it an effective tool for the task. It operates at a speed of 21.2 kg/min, ensuring a relatively high throughput for processing large volumes of compost (Lin et al., 2021).

4 Chemical and Physical Effects on Composting and Vermitechnology

Composting is a carefully managed process that involves the decomposition of organic materials by microorganisms in an aerobic environment. This controlled process transforms the initial materials into a sanitary, odorless, and humus-like substance, which is highly beneficial for soil health and plant growth. Through composting, organic materials are broken down and stabilized, resulting in products that can be used as soil conditioners and organic fertilizers (Negro et al., 1999). Assessing compost's maturity, stability, and quality is difficult due to variable factors like raw materials and composting methods (Bustamante et al., 2010). The challenge is more apparent in materials like winery-distillery residues, as they showcase unusual patterns in humification aspects during composting (Bustamante et al., 2008; Meena et al., 2021; Bernal et al., 2009; Raj & Antil, 2011) (Table 16.3). To assess compost maturity, monitoring the ratio of humified to non-humified carbon serves as an indicator. Analyzing organic matter evolution may help solve the compost maturity problem (Castaldi et al., 2005). Research suggests that the quantity and chemical composition of humic substances, particularly those formed during composting, are crucial indicators for evaluating compost maturity and stability (Tomati et al., 2000).

Table 16.3 A summary of the various parameters that have been identified in the literature as critical in determining compost or vermicompost maturity

Maturity parameter	Chemical properties
Total C:N ratio	<20 and preferably <10
Humic acid to fulvic acid ratio	>1.9
Water soluble organic-C	≤4 g/kg
Water soluble carbon: Organic nitrogen ratio	<0.55
Germination index	50–70%
Organic matter loss	>42%
Humification index	>30%
NH_4-N to NO_3-N ratio	<0.16
Polymerization index	>1
Humification ratio	>7

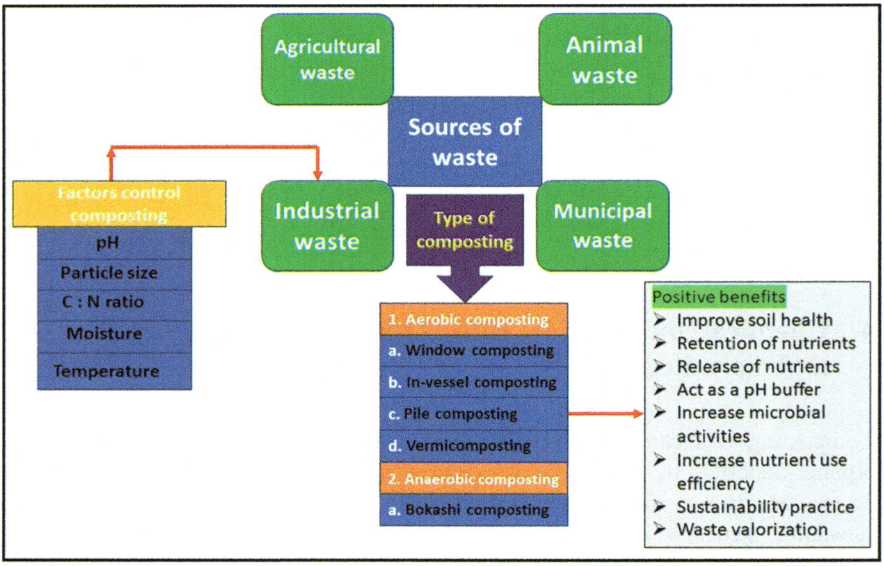

Fig. 16.2 Types of composting, factor influences, and its positive benefits

Evaluate the maturity and stability of compost, various methods are employed, categorized into physical, chemical, and biological parameters (Tiquia, 2005). These methods for assessing the maturity and stability of compost are further categorized into two groups: "maturity evaluation techniques" and "stability evaluation techniques." Controlling a composting process involves managing key factors, such as environmental parameters (depicted in Fig. 16.2) like temperature, moisture content, pH, and aeration, alongside substrate nature parameters such as the C/N ratio, particle size, and nutrient content (Diaz et al., 2002).

4.1 pH

Changes in compost pH can be attributed to various factors, including the breakdown of organic acids and the release of ammonia, as well as the release of organic acids, carbon dioxide (CO_2), and ammonia (NH_3) due to microbial activity (Yang et al., 2013). An initial pH of 6–8 is considered suitable for initiate the composting process (Bustamante et al., 2008). Wang et al. (2017) found that treatment to attain an initial pH of 5–6 can aid in the maturation of compost derived from distilled grain waste. Environmental acidity or alkalinity plays a significant role in the formation, size, and distribution of humic substances (HSs), which are complex organic compounds derived from the decomposition of plant and animal matter. The pH level influences the stability of HSs and their molecular composition (Pédrot et al., 2010).

Fluctuations in pH during the composting period are essential indicators of biological activity, reflecting the dynamic nature of the composting process. The pH level in composting systems is influenced by several factors, including the rate of decomposition and the characteristics of the initial materials being composted. Edwards (1998) has reported that earthworms are sensitive to pH and survive within the range of 4.5 to 9.

4.2 C:N

The carbon-to-nitrogen (C:N) ratio is a crucial factor in determining the efficiency and quality of composting, as well as the characteristics of the resulting compost and its associated products, such as humic substances (Wang et al., 2015). An increased initial C/N ratio encourages the oxidation of organic compounds, resulting in faster stabilization. Conversely, a lower initial C/N ratio may cause the accumulation of partially oxidized organic matter, leading to immature final compost and potential nitrogen losses (Silva et al., 2014). In general, a C/N ratio between 25 and 30 is considered ideal for aerobic composting processes. This balance allows for efficient decomposition of organic materials, leading to faster stabilization and a higher quality final compost product (Silva et al., 2014). Wu et al. (2016) reported that an initial C/N ratio of 25 could favor the formation of high-quality and low-toxicity compost. Aerobic composting occurs at a broad pH range from 5.5 to 9.0, however, the preferred range of pH is 6.5–8.0 (Rynk et al., 1992a). For active composting, ideal bulk C:N ratio is considered from 25:1 to 30:1 with acceptable low ranges is 20:1 (Rynk et al., 1992a). When the C/N ratio exceeds 40:1, a longer composting period becomes necessary, and there is a risk of nitrogen immobilization upon soil application. Conversely, if the ratio is too low, nitrogen may be lost as ammonia, leading to unpleasant odors and potentially affecting the overall compost quality. Maintaining an optimal C/N ratio is crucial for successful aerobic composting.

4.3 Temperature

In conventional compost production for organic agriculture, one primary objective is to deactivate weed seeds and reduce potential pathogens. Temperature plays a vital role in any composting operation, as it influences the growth and activity of microorganisms responsible for breaking down organic materials. Maintaining optimal temperatures ensures effective decomposition and contributes to the overall quality of the final compost product in vermicomposting, which involves the use of earthworms to break down organic materials, temperature has a significant impact on various aspects of the earthworms' life processes (Reinecke et al., 1992). To achieve optimal conditions for earthworm activity and efficient composting in vermicomposting systems, it is essential to manage the compost temperature to reach around 60 °C. Temperature plays a crucial role in controlling the population of microorganisms involved in the composting process, as their activities are generally temperature-dependent. The preferred temperature range for these microorganisms is typically between 54 and 60 °C, as it allows for efficient decomposition and supports the growth and activity of the beneficial microbes responsible for breaking down organic materials (Rynk et al., 1992b). While maintaining a temperature of around 60 °C is ideal for optimizing microbial activities in the composting process, vermicomposting systems generally aim to maintain mesothermic temperatures. This means keeping the composting temperatures within the optimal range for earthworms, which ensures their well-being and contributes to the overall effectiveness of the system. For red wiggler or California worms (Eisenia fetida), a common earthworm species used in vermicomposting in the USA, the ideal temperature range is between 15 to 25 °C (Rostami et al., 2010). In some cases, vermicomposting may be preceded by a thermophilic composting cycle, especially to meet regulatory requirements for organic certification. Although both vermicomposting and thermophilic composting systems use similar feedstocks, they foster different microbial communities due to the distinct conditions and processes involved. Vermicompost, which results from the vermicomposting process, generally exhibits higher microbial activity compared to thermophilic compost (Domínguez et al., 2019).

4.4 Moisture

In the composting process, water plays a crucial role and is involved in various aspects. It enters the system through multiple pathways, including microbial activities like metabolism and the addition of forced air (Cai et al., 2012). Moisture content plays a vital role in determining the physical and biological characteristics of compost. It significantly impacts various properties, including bulk density, free air space, air permeability, thermal conductivity, and biological activity (Kulikowska, 2016). Moisture content can be modified by bulking agents, and generally mixture of ideal moisture should be between 50 and 60% by weight (Bernal et al., 2009).

Usually, the moisture content in composting is maintained between 40 and 65% (Misra et al., 2003). Moisture is not only necessary to support microbial activity and metabolism but also for nutrient transport (Shi et al., 1999). A moisture level below 45% results in slowed microbial activity, whereas moisture levels above 65% occupied pore spaces causing the process anaerobic (Rynk et al., 1992b). Post-production, compost should not be dried excessively (i.e., less than 45% moisture) as microbial activity is lost and generates dust while handling, and it makes it challenging to rewater the compost (Moon, 1997).

4.5 *Particle Size*

The size and shape of particles in compost significantly influence both the operational costs associated with sieving and the effectiveness of the compost when applied to soil (Wang & Ai, 2016). Sizing-wise, large particles exhibit a reduced overall surface area, making them less accessible to microorganisms compared to their finer counterparts (Verma & Marschner, 2013). According to research findings, finer composts tend to release higher amounts of nitrogen (N) and phosphorus (P) compared to coarser composts (Duong et al., 2012). This widespread observation implies that the overall quality of vermicompost is frequently deemed better than that of conventional compost (Sinha et al., 2010). While it is widely believed that vermicompost often surpasses regular compost in terms of quality, there is a lack of comprehensive scientific literature comparing both products when prepared from identical particle size fractions derived from various feedstocks.

4.6 *Aeration*

Aerobic composting is a process where organic materials, such as plant and animal waste, undergo decomposition in the presence of oxygen (Liang et al., 2003). During the aerobic composting process, the primary byproducts generated are carbon dioxide (CO_2), water (H_2O), and heat (Bari & Koenig, 2001). Aeration plays a vital role in the aerobic composting process, as it ensures an adequate supply of oxygen for the microorganisms involved in breaking down the organic materials. There are several methods to achieve proper aeration, and the choice of method depends on factors such as the scale of the composting operation and available resources.

Physical turning: this method involves manually or mechanically turning the compost pile to mix the materials and introduce fresh oxygen. Turning also helps to redistribute heat and moisture, ensuring uniform conditions throughout the pile. This is the most common and primary aeration technique used in composting (Diaz et al., 2002). If the aeration rate is too high, it can lead to a decrease in temperature due to increased air circulation (Rynk et al., 1992a). Insufficient aeration can lead to anaerobic conditions during composting (Brodie et al., 2000).

To maintain optimal aerobic conditions during composting, a minimum oxygen concentration of 5% within the pore space of the compost pile is required (Haung, 1993). In recent decades, numerous studies have been carried out to assess the effectiveness of composting using various activity parameters. This has led to the development of specific methods for evaluating the quality and maturity of compost derived from various types of organic waste (Melis & Castaldi, 2004). The key distinction between anaerobic and aerobic decomposition lies in the gases produced and the energy released during composting. Aerobic decomposition, which involves exothermic reactions, releases a higher amount of energy compared to anaerobic decomposition, which involves chemical reduction. During aerobic composting, microorganisms produce gases such as CO_2, NO_2, and NO_3. In contrast, anaerobic decomposition results in the production of CH_4 and CO_2 (Talyan et al., 2008). The composting of biodegradable solid wastes produces a valuable product that can be used as nutrient-rich organic fertilizers or for land application. This practice is recognized as beneficial by the United Nations (United Nations, 2000).

5 Current and Recent Study

Composting is a beneficial method that speeds up the biodegradation and transformation of organic matter into a stable and mature product under controlled conditions. Numerous studies have investigated composting technology for managing waste from livestock, poultry, kitchen, urban, and courtyard sources, and have applied it to organic agriculture production. Composting and vermicomposting produce valuable organic amendments that play a crucial role in enhancing soil health and sustainability. Applying different types of waste-derived compost as a soil enhancer offers numerous benefits, including reducing soil bulk density, improving infiltration rate and hydraulic conductivity, enhancing water retention, and boosting overall plant growth and productivity. Additionally, compost application increases soil microbial diversity, suppresses soil pathogens, and helps inhibit fungal diseases like *Fusarium oxysporum* and *Phytophthora drechsleri* (Kranz et al., 2020). Research has shown that the combined use of vermicompost and water hyacinth drum compost has a beneficial impact on soil quality and plant growth in agroecosystems. Additionally, combining vermicompost with zinc oxide nanoparticles has been found to have positive effects on the root and shoot length of *Arachis hypogaea L.,* as well as increased sugar and chlorophyll contents and dry weight (Chia et al., 2020). Furthermore, the combined application of vermicompost and biochar enhanced the quality and yield of cucumber plants (increasing by 29.2%–56%), improved soil properties, and helped address issues related to continuous cropping (Wang et al., 2021). Shrimal and Khan (2017) found that vermicomposts have beneficial effects on the yield of various commercial crops, including cereals, legumes, ornamental plants, and vegetables. The nutrient composition of vermicomposts varies depending on the raw materials used (type of waste and co-additives), and its impact varies from plant to plant. Vermicomposting is recognized as an efficient and environmentally friendly method for waste stabilization by introducing earthworms and microorganisms (Sharma & Garg, 2018). Vermicomposting is known to enhance the decomposition process, increase the activities of enzymes like lipases, proteases, amylases, and celluloses, and enrich the soil with nitrogen (Usmani et al., 2017). Vermicomposting converts waste into valuable matter, addressing waste management issues and supporting sustainable agriculture practices (Bhattacharya & Chattopadhyay, 2006). The impact of date palm waste compost on soil microbial communities is not well understood, despite its benefits for barley crop growth and soil properties. Ghouili et al. (2023b) found that compost treatment in barley crops led to an increased abundance of bacteria and fungi, resulting in significant changes to the richness and diversity of both fungal and bacterial communities. Ghouili et al. (2023b) observed that incorporating compost could be viewed as a sustainable method to promote a beneficial soil microbiome, leading to enhanced soil quality and increased barley yield. During abiotic stress conditions, the use of compost has been shown to have beneficial impacts on the growth and yield of canola (Rashtbari et al., 2020) and wheat (Omara et al., 2022), while also enhancing the water-holding capacity of the soil in both cases. Research has indicated that applying compost has enhanced the growth and development of radishes by increasing soil microbial biomass, total soil carbon, and nitrogen levels (Heisey et al., 2022). Likewise, compost has been shown to have a positive impact on the growth and development of grapevines (Gaiotti et al., 2017) and barley roots (Ghouili et al., 2023a). Moreover, the germination of green bean seeds has been shown to benefit from the incorporation of compost derived from municipal solid waste at varying ratios. This suggests that such compost can enhance biological activity in the soil (Soobhany et al., 2017). In a study by Ghouili et al. (2022), it was found that compost made from date palm waste significantly boosted the presence of proteins in barley related to key metabolic pathways like photosynthesis, energy metabolism, and secondary metabolism. Furthermore, the research identified specific gene-coding proteins in barley that may be influenced by compost application, potentially enhancing the plant's defense mechanisms against both biotic and abiotic stressors. In vermicomposts, inorganic nitrogen is mostly present in nitrate form. It is found that rabbit manure vermicompost contained 1303 mg kg^{-1} N-NO_3 and only 15 mg kg^{-1} N-NH_4^+ (Lazcano et al., 2010), while cow manure had 220 mg kg^{-1} N-NO_3 and 53 mg kg^{-1} N-NH_4^+ (Yue et al., 2017). In contrast, pig manure vermicompost contained 297 mg kg^{-1} N-NH_4^+ and 104 mg kg^{-1} N-NO_3 (Lazcano et al., 2009). There is evidence to suggest that vermicomposts enhance the conversion of ammonium into nitrate through the process of nitrification (Cáceres et al., 2018). Thus, raw cattle manure contained 610 mg kg^{-1} N-NH_4^+ and 19 mg kg^{-1} N-NO_3, which increased up to 1235 mg kg^{-1} N-NH_4^+ and 721 mg kg^{-1} N-NO_3 in composted manure (Lazcano et al., 2008).

In addition, Lukashe et al. (2020) conducted a study assessing the use of fly ash enriched with vermicompost to enhance the growth of *C. Gayana* by increasing available phosphorus levels. The improved growth of *C. Gayana* in vermicompost-amended soils was linked to enhanced nutrient availability and reduced phytotoxicity of trace elements. Therefore, the reclamation of mine waste-contaminated soils can be facilitated by incorporating fly ash-enriched vermicompost at an optimal rate of 40 mg-P/kg FV. In summary, these studies highlight the beneficial effects of compost and vermicompost in boosting soil quality, promoting crop growth, and efficiently managing waste. This underscores their importance in sustainable agricultural practices.

6 Way Forward

The rise in population has led to a significant increase in organic waste, placing strain on waste management systems. Implementing new strategies in organic waste management is a key solution to tackle this challenge. It is essential to enhance current waste management practices, streamline processes, and promote the adoption of sustainable techniques by urban areas, local governments, and large organic waste generators. The combination of rotary drum composting and vermicomposting shows promise for effective utilization. In-vessel composting can handle large organic loads in limited space near waste generation hotspots, reducing landfill pressure. Vermicomposting enhances product consistency but increases compost production costs. In summary, the combination effectively manages larger waste loads and produces marketable compost products. Advancements in composting methods and the use of additives to improve compost quality offer hope for high-quality humus from diverse organic residues in the future. This is expected to lead to increased recognition and attention towards composting practices.

7 Conclusion

Composts and vermicomposts are increasingly valued as nutrient sources for soil fertility, acknowledging the limitations of inorganic synthetic fertilizers in maintaining soil quality. These end-products of composting and vermicomposting are commercially referred to as humus or earthworm humus. These substances have a fine and porous texture, are biologically stable, and have a high water-retention capacity. They also have a low C/N ratio and high nutrient content in forms readily available for plant absorption. Rich in a complex microbial community, these organic materials offer various beneficial effects to soil–plant systems. Compost and vermicompost help enhance soil health and fertility, increase nutrient and microbial content in soils, improve water retention, and reduce reliance on fertilizers and pesticides.

References

Ahmed, S., Matiullah, K. H. A. N., Taqi, R. A. Z. A., Ahmad, R., & Iqbal, J. (2022). Integrated use of bio-organic and chemical fertilizer to enhance yield and nutrients content of tomato. *Eurasian Journal of Soil Science, 11*(2), 126–132.

Alavi, N., Daneshpajou, M., Shirmardi, M., Goudarzi, G., Neisi, A., & Babaei, A. A. (2017). Investigating the efficiency of co-composting and vermicomposting of vinasse with the mixture of cow manure wastes, bagasse, and natural zeolite. *Waste Management, 69*, 17–126. https://doi.org/10.1016/j.wasman.2017.07.039

Atiyeh, R. M., Arancon, N., Edwards, C. A., & Metzger, J. D. (2000). Influence of earthworm processed pig manure on the growth and yield of greenhouse tomatoes. *Bioresource Technology, 75*(3), 175–180. https://doi.org/10.1016/S0960-8524(00)00064-X

Awasthi, M. K., Duan, Y., Liu, T., Awasthi, S. K., & Zhang, Z. (2020). Relevance of biochar to influence the bacterial succession during pig manure composting. *Bioresource Technology, 304*, 122962.

Bari, Q. H., & Koenig, A. (2001). Effect of air recirculation and reuse on composting of organic solid waste. *Resources Conservation and Recycling, 33*, 93–111.

Bartolozzi, I., Baldereschi, E., Daddi, T., & Iraldo, F. (2018). The application of life cycle assessment (LCA) in municipal solid waste management: A comparative study on street sweeping services. *Journal of Cleaner Production, 182*, 455–465. https://doi.org/10.1016/j.jclepro.2018.01.230

Bellitürk, K., Fang, L., & G¨orres, J. H. (2023). Effect of post-production vermicompost and thermophilic compost blending on nutrient availability. *Waste Management, 155*(2023), 146–152.

Bernal, M. P., Alburquerque, J. A., & Moral, R. (2009). Composting of animal manures and chemical criteria for compost maturity assessment. A review. *Bioresource Technology, 100*, 5444–5453. https://doi.org/10.1016/jbiortech.2008.11.027

Bhatia, A., Ali, M., Sahoo, J., Madan, S., Pathania, R., Ahmed, N., & Kazmi, A. A. (2012). Microbial diversity during rotary drum and windrow pile composting. *Journal of Basic Microbiology, 52*, 5–15. https://doi.org/10.1002/jobm.201100077

Bhattacharya, S., & Chattopadhyay, G. (2006). Effect of vermicomposting on the transformation of some trace elements in fly ash. *Nutrition Cycle Agroecosystem., 75*, 223–231.

Borker, S. S., Prashant, S. P., Kumari, A., Kumari, S., Devi, R., & Kumar, R. (2021). Metabolic pathways in biodegrading psychrotrophic bacteria under cold environments. In *Extreme environment* (pp. 1–17). First CRC Press.

Brodie, H. L., Carr, L. E., & Condon, P. (2000). A comparison of static pile and turned windrow methods for poultry litter compost production. *Compost Science and Utilization, 8*, 178–189.

Bustamante, M. A., Paredes, C., Marhuenda-Egea, F. C., Pérez-Espinosa, A., Bernal, M. P., & Moral, R. (2008). Co-composting of distillery wastes with animal manures: Carbon and nitrogen transformations in the evaluation of compost stability. *Chemosphere, 72*, 551–557. https://doi.org/10.1016/j.chemosphere.2008.03.030

Bustamante, M. A., Suárez-Estrella, F., Torrecillas, C., Paredes, C., Moral, R., & Moreno, J. (2010). Use of chemometrics in the chemical and microbiological characterization of composts from agroindustrial wastes. *Bioresource Technology, 101*, 4068–4074. https://doi.org/10.1016/j.biortech.2010.01.085

Cáceres, R., Malińska, K., & Marfà, O. (2018). Nitrification within composting: A review. *Waste Management, 72*, 119–137.

Cai, L., Gao, D., Chen, T. B., Liu, H. T., Zheng, G. D., & Yang, Q. W. (2012). Moisture variation associated with water input and evaporation during sewage sludge bio-drying. *Bioresource Technology, 117*, 13–19.

Cao, Y. N., Tian, Y. Q., Wu, Q., Li, J. S., & Zhu, H. Y. (2021). Vermicomposting of livestock manure as affected by carbonrich additives (straw, biochar and nanocarbon): A comprehensive evaluation of earthworm performance, microbial activities, metabolic functions and vermicompost quality. *Bioresource Technology, 320*(Part B), 124404.

Castaldi, P., Alberti, G., Merella, R., & Melis, P. (2005). Study of the organic matter evolution during municipal solid waste composting aimed at identifying suitable parameters for the evaluation of compost maturity. *Waste Management, 25*, 209–213. https://doi.org/10.1016/j.wasman.2004.12.011

Chen, H., Yan, S. H., Ye, Z. L., Meng, H. J., & Zhu, Y. G. (2012). Utilization of urban sewage sludge: Chinese perspectives. *Environment Science Pollutant Research, 19*, 1454–1463.

Chia, W., Chew, K., Le, C., Lam, S., Chee, C., Ooi, M., & Show, P. (2020). Sustainable utilization of bio-waste compost for renewable energy and soil amendments. *Environment Pollutant, 267*, 115662.

De Bertoldi, M. D., Vallini, G. E., & Pera, A. (1983). The biology of composting: A review. *Waste Management Research, 1*, 157–176. https://doi.org/10.1016/0734-242X(83)90055-1

Diaz, M. J., Madejon, E., Lopez, F., Lopez, R., & Cabrera, F. (2002). Optimization of the rate vinasse/grape marc for co-composting process. *Process Biochemistry, 37*, 1143–1150.

Dominguez, J., & Edwards, C. A. (2011). Relationship between composting and vermicomposting. In *Vermiculture technology-earthworms. Organic wastes and environmental management* (pp. 11–26). CRC Press.

Domínguez, J., Aira, M., Kolbe, A. R., G'omez-Brand'on, M., & P'erez-Losada, M. (2019). Changes in the composition and function of bacterial communities during vermicomposting may explain beneficial properties of vermicompost. *Scientific Reports, 9*(1), 9657. https://doi.org/10.1038/s41598-019-46018-w

Duan, Y. M., Pandey, A., Zhang, Z. Q., Awasthi, M. K., Bhatia, S. K., & Taherzadeh, M. J. (2020). Organic solid waste bio refinery: Sustainable strategy for emerging circular bio economy in China. *Industrial Crops and Products, 153*, 112568.

Duong, T. T. T., Penfold, C., & Marschner, P. (2012). Differential effects of composts on properties of soils with different textures. *Biology Fertility Soils, 48*, 699–707.

Edwards, C. A. (1995). Historical overview of vermicomposting. *Bio Cycle, 36*(6), 56–58. Epa, US.

Edwards, C. A. (Ed.). (1998). *The use of earthworms in the breakdown and management of organic wastes, in earthworm ecology*. St. Lucie Press.

FAO. (2011). *Global food losses and food waste*. www.fao.org/docrep/014/mb060e/mb060e00.pdf

Gabhane, J., William, S. P., Bidyadhar, R., Bhilawe, P., Anand, D., Vaidya, A. N., & Wase, S. R. (2012). Additives aided composting of green waste: Effects on organic matter degradation, compost maturity, and quality of the finished compost. *Bioresource Technology, 114*, 382–388.

Gaiotti, F., Marcuzzo, P., Belfiore, N., Lovat, L., Fornasier, F., & Tomasi, D. (2017). Influence of compost addition on soil properties, root growth and vine performances of Vitis vinifera cv. Cabernet sauvignon. Science. *Horticulture, 225*, 88–95. [Google Scholar] [CrossRef].

Ghouili, E., Abid, G., Jebara, M., Nefissi Ouertani, R., de Oliveira, A. C., El Ayed, M., & Muhovski, Y. (2022). Proteomic analysis of barley (*Hordeum vulgare* L.) leaves in response to date palm waste compost application. *Plants, 11*, 3287. https://doi.org/10.3390/plants11233287

Ghouili, E., Sassi, K., Hidri, Y., M'Hamed, H. C., Somenahally, A., Xue, Q., Jebara, M., Nefissi Ouertani, R., Riahi, J., & de Oliveira, A. C. (2023a). Effects of date palm waste compost application on root proteome changes of barley (*Hordeum vulgare* L.). *Plants, 12*, 526. https://doi.org/10.3390/plants12030526

Ghouili, E., Abid, G., Hogue, R., Jeanne, T., D'Astous-Pagé, J., Sassi, K., Hidri, Y., M'Hamed, H. C., Somenahally, A., & Xue, Q. (2023b). Date palm waste compost application increases soil microbial community diversity in a cropping barley (*Hordeum vulgare* L.) field. *Biology, 12*, 546. https://doi.org/10.3390/biology12040546

Gong, X. Q., Zhang, Z. T., & Wang, H. (2021). Effects of Gleditsia sinensis pod powder, coconut shell biochar and rice husk biochar as additives on bacterial communities and compost quality during vermicomposting of pig manure and wheat straw. *Journal Environment Management, 295*, 113136. https://doi.org/10.1016/j.jenvman.2021.113136

Haung, R. T. (1993). *The practical handbook of compost engineering*. Lewis Publishers.
Hayawin, Z. N., et al. (2014). Study on the effect of adding zeolite and other bio-adsorbance in enhancing the quality of the palm based vermicompost. *Applied Mechanics and Materials, 472*, 926–934. https://doi.org/10.4028/www.scientific.net/AMM.472.926
Heisey, S., Ryals, R., Maaz, T. M., & Nguyen, N. H. (2022). A single application of compost can leave lasting impacts on soil microbial community structure and alter cross-domain interaction networks. *Frontiers in Soil Science, 2*, 749212.
Joshiand, R., & Vig, A. P. (2010). Effect of vermicompost on growth, yield and quality of tomato *(Lycopersicum esculentum L). African Journal of Basic and Applied Sciences, 2*(3–4), 117–123.
Kaza, S., Yao, L., Bhada-Tata, P., & Van Woerden, F. (2018). *What a waste 2.0: A global snapshot of solid waste management to 2050*. World Bank Publications.
Khan, K. S., Qadir, M. F., Ahmad, A., Naveed, M., Raza, T., & Ditta, A. (2022). Efficacy of different endophytic bacterial strains in enhancing growth, yield, and physiological and biochemical attributes of Linum usitatissimum L. *Journal of Soil Science and Plant Nutrition, 22*(4), 4365–4376. https://doi.org/10.1007/s42729-022-01035-z
Kranz, C., Mclaughlin, R., Johnson, A., Miller, G., & Heitman, J. (2020). The effects of compost incorporation on soil physical properties in urban soils-a concise review. *Journal Environment Management, 261*, 110209.
Kulikowska, D. (2016). Kinetics of organic matter removal and humification progress during sewage sludge composting. *Waste management, 49*, 196–203. https://doi.org/10.1371/journal.pone.0146237
Kulikowska, D., & Bernat, K. (2021). Waste willow-bark from salicylate extraction successfully reused as an amendment for sewage sludge composting. *Sustainability, 13*, 6771.
Lavelle, P., Decaens, T., Aubert, M., Barot, S., Blouin, M., & Bureau, F. (2006). Soil invertebrates and ecosystem services. *European Journal Soil Biology, 42*, 3–15.
Lazcano, C., & Domínguez, J. (2011). The use of vermicompost in sustainable agriculture: Impact on plant growth and soil fertility. In M. Miransari (Ed.), *Soil nutrients* (pp. 230–254). Nova Science Publishers.
Lazcano, C., Gomez-Brandon, M., & Dominguez, J. (2008). Comparison of the effectiveness of composting and vermicomposting for the biological stabilization of cattle manure. *Chemosphere, 72*, 1013–1019.
Lazcano, C., Arnold, J., Tato, A., Zaller, J. G., & Domínguez, J. (2009). Compost and vermicompost as nursery pot components: Effects on tomato plant growth and morphology. *Spanish Journal Agriculture Research, 7*, 944–951.
Lazcano, C., Sampedro, L., Zas, R., & Domínguez, J. (2010). Vermicompost enhances germination of the maritime pine (*Pinus pinaster* Ait.). *New Forest, 39*, 387–400.
Liang, C., Das, K. C., & McClendon, R. W. (2003). The influence of temperature and moisture contents regimes on the aerobic microbial activity of a bio solids composting blend. *Bio/Technology, 86*, 131–137.
Lin, J. C., Zhao, S., & Yuan, Q. X. (2021). A novel technology for separating live earthworm from vermicompost: Experiment, mechanism analysis, and simulation. *Waste Management, 131*, 50–60. https://doi.org/10.1016/j.wasman.2021.05.031
Lukashe, S. N., Mnkeni, P. N. S., & Mupambwa, H. A. (2020). Growth and elemental uptake of Rhodes grass *(Chloris gayana)*grown in a mine waste-contaminated soil amended with fly ash-enriched vermicompost. *Environmental Science Pollutant Research, 27*(16), 19461–19472. https://doi.org/10.1007/sl11356-020-08354-7
Manzoor, R., Akhtar, M. S., Khan, K. S., Raza, T., Rehmani, M. I. A., Rosen, C., & Sabag, A. E. (2022). Diagnosis and recommendation integrated system assessment of the nutrients limiting and nutritional status of tomato. *Phyton, 12*(91), 2759–2774. https://doi.org/10.32604/phyton.2022.022988
Meena, M. D., Dotaniya, M. L., Meena, M. K., Meena, K. N., Doutaniya, R. K., Meena, H. S., Moharana, P. C., & Rai, P. K. (2021). Maturity indices as an index to evaluate of Sulphur. *Waste Management, 126*, 180–190.

Melis, P., & Castaldi, P. (2004). Thermal analysis of the evaluation of the organic matter during municipal solid waste aerobic composting process. *Thermochimica Acta, 41*, 209–214.
Misra, R.V., Roy, R. N., & Hiraoka, H. (2003). *On-farm composting methods.*
Mohee, R. (2002). Assessing the recovery potential of solid waste in Mauritius. *Resource Conservation Recycle, 36*, 33–43.
Mohee, R., Karagiannidis, A., Themelis, N., & Kontogianni, S. T. (2010). The mounting problems in managing waste in rapidly developing islands: The Mauritius case. In *3rd International symposium on energy from biomass and waste.* 8–11 November. CISA, Environmental Sanitary Engineering Centre.
Moon, P. (1997). *Final report on basic on-farm composting manual. CM-97-3.*
Mupambwa, H. A., & Mnkeni, P. N. S. (2018). Optimizing the vermicomposting of organic wastes amended with inorganic materials for production of nutrient-rich organic fertilizers. A review. *Environmental Science Pollutant Research, 25*, 10577–10595.
Mupambwa, H. A., Nciizah, A. D., & Nyambo, P. (2022). Can organic soil fertility management sustain farming and increase food security among African smallholder farmers? In *Food security for African smallholder farmers* (Sustainability sciences in Asia and Africa). Springer, Singapore. https://doi.org/10.1007/078-981-16-6771-8-6
Negro, M. J., Solano, P. C., & Carrasco, J. (1999). Composting of sweet sorghum bagasse with other wastes. *Bioresource Technology, 67*, 89–92.
Nurhidayati, N., Machfudz, M., & Murwani, I. (2018). Direct and residual effect of various vermicompost on soil nutrient and nutrient uptake dynamics and productivity of four mustard Pak-Coi (*Brassica rapa* L.) sequences in organic farming system. *International journal of recycling of organic waste in agriculture, 7*, 173–181. https://doi.org/10.1007/s40093-018-0203-0
Omara, A. E. D., Hafez, E. M., Osman, H. S., Rashwan, E., El-Said, M. A. A., Alharbi, K., Abd El-Moneim, D., & Gowayed, S. M. (2022). Collaborative impact of compost and beneficial rhizobacteria on soil properties, physiological attributes, and productivity of wheat subjected to deficit irrigation in salt affected soil. *Plants., 11*, 877.
Onwosi, C. O., Igbokwe, V. C., Odimba, J. N., Eke, I. E., Nwankwoala, M., Iroh, I. N., & Ezeogu, L. I. (2017). Composting technology in waste stabilization: On the methods, challenges and future prospects. *Journal Environment Management, 190*, 140–157.
Papathanasiou, F., Papadopoulos, I., Tsakiris, I., & Tamoutsidis, E. (2012). Vermicompost as a soil supplement to improve growth, yield and quality of lettuce (*Lactuca saliva* L.). *Journal of Food Agriculture and Environment, 10*, 677–682.
Pédrot, M., Dia, A., & Davranche, M. (2010). Dynamic structure of humic substances: Rare earth elements as a fingerprint. *Journal Colloid Interface Science, 345*, 206–213.
Qadir, M. F., Naveed, M., Khan, K. S., Mumtaz, T., Raza, T., Mohy-Ud-Din, W., & Mustafa, A. (2024). Divergent responses of phosphorus solubilizing bacteria with P-laden biochar for enhancing nutrient recovery, growth, and yield of canola (Brassica napus L.). *Chemosphere*, 141565. https://doi.org/10.1016/j.chemosphere.2024.141565
Ramnarain, Y. I., Ansari, A. A., & Ori, L. (2019). Vermicomposting of different organic materials using the epigeic earthworm *Eisenia foetida. International Journal of Recycling of Organic Waste in Agriculture, 8*, 23–36. https://doi.org/10.1007/s40093-018-0225-7
Raj, D., & Antil, R. S. (2011). Evaluation of maturity and stability parameters of composts prepared from agro-indastrial wastes. *Bioresource Technology, 102*, 2868–2873. https://doi.org/10.1016/g.biotech.2010.10.077
Rashtbari, M., Ali, A. H., & Ghorchiani, M. (2020). Effect of vermicompost and municipal solid waste compost on growth and yield of canola under drought stress conditions. *Community Soil Science Plant Analysis, 51*, 2215–2222. [Google Scholar] [CrossRef].
Raza, T., Qadir, M. F., Khan, K. S., Eash, N. S., Yousuf, M., Chatterjee, S., & Oetting, J. N. (2023). Unrevealing the potential of microbes in decomposition of organic matter and release of carbon in the ecosystem. *Journal of Environmental Management, 344*, 118529. https://doi.org/10.1016/j.jenvman.2023.118529

Reinecke, A. J., Viljoen, S. A., & Saayman, R. J. (1992). The suitability of *Eudrilus Eugeniae*. *Soil Biology and Biochemistry, 24*(12), 1295–1307.

Rostami, R., Nabaei, A., & Eslami, A. (2010). Survey of optimal temperature and moisture for worm's growth and operating vermicompost production of food wastes. *Journal Environment Health, 1*, 105–112.

Rynk, R. F., van de Kamp, M., Willson, G. B., Singley, M. E., Richard, T. L., Kolega, J. J., Gouin, F. R., Laliberty, L. L., Kay, D., Murphy, D. W., Hoitink, H. A. J., & Brinton, W. F. (1992a). *On-farm composting handbook*. Natural Resource, Agriculture, and Engineering Service (NRAES).

Rynk, R., Kamp, M., Willson, G. B., Singley, M. E., Richard, T. L., Kolega, J. J., Gouin, F. R., Laliberty, L., Kay, D., Murpy, D. W., Hoitink, H. A. J., & Brinton, W. F. (1992b). *On-farm composting handbook, NRAES-54*. Cooperative Extension Service, North Regional Agricultural Engineering Service.

Sanchez-Monedero, M. A., Cayuela, M. L., Roig, A., Jindo, K., Mondini, C., & Bolan, N. (2018). Role of biochar as an additive in organic waste composting. *Bioresource Technology, 247*, 1155–1164.

Saranraj, P., Anandaraj, K., Raja, R. E., Sivasakthi, S., Nadu, T., Arts, S., & Nadu, T. (2018). Vermicomposting of market wastes and analysis of its physico-chemical and biological properties. *International Journal Research Analysis Review, 5*, 287–296.

Sharma, S., Pradhan, K., Satya, S., & Vasudevan, P. (2005). Potentiality of earthworms for waste management and in other uses–A review. *The Journal of American Science, 1*(1), 4–16.

Sharma, K., & Garg, V. (2018). Comparative analysis of vermicompost quality produced from rice straw and paper waste employing earthworm *Eisenia fetida* (Sav.). *Bioresource Technology, 250*, 708–715.

Shi, W., Norton, J. M., Miller, B. E., & Pace, M. G. (1999). Effects of aeration and moisture during windrow composting on the nitrogen fertilizer values of dairy waste composts. *Applied Soil Ecology, 11*, 17–28.

Shrimal, P., & Khan, T. I. (2017). Studies on the effects of vermicompost on growth parameters and chlorophyll content of bengal gram (*Cicer arietinum* L.) var. RSG-896. *IOSR Journal Environmental Science Toxicology Food Technology, 11*, 12–16.

Silva, M. E. F., de Lemos, L. T., Nunes, O. C., & Cunha-Queda, A. C. (2014). Influence of the composition of the initial mixtures on the chemical composition, physicochemical properties and humic-like substances content of composts. *Waste Management, 34*, 21–27.

Sindhu, R., Gnansounou, E., Rebello, S., Binod, P., Varjani, S., Thakur, I. S., Nair, R. B., & Pandey, A. (2019). Conversion of food and kitchen waste to value-added products. *Journal Environment Management, 241*, 619–630.

Sinha, R. K., Agarwal, S., Chauhan, K., & Valani, D. (2010). The wonders of earthworms & its vermicompost in farm production: Charles Darwin's friends of farmers, with potential to replace destructive chemical fertilizers from agriculture. *Agricultural Sciences, 1*, 76–94. https://doi.org/10.4236/as.2010.12011

Soobhany, N., Mohee, R., & Garg, V. K. (2017). A comparative analysis of composts and vermicomposts derived from municipal solid waste for the growth and yield of green bean (*Phaseolus vulgaris*). *Environmental Science and Pollution Research, 24*, 11228–11239.

Stevenson, F. J. (1994). *Humic chemistry: Genesis, composition, reactions*. Wiley.

Suthar, S. (2006). Potential utilization of guar gum industrial waste in vermicompost production. *Bioresource technology, 97*(18), 2474–2477. https://doi.org/10.1016/j.biortech.2005.10.018

Suthar, S., & Singh, S. (2008). Vermicomposting of domestic waste by using two epigeic earthworms (*Perionyx excavatus* and *Perionyx sansibaricus*). *International Journal Environment Science Technology, 5*, 99–106. https://doi.org/10.1007/BF03326002

Syarifinnur, N. Y., Prasetya, B., & Handayanto, E. (2020). Effectiveness of compost and vermicompost from market organic waste to improve soil chemical properties. *IOP Conference Series: Materials Science and Engineering, 980*(1), 012068.

Syed, A., Raza, T., Bhatti, T. T., & Eash, N. S. (2022). Climate impacts on the agricultural sector of Pakistan: Risks and solutions. *Environmental Challenges, 6*, 100433.
Talyan, V., Dahiya, R. P., & Sreekrishnan, T. R. (2008). State of municipal solid waste management in Delhi, the capital of India. *Waste Management, 28*, 1276–1287.
Tiquia, S. M. (2005). Microbiological parameters as indicators of compost maturity. *Journal Application Microbiology., 99*, 816–828.
Tomati, U., Madejon, E., & Galli, E. (2000). Evolution of humic acid molecular weight as an index of compost stability. *Compost Science Utilization, 8*, 108–115. https://doi.org/10.1080/1065657x.2000.10701756
United Nations. (2000). *Waste, State of the Environment in Asia and the Pacific*. United Nations Publication.
Ur Rehman, S., Ijaz, S. S., Raza, M. A., Din, A. M. U., Khan, K. S., Fatima, S., Raza, T., Mehmood, S., Saeed, A., & Ansar, M. (2023). Soil organic carbon sequestration and modeling under conservation tillage and cropping systems in a rainfed agriculture. *European Journal of Agronomy, 147*, 126840.
Usmani, Z., Kumar, V., & Mritunjay, S. K. (2017). Vermicomposting of coal fly ash using epigeic and epi-endogeic earthworm species: Nutrient dynamics and metal remediation. *RSC Advances, 7*, 4876–4890.
Verma, S. L., & Marschner, P. (2013). Compost effects on microbial biomass and soil P pools as affected by particle size and soil properties. *Journal Soil Science Plant Nutrition, 13*, 313–328.
Vukovic, A., Velki, M., Ecimovic, S., Vukovic, R., Štolfa Camagajevac, I., & Loncaric, Z. (2021). Vermicomposting-facts, benefits and knowledge gaps. *Agronomy, 11*, 1952.
Wang, T. T., Wang, S. P., Zhong, X. Z., Sun, Z. Y., Huang, Y. L., Tan, L., & Kida, K. (2017). Converting digested residue eluted from dry anaerobic digestion of distilled grain waste into value-added fertilizer by aerobic composting. *Journal of cleaner production, 166*, 530–536. https://doi.org/10.1016/j.jclepro.2017.08.075
Wang, Y., & Ai, P. (2016). Integrating particle physical geometry into composting degradation kinetics. *Bioresource Technology, 200*, 514–520.
Wang, X., Cui, H., Shi, J., Zhao, X., Zhao, Y., & Wei, Z. (2015). Relationship between bacterial diversity and environmental parameters during composting of different raw materials. *Bioresource Technology, 198*, 395–402.
Wang, F., Wang, X., & Song, N. (2021). Biochar and vermicompost improve the soil properties and the yield and quality of cucumber (*Cucumis sativus l.*) grown in plastic shed soil continuously cropped for different years. Agriculture ecosystem. *Environment, 315*, 107425.
Wu, S., Shen, Z., Yang, C., Zhou, Y., Li, X., Zeng, G., Ai, S., & He, H. (2016). Effects of C/N ratio and bulking agent on speciation of Zn and cu and enzymatic activity during pig manure composting. *International Biodeterior Biodegrade, 119*, 429–436.
Wurff, A. W. G., van der Fuchs, J. G., Raviv, M., & Termorshuizen, A. (2016). *Handbook for composting and compost use in organic horticulture*. BioGreenhouse. https://doi.org/10.18174/375218
Yang, F., Li, G. X., Yang, Q. Y., & Luo, W. H. (2013). Effect of bulking agents on maturity and gaseous emissions during kitchen waste composting. *Chemosphere, 93*, 1393–1399.
Yang, Y. J., Awasthi, M. K., Ren, X. N., Guo, H. H., & Lv, J. L. (2019). Effect of bean dregs on nitrogen transformation and bacterial dynamics during pig manure composting. *Bioresource Technology, 288*, 121430. https://doi.org/10.1016/j.biortech.2019.121430
Yue, Y., Lin, Q., Xu, Y., Li, G., & Zhao, X. (2017). Slow pyrolysis as a measure for rapidly treating cow manure and the biochar characteristics. *Journal of analytical and applied pyrolysis, 124*, 355–361. https://doi.org/10.1016/j.jaap.2017.01.008
Zandvakili, O. R., Barker, A. V., Hashemi, M., & Etemadi, F. (2019). Biomass and nutrient concentration of lettuce grown with organic fertilizers. *Journal of Plant Nutrition, 42*(5), 444–457. https://doi.org/10.1080/01904167.2019.1567778
Zhao, X., Tan, W., Peng, J., Dang, Q., Zhang, H., & Xi, B. (2020). Bio-waste-source-dependent synthetic pathways of redox functional groups within humic acids favoring pentachlorophenol dechlorination in composting process. *Environment International, 135*, 105380.

Chapter 17
Advancing Phytoremediation from Lab Research to Field Applications

Niti Chawla, Sanjeev Kumar, and Lalita Gupta

1 Introduction to Phytoremediation

The desire to keep up with a perplexedly growing population has led to progress towards rapid, innovative, and advanced but disruptive technologies. Intensive agricultural practices, industrial activities, urbanization, and construction activities largely rely on these modern technologies, which exploit the environment and natural resources around the world (Kamran et al., 2021). As a result, soil and water ecosystems have been contaminated with harmful substances such as agro-chemicals, heavy metals, metalloids, radionuclides, persistent organic compounds, and oil spills (Malik et al., 2017; Kaur et al., 2021; Kumar & Verma, 2023; Prakash et al., 2013). These contaminated sites should be replenished immediately. Otherwise, the existence of life on Earth will be threatened in the near future. Most conventional remediation approaches to the contaminated sites are limited because they involve some form of environmental degradation (DalCorso et al., 2019). Some examples of these approaches include incineration, thermal desorption, excavation, landfilling, electrical remediation, and photo-oxidation. These approaches mainly rely on high temperatures, radiation, or chemicals and are less efficient, expensive, and damaging to the environment.

As an alternative to the conventional approaches, phytoremediation has several benefits, such as sustainability, manageability, environmental friendliness, and affordability, that make it a highly promising approach for combating polluted environments. It has been found to be effective in the remediation of varied kinds of pollutants, such as pesticides and heavy metals (Chandra & Kumar, 2017; Bhat et al., 2020; Sabreena et al., 2022; Shao et al., 2022). During phytoremediation, plant

N. Chawla · S. Kumar
Department of Biotechnology, Chaudhary Bansi Lal University, Bhiwani, Haryana, India

L. Gupta (✉)
Department of Zoology, Chaudhary Bansi Lal University, Bhiwani, Haryana, India

V. Kumar et al. (eds.), *Environmental Engineering and Waste Management*,
https://doi.org/10.1007/978-3-031-58441-1_17

species employ different strategies to rejuvenate the contaminated sites (Ashraf et al., 2019; Sabreena et al., 2022). These strategies include phytoextraction (plant accumulation), phytotransformation/phytodegradation, phytovolatilization, phytostabilization, and rhizofiltration (Pivetz, 2001). Various mechanisms involved in these strategies include immobilization or extraction of contaminants from contaminated sites and their accumulation in plant tissues, conversion of contaminants to less toxic substances, volatilizing contaminants in non-toxic form into the atmosphere, stabilizing contaminants on root surfaces or in the rhizosphere soil, or filtering contaminants from contaminated water by the plant roots, etc. Plants remediate different types of pollutants at variable rates depending on their physiology and soil characteristics (Shao et al., 2022). Site-specific phytoremediation approaches are also available for highly contaminated soil, sludge, and water bodies (Chandra et al., 2018a; Kumar et al., 2021a, b). These are called riparian buffer strips or plant caps. Moreover, the competence of these conventional phytoremediation strategies can be enhanced using a variety of tools, including the use of natural or chemical additives, microbial inoculation, and genetic engineering of plants (Sabreena et al., 2022).

The term "phytoremediation"was coined by Ilya Raskin of Rutgers University. The earliest phytoremediation was observed in the sixteenth century as metal resistance and in the 1940s as the breakdown of organic pollutants in some plants. Extensive research on phytoremediation began in the early 1990s. Scientists have so far studied many potential phytoremediation facilities that can detoxify various contaminants from the environment, such as pesticides, heavy metals, radionuclides, explosives, and oil spills. However, it is not yet practically feasible due to the fundamental challenges (Shao et al., 2022) that still remain unanswered. The main challenge observed is the lengthy period (several years) required for the complete removal of contaminants from the sites. The sustainability of phytoremediation primarily relies on climate and weather conditions. At field/large scale these climatic or weather conditions pose many types of stresses on plants. As a result, low biomass and slow growth rates of plants usually occur in the field when compared to the laboratory scale (Huang et al., 2004). The accumulative capacity of some plants can also be reduced or even completely lost due to the spread of pests and diseases (Nedunuri et al., 2000). It is also difficult to enhance the bioavailability of firmly bound contaminants in soil on a large scale. Agricultural practices and soil additives can also influence the availability of pollutants.

The introduction of exotic plant species as hyperaccumulators can also be seen as a challenge because that can affect the diversity of native plants (Gressel & Al-Ahmad, 2005). Moreover, phytoremediation applies to sites with a low level of contamination due to unmanageable plant growth on heavily contaminated soil. Therefore, it is also difficult to prevent the risk of accumulated pollutants entering the food chain (Kumar et al., 2022a, b). It is also difficult to treat or dispose of the large amounts of contaminated biomass produced on a field scale. Spatial heterogeneity under field conditions can also lead to data variability during large-scale trials.

Coordinated efforts by the government, industry, and researchers have proposed solutions to the challenges observed at the field scale. Some of these strategies include inoculation with rhizobia or beneficial endophytes to enhance plant growth and the use of native plant species that can easily adapt to contaminated sites and require little maintenance. Therefore, this chapter summarizes the successes achieved so far and the challenges faced in the large-scale application of phytoremediation strategies. The existing information on conventional and improved phytoremediation strategies, including their principles, mechanisms, possibilities, advantages, and limitations, are also discussed.

2 Need for Phytoremediation

In recent decades, a variety of industrial, technological, agricultural, and therapeutic sources have contributed to an alarming increase in waste and global pollution (Gudeta et al., 2023; Sharma et al., 2023). Additionally, inadequate waste disposal and poor management strategies caused the accumulation of hazardous organic chemicals, heavy metals, and radioactive materials in the environment impacting natural resources, ecosystems, and public health (Sabreena et al., 2023). Therefore, efficient repair techniques are highly recommended. Many of the conventional synthetic remedial techniques, such as physicochemical, photochemical, or photocatalytic, rely on hazardous radiations, chemicals, or high temperatures that are already in practice. They are very expensive, energy-intensive, less efficient, have suboptimal performance, produce a lot of sludge, and are technically difficult to implement. They also produce secondary pollutants that, in turn, degrade the soil, impair its biological activity, and make it barren. Furthermore, these approaches cannot handle contaminants at the release site and require transport to the site of action. This entails transportation and handling hazards and exposes the surrounding area to exposure risks. Additionally, remediated or contamination-free soil is typically not brought to its original site after treatment and is disposed of somewhere else as waste. Above all, the limitations of conventional remediation techniques call for eco-friendly, cost-effective, and efficient alternatives that can effectively cleanse the environment.

Phytoremediation is one of the emerging alternative approaches that use plants and their associated microbes to decontaminate the environment. It is an increasingly developing technology that has the potential to clean a wide range of organic and inorganic wastes in an efficient- and cost-effective manner. The protocol is relatively easy to implement and also has minimal maintenance costs after initial site preparation and planting. It is based on natural systems and has received significant scientific attention because it performs on-site remediation of the contaminated sites. It is essentially self-sustaining because it supplies nutrients, organic materials and oxygen to the soil via plant and microbial metabolic activities. Additionally, the plants and their associated microbes improve the overall soil quality and texture of the remedial site. Moreover, phytoremediation can be employed on remedial sites of any size and location that can support plant growth. This process also has high public acceptance thus, making it an attractive option for remediation. Hence, there is no uncertainty in accepting phytoremediation as a consistent, sustainable, and economically favorable approach applicable to contaminated sites.

3 Conventional Phytoremediation Strategies

In phytoremediation, plants employ different strategies to remediate contaminated sites. These strategies include phytoextraction, phytostabilization, rhizofiltration, phytotransformation/phytodegradation, rhizodegradation, and phytovolatilization (Fig. 17.1). The important features of these strategies are discussed below.

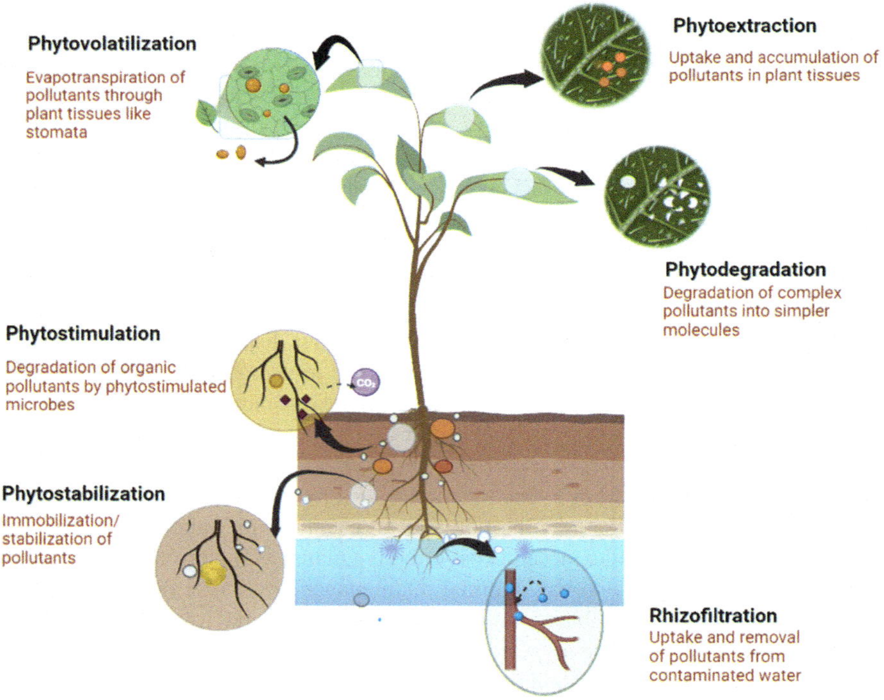

Fig. 17.1 Phytoremediation strategies of plant for remediation of contaminated sites. (This figure was created using BioRender available at https://app.biorender.com/illustrations/6325e71232c4984 6faebee40)

3.1 Phytoextraction

Phytoextraction is also known as phytoabsorption, phytosequestration, phytoaccumulation, or phytoconcentration. In this technique, plants absorb pollutants from the contaminated sites (soil or water) via their roots and, with the help of translocation proteins, translocate them to easily harvestable aboveground aerial parts. The root-to-shoot migration system keeps the absorption process in continuation. Generally, plants do not break down the contaminants, however, they accumulate them in their cellular components, such as vacuoles, cell walls, and cell membranes. Plants resist or protect themselves from the harmful effects of these contaminants during transportation and storage. Such plants are called hyperaccumulators, for example, Indian mustard, sunflower, barley, serpentine, and dandelions (Qadir et al., 2021). Hyperaccumulators are ideal for plant extraction processes because they possess certain characteristics, such as growing rapidly, attaining high biomass in a short period, and having extensive root and shoot systems that can accumulate contaminants even up to a hundred times higher than those of normal plants (Chandra et al., 2018b). Additionally, they are (a) capable of absorbing and accumulating contaminants very effectively in easily harvestable aerial parts; (b) highly adaptable to varied environmental conditions; (c) easy to cultivate; (d) highly resistant to pathogens and pests; and (e) repel herbivores to prevent contaminants from entering the food chain (Shabani & Sayadi, 2012). Typical hyperaccumulating plants are reported in the plant families *Brassicaceae*, *Asteraceae*, *Labiatae*, and *Euphorbiaceae*.

The efficiency of phytoextraction depends upon several factors, such as the plant type, pollutant concentration, and climate. It is primarily used to remove heavy metals, metalloids, or radionuclides from contaminated soils, sediments, and sludges. To a lesser extent, it is applicable to polluted water as well. Several studies have been carried out to explore the phytoextraction potential of different plant species. Jacobs and his co-workers observed a significant absorption of zinc (Zn) and cadmium (Cd) by *Noccaea caerulescens* leaves under field conditions. On extending the growth period from 6 months to 12 months, they further observed 3-fold and 1.5-fold absorption for Zn and Cd, respectively, and concluded *N. caerulescens* as an ideal candidate for phytoextraction (Jacobs et al., 2018). Ghazaryan et al. compared the phytoextraction potential of *Melilotus officinalis* and *Amaranthus retroflexus* growing on copper- and molybdenum-contaminated soils. Their results revealed that *M. officinalis* preferred to accumulate zinc in roots, while *A. retroflexus* favored building up copper in the shoot region (Ghazaryan et al., 2021). Interestingly, the cadmium and zinc remediation abilities of three varieties of Napier grass (*Pennisetum purpureum*) revealed that 197.5 g ha^{-1} and 5023.9 g ha^{-1} of cadmium and zinc, respectively, were removed by *P. purpureum* cv. Guiminyin (Yang et al., 2020). *Helianthus annuus* has also been found to be effective in the extraction and accumulation of radionuclides and certain heavy metals. Miller planted it in a pond contaminated with the radionuclides Cesium-137 and Strontium-90 in the Chernobyl area of Ukraine. Radionuclides were reported to accumulate in *H. annuus* roots and resulted in a 90% reduction of their concentration in water within two weeks (Miller, 1996).

The process of phytoextraction is carried out in steps. In the first step, the corresponding plant or crop species are grown on the contaminated site, and once they reach the desired growth or pollutant accumulation in the tissues, they are recycled or disposed of. Valuable pollutants are reclaimed from them by phytomining, and the rest of the biomass is disposed of after certain treatments like incineration or pyrolysis which produce ash that is further stabilized first and disposed of thereafter. Biomass is also disposed of in a landfill, but anaerobic decomposition can solubilize contaminants and leach them into groundwater. Therefore, thermal treatment is proven to be more promising than landfills. Phytoextraction can be natural or induced. In natural phytoextraction, plants use their inherent potential to extract contaminants from the site. Contaminants must be water soluble then only they are absorbed by the plants. Moreover, the plants also produce siderophores (Iron chelating compounds), organic acids, and many other chelating agents to chelate and absorb them if they are not soluble (Padmavathiamma & Li, 2012). Sometimes chelating agents are also added artificially to increase the availability of contaminants to plants. This is known as induced phytoextraction. These chemicals are added throughout the growth cycle of the plants and therefore, this method can have the potential risk of excessive doses over the years (Lombi et al., 2001).

There are associated advantages as well as limitations to the phytoextraction strategies. The main advantage of this technique is that it is less expensive compared to other known restoration techniques. This is mainly because of the plant's ability to grow in the presence of sunlight without requiring any other source of energy. The contaminants with high economic value can also be easily recovered from the plant tissues. Plants with high biomass content can also be used for bioenergy production after pollutant recovery. Material handling is also limited, which reduces the transportation cost when compared with other approaches such as excavation and landfilling. Another advantage of this technology is its minimal impact on the site of action, which makes it more acceptable. Regarding limitations, this technology is still in its developing phase. It is slow in action as decontamination can take a relatively long time, even years, depending on the availability of pollutants, therefore, repeated cultivation of hyper-accumulating species may be required to reach the desired level of decontamination. It can also be used most effectively only on lightly to moderately contaminated soil. It can address a specific set of contaminants and conditions in a way similar to other technologies (therefore, it is not considered a silver bullet). The absolute performance of plant extraction is limited by metal speciation. Most plants that are considered suitable for this technique do not thrive in submerged (wetland) conditions. It removes metals, metalloids, and radionuclides from the contaminated sites with limitations to the organic contaminants. Its decontaminating capability is usually confined to the immediate vicinity of the plant roots which are mostly limited to the top soil. Therefore, root depth determines the effectiveness of plant-mediated extraction. Potential leaching of metals from original sources and dilution of contamination by cultivation might be a challenge to this methodology. The accumulation of contaminants on the above-ground aerial parts of plants can pose a risk to grazing animals.

Exploring the potent plant species that can extract contaminants from the wetlands is a prerequisite to enhance the applicability of phytoextraction. The use of native plants is mostly preferred at the field scale to reduce the number of trials that are required with exotic species. If non-native plants are selected, steps should be taken to prevent further infestation such as using sterile plants or harvesting plants prior to planting.

3.2 *Phytostabilization/Phytoimmobilization*

Phytostabilization is the in-situ stabilization or immobilization of potentially toxic contaminants by plants. The stabilization might occur through immobilization in rhizosphere soil, adsorption to the root surface, or accumulation in root cell vacuoles. Phytochemicals released by plant roots play an important role in the stabilization process. Plants fix pollutants using a variety of mechanisms. They can physically, chemically, or biologically alter the rhizosphere area compared to the bulk soil. This change alters the chemical form of contaminants, making them slightly more complex and immobile in the rhizosphere region. Hammond and his co-workers evaluated the arsenic stabilization potential of *Prosopis juliflora* and found that this metalloid is anchored to the epidermis of plant root. Arsenic binds to iron sulfide in the vacuole and forms arsenic-tris-thiolate in the rhizosphere (Hammond et al., 2018). In addition, the rhizosphere region is naturally rich in several organic molecules, such as organic acids, amino acids, sugars, proteins, lipids, and flavonoids, which fix impurities and render them insoluble by redox or precipitation reactions. In addition, microorganisms that live in the rhizosphere, such as bacteria and mycorrhizae, as well as endophytes, also increase the efficiency of plant stabilization. Recent studies revealed a strong phytostabilization in uranium and lead-contaminated soil by *Leptochloa fusca* that was, in fact, due to three bacterial endophytes namely, *Pantoea stewartii* ASI11, *Enterobacter* sp. HU38 and *Microbacterium arborescens* HU33 in the plant (Ahsan et al., 2017). These microorganisms may release certain chemicals that immobilize the contaminants. Some contaminants, mainly heavy metals, can also bind to the cell wall of microorganisms and thus, become unavailable in the soil. The fibrous root system also favors the plant stabilization process as it provides more surface area for the extensively branched microscopic root processes to contact the contaminated site. However, root depth varies greatly with plant type, water content, and soil conditions.

The stabilization process can be enhanced by the artificial addition of additives or chemicals, called aided phytostabilization. These additives increase the levels of organic matter and essential nutrients in the soil, improve physicochemical and biological properties, promote plant colonization, and improve soil water retention. Some of the benefits associated with this technique include reducing the bioavailability of these toxic contaminants and limiting their toxicity to plants. It protects the surrounding ecosystem by preventing the leaching of contaminants into groundwater and the diffusion of contaminants by wind and running water. It is also suitable for replanting, as it reduces the risk of contaminated soil. Ease of use, simplicity, safety, friendliness, and low-cost permit its usage in a wide variety of polluted sites. This technology produces no secondary waste that requires further treatment, making it highly effective where rapid immobilization is required to protect ground- and surface-water. It is suitable for large areas with low levels of contaminants and is probably more tolerable than alternatives such as soil washing that does not interfere with ongoing site activity. Moreover, it gives effective results when used in combination with other technologies. It also provides an opportunity to generate income from the vegetation used for crop stabilization. Plant stabilization sites can be managed to produce valuable products from plant biomass. This could be non-edible products such as wood, bioenergy, biochar, essential oils, or other plant products. In addition, plants play an important role in environmental clean-up as carbon sinks (Susarla et al., 2002). Besides the numerous advantages, some limitations have also been observed with this technique.

3.3 Phytodegradation/Phytotransformation

Phytodegradation is the breakdown of pollutants inside or outside the plants through intra- or extra-metabolic processes. Therefore, these two major mechanisms of phytodegradation are either uptake and subsequent degradation of contaminants or degradation outside by the released enzymes. Doty et al., 2003 provided the first evidence of a tropical tree (*Leucaena leucocephala*) that effectively metabolizes organic pollutants such as ethylene dibromide and trichloroethylene from contaminated soil and groundwater. de Farias et al., 2009 observed the phytodegradation of petroleum components by *Erythrina crista-galli* L. The main pollutants susceptible to phytodegradation are organic compounds, which are broken down into simpler forms and incorporated into plant tissues. As primary photosynthetic producers, plants usually derive anabolic precursors from inorganic chemicals and incorporate them into organic compounds. However, in this process, the absorption of organic compounds takes place directly. The absorption depends on the solubility and hydrophobicity/hydrophilicity of organic compounds. Strongly hydrophobic compounds tend to bind to or distribute to the root surface but are unable to move further within the plant. Moderately hydrophobic compounds are most readily taken up and translocated into the plants, whereas hydrophilic or highly soluble compounds do not translocate easily.

Phytodegradation can be applied to soil, sediment, sludge, groundwater, or surface water to remove organic contaminants such as chlorinated solvents, herbicides, pesticides, inorganic nutrients, explosives, and many more. Plants that can undergo phytodegradation are poplar, willow, cottonwood, rye, Bermuda, sorghum, clover, alfalfa, cowpea, etc. Examples of enzymes produced by plants include oxidoreductases, nitrilases, dehalogenases, peroxidases, laccases, and phosphatases

(McCutcheon et al., 2003). Phytodegradation is more efficient than rhizodegradation because microorganisms cannot survive in highly polluted areas, but plants can survive and degrade the contaminants into simple substances. Thus, phytodegradation can even occur in soils where microbial degradation is not possible. Microbes inhabiting the rhizosphere apply positive effects on plant growth and also assist them in the degradation of pollutants. Therefore, a fine balance must be maintained among rhizosphere microbes, plants, and their respective tolerances to various contaminants in order to achieve sustainable plant degradation of organic pollutants. Some limitations are associated with phytodegradation. The toxic intermediates or degradation products may form in phytodegradation, which can affect the health of soil, plants, and microbes as well. Sometimes it becomes difficult to detect the presence of pollutants or their degradation products in plants and also confirm the disappearance of contamination from the sites.

3.4 *Rhizodegradation*

Rhizodegradation is also known as plant-assisted degradation, plant-assisted in-situ biodegradation, or rhizosphere biodegradation. In this technique, organic contaminants are degraded by the microorganisms present in the rhizospheric region. Plants secrete about one-fifth of all photosynthetic products into the rhizosphere region, including sugars, organic acids, amino acids, fatty acids, sterols, growth factors, nucleotides, enzymes, and many other compounds. As a result, microbial growth in the rhizosphere is much greater (approximately 3–4 times) compared to the bulk soil (Weyens et al., 2009). Microorganisms co-metabolically degrade pollutants and break them down into simple compounds. In addition to releasing root exudates, plants improve the physical and chemical properties of the soil. They aerate the soil and improve the water content of the soil. In addition, plants perform better over time as new roots penetrate nearby soil areas. If the root exudate is water-soluble, root action can reach slightly deeper than the root zone. If polluted groundwater is within the root's area of influence, it can be treated as well. The rate of rhizodegradation depends on the type, amount, and intensity of exudates and enzymes produced by plant roots. There are many contaminants that are subjected to rhizodegradation. These include petroleum hydrocarbons, polyaromatic hydrocarbons, pesticides, etc. Some examples of plants capable of rhizodegradation are mulberry, apple, orange, spearmint, alfalfa, soybean, rye, Bermuda, etc.

In rhizodegradation, the destruction of contaminants is done on the spot. Dams et al., 2007 found enhanced degradation of pentachlorophenol by a strain of *Sphingobium chlorophenolicum* in the field of wheat crops. Accumulation of contaminants either in the plants or release into the atmosphere is less likely because microbial-mediated degradation occurs at the source under the influence of plants. Low installation and maintenance costs are required for this technique in comparison to other remedies with some drawbacks, such as the fact that plants can take a longer time to develop an extensive root system. Physical structure or moisture conditions of the soil also affect root penetration into the soil, and fertilizers may be needed to fulfill the nutritional requirements of microbes. In addition, to support degrading microbes, exudates can stimulate non-degrading microorganisms as well, at the expense of degraders. Moreover, the interpretation of the results of rhizodegradation is also difficult because it takes place collectively with phytodegradation.

3.5 Rhizofiltration

Rhizofiltration is the absorption, concentration, and precipitation of pollutants from contaminated water by the rapidly growing roots of plants. The fibrous root system and large root surfaces play an important role here. A plant is first grown hydroponically in contaminated water and allowed to acclimatize. These plants are then transported to contaminated sites to collect pollutants and harvested once the roots are saturated. Rai (2019) investigated the simultaneous removal of six toxic metals (Fe, Cu, Cd, Cr, Zn) by three wetland plants, *Pistia stratiotes* (water lettuce), *Spirodela polyrhiza* (duckweed), and *Eichhornia crassipes* (water hyacinth), from a Ramsar site of a biodiversity hotspot (Rai, 2019). Pratas et al., 2012 evaluated the potential of aquatic plants for the rhizofiltration of uranium (U) from contaminated water. They observed that *Callitriche stagnalis*, *Lemna minor*, and *Fontinalis antipyretica*, growing in the uraniferous geochemical provinces of Central Portugal, can accumulate significant amounts of uranium. The highest concentration of uranium was found in *Callitriche stagnalis* (1948.41 mg/kg), followed by *Fontinalis antipyretica* (234.79 mg/kg), and *Lemna minor* (52.98 mg/kg).

Rhizofiltration is highly flexible and applicable to the treatment of surface and groundwater, industrial and domestic wastewater, rainwater, acid mine effluent, agricultural effluent, diluted sludge, and solutions contaminated with radionuclides. It is an eco-friendly method because wastewater is treated on-site, resulting in minimal impact on the environment. It has been found to be very efficient in the removal of even very small concentrations of contaminants from large volumes of water. Removed contaminants remain confined to roots only and do not get transported to shoot, so contaminated biomass debris is produced in small amounts and its disposal can be done easily. Harvested biomass can be used as an alternative to fossil fuels. In addition, this technology also prevents contaminant runoff in nearby regions.

Rhizofiltration is relatively inexpensive but potentially more effective than comparable techniques. It shares similarities with phytoextraction as it mainly removes the metals, metalloids, and radionuclides from the contaminated sites and accumulates them in tissues. Significant removal of heavy metals such as lead, copper, zinc, cadmium, nickel, and radioactive contaminants from liquid streams has been reported by several aquatic as well as terrestrial plant species. Some examples of such plants are sunflower, Indian mustard, tobacco, rye, spinach, maize, iris leaf firefly, and many more. Rhizofiltratiom has some limitations, too. Contaminants below the rooting depth cannot be extracted. Plants used for rhizofiltration may not grow in heavily polluted areas. Most importantly, it may take years to reach regulatory levels. This leads to long-term maintenance. There may be a combination of metals and organics that cannot be processed by this method. Crops grown in polluted water and soil can easily create a risk to human and animal health. Therefore, special attention should be paid to the harvesting process and only non-forage crops should be selected for the root filtration recovery method.

3.6 Phytovolatilization

Phytovolatilization is the absorption of pollutants by plant roots, their subsequent transformation into less toxic volatile forms, and then releasing them into the atmosphere via transpiration. Thus, phytodegradation and phytotransformation accompany phytovolatilization. Phytovolatilization is applicable mainly to remove organic pollutants from groundwater, soil, sediment, and sludge. This technique works well when volatilized pollutants have a low toxic effect and diffused from the ground into the atmosphere. Guarino et al., 2020 studied the phytoremediation of arsenic using the *Arundo donax* L. plant and its associated microbial strains *Stenotrophomonas maltophilia* sp. and *Agrobacterium* sp. They grew the plants in sterile garden soil amended with 2, 10, or 20 mgL^{-1} of sodium arsenite ($NaAsO_2$). They observed that 75% of arsenic was volatilized by transpiration, 25% remained in the soil, and 0.15% accumulated in the plant. Volatilization directly takes place from stems and leaves, or through cuts, the outermost layer epidermis, suberin, and other surface layers. Also, some compounds in plant systems can travel up the transpiration stream and be lost to the atmosphere during the transpiration process through stomata. The advantage of phytovolatilization compared to other phytoremediation strategies is that heavy metal contaminants (metalloids) are removed from the field and dispersed as gaseous compounds without the need for crop harvesting or disposal. However, as a remedial strategy, plant volatilization does not completely remove contaminants from the environment. It just transfers pollutants from the ground into the atmosphere, polluting the surrounding air with toxic, volatile compounds. Further, there is a risk of their redeposition in soils by precipitation (Vangronsveld et al., 2009). Some compounds, after they are released into the atmosphere, can react with hydroxyl radicals and remain in the atmosphere as serious pollutants. Therefore, a risk assessment is required before applying this technique in the field.

Pollutants or metabolites released into the atmosphere may undergo more effective or faster natural degradation processes, such as photodegradation. Phytovolatilization is a less expensive method, and there is no need to install a technical pump-and-treat system. A matter of consideration is the effect of climatic and seasonal conditions on the efficiency of phytovolatilization. The water absorption rate does not remain constant all the time. For example, water absorption by deciduous plants is significantly reduced in winter. Besides that, root depth also affects the efficiency of extracting pollutants from groundwater. The deeper the roots, the more pollutants can be extracted from the groundwater. Although in phytovolatilization, metabolites (transformed pollutants) are released into the atmosphere, however, a small amount might remain accumulated in the plant tissues and can pass to the food chain.

Contaminants amenable to phytovolatilization are chlorinated solvents (trichloroethane, carbon tetrachloride), aromatics (BTEX), explosives (tri-nitrotoluene, RDX), Inorganic impurities (Se, Hg), herbicides, nutrients, etc., and the plants used for volatilization include poplar, rye, Bermuda, sorghum, fescue, alfalfa, mustard, and canola.

4 Improved Phytoremediation Technologies

Phytoremediation can effectively remove contaminants from different sites; however, there are ways to make it even more efficient. Addition of synthetic and natural additives, the introduction of microorganisms that can assist plants or genetically modify plants to make the changes needed for efficient phytoremediation (Fig. 17.2).

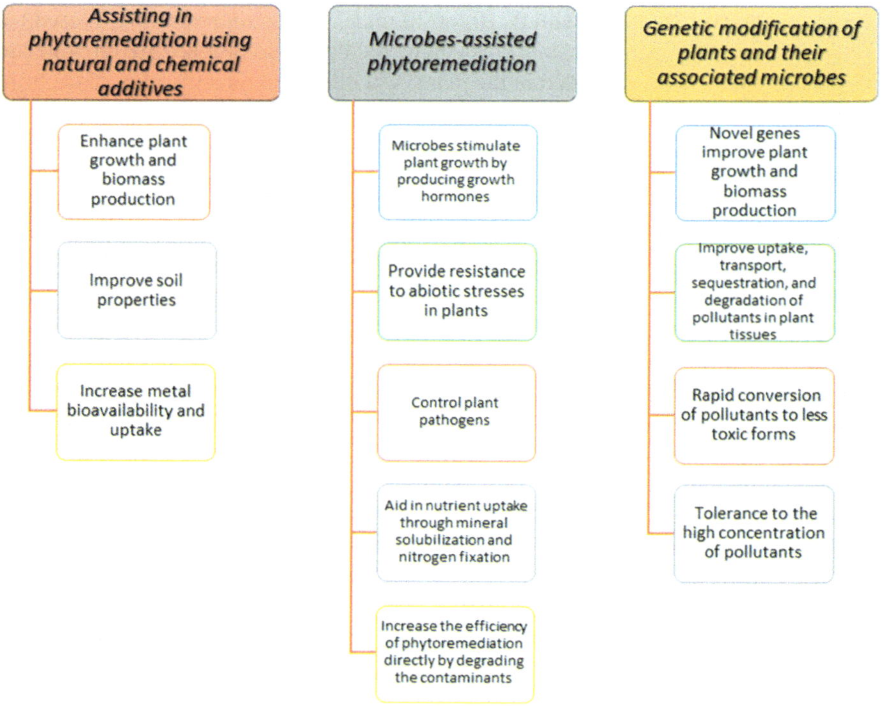

Fig. 17.2 Phytoremediation improving technologies and their effects

Adjuvants promote the formation of metal-chelator complexes, increase plant tolerance to contaminants, and also facilitate contaminant mobility and accumulation within plants, thereby increasing the overall efficiency of phytoremediation. Microorganisms promote the growth of plants and their ability to accumulate harmful substances (Agrawal et al., 2022). Microorganisms can also produce chelating agents that make pollutants bioavailable to plants. Details of these improved phytoremediation strategies are discussed below.

4.1 Assisting in Phytoremediation Using Natural and Chemical Additives

In practice, various additives are used to support phytoremediation. These additives enhance plant growth and biomass production. In addition, they also improve soil properties such as texture, moisture, and nutrient retention. Occasionally, these additives act as chelating agents, increasing the solubility, mobility, and thus bioavailability of nutrients and other organic or inorganic substances, mainly metals, metalloids, and radionuclides, thereby increasing their extraction by plants. As a result, nutrients promote plant growth and pollutants accumulate in tissues and are removed from the environment.

These additives are of two main types: natural additives and synthetic additives. Natural additives are organic/inorganic natural substances used as such or resulting from the processing of natural substances. The most commonly reported additives are the wastes from the paper industry, distilleries, poultry litter, biochar, compost, organic acids, fly ash, sugar foam, lime, zero-valent iron, clay, etc. *Trifolium repens* L. has been reported to enhance the extraction of heavy metals such as cadmium, chromium, and nickel from the soil when the soil is amended with sugar beet residues (agricultural industrial waste) (Azc'on et al., 2009). Composted sewage sludge has also been reported to cause the solubilization of cadmium and nickel in soil and enhance their absorption by sunflower plants. Chelating agents in compost also dissolve these metals and make them bioavailable to sunflower plants (Liphadzi & Kirkham, 2006).

Natural additives are readily biodegradable in soil and therefore may pose less risk to the environment than synthetic additives. The use of natural additives to support phytoremediation is therefore more promising. Synthetic additives include ethylene diamine tetra-acetic acid (EDTA), ethylene glycol tetra-acetic acid (EGTA), diethylenetriamine acid (DTPA), S, S-ethylenediamine disuccinic acid (EDDS), and sodium dodecyl sulfate (SDS). These chemicals act primarily as chelators, increasing the bioavailability of substances, mainly heavy metals, and radionuclides, and facilitating their uptake and translocation into plant tissues (Sarwar et al., 2017). EDTA is the most studied synthetic chemical additive, with great potential for heavy metal mobilization. It improves the uptake of heavy metals by roots after the formation of soluble complexes with metals and also aids their loading in the xylem (Awokunmi et al., 2012; Dipu et al., 2012). In one experiment, Schaider et al., 2006 showed the presence of EDTA-complexed cadmium, lead, and iron in the xylem sap of *Brassica juncea* cultured in solution as well as sand. EDTA can even enhance the uptake of some heavy metals up to two hundred times, depending on the plant species and type of soil. In other cases, EDTA also enhanced the uptake of copper, lead, zinc, nickel, and cadmium in *Lolium perenne* and *Zea mays* (Zhao et al., 2010), lead in *Sedum alfredii, Vetiveria zizanioides,* and *Vicia faba* (Andra et al., 2009; Sun et al., 2009; Shahid et al., 2014) and cadmium in *Helianthus annuus* (Chen & Cutright, 2001).

EGTA is a very potent and widely used chelating agent. Like EDTA, it also increases the absorption of pollutants, especially heavy metals, by plants. Increased cadmium accumulation has been reported in the plant species *Calendula officinalis* and *Mirabilis Jalapa* in the presence of EGTA (Liu et al., 2010; Wang & Liu, 2014). An increase in Pb remediation was also seen in case of *Cicer arietinum* in the presence of EGTA (Sakouhi et al., 2016). SDS is also used to remove heavy metals and other contaminants from the soil. SDS improves the solubility of various hydrocarbons and heavy metals such as zinc, cadmium, lead, and copper and facilitates their removal by phytoremediation (Messina et al., 2014). Application of SDS reported increased cadmium accumulation in *Althea rosea* (Liu et al., 2009) and *Calendula officinalis* (Liu et al., 2010), and zinc accumulation in *Populus alba* (Pierattini et al., 2018).

There are some limitations of chemical additives. These are of environmental concern due to their toxic effects on microorganisms as well as plant species. They can also disrupt soil's chemical and physical structure by dissolving minerals. Also, they are not biodegradable and can last for a longer time leading to the leaching of metals.

4.2 Microbe-Assisted Phytoremediation

Microbe-assisted phytoremediation harnesses the collective potential of plants and associated microbes to remove various contaminants. Microorganisms are prime candidates for bioremediation due to their unique ability to degrade various xenobiotic compounds through specific metabolic processes. They use pollutants as a source of carbon and energy. They can adapt to heavily polluted environments easily. However, the efficiency of microbial remediation may be limited in soil contaminated with persistent organic pollutants. In such cases, sharing microbes and plants can increase efficiency. Plant roots provide favorable conditions and their exudate accelerates the bioavailability and degradability of organic xenobiotics by providing substrates for microbial metabolism. Some important microbial genera that facilitate phytoremediation are *Pseudomonas*, *Burkholderia*, *Alcaligenes*, *Achromobacter*, *Rhodococcus*, *Sphingomonas*, and *Ralstonia* (Hussain et al., 2018).

Plant-supporting microorganisms fall into two broad categories: plant growth-promoting rhizobia (PGPR) and endophytes. PGPR produces growth hormones such as auxins, cytokinins, gibberellins, and ethylene that stimulate plant growth and induce stress tolerance. PGPR also produces bioactive compounds such as various enzymes, organic acids, osmolytes, biosurfactants, siderophores, and antibiotics. These compounds help plants by controlling pathogens, providing resistance to abiotic stresses, and aiding in nutrient uptake through mineral solubilization and nitrogen fixation. Thus, microbes increase the efficiency of phytoremediation either directly by degrading the contaminants or indirectly by promoting plant growth. Sampaio et al., 2019 evaluated the association of *Rhizophora mangle* L. (mangrove plant) and plant growth-promoting bacteria (*Pseudomonas aeruginosa* and *Bacillus* sp.) for their ability to degrade polycyclic aromatic hydrocarbons (PAH) in diesel oil-contaminated sediments.

In addition to PGPR, endophytes also support phytoremediation. They are not obligate biotrophs. They live in plant tissues and utilize nutrients from the host. They coexist with PGPR to promote plant growth and increase the efficiency of phytoremediation. They promote plant growth and development even under biotic or abiotic stress conditions. The association of endophytes can synergistically alter plant physiology and metabolism. This improves the removal of contaminants from contaminated sites and offers great potential for environmental protection. They also confer better adaptability to plants through biological control mechanisms and induce systemic resistance to plant pathogens. They produce plant hormones, organic acids, siderophores, enzymes, and growth regulators involved in plant growth. Therefore, inoculating plants with PGPR and endophytes could be a good strategy to enhance the phytoremediation of contaminated sites.

Fungi have also been reported to facilitate phytoremediation by directly or indirectly interacting with contaminants. Mycorrhizal fungi influence the bioavailability of heavy metals by causing physicochemical changes in soil (Bhantana et al., 2021). Arbuscular mycorrhizal fungi promote plant growth through the production of plant hormones and increase the surface area of plant roots and enhance nutrient, water as well and heavy metals uptake (Vamerali et al., 2010).

4.3 Genetic Approaches

Genetic manipulation of plants is a promising tool to enhance their efficiency of phytoremediation. The genetically modified plants that we obtain are also called transgenic plants. Novel genes that may not naturally present and encode factors for efficient phytoremediation are incorporated into the plants (Fig. 17.3) or sometimes the overexpression of native genes is also performed using genetic engineering tools. These genes are responsible for the rapid growth rate, high biomass production, rapid uptake, transport, sequestration, and degradation of pollutants in plant tissues. In addition, properties such as the conversion of pollutants to less toxic forms, resistance to biotic and abiotic stresses, and tolerance to high concentrations of pollutants may also be genetically induced in plants. The source of the genes may be microorganisms (such as bacteria and fungi), plants, or animals from which genes

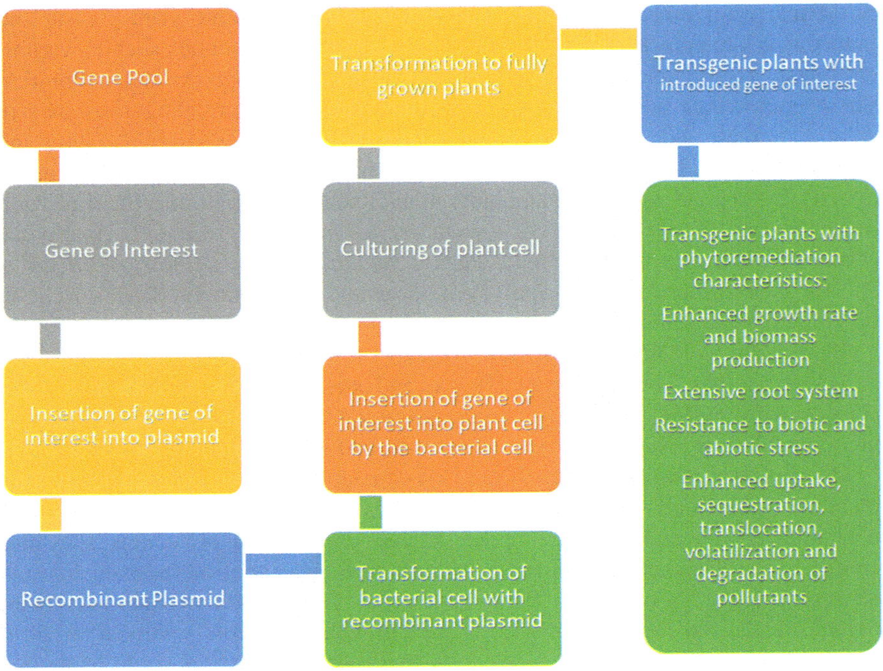

Fig. 17.3 Genetic modification of plants for introducing potential phytoremediation characteristics

are transferred to the target plants. Hannink et al., 2007 created transgenic tobacco (*Nicotiana tabacum*) by constitutively expressing the nsfI nitroreductase gene from the bacteria *Enterobacter cloacae*. Transgenic tobacco showed enhanced 2,4,6-trinitrotoluene (TNT) degradation.

Genetic engineering tools have certain advantages over traditional methods. Genetic engineering tools introduce desired genes into plants in a short period of time compared to traditional breeding methods. Moreover, desirable genes can be transferred between sexually incompatible plant species (Marques et al., 2009). Apart from that, genetic engineering also has some limitations. Sometimes phytoremediation process is complex and controlled by multiple genes. In such cases, it is not easy to study gene regulation mechanisms and the integration of multiple genes. Moreover, transgenic plants have shown promising results at a small or laboratory scale under controlled environmental conditions. However, the potential of transgenic plants should also be analyzed under field conditions. It is also difficult to obtain permits to grow genetically modified organisms (GMOs) at the field scale due to the risks to food and ecosystem security. Therefore, whenever the use of genetic engineering is impractical, alternative approaches should be used.

5 Phytoremediation: From Lab-Scale to Field-Scale Transition

The transition from laboratory experimentation to field-scale applications for phytoremediation requires a stepwise approach. The first step is an initial investigation carried out to determine the type and limits of a contaminated site. Next in line is choosing plant species based on their ability to grow and cleanse specific pollutants. Lab testing enables optimal plant growth and pollutant extraction. That may help to narrow the gap between controlled greenhouse conditions and field conditions as well as field-site selection (Rheay et al., 2020; Marchiol & Fellet, 2011; Abdelhafeez et al., 2022). Pilot studies assess viability and effectiveness in miniature, measuring the effect of plants on their surroundings and how much pollution they are eliminating. Periodic observation and care are needed to keep it in effect. Twine et al. (2019) emphasized the importance of regulatory compliance and engaging with the public, data collection and analysis, inform adaptive management, and making necessary changes. This is where long-term planning comes into practice—making provision for re-contaminations in future and ensuring sustainability. Regular records and reports facilitate transparency and responsibility at every step of the process (Rheay et al., 2020; Ehiomogue et al., 2021; Cantamessa et al., 2020).

The overall in-field success of phytoremediation is dependent upon interdisciplinary teamwork and flexibility to local conditions. Although phytoremediation has proven its potential in laboratory and greenhouse experiments, however, due to hurdles along the way, limited success has been achieved in field conditions. Nevertheless, promising results provide justification for its continued usage in field conditions. For example, Doty and his co-workers 2017 conducted a 3-year field trial of endophyte-assisted phytoremediation of trichloroethane (TCE) by poplar in the research area of Silicon Valley, California. TCE concentration decreased from 300 μg/L to less than 5 μg/L. TCE derivatives were similarly removed, with cis-1,2-dichloroethene from 160 μg/L to <5 μg/L and trans-1,2-dichloroethene from 3.1 μg/L to <0.5 μg/L. Siciliano et al., 2003 observed a 30% reduction in polyaromatic hydrocarbon concentration in a phytoremediation field trial over a period of two years. A study by Nedunuri et al., 2000 demonstrated the degradation of petroleum hydrocarbons at a field scale by ryegrass (*Lolium annual*) and St. Augustine grass (*Stenotaphrum secundatum*) over a period of 21 months. The results showed 42% and 50% reductions in the petroleum hydrocarbon concentration, respectively due to phytoremediation by these two types of grasses. Haa et al. (2019) evaluated the phytoextraction and phytostabilization potential of 21 naturally growing plant species in the largest tungsten mine in Vietnam. Their results showed a hyperaccumulation of arsenic in *Pityrogramma calomelanos* (2400 mg/kg) and *Pteris vittata* L (1860 mg/kg). So, these native plant species were considered promising for the phytoextraction of arsenic-contaminated soils. Another successful field experiment was conducted to remediate a BTEX (Benzene, toluene, ethylbenzene, xylene)-contaminated groundwater plume on the premises of an automobile factory in Belgium by the collective action poplar and associated rhizosphere and endophytic microorganisms (Barac et al., 2009). Monitoring conducted over a period of six years showed that poplar and its associated microbes played an active role in the complete containment of contamination of BTEX. In a 60-day field trial, 96% of 2,4,6-trinitrotoluene (TNT) was removed by maize (*Zea mays*), from the test plots and neither spontaneous attenuation nor bioaugmentation with *Pseudomonas putida* JLR11 had any significant effect on TNT levels (Van Dillewijn et al., 2007). They concluded that the phytoremediation by maize masked the effects of rhizoremediation by *P. putida* JLR11 under these field conditions. White Jr. et al. (2006) observed that the phytoremediation using a combination of grass and fertilizer reduced polyaromatic hydrocarbon in oil-spilled sites. Further examples of successful field-scale phytoremediations are summarized in Table 17.1.

6 Challenges Behind the Lab-Scale to Field-Scale Transition

Past advantages have increased interest in large-scale applications of phytoremediation strategies. However, many challenges (Fig. 17.4) limit the transition from laboratory to field scale. The most common issue is the much slower speed of phytoremediation in the field. Plants can experience stress under field conditions due to a variety of physical, chemical, and biological barriers that are not encountered under lab conditions. These include variations in temperature, humidity, nutrients, herbivores, pests, diseases, and weeds. These factors inhibit or prevent plant growth and the ability to accumulate or degrade pollutants under field conditions, thus adversely affecting the phytoremediation process. Phytoremediation capacity is also compromised if the contaminated soil is deeper than the root zone. Furthermore, tightly bound contaminants have limited bioavailability and are difficult to remove from the soil on a large scale. Changes in farming practices and soils can also affect contaminant transfer. Furthermore, phytoremediation is only applied in locations with low to moderate concentrations of contaminants, as there is no sustained plant growth in heavily contaminated soils. It is also difficult to prevent grazing in large growing regions, and so there is a risk of accumulating contaminants in the food chain. It is also difficult to handle and dispose of a large amount of contaminated biomass generated on a field scale.

Many other factors affecting the phytoremediation performance on large scales deserve significant consideration. Contaminants distributed across large contaminated areas generate regions with low or high contaminant levels, in contrast to lab experimental set-ups where contaminant concentrations are evenly distributed. This is because, in labs, it is feasible to properly mix the contaminated samples to achieve homogeneity. Spatial heterogeneity in initial contaminant concentrations can lead to data variability and make it difficult to statistically demonstrate significant treatment effects in field trials. This is also an issue when it comes to regulatory standards. This is because remediation success is often evaluated selectively rather than averaging

Table 17.1 Field-scale phytoremediations

Plants	Contaminants phytoremediated	Phytoremediation mechanism involved	Location of field site	References
Trifolium repens, *Lolium perenne*	Cd, Pb, Zn	Phytostabilization	Northern France	Bidar et al. (2007)
Quercus ilex	As, Bi, Cd, Cu, Pb, Sb, Tl, Zn	Phytostabilization	Aznalcollar, Spain	Dominguez et al. (2008)
Zea mays, *Nicotiana tabacum*, *Salix* species, *Brassica napus*, *Hybrid poplar*	Cd, Zn	Phytoextraction	Lommel, Belgium	Vangronsveld et al. (2009)
Medicago sativa L.	Polychlorinated biphenyls	Phytodegradation	Taizhou, Zhejiang Province, China	Tu et al. (2011)
Populus alba L.	Cd, Cu, Pb, Zn	Phytoextraction and Phytostabilization	Shapogang, Gansu Province, China	Hu et al. (2013)
Populus nigra, *Paulownia tomentosa*, *Cytisus scoparius* L.	Polyaromatic hydrocarbons	Phytoextraction and Phytodegradation	Pisa, Italy	Macci et al. (2013)
Festuca rubra L.	Pb, Zn	Phytostabilization	Province of Biscay, Spain	Galende et al. (2014)
Lolium perenne L., *Trifolium repens* L.	Petroleum hydrocarbons	Phytodegradation	Republic of Ireland	Germaine et al. (2015)
Zea mays L., *Sorghum bicolor* L.	Cd, Zn	Phytoextraction	Hangzhou, Zhejiang Province, China	Deng et al. (2016)
Pteris vittata L., *Sedum alfredii*, Sugar cane, Mulberry tree	As, Cd, Pb	Phytoextraction	Huan Jiang Maonan county, China	Wan et al. (2016)
Canavalia ensiformis and *Crotalaria juncea*	Sulfentrazone herbicide	Phytoaccumulation	Espírito Santo University Centre, Brazil	Belo et al. (2016)
Cannabis sativa L.	Cd, Pb and Zn	Phytoaccumulation	Sardinia, Italy	Canu et al. (2022)
Corn	Pb	Phytoaccumulation	Taiwan	Chang et al. (2019)
Eucalyptus globulus	Multi-metal-polluted sites	Phytoaccumulation	Guangdong, Province, China	Luo et al. (2018)

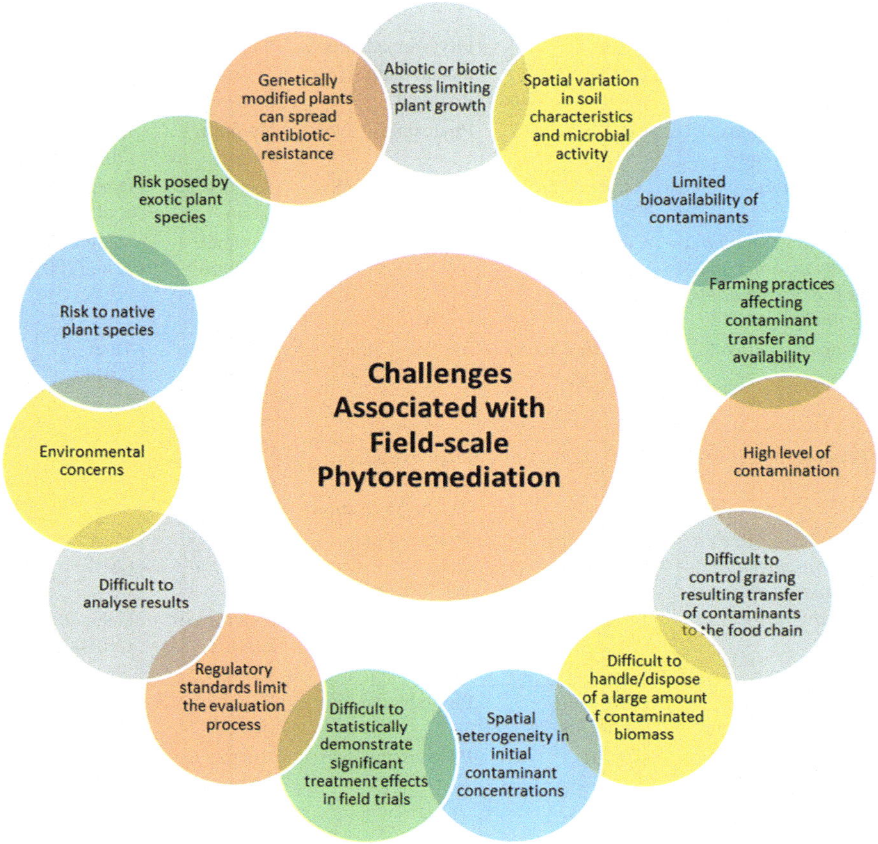

Fig. 17.4 Challenges associated with the field-scale application of phytoremediation

data points across sites. If one point fails to meet regulatory requirements, the entire site will fail. In addition to contaminant concentrations, factors such as root structure, soil structure, soil composition, soil pH, water content, and microbial activity also often change over time at a particular location and show spatial variation. This can affect plant and microbial growth and, thus, the potential for remediation.

Agricultural techniques such as tillage and the addition of nutrients and organic matter are often performed before planting. These practices are generally designed to improve soil texture and composition to promote plant and microbial growth. It can affect availability and cause the degradation of contaminants, even in bulk soils. Therefore, it becomes difficult to conclude whether the process of phytoremediation is superior to natural physical and biological degradation. Analysis of results obtained in the field is also difficult, as it is challenging to independently determine the degree of restoration by early rootlets, the soils associated with these rootlets, the rhizosphere soil, or large bulk soils. Regulatory restrictions are also relevant to the phytoremediation of hazardous waste due to environmental concerns. Bioaccumulation of contaminants in plants is considered. What will happen if the contaminated plants are eaten by the animals, and what will be the fate of decomposed and volatilized compounds? These compounds, in fact, can diffuse from the phytoremediation sites and affect native plant species.

The introduction of exotic plants is also another issue to consider. Additionally, the use of genetically modified plants for phytoremediation may be of concern as they contain antibiotic-resistant genes. Genetically engineered microbial strains, when used, may not be able to compete with the native microorganisms in the rhizosphere and reduce their numbers to levels where they cannot effectively support repair.

7 Devastating the Challenges

The efficiency of phytoremediation mainly depends on plant growth rate and pollutant accumulation. However, under field conditions, plants are exposed to the direct effects of various physical, chemical, and biological stresses that limit their growth and pollutant accumulation capacity. A successful strategy is to use rhizosphere bacteria (PGPR) to promote plant growth. These bacteria promote plant growth in different ways. They increase the uptake of nutrients by plants, produce plant hormones, act as biocontrol agents, and mitigate the effects of pathogens. Under stress, plant ethylene biosynthesis is triggered, and plant growth is inhibited. However, PGPR slows down ethylene biosynthesis. As a result, PGPR improves seed germination and growth rates under stressful conditions. They show more potent effects, especially when used in combination with hyperaccumulating plant species. Many PGPRs, such as *Pseudomonas*, act as both plant growth promoters and pollutant degraders. Therefore, PGPR also supports plants in degrading pollutants. In addition to PGPR, endophytes are also a solution. In contrast to PGPRs, which colonize the surface of the root and rhizosphere, endophytes colonize the internal tissues of plants. Endophyte is a better solution than PGPR because there are several factors that limit the effectiveness of PGPR, such as soil pH, temperature, water content, and competing pressures in the rhizosphere. Endogenous bacteria have also been reported to degrade organic pollutants.

In addition, phytoremediation is a time-consuming process that may be effective in years. Genetic modification of plants and their associated microbes can accelerate this process. However, genetic modifications increase the ecological risk of gene transfer to other plant species, especially antibiotic-resistant genes. The risk of gene flow can be minimized by using a biological containment system (Gressel & Al-Ahmad, 2005). Instead of introducing one or two genes, multiple transgenes can prevent gene flow between the introduced species and other species in the environment. Moreover, mitigator genes linked to the primary transgene can be added to strengthen the containment system. Mitigation genes confer harmless traits to phytoremediation species but are harmful to related species if gene transfer occurs. Alternatively, they may prevent phytoremediation species from successfully competing outside the contaminated site. Another obstacle is the lack of understanding among remediation workers. Exploring more phytoremediation plants and understanding the phytoremediation processes can help us to increase the phytoremediation potential of plants through genetic modifications.

Another challenge in field conditions is inconsistency in sampling methods, monitoring protocols, and data analysis. Government agencies, industrialists, and researchers have developed protocols to address these issues (Remediation Technologies Development Forum, EPAJ542/F-99/023, Nov. 1999). Comparative testing of remediation strategies should be conducted in different geographic locations according to these protocols. In addition to using pooled samples for analysis, it is useful to view the data as changes in contaminant concentration over time at approximately the same location on a given graph over time. This means that if we sample 10 points in the field at each time interval, we need to compare the remediation at each point over time, not just the average across the site. Appropriate use of trends in time series data can also solve statistical problems due to data dispersion. Assessing the total intake of contaminants in crops can be used as a measure of success in the field, rather than trying to quantify the reduction in soil contamination at the end of the growing season. Consideration of 10% as the significance level for statistical analysis of field data instead of the 5% standard accepted for lab studies. This is because the inherent variability in field studies is greater than in laboratory studies. Additionally, a 3-year endpoint is considered more realistic than expecting a complete or significant recovery in one growing season. A longer field trial period allows for a better evaluation of the phytoremediation system and gives time to make improvements. If a problem occurs in the field, a lab or greenhouse experiment can be performed to fix the problem before the next field season begins.

Another strategy for successful field applications of phytoremediation is to use only native plants. The use of native plant species can serve the dual purpose of restoring native habitats. In addition to the recognized ecological benefits, this appears to be one of the best strategies for microbial-assisted phytoremediation, as native species are already adapted to local conditions.

8 Conclusions

Phytoremediation is a very effective technology in comparison to conventional environment-depleting and costly technologies. We expect significant applications of this technology. However, constraints related to the field scale applications confine its usage to a small scale. There is the need to be completely privy to the constraints of this remediation approach. We should not be discouraged due to the challenges encountered in bringing this technology into recurring practices. We are progressing towards fixing the problems faced, specifically in dealing with crop stress inside the field. It continues to be modified and developed through experiments and skilled work in greenhouses and laboratories to cope with increasing problems. New protocols are also being developed to sample, evaluate, and interpret results from field scales to demonstrate the efficacy of phytoremediation. More integrated efforts and research on the applicability of phytoremediation will take us toward sustainability in the near future.

References

Abdelhafeez, I. A., El-Tohamy, S. A., Ul-Malik, M., Abdel-Raheem, S.M., & El-Dars, F.M.S.E. (2022). *A review on green remediation techniques for hydrocarbons and heavy metals contaminated soil*. https://scite.ai/reports/10.5267/j.ccl.2021.9.006

Agrawal, S., Kumar, V., Kumar, S., & Shahi, S. K. (2022). Plant development and crop protection using phytonanotechnology: A new window for sustainable agriculture. *Chemosphere, 299*, 134465.

Ahsan, M. T., Najam-ul-haq, M., Idrees, M., Ullah, I., & Afzal, M. (2017). Bacterial endophytes enhance phytostabilization in soils contaminated with uranium and lead. *International Journal of Phytoremediation., 19*, 937–946.

Andra, S. S., Datta, R., Sarkar, D., Saminathan, S. K. M., Mullens, C. P., & Bach, S. B. H. (2009). Analysis of phytochelatin complexes in the lead tolerant vetiver grass *Vetiveria zizanioides* (L.) using liquid chromatography and mass spectrometry. *Environmental Pollution, 157*, 2173–2183.

Ashraf, S., Ali, Q., Zahir, Z. A., Ashraf, S., & Asghar, H. N. (2019). Phytoremediation: Environmentally sustainable way for reclamation of heavy metal polluted soils. *Ecotoxicology and Environmental Safety, 174*, 714–727.

Awokunmi, E., Asaolu, S., Ajayi, O., & Adebayo, O. (2012). The role of EDTA on heavy metals phytoextraction by *Jatropha gossypifolia* grown on soil collected from dumpsites in Ekiti state Nigeria. *British Journal of Environment and Climate Change, 2*, 153.

Azc'on, R., Medina, A., Rold'an, A., Bir'o, B., & Vivas, A. (2009). Significance of treated agrowaste residue and autochthonous inoculates (arbuscular mycorrhizal fungi and *Bacillus cereus*) on bacterial community structure and phytoextraction to remediate soils contaminated with heavy metals. *Chemosphere, 75*, 327–334.

Barac, T., Weyens, N., Oeyen, L., Taghavi, S., van der Lelie, D., Dubin, D., Spliet, M., & Vangronsveld, J. (2009). Field note: Hydraulic containment of a BTEX plume using poplar trees. *International Journal of Phytoremediation, 11*, 416–424.

Belo, A F., Pires, F R., Bonomo, R., Filho, A C., & Tenis, L H O. (2016). *Sulfentrazone phytoremediation under field conditions*. https://scite.ai/reports/10.1590/1983-21252016v29n114rc

Bhantana, P., Rana, M. S., Sun, X., Moussa, M. G., Saleem, M. H., Syaifudin, M., Shah, A., Poudel, A., Pun, A. B., & Bhat, M. A. (2021). Arbuscular mycorrhizal fungi and its major role in plant growth, zinc nutrition, phosphorous regulation, and phytoremediation. *Symbiosis, 84*, 19–37.

Bhat, S. A., Qadri, H., Cui, G., & Li, F. (2020). Remediation of pesticides through microbial and phytoremediation techniques. In H. Qadri, R. Bhat, M. Mehmood, & G. Dar (Eds.), *Fresh water pollution dynamics and remediation*. Springer. https://doi.org/10.1007/978-981-13-8277-2_13

Bidar, G., Garcon, G., Pruvot, C., Dewaele, D., Cazier, F., Douay, F., & Shirali, P. (2007). Behavior of *Trifolium repens* and *Lolium perenne* growing in a heavy metal contaminated field: Plant metal concentration and phytotoxicity. *Environmental Pollution, 147*, 546–553.

Cantamessa, S., Gamalero, E., & Berta, G. (2020). *Phytoremediation of a highly arsenic polluted site, Using Pteris vittata L. and Arbuscular Mycorrhizal Fungi*. https://scite.ai/reports/10.3390/plants9091211

Canu, M., Mulè, P., Spanu, E., Fanni, S., Marrone, A., & Carboni, G. (2022). Hemp cultivation in soils polluted by Cd, Pb and Zn in the Mediterranean area: Sites characterization and phytoremediation in real scale settlement. *Applied Sciences, 12*, 3548. https://doi.org/10.3390/app12073548

Chandra, R., & Kumar, V. (2017). Phytoextraction of heavy metals by potential native plants and their microscopic observation of root growing on stabilised distillery sludge as a prospective tool for In-situ phytoremediation of industrial waste. *Environmental Science and Pollution Research, 24*, 2605–2619. https://doi.org/10.1007/s11356-016-8022-1

Chandra, R., Dubey, N. K., & Kumar, V. (2018a). *Phytoremediation of environmental pollutants*. CRC Press. https://doi.org/10.1201/9781315161549

Chandra, R., Kumar, V., & Singh, K. (2018b). Hyperaccumulator versus nonhyperaccumulator plants for environmental waste management. In R. Chandra, N. Dubey, & V. Kumar (Eds.), *Phytoremediation of environmental pollutants*. CRC Press.

Chang, J. H., Dong, C. D., & Shen, S. Y. (2019). The lead contaminated land treated by the circulation-enhanced electrokinetics and phytoremediation in field scale. *Journal of Hazardous Materials, 368*, 894–898. https://doi.org/10.1016/j.jhazmat.2018.08.085

Chen, Y., & Cutright, T. (2001). EDTA and HEDTA effects on Cd, Cr, and Ni uptake by *Helianthus annuus*. *Chemosphere, 45*, 21–28.

DalCorso, G., Fasani, E., Manara, A., Visioli, G., & Furini, A. (2019). Heavy metal pollutions: State of the art and innovation in phytoremediation. *International Journal of Molecular Sciences, 20*, 3412. https://doi.org/10.3390/ijms20143412

Dams, R. I., Paton, G. I., & Killham, K. (2007). Rhizoremediation of pentachlorophenol by *Sphingobium chlorophenolicum* ATCC 39723. *Chemosphere, 68*, 864–870.

Deng, L., Li, Z., Wang, J., Liu, H., Li, N., Wu, L., et al. (2016). Long-term field phytoextraction of zinc/cadmium contaminated soil by *Sedum plumbizincicola* under different agronomic strategies. *International Journal of Phytoremediation, 18*, 134–140.

Dipu, S., Kumar, A. A., & Thanga, S. G. (2012). Effect of chelating agents in phytoremediation of heavy metals. *Remediation Journal, 22*, 133–146.

Dominguez, M. T., Maranon, T., Murillo, J. M., Schulin, R., & Robinson, B. H. (2008). Trace element accumulation in woody plants of the Guadiamar Valley, SW Spain: A large-scale phytomanagement case study. *Environmental Pollution, 152*, 50–59.

Doty, S. L., Shang, T. Q., Wilson, A. M., Moore, A. L., Newman, L. A., Strand, S. E., & Gordon, M. P. (2003). Metabolism of the soil and groundwater contaminants, ethylene dibromide and trichloroethylene, by the tropical leguminous tree, *Leuceana leucocephala*. *Water Research, 37*, 441–449.

Doty, S. L., Freeman, J. L., Cohu, C. M., Burken, J. G., Firrincieli, A., Simon, A., Khan, Z., Isebrands, J. G., Lukas, J., & Blaylock, M. J. (2017). Enhanced degradation of TCE on a superfund site using endophyte-assisted poplar tree phytoremediation. *Environmental Science and Technology, 51*(17), 10050–10058. https://doi.org/10.1021/acs.est.7b01504

Ehiomogue, O. P., Ahuchaogu, I. I., & Ahaneku, I. E. (2021). *A review of bioremediation of hydrocarbon contaminated soils in Niger Delta area of Nigeria*. https://scite.ai/reports/10.5937/poljteh2103023p

de Farias, V., Maranho, L. T., de Vasconcelos, E. C., da Silva Carvalho Filho, M. A., Lacerda, L. G., Azevedo, J. A., Pandey, A., & Soccol, C. R. (2009). Phytodegradation potential of *Erythrina crista-galli* L., Fabaceae, in petroleum-contaminated soil. *Applied Biochemistry and Biotechnology, 157*(1), 10–22. https://doi.org/10.1007/s12010-009-8531-1

Galende, M. A., Becerril, J. M., Barrutia, O., Artetxe, U., Garbisu, C., & Herna'ndez, A. (2014). Field assessment of the effectiveness of organic amendments for aided phytostabilization of a Pb-Zn contaminated mine soil. *Journal of Geochemical Exploration, 145*, 181–189.

Germaine, K. J., Byrne, J., Liu, X., Keohane, J., Culhane, J., Lally, R. D., et al. (2015). Ecopiling: A combined phytoremediation and passive biopiling system for remediating hydrocarbon impacted soils at field scale. *Frontiers in Plant Science, 5*, 756. https://doi.org/10.3389/fpls.2014.00756

Ghazaryan, K. A., Movsesyan, H. S., Minkina, T. M., Sushkova, S. N., & Rajput, V. D. (2021). The identification of phytoextraction potential of *Melilotus officinalis* and *Amaranthus retroflexus* growing on copper- and molybdenum-polluted soils. *Environmental Geochemistry and Health, 43*. https://doi.org/10.1007/s10653-019-00338-y

Gressel, J., & Al-Ahmad, H. (2005). Assessing and managing biological risks of plants used for bioremediation, including risks of transgene flow. *Zeitschrift für Naturforschung – Section C Journal of Biosciences, 60*(3–4), 154–165. https://doi.org/10.1515/znc-2005-3-402

Guarino, F., Miranda, A., Castiglione, S., & Cicatelli, A. (2020). Arsenic phytovolatilization and epigenetic modifications in *Arundo donax* L. assisted by a PGPR consortium. *Chemosphere, 251*, 126310. https://doi.org/10.1016/j.chemosphere.2020.126310

Gudeta, K., Kumar, V., Bhagat, A., Julka, J. M., Bhat, S. A., Ameen, F., Qaqdri, H., Singh, S., & Amarowicz, R. (2023). Ecological adaptation of earthworms to cope with plant polyphenols, heavy metals, and microplastics in the soil: A critical review. *Heliyon, 9*(3), e14572. https://doi.org/10.1016/j.heliyon.2023.e14572

Haa, N. T. H., Hab, N. T., Ngaa, T. T. H., Minha, N. N., Anhc, B. T. K., Hangd, N. T. A., Duca, N. A., Nhuana, M. T., & Kimb, K. W. (2019). Uptake of arsenic and heavy metals by native plants growing near Nui Phao multi-metal mine, northern Vietnam. *Applied Geochemistry, 108*, 104368.

Hammond, C. M., Root, R. A., Maier, R. M., & Chorover, J. (2018). Mechanisms of arsenic sequestration by *Prosopis juliflora* during the phytostabilization of metalliferous mine tailings. *Environmental Science and Technology, 52*, 1156–1164.

Hannink, N., Subramanian, N., Rosser, S. J., Basran, A., Murray, J. A. H., & Shanks, J. V. (2007). Enhanced transformation of TNT by tobacco plants expressing a bacterial nitroreductase. *International Journal of Phytoremediation, 9*, 385–401. https://doi.org/10.1201/9781315161549-1

Hu, Y., Nan, Z., Su, J., & Wang, N. (2013). Heavy metal accumulation by poplar in calcareous soil with various decrees of multi-metal contamination: Implications for phytoextraction and phytostabilization. *Environmental Science and Pollution Research, 20*, 7194–7203.

Huang, X. D., El-Alawi, Y. S., Penrose, D., Glick, B. R., & Greenberg, B. M. (2004). Responses of three grass species to creosote during phytoremediation. *Environmental Pollution, 130*, 453–463.

Hussain, F., Hussain, I., Khan, A. H. A., Muhammad, Y. S., Iqbal, M., Soja, G., Reichenauer, T. G., Zeshan, & Yousaf, S. (2018). Combined application of biochar, compost, and bacterial consortia with Italian ryegrass enhanced phytoremediation of petroleum hydrocarbon contaminated soil. *Environmental and Experimental Botany, 153*, 80–88. https://doi.org/10.1016/j.envexpbot.2018.05.012

Jacobs, A., Drouet, T., & Noret, N. (2018). Field evaluation of cultural cycles for improved cadmium and zinc phytoextraction with *Noccaea caerulescens*. *Plant and Soil, 430*, 381–394.

Kamran, M., Danish, M., Saleem, M. H., Malik, Z., Parveen, A., Abbasi, G. H., Jamil, M., Ali, S., Afzal, S., & Riaz, M. (2021). Application of abscisic acid and 6-benzylaminopurine modulated morpho-physiological and antioxidative defense responses of tomato (*Solanum lycopersicum* L.) by minimizing cobalt uptake. *Chemosphere, 263*, 128169.

Kaur, J., Bhatti, S. S., Bhat, S. A., Nagpal, A. K., Kaur, V., & Katnoria, J. K. (2021). Evaluating potential ecological risks of heavy metals of textile effluents and soil samples in the vicinity of textile industries. *Soil Systems, 5*, 63. https://doi.org/10.3390/soilsystems5040063

Kumar, V., & Verma, P. (2023). A critical review on environmental risk and toxic hazards of refractory pollutants discharged in chlorolignin waste of pulp and paper mills and their remediation approaches for environmental safety. *Environmental Research, 236*, 116728.

Kumar, V., Ferreira, L. F. R., Sonkar, M., & Singh, J. (2021a). Phytoextraction of heavy metals and ultrastructural changes of *Ricinus communis* L. grown on complex organometallic sludge discharged from alcohol distillery. *Environmental Technology & Innovation, 22*, 101382.

Kumar, V., Shahi, S. K., Ferreira, L. F. R., Bilal, M., Biswas, J. K., & Bulgariu, L. (2021b). Detection and characterization of refractory organic and inorganic pollutants discharged in biomethanated distillery effluent and their phytotoxicity, cytotoxicity, and genotoxicity assessment using *Phaseolus aureus* L. and *Allium cepa* L. *Environmental Research, 201*, 111551.

Kumar, V., Agrawal, S., Bhat, S. A., Américo-Pinheiro, J. H. P., Shahi, S. K., & Kumar, S. (2022a). Environmental impact, health hazards, and plant-microbes synergism in remediation of emerging contaminants. *Cleaner Chemical Engineering, 2*, 100030.

Kumar, V., Ameen, F., Islam, M. A., Agrawal, S., Motghare, A., Dey, A., Shah, M. P., Américo-Pinheiro, J. H. P., Singh, S., & Ramamurthy, P. C. (2022b). Evaluation of cytotoxicity and genotoxicity effects of refractory pollutants of untreated and biomethanated distillery effluent using *Allium cepa*. *Environmental Pollution, 300*, 118975.

Liphadzi, M. S., & Kirkham, M. B. (2006). Chelate-assisted heavy metal removal by sunflower to improve soil with sludge. *Journal of Crop Improvement, 16*, 153–172.

Liu, J. N., Zhou, Q. X., Wang, S., & Sun, T. (2009). Cadmium tolerance and accumulation of *Althaea rosea* Cav. and its potential as a hyperaccumulator under chemical enhancement. *Environment Monitoring and Assessment, 149*, 419–427.

Liu, J., Zhou, Q., & Wang, S. (2010). Evaluation of chemical enhancement on phytoremediation effect of cd-contaminated soils with *Calendula officinalis* L. *International Journal of Phytoremediation, 12*, 503–515.

Lombi, E., Zhao, F. J., Dunham, S. J., & McGrath, S. P. (2001). Phytoremediation of heavy metal contaminated soils: Natural hyperaccumulation versus chemically enhanced phytoextraction. *Journal of Environmental Quality, 30*, 1919–1926.

Luo, J., He, M., Qi, S., et al. (2018). Effect of planting density and harvest protocol on field-scale phytoremediation efficiency by Eucalyptus globulus. *Environmental Science and Pollution Research, 25*, 11343–11350. https://doi.org/10.1007/s11356-018-1427-2

Macci, C., Doni, S., Peruzzi, E., Bardella, S., Filippis, G., Ceccanti, B., et al. (2013). A real-scale soil phytoremediation. *Biodegradation, 24*, 521–538.

Malik, Z., Ahmad, M., Abassi, G. H., Dawood, M., Hussain, A., & Jamil, M. (2017). Agrochemicals and soil microbes: Interaction for soil health. In *Xenobiotics in the soil environment* (pp. 139–152). Springer.

Marchiol, L., & Fellet, G. (2011). *Agronomy towards the green economy. Optimization of metal phytoextraction.* https://scite.ai/reports/10.4081/ija.2011.e30

Marques, A. P., Rangel, A. O., & Castro, P. M. (2009). Remediation of heavy metal contaminated soils: Phytoremediation as a potentially promising clean-up technology. *Critical Reviews of Environment, Science, and Technology, 39*, 622–654.

McCutcheon, L. E., Ashe, D. D., Houran, J., & Maltby, J. (2003). A cognitive profile of individuals who tend to worship celebrities. *The Journal of Psychology, 137*, 309–322.

Messina, C. M., Faggio, C., Laudicella, V. A., Sanfilippo, M., Trischitta, F., & Santulli, A. (2014). Effect of sodium dodecyl sulfate (SDS) on stress response in the Mediterranean mussel (*Mytilus Galloprovincialis*): Regulatory volume decrease (Rvd) and modulation of biochemical markers related to oxidative stress. *Aquatic Toxicology, 157*, 94–100.

Miller, R. R. (1996). Phytoremediation, technology overview report. *Ground Water Remediation Technologies Analysis Center, 39*, 80.

Nedunuri, K. V., Govindaraju, R. S., Banks, M. K., Schwab, A. P., & Chen, Z. (2000). Evaluation of phytoremediation for field-scale degradation of total petroleum hydrocarbons. *Journal of Environment and Engineering, 126*, 483–490.

Padmavathiamma, P. K., & Li, L. Y. (2012). Rhizosphere influence and seasonal impact on phytostabilisation of metals—A field study. *Water, Air and Soil Pollution, 223*, 107–124. https://doi.org/10.1007/s11270-011-0843-4

Pierattini, E. C., Francini, A., Raffaelli, A., & Sebastiani, L. (2018). Surfactant and heavy metal interaction in poplar: A focus on SDS and Zn uptake. *Tree Physiology, 38*, 109–118.

Pivetz, B. E. (2001). *Phytoremediation of contaminated soil and groundwater at hazardous waste sites*. US Environmental Protection Agency.

Prakash, D., Gabani, P., Chandel, A. K., Ronen, Z., & Singh, O. V. (2013). Bioremediation: A genuine technology to remediate radionuclides from the environment. *Microbial Biotechnology, 6*, 349–360.

Pratas, J., Favas, P. J., Paulo, C., Rodrigues, N., & Prasad, M. (2012). Uranium accumulation by aquatic plants from uranium-contaminated water in Central Portugal. *International Journal of Phytoremediation, 14*, 221–234.

Qadir, A., Hameed, M. A., Zafar, M. S. B., Farooqi, Z. U. R., Younas, F., Hussain, M. M., & Mohy-Ud-Din, W. (2021). Phytoremediation of inorganic pollutants: An eco-friendly approach, its types and mechanisms. *Plant and Environment, 1*(2), 110–129.

Rai, P. K. (2019). Heavy metals/metalloids remediation from wastewater using free-floating macrophytes of a natural wetland. *Environmental Technology and Innovation, 15*, 100393.

Remediation Technologies Development Forum Phytoremediation Action Team Field Study Protocol. (1999, July). www.rtdf.org/public/phyto/protocol/protocol99.htm (July 22, 2008).

Rheay, H. T., Omondi, E. C., & Brewer, C. E. (2020). *Potential of hemp (Cannabis sativa L.) for paired phytoremediation and bioenergy production*. https://scite.ai/reports/10.1111/gcbb.12782

Sabreena, Hassan, S., Bhat, S. A., Kumar, V., Ganai, B. A., & Ameen, F. (2022). Phytoremediation of heavy metals: An indispensable contrivance in green remediation technology. *Plants, 11*, 1255. https://doi.org/10.3390/plants11091255

Sabreena, Hassan, S., Kumar, V., Bhat, S. A., & Ganai, B. A. (2023). Unravelling the microbes as potential proxies for remediation of heavy metal and pesticide contamination: A state-of-the art review. *International Journal of Environmental Research, 17*, 55. https://doi.org/10.1007/s41742-023-00544-8

Sakouhi, L., Rahoui, S., Massoud, M. B., Munemasa, S., Ferjani, E. F., Murata, Y., & Chaou, A. (2016). Calcium and EGTA alleviate cadmium toxicity in germinating chickpea seeds. *Journal of Plant Growth Regulation, 35*, 1064–1073.

Sampaio, J. S., de Souza, R. B., Damião, A. O., Bahiense, T. C., & Roque, R. A. (2019). Biodegradation of polycyclic aromatic hydrocarbons (PAHs) in a diesel oil-contaminated mangrove by plant growth-promoting rhizobacteria. *3 Biotech, 9*(4), 155. https://doi.org/10.1007/s13205-019-1686-8

Sarwar, N., Imran, M., Shaheen, M. R., Ishaque, W., Kamran, M. A., Matloob, A., Rehim, A., & Hussain, S. (2017). Phytoremediation strategies for soils contaminated with heavy metals: Modifications and future perspectives. *Chemosphere, 171*, 710–721.

Schaider, L., Parker, D., & Sedlak, D. (2006). Uptake of EDTA-complexed Pb, Cd and Fe by solution-and sand-cultured *Brassica juncea*. *Plant and Soil, 286*, 377–391.

Shabani, N., & Sayadi, M. (2012). Evaluation of heavy metals accumulation by two emergent macrophytes from the polluted soil: An experimental study. *Environment, 32*, 91–98.

Shahid, A., Austruy, G., Echevarria, M., Arshad, M., Sanaullah, M., Aslam, M., Nadeem, M., Nasim, W., & Dumat, C. (2014). EDTA-enhanced phytoremediation of heavy metals: A review. *Soil and Sediment Contamination: An International Journal, 23*, 389–416.

Shao, H., Cui, G., & Bhat, S. A. (2022). Phytoremediation: A green technology for treating heavy metal-contaminated soil. In A. Malik, M. K. Kidwai, & V. K. Garg (Eds.), *Bioremediation of toxic metal(loid)s* (1st ed.). CRC Press. https://doi.org/10.1201/9781003229940

Sharma, U., Sharma, S., Rana, V. S., Rana, N., Kumar, V., Sharma, S., Qadri, H., Kumar, V., & Bhat, S. A. (2023). Assessment of microplastics pollution on soil health and eco-toxicological risk in horticulture. *Soil Systems, 7*(1), 7. https://doi.org/10.3390/soilsystems7010007

Siciliano, S. D., Germida, J. J., Banks, K., & Greer, C. W. (2003). Changes in microbial community composition and function during a polyaromatic hydrocarbon phytoremediation field trial. *Applied and Environmental Microbiology, 69*, 483–489.

Sun, Y., Zhou, Q., Wang, L., & Liu, W. (2009). The influence of different growth stages and dosage of EDTA on cd uptake and accumulation in Cd-hyperaccumulator (*Solanum nigrum* L.). *Bulletin of Environmental Contamination and Toxicology, 82*, 348–353.

Susarla, S., Medina, V. F., & McCutcheon, S. C. (2002). Phytoremediation: An ecological solution to organic chemical contamination. *Ecological Engineering, 18*, 647–658.

Tu, C., Teng, Y., Luo, Y., Sun, X., Deng, S., Li, Z., et al. (2011). PCB removal soil enzyme activities, and microbial community structures during the phytoremediation by alfalfa in field soils. *Journal of Soils and Sediments, 11*, 649–656.

Twine, R., Hundt, G. L., & Kahn, K. (2019). *Dilemmas of ethics in practice in longitudinal health research: Identifying opportunities for widening participation of residents*. https://scite.ai/reports/10.3389/fsoc.2019.00033

Vamerali, T., Bandiera, M., & Mosca, G. (2010). Field crops for phytoremediation of metal-contaminated land. A review. *Environmental Chemistry Letters, 8*, 1–17.

Van Dillewijn, P., Caballero, A., Paz, J.A, Gonza'lez-Pe'rez, M.M., Oliva, J.M. and. Ramos, J.L. 2007. Bioremediation of 2,4,6-trinitrotoluene under field conditions. *Environmental Science and Technology,* 41:1378–1383.

Vangronsveld, J., Herzig, R., Weyens, N., Boulet, J., Adriaensen, K., Ruttens, A., et al. (2009). Phytoremediation of contaminated soils and groundwater: Lessons from the field. *Environmental Science and Pollution Research, 16*, 765–794.

Wan, X., Lei, M., & Chen, T. (2016). Cost-benefit calculation of phytoremediation technology for heavy-metal contaminated soil. *Science of the Total Environment, 563–564*, 796–802.

Wang, S., & Liu, J. (2014). The effectiveness and risk comparison of EDTA with EGTA in enhancing Cd phytoextraction by *Mirabilis jalapa* L. *Environmental Monitoring and Assessment, 186*, 751–759.

Weyens, N., van der Lelie, D., Taghavi, S., Newman, L., & Vangronsveld, J. (2009). Exploiting plant-microbe partnerships to improve biomass production and remediation. *Trends in Biotechnology, 27*, 591–598.

White, P. M., Jr., Wolf, D. C., Thoma, G. J., & Reynolds, C. M. (2006). Phytoremediation of alkylated polycyclic aromatic hydrocarbons in a crude oil-contaminated soil. *Water, Air and Soil Pollution, 169*, 207–220.

Yang, W., Gu, J., Zhou, H., Huang, F., Yuan, T., Zhang, J., Wang, S., Sun, Z., Yi, H., & Liao, B. (2020). Effect of three Napier grass varieties on phytoextraction of Cd- and Zn-contaminated cultivated soil under mowing and their safe utilization. *Environmental Science and Pollution Research, 27*, 16134–16144.

Zhao, Z., Xi, M., Jiang, G., Liu, X., Bai, Z., & Huang, Y. (2010). Effects of IDSA, EDDS and EDTA on heavy metals accumulation in hydroponically grown maize (*Zea mays*, L.). *Journal of Hazardous Materials, 181*, 455–459.

Chapter 18
Advancements in Biomass Degradation: The Role of Cross-linked Endocellulase in Agroindustrial Waste Processing

Dileep Dasari, Vivek Manyapu, and Ashootosh Mandpe

1 Introduction

Many new agricultural technologies, with the intention of increasing yield, are resulting in the global challenge confronting recycling farm waste. A large amount of agroindustrial waste (AIW) is causing health and environmental issues. More than two billion tons of agricultural waste are generated each year, with the vast majority of junk not recycled (World Bank, 2023). Sources like industries, post-harvesting, livestock, hospitals, and households (Koul et al., 2022; Millati et al., 2019) contribute to the high production of AIW. Improperly managed AIW can cause water pollution through the unwanted leaching of pesticides, fertilizers, and liquid waste from farms (Speight, 2020). The anaerobic decomposition of plant waste and manure releases methane and nitrous oxide, contributing to the greenhouse effect (Lamb et al., 2021). Antiquated methods of waste disposal, such as burning, can lead to air pollution through the release of volatile compounds and gases (Singh & Chauhan, 2022). The aforementioned malpractices in AIW management can also lead to soil degradation as a result of the deterioration of beneficial microorganisms and

D. Dasari
Department of Biological Science and Technology, National Pingtung University of Science and Technology, Pingtung, Taiwan

V. Manyapu
Department of Biological Science and Technology, National Pingtung University of Science and Technology, Pingtung, Taiwan

Department of Tropical Agriculture and International Cooperation, National Pingtung University of Science and Technology, Pingtung, Taiwan

A. Mandpe (✉)
Department of Civil Engineering, Indian Institute of Technology Indore, Indore, Madhya Pradesh, India
e-mail: as_mandpe@iiti.ac.in

V. Kumar et al. (eds.), *Environmental Engineering and Waste Management*,
https://doi.org/10.1007/978-3-031-58441-1_18

earthworms (Lin & Begho, 2022). Chemical treatment of AIW can also hamper soil quality and reduce crop growth (Bai et al., 2018; Matsagar & Wu, 2022). People in underdeveloped and developing countries are less likely to consider recycling agroindustrial waste and its benefits due to the lack of technology and strict government policies. Agricultural wastes can be used to make biofuels (Ali et al., 2019), papers (Gonzalo et al., 2017), bioproducts (Ginni et al., 2021), and fertilizers (Manyapu et al., 2018).

Mainly, AIW is composed of green plants and plant leftovers, as well as plant-based industrial wastes such as rotten vegetables, rotten fruits, crop residues, wood wastes, etc. Most of the AIW is made of high lingo-cellulose content, out of which cellulose is prominent, ranging from 11% to 47% (Huzir et al., 2018). Cellulose breakdown is highly challenging because of its complex and crystalline structure. Moreover, cellulose is bound with lignin and hemicellulose, thus making it highly difficult to degrade (Urbina et al., 2021). Therefore, the endocellulase enzyme is highly effective in addressing the challenges of breaking down and producing significant by-products.

Cellulase is a unique mixture of enzyme proteins with distinct glycosidic bond hydrolysis specificities. Recent assessments show that pulp and paper, detergent, textile, healthcare, food, and beverages are the major sectors of the industry where cellulase enzyme is profusely used (Jayasekara et al., 2019). The food and beverage industry is connected to its extensive use in the manufacturing of fruit juice, wine, and coffee (Kumar et al., 2019). Cellulases are most commonly and effectively utilized within the textile industry, where they play a pivotal role in processes like biostoning for jeans, biopolishing fabrics, and enhancing the overall appearance of cotton and various textiles. In addition to the textile sector, a significant portion (20–30%) of enzyme sales is attributed to their application in household laundry detergents. In this specific domain, proteases and lipases take precedence as the primary enzymatic components, working alongside cellulase. Within the detergent industry, the compatibility of cellulases with alkaline conditions and other detergent formulation ingredients, coupled with their inherent thermostability, is of paramount importance. Alkaline cellulases are particularly valuable, as they contribute significantly to the enhancement of colour brightness and the effective removal of dirt and stains from fabrics, making them indispensable components in the formulation of laundry detergents (Ejaz et al., 2021). There are three main cellulase enzymes: endocellulase, exocellulase, and β-glucosidase. Endocellulase has been an important biocatalyst for many decades. Although it has more capabilities than other cellulase enzymes, these characteristics are shown to have endocellulase potentiality in many industries (Sohail et al., 2022). Microbial endocellulase is applied in a variety of industries, including agriculture, food and feed, textiles, paper and pulp, medicine, laundry, and detergent (Jayasekara et al., 2019). Furthermore, an appropriate strategy for immobilizing the endocellulase will be aiding in fulfilling the demand and succeeding in the production of required by-products in a way easier than traditional immobilization techniques. Therefore, this chapter discusses the application of the cross-linking immobilization technique to the endocellulase enzyme and its utilization for the bioconversion of agroindustrial waste. Moreover, an overview of the degradation pathway of cellulose by endocellulase has also been elucidated.

2 High Cellulose-Containing Agroindustrial Wastes and Their Sources

Cellulose in plants and their products is significantly high, making them the most significant organic carbon source worldwide. This makes cellulose the most abundant and renewable polymer resource currently available (Seddiqi et al., 2021). Cellulose plays a vital role as a structural component of the primary cell walls in green plants, algae, and oomycetes (Rongpipi et al., 2019). It is typically obtained from wood pulp and cotton and has a variety of industrial applications. While it is most commonly known for being used in the manufacture of paper products, it can also be processed into a range of other materials, including cellophane and rayon (Bajpai, 2018). Furthermore, in order to address concerns about sustainable energy sources, researchers are exploring the use of cellulose obtained from energy crops to produce biofuels like ethanol (de Almeida et al., 2023).

The primary source of cellulose is plants and their products. Agricultural companies, farms, and orchards generate cellulosic residues in various forms, such as straws, stems, leaves, chaffs, stalks, bunches, stumps, and stubbles, along with damaged grains and waste from fruits and vegetables. Similarly, the food industry also generates cellulosic waste in the form of post-processing residues like skins, bunches, seeds, leaves, bagasse, vinasse, pulps, peels, and similar materials, in addition to damaged fruits and vegetables (Heinze et al., 2018). Some of the common agroindustrial wastes, along with their cellulosic content, have been enlisted in Table 18.1. These agroindustrial wastes are rich in cellulose, making them valuable resources for biofuel production, papermaking, and other applications where cellulose-derived products are essential.

Cellulose, a versatile material, originates from four distinct sources. The primary source encompasses fibres from textiles, paper, wood, and industrial crops. The secondary source includes unprocessed waste materials from agriculture, forestry, and the food industry, like leaves, bark, husks, and straws. The tertiary source comprises processed waste generated during the utilization, modification, and conversion of cellulosic biomass, such as pulp, bagasse, and food residues. Lastly, the quaternary source encompasses a limited range of sources, notably cellulose extracted from marine animals, bacteria, and fungi (Ahmad Khorairi et al., 2023).

Table 18.1 Common agroindustrial wastes with their cellulosic content

S. no	Agroindustrial waste	Source	Cellulose content (%)	References
1	Pineapple leaves	Pineapple plantation, Malaysia	40.51	Abdul Karim et al. (2022)
2	Palm fibre	Local farm, Oman	33.6	Al kamzari et al. (2021)
3	Palm fronds	Local farm, Oman	18.2	
4	Coffee silverskin	Multi-Café Company, Iran	23.8	Alghooneh et al. (2017)
5	Banana peduncle	Local plantation of the Napaam area, India	63.12	Baruah et al. (2022)
6	Wheat husks	Agriculture farm, Germany	36	Bledzki et al. (2010)
7	Potato pulp	Huaou Group Co. Ltd, China	81.34	Cheng et al. (2019)
8	Pineapple peel	Local pineapple industry, Thailand	21.98	Choonut et al. (2014)
9	Babassu flour	TOBASA Bioindustrial de Babaçu S. A., Brazil	13.5	Cinelli et al. (2014)
10	Cassava flour waste	Mixed organic waste, Japan	16	Daengprasert et al. (2011)
11	Horticultural waste	Ecowise Solution Pvt. Ltd Singapore	34.5	Geng et al. (2012)
12	Sponge gourd fibres	Local market, Brazil	66.59	Guimarães et al. (2009)
13	Sugarcane bagasse	Alcohol plant, Brazil	48.4	
14	Rice straw	Suburb of Deyang city of Sichuan Province, China	35.06	Jiang et al. (2011)
15	Cotton stalk	Reliance Industries Ltd., India	41.6	Karthyani et al. (2017)
16	Barley hull	Agriculture land, Virginia	34	Kim et al. (2008)
17	Empty fruit bunches	Palm oil mill, Malaysia	28.7	Kim et al. (2012)
18	Wheat straw	Agricultural University, Cracow	40.8	Kopania et al. (2012)
19	Carrot leaves	Agricultural University, Cracow	31.58	
20	Corn straw	Agricultural University, Cracow	38.83	
21	Hemp straw	Agricultural University, Cracow	60.09	
22	Bean straw	Agricultural University, Cracow	40.2	
23	Flax straw-oil variety	Agricultural University, Cracow	41.86	
24	Rye straw	Agricultural University, Cracow	45.07	
25	Rape straw	Agricultural University, Cracow	40.85	
26	Burley stalks	Agricultural land, Srem	34.15	Kulic and Radojicic (2011)
27	Virginia stalks	Agriculture land of srem, Srem	30.35	
28	Pulp and paper sludge	Pulp & Paper plant, China	23.4	Lin et al. (2012)

(continued)

Table 18.1 (continued)

S. no	Agroindustrial waste	Source	Cellulose content (%)	References
29	Water lettuce biomass	Effluent inflow pond, China	53	Luo et al. (2011)
30	Almond shell	MIMSA S.A, Spain	50.7	Nabarlatz et al. (2007)
31	Coconut residue	Grocery store, Malaysia	72.6	Ng et al. (2010)
32	Garlic skin	Local food processing company, Korea	41.7	Prasad Reddy and Rhim (2014)
33	Onion skin	Local food processing company	41.1	Rhim et al. (2015)
34	Corn stower	Auburn University, National renewable energy laboratory, Colorado	34.5	Kumar et al. (2011b)
35	Oil palm fronds	Malaysian Palm Oil Board, Malaysia	44	Sabiha-Hanim et al. (2011)
36	Douglas fir	Colorado State Forest Service, USA	35–48	Schell et al. (1999)
37	Corn cover	Local market, India	25.9	Singh et al. (2020)
38	Corn cob	Local market, India	29.6	
39	Nut shells	Mixed organic waste, Germany	25–30	Sinner et al. (1979)
40	Corn straw	Farm of Qiqihar City, China	30–40	Song et al. (2020)
41	Beech wood	Mixed organic waste, Turkey	45.1	Tekin et al. (2012)
42	Cow manure	Dairy farm, China	16.62	Yan et al. (2018)
43	Rice husk	Maccopa rice mill, Indonesia	58.73	Yunus (2019)

3 Bioconversion of the Agroindustrial Wastes Using Endocellulase

A multistep process is needed to bio-convert lignocellulosic material into valuable products, including pretreatment of the material to remove lignin and make the polysaccharides more accessible for subsequent hydrolysis of the carbohydrates and fermentation of the resulting monosaccharides. Microbial hydrolytic enzymes are highly environmentally friendly for saccharification. They provide a sustainable and green approach to biomass conversion, minimizing environmental impact. This technology leverages microorganisms to efficiently break down biomass into valuable sugars (Thapa et al., 2020).

Plants are the most abundant polymer, generating around 200 billion tons of cellulose annually (Antony & Vasudevan, 2018). Cellulose, a key sugar found in plant materials, is transformed into glucose with the help of three enzymes: endoglucanases, exoglucanases, and β-glucosidases. Endoglucanases break cellulose bonds randomly, creating small sugar pieces. Exoglucanases further break down these pieces, forming cellobiose. Finally, β-glucosidases convert cellobiose into glucose (Bhardwaj et al., 2021). Endocellulase is a type of enzyme that catalyses cellulose hydrolysis, breaking it down into tinny molecules such as cellobiose and glucose (Morais Junior et al., 2021). Endocellulases are classified in glycoside hydrolase families 5–10, 12, 44, 45, 48, 51, 74, and 124, while exocellulases are classified in GH families 5, 6, and 9 (http://www.cazy.org/Glycoside-Hydrolases.html). Endocellulase has been extensively used in bioconversion processes to degrade cellulose and generate valuable products and has greater potential for reducing agricultural and industrial waste (Siu-Rodas et al., 2018; Theradimani & Kumar, 2015; Ventorino et al., 2016).

A wide range of cellulolytic enzymes are available in different biotic sources. This characteristic is ubiquitous across a broad spectrum of organisms, spanning bacteria, fungi, and certain plant species. Due to the widespread distribution of plant biomass in both terrestrial and aquatic environments, the breakdown of lignocellulose has been observed in numerous taxonomic groups. The carbon cycle and nitrogen cycle are heavily influenced by microbial degradation, with fungi and bacteria playing a particularly pivotal role in this process. A comprehensive analysis of microbial genomes at a large scale has revealed that the distribution and abundance of around 60,000 possible cellulases from 13 glycoside hydrolase families vary across different microbial lineages (Berini et al., 2018; Bhattacharya et al., 2015; Cragg et al., 2015).

Cellulases have found an intriguing niche within the wine industry, where their application holds promise for elevating the quality of wine production. In the natural world, sugars are predominantly locked within the intricate structures of hemicellulose and cellulose. This inherent abundance of cellulosic biomass offers a unique avenue for alcohol production. Cellulase enzymes play a main role in the conversion of these cellulosic biomaterials into sugars suitable for fermentation. These sugars are then skillfully transmuted into alcohol through the action of yeast, ultimately contributing to the production of wines of exceptional quality (Kumar et al., 2019). The fruit juice industry is facing challenges with cellulosic polysaccharides during fruit juice extraction. To address this, cellulases are employed to break down cell walls and dissolve these polysaccharides, facilitating juice extraction and improving the viscosity of puree and nectar from various fruits (Raveendran et al., 2018). These enzymes also play a role in extracting flavonoids from seeds and flowers. Moreover, cellulases assist in eliminating insoluble and dense fibres in juices, preventing manufacturing line blockages and effectively reducing turbidity, acidity, and viscosity in orange juice by utilizing a multienzyme preparation that includes cellulase and xylanase. Notably, cellulases are recognized for enhancing the aroma and taste of citrus fruits (Shariq & Sohail, 2019).

3.1 Endocellulase from Microorganisms and Insects

Endocellulase is required for cellulose degradation, but its availability is limited. For the treatment of large amounts of cellulosic waste, high production of enzymes is needed, thus making it a challenging and expensive process. Cellulolytic enzymes are produced by various microorganisms; for example, fungi and yeast produce cellulolytic enzymes in the rumen, and insects degrade cellulose. Therefore, by the application of microorganisms and with the aid of a cross-linking immobilization approach, the required amount of enzyme can be helpful for the bioconversion of cellulolytic agroindustrial waste. The potential microbial and insect sources encoding cellulolytic properties have been discussed in the following.

3.1.1 Cellulolytic Fungi

Fungi are well known for their ability to degrade plant biomass. It indicates that naturally occurring fungi have evolved various enzymes for decomposing plant components. Fungi have been the most widely investigated and characterized microorganisms under the plant biomass-degrading group due to their extraordinary enzyme production capabilities and the long-lasting nature of their degradation enzymes (Juturu & Wu, 2014). More than a few hundred fungal genomes have been scrutinized, and an overwhelming 98% of the potential cellulase sequences that have been detected are derived from the sub-kingdom Dikarya. This grouping includes the two phyla that produce dikaryons, namely Ascomycota and Basidiomycota. Ascomycota, in particular, stands out as the most diverse fungal phylum, boasting an enormous range of morphological and ecological variation. Additionally, the subkingdom Dikarya encompasses the majority of soft rot fungi, a heterogeneous group recognized for their ability to induce cavities and deterioration in plant cell walls (Beimforde et al., 2014). Fungal species belonging to the *Ascomycota, Basidiomycota,* and *Deuteromycota phyla* are primarily known for their cellulolytic capabilities, with some chytrids also found in the rumens of certain animals. Fungi are favoured for cellulase production due to their ability to secrete large amounts of highly adaptable extracellular cellulase. A broad variety of hydrolytic and oxidative enzymes are secreted, and fungal cellulases can efficiently hydrolyse lignocellulosic biomass (Kaur et al., 2020). Fungi that break down plant material are categorized as brown rot, soft rot, or white rot based on visible changes in the degradation process. These fungi inhabit distinct environmental niches and produce enzymes that degrade various plant biomass components. Whole genome sequences of fungi like *Aspergillus niger*, *Trichoderma reecei*, and *Postia placenta* reveal different enzyme combinations linked to degradation patterns. Advancements in sequencing technology have exposed a wide array of lignocellulolytic strategies in various fungal species, challenging the conventional classification. Comparative genomic studies indicate that the natural plant material degradation mechanisms in fungi are more diverse than previously thought (Liu et al., 2021; Nagy et al., 2016).

3.1.2 Cellulolytic Bacteria

Cellulolytic bacteria are microorganisms that can degrade cellulose, the most abundant biopolymer on earth, into glucose and other products. Cellulolytic bacteria have a high biotechnological potential for various applications, such as biofuel production, bioremediation, animal feed, and waste management. Recent studies have explored the diversity, ecology, genetics, and enzymatic mechanisms of cellulolytic bacteria from different sources, such as soil, rumen, compost, and hot springs. The focus on discovering and characterizing novel lignocellulolytic enzymes has shifted towards bacteria, even though the widely used commercial cellulases are generated from fungi. This is due to the better specific activity and efficacy of bacterial systems in heterologous synthesis (Obeng et al., 2017). The fast growth rate and diverse habitat of bacterial species have drawn considerable attention to their cellulolytic activities. This is due to their ability to produce cellulolytic enzymes that are resistant to environmental stress, distinguishing them from other microorganisms. While most bacterial species may catalyse the conversion of soluble derivatives or amorphous areas of crystalline cellulose, only a few, known as "true cellulolytic," can substantially hydrolyse crystalline cellulose. Conversely, bacterial species lacking a complete cellulase system are referred to as "pseudo cellulolytic." The well-researched aerobic cellulolytic bacteria comprise *Microbispora bispora*, *Bacillus* sp., and *Cellulomonas biazotea.* In contrast, the extensively researched anaerobic cellulolytic bacteria include *Ruminococcus albus*, *Clostridium thermocellum*, *Acetivibrio cellulolyticus,* and *Fibrobacter succinogenes*. Moreover, a multitude of marine cellulolytic bacterial genera has been identified as well. Thermophilic bacteria, such as *Anoxybacillus* sp., *Geobacillus* sp., and *Bacteroides*, have also been discovered to possess cellulase activity (Dhillon et al., 1985; Kuhad et al., 2011).

3.1.3 Cellulolytic Actinomycetes

Actinomycetes are unique Gram-positive bacteria with fungal-like morphological features. They thrive in various environments and are known for their ability to produce a wide range of biologically active compounds, including lignocellulolytic enzymes. These enzymes, which include cellulases, hemicellulases, and lignin-degrading enzymes, are crucial for breaking down tough plant biomass. They find applications in industries such as bioethanol production, paper manufacturing, and wastewater treatment. Actinomycetes are a promising source of these valuable enzymes, making them a focus of research in lignocellulose production. The cellulose-degrading abilities of actinomycetes have been investigated extensively as an alternative to other microorganisms for cellulase production. Numerous research studies have consistently demonstrated the significant cellulose degradation potential of actinomycetales-derived microbes. A diverse array of actinomycetes *Streptomyces* sp., *Cellulomonas fimi, Micromonospora* sp., *Thermomonospora* sp., *Pseudonocardia thermophila, Actinopolyspora halophila, Streptomyces drozdowiczii,* etc., exhibits the capability of producing cellulase enzymes (Saini et al., 2015).

3.1.4 Cellulolytic Insects

Insects are a hugely wide range of organisms, with approximately ten million different species living on the planet. Insects with cellulolytic abilities possess the capacity to break down cellulose, which is a complex carbohydrate providing the structural framework of plant cell walls. Through the evolution of specialized digestive systems, these insects have developed enzymes cellulases, and hemicellulases are enzymes that aid in the degradation of cellulose into simpler carbohydrates. The resulting sugars serve as a viable source of energy for the insects. Cellulolytic insects include termites, wood-boring beetles, as well as certain species of ants and cockroaches. These insects are influential in nutrient cycling and decomposition in terrestrial ecosystems. The digestive tracts of termites are abundant in microorganisms, such as fungi, bacteria, and protozoa, which are capable of degrading cellulose and hemicellulose effectively (Chatterjee et al., 2015). Additionally, cellulase-producing genes have been discovered in Hemiptera, Hymenoptera, and Phthiraptera. Numerous publications have described the cellulolytic activity exhibited by various insects, such as *Tenebrio molitor, Reticulitermes flavipes, Pachyda marginata, Anoplophora glabripennis, Zootermopsis angusticollis*, among others (Ferreira et al., 2001).

Degradation of cellulose by microbial cellulase varies significantly with various physicochemical parameters of the substrate (Petre et al., 1999): (i) the permeability and size of cellulases and other reagents involved with respect to the size and surface characteristics of the grown fibrils and the gap between microfibrils and cellulose molecules from the amorphous area, (ii) crystallinity degree of cellulose, (iii) unit cellular dimension of cellulose, (iv) rigidity and stereoscopic conformation of units of anhydroglucose, (v) polymerization degree of cellulose units, (vi) characteristics of celluloses-associated components, and (vii) the distribution, concentration, and nature of groups substituted.

The endocellulase sourced from various microorganisms and insects has wide applications for producing different by-products.

3.2 Aerobic and Anaerobic Degradation of Cellulose

Cellulase enzyme production by fungi, bacteria, and actinomycetes is known to be quite effective in the natural environment. The microorganisms produce different cellulases that can be either free or bound to their cell surface, with varying levels of efficiency and complexity in their enzyme production. Moreover, cellulases can be produced both by aerobic and anaerobic microorganisms. Breakdown of the β-1,4-linkages in the cellulose polymer by cellulases releases sugar subunits like glucose. This process is applied in industries where cellulose is utilized as a raw material or where cellulose degradation is necessary. *Trichoderma viride* and *T. reesei*, which are aerobic cellulolytic fungi, have been the subject of extensive research (Jayasekara et al., 2019).

Aerobic cellulolysis occurs in the presence of oxygen, leading to the conversion of cellulose into bioethanol (Lu & Imlay, 2021), and is carried out by aerobic microorganisms such as bacteria and fungi. Aerobic cellulolysis is a process where cellulose is broken down into glucose units by the action of three types of enzymes: endoglucanases, exogluconases, and β-glucosidases. Endoglucanases, also known as 1,4-β-D-glucan 4-glucanohydrolases, break the internal bonds within cellulose chains, creating shorter chains called cellodextrins. Exogluconases, on the other hand, work on the ends of these shorter chains, breaking them down into even smaller glucose units. Finally, β-glucosidases, or β-D-glucoside glucohydrolases, convert the remaining cellodextrins and other glycosides into individual glucose units, which can be used by cells for energy. Overall, these three types of enzymes work together to break down cellulose into its constituent glucose units, allowing cells to use this abundant polysaccharide as an energy source (Sapkota, 2022).

Biodegradation of agroindustrial waste products can be carried out both aerobically and anaerobically. Aerobic biodegradation leads to carbon dioxide and water (Manyapu et al., 2018), while anaerobic biodegradation may result in the production of methane, which can be used as biogas (Manyapu et al., 2021).

In contrast, anaerobic cellulolysis involves a more complex mechanism that likely involves the concerted action of several enzymes. One proposed mechanism involves the binding of cellulose to a multienzyme complex, known as a cellulosome, that is anchored to the bacterial cell surface. This complex contains several types of enzymes, including endoglucanases, exogluconases, and other enzymes, that modify the cellulose substrate. Endoglucanases initiate the attack on amorphous regions, followed by exogluconases attacking newly exposed ends of cellulose chains. Other enzymes modify the substrate for further enzymatic attack (Sapkota, 2022).

Many microorganisms efficiently produce endocellulase during solid-state fermentation under anaerobic conditions, for instance, *Aspergillus niger*, *Penicillium* sp., *Saccharomyces cerevisiae, Thermoascus aurantiacus*, *Trichoderma reesei*, etc. (Shah et al., 2019). In the case of *T. aurantiacus* and *A. niger*, the endoglucanase activity was found to be 987 U/g (Kalogeris et al., 1999) and 215 U/g (Yang et al., 2004), respectively.

Both aerobic and anaerobic processing of cellulose can result in the production of biofuels like ethanol and biogas. The use of cellulolytic microorganisms, such as yeast and bacteria, in aerobic treatment includes the production of ethanol from cellulose. There have been a few reports of cellulolytic facultative anaerobes, such as *Cellulomonas uda* (Poulsen et al., 2016) and *Caulobacter* sp. FMC1 (Song et al., 2013). Research by Kim et al. (2018) found that *Paenibacillus* sp. CAA11, a bacterium that can digest cellulose both aerobically and anaerobically, produced ethanol from cellulose. The bacteria were modified to express heterologous endoglucanase, increasing its capacity for cellulolysis and ethanol production.

Utilizing methanogenic bacteria like *Thermoanaerobacterium saccharolyticum*, *Clostridium* sp., and *Ruminococcus* sp., which primarily produce methane and carbon dioxide from cellulose as part of anaerobic treatment, results in the production of biogas (Sasaki et al., 2021).

The majority of the cellulose is degraded aerobically, while only 5–10% is degraded anaerobically (Ward, 2015). However, anaerobic and aerobic cellulolysis occur under different conditions, and as such, optimizing the conditions for each method is crucial to enhancing productivity. This is due to the fact that different bacteria and fungi are involved in the production of cellulolytic enzymes for each method. Consequently, the application of cellulolytic enzymes produced by microorganisms provides an excellent solution for enhancing productivity, as each method has different productivity levels due to the unique conditions involved (Leschine, 1995).

3.3 Conversion of Cellulose into Bioethanol

Modern bioenergy applications are based on practical solid, liquid, and gaseous energy carriers, with pellets, bioethanol, and methane as common examples. Bioethanol is now the biofuel produced in the largest volume. The economic viability of bioethanol production from lignocellulosic biomass is attributed to the ample presence of cellulose and hemicellulose sugars in comparison to starch or sugar-based biomass (Periyasamy et al., 2023).

The complexity of the plant cell wall's structure contributes to a rise in the production cost of bioethanol. The conversion of lignocellulosic materials into bioethanol involves two essential steps. The first step requires the pretreatment of the biomass to separate individual components and subsequently hydrolyse them to produce fermentable sugars. In the second step, the fermentable sugars are then transformed into bioethanol. In biologically mediated processes for the production of bioethanol, cellulase enzymes are typically utilized for hydrolysis, while yeast or bacterial strains are employed for fermentation. While cellulase enzymes possess the ability to access lignin and hemicellulose present in lignocellulosic biomass, this can result in decreased hydrolysis efficiency. The removal of lignin and hemicellulose from lignocellulosic biomass results in a reduction of cellulose crystallinity and an increase in porosity. Employing diverse pretreatment techniques can be advantageous in enhancing the accessibility of cellulose to cellulolytic enzymes during the hydrolysis process. These changes significantly improve hydrolysis performance, ultimately enhancing the bioconversion efficiency of lignocellulosic materials to bioethanol (Chen et al., 2017).

There exist various strategies that can be employed to enhance hydrolysis efficiency in bioethanol production from lignocellulosic biomass. These approaches comprise the optimization of the pretreatment conditions, utilization of more efficient cellulase enzymes, and the adoption of genetically modified microorganisms to improve bioconversion.

Narendra Kumar et al. (2023) uncovered a highly efficient cellulase-producing bacterium, identified as *Chryseobacterium culicis* strain Bp16, isolated from soil within the Bandipur forest. Under carefully optimized conditions, including a pH level of 7, a temperature of 30 °C, and a 72-hour incubation period, this bacterium exhibited impressive cellulase enzyme activity. Agricultural waste materials like coconut husk and rice bran were harnessed as substrates, facilitating cellulase enzymatic activity and the subsequent generation of reducing sugars through submerged fermentation. Notably, these reducing sugars were successfully transformed into ethanol, achieving a yield of 0.53%, employing both baker's yeast and specialized wine starter cultures. This breakthrough presents a promising avenue for sustainable ethanol production from underutilized biomass resources.

Jugwanth et al. (2020) explored the potential of sugarcane bagasse (SCB), an often-overlooked residual by-product of the sugar industry, as a valuable resource for ethanol production. The research encompassed a comprehensive study involving a simultaneous saccharification and fermentation (SSF) process and systematically fine-tuned the conditions for optimal results. This included setting the yeast titre to 1 time, utilizing an enzyme loading of 100 U/g with the commercial cellulase enzyme sourced from *Trichoderma reesei*, and maintaining a temperature of 39 ° C. The culmination of these efforts yielded an impressive bioethanol concentration of 4.88 g/L. These findings not only hold promise for sustainable biofuel production but also offer valuable insights for the effective scaling-up of industrial processes in this field.

The engineered GH1 β-glucosidase highly expressed by *Trichoderma harzianum* was tolerant to glucose inhibition and effective in saccharification of sugarcane baggage, thus boosting bioethanol production by twofolds (Santos et al., 2019). Out of 19 isolates of white-rot fungi, *Cotylidia pannosa* revealed the highest potential for endocellulase production with 7.1 U/ml with wheat bran as substrate at 30°°C. *Cotylidia pannosa* produced 4.2 g/L of ethanol after 56 hours of incubation from wheat bran using separate hydrolysis and fermentation technique. The synergistic action of *C. pannosa* and *S. cerevisiae* enhanced bioethanol production.

3.4 Conversion of Cellulose into Biocompost

The breakdown of lignocellulose by a variety of microorganisms results in the production of carbon dioxide and water under an aerobic environment as well as carbon dioxide, methane, and water under an anaerobic environment. Although anaerobic lignocellulose-degrading protozoa have also been discovered in the rumen, bacteria and fungi still make up the bulk of lignocellulolytic microorganisms (Gupta et al., 2012).

The most essential biological step in composting is the breakdown of cellulose, which is mostly carried out by microorganisms that produce cellulase enzymes. The rate and efficiency of the composting process are determined by the ability of these bioagents to produce cellulase. During composting, these bioagents play a vital role in the decomposition of cellulose and lignin components in the waste material. The waste's indigenous microflora is capable of producing large levels of enzymes like cellulase and lignin-modifying enzymes, which speed up the breakdown of these complicated compounds. This enzymatic activity eventually leads to a more rapid and efficient composting process. However, a limited number of microbes are capable of efficiently utilizing cellulose as a substrate. According to certain searches, the *Aspergillus fumigatus* F12 strain is exceptionally successful at degrading cellulose when compared with other biopolymers, which indicates its potential as an efficient cellulose-degrading microbe (Cao et al., 2013; Kumar et al., 2010).

A significant number of microorganisms have demonstrated the capacity to generate cellulolytic enzymes, which are utilized in the degradation of waste during composting. Nonetheless, optimizing the conditions of the composting process can enhance microbial diversity and activity. The identification and utilization of efficient microorganism strains can be highly advantageous in the composting process.

Yin et al. (2023) tested three treatments in the composting experiments: no added biochar in control (CK), addition of 10% (dry weight) corn stalk biochar (CSB), and 10% (dry weight) rape straw biochar (RSB). The primary objective was to elucidate the roles played by cellulase enzymes and fungal communities in the intricate process of organic matter conversion. The findings of this research uncovered a significant revelation. Within the RSB-treated compost, there was a notable enhancement in both organic matter transformation and cellulase activity. In comparison to the other treatment groups, the fungal community structure in the RSB treatment remained stable throughout the middle and late phases of composting. Most intriguingly, RSB demonstrated a unique capacity to augment the abundance of fungi intricately associated with the degradation and conversion of organic matter during the composting process. The research findings pave the way for the effective use of biochar in improving composting processes.

Zhang and Dong (2022) isolated seven microbial strains from the compost pile of rabbit faeces and sesame oil by using the dilution plate method. Seven powerful cellulose-degrading bacteria were grouped into six teams. These teams were used to improve the composting of agricultural waste. Notably, consortium VI, encompassing all seven strains, demonstrated exceptional cellulase activity, recording values of 141.89 U/ml for CMCase (carboxymethyl cellulase), 104.56 U/ml for FPase (filter paper cellulase), and 131.18 U/ml for β-Gase (β-glucosidase). Researchers optimized CMCase production in Consortium VI using a single-factor approach and then fine-tuned it with a more advanced method called RSM. The best conditions they found were a 4.25-day culture time, a temperature of 35.5 °C, pH 6.6, and an inoculum volume of 5% (v/v). This improved the CMCase activity to 170.83 U/ml, and Consortium VI outperformed another microbe group in breaking down organic materials (rabbit faeces and sesame oil cake), enhancing composting, and promoting sustainable agriculture.

Bhupathi and Subbaiah (2020) experimented on bioconversion of different agricultural residues, viz. paddy straw, sorghum stalks, sugarcane bagasse, maize stalks, soybean hay, black gram hay, ground nut haulms, sawdust, paddy straw compost, and coir pith compost, into protein-rich biomass using *Calocybe indica* var. Apk2. It was found that the mushroom produced the highest endocellulase of 2.4% in the case of paddy straw conversion. Certainly, the paddy straw produced the highest yield of mushrooms, 356.5 g per bed, and also the highest level of protein, 5.858 g per kg of substrate.

3.5 Conversion of Cellulose into Single-Cell Protein

Single-cell protein is a dried microbial cell or protein extracted from bacteria, yeast, fungi, or algae. These microorganisms are grown in large-scale fermentation systems and serve as a source of protein for food or feed supplements (Abdel-Azeem & Sheir, 2020; Zepka et al., 2010).

The production of single-cell protein (SCP) from lignocellulosic wastes is a complex yet sustainable process. It begins with the collection and preparation of lignocellulosic materials, which are often abundant agricultural or forestry residues. These materials undergo a series of pretreatments, including physical, chemical, physicochemical, and biological methods, to break down their complex structures. Hydrolysis is then employed to release sugars from cellulose and hemicellulose, which are essential for the subsequent fermentation stage. Microorganisms are introduced to the hydrolysed solution to consume these sugars and produce SCP. After fermentation, the biomass is harvested and dried, resulting in a valuable single-cell protein product. This process not only addresses the challenge of lignocellulosic waste disposal but also contributes to sustainable protein production for various applications, making it a promising approach for the future. Cellulolytic enzymes also played an important part in hydrolysis throughout this phase. Enzymatic hydrolysis necessitates the use of particular enzymes, such as cellulases and hemicellulases, to convert complex polysaccharides into fermentable sugars. The production of single-cell protein (SCP) from lignocellulosic wastes is a complex yet sustainable process. It begins with the collection and preparation of lignocellulosic materials, which are often abundant agricultural or forestry residues. These materials undergo a series of pretreatments, including physical, chemical, physicochemical, and biological methods, to break down their complex structures. Hydrolysis is then employed to release sugars from cellulose and hemicellulose, which are essential for the subsequent fermentation stage. Microorganisms are introduced to the hydrolysed solution to consume these sugars and produce SCP. After fermentation, the biomass is harvested and dried, resulting in a valuable single-cell protein product. The production of single-cell protein (SCP) from lignocellulosic waste not only addresses waste disposal challenges but also offers a sustainable source of protein for various applications. This makes it a promising and environmentally friendly approach for the future (Bajpai, 2016).

Agricultural waste, due to its high content of cellulose, hemicellulose, and lignin, serves as a primary source for the production of single-cell protein. Prior to utilization, agricultural waste undergoes pretreatment and hydrolysis. Endocellulase and exocellulase, which are cellulolytic enzymes, are commonly used in the enzymatic hydrolysis process. These enzymes are selected due to their ability to efficiently break down agricultural waste materials.

Gu et al. (2022) experiment was carried out on paper mulberry to explore its conversion into high-value products. This involved employing autohydrolysis and enzymatic hydrolysis techniques. The outcomes of the experiment indicated the recovery of 8.3 g of xylo-oligosaccharides, 5.7 of low-degree xylo-oligosaccharides, and 1.8 g of amino acids. Furthermore, when cellulase was used to hydrolyse the autohydrolysed paper mulberry, it resulted in the generation of 20.1 g of fermentable monosaccharides. These monosaccharides were subsequently employed in a fermentation process with *Candida utilis*, leading to the production of 4.8 grams of single-cell protein (SCP).

Tropea et al. (2022) conducted a study where they developed a single-cell protein from multi food waste using the *Saccharomyces cerevisiae* and enzyme treatment. The study involved the fermentation of animal waste, pineapple, banana, apple, and citrus peels, etc. The substrate's composition was assessed after the fermentation process, with a focus on crude lipid, ash, and lignin, as well as soluble and insoluble sugars. The results indicated that the substrate had 14.46% crude lipid, 1.08% ash, and 6.29% lignin. The substrate contained an impressive 40.19% protein, making it an excellent choice for animal feed. The study showcased the feasibility of using mixed animal and plant food waste for producing single-cell protein.

3.6 Conversion of Cellulose into Protein-Rich Feed

In a bioconversion process, cellulose waste was converted into useful products using two strains of fungi called *P. ostreatus* and *P. florida*. These fungi were attached to tiny hydrogel pellets and placed into a special container called a fluidized mobile bed bioreactor. This container was designed to move the pellets around while also allowing liquid to flow through them. The goal of this process was to break down the cellulose waste into useful products with the help of fungi. As a single-cell protein production process, this substrate was pretreated. Enzymatic hydrolysis is the penultimate phase, in which cellulosic enzymes break down the cellulose into simple sugars. Later, fungal and bacterial energy is used to make rich protein, and biomass is extracted. Rich protein can be obtained by optimizing the enzymatic hydrolysis condition in this process (Baker et al., 1998; Bommarius et al., 2008).

Jabri et al. (2023a) investigated the enzymatic valorization of alkali-treated chickpea straw and sunflower stalks as a source of digestible fibre for ruminants. The research showed that using NaOH, urea, and external fibrolytic enzymes improved the nutritional value and digestibility of chickpea straw and sunflower stalks in ruminants, and with the addition of exogenous fibrolytic enzymes, their digestibility was enhanced. Jabri et al. (2023b) conducted a study on faba bean straw supplemented with exogenous fibrolytic enzymes. Proved that adding external fibrolytic enzymes to faba bean straw can result in a bio-converted by-product that is rich in both energy and protein, making it a better option than traditional cereal straw in ruminant nutrition.

Application of microalgae, such as *Chlorella*, *Spirulina*, *Scenedesmus*, and *Dunaliella*, can grow on a variety of carbon sources, including cellulose hydrolysates (Jones et al., 2020). By producing cellulases, these microalgae can be used to produce a high protein-rich feed (up to 70% of dry weight) (Ritala et al., 2017). Microalgae may also produce vital amino acids, vitamins, minerals, and antioxidants.

3.7 Conversion of Cellulose into Industrial Reagents (Ethanol, Isopropanol, Butanol) and Fuels

The use of lignocellulosic biomass, including agricultural residues, forestry wastes, crop wastes, and municipal solid wastes, and manures, is a promising option for producing industrial reagents and biofuels due to their large availability and non-food use (Chundawat et al., 2011).

The process of producing valuable products from lignocellulosic biomass involves four main steps, namely pretreatment, hydrolysis of carbohydrates to sugar monomers, fermentation of sugars, and recovery. Before enzymatic hydrolysis, a pretreatment step is necessary to open up the biomass matrix, thus enhancing enzyme accessibility to cellulose. During cellulose hydrolysis, lengthy glucose molecule chains are broken down into glucose monomers, which are then fermented to yield (Wang et al., 2012).

Glucose is a monosaccharide that is utilized in the synthesis of many industrial chemicals, the production of biofuels, and the manufacturing of food and feed. In a study by Kumar et al. (2011a), glucose was extracted from orange peel waste using a rapid extraction method with the mixture of methanol–dichloromethane–water in the proportions 0.3:4:1 by volume, an agroindustrial waste, utilizing endocellulase and sulphite digestion reagents. Apart from glucose, other industrially significant sugars like fructose, galactose, arabinose, and xylose can also be extracted from orange peel waste.

An experiment conducted by Khantibongse and Ratanatamskul (2023) investigated the enzymatic hydrolysis of rice straw. Various enzymatic hydrolysis approaches were employed in the study, including hot water pretreatment with enzymes, alkaline pretreatment with enzymes, and acid pretreatment with enzymes. Two types of cellulase enzymes were used: laboratory-grade cellulase from *Aspergillus niger* and a commercial cellulase enzyme called Atres Zymix Plus. Research showed that the most effective method for obtaining monosaccharides from rice straw was boiling it with water and using a commercial cellulase enzyme. This approach primarily yielded glucose, making up 78–86% of the total. When using acid pretreatment, glucose remained the dominant monosaccharide. In contrast, alkali pretreatment resulted in xylose as the primary monosaccharide, ranging from 48% to 85%, with glucose at 5–49%. Furthermore, boiling rice straw with the commercial cellulase enzyme produced the highest glucose yield compared with other methods. The analysis also detected up to 62 different phenolic compounds and aromatic and aliphatic hydrocarbons during the enzymatic hydrolysis process. The monosaccharides produced from the degradation of cellulose and hemicellulose were transformed into biofuels such as ethanol and biogas.

The treated coffee husks with mild acid, alkali, and high-temperature water treatments after the enzymatic saccharification using the Cellic CTec2 enzyme cocktail (mixture of cellulase enzymes) was found to be the best conditions to convert them into glucose, a sugar used in the production of ethanol (de Almeida et al., 2023). The optimal conditions for saccharification included maintaining a pH level of 5.5 and keeping the temperature within the range of 30 °C to 75 °C, resulting in a conversion yield of 16.2%. Further enhancements were achieved by adjusting the biomass loading to fall between 11% and 14% and maintaining an enzyme concentration of 20–25 FPU/g. This adjustment led to the production of 5 g/L of glucose, and the biomass conversion yield improved to an impressive 24.46%. These findings underscore the potential of coffee husks as a sustainable source for ethanol production, in alignment with the principles of the bio-economy, where renewable resources are maximized while minimizing waste.

The cellulose and hemicellulose can be converted into aromatic compounds in two steps (Yamaguchi et al., 2019). In the first step, the cellulose and hemicellulose content of sugarcane bagasse was hydrolysed into sugar alcohols such as sorbitol and xylitol using a carbon-supported platinum catalyst, and in the second step, the sugar alcohols with 64% yield were converted into aromatic compounds, viz., erythritol, glycerin, propylene glycol, and ethylene glycol using hydrogen catalyst. The batch reaction took place in the water at a temperature of around 190 °C for 16 hours (Yamaguchi et al., 2019).

Zhao et al. (2019) achieved the development of a cost-effective approach for enhancing ethanol production by concurrently utilizing engineered strains of lignocellulase (endocellulase, exocellulase)-producing *S. cerevisiae* and the high-efficiency ethanol-producing *S. cerevisiae* strain, WXY12. In this study, pretreated corn stalks were subjected to enzymatic hydrolysis using fermentation broth obtained from *Saccharomyces cerevisiae* strains that had been engineered to produce lignocelluloses. The application of this approach resulted in a reduction of costs and an increase in ethanol production from corn stalks.

For carrying out the bioconversion process in an economical and sustainable approach, there is a need for a simple and effective immobilization technique.

4 Immobilization of Endocellulase

The selection of an appropriate enzyme immobilization method is pivotal, as it profoundly influences enzyme performance. Key factors to weigh include enzyme activity, especially its effect on reaction efficiency, and the role of lipase in specific applications. The method's ability to preserve and regenerate enzyme properties over time is essential. Cost-effectiveness is another crucial consideration, and the potential toxicity of immobilization compounds should be assessed to avoid unwanted side effects. Lastly, the final properties of immobilized enzymes, such as activity, stability, and reusability, must align with the specific needs of the application. This comprehensive evaluation process ensures a well-informed choice for enzyme immobilization (Maghraby et al., 2023). Enzyme immobilization methods fall into two main categories: physical and chemical. Physical immobilization uses non-covalent interactions like hydrogen bonds, van der Waals forces, and more to attach enzymes to a substrate. In contrast, chemical immobilization forms covalent bonds between the enzyme and support material using linkages like amide or carbamate. The choice of immobilization method is critical, as it profoundly affects enzyme performance during specific reactions. The decision hinges on the application's requirements and the desired properties of the immobilized enzyme, offering flexibility and tailored solutions to meet distinct goals (Mohamad et al., 2015). An enzyme works as a biological catalyst in an enzymatic reaction, promoting the reaction rate while being unaffected by the reactions. As a result, as long as the enzyme is active, it can be used repeatedly. Thus, several immobilization strategies for immobilizing enzymes on a solid surface have been established. Adsorption, entrapment, covalent bonding, affinity bonding, and cross-linking methods can be used to immobilize enzymes. However, due to the complex structure of proteins, no single strategy is universally effective for all enzymes and applications (Nguyen & Kim, 2017; Guisan, 2006).

4.1 Limitations of Different Immobilization Techniques

In the adsorption method, enzymes are attached to a carrier material through weak, non-specific forces such as van der Waals forces, hydrogen bonds, and hydrophobic interactions. This method allows enzymes to adhere to the surface of the carrier without the formation of strong chemical bonds. Adsorption is a commonly used technique for enzyme immobilization due to its simplicity and effectiveness in various applications (Imam et al., 2021). Ionic bonding enables enzymes to be linked together through salt bonds. These are relatively weak and non-specific forces that can have drawbacks, including the potential for enzyme leakage from the carrier material. While ionic bonding can be used for enzyme immobilization, the risk of enzyme detachment and loss of activity may be a concern in certain applications, making it necessary to consider alternative immobilization methods (Braham et al., 2021). The surface charge of an enzyme is modulated by the pH value of the surrounding liquid solution and the isoelectric point. Advanced methodologies enable precise calculation and visualization of the allocation of charges on the enzyme surface (Hanefeld et al., 2009). Ionic bonding for enzyme immobilization has its drawbacks. It is highly sensitive to changes in buffer pH, temperature, and ionic strength. Additionally, the physical link established through ionic bonding may not always guarantee a secure enzyme attachment to the carrier, potentially resulting in enzyme leakage. In industrial settings with high concentrations of reactants and products, as well as elevated ionic strength, the risk of enzyme leaching increases. This could lead to substrate contamination, which is a significant concern when implementing enzymatic processes in such environments (Wahab et al., 2020).

Enzyme immobilization via entrapment refers to a nonreversible technique where enzymes are confined within a carrier, such as a material's lattice structure, polymer membranes, or the inside of fibres (Bhattacharjee et al., 2023; Imam et al., 2021). The key element in this process is enabling the enzyme to be retained while allowing the diffusion of products. This balance between immobilizing the enzyme and facilitating the movement of reaction products is crucial for efficient and effective enzymatic reactions. It ensures that the enzyme remains in place to catalyse the reaction while allowing the products to move away from the active site, preventing any inhibitory effects that product buildup might have on the enzyme's activity (Imam et al., 2021). The usefulness of the entrapment technique is restricted by mass transfer constraints that affect substrate or analyte delivery to the enzyme's active site (Górecka & Jastrzębska, 2011). There are several important limitations to consider in this process. These include the potential for enzyme leakage if the support matrix has large enough holes, the possibility of enzyme deactivation during immobilization, limitations in weight-bearing capacity, and potential deterioration of the carrier material during use. Another crucial factor is the bonding between the particle size of immobilized enzymes and the pore size of the support material, which plays a vital role in determining the efficiency of the produced probes or immobilized enzyme systems. Addressing these limitations is essential to achieving effective and reliable enzymatic processes (Mohamad et al., 2015).

Covalent bonding immobilization offers the advantage of creating a strong and stable bond between the carrier and the enzyme, preventing enzyme leakage and leading to improved enzyme stability and enhanced stereospecificity (Zhang et al., 2015). However, this process is more difficult and requires longer incubation durations than adsorption immobilization. To acquire functional groups for covalent bonding, the enzyme molecule may need to be chemically modified. This functionalization of the enzyme may result in its denatured proteins. Also, the quantity of the enzyme capable of being immobilized on the carrier is extremely low (Marrazza, 2014). Furthermore, covalent bonding is not suitable because its matrix is not regenerable, and there is a greater chance of coupling with toxicants (Sassolas et al., 2012).

Affinity bonds between an affinity tag (such as biotin) on a protein sequence and a functional group (such as avidin) on support, but it needs the presence of specific groups on enzymes like biotin, without which immobilization cannot be carried out (Sassolas et al., 2012).

4.2 Cross-Linking Technique for Endocellulase Immobilization

Above all the aforementioned immobilization techniques, cross-linking is the technique with minimum drawbacks, which can be eliminated with improvements in the strategy. Cross-linking has various advantages over other techniques of immobilizing enzymes. For example, cross-linking can produce a more stable and robust enzyme-support complex that can endure harsh reaction conditions like high temperature, pH, or pressure. Cross-linking can improve enzyme reusability and shelf-life, lowering the cost and waste associated with enzyme synthesis. Moreover, cross-linking can improve enzyme catalysis and selectivity, allowing them to perform specific processes more effectively and with larger yields (Govardhan, 1999). Overall, the cross-linking technique for enzyme immobilization has a lot of potential in sectors including bio-catalysis, biosensors, bioremediation, and drug development.

The process of cross-linking represents an irreversible method for the immobilization of enzymes that is distinct from other methods because other procedures necessitate the use of carriers, whereas this method does not (Shi et al., 1995). Enzymes simply serve as carriers in this method. Thus, this technique is commonly referred to as carrier-free immobilization (Cao et al., 2003; Sheldon, 2007). Cross-linking is a process that entails the creation of covalent bonds between enzyme molecules, resulting in the formation of intermolecular cross-links. This is accomplished by using a multifunctional reagent as a linker to bind enzyme molecules together, creating three-dimensional cross-linked aggregates. The enzyme that has been immobilized is not affixed to any substrate or carrier and remains present within the solution mixture. Cross-linking immobilization can be achieved through two techniques: cross-linking enzyme aggregate and cross-linking enzyme crystal. In these techniques, a cross-linking agent, such as glutaraldehyde, is employed to link enzyme molecules together via a reaction with the free amino groups present on the reactive sites of adjacent enzyme molecules. The cross-linking method is a sample enzyme immobilization method where enzyme leakage is exclusively determined by the degree of chemical binding of enzyme molecules. One of the key advantages of this methodology is its capacity to modulate the microenvironment for enzymes through the utilization of specific stabilizing compounds via surface complementarity, resulting in enhanced stability. However, the use of glutaraldehyde may, at times, perturb the enzymes, leading to alterations in their performance. To circumvent enzyme modifications, gelatin and bovine serum albumin are examples of inert proteins incorporated into the immobilization process (Broun, 1976; Chang & Mahoney, 1995).

In the cross-linked enzyme aggregation (CLEA) method, a cross-linking agent is used to bind numerous enzymes together, forming a solid aggregate. This aggregate is then extracted from the solution and can be utilized in industrial operations like biocatalysis or biotransformation. The CLEA method is an inexpensive and simple technique that can be used for a wide range of enzymes. A cross-linking agent is involved in binding the enzymes together to generate a crystal structure. These crystals have the potential to be significantly larger and more stable than CLEAs, allowing for greater control of enzyme processes. CLEC, on the other side, is a more complicated and costly approach that necessitates particular conditions for the creation of enzyme crystals. CLEC is a more stable and predictable structure for enzymatic processes, but it requires more specialized conditions to produce. CLEA is a simpler and less expensive strategy for cross-linking enzymes. Both procedures have advantages and disadvantages, and which one to use depends on the unique needs of the industrial process (Velasco-Lozano et al., 2016).

4.2.1 Challenges and Current Scenario of Cross-Linking Technique

Cross-linked methods can improve the stability, reusability, and activity of the enzymes, as well as reduce the cost of enzyme production. However, there are some challenges to cross-linked endocellulase, such as

- Finding the optimal conditions for cross-linked endocellulase formation, such as the type and concentration of precipitant, cross-linker, and additives (Barbosa et al., 2020).
- Maintaining the balance between enzyme activity and stability, as excessive cross-linking can reduce enzyme accessibility and catalytic efficiency (Noori et al., 2019).
- Avoid enzyme leaching and aggregation during storage and operation (Bhardwaj et al., 2021).
- Enhancing the mass transfer and diffusion of substrates and products through the cross-linked endocellulase matrix.

These challenges require further research and development to optimize the cross-linked endocellulase technology and expand its applications in various fields.

The current scenario of research on cross-linked endocellulase is focused on exploring the diversity, ecology, genetics, and enzymatic mechanisms of cellulolytic bacteria from different sources, such as soil, rumen, compost, and hot springs. A review article by Bhardwaj et al. (2021) provides an overview of the current status and future prospects of microbial cellulases production and applications in various industries. It also discusses the strategies to enhance the yield, efficiency, and specificity of cellulases by using waste feedstock, process parameters control, genetic manipulations, and immobilization techniques. Jayasekara et al. (2019) described the different methods and applications of immobilizing enzymes by entrapment using various materials, such as alginate, gelatin, polyacrylamide, and silica. It also provides some examples of entrapped enzymes for biocatalysis and biosensing.

4.2.2 Parameters to Be Considered in Cross-Linking Enzyme Aggregation

1. Physical properties: In industrial applications, particle size and morphology are crucial for cross-linked enzyme aggregates (CLEAs) and their impact on mass transfer and filterability. Mass transfer limitations can hinder CLEA catalytic potential, especially with enzymes acting on large substrates. CLEA particle size is vital; larger particles risk reduced enzyme–substrate interaction and potential activity loss, while smaller particles enhance mass transfer. CLEAs are categorized as type 1 and type 2 aggregates based on morphology. Type 1 CLEAs have fine-grained structures with numerous cavities, typically promoting better mass transfer. Particle size and morphology control are achievable through various synthesis parameters. Smaller CLEA particles often deliver optimal mass transfer, but practical concerns may require larger particles for ease of handling. Encapsulation methods can increase particle size for large-scale applications, balancing practicality with high activity levels. Therefore, characterizing CLEA particle size before large-scale use remains crucial (Talekar et al., 2013).
2. Stability: Enzyme immobilization offers the advantage of increased resistance to high temperatures and organic solvents. This resilience is primarily due to a reduction in the enzyme's flexibility. When enzymes are immobilized, their structures are less likely to undergo significant changes or unfolding, which could compromise their three-dimensional or tertiary structure. This tertiary structure is crucial for the enzyme's proper functioning. In essence, immobilization helps enzymes maintain their integrity and activity under harsh conditions, making them more versatile and robust in various industrial applications (Patel, 2006).
3. Recyclability: Another notable advantage of enzyme immobilization is its ability to simplify the collection and recycling of enzymes from the reaction medium. This streamlined approach leads to more efficient downstream processing and reduces the overall cost of enzymes as a component in the final product. By immobilizing enzymes, it becomes easier to separate them from the reaction mixture, enabling their reuse in subsequent reactions. This not only conserves valuable enzymes but also contributes to cost savings in industrial applications, making enzyme-based processes more economically sustainable (van Langen et al., 2005).
4. Reduced processing time: Cross-linking, facilitated by enzymes like endocellulase, significantly accelerates various industrial processes, such as pulp and paper manufacture. In traditional paper production, which can take several hours or even days, the use of endocellulase cross-linking enzymes to pretreat cellulose fibres enhances their resistance to degradation. This pretreatment reduces the time required for subsequent processing or treatments. By reinforcing cellulose fibres and simplifying their processing, enzymes play a crucial role in enhancing productivity and cost efficiency throughout the production process. This streamlined approach is particularly valuable in industries

where time and resource efficiency are paramount, such as paper manufacturing (Li et al., 2022).

It can be inferred that cellulose degradation using a normal cellulase enzyme with the usual method of treatment is extremely challenging. The need for immobilization of the enzyme is highly preferable for enhancing the enzyme and its characteristics. The cross-linking technique is reasonable for the immobilization of cellulase.

5 Cellulose Breakdown Mechanism

'Elementary fibrils', which are longer slander bundles of long chains of β-D-glucopyranose residues connected by 1–4 glycosidic linkages, make up the cellulose molecules. The cellulose chains are reported to consist of 24–36 chains on the basis of scattering data (Fernandes et al., 2011) and the cellulose synthase data (Endler & Persson, 2011). The cellulose molecules are laterally bonded inside each primary fibril, and the nearby molecules run parallel but in opposite directions with varying degrees of orientation (Horn et al., 2012). Cellulose molecules consist of two regions: crystalline cellulose, comprising highly oriented molecules, and amorphous cellulose, consisting of less oriented molecules. The basic fibrils are combined to create microfibrils, and multiple microfibrils form a macrofibril. Enzymatic saccharification of these crystalline, non-soluble microfibrils is challenging. Cellobiose, a disaccharide, and the repeating unit is rotated 180 degrees by successive sugars along the chains of crystalline cellulose. Cellulose often exhibits both well-organized crystalline portions and disorganized, amorphous regions. Naturally occurring crystalline cellulose can exist in various forms, such as isomers I (Iα and Iβ) to IV, with Iβ being the more common form in plants. Different types, including Type II, III, and Type IV crystalline cellulose, may develop due to pretreatment processes. These structural variations impact the accessibility of cellulose to enzymatic breakdown, influencing its suitability for various applications, including biofuel production and paper manufacturing (Seddiqi et al., 2021). Although its resistance to enzymatic breakdown may present issues, cellulose has several benefits, including uniformity. Only glucose remains after completely depolymerizing cellulose.

Cellulases are important in both the natural and industrial realms by breaking down insoluble cellulose into soluble sugars, thereby contributing significantly to the global carbon cycle. Each of the three types of cellulases, i.e. β-1,4-endoglucanase (endocellulase), cellobiohydrolase, and β-glucosidase, has a unique mode of operation and structure. Endoglucanases, a type of endo cellulase, cleave the cellulose molecule at the internal bonds of the molecule's non-crystalline surface, attacking the cellulose chain at random and splitting the β-1,4-glucosidic linkages present within the molecule (Fig. 18.1). The primary function of endoglucanases is to reduce the length of the cellulose molecule, making it more accessible to other enzymes for further breakdown. Endoglucanase is critical in this

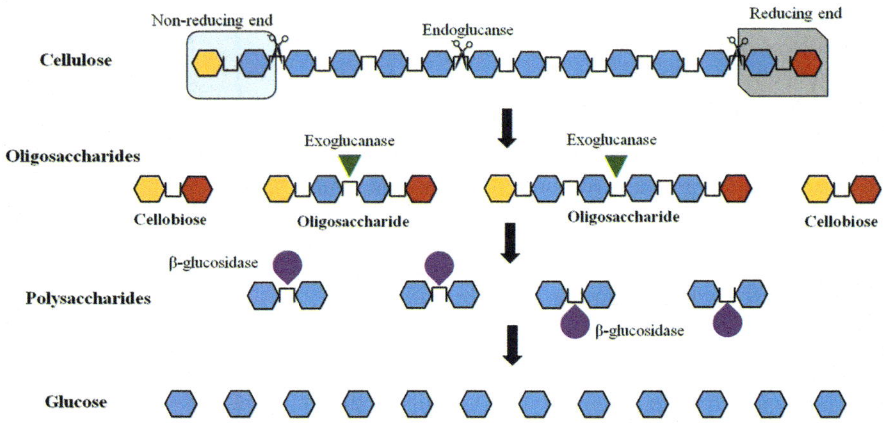

Fig. 18.1 Mechanism of bioconversion of cellulose using endocellulase and other cellulases

scenario because it is the only enzyme capable of breaking down long cellulose chains. Once endoglucanase has reduced the length of the chains, other cellulases can then break down the cellulose further. Exoglucanases are responsible for digesting the smaller tetrasaccharides and disaccharides created by endoglucanases, while cellobiases are enzymes that convert cellobiose units (disaccharides, trisaccharides, and tetrasaccharides) into individual glucose units. Cellobiases are also referred to as β-glucosidases since they break down cellobiose into monomeric glucose units. Together, these enzymes work in a degradation process to break down cellulose (Uzuner & Cekmecelioglu, 2019; Wilson, 2009).

Cellulose is degraded by endoglucanase through the β -1,4 glycosidic bonds' hydrolysis that connects glucose units. Endoglucanases (EC 3.2.1.4) arbitrarily target the amorphous regions within the chains of cellulose, causing glycosidic linkages' hydrolysis and producing new ends for additional hydrolysis to begin the cellulose breakdown process. Afterward, exoglucanases, often referred to as cellobiohydrolases, react with long-chain oligosaccharides produced from either reducing (EC 3.2.1.176) or non-reducing (EC 3.2.1.91) endings. Some endocellulases are referred to as processive endoglucanases since they travel in a single direction and progressively along cellulose chains from one end. Exoglucanases hydrolyse short-chain oligosaccharides, primarily cellobiose, a disaccharide that may be converted into glucose by the action of β-glucosidases (EC: 3.2.1.21) (Horn et al., 2012; Lakhundi et al., 2015). This process is repeated until the cellulose molecule is entirely broken down into its constituent sugars. Endoglucanases work together with other cellulase enzymes, including cellobiohydrolases and β-glucosidase, to fully break down cellulose into glucose (Meng et al., 2021). Endoglucanase hydrolysis of cellulose includes various stages, such as substrate binding, catalysis, and product release, all of which are controlled by particular amino acid residues in the active site of the enzyme (Kumar & Murthy, 2013; Vermaas et al., 2019).

Endocellulase belongs to the glycoside hydrolases family. Based on how similar their amino acid sequences are, glycoside hydrolysases have been classified into 165 families (Liu et al., 2021). Endoglucanase activity was discovered in a number of families that exhibit distinctive protein fold structures, such as the (α/α) 6 barrel (GH8, GH9, and GH48), (β/α) 8-barrel fold (GH5, GH44, and GH51), modified α/β barrel (GH6), modified β barrel (GH45), seven-fold β propeller (GH74), β-jelly roll (GH7 and GH12). Given that several GH families exhibit different protein fold classes and have evolved to bind and break down the same substrate, it is possible that cellulases originated separately from various evolutionary sources but functionally converged.

There are a few factors that can change the activity of endocellulase, viz. pretreatment of the substrate, incubation time, age and size of the inoculum, moisture content and moistening agent, temperature and pH, nitrogen source, etc. (Acharya et al., 2014).

(a) Pretreatment of substrate: Substrate pretreatment with some acid or base can improve the endocellulase activity.
(b) Incubation time: Duration of action by microorganisms also plays a significant role in the change in endocellulase activity. The maximum endocellulase activity was observed in 4 days of incubation in the case of *Aspergillus niger* (Acharya et al., 2014).
(c) Age and size of inoculum: There is a need for inoculating microbes with optimized age and size. The activity of endocellulase can be at maximum when the age of *A. niger* is 96 hours incubated. Furthermore, an increase in the size of the microbial population can improve the enzyme activity but could reduce the activity if overpopulated and nutrients are limited.
(d) Moisture content: Every enzyme activity needs favourable moisture. The moisture content deals with aeration and nutrient availability. An optimum endocellulase of around 4 IU/g of substrate was observed at 65% moisture for *Aspergillus niger* (Acharya et al., 2014). Thus, the selection of substrate must be considered at prior importance.
(e) Temperature and pH: Like every enzyme, endocellulase is also dependent on suitable temperature and pH. Incubation temperature of 28 °C for *A. niger* to produce the maximum endocellulase of more than 4 IU/g (Acharya et al., 2014). A similar temperature was found for the maximum cellulase activity in the case of *Trichoderma harzianum* (Haq et al., 2005).
(f) Nitrogen source: Sources like urea, ammonium sulphate, yeast extract, and peptone have been analysed for quantifying the endocellulase activity, where Acharya et al. (2014) found that urea produced the highest endocellulase activity of 2.129 IU/g for *A. niger* (Acharya et al., 2014).

6 By-products

The by-products generated during cellulose degradation are influenced by the precise reaction conditions and the particular endoglucanase enzyme employed. Generally, the primary by-products consist of oligosaccharides. These oligosaccharides are liberated from the interior regions of cellulose's amorphous structure as a consequence of the hydrolysis reaction. The composition and quantity of these by-products can vary depending on the specific enzymatic process and environmental factors (Rahman et al., 2018). These oligosaccharides are namely cellobiose, cellotriose, cellotetraose, and cello-oligosaccharides.

(a) Cellobiose is a zero-calorie sweetener that may be used in place of sucrose or lactose in food products (Zhong et al., 2017). Various foods and beverages can employ it as a filler and transporter for flavours or highly concentrated sweeteners. Cellobiose has a long history of use in medical applications, particularly in the cellobiose/mannitol (Ce/Ma) test. This non-invasive test assesses intestinal permeability and provides valuable insights into digestive health and potential issues (EFSA Panel on Nutrition, Novel Foods and Food Allergens (NDA) et al., 2022).
(b) Cellotriose is a substrate for the enzyme cellodextrin phosphorylase, which catalyses the reversible phosphorolysis of cello-oligosaccharides. Using engineered cellobiose phosphorylase, cellobiose and glucose-1-phosphate may be converted into cellotriose (Ubiparip et al., 2020), which can be used to produce potential prebiotic cellodextrins. Cellulosic bacteria may also use cellotriose as a source of carbohydrates (Tsuji et al., 2013).
(c) Cellotetraose can serve as a substrate for cellulases, which accelerates the breakdown of cellulose. By using cellotetraose, X-ray crystallography may be utilized to investigate the binding and catalytic processes of cellulases (Cheng et al., 2014; Ferrara et al., 2014; Prates et al., 2013). Glucosidases, enzymes that hydrolyse glycosidic linkages in glucans, can also be controlled by the use of cellotetraose (Ferrara et al., 2014).
(d) Cello-oligosaccharides have the potential to function as prebiotics, encouraging the development and activity of favourable bacteria in the human digestive tract (Ávila et al., 2021). Cello-oligosaccharides can also improve the immune system and metabolic health by modifying the composition and function of the gut microbiota (Ávila et al., 2021). Additionally, they can be added to ruminant feed to enhance rumen fermentation and animal performance (Karnaouri et al., 2019).

These oligosaccharides can be further degraded into glucose by other cellulase enzymes, like cellobiohydrolases and β-glucosidase. Glucose monomers can then be fermented to create biofuels, such as ethanol or butanol, which can generate renewable energy. However, several factors, such as enzyme concentration, reaction temperature, pH, and cellulose substrate, can significantly impact the efficiency of the cellulose degradation process (Geem et al., 2021; Vidhya, 2022). Moreover, endoglucanase plays a crucial role in cellulose degradation as it targets the amorphous areas of cellulose fibres, which are more challenging to break down compared to the crystalline regions. By cleaving 1,4-glycosidic bonds in these amorphous regions, endoglucanases create new chain ends and release minor oligosaccharides (Watson et al., 2009). This process enhances the accessibility of cellulose to other cellulase enzymes, such as cellobiohydrolases and β-glucosidase, that can further degrade cellulose into glucose. Moreover, endoglucanases are considered the most beneficial enzyme for fibrillating fibres and increasing freeness during the papermaking process without compromising the fibre properties. The use of endoglucanase enzymatic refining in paper production can reduce the requirement for chemical additives, resulting in a more environmentally friendly process (Lakhundi et al., 2015; Nagl et al., 2021).

7 Challenges

Employing cross-linked endocellulase as an eco-friendly method for the bioconversion of agroindustrial residues gives rise to numerous hurdles that require attention for its successful application. These challenges comprise:

(a) Enzyme Stability: To ensure continuous and efficient cellulose degradation, it is essential that cross-linked endocellulases maintain their stability when exposed to diverse conditions, such as fluctuations in temperature and pH levels.
(b) Enzyme Immobilization: Immobilizing cross-linked endocellulase on appropriate supports in a stable manner poses a considerable challenge, given its significant influence on both enzyme activity and durability.
(c) Cost-Effectiveness: Simplifying the production and immobilization processes for cross-linked endocellulases is crucial. The development of cost-efficient methods for enzyme production and immobilization is essential to enhance the overall commercial viability of this sustainable approach.
(d) Substrate Specificity: It is vital to guarantee that cross-linked endocellulases exhibit effective cellulose substrate targeting across a diverse spectrum of agroindustrial waste materials. This versatility is essential for their practical application.
(e) Yield and Productivity: One of the central challenges lies in optimizing the production of valuable products, such as biofuels or bio-products, through cellulose bioconversion. Balancing high yields with efficient productivity is critical.

Overcoming these challenges is essential to the effective utilization of cross-linked endocellulases as a sustainable solution for transforming agroindustrial waste into valuable resources. This contributes to promoting eco-friendly and sustainable waste management strategies (Cui & Jia, 2015).

8 Future Scope

The study's future scope involves continued research and development of the cross-linked endocellulase enzyme for sustainable bioconversion of agroindustrial waste. This includes optimizing immobilization methods and exploring new applications for the enzyme, like exploring state-of-the-art methods for immobilizing enzymes and using nanomaterial-based frameworks or modifying enzymes through engineering to improve both their stability and activity. Engage in interdisciplinary collaborations to unlock the full potential of enzymes across diverse sectors, ranging from biofuel manufacturing to pharmaceuticals. The priority is to create affordable production methods for this eco-friendly technology, making it accessible to a broader audience. By doing so, one can optimize both the economic and environmental benefits of enzyme applications. Looking ahead, the central focus of the researcher's efforts will be to overcome the hurdle of scaling up enzyme production from the laboratory to industrial levels while ensuring the enzyme's continued effectiveness. The research may also inspire further investigation into other enzyme systems for cellulose degradation and their potential for sustainable waste management and renewable energy. In summary, this study offers promising opportunities for advancing biotechnology to support sustainability efforts and tackle environmental challenges.

Proposed Solutions for Future Research:

- Sustainable Production: Investigate eco-friendly production techniques.
- Synergy Studies: Explore enzyme combinations for enhanced cellulose degradation.
- Environmental Assessment: Carry out comprehensive assessments to evaluate the environmental impacts thoroughly.
- Industrial-Scale Optimization: Create procedures for scaling up enzyme production to an industrial level.
- Genomic exploration: Study natural microbial communities to find new cellulase enzymes with unique properties for sustainable bioconversion.
- Biological sensors: Create biological sensors capable of real-time monitoring and controlling enzyme activity to ensure peak performance in different applications.

9 Conclusion

The need for endocellulase is highly inevitable in the cellulose degradation process; thus, sustainable and cost-effective immobilization approaches like cross-linking can be amalgamated with the endocellulase. By comparing the drawbacks of different immobilization techniques, the CLEA strategy, a carrier-free technology, is found to be more appropriate. Cross-linked endocellulase has much potential as an environmentally friendly way of recycling agroindustrial waste. The application of cross-linked endocellulase can aid in overcoming the challenges connected with high cellulose agroindustrial waste degradation. Cross-linked endocellulase can produce renewable energy and aid in eliminating waste by breaking down cellulose into simple sugars. These sugars can be helpful for various microflora to produce various valuable by-products like bioethanol. The efficacy and durability of cross-linked endocellulase make it a promising route for future biotechnology research and development.

References

Abdel-Azeem, A. M., & Sheir, D. H. (2020). Bioconversion of lignocellulosic residues into single-cell protein (SCP) by *Chaetomium*. In A. M. Abdel-Azeem (Ed.), *Recent developments on genus Chaetomium, fungal biology* (pp. 343–375). Springer International Publishing. https://doi.org/10.1007/978-3-030-31612-9_13

Abdul Karim, M. S., Zainol, N., Abu Hassan As'ari, N. I. A., Abu Talip Yusof, N., & Aziz, N. H. (2022). Effect of processing parameters on cellulose content extracted from pineapple leaf. *Biocatalysis and Agricultural Biotechnology, 42*, 102339. https://doi.org/10.1016/j.bcab.2022.102339

Acharya, D. K., Bhojani, G. R., Prajapati, P. R., Patel, H. M., & Shukla, R. M. (2014). Bioprocessing of bajra straw using locally isolated *Aspergillus niger* HD-6 for endocellulase production. *Applied Biological Research, 16*, 223. https://doi.org/10.5958/0974-4517.2014.00014.7

Ahmad Khorairi, A. N. S., Sofian-Seng, N.-S., Othaman, R., Abdul Rahman, H., Mohd Razali, N. S., Lim, S. J., & Wan Mustapha, W. A. (2023). A review on agro-industrial waste as cellulose and nanocellulose source and their potentials in food applications. *Food Review International, 39*, 663–688. https://doi.org/10.1080/87559129.2021.1926478

Al kamzari, S. M. A., Nageswara Rao, L., Lakavat, M., Gandi, S., Reddy, P. S., & Kavitha Sri, G. (2021). Extraction and characterization of cellulose from agricultural waste materials. *Materials Today: Proceedings*. https://doi.org/10.1016/j.matpr.2021.07.030

Alghooneh, A., Mohammad Amini, A., Behrouzian, F., & Razavi, S. M. A. (2017). Characterisation of cellulose from coffee silverskin. *International Journal of Food Properties, 20*, 2830–2843. https://doi.org/10.1080/10942912.2016.1253097

Ali, M., Saleem, M., Khan, Z., & Watson, I. A. (2019). 16 – The use of crop residues for biofuel production. In D. Verma, E. Fortunati, S. Jain, & X. Zhang (Eds.), *Biomass, biopolymer-based materials, and bioenergy, Woodhead publishing series in composites science and engineering* (pp. 369–395). Woodhead Publishing. https://doi.org/10.1016/B978-0-08-102426-3.00016-3

Antony, A. S., & Vasudevan, R. T. (2018). A review on cellulose and its utilization from agro-industrial waste. *Drug Invention Today, 10*, 89–94.

Ávila, P. F., Silva, M. F., Martins, M., & Goldbeck, R. (2021). Cello-oligosaccharides production from lignocellulosic biomass and their emerging prebiotic applications. *World Journal of Microbiology and Biotechnology, 37*, 73. https://doi.org/10.1007/s11274-021-03041-2

Bai, Z., Caspari, T., Gonzalez, M. R., Batjes, N. H., Mäder, P., Bünemann, E. K., de Goede, R., Brussaard, L., Xu, M., Ferreira, C. S. S., Reintam, E., Fan, H., Mihelič, R., Glavan, M., & Tóth, Z. (2018). Effects of agricultural management practices on soil quality: A review of long-term experiments for Europe and China. *Agriculture, Ecosystems and Environment, 265*, 1–7. https://doi.org/10.1016/j.agee.2018.05.028

Bajpai, P. (2016). *Pretreatment of lignocellulosic biomass for biofuel production*. Springer Briefs in Molecular Science. https://doi.org/10.1007/978-981-10-0687-6

Bajpai, P. (2018). Chapter 2 – Wood and fibre fundamentals. In P. Bajpai (Ed.), *Biermann's handbook of pulp and paper (Third Edition)* (pp. 19–74). Elsevier. https://doi.org/10.1016/B978-0-12-814240-0.00002-1

Baker, J. O., Ehrman, C. I., Adney, W. S., Thomas, S. R., & Himmel, M. E. (1998). Hydrolysis of cellulose using ternary mixtures of purified cellulases. *Applied Biochemistry and Biotechnology, 70–72*, 395–403. https://doi.org/10.1007/BF02920154

Barbosa, F. C., Silvello, M. A., & Goldbeck, R. (2020). Cellulase and oxidative enzymes: New approaches, challenges and perspectives on cellulose degradation for bioethanol production. *Biotechnology Letters, 42*, 875–884. https://doi.org/10.1007/s10529-020-02875-4

Baruah, J., Bardhan, P., Mukherjee, A. K., Deka, R. C., Mandal, M., & Kalita, E. (2022). Integrated pretreatment of banana agrowastes: Structural characterization and enhancement of enzymatic hydrolysis of cellulose obtained from banana peduncle. *International Journal of Biological Macromolecules, 201*, 298–307. https://doi.org/10.1016/j.ijbiomac.2021.12.179

Beimforde, C., Feldberg, K., Nylinder, S., Rikkinen, J., Tuovila, H., Dörfelt, H., Gube, M., Jackson, D. J., Reitner, J., Seyfullah, L. J., & Schmidt, A. R. (2014). Estimating the Phanerozoic history of the Ascomycota lineages: Combining fossil and molecular data. *Molecular Phylogenetics and Evolution, 78*, 386–398. https://doi.org/10.1016/j.ympev.2014.04.024

Berini, F., Katz, C., Gruzdev, N., Casartelli, M., Tettamanti, G., & Marinelli, F. (2018). Microbial and viral chitinases: Attractive biopesticides for integrated pest management. *Biotechnology Advances, 36*, 818–838. https://doi.org/10.1016/j.biotechadv.2018.01.002

Bhardwaj, N., Kumar, B., Agrawal, K., & Verma, P. (2021). Current perspective on production and applications of microbial cellulases: A review. *Bioresources and Bioprocessing, 8*, 95. https://doi.org/10.1186/s40643-021-00447-6

Bhattacharjee, N., Alonso-Cotchico, L., & Lucas, M. F. (2023). Enzyme immobilization studied through molecular dynamic simulations. *Frontiers in Bioengineering and Biotechnology, 11*, 1–15. https://doi.org/10.3389/fbioe.2023.1200293

Bhattacharya, A. S., Bhattacharya, A., & Pletschke, B. I. (2015). Synergism of fungal and bacterial cellulases and hemicellulases: A novel perspective for enhanced bio-ethanol production. *Biotechnology Letters, 37*, 1117–1129. https://doi.org/10.1007/s10529-015-1779-3

Bhupathi, P., & Subbaiah, K. A. (2020). Conversion of agricultural residues into protein biomass by milky mushroom fungus *Calocybe indica* var. Apk2 through solid state fermentation. *Madras Agricultural Journal, 107*. https://doi.org/10.29321/MAJ.S.000438

Bledzki, A. K., Mamun, A. A., & Volk, J. (2010). Physical, chemical and surface properties of wheat husk, rye husk and soft wood and their polypropylene composites. *Composites Part A: Applied Science and Manufacturing, 41*, 480–488. https://doi.org/10.1016/j.compositesa.2009.12.004

Bommarius, A. S., Katona, A., Cheben, S. E., Patel, A. S., Ragauskas, A. J., Knudson, K., & Pu, Y. (2008). Cellulase kinetics as a function of cellulose pretreatment. *Metabolic Engineering, 10*, 370–381. https://doi.org/10.1016/j.ymben.2008.06.008

Braham, S. A., Siar, E.-H., Arana-Peña, S., Carballares, D., Morellon-Sterling, R., Bavandi, H., de Andrades, D., Kornecki, J. F., & Fernandez-Lafuente, R. (2021). Effect of concentrated salts solutions on the stability of immobilized enzymes: Influence of inactivation conditions and immobilization protocol. *Molecules, 26*, 968. https://doi.org/10.3390/molecules26040968

Broun, G. B. (1976). Chemically aggregated enzymes. *Methods in Enzymology, 44*, 263–280. https://doi.org/10.1016/s0076-6879(76)44022-3

Cao, L., van Langen, L., & Sheldon, R. A. (2003). Immobilised enzymes: Carrier-bound or carrier-free? *Current Opinion in Biotechnology, 14*, 387–394. https://doi.org/10.1016/S0958-1669(03)00096-X

Cao, Y., Chang, Z., Wang, J., Ma, Y., & Fu, G. (2013). The fate of antagonistic microorganisms and antimicrobial substances during anaerobic digestion of pig and dairy manure. *Bioresource Technology, 136*, 664–671. https://doi.org/10.1016/j.biortech.2013.01.052

Chang, B. S., & Mahoney, R. R. (1995). Enzyme thermostabilization by bovine serum albumin and other proteins: Evidence for hydrophobic interactions. *Biotechnology and Applied Biochemistry, 22*, 203–214.

Chatterjee, S., Meghvansi, M. K., & Veer, V. (2015). EDITORIAL: Special Issue on "Recent Advances in Biodegradation, Sanitation, and Bioremediation". *South Asian Journal of Experimental Biology, 5*(6), 194–196. https://doi.org/10.38150/sajeb.5(6).p194-196

Chen, H., Liu, J., Chang, X., Chen, D., Xue, Y., Liu, P., Lin, H., & Han, S. (2017). A review on the pretreatment of lignocellulose for high-value chemicals. *Fuel Processing Technology, 160*, 196–206. https://doi.org/10.1016/j.fuproc.2016.12.007

Cheng, Y.-S., Huang, C.-H., Chen, C.-C., Huang, T.-Y., Ko, T.-P., Huang, J.-W., Wu, T.-H., Liu, J.-R., & Guo, R.-T. (2014). Structural and mutagenetic analyses of a 1,3-1,4-β-glucanase from Paecilomyces thermophila. *Biochimica et Biophysica Acta, 1844*, 366–373. https://doi.org/10.1016/j.bbapap.2013.11.005

Cheng, L., Hu, X., Gu, Z., Hong, Y., Li, Z., & Li, C. (2019). Characterization of physicochemical properties of cellulose from potato pulp and their effects on enzymatic hydrolysis by cellulase. *International Journal of Biological Macromolecules, 131*, 564–571. https://doi.org/10.1016/j.ijbiomac.2019.02.164

Choonut, A., Saejong, M., & Sangkharak, K. (2014). The production of ethanol and hydrogen from pineapple peel by *Saccharomyces cerevisiae* and *Enterobacter aerogenes*. In *Energy Procedia, 2013 international conference on Alternative Energy in Developing Countries and Emerging Economies (2013 AEDCEE)* (Vol. 52, pp. 242–249). https://doi.org/10.1016/j.egypro.2014.07.075

Chundawat, S. P. S., Beckham, G. T., Himmel, M. E., & Dale, B. E. (2011). Deconstruction of lignocellulosic biomass to fuels and chemicals. *Annual Review of Chemical and Biomolecular Engineering, 2*, 121–145. https://doi.org/10.1146/annurev-chembioeng-061010-114205

Cinelli, B. A., López, J. A., Castilho, L. R., Freire, D. M. G., & Castro, A. M. (2014). Granular starch hydrolysis of babassu agro-industrial residue: A bioprocess within the context of biorefinery. *Fuel, 124*, 41–48. https://doi.org/10.1016/j.fuel.2014.01.076

Cragg, S. M., Beckham, G. T., Bruce, N. C., Bugg, T. D., Distel, D. L., Dupree, P., Etxabe, A. G., Goodell, B. S., Jellison, J., McGeehan, J. E., McQueen-Mason, S. J., Schnorr, K., Walton, P. H., Watts, J. E., & Zimmer, M. (2015). Lignocellulose degradation mechanisms across the tree of life. *Current Opinion in Chemical Biology, 29*, 108–119. https://doi.org/10.1016/j.cbpa.2015.10.018

Cui, J. D., & Jia, S. R. (2015). Optimization protocols and improved strategies of cross-linked enzyme aggregates technology: Current development and future challenges. *Critical Reviews in Biotechnology, 35*, 15–28. https://doi.org/10.3109/07388551.2013.795516

Daengprasert, W., Boonnoun, P., Laosiripojana, N., Goto, M., & Shotipruk, A. (2011). Application of sulfonated carbon-based catalyst for solvothermal conversion of cassava waste to hydroxymethylfurfural and furfural. *Industrial and Engineering Chemistry Research, 50*, 7903–7910. https://doi.org/10.1021/ie102487w

de Almeida, M. N., Halfeld, G. G., da Costa, I. B., de Lima Guimarães, L. G., Cordeiro, B., & Guimarães, V. M. (2023). Exploring the potential of coffee husks as a raw material for second-generation ethanol production. *Bioenergy Research, 17*, 281. https://doi.org/10.1007/s12155-023-10655-x

Dhillon, N., Chhibber, S., Saxena, M., Pajni, S., & Vadehra, D. V. (1985). A constitutive endoglucanase (CMCase) from *Bacillus licheniformis*-1. *Biotechnology Letters, 7*, 695–697. https://doi.org/10.1007/BF01040212

EFSA Panel on Nutrition, Novel Foods and Food Allergens (NDA), Turck, D., Bohn, T., Castenmiller, J., De Henauw, S., Hirsch-Ernst, K. I., Maciuk, A., Mangelsdorf, I., McArdle, H. J., Naska, A., Pelaez, C., Pentieva, K., Siani, A., Thies, F., Tsabouri, S., Vinceti, M., Cubadda, F., Frenzel, T., Heinonen, M., Prieto Maradona, M., Marchelli, R., Neuhäuser-Berthold, M., Poulsen, M., Schlatter, J. R., van Loveren, H., Albert, O., & Knutsen, H. K.

(2022). Safety of cellobiose as a novel food pursuant to regulation (EU) 2015/2283. *EFSA Journal, 20*, e07596. https://doi.org/10.2903/j.efsa.2022.7596
Ejaz, U., Sohail, M., & Ghanemi, A. (2021). Cellulases: From bioactivity to a variety of industrial applications. *Biomimetics, 6*, 44. https://doi.org/10.3390/biomimetics6030044
Endler, A., & Persson, S. (2011). Cellulose synthases and synthesis in Arabidopsis. *Molecular Plant, 4*, 199–211. https://doi.org/10.1093/mp/ssq079
Fernandes, A. N., Thomas, L. H., Altaner, C. M., Callow, P., Forsyth, V. T., Apperley, D. C., Kennedy, C. J., & Jarvis, M. C. (2011). Nanostructure of cellulose microfibrils in spruce wood. *Proceedings of the National Academy of Sciences, 108*, E1195–E1203. https://doi.org/10.1073/pnas.1108942108
Ferrara, M. C., Cobucci-Ponzano, B., Carpentieri, A., Henrissat, B., Rossi, M., Amoresano, A., & Moracci, M. (2014). The identification and molecular characterization of the first archaeal bifunctional exo-β-glucosidase/N-acetyl-β-glucosaminidase demonstrate that family GH116 is made of three functionally distinct subfamilies. *Biochimica et Biophysica Acta, 1840*, 367–377. https://doi.org/10.1016/j.bbagen.2013.09.022
Ferreira, A. H., Marana, S. R., Terra, W. R., & Ferreira, C. (2001). Purification, molecular cloning, and properties of a beta-glycosidase isolated from midgut lumen of *Tenebrio molitor* (Coleoptera) larvae. *Insect Biochemistry and Molecular Biology, 31*, 1065–1076. https://doi.org/10.1016/s0965-1748(01)00054-6
Geem, K. R., Song, Y., Hwang, I., Bae, H.-J., & Lee, D. W. (2021). Production of *Gloeophyllum trabeum* endoglucanase Cel12A in *Nicotiana benthamiana* for cellulose degradation. *Frontiers in Plant Science, 12*, 1–9. https://doi.org/10.3389/fpls.2021.696199
Geng, A., Xin, F., & Ip, J. (2012). Ethanol production from horticultural waste treated by a modified organosolv method. *Bioresource Technology, 104*, 715–721. https://doi.org/10.1016/j.biortech.2011.10.076
Ginni, G., Kavitha, S., Kannah, Y., Bhatia, S. K., Kumar, A., Rajkumar, M., Kumar, G., Pugazhendhi, A., & Chi, N. T. L. (2021). Valorization of agricultural residues: Different biorefinery routes. *Journal of Environmental Chemical Engineering, 9*, 105435. https://doi.org/10.1016/j.jece.2021.105435
Gonzalo, A., Bimbela, F., Sánchez, J. L., Labidi, J., Marín, F., & Arauzo, J. (2017). Evaluation of different agricultural residues as raw materials for pulp and paper production using a semichemical process. *Journal of Cleaner Production, 156*, 184–193. https://doi.org/10.1016/j.jclepro.2017.04.036
Górecka, E., & Jastrzębska, M. (2011). Immobilization techniques and biopolymer carriers. *Biotechnology and Food Science, 75*(1), 65–86. https://doi.org/10.34658/bfs.2011.75.1.65-86
Govardhan, C. P. (1999). Crosslinking of enzymes for improved stability and performance. *Current Opinion in Biotechnology, 10*, 331–335. https://doi.org/10.1016/S0958-1669(99)80060-3
Gu, Y., Hu, Y., Huang, C., Lai, C., Ling, Z., & Yong, Q. (2022). Co-production of amino acid-rich xylooligosaccharide and single-cell protein from paper mulberry by autohydrolysis and fermentation technologies. *Biotechnology for Biofuels and Bioproducts, 15*, 1. https://doi.org/10.1186/s13068-021-02095-6
Guimarães, J. L., Frollini, E., da Silva, C. G., Wypych, F., & Satyanarayana, K. G. (2009). Characterization of banana, sugarcane bagasse and sponge gourd fibres of Brazil. *Industrial Crops and Products, 30*, 407–415. https://doi.org/10.1016/j.indcrop.2009.07.013
Guisan, J. M. (Ed.). (2006). *Immobilization of enzymes and cells, methods in biotechnology™*. Humana Press. https://doi.org/10.1007/978-1-59745-053-9
Gupta, P., Samant, K., & Sahu, A. (2012). Isolation of cellulose-degrading bacteria and determination of their cellulolytic potential. *International Journal of Microbiology, 2012*, e578925. https://doi.org/10.1155/2012/578925
Hanefeld, U., Gardossi, L., & Magner, E. (2009). Understanding enzyme immobilisation. *Chemical Society Reviews, 38*, 453–468. https://doi.org/10.1039/B711564B

Haq, I., Hameed, U., Shahzadi, K., Javed, M. M., Ali, S., & Qadeer, M. A. (2005). Cotton saccharifying activity of cellulases by *Trichoderma harzianum* UM-11 in shake flask. *International Journal of Botany, 1*, 19–22. https://doi.org/10.3923/ijb.2005.19.22

Heinze, T., El Seoud, O. A., & Koschella, A. (2018). Production and characteristics of cellulose from different sources. In T. Heinze, O. A. El Seoud, & A. Koschella (Eds.), *Cellulose derivatives: Synthesis, structure, and properties, springer series on polymer and composite materials* (pp. 1–38). Springer International Publishing. https://doi.org/10.1007/978-3-319-73168-1_1

Horn, S. J., Vaaje-Kolstad, G., Westereng, B., & Eijsink, V. (2012). Novel enzymes for the degradation of cellulose. *Biotechnology for Biofuels, 5*, 45. https://doi.org/10.1186/1754-6834-5-45

Huzir, N. M., Aziz, M. M. A., Ismail, S. B., Abdullah, B., Mahmood, N. A. N., Umor, N. A., & Syed Muhammad, S. A. F. (2018). Agro-industrial waste to biobutanol production: Eco-friendly biofuels for next generation. *Renewable and Sustainable Energy Reviews, 94*, 476–485. https://doi.org/10.1016/j.rser.2018.06.036

Imam, H. T., Marr, P. C., & Marr, A. C. (2021). Enzyme entrapment, biocatalyst immobilization without covalent attachment. *Green Chemistry, 23*, 4980–5005. https://doi.org/10.1039/D1GC01852C

Jabri, J., Abid, K., Yaich, H., Malek, A., Rekhis, J., & Kamoun, M. (2023a). Enzymatic valorization of alkali-treated chickpea straw and sunflower stalks as high fibrous agricultural wastes for sustainable ruminant nutrition. *Biomass Conversion and Biorefinery*. https://doi.org/10.1007/s13399-023-04659-y

Jabri, J., Abid, K., Yaich, H., Malek, A., Rekhis, J., & Kamoun, M. (2023b). Fermentative profile and nutritional value of untreated and alkali-treated faba bean (*Vicia faba* L.) straw supplemented with exogenous fibrolytic enzymes derived from *Trichoderma longibrachiatum*, *Aspergillus* strains, and *Neurospora intermedia*. *Biomass Conversion and Biorefinery*. https://doi.org/10.1007/s13399-023-04251-4

Jayasekara, S., Ratnayake, R., Jayasekara, S., & Ratnayake, R. (2019). Microbial cellulases: An overview and applications, cellulose. *IntechOpen*. https://doi.org/10.5772/intechopen.84531

Jiang, M., Zhao, M., Zhou, Z., Huang, T., Chen, X., & Wang, Y. (2011). Isolation of cellulose with ionic liquid from steam exploded rice straw. *Industrial Crops and Products, 33*, 734–738. https://doi.org/10.1016/j.indcrop.2011.01.015

Jones, S. W., Karpol, A., Friedman, S., Maru, B. T., & Tracy, B. P. (2020). Recent advances in single cell protein use as a feed ingredient in aquaculture. *Current Opinion in Biotechnology, 61*, 189–197. https://doi.org/10.1016/j.copbio.2019.12.026

Jugwanth, Y., Sewsynker-Sukai, Y., & Gueguim Kana, E. B. (2020). Valorization of sugarcane bagasse for bioethanol production through simultaneous saccharification and fermentation: Optimization and kinetic studies. *Fuel, 262*, 116552. https://doi.org/10.1016/j.fuel.2019.116552

Juturu, V., & Wu, J. C. (2014). Microbial cellulases: Engineering, production and applications. *Renewable and Sustainable Energy Reviews, 33*, 188–203. https://doi.org/10.1016/j.rser.2014.01.077

Kalogeris, E., Fountoukides, G., Kekos, D., & Macris, B. J. (1999). Design of a solid-state bioreactor for thermophilic microorganisms. *Bioresource Technology, 67*, 313–315. https://doi.org/10.1016/S0960-8524(98)00124-2

Karnaouri, A., Matsakas, L., Krikigianni, E., Rova, U., & Christakopoulos, P. (2019). Valorization of waste forest biomass toward the production of cello-oligosaccharides with potential prebiotic activity by utilizing customized enzyme cocktails. *Biotechnology for Biofuels, 12*, 285. https://doi.org/10.1186/s13068-019-1628-z

Karthyani, S., Pandey, A., & Devendra, L. (2017). Delignification of cotton stalks using sodium cumene sulfonate for bioethanol production. *Biofuels, 11*, 1–10. https://doi.org/10.1080/17597269.2017.1370884

Kaur, P., Taggar, M. S., & Sidhu, J. (2020). Cellulolytic microorganisms: Diversity and role in conversion of rice straw to bioethanol. *Cellulose Chemistry and Technology, 54*, 613–634. https://doi.org/10.35812/CelluloseChemTechnol.2020.54.61

Khantibongse, P., & Ratanatamskul, C. (2023). Insight into pathway of monosaccharide production from integrated enzymatic hydrolysis of rice straw waste as feed stock for anaerobic digestion. *Scientific Reports, 13*, 148. https://doi.org/10.1038/s41598-023-27398-6

Kim, T. H., Taylor, F., & Hicks, K. B. (2008). Bioethanol production from barley hull using SAA (soaking in aqueous ammonia) pretreatment. *Bioresource Technology, 99*, 5694–5702. https://doi.org/10.1016/j.biortech.2007.10.055

Kim, J. S., Choi, W. I., Kang, M., Park, J. Y., & Lee, J.-S. (2012). Kinetic study of empty fruit bunch using hot liquid water and dilute acid. *Applied Biochemistry and Biotechnology, 167*, 1527–1539. https://doi.org/10.1007/s12010-011-9528-0

Kim, E. S., Kim, B.-S., Kim, K.-Y., Woo, H.-M., Lee, S.-M., & Um, Y. (2018). Aerobic and anaerobic cellulose utilization by *Paenibacillus* sp. CAA11 and enhancement of its cellulolytic ability by expressing a heterologous endoglucanase. *Journal of Biotechnology, 268*, 21–27. https://doi.org/10.1016/j.jbiotec.2018.01.007

Kopania, E., Wietecha, J., & Ciechanska, D. (2012). Studies on isolation of cellulose fibres from waste plant biomass. *Fibres & Textiles in Eastern Europe, 96*, 167–172.

Koul, B., Yakoob, M., & Shah, M. P. (2022). Agricultural waste management strategies for environmental sustainability. *Environmental Research, 206*, 112285. https://doi.org/10.1016/j.envres.2021.112285

Kuhad, R. C., Gupta, R., & Singh, A. (2011). Microbial cellulases and their industrial applications. *Enzyme Research, 2011*, 280696. https://doi.org/10.4061/2011/280696

Kulic, G., & Radojicic, V. (2011). Analysis of cellulose content in stalks and leaves of large leaf tobacco. *Journal of Agricultural Sciences (Belgrade), 56*, 207–215. https://doi.org/10.2298/JAS1103207K

Kumar, D., & Murthy, G. S. (2013). Stochastic molecular model of enzymatic hydrolysis of cellulose for ethanol production. *Biotechnology for Biofuels, 6*, 63. https://doi.org/10.1186/1754-6834-6-63

Kumar, M., Ou, Y.-L., & Lin, J.-G. (2010). Co-composting of green waste and food waste at low C/N ratio. *Waste Management, 30*, 602–609. https://doi.org/10.1016/j.wasman.2009.11.023

Kumar, C. S. C., Mythily, R., & Chandraju, S. (2011a). Extraction of carbohydrate from sweet orange peels (*Citrus sinensis* L.) and their identification via LC /MS & Thin Layer Chromatographic Analysis. *Biosciences, Biotechnology Research Asia, 8*, 709–715. https://doi.org/10.13005/bbra/923

Kumar, S., Kothari, U., Kong, L., Lee, Y. Y., & Gupta, R. B. (2011b). Hydrothermal pre-treatment of switchgrass and corn stover for production of ethanol and carbon microspheres. *Biomass and Bioenergy, 35*, 956–968. https://doi.org/10.1016/j.biombioe.2010.11.023

Kumar, V. A., Kurup, R. S. C., Snishamol, C., & Prabhu, G. N. (2019). Role of cellulases in food, feed, and beverage industries. In B. Parameswaran, S. Varjani, & S. Raveendran (Eds.), *Green bio-processes: Enzymes in industrial food processing, energy, environment, and sustainability* (pp. 323–343). Springer. https://doi.org/10.1007/978-981-13-3263-0_17

Lakhundi, S., Siddiqui, R., & Khan, N. A. (2015). Cellulose degradation: A therapeutic strategy in the improved treatment of *Acanthamoeba* infections. *Parasites & Vectors, 8*, 23. https://doi.org/10.1186/s13071-015-0642-7

Lamb, W. F., Wiedmann, T., Pongratz, J., Andrew, R., Crippa, M., Olivier, J. G. J., Wiedenhofer, D., Mattioli, G., Khourdajie, A. A., House, J., Pachauri, S., Figueroa, M., Saheb, Y., Slade, R., Hubacek, K., Sun, L., Ribeiro, S. K., Khennas, S., de la Rue du Can, S., Chapungu, L., Davis, S. J., Bashmakov, I., Dai, H., Dhakal, S., Tan, X., Geng, Y., Gu, B., & Minx, J. (2021). A review of trends and drivers of greenhouse gas emissions by sector from 1990 to 2018. *Environmental Research Letters, 16*, 073005. https://doi.org/10.1088/1748-9326/abee4e

Leschine, S. (1995). Cellulose degradation in anaerobic environments. *Annual Review of Microbiology, 49*, 399–426. https://doi.org/10.1146/annurev.mi.49.100195.002151

Li, T., Gong, X., Yang, G., Li, Q., Huang, J., Zhou, N., & Jia, X. (2022). Cross-linked enzyme aggregates (CLEAs) of cellulase with improved catalytic activity, adaptability and reusability. *Bioprocess and Biosystems Engineering, 45*, 865–875. https://doi.org/10.1007/s00449-022-02704-0

Lin, M., & Begho, T. (2022). Crop residue burning in South Asia: A review of the scale, effect, and solutions with a focus on reducing reactive nitrogen losses. *Journal of Environmental Management, 314*, 115104. https://doi.org/10.1016/j.jenvman.2022.115104

Lin, Y., Wang, D., & Wang, T. (2012). Ethanol production from pulp & paper sludge and monosodium glutamate waste liquor by simultaneous saccharification and fermentation in batch condition. *Chemical Engineering Journal, 191*, 31–37. https://doi.org/10.1016/j.cej.2011.09.040

Liu, L., Huang, W.-C., Liu, Y., & Li, M. (2021). Diversity of cellulolytic microorganisms and microbial cellulases. *International Biodeterioration and Biodegradation, 163*, 105277. https://doi.org/10.1016/j.ibiod.2021.105277

Lu, Z., & Imlay, J. A. (2021). When anaerobes encounter oxygen: Mechanisms of oxygen toxicity, tolerance and defence. *Nature Reviews. Microbiology, 19*, 774–785. https://doi.org/10.1038/s41579-021-00583-y

Luo, G., Shi, W., Chen, X., Ni, W., Strong, P. J., Jia, Y., & Wang, H. (2011). Hydrothermal conversion of water lettuce biomass at 473 or 523 K. Biomass bioenergy, land use impacts of bioenergy. Selected papers from the IEA bioenergy task 38 meetings in Helsinki, 2009 and Brussels, 2010, vol. 35, pp. 4855–4861. https://doi.org/10.1016/j.biombioe.2011.10.002

Maghraby, Y. R., El-Shabasy, R. M., Ibrahim, A. H., & Azzazy, H. M. E.-S. (2023). Enzyme immobilization technologies and industrial applications. *ACS Omega, 8*, 5184–5196. https://doi.org/10.1021/acsomega.2c07560

Manyapu, V., Mandpe, A., & Kumar, S. (2018). Synergistic effect of fly ash in in-vessel composting of biomass and kitchen waste. *Bioresource Technology, 251*, 114–120. https://doi.org/10.1016/j.biortech.2017.12.039

Manyapu, V., Kumar, A., & Kumar, R. (2021). Psychrophilic biomethanation for enhanced bioenergy production in cold regions. *Clean Technologies and Environmental Policy*. https://doi.org/10.1007/s10098-021-02223-8

Marrazza, G. (2014). Piezoelectric biosensors for organophosphate and carbamate pesticides: A review. *Biosensors, 4*, 301–317. https://doi.org/10.3390/bios4030301

Matsagar, B. M., & Wu, K. C.-W. (2022). Chapter 1 – Agricultural waste-derived biochar for environmental management. In D. C. W. Tsang & Y. S. Ok (Eds.), *Biochar in agriculture for achieving sustainable development goals* (pp. 3–13). Academic Press. https://doi.org/10.1016/B978-0-323-85343-9.00026-4

Meng, J., Mäkelä, M. R., & de Vries, R. P. (2021). Chapter 3 – Molecular engineering to improve lignocellulosic biomass based applications using filamentous fungi. In G. M. Gadd & S. Sariaslani (Eds.), *Advances in applied microbiology* (pp. 73–109). Academic Press. https://doi.org/10.1016/bs.aambs.2020.09.001

Millati, R., Cahyono, R. B., Ariyanto, T., Azzahrani, I. N., Putri, R. U., & Taherzadeh, M. J. (2019). Chapter 1 – Agricultural, industrial, municipal, and Forest wastes: An overview. In M. J. Taherzadeh, K. Bolton, J. Wong, & A. Pandey (Eds.), *Sustainable resource recovery and zero waste approaches* (pp. 1–22). Elsevier. https://doi.org/10.1016/B978-0-444-64200-4.00001-3

Mohamad, N. R., Marzuki, N. H. C., Buang, N. A., Huyop, F., & Wahab, R. A. (2015). An overview of technologies for immobilization of enzymes and surface analysis techniques for immobilized enzymes. *Biotechnology and Biotechnological Equipment, 29*, 205–220. https://doi.org/10.1080/13102818.2015.1008192

Morais Junior, W. G., Pacheco, T. F., Gao, S., Martins, P. A., Guisán, J. M., & Caetano, N. S. (2021). Sugarcane bagasse saccharification by enzymatic hydrolysis using endocellulase and β-glucosidase immobilized on different supports. *Catalysts, 11*, 340. https://doi.org/10.3390/catal11030340

Nabarlatz, D., Ebringerová, A., & Montané, D. (2007). Autohydrolysis of agricultural by-products for the production of xylo-oligosaccharides. *Carbohydrate Polymers, 69*, 20–28. https://doi.org/10.1016/j.carbpol.2006.08.020

Nagl, M., Haske-Cornelius, O., Skopek, L., Pellis, A., Bauer, W., Nyanhongo, G. S., & Guebitz, G. (2021). Biorefining: The role of endoglucanases in refining of cellulose fibres. *Cellulose, 28*, 7633–7650. https://doi.org/10.1007/s10570-021-04022-2

Nagy, L. G., Riley, R., Tritt, A., Adam, C., Daum, C., Floudas, D., Sun, H., Yadav, J. S., Pangilinan, J., Larsson, K.-H., Matsuura, K., Barry, K., Labutti, K., Kuo, R., Ohm, R. A., Bhattacharya, S. S., Shirouzu, T., Yoshinaga, Y., Martin, F. M., Grigoriev, I. V., & Hibbett, D. S. (2016). Comparative genomics of early-diverging mushroom-forming fungi provides insights into the origins of lignocellulose decay capabilities. *Molecular Biology and Evolution, 33*, 959–970. https://doi.org/10.1093/molbev/msv337

Narendra Kumar, H. K., Chandra Mohana, N., Rakshith, D., Abhilash, M. R., Harini, B. P., & Sathish, S. (2023). Bioprocessing of cellulosic waste biomass for ethanol production by *Chryseobacterium culicis* Bp16. *Sustainable Chemistry and Pharmacy, 33*, 101081. https://doi.org/10.1016/j.scp.2023.101081

Ng, S., Tan, C., Lai, O., Long, K., & Mirhosseini, H. (2010). Extraction and characterization of dietary fibre from coconut residue. *Journal of Food, Agriculture and Environment, 8(2): 172-177.*

Nguyen, H. H., & Kim, M. (2017). An overview of techniques in enzyme immobilization. *Applied Science and Convergence Technology, 26*, 157–163. https://doi.org/10.5757/ASCT.2017.26.6.157

Noori, R., Perwez, M., & Sardar, M. (2019). Cross-linked enzyme aggregates: Current developments and applications. In Q. Husain & M. F. Ullah (Eds.), *Biocatalysis: Enzymatic basics and applications* (pp. 83–112). Springer International Publishing. https://doi.org/10.1007/978-3-030-25023-2_5

Obeng, E. M., Adam, S. N. N., Budiman, C., Ongkudon, C. M., Maas, R., & Jose, J. (2017). Lignocellulases: A review of emerging and developing enzymes, systems, and practices. *Bioresources and Bioprocessing, 4*, 16. https://doi.org/10.1186/s40643-017-0146-8

Patel, R. N. (Ed.). (2006). *Biocatalysis in the pharmaceutical and biotechnology industries*. CRC Press. https://doi.org/10.1201/9781420019377

Periyasamy, S., Beula Isabel, J., Kavitha, S., Karthik, V., Mohamed, B. A., Gizaw, D. G., Sivashanmugam, P., & Aminabhavi, T. M. (2023). Recent advances in consolidated bioprocessing for conversion of lignocellulosic biomass into bioethanol—A review. *Chemical Engineering Journal, 453*, 139783. https://doi.org/10.1016/j.cej.2022.139783

Petre, M., Zarnea, G., Adrian, P., & Gheorghiu, E. (1999). Biodegradation and bioconversion of cellulose wastes using bacterial and fungal cells immobilized in radiopolymerized hydrogels. *Resources, Conservation and Recycling, 27*, 309–332. https://doi.org/10.1016/S0921-3449(99)00028-2

Poulsen, H. V., Willink, F. W., & Ingvorsen, K. (2016). Aerobic and anaerobic cellulase production by *Cellulomonas uda*. *Archives of Microbiology, 198*, 725–735. https://doi.org/10.1007/s00203-016-1230-8

Prasad Reddy, J., & Rhim, J.-W. (2014). Isolation and characterization of cellulose nanocrystals from garlic skin. *Materials Letters, 129*, 20–23. https://doi.org/10.1016/j.matlet.2014.05.019

Prates, É. T., Stankovic, I., Silveira, R. L., Liberato, M. V., Henrique-Silva, F., Jr., Pereira, N., Polikarpov, I., & Skaf, M. S. (2013). X-ray structure and molecular dynamics simulations of endoglucanase 3 from *Trichoderma harzianum*: Structural organization and substrate recognition by endoglucanases that lack cellulose binding module. *PLoS One, 8*, e59069. https://doi.org/10.1371/journal.pone.0059069

Rahman, M. S., Fernando, S., Ross, B., Wu, J., & Qin, W. (2018). Endoglucanase (EG) activity assays. *Methods in Molecular Biology Clifton NJ, 1796*, 169–183. https://doi.org/10.1007/978-1-4939-7877-9_13

Raveendran, S., Parameswaran, B., Ummalyma, S. B., Abraham, A., Mathew, A. K., Madhavan, A., Rebello, S., & Pandey, A. (2018). Applications of microbial enzymes in food industry. *Food Technology and Biotechnology, 56*, 16–30. https://doi.org/10.17113/ftb.56.01.18.5491

Rhim, J.-W., Reddy, J. P., & Luo, X. (2015). Isolation of cellulose nanocrystals from onion skin and their utilization for the preparation of agar-based bio-nanocomposites films. *Cellulose, 22*, 407–420. https://doi.org/10.1007/s10570-014-0517-7

Ritala, A., Häkkinen, S. T., Toivari, M., & Wiebe, M. G. (2017). Single cell protein—State-of-the-art, industrial landscape and patents 2001–2016. *Frontiers in Microbiology, 8*. https://doi.org/10.3389/fmicb.2017.02009

Rongpipi, S., Ye, D., Gomez, E. D., & Gomez, E. W. (2019). Progress and opportunities in the characterization of cellulose—An important regulator of cell wall growth and mechanics. *Frontiers in Plant Science, 9*(1894), 1–28. https://doi.org/10.3389/fpls.2018.01894

Sabiha-Hanim, S., Noor, M. A. M., & Rosma, A. (2011). Effect of autohydrolysis and enzymatic treatment on oil palm (*Elaeis guineensis* Jacq.) frond fibres for xylose and xylooligosaccharides production. *Bioresource Technology, 102*, 1234–1239. https://doi.org/10.1016/j.biortech.2010.08.017

Saini, A., Aggarwal, N. K., Sharma, A., & Yadav, A. (2015). Actinomycetes: A source of lignocellulolytic enzymes. *Enzyme Research, 2015*, 279381. https://doi.org/10.1155/2015/279381

Santos, C. A., Morais, M. A. B., Terrett, O. M., Lyczakowski, J. J., Zanphorlin, L. M., Ferreira-Filho, J. A., Tonoli, C. C. C., Murakami, M. T., Dupree, P., & Souza, A. P. (2019). An engineered GH1 β-glucosidase displays enhanced glucose tolerance and increased sugar release from lignocellulosic materials. *Scientific Reports, 9*, 4903. https://doi.org/10.1038/s41598-019-41300-3

Sapkota, A. (2022). Microbial degradation of cellulose (enzymes, steps, mechanisms) [WWW Document]. URL https://microbenotes.com/microbial-degradation-of-cellulose/ (accessed 4.3.23).

Sasaki, K., Sasaki, D., Tsuge, Y., Morita, M., & Kondo, A. (2021). Enhanced methane production from cellulose using a two-stage process involving a bioelectrochemical system and a fixed film reactor. *Biotechnology for Biofuels, 14*, 7. https://doi.org/10.1186/s13068-020-01866-x

Sassolas, A., Blum, L. J., & Leca-Bouvier, B. D. (2012). Immobilization strategies to develop enzymatic biosensors. *Biotechnology Advances, 30*, 489–511. https://doi.org/10.1016/j.biotechadv.2011.09.003

Schell, D., Ruth, M., & Tucker, M. (1999). Modeling the enzymatic hydrolysis of dilute-acid pretreated Douglas fir. *Applied Biochemistry and Biotechnology, 77*, 67–81. https://doi.org/10.1385/ABAB:77:1-3:67

Seddiqi, H., Oliaei, E., Honarkar, H., Jin, J., Geonzon, L. C., Bacabac, R. G., & Klein-Nulend, J. (2021). Cellulose and its derivatives: Towards biomedical applications. *Cellulose, 28*, 1893–1931. https://doi.org/10.1007/s10570-020-03674-w

Shah, F., Ranawat, B., & Mishra, S. (2019). *An approach toward cellulase production, bioconversion, and utilization, in: Advanced bioprocessing for alternative fuels, biobased chemicals, and bioproducts* (pp. 207–223). Elsevier. https://doi.org/10.1016/B978-0-12-817941-3.00011-5

Shariq, M., & Sohail, M. (2019). *Citrus limetta* peels: A promising substrate for the production of multienzyme preparation from a yeast consortium. *Bioresources and Bioprocessing, 6*, 43. https://doi.org/10.1186/s40643-019-0278-0

Sheldon, R. A. (2007). Cross-linked enzyme aggregates (CLEAs): Stable and recyclable biocatalysts. *Biochemical Society Transactions, 35*, 1583–1587. https://doi.org/10.1042/BST0351583

Shi, Y.-F., Fengxie, J., Yiying, W., Fu, Y., Xingju, Y., & Quan, Y. (1995). Improvement of immobilized cells through permealizing and crosslinking. *Chinese Journal of Biotechnology, 12*, 111–113.

Singh, R. P., & Chauhan, A. (2022). Chapter 1 – Sources of atmospheric pollution in India. In R. P. Singh (Ed.), *Asian atmospheric pollution* (pp. 1–37). Elsevier. https://doi.org/10.1016/B978-0-12-816693-2.00029-9

Singh, H. K., Patil, T., Vineeth, S. K., Das, S., Pramanik, A., & Mhaske, S. T. (2020). Isolation of microcrystalline cellulose from corn stover with emphasis on its constituents: Corn cover and corn cob. In *Materials Today Proceedings, First international conference on recent advances in materials and manufacturing 2019* (Vol. 27, pp. 589–594). https://doi.org/10.1016/j.matpr.2019.12.065

Sinner, M., Puls, J., & Dietrichs, H. (1979). Carbohydrate composition of nut shells and some other agricultural residues. *Starch-Stärke, 31*, 267–269. https://doi.org/10.1002/star.19790310807

Siu-Rodas, Y., de los Angeles Calixto-Romo, M., Guillén-Navarro, K., Sánchez, J. E., Zamora-Briseño, J. A., & Amaya-Delgado, L. (2018). *Bacillus subtilis* with endocellulase and exocellulase activities isolated in the thermophilic phase from composting with coffee residues. *Revista Argentina de Microbiología, 50*, 234–243. https://doi.org/10.1016/j.ram.2017.08.005

Sohail, M., Barzkar, N., Michaud, P., Tamadoni Jahromi, S., Babich, O., Sukhikh, S., Das, R., & Nahavandi, R. (2022). Cellulolytic and xylanolytic enzymes from yeasts: Properties and industrial applications. *Molecules, 27*, 3783. https://doi.org/10.3390/molecules27123783

Song, N., Cai, H.-Y., Yan, Z.-S., & Jiang, H.-L. (2013). Cellulose degradation by one mesophilic strain *Caulobacter* sp. FMC1 under both aerobic and anaerobic conditions. *Bioresource Technology, 131*, 281–287. https://doi.org/10.1016/j.biortech.2013.01.003

Song, H., Jia, H., Wang, Q., Zhao, X., Yang, G., Zhang, M., Zhou, H., Xu, S., Zang, Y., Wang, Y., & Ma, L. (2020). A new environmentally-friendly system for extracting cellulose from corn straw: The low temperature laccase system. *Materials, 13*, 437. https://doi.org/10.3390/ma13020437

Speight, J. G. (2020). 5 – Sources of water pollution. In J. G. Speight (Ed.), *Natural water remediation* (pp. 165–198). Butterworth-Heinemann. https://doi.org/10.1016/B978-0-12-803810-9.00005-X

Talekar, S., Joshi, A., Joshi, G., Kamat, P., Haripurkar, R., & Kamble, S. (2013). Parameters in preparation and characterization of cross linked enzyme aggregates (CLEAs). *RSC Advances, 3*, 12485–12511. https://doi.org/10.1039/C3RA40818C

Tekin, K., Karagöz, S., & Bektaş, S. (2012). Hydrothermal liquefaction of beech wood using a natural calcium borate mineral. *Journal of Supercritical Fluids, 72*, 134–139. https://doi.org/10.1016/j.supflu.2012.08.016

Thapa, S., Mishra, J., Arora, N., Mishra, P., Li, H., O'Hair, J., Bhatti, S., & Zhou, S. (2020). Microbial cellulolytic enzymes: Diversity and biotechnology with reference to lignocellulosic biomass degradation. *Reviews in Environmental Science and Biotechnology, 19*, 621–648. https://doi.org/10.1007/s11157-020-09536-y

Theradimani, M., & Kumar, S. R. (2015). Bioconversion of cashew nut shell into organic manure by white rot fungus. *Acta Horticulturae*, 165–170. https://doi.org/10.17660/ActaHortic.2015.1080.19

Tropea, A., Ferracane, A., Albergamo, A., Potortì, A., Lo Turco, V., & Di Bella, G. (2022). Single cell protein production through multi food-waste substrate fermentation. *Fermentation, 8*, 91. https://doi.org/10.3390/fermentation8030091

Tsuji, A., Tominaga, K., Nishiyama, N., & Yuasa, K. (2013). Comprehensive enzymatic analysis of the cellulolytic system in digestive fluid of the sea hare *Aplysia kurodai*. Efficient glucose release from sea lettuce by synergistic action of 45 kDa endoglucanase and 210 kDa ß-Glucosidase. *PLoS One, 8*, e65418. https://doi.org/10.1371/journal.pone.0065418

Ubiparip, Z., Moreno, D. S., Beerens, K., & Desmet, T. (2020). Engineering of cellobiose phosphorylase for the defined synthesis of cellotriose. *Applied Microbiology and Biotechnology, 104*, 8327–8337. https://doi.org/10.1007/s00253-020-10820-8

Urbina, L., Corcuera, M. Á., Gabilondo, N., Eceiza, A., & Retegi, A. (2021). A review of bacterial cellulose: Sustainable production from agricultural waste and applications in various fields. *Cellulose, 28*, 8229–8253. https://doi.org/10.1007/s10570-021-04020-4

Uzuner, S., & Cekmecelioglu, D. (2019). Chapter 3 – Enzymes in the beverage industry. In M. Kuddus (Ed.), *Enzymes in food biotechnology* (pp. 29–43). Academic Press. https://doi.org/10.1016/B978-0-12-813280-7.00003-7

van Langen, L., Selassa, R., van Rantwijk, F., & Sheldon, R. (2005). Cross-linked aggregates of R-oxynitrilase: A stable, recyclable biocatalyst for enantioselective hydrocyanation. *Organic Letters, 7*, 327–329. https://doi.org/10.1021/ol047647z

Velasco-Lozano, S., López-Gallego, F., Mateos-Díaz, J., & Favela-Torres, E. (2016). Cross-linked enzyme aggregates (CLEA) in enzyme improvement—A review. *Biocatalysis, 1*. https://doi.org/10.1515/boca-2015-0012

Ventorino, V., Ionata, E., Birolo, L., Montella, S., Marcolongo, L., de Chiaro, A., Espresso, F., Faraco, V., & Pepe, O. (2016). Lignocellulose-adapted endo-cellulase producing streptomyces strains for bioconversion of cellulose-based materials. *Frontiers in Microbiology, 7*(2061), 1–15. https://doi.org/10.3389/fmicb.2016.02061

Vermaas, J. V., Kont, R., Beckham, G. T., Crowley, M. F., Gudmundsson, M., Sandgren, M., Ståhlberg, J., Väljamäe, P., & Knott, B. C. (2019). The dissociation mechanism of processive cellulases. *Proceedings of the National Academy of Sciences, 116*, 23061–23067. https://doi.org/10.1073/pnas.1913398116

Vidhya, C. V. (2022). Chapter 18 – Microalgae—The ideal source of biofuel. In B. Gurunathan, R. Sahadevan, & Z. A. Zakaria (Eds.), *Biofuels and bioenergy* (pp. 389–405). Elsevier. https://doi.org/10.1016/B978-0-323-85269-2.00007-1

Wahab, R. A., Elias, N., Abdullah, F., & Ghoshal, S. K. (2020). On the taught new tricks of enzymes immobilization: An all-inclusive overview. *Reactive and Functional Polymers, 152*, 104613. https://doi.org/10.1016/j.reactfunctpolym.2020.104613

Wang, M., Li, Z., Fang, X., Wang, L., & Qu, Y. (2012). Cellulolytic enzyme production and enzymatic hydrolysis for second-generation bioethanol production. In F.-W. Bai, C.-G. Liu, H. Huang, & G. T. Tsao (Eds.), *Biotechnology in China III: Biofuels and bioenergy, advances in biochemical engineering biotechnology* (pp. 1–24). Springer, Berlin, Heidelberg. https://doi.org/10.1007/10_2011_131

Ward, B. (2015). Chapter 11 – Bacterial energy metabolism. In Y.-W. Tang, M. Sussman, D. Liu, I. Poxton, & J. Schwartzman (Eds.), *Molecular medical microbiology (Second Edition)* (pp. 201–233). Academic Press. https://doi.org/10.1016/B978-0-12-397169-2.00011-1

Watson, B. J., Zhang, H., Longmire, A. G., Moon, Y. H., & Hutcheson, S. W. (2009). Processive endoglucanases mediate degradation of cellulose by *Saccharophagus degradans*. *Journal of Bacteriology, 191*, 5697–5705. https://doi.org/10.1128/JB.00481-09

Wilson, D. B. (2009). Cellulases. In M. Schaechter (Ed.), *Encyclopedia of microbiology (Third Edition)* (pp. 252–258). Academic Press. https://doi.org/10.1016/B978-012373944-5.00138-3

World Bank. (2023). Trends in solid waste management in "What a waste 2.0" [WWW Document]. URL https://datatopics.worldbank.org/what-a-waste/ (accessed 4.10.23).

Yamaguchi, A., Mimura, N., Shirai, M., & Sato, O. (2019). Cascade utilization of biomass: Strategy for conversion of cellulose, hemicellulose, and lignin into useful chemicals. *ACS Sustainable Chemistry & Engineering, 7*, 10445–10451. https://doi.org/10.1021/acssuschemeng.9b00786

Yan, Q., Liu, X., Wang, Y., Li, H., Li, Z., Zhou, L., Qu, Y., Li, Z., & Bao, X. (2018). Cow manure as a lignocellulosic substrate for fungal cellulase expression and bioethanol production. *AMB Express, 8*, 190. https://doi.org/10.1186/s13568-018-0720-2

Yang, Y. H., Wang, B. C., Wang, Q. H., Xiang, L. J., & Duan, C. R. (2004). Research on solid-state fermentation on rice chaff with a microbial consortium. *Colloids and Surfaces. B, Biointerfaces, 34*, 1–6. https://doi.org/10.1016/j.colsurfb.2003.10.009

Yin, Y., Li, M., Tao, X., Yang, C., Zhang, W., Li, H., Zheng, Y., Wang, X., & Chen, R. (2023). Biochar enhanced organic matter transformation during pig manure composting: Roles of the cellulase activity and fungal community. *Journal of Environmental Management, 333*, 117464. https://doi.org/10.1016/j.jenvman.2023.117464

Yunus, M. (2019). Extraction cellulose from rice husk. *Jurnal Akta Kimia Indonesia (Indonesia Chimica Acta), 12*, 79. https://doi.org/10.20956/ica.v12i2.6559

Zepka, L. Q., Jacob-Lopes, E., Goldbeck, R., Souza-Soares, L. A., & Queiroz, M. I. (2010). Nutritional evaluation of single-cell protein produced by *Aphanothece microscopica* Nägeli. *Bioresource Technology, 101*, 7107–7111. https://doi.org/10.1016/j.biortech.2010.04.001

Zhang, G., & Dong, Y. (2022). Design and application of an efficient cellulose-degrading microbial consortium and carboxymethyl cellulase production optimization. *Frontiers in Microbiology, 13*, 957444.

Zhang, Y., Ge, J., & Liu, Z. (2015). Enhanced activity of immobilized or chemically modified enzymes. *ACS Catalysis, 5*, 4503. https://doi.org/10.1021/acscatal.5b00996

Zhao, W., Zhao, F., Zhang, S., Gong, Q., & Chen, G. (2019). Ethanol production by simultaneous saccharification and cofermentation of pretreated corn stalk. *Journal of Basic Microbiology, 59*, 744–753. https://doi.org/10.1002/jobm.201900117

Zhong, C., Wei, P., & Zhang, Y.-H. P. (2017). A kinetic model of one-pot rapid biotransformation of cellobiose from sucrose catalyzed by three thermophilic enzymes. *Chemical Engineering Science, 161*, 159–166. https://doi.org/10.1016/j.ces.2016.11.047

Chapter 19
Nano-bioremediation for Sustainable Treatment of Waste: Applications, Advancements, and Challenges

Shradhanjali Sethi, Ankita Bharshankh, Rojalin Sahu, and Rima Biswas

1 Introduction

Nanotechnology is a dynamic research field bridging atomic and molecular structures with bulk materials, initially coined by Professor Norio Taniguchi of Tokyo Science University and introduced by Richard Feynman. This area focuses on understanding, designing, and applying nanomaterials (NMs) due to their unique porosity, surface area-to-volume ratio, and customization potential for diverse scientific, engineering, and technological domains (Sadri et al., 2018; Portela et al., 2020). Consequently, nanomaterials find applications in various industries like food, pharmaceuticals, agriculture, healthcare, energy, and personal care products (Fig. 19.1) (Sarno, 2020; Pelaz et al., 2017; Dimkpa & Bindraban, 2018; Prasad et al., 2017).

Nanomaterials are synthesized via diverse methods: (i) physical techniques (top-down), such as sonication, pulse wire discharge, radiation, and spray pyrolysis; (ii) chemical methods (bottom-up), including hydrothermal, co-precipitation, microwave, and photochemical; and (iii) biological methods (alternative bottom-up), involving bacteria, algae, plants, yeast, fungi, and biopolymers (Table 19.1) (Rafique et al., 2017a; Marslin et al., 2018). The synthesis process significantly impacts nanomaterial stability, morphology, and physicochemical properties, leading to a growing emphasis on precise control of nanomaterial morphologies (Baig et al., 2021). Traditional approaches are costly, dependent on skilled operators, lack fundamental understanding and modeling, and yield unstable nanomaterials in

S. Sethi · A. Bharshankh · R. Sahu · R. Biswas (✉)
Academy of Scientific and Innovative Research (AcSIR), CSIR-Human Resource Development Centre (CSIR-HRDC), Ghaziabad, India

Wastewater Management Sub-vertical, CSIR-National Environmental Engineering Research Institute, Nagpur, Maharashtra, India
e-mail: ra_biswas@neeri.res.in

V. Kumar et al. (eds.), *Environmental Engineering and Waste Management*,
https://doi.org/10.1007/978-3-031-58441-1_19

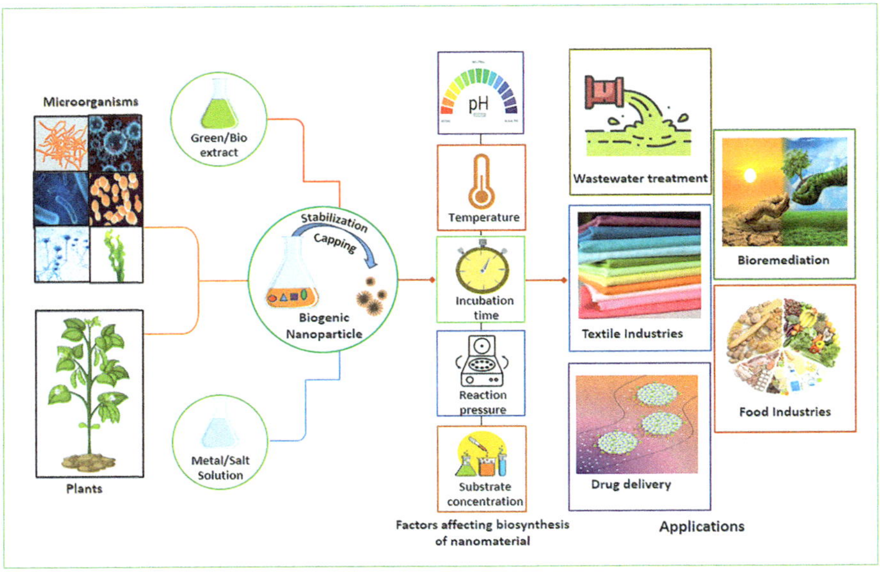

Fig. 19.1 Schematic depiction of the diverse biogenic components participating in nanoparticle synthesis, the influential factors impacting nanomaterial synthesis, and their potential applications

harsh environments, potentially causing bioaccumulation or toxicity and posing environmental challenges (Rafique et al., 2017b; Singh et al., 2018). Despite nanomaterials' extensive utility, their unregulated use and improper disposal in industrial waste and wastewater sludge raise ecological toxicity concerns, especially in aquatic ecosystems.

Addressing concerns regarding conventional approaches, recent nanomaterials research and development have embraced green synthesis, emphasizing sustainability. Green synthesis aligns with green chemistry principles, aiming to design safer chemical processes and products while minimizing the use of hazardous substances (Collins, 2017). The objectives of green nanomaterial synthesis encompass (i) incorporating nanomaterials into environmental technologies for pollution remediation and (ii) creating nanomaterials that mitigate the adverse effects of human activities on public health and the ecosystem (Sivaraj et al., 2015). In comparison to traditional methods, green synthesis offers several advantages: it is safe and environmentally friendly, avoiding toxic substances; it is cost-effective, requiring fewer resources and less energy; and it employs biological components as reducing agents, yielding eco-friendly end products (Singh et al., 2018).

Green synthesis of nanomaterials involves redox reactions, wherein organisms or their metabolic products utilize nicotinamide adenine dinucleotide phosphate hydrogen (NADPH)-dependent reductase enzymes to reduce metal ions, resulting in stable nanomaterials through an enzymatic electron-shuttle process for metal reduction

Table 19.1 Different biogenic components for nanoparticle synthesis and their potential applications

Biogenic material type	Name	Nanoparticles	Applications	References
Algae	*Scenedesmus* sp.	Ag	Para-nitrophenol degradation, antibacterial and antifungal activity	Priyadarshini et al. (2023)
	Chlorella vulgaris	Ag	Treatment of wastewater	Karthikeyan et al. (2015)
	Scenedesmus quadricauda	Au, Ag, Cu	Degradation of PAHs	Agarwal et al. (2019)
	Gracilaria edulis	Ag	Exhibiting anticancer properties against human PC3 cell lines	Priyadharshini et al. (2014)
	Lemanea fluviatilis (L.)	Au	Antioxidant property	Sharma et al. (2014)
	Turbinaria conoides	Au	Enhanced catalytic performance in reducing organic dye molecules and aromatic nitro compounds	Ramakrishna et al. (2016)
	Chlorella vulgaris	Ag	Catalytic property	Eroglu et al. (2013)
Fungi	*Fusarium solani*	Ag	Resistance to Ag ions and biotransformati on property	Sayed and El-Sayed., (2020)
	Pleurotus djamor	Ag	Anticancerous activity	Raman et al. (2015)
	Trichoderma viride	Au	Para-nitrophenol degradation	Mishra et al. (2014)
	Trichoderma koningiopsis	Cu	Treatment of wastewater	Salvadori et al. (2014)
	Ganoderma sp.	Ag, Au	Biocompatibility property	Musarrat et al. (2010)
	Fusarium oxysporum	CdS	Cellular imaging in real time and diagnostic applications	Raheman et al. (2011)
	Bipolaris nodulosa	Ag	Antifungal property	Chen et al. (2014)
Bacteria	*Bacillus licheniformis*	ZnO nanoflowers	Enhanced photocatalysis of methylene blue dye	Agarwal et al. (2017)
	Leptothrix	Au	Biomedicine and catalysis	Kunoh et al. (2018)
	Pseudomonas veronii	Au	Antibacterial activity	Baker and Satish (2015)
	Streptomyces coelicolor	MgO	Agent against multidrug-resistant pathogens	El-Moslamy (2018)
	Sphingobacteriu m thalpophilum	ZnO	Antimicrobial activity	Rajabairavi et al. (2017)
	Enterobacter hormaechei	Ag	Antimicrobial activity	Monowar et al. (2021)
	Cytobacillus firmus	Ag	Antimicrobial activity	Sudarsan et al. (2021)

(continued)

Table 19.1 (continued)

Biogenic material type	Name	Nanoparticles	Applications	References
Actinobacteria	*Streptomyces naganishii*	Ag	Showing antioxidant, anticancerous and anti biofouling activities	Shanmugasundaram et al. (2013)
	Streptomyces viridogens	Au	Antibacterial activity	Balagurunathan et al. (2011)
	Streptomyces sp.	Au	Antimalarial activity	Karthik et al. (2013)
	Thermomonospora sp.	Au	Biosensor	Torres-Chavolla et al. (2010)
	Streptomyces sp.	Au	Fungal inhibiting property	Vinay Gopal et al. (2013)
	Streptomyces hygroscopicus	Au	Antibacterial activity	Sadhasivam et al. (2012)
Plant extracts	*Tragia involucrata*	Pt	Pharmaceutical and biomedical	Selvi et al. (2020)
	Calotropis gigantea	ZnO	Sensing of nitrite, photocatalytic action, and bacterial inhibiting properties	Kumar et al. (2020)
	Lemon peel extract	TiO2	Photocatalytic property	Nabi et al. (2020)
	Pimenta dioica	Au	Anticancerous activity	Kharey et al. (2020)
	Litsea cubeba	Au	Degradation of Para-nitrophenol	Doan et al. (2020)
	Orobanche aegyptiaca	Cu	Nematode killing property	Akhter et al. (2020)
Yeast	*Yarrowia lipolytica*	Ag	Biofilm inhibiting property	Apte et al. (2013)
	Candida utilis	Ag	Microbial inhibiting property	Waghmare et al. (2015)
	Candida glabrata	CdS	Detoxification	Dameron et al. (1989)
	Yarrowia lipolytica	Au	Antimicrobial activity	Agnihotri et al. (2009)

(El-Seedi et al., 2019). Additionally, the utilization of biological constituents for generating metal oxide nanoparticles has gained significant attention due to their unique characteristics. While microorganism-mediated nanomaterial synthesis has been extensively explored (AlNadhari et al., 2021; González-Ballesteros et al., 2017; Molnár et al., 2018; Clarance et al., 2020; Hassan et al., 2019), the use of plant extracts for nanomaterial synthesis is on the rise, favored for its simplicity, cost-effectiveness, environmental compatibility, and scalability (Priyadarshini et al., 2023; Teimuri-mofrad et al., 2017). Moreover, research is exploring solvent-free biosynthesis of nanomaterials (Priyadarshini et al., 2023; Rostami-Vartooni et al., 2015; Oliveira et al., 2019) to strengthen and advance a core tenet of green chemistry.

To address knowledge gaps hindering progress in nanomaterial biosynthesis, this chapter conducts a thorough review of green synthesis methods using various biological capping agents. It begins with a brief examination of nanomaterial stability and environmental toxicity. It then critically assessed commonly used solvents in nanomaterial biosynthesis. After discussing the fundamental requirements for biosynthesis, the chapter provides a comprehensive survey of nanomaterial applications across various sectors. The final section offers a concise overview of the benefits and challenges of using specific biological agents in green nanomaterial synthesis and outlines a roadmap for future research. By highlighting insights from cutting-edge research in nanomaterial biosynthesis, this chapter aims to equip key stakeholders in the field with essential knowledge to guide future investigations in green nanotechnology synthesis.

2 Influential Factors Affecting the Synthesis of Nanomaterials

Nanomaterial synthesis is a complex and dynamic process influenced by numerous factors. Key factors impacting synthesis include precursor materials, reaction conditions, and chosen methods. Temperature, pressure, and the reaction environment composition also play roles, affecting particle size, shape, and crystallinity (Fig. 19.1). The choice of synthesis method, such as chemical vapor deposition, sol–gel, or ball milling, can significantly alter outcomes. Additionally, nanomaterial synthesis advances continuously as researchers explore novel techniques and gain deeper insights into these factors' interactions, shaping nanomaterial properties, and applications.

2.1 pH

In the synthesis of metal nanoparticles (NPs), pH plays a pivotal role, exerting influence on NP size and shape by affecting the biomolecules responsible for capping and stabilizing them, as noted by Verma and Mehata, in 2016 and more recently by Tripathy et al. in 2023. Additionally, pH indirectly impacts NP synthesis by regulating functional groups. Research indicates that lower pH tends to result in rod-like NPs, while higher pH leads to spherical particles, as reported by Ahmed et al. in 2022. Jameel et al. (2020) compared two studies: one by Al-Radadi, 2019, which observed rapid platinum NP synthesis in an alkaline medium within a pH range of 1.5–8.5, and another by Karim et al. in 2019, demonstrating uniform particle distribution at higher pH levels (9–11) with a slight size reduction (up to 2 nm). These findings underscore the substantial impact of pH on customizing metal NP characteristics. Ibrahim, in 2015 further supports this, where it was found that at a highly acidic pH of 2.0, no nanoparticles were detected. However, when pH was raised to 11, highly dispersed nanomaterials with an average size of 23 nm were observed. This significant disparity in outcomes led to the conclusion that under highly acidic conditions, the biomolecules involved in synthesis lose their functionality, impeding nanoparticle formation.

2.2 *Reaction Pressure*

Pressure plays a crucial role in the environmentally friendly synthesis of nanomaterials, impacting both nanoparticle size and shape. Moreover, pressure governs the reduction rate, with optimal pressure conditions leading to accelerated reaction rates, as emphasized by Patra and Baek, in 2014. This relationship is exemplified in the synthesis of $Dy_2Ce_2O_7$ nanoparticles, where CO_2 pressure proves highly significant. The study reveals that the highest yield percentage is achieved within the 2.0–2.5 MPa pressure range. Consequently, elevating the reaction pressure is beneficial for enhancing the synthesis rate, as long as CO_2 pressure remains below 2.5 MPa. Therefore, 2.5 MPa of CO_2 pressure emerges as the ideal reaction pressure for the environmentally friendly synthesis of $Dy_2Ce_2O_7$ nanoparticles, as concluded by in Feng et al. in 2020.

2.3 *Reaction Temperature*

Reaction temperature is a pivotal factor in shaping nanomaterial properties, influencing their size and shape significantly. Lower temperatures tend to result in rod- or plate-shaped Au nanoparticles, whereas higher temperatures favor spherical ones (Ahmed et al., 2022). Zhang et al. (2020) reported that nanoparticles produced from *Citrus sinensis* were 35 nm in size at 25 °C but decreased to 10 nm at 60 °C, with higher temperatures promoting more extensive formation of Au nanoparticles and faster conversion rates. This enables rapid absorption of metal ions during nucleus formation, initially resulting in reduced nanoparticle (NP) aggregation and smaller size (Tripathy et al., 2023). Vijayaraghavan and Ashokkumar (2017) observed that NPs formed at high temperatures (e.g., 100 °C) exhibited greater stability and size increments compared to those formed at lower temperatures (e.g., 25 °C). Elevated reaction temperatures increase molecular kinetic energy, accelerating silver ion consumption and limiting particle growth. An ideal temperature for synthesizing silver nanoparticles was found to be 40 °C, as noted by Bala and Rani (2020). In summary, reaction temperature is a critical factor in determining nanomaterial properties, influencing size, shape, stability, and conversion rates.

2.4 Reaction Time

The impact of reaction time on nanomaterial synthesis is diverse and contingent on the specific material under investigation. For instance, in the production of AgNPs from *Ananas comosus*, rapid reduction of $AgNO_3$ leads to nanoparticle formation within 2 min, with significantly reduced production observed following the initial 2 min (Shah et al., 2015). The duration of the reaction as well influences size of nanoparticle, as demonstrated by a synthesis involving *Azadirachta indica* leaf extract and $AgNO_3$ solution, where varying the reaction time between 30 min and 4 h resulted in a shift from 10 to 35 nm in average particle size (Prathna et al., 2011). Additionally, in the synthesis of magnetic NPs using green tea extract, Singh et al. (2020) noted that increasing the reaction time from 7.5 to 12 nm caused particle size to increase and improved the saturation magnetization. In summary, reaction time is a critical parameter in nanomaterial synthesis, impacting both production yield and nanoparticle size, while also influencing specific properties like magnetization.

2.5 Reactant Concentration

The concentration of metal ions and biomolecules extracts derived from plants plays a crucial part in determining the size, morphology, and characteristics of NPs. Reactant concentration is directly linked to their effectiveness against specific microbial species, such as Ag NPs, enhanced antibacterial properties against *E. coli* compared to *S. aureus*. Altering ion concentration affects absorption peaks and medium-dependent nanoparticle shape. For instance, Aloe vera leaf extract changes the proportion of gold triangular plates to spherical particles. Different *Plectranthus amboinicus* concentrations result in diverse Ag nanoparticle shapes and sizes. Additionally, plant extract concentration significantly impacts synthesis, with even low concentrations altering nanoparticle size, as seen in the case of *Gomphrena globosa* leaf extract (Shah et al., 2015; Bala & Rani, 2020). In nanomaterial research, these concentrations are pivotal determinants worth careful consideration.

2.6 Preparation Cost

Besides the physical and chemical factors, several other parameters play a crucial role in governing the synthesis of nanomaterials, including the cost of preparation. While significant advancements have been made in refining the chemical and physical processes involved in nanoparticle production, these methods often come at a high price tag. In response to this, researchers have increasingly turned their attention toward green synthesis. Green synthesis, also known as biological synthesis, offers a more cost-effective alternative to traditional physical and chemical methods, albeit with a slightly lower yield (Ijaz et al., 2020). These eco-friendly synthesis techniques not only promote environmental sustainability but also attest cost efficient by utilizing natural resources such as microorganisms and plants. Moreover, they boast minimal operating costs since they do not necessitate elevated energy consumption, heavy machinery, or toxic chemicals (Nagababu & Umamaheswara Rao, 2016).

3 Exploring Solvent Selection Strategies for Nanomaterial Synthesis: Enhancing Performance and Sustainability

By navigating the complex landscape of solvent options, the research aims to uncover strategies that not only enhance the performance of nanomaterial synthesis but also align with the principles of sustainability. This investigation holds the promise of shedding light on solvent choices that optimize both the technical outcomes of nanomaterial synthesis processes and their environmental impact, thereby contributing to the advancement of efficient and eco-friendly nanomaterial fabrication techniques.

3.1 Water

Water's remarkable physicochemical properties make it an exceptional solvent for nearly all inorganic compounds and certain organic molecules like sugars and proteins. Its superior solvency remains unmatched by other solvents, irrespective of factors such as temperature and pressure, further enhanced by its widespread availability (Ali et al., 2020). Most notably, many ionic compounds can undergo dissociation in aqueous solutions, weakening electrostatic forces between ions and allowing them to move freely. Numerous experiments utilize extracts from plants, animals, and microbes, primarily extracted with water as a leading eco-friendly solvent, designating the resulting extracts as secondary eco-friendly solvents.

For instance, environmentally friendly silver nanoparticles were synthesized using the microalga *Scenedesmus* sp. and $AgNO_3$, with the reduction process occurring in an aqueous medium and assisted by sodium borohydride ($NaBH_4$)

under standard ambient conditions (Priyadarshini et al., 2023). In another experiment, graphene oxide was synthesized in an aqueous medium through graphite subjected to water electrolytic oxidation. This electrochemical oxidation reaction rapidly and thoroughly oxidized the graphite lattice in seconds. However, it is important to note that while water is the optimal solvent, the production process demands significant energy input due to its high heat capacity compared with other substances (Duan et al., 2015).

3.2 *Ionic Fluid*

Ionic fluids (IFs) are emerging as eco-friendly alternatives to toxic organic solvents due to their low vapor pressures. They solvate compounds based on hydrogen bonding and polarity, determining solubility (Kharissova et al., 2019). IFs consist of ions with melting points below 100 °C and are nonaqueous, nonvolatile, and polar (Singh et al., 2018). IFs' hydrophobic or hydrophilic nature depends on cations and anions. Common cations include N-alkyl pyridinium, 1-alkyl-3-methylimidazolium, and NR_{4+} (Kharissova et al., 2019), with common anions including [PF6] $^{-}$, [CH3COO^{-}], alkyl sulfate, Cl^{-}, NO3^{-}, and more.

IFs provide a versatile medium for synthesizing various nanoparticles through chemical reduction, using gases, organic, or inorganic compounds as reducing agents (Duan et al., 2015). IFs can absorb microwave energy due to high dielectric constants, ionic charges, and polarities, making them compatible with eco-friendly heating technologies like ultrasound and microwave, enhancing reaction efficiency in terms of rate, product quality, and time. For example, *Couroupita guianensis Abul* leaves extract was used to synthesize ytterbium oxide (Yb_2O_3) and ionic fluid-assisted ytterbium oxide (Yb_2O_3-IF) nanoparticles, with Yb_2O_3-IF exhibiting superior antibacterial, anticancer, and anti-inflammatory activity compared with Yb_2O_3 nanoparticles (Muthulakshmi & Sundrarajan, 2020). In another study, Ag–Au/Y_2O_3 NPs synthesized using the ionic liquid [BMIM-PF_6] showed enhanced anticancer activity against the HeLa cell line and stronger antibacterial properties against *S. aureus* and *E. coli* (Pandiyan et al., 2021).

3.3 *Supercritical Fluids*

Ordinary solvents become supercritical fluids (SCFs) under high temperatures and pressures exceeding their critical points (Singh et al., 2018). SCFs bridge the gap between liquids and gases, offering enhanced chemical reactivity and the ability to dissolve solutes, all while retaining gas-like properties such as improved diffusivity, no surface tension, and low viscosity (Ahmed et al., 2022). Notable SCFs include supercritical carbon dioxide (SC-CO_2) and supercritical water (SC-W), as well as eco-friendly bio-based solvents derived from microbial cells, plants, animal tissues, or explants. SC-CO_2, with a critical pressure of 74 bar and a critical temperature of 304 K, efficiently penetrates nanopores, making it ideal for depositing metal nanoparticles into porous nanomaterials. Moreover, it is non-flammable, environmentally safe, nontoxic, chemically inert, and cost effective (Siril & Türk, 2020; Ahmed et al., 2022).

The bottom-up approach called supercritical fluid reactive deposition (SFRD) can produce metal nanostructures in large quantities, and SC-CO_2's high wetting ability makes it suitable for depositing nanoparticles on extremely hydrophobic surfaces like graphene and polymers (Moreno et al., 2018). In another experiment, tungsten oxide and tungsten blue oxide were synthesized using subcritical and supercritical water and methanol (Kim et al., 2016b). Subcritical and supercritical methanol proved more effective than glycerol hydrothermal reduction for reducing tungsten oxides. SC-W, with a critical pressure of 22.1 MPa and a critical temperature of 646 K, can interact with both inorganic and organic substances to form a homogenous phase (Kharissova et al., 2019; Ahmed et al., 2022).

4 Utilization of Various Biogenic Components in Green Synthesis of Nanomaterials

4.1 Bacteria

Bacteria are well known for their ability to produce metallic nanoparticles like silver and gold, both intracellularly and extracellularly (Rónavári et al., 2018). Intracellular synthesis transports ions into microbial cells for nanoparticle formation with enzymes, whereas extracellular synthesis traps metal ions on cell surfaces, followed by enzymatic reduction (Fig. 19.2) (Llorens et al., 2012). To ensure successful nanoparticle synthesis, it is crucial to determine the minimum inhibitory concentration of metals beforehand, as these mechanisms have evolved to counteract the harmful effects of metals. Silver nanoparticle production primarily relies on the enzyme nitrate reductase and converts NO_3 to NO_2. The NADPH-dependent nitrate reductase, using alpha-NADPH, streamlines the process, eliminating the need for extra downstream steps. This method stands out when compared to other in vitro approaches discussed by Bala and Rani, in 2020.

Furthermore, it is worth noting that bacterial cells can bind substantial quantities of metallic cations. For example, non-magnetotactic bacteria like *Geobacter metallireducens GS-15* were used in anaerobic conditions to create magnetite nanoparticles sized between 10 and 50 nm (Lovley et al., 1987). Similarly, John et al. in 2020 reported in that *Bacillus thuringiensis* was used to produce silver nanoparticles Ag-NPs with diameters ranging from 43.51 to 142.96 nm. Gold nanoparticles were successfully generated using various bacterial strains, including *Pseudomonas aeruginosa, E. coli, Rhodopseudomonas capsulata, DH5, Bacillus* species like *subtilis and licheniformis*. The production of silver nanoparticles with

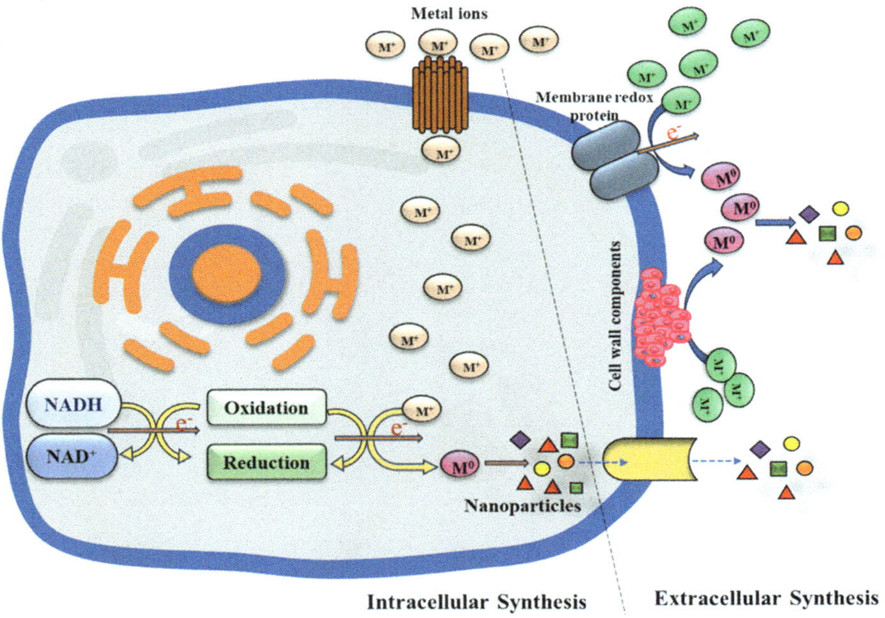

Fig. 19.2 Mechanisms of intracellular and extracellular synthesis of bacterial nanoparticles

distinct size and shape morphologies primarily relied on bacterial strains like *Lactobacillus casei, Escherichia coli SH10, Bacillus cereus, Aeromonas* sp., *Antarctica Bacillus amyloliquefaciens, Bacillus indicus, Bacillus cecembensis, Enterobacter cloacae, Geobacter* spp., *Arthrobacter gangotriensis, Corynebacterium* sp. *SH09,* and *Shewanella oneidensis.* Cadmium nanoparticles were previously synthesized using bacteria such as *Escherichia coli, Clostridium thermoaceticum, and Rhodopseudomonas palustris* (Singh et al., 2018).

4.2 Fungi

Fungi are highly efficient agents for producing well-defined, monodispersed nanoparticles, because of their remarkable metal-binding capacity, resilience in the face of metals, and ability to accumulate them internally, a capability comparable to that of bacteria (Ahmad et al., 2003). Fungi offer distinct advantages over other organisms due to the presence of enzymes, proteins, and reducing agents on their cell surfaces. Their unique approach to nanoparticle synthesis involves producing copious amounts of enzymes that break down silver ions and facilitate metallic nanoparticle growth. Researchers worldwide harness fungi for nanoparticle biosynthesis, achieving both intracellular and extracellular production (Fig. 19.3) (Tripathy et al., 2023).

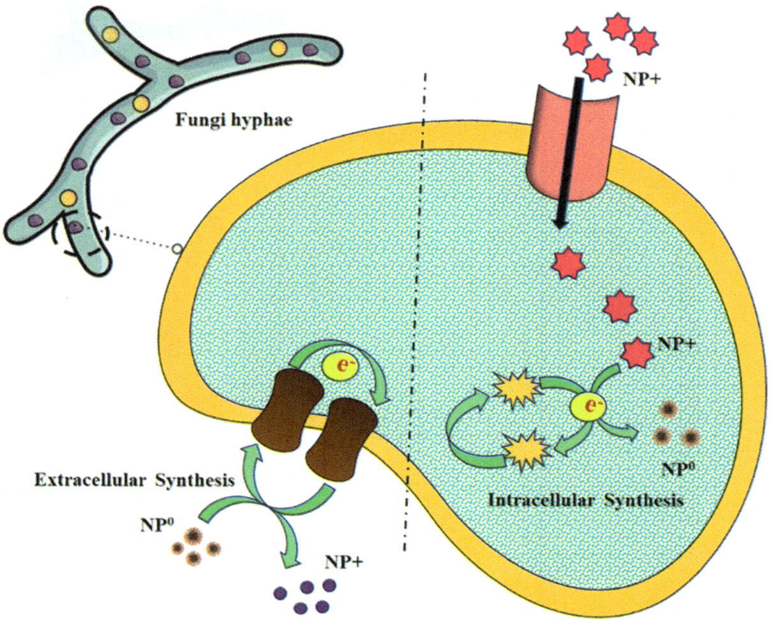

Fig. 19.3 Mechanisms of fungal intracellular and extracellular synthesis of nanoparticles

Intracellularly, *Aspergillus flavus* generates silver nanoparticles, while *Verticillium* spp. can create both gold and silver nanoparticles. Gold nanoparticles produced by *Trichothecium* species and *Verticillium luteoalbum* are among the most common, exhibiting spherical, quasi-hexagonal, and rod morphologies within the size range of 8.92–25 nm (Mukherjee et al., 2001; Vigneshwaran et al., 2007; Gericke & Pinches, 2006; Ahmad et al., 2003). Extracellularly, *F. oxysporum, Trichothecium* species, and *Colletotrichum species* are responsible for generating Au-NP ranging in size from 5 to 200 nm, features spherical, triangular, and hexagonal shapes. Additionally, silver nanoparticles are produced by a variety of fungi, including *T. asperellum, T. viride, F. oxysporum, A. niger, F. solani, F. semitectum, A. fumigatus, P. chrysosporium, C. versicolor, P. glomerata, P. brevicompactum, and C. cladosporioides* (Bhatt et al., 2022).

4.3 Yeast

Yeasts, classified as single-celled eukaryotic microorganisms, represent a diverse group with over 1500 different species. Scientists have effectively utilized yeast in the synthesis of nanoparticles and various nanomaterials, as documented by Singh et al. in 2018. Remarkably, a yeast strain tolerant to silver, along with a solution of

Saccharomyces cerevisiae, has shown the capability to biologically produce both silver and gold nanoparticles. Several studies have showcased yeast's capacity to produce AgNPs, with strains like *Saccharomyces boulardii*, *Candida albicans*, and *Candida utilis* being notable examples. *Schizosaccharomyces pombe* and *Candida glabrata* have been shown to generate wurtzite hexagonal CdS crystals, sized at 1–1.5 nm. These crystals are spherical, monodispersed, and peptide-bound CdS quantum crystallites, with a size of 20 (Dameron et al., 1989).

Similarly, *Turulopsis* sp. was employed to construct PbS nanocrystallites within its vacuoles, with dimensions ranging from 2 to 5 nm. As described by Khati et al., in 2018, these nanoparticles were later used to generate an optimal diode. Additionally, the yeast strain MKY3, known for its ability to tolerate silver, has been used to produce silver nanoparticles externally, as shown by (Soliman et al., 2018). Additionally, as noted by Rónavári et al. in 2018, extremophilic yeasts found in acidic spring outflows function as biocatalysts in the creation of both gold and silver nanoparticles.

4.4 Viruses

The utilization of viruses for nanoparticle biosynthesis represents a unique and innovative approach to producing various inorganic nanomaterials, including zinc sulfide, cadmium sulfide, iron oxide, and silicon dioxide (Zhao et al., 2013). This method employs a range of biological components, such as DNA, biolipid cylinders, protein cages, viroid capsules, S-layers, bacterial replisomes, and multicellular superstructures, as templates for the synthesis of inorganic nanoparticles and microstructures (Alivisatos et al., 1996; CA Mirkin et al., 2020; Braun et al., 1998). Additionally, various biological substances like fatty acids, amino acids, and polyphosphates are employed as templates in the formation of semiconductor nanocrystals (Shah et al., 2015). Notably, the shapes of cadmium sulfide, cadmium selenide, and cadmium telluride nanocrystals can be modified via adjusting the proportions of different fatty acids (Henglein, 1989).

One intriguing feature of this approach is the intricate surface of viruses, which safeguards the capsid protein structure and creates a highly sensitive surface when combined with metal ions (Makarov et al., 2014). The inclusion of TMV, composed of approximately 2130 capsid proteins, has a significant influence on the quantity, size, and shape of the resulting silver and gold nanoparticles. TMV is utilized as a template for processes such as the oxidative hydrolysis synthesis of iron oxides, the co-crystallization of CdS and PbS, and the sol–gel condensation synthesis of SiO_2 (Shenton et al., 1999). Additionally, M13 bacteriophages, capable of forming ZnS and CdS nanocrystals, produce peptides A7 and J140 as pVIII fusion proteins within their capsids (Singh et al., 2015).

4.5 Plant

Various plant species have displayed remarkable capabilities in detoxifying and sequestering heavy metals. Plant-assisted synthesis of nanomaterials is preferred due to the stability of the produced nanomaterials and its advantages, such as speed, non-pathogenicity, cost-effectiveness, environmental safety, and a streamlined, single-step biosynthetic process (Singh, 2016, 2020). In this process, the photosynthetic apparatus of the thylakoid membrane, including photosystem I and photosystem II, utilizes electrons derived from water to reduce $NADP^+$ to NADPH when H^+ ions are transported across the thylakoid membrane (Kim et al., 2016a). This proton gradient drives ATP production and the release of oxygen, making plants a promising candidate for nanoparticle biosynthesis (Fig. 19.4).

Copper and copper oxide nanoparticles have been synthesized using plant extracts from species such as Aloe vera, Soy, and Tridax daisy leaf (Gunalan et al., 2012; Rafique et al., 2017a, b). An economical and uncomplicated method for synthesizing silver nanoparticles entails reducing and stabilizing silver ions with diverse biomolecules found within plant extracts. These biomolecules encompass amino acids, polysaccharides, proteins, vitamins terpenoids, alkaloids, saponins, and phenolic compounds as elucidated by Stampoulis et al. in 2009. As an illustration, Ahmed et al. (2016) effectively generated silver nanoparticles via an aqueous extract derived from *Azadirachta indica* leaves, employing a straightforward, rapid,

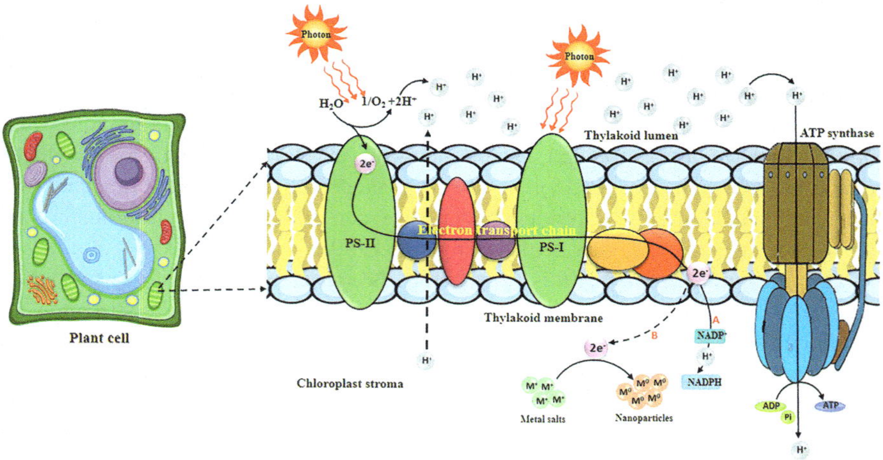

Fig. 19.4 Plant cells stimulate nanoparticle synthesis through the activation of the electron transport chain within the photosynthetic machinery located in the thylakoid membrane of chloroplasts. Light energizes both PSI and PSII, causing them to draw electrons from water and ultimately resulting in the reduction of $NADP^+$ to NADPH (**a**) Simultaneously, the process creates a H^+ gradient across the thylakoid membrane, which serves as the driving force for ATP synthesis (**b**) Moreover, this system has the capability to reduce metal ions into nanoparticles

environmentally benign, and nontoxic procedure. Additionally, aqueous leaf cell extracts from plants like *Eucalyptus, Turnera ulmifolia, and Couroupita* guianensis have been utilized for the production of Fe-NPs and composites (Salem & Fouda, 2021). Flavonoids, with their functional groups, boost metal ion reduction. The enol-to-keto conversion in *Ocimum basilicum* extracts is pivotal in biogenic silver nanoparticle synthesis (Ahmed et al., 2022). Another case involves eugenol, a key terpenoid in cinnamon extracts, which is essential for bioconversion of metal salts into corresponding nanoparticles (Singh et al., 2016).

4.6 Algae

Algae, which are marine microorganisms, not only absorb heavy metals from the environment but also produce metallic nanoparticles. They can be multicellular (like brown algae) or unicellular (such as chlorella) photosynthetic microorganisms (Shah et al., 2021). Algae are commonly used for zinc oxide nanoparticle production and occasionally for silver and gold nanoparticles (Agarwal et al., 2017). *Spirulina platensis*, a blue-green alga, has been used to produce 5 nm gold nanoparticles (Uma Suganya et al., 2015). Agarwal et al. used *Sargassum myriocystum and Sargassum muticum microalgae* to create 36 nm zinc oxide nanoparticles. Phytoplanktonic alga *Phaeodactulum tricornatum* produced cadmium-exposed phytochelatin-coated CdS nanocrystals.

Extracellular biosynthesis of Au-NP has been achieved rapidly using a marine alga called *Sargassum wightii Greville and Turbinaria conoides* (Scarano & Morelli, 2003; Vijayan et al., 2014). In contrast, four marine macroalgae—*Colpmenia sinusa, Jania rubins, Pterocladia capillacae,* and *Ulva faciata*—have been employed in the production of silver nanoparticles. Diatoms like *Navicula atomus* and *Diadesmis gallica* have been used to produce extracellular polymeric compounds, resulting in gold, silica-gold bio-nanocomposites, and gold nanoparticles.

4.7 Actinomycetes

Actinomycetes are prokaryotic microorganisms belonging to the Actinobacteria phylum, and they share several characteristics with fungi. They are sometimes referred to as "ray fungus" (Ahmad et al., 2003). Actinobacteria are capable of producing metallic nanoparticles, either extracellularly or intracellularly (Kumari et al., 2021). Research on alkalo-tolerant actinomycetes, especially *Rhodococcus* species, found intracellular gold nanoparticles ranging from 5 to 15 nm. These nanoparticles accumulate as Ag^0 on cell walls and membranes (Pal et al., 2019). Gold nanoparticle formation is also documented in various actinomycetes, including

Streptomyces viridogens, *S. hygroscopicus*, *Nocardia farcinica*, *Thermoactinomyces* sp., and *Thermomonospora* (Składanowski et al., 2017). Significantly, *Thermomonospora* sp., an extremophilic actinomycete, successfully synthesized spherical, monodisperse 8 nm spherical gold nanoparticles that remain stable for 6 months (Lu et al., 2017). *Streptomyces* species are known for producing metallic nanoparticles, including Ag, Zn, Cu, and Mn nanoparticles (Salem & Fouda, 2021). Similarly, actinomycetes like *Rhodococcus* sp., *Thermoactinomyces* sp., and *Streptomyces* sp. have been reported to produce silver nanoparticles (Alani et al., 2012; Singh, 2016; Otari et al., 2012).

5 Toxicity and Stability Assessment of Nanomaterial

Nanoparticle (NP) toxicity in the human body often results from excessive reactive oxygen species (ROS) generation (Nel et al., 2006; Møller et al., 2010; Unfried et al., 2007). For example, iron-based NPs catalyze ROS production through the Fenton reaction, leading to oxidative stress (Bhatt et al., 2022). TiO_2 NPs induce proinflammatory responses in human endothelial cells, while carbon NPs cause oxidative stress in fish brain cells and pulmonary inflammation in rats (Jiang et al., 2009; Warheit, 2003). The cytotoxicity of Ag-NPs has been extensively studied, especially in relation to the respiratory, cardiovascular systems, embryonic development, and more (Albers et al., 2013). Some researchers suggest that AgNPs can convert into Ag^+ ions intracellularly, elevating Ag^+ concentration within cells (Maiorano et al., 2010; Mudunkotuwa & Grassian, 2015).

Gold nanoparticles accumulating around maize roots have raised toxicity concerns, blocking uptake pathways, reducing plant growth, and nutrient absorption (Wang et al., 2020). In the case of metals like copper, nanometal oxides release soluble metal ions, with CuO-NPs being more toxic than bulk CuO. Even low ${Cu_2}^+$ levels can harm algal cells, suggesting ion dissolution plays a role in Cu-NPs toxicity (Wang et al., 2008).

Stabilization is crucial for maintaining the stability of colloidal metal nanoparticles, involving the creation of an electrical double layer on nanoparticle surfaces through electrostatic mechanisms. Nanoparticles tend to aggregate due to electrostatic, magnetic, and van der Waals interactions in solutions (Cosgrove, 2010). Ligands used for surface capping or stabilization during nanoparticle synthesis can impact the final nanoparticle size (Brust et al., 1995). Colloidal nanoparticles stabilize by adsorbing nearby molecules or coagulating to minimize surface area (Derjaguin & Churaev, 1974). Stabilization can involve the adsorption of large molecules containing surfactants or polymers onto nanoparticle surfaces, preventing aggregation. Steric stabilization through adsorbed molecules increases osmotic repulsive force, enhancing nanostructured particle stability (Sharma et al., 2015).

6 Potential Applications of Green Nanoparticles

6.1 Antimicrobial Effects

The emergence of microbial resistance to standard antiseptics and antibiotics has prompted numerous studies to enhance antimicrobial activity. Metallic nanoparticles have shown the potential to effectively inhibit various bacteria in vitro antimicrobial studies, making them a promising avenue to combat microbial drug resistance (Fouda et al., 2020). Nanoparticles can prevent biofilm formation and antibiotic resistance through several mechanisms (Fig. 19.5). The substance used for nanoparticle synthesis and their particle size are crucial factors that determine the effectiveness of metallic nanoparticles against microorganisms (Mba & Nweze, 2021). Researchers have successfully generated various types of nanoparticles (NPs) for medical applications, including metals like Au and Ag, as well as metal oxides such as Ag_2O, CaO, TiO_2, ZnO, MgO, CuO, and SiO_2 (Table 19.1) (Dikshit et al., 2021). Silver nanoparticles (AgNPs) are particularly valuable in the medical field as potent antibacterial agents. Both Au and Ag nanoparticles display significant antibacterial activity against Gram-negative pathogens, including *E. coli, Salmonella typhimurium, Enterobacter aerogenes, Shigella dysenteriae, Pseudomonas aeruginosa, Klebsiella pneumonia, Citrobacter* sp., *and Proteus mirabilis* (Raghunath & Perumal, 2017).

Biosynthesized Au-NPs and Ag-NPs exhibit antimicrobial activity against Gram-positive pathogenic bacteria, including *Staphylococcus epidermidis, Staphylococcus aureus (including MRSA), Streptococcus pyogenes, Enterococcus faecalis, and Bacillus subtilis* (Salem & Fouda, 2021). Furthermore, biosynthesized Ag-NPs

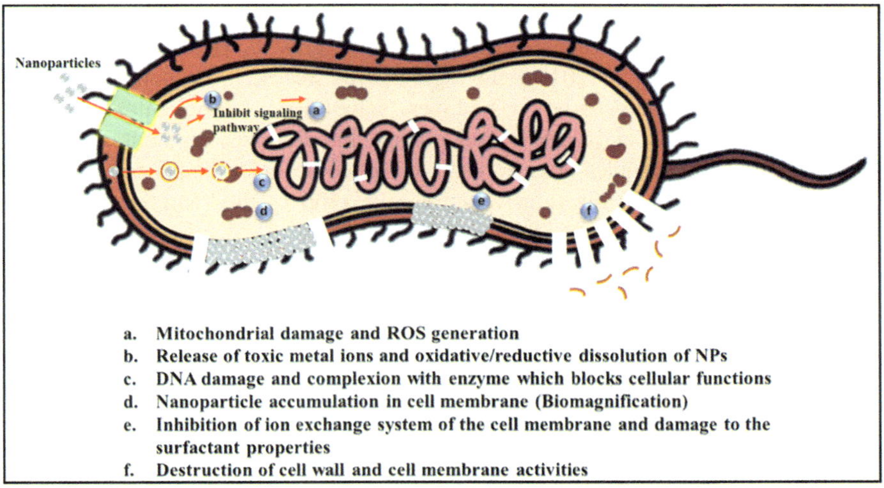

Fig. 19.5 Green nanoparticles display antimicrobial activity through various distinct mechanisms

exhibit antifungal properties against a variety of fungi, including *Aspergillus flavus, Aspergillus fumigatus, Candida species such as glabrata, albicans, parapsilosis, krusei*, and *tropicalis*, as well as *Cryptococcus* species like *gattii* and *neoformans*, along with *Trichophyton species* such as *rubrum* and *mentagrophytes, Epidermophyton floccosum, Mucor hiemalis,* and *Sporothrix schenckii* (Huang et al., 2023; Romero et al., 2018).

6.2 Agriculture and Nanoparticle

Agriculture serves a crucial role in supplying essential elements such as animal feed, human food, and raw materials for diverse industries, including textiles, leather, and chemicals like sugars, starch, and xylan (Salem & Fouda, 2021). In recent times, agriculture nanotechnology has gained significant attention, focusing on research and applications to address concerns related to crop enhancement, agricultural sustainability, increased production, and environmental considerations (Bhatt et al., 2022). Nanozeolite, nanosilicon dioxide, and titanium oxide nanoparticles have positively impacted maize crop yields and soil conditions (Khati et al., 2018). Gypsum and nanogypsum have shown the ability to enhance bacterial cultures in nutrient broth and improve agronomic characteristics in maize compared to control samples (Chaudhary & Sharma, 2018). Silica nano-fertilizers demonstrated similar positive effects on basil, improving specific growth characteristics (Kalteh et al., 2018). Similarly, zinc oxide nano-fertilizers have shown growth-promoting effects in various crops, including moringa, tomatoes, pearl millet, and coffee (Rehmanullah et al., 2020; Nongbet et al., 2022).

Various nanoparticles, including iron, zinc oxide, urea hydroxyapatite, carboxymethyl, titanium oxide, and cellulose-based hydroxyapatite, can be used to enhance secondary metabolites, germination rates, growth, yield, and grain nutritional quality (Chhipa, 2019). In agricultural research, nanomaterials and nanostructures such as carbon nanotubes, nanofibers, and quantum dots are employed as biosensors to assess soil quality and fertilizer distribution (Sharon et al., 2010). Nanoparticles are also used to reduce pesticide dispersion, minimize nutrient loss during fertilization, and improve crop quality and production by delivering nutrients efficiently (Sangeetha et al., 2021).

6.3 Wastewater Treatment

Water, an essential element for life, faces challenges due to population growth, resulting in increased wastewater generation. Traditional wastewater treatment methods, limited by time constraints and the presence of hazardous compounds, often struggle to completely remove contaminants (Ahmed et al., 2022). Nanotechnology offers an innovative solution to address the pressing issues related to water quality and scarcity (Madhura et al., 2019). Nanoscale metals like Ag, Au, CuO, TiO_2, ZnO, and carbon nanotubes show great promise in wastewater and water treatment due to their effectiveness as catalysts (Das et al., 2015).

Silver nanoparticles have proven effective in removing pathogens, phosphorus, sulfur, and various contaminants from wastewater (Singh et al., 2016). TiO_2 nanoparticles demonstrate efficiency in the photocatalytic treatment of wastewater, while magnetic nanoparticles are successful in eliminating harmful metal ions like mercury (II) and cadmium (II) from wastewater (Madhura et al., 2019). FeNp obtained from eucalyptus leaves has shown promise in swine effluent treatment. Fe/Pd and Fe bimetallic NPs are employed for eco-friendly dye degradation (Ahmed et al., 2022). Furthermore, magnesium oxide and hematite nanoparticles have demonstrated significant efficiency in eliminating chromium from textile industry effluents (Fouda et al., 2021). Copper oxide nanoparticles (Cu_2O-NPs) show promise in the (Ajith et al., 2021).

6.4 Food Industry

Nanotechnology has made significant advancements in the food processing sector, offering various applications to enhance food quality, safety, and packaging. Nanomaterials are utilized to improve the mechanical strength and stability of packaging materials, serving as antibacterial agents, anticaking agents, food additives, and nutrient transporters (Ezhilarasi et al., 2013). Innovative food packaging is essential for preserving food safety and quality by protecting it from harmful bacteria. Nanoparticles have an important role in protecting the quality and safety of perishable and semi-perishable food goods. Silver and zeolite nanoparticles are employed to reduce the microbial load of bacteria like *Staphylococcus aureus* by releasing silver ions, making them valuable in robust food packaging systems (Llorens et al., 2012). Silver nanoparticles, known for their antibacterial properties, are used as natural antibacterial agents in food processing to inhibit bacterial and pathogenic growth (Verma & Mehata, 2016).

Zinc oxide and titanium dioxide nanoparticles are two of the most often employed antimicrobial metal oxides in active food packaging. TiO_2 nanoparticles, approved by the FDA in the United States, enhance food product quality due to their properties such as high refractive index, whiteness, coverage, and neutral taste (Silvestre et al., 2011). Amorphous silica (SiO_2) is another FDA-approved food ingredient used as an anticaking agent in powdered or granulated foods, improving powder flow and preventing lumps. Iron oxide (Fe_2O_3) nanomaterials are employed as coloring agents in the food industry, while chitosan is used to coat fresh fruits (Jafarizadeh-Malmiri et al., 2019). Nanoparticles also play a crucial role in enhancing the sensitivity and specificity of food sensors, enabling the detection of a wide range of signals (Sharma et al., 2015). Common antibacterial agents used in food processing and packaging include Ti, Ag, Se, Cu, ZnO, $Mg(OH)_2$, and Au nanoparticles (Hoseinnejad et al., 2018).

6.5 Textile Industry

In the textile industry, approximately 60% of dyes are used in the coloring process for various materials. However, the persistent nature of dyes presents a significant environmental challenge, with approximately 15% of dyes being wasted and discharged into the hydrosphere after fabric processing (Carmen & Daniela, 2012). Nanoparticles (NPs) have emerged as superior alternatives to traditional textile finishing methods. They do not compromise the esthetic appeal of textile materials and offer enhanced durability (Ali et al., 2023). Various types of NPs, including carbon nanofibers, carbon nanotubes, nano-clay, nanocomposites, hybrids, and polymeric nanoparticles, are revolutionizing the textile industry (Perera et al., 2022). Silver nanoparticles have been used in finished textiles to enhance their antibacterial properties, impart self-cleaning attributes, and provide UV-blocking capabilities (Mowafi et al., 2017). Additionally, zinc oxide nanoparticles (ZnO-NPs) are applied in the textile sector to enhance UV blocking and antibacterial properties (Fouda et al., 2021). Research has also explored the use of nanoparticles like ZnO, TiO_2, and SiO_2 to enhance fiber properties, ensuring they remain chemically stable and nontoxic when exposed to UV radiation and high temperatures. These nanoparticles, such as nanosilver, antimony-doped tin oxide, and silane nanosols, provide antistatic properties to synthetic fibers.

Furthermore, extensive studies have been conducted on the catalytic activity of nanoparticles in dye degradation. Nanoparticles like cadmium sulfide (CdS)-TiO_2-gold (Au), iron (Fe) doped TiO_2, and TiO_2-coated silica gel beads are widely employed for the degradation of dyes such as methyl orange (MO) (Shnoudeh et al., 2019). Silica nanoparticles (SNPs) are preferred due to their large surface area and ease of application to fabric surfaces, making them a common choice in textile curing processes (Shnoudeh et al., 2019). The integration of nanoparticles in the textile industry demonstrates the potential to address environmental concerns and enhance the functionality and sustainability of textile products.

6.6 Drug Delivery and Cancer Therapy

Nanoparticles (NPs) are finding increasing applications in various biological areas, including drug delivery, gene transfer, and cancer therapy, thanks to their unique properties (Yusuf et al., 2023). NP-based therapies are gaining popularity due to their enhanced efficacy, reduced side effects, and targeted delivery to specific cells, particularly in cancer treatment. The small size of nanoparticles enables them to easily penetrate tiny capillaries, facilitating improved cell uptake and effective drug accumulation (Bandopadhyay et al., 2020). In drug delivery, nanoparticles play a crucial role, providing a platform for precise and efficient medication delivery to specific body regions (Bhatt et al., 2022). Various types of nanoparticles, such as zinc oxide, silver, mesoporous silica, among others, have shown promise in drug delivery applications (Kalpana & Devi Rajeswari, 2018). For instance, silver nanoparticles are integrated into implants, bone cement, and wound dressings due to their proven antibacterial properties (Schröfel et al., 2014). Copper nanoparticles (CuNPs) are being explored for targeted drug delivery to combat bacterial infections, leveraging their antibacterial properties (Mba & Nweze, 2021). Coating nanoparticles with polyethylene glycol (PEG) enhances the effectiveness of nanoparticle-based drug formulations, enabling more precise targeting of bacterial infections within the body (Suk et al., 2016). Zinc oxide (ZnO) and silicon (Si) nanoparticles have demonstrated potential as drug carriers, enhancing cell uptake and transfection efficiency (Kalpana & Devi Rajeswari, 2018).

In the field of cancer therapy, nanoparticles, including bimetallic composite nanoparticles, have shown potential to induce mitochondrial stress and cell death, leading to p53-independent apoptosis mediated by reactive oxygen species (ROS) (Yusuf et al., 2023). Magnesium oxide nanoparticles produced through bio methods have exhibited a lethal effect against lung cancer cell lines by increasing ROS production (Kapinder et al., 2021; Ahmed et al., 2022). Inorganic nanomaterials, such as metal oxides and metal particles like zinc oxide, iron oxide, titanium dioxide, gold, silver, and nickel, hold promise in various medical applications, including cell imaging, biosensing, gene or drug delivery, and cancer treatment (Khalafi et al., 2019; Sekhon & Kamboj, 2010). The field of nanomedicine is continually evolving, presenting new opportunities for enhanced drug delivery and targeted therapies across various biomedical applications.

6.7 Bioimaging

In the field of medicine, imaging tools play a crucial role in drug development, disease diagnosis, and therapy. Over the past decade, there has been significant progress in fluorescent photography, enabling the observation of biological activities at various scales, including molecular, cellular, organellar, and organism-wide scales. Nanoparticles have gained widespread use in bioimaging, contributing to advancements in molecular imaging methods such as magnetic resonance imaging (MRI), positron emission tomography (PET), ultrasound imaging (USI), optical imaging (OI), and more, for imaging biological specimens both in vitro and in vivo (Margolis et al., 2007; Weissleder, 2002). Nanoparticle-based optical imaging agents can generally be categorized into quantum dots and dye-doped nanoparticles, each with its distinct properties and advantages (Alivisatos et al., 1996). Bioimaging systems often utilize quantum dot (QD)-peptide conjugates, as described by Chang et al., which present luminescent probes with intrinsic signal amplification upon interaction with proteolytic enzymes (Zhao et al., 2013). Additionally, nanoparticles like cross-linked iron oxide (CLIO) nanoparticles, known for their stability and practicality, have been utilized for targeted imaging with excellent cellular uptake (Zhao et al., 2013).

In the context of cancer, targeted detection and treatment are critical. Nanoparticles, such as magnetic poly-(D,L-lactide-co-glycolide), nanoparticles loaded with doxorubicin (DOX), have been developed for the simultaneous targeted detection and treatment of breast cancer (Wang et al., 2008). Zinc oxide nanoparticles, often used due to their emission of near-UV light and essential for excitonic blue, result in green luminescence according to Kalpana and Devi Rajeswari (2018).

6.8 Bioremediation

The application of nanomaterials in bioremediation offers a promising avenue for addressing environmental pollution. Nanoparticles possess distinct properties that make them well suited for this purpose. Moreover, the quantum effects of nanomaterials lower activation energy, facilitating reactions vital for pollutant degradation. Additionally, the ability to utilize nanomaterials of varying sizes and shapes enhances their applicability in environmental remediation (Anyanwu et al., 2021). Pollutants like azo dyes, acid dyes, cationic dyes, and others pose a significant threat to the environment and human health. Bioremediation involving nanomaterials has gained attention as a promising approach to address these issues (Sivaraj et al., 2015). Nano-bioremediation involves utilizing various nanoparticles in conjunction with microbes to develop eco-friendly strategies for cleaning hazardous environmental pollutants, presenting an advancement in bioremediation (Chaudhary & Sharma, 2018). For instance, a study demonstrated the breakdown of phenol from groundwater using a bacterial consortia and calcium peroxide nanoparticles, showcasing the efficacy of this approach (Lu et al., 2017).

7 Environmental Influence of Nanoparticles

The presence of organic and inorganic contaminants in nanoparticles directs significant warnings, impacting not only human health but also the environment. Nano-bioremediation emerges as an efficient technology for mitigating these contaminants with reduced toxicity. Its effectiveness in removing harmful chemicals contributes to the creation of a more sustainable environment. In nanoparticles, toxicity primarily arises from the generation of reactive oxygen species (ROS), which disrupt electron transport processes in mitochondria and chloroplasts, as observed in a study by Yan and Chen, in 2019. Moreover, ROS can lead to alterations in lipid and protein peroxidation and ultimately result in DNA impairment (Wang et al., 2020).

The implication of nanoparticles in the environs has led to their penetration into plants through both roots and leaves, resulting in alterations in the morphological and physiological functions of the plants (Khati et al., 2018; Bhatt et al., 2022). The impact of NPs on plant productivity and health can vary, with both positive and negative effects, contingent upon the type and concentration of the NPs (Bhatt et al., 2022). For instance, aluminum oxide and zinc NPs have been shown to have toxic effects on green pea plants (Mukherjee et al., 2016). Similarly, zinc NPs have been found to negatively impact fruit yield in cucumber (Zhao et al., 2013). In zucchini plants, the combined presence of AgNPs and CuNPs reduces root growth and development, with AgNPs also inhibiting transpiration rates by up to 41–57% (Stampoulis et al., 2009). Higher concentrations of titanium dioxide (TiO_2) NPs have been observed to have a detrimental effect on the growth of wheat plants (Feizi et al., 2012). Sillen et al. (2015) documented the impact of nanosilver at a concentration of 100 $mg{\cdot}kg^{-1}$ on FDA (fluorescein diacetate activity) within the maize rhizosphere, revealing a decrease in FDA activity in the treated soil (McGee et al., 2017). Several studies have indicated that reductions in cytokinins, auxins, and hormonal imbalances can disrupt the whole metabolic processes in plants also (Vankova et al., 2017).

8 Pros and Cons in the Fabrication of Nanomaterials

Biogenic synthesis of nanomaterials is considered environmentally friendly in contrast to chemical and physical methods, offering the ability to produce highly stable, monodispersed nanomaterials through adjustments in parameters like pH, temperature, mixing ratios, and incubation period (Singh, 2016). The use of microbes, particularly in extracellular nanoparticle synthesis, reduces downstream processing needs, exhibits high metal tolerance (Rana et al., 2020), and simplifies nanomaterial harvesting compared to intracellular methods. Additionally, green synthesis provides numerous advantages, including cost-effectiveness, energy efficiency, process and product safety, reduced waste generation, and potential applications in biomedicine and pharmaceuticals due to the inherent pharmacological properties of the synthesized materials (Ijaz et al., 2020; Hassan et al., 2019). A comparative details of greener components and solvents used in nanomaterial biosynthesis are summarized in Table 19.2.

The utilization of nanotechnologies for environmental contaminant removal has primarily been confined to laboratory settings, with industrial-scale implementation remaining a significant challenge. While nanoparticles have shown promise in remediating contaminants, they come with certain limitations in field applications, including loss of reactivity over time, and their potential impact on microorganisms (Das et al., 2018). Microbe-driven nanoparticle synthesis, though effective, is slow and demands intricate maintenance procedures involving microbial cultivation, isolation, and incubation. Additionally, a lack of comprehensive knowledge exists regarding the effects of nanoparticles on microbial communities, with conflicting results in studies (He et al., 2018). Some research has indicated inhibitory effects on microorganisms like *Staphylococcus aureus* and *Escherichia coli* (Babaei-Ghazvini et al., 2021), while others have shown stimulatory effects of nanoparticles as electron donors for bacteria and methanogens (Ahmed et al., 2022). Generally, various

Table 19.2 Pros and Cons associated with various synthesis components and solvents

Synthesis ingredients	Pros	Cons
Algae	Easily available	Intensive upkeep
	Effective in heavy metals removal	Temperature sensitive
	No capping agent	Prolonged incubation period
	Quick synthesis	
Fungi	Generate ample amounts of enzymes with numerous beneficial practices	Require frequent maintenance
	Easy to yield mycelial cells	Extended incubation period
	Monodispersed nanoparticles with clearly visible morphology	
	Metal tolerance and the capacity for bioaccumulation	
	Facilitates scaling-up, downstream processing, and large-scale production	
Actinomycetes	Enhanced sizing and polydispersed nanomaterials	Extensive maintenance
	Production of valuable secondary metabolites, antibiotics	Prolonged incubation duration
	simplifies scalability	
Bacteria	Monodispersed nanoparticle with high purity	Metal tolerance in comparison to fungi is relatively lower
	Excellent choices for extracellular synthesis	Prolonged incubation period
		Demands extensive maintenance
Yeast	Generates a variety of nanomaterials	High maintenance
	Efficient nanoparticles production	Long incubation period
Synthesis facilitated by microwaves	Enables the production of a broad spectrum of nanoparticles	Direct use of nonpolar solvents is not feasible
	Accelerated reaction rate	Constraining the practical industrial utilization
Sonochemical synthesis	Reduced reaction time and fast synthesis of NPs	Restricted industrial applications
	No requirement of additional capping agents	
	Synthesize NPs with high purity	
	Production of different variety of nanomaterials	
Biopolymers	Significant bioactivity and biocompatibility of NPs	Challenging to extract biopolymers
	High stability of NPs	Difficulty in maintaining exact size and shape of NPs
	Low toxicity non-antigenicity of synthesized NPs	

(continued)

Table 19.2 (continued)

Synthesis ingredients	Pros	Cons
Water	Environmentally friendly solvent	Energy consuming solvent because of high heat capacity
	Innocuous and non-flammable in nature	
	Financially viable	
	Available in plenty	
Supercritical fluids	Environmentally favorable	Deterioration of equipment due to corrosion
	Cost effective	
	Innocuous and non flammable in nature	
	Energy efficient	
	Enables the alteration of nanomaterial (NM) properties by adjusting operational parameters	
Ionic solvents	Nonhazardous in nature	Noncoordinative
	Reduced reaction duration and energy consumption No extra reducing agents Gases, polar organics, and metal catalysts dissolve readily Higher thermal stability	Nonbiodegradable

mechanisms underlie nanoparticle toxicity, including the disruption of cell membranes through the generation of reactive oxygen species and hindrance of nutrient absorption, ultimately inhibiting microbial growth. Addressing bacterial inhibition could involve exploring the coating of nanoparticles with organic polymers to minimize their toxic effects, an area yet to be extensively explored in nanotechnology.

9 Concluding Thoughts and Future Directions

The realm of nanomaterials, encompassing their study, production, and utilization, is experiencing rapid expansion and remains apt for extensive exploration. This domain holds particular promise in terms of advancing eco-friendly and efficient synthesis methods, presenting a significant avenue for further research. This chapter delves into a comprehensive investigation of environmentally sustainable approaches to synthesizing nanomaterials, including biological routes, while scrutinizing the associated process variables, as well as the resulting stability and potential toxicity of these materials. Biologically derived nanomaterials claim an array of advantages, particularly in terms of biocompatibility and sensitivity, which positions them as valuable assets across various fields, encompassing consumer industries, biomedicine, and environmental remediation efforts.

Acknowledgments The authors are grateful to the Director, CSIR-NEERI, Nagpur, India, for his guidance. Authors acknowledge the efforts taken by Knowledge Resource Centre, CSIR-NEERI, Nagpur, India, for plagiarism check of the article using iThenticate software. SS, AB, and RS acknowledge the Academy of Scientific and Innovative Research (AcSIR), India, for facilitating doctoral research. AB and RS thankfully acknowledge the financial assistance granted for the JRF 201610084213 and 190520856226 by the University Grants Commission (UGC) India, respectively.

References

Agarwal, H., Venkat Kumar, S., & Rajeshkumar, S. (2017). A review on green synthesis of zinc oxide nanoparticles – An eco-friendly approach. *Resource-Efficient Technologies, 3*(4), 406–413. https://doi.org/10.1016/j.reffit.2017.03.002

Agarwal, P., Gupta, R., & Agarwal, N. (2019). Advances in synthesis and applications of microalgal nanoparticles for wastewater treatment. *Journal of Nanotechnology, 2019*, 1–9. https://doi.org/10.1155/2019/7392713

Agnihotri, M., Joshi, S., Kumar, A. R., Zinjarde, S., & Kulkarni, S. (2009). Biosynthesis of gold nanoparticles by the tropical marine yeast *Yarrowia lipolytica* NCIM 3589. *Materials Letters, 63*(15), 1231–1234. https://doi.org/10.1016/j.matlet.2009.02.042

Ahmad, A., Mukherjee, P., Senapati, S., Mandal, D., Khan, M. I., Kumar, R., & Sastry, M. (2003). Extracellular biosynthesis of silver nanoparticles using the fungus *Fusarium oxysporum*. *Colloids and Surfaces B: Biointerfaces, 28*(4), 313–318. https://doi.org/10.1016/S0927-7765(02)00174-1

Ahmed, S., Ahmad, M., Swami, B. L., & Ikram, S. (2016). A review on plants extract mediated synthesis of silver nanoparticles for antimicrobial applications: A green expertise. *Journal of Advanced Research, 7*(1), 17–28. https://doi.org/10.1016/j.jare.2015.02.007

Ahmed, S. F., Mofijur, M., Rafa, N., Chowdhury, A. T., Chowdhury, S., Nahrin, M., Islam, A. B. M. S., & Ong, H. C. (2022). Green approaches in synthesising nanomaterials for environmental nanobioremediation: Technological advancements, applications, benefits and challenges. *Environmental Research, 204*, 111967. https://doi.org/10.1016/j.envres.2021.111967

Ajith, M. P., Aswathi, M., Priyadarshini, E., & Rajamani, P. (2021). Recent innovations of nanotechnology in water treatment: A comprehensive review. *Bioresource Technology, 342*, 126000. https://doi.org/10.1016/j.biortech.2021.126000

Akhter, G., Khan, A., Ali, S. G., Khan, T. A., Siddiqi, K. S., & Khan, H. M. (2020). Antibacterial and nematicidal properties of biosynthesized Cu nanoparticles using extract of holoparasitic plant. *SN Applied Sciences, 2*(7), 1268. https://doi.org/10.1007/s42452-020-3068-6

Alani, F., Moo-Young, M., & Anderson, W. (2012). Biosynthesis of silver nanoparticles by a new strain of *Streptomyces* sp. Compared with *Aspergillus fumigatus*. *World Journal of Microbiology and Biotechnology, 28*(3), 1081–1086. https://doi.org/10.1007/s11274-011-0906-0

Albers, C. E., Hofstetter, W., Siebenrock, K. A., Landmann, R., & Klenke, F. M. (2013). *In vitro* cytotoxicity of silver nanoparticles on osteoblasts and osteoclasts at antibacterial concentrations. *Nanotoxicology, 7*(1), 30–36. https://doi.org/10.3109/17435390.2011.626538

Ali, K., Cherian, T., Fatima, S., Saquib, Q., Faisal, M., Alatar, A. A., Musarrat, J., & Al-Khedhairy, A. A. (2020). Role of solvent system in green synthesis of nanoparticles. In Q. Saquib, M. Faisal, A. A. Al-Khedhairy, & A. A. Alatar (Eds.), *Green synthesis of Nanoparticles:*

Applications and prospects (pp. 53–74). Springer Singapore). https://doi.org/10.1007/978-981-15-5179-6_3

Ali, T., Najam-ul-Haq, M., Mohyuddin, A., Musharraf, S. G., & Hussain, D. (2023). Next-generation functional nanotextiles—Prospects and challenges. *Sustainable Materials and Technologies, 36*, e00640. https://doi.org/10.1016/j.susmat.2023.e00640

Alivisatos, A. P., Johnsson, K. P., Peng, X., Wilson, T. E., Loweth, C. J., Bruchez, M. P., & Schultz, P. G. (1996). Organization of "nanocrystal molecules" using DNA. *Nature, 382*(6592), 609–611. https://doi.org/10.1038/382609a0

AlNadhari, S., Al-Enazi, N. M., Alshehrei, F., & Ameen, F. (2021). A review on biogenic synthesis of metal nanoparticles using marine algae and its applications. *Environmental Research, 194*, 110672. https://doi.org/10.1016/j.envres.2020.110672

Al-Radadi, N. S. (2019). Green synthesis of platinum nanoparticles using Saudi's Dates extract and their usage on the cancer cell treatment. *Arabian Journal of Chemistry, 12*(3), 330–349. https://doi.org/10.1016/j.arabjc.2018.05.008

Anyanwu, N. G., Stanley, H. O., Okpokwasili, G. C., & Akaranta, O. (2021). Biogenic nanoparticles and their environmental applications in bioremediation and pollution control. *International Journal of Innovative Science and Research Technology, 6*(10), 113–121.

Apte, M., Sambre, D., Gaikawad, S., Joshi, S., Bankar, A., Kumar, A. R., & Zinjarde, S. (2013). Psychrotrophic yeast *Yarrowia lipolytica* NCYC 789 mediates the synthesis of antimicrobial silver nanoparticles via cell-associated melanin. *AMB Express, 3*(1), 32. https://doi.org/10.1186/2191-0855-3-32

Babaei-Ghazvini, A., Acharya, B., & Korber, D. R. (2021). Antimicrobial biodegradable food packaging based on chitosan and metal/metal-oxide bio-nanocomposites: A review. *Polymers, 13*(16), 2790. https://doi.org/10.3390/polym13162790

Baig, N., Kammakakam, I., & Falath, W. (2021). Nanomaterials: A review of synthesis methods, properties, recent progress, and challenges. *Materials Advances, 2*(6), 1821–1871. https://doi.org/10.1039/D0MA00807A

Baker, S., & Satish, S. (2015). Biosynthesis of gold nanoparticles by *Pseudomonas veronii* AS41G inhabiting Annona squamosa L. *Spectrochimica Acta Part A: Molecular and Biomolecular Spectroscopy, 150*, 691–695. https://doi.org/10.1016/j.saa.2015.05.080

Balagurunathan, R., Radhakrishnan, M., Rajendran, R. B., & Velmurugan, D. (2011). Biosynthesis of gold nanoparticles by actinomycete *Streptomyces viridogens* strain HM10. I*ndian Journal of Biochemical and Biophysics 48(5),* 331-5

Bala, A., & Rani, G. (2020). A review on phytosynthesis, affecting factors and characterization techniques of silver nanoparticles designed by green approach. *International Nano Letters, 10*(3), 159–176. https://doi.org/10.1007/s40089-020-00309-7

Bandopadhyay, S., Manchanda, S., Chandra, A., Ali, J., & Deb, P. K. (2020). Overview of different carrier systems for advanced drug delivery. In *Drug delivery systems* (pp. 179–233). Elsevier. https://doi.org/10.1016/B978-0-12-814487-9.00005-3

Bhatt, P., Pandey, S. C., Joshi, S., Chaudhary, P., Pathak, V. M., Huang, Y., Wu, X., Zhou, Z., & Chen, S. (2022). Nanobioremediation: A sustainable approach for the removal of toxic pollutants from the environment. *Journal of Hazardous Materials, 427*, 128033. https://doi.org/10.1016/j.jhazmat.2021.128033

Braun, E., Eichen, Y., Sivan, U., & Ben-Yoseph, G. (1998). DNA-templated assembly and electrode attachment of a conducting silver wire. *Nature, 391*(6669), 775–778. https://doi.org/10.1038/35826

Brust, M., Fink, J., Bethell, D., Schiffrin, D. J., & Kiely, C. (1995). Synthesis and reactions of functionalised gold nanoparticles. *Journal of the Chemical Society, Chemical Communications, 16*, 1655. https://doi.org/10.1039/c39950001655

Carmen, Z., & Daniela, S. (2012). Textile organic dyes-characteristics, polluting effects and separation/elimination procedures from industrial effluents-A critical overview. In *Organic pollutants ten years after the stockholm convention—environmental and analytical update*. IntechOpen.

Chaudhary, P., & Sharma, A. (2018). Response of nanogypsum on the performance of plant growth promotory bacteria recovered from nanocompound infested agriculture field. *Environment and Ecology, 37*(1B), 363–372.
Chen, J., Peng, H., Wang, X., Shao, F., Yuan, Z., & Han, H. (2014). Graphene oxide exhibits broad-spectrum antimicrobial activity against bacterial phytopathogens and fungal conidia by intertwining and membrane perturbation. *Nanoscale, 6*(3), 1879–1889. https://doi.org/10.1039/C3NR04941H
Chhipa, H. (2019). Applications of nanotechnology in agriculture. *Methods in Microbiology, 46*, 115–142.
Clarance, P., Luvankar, B., Sales, J., Khusro, A., Agastian, P., Tack, J.-C., Al Khulaifi, M. M., Al-Shwaiman, H. A., Elgorban, A. M., Syed, A., & Kim, H.-J. (2020). Green synthesis and characterization of gold nanoparticles using endophytic fungi *Fusarium solani* and its in-vitro anticancer and biomedical applications. *Saudi Journal of Biological Sciences, 27*(2), 706–712. https://doi.org/10.1016/j.sjbs.2019.12.026
Collins, T. J. (2017). Review of the twenty-three year evolution of the first university course in green chemistry: Teaching future leaders how to create sustainable societies. *Journal of Cleaner Production, 140*, 93–110. https://doi.org/10.1016/j.jclepro.2015.06.136
Cosgrove, T. (2010). *Colloid science: Principles, methods and applications*. Wiley.
Dameron, C. T., Reese, R. N., Mehra, R. K., Kortan, A. R., Carroll, P. J., Steigerwald, M. L., Brus, L. E., & Winge, D. R. (1989). Biosynthesis of cadmium sulphide quantum semiconductor crystallites. *Nature, 338*(6216), 596–597. https://doi.org/10.1038/338596a0
Das, S., Sen, B., & Debnath, N. (2015). Recent trends in nanomaterials applications in environmental monitoring and remediation. *Environmental Science and Pollution Research, 22*(23), 18333–18344. https://doi.org/10.1007/s11356-015-5491-6
Das, P., Ghosh, S., Ghosh, R., Dam, S., & Baskey, M. (2018). Madhuca longifolia plant mediated green synthesis of cupric oxide nanoparticles: A promising environmentally sustainable material for waste water treatment and efficient antibacterial agent. *Journal of Photochemistry and Photobiology B: Biology, 189*, 66–73. https://doi.org/10.1016/j.jphotobiol.2018.09.023
Derjaguin, B. V., & Churaev, N. V. (1974). Structural component of disjoining pressure. *Journal of Colloid and Interface Science, 49*(2), 249–255. https://doi.org/10.1016/0021-9797(74)90358-0
Dikshit, P., Kumar, J., Das, A., Sadhu, S., Sharma, S., Singh, S., Gupta, P., & Kim, B. (2021). Green synthesis of metallic nanoparticles: Applications and limitations. *Catalysts, 11*(8), 902. https://doi.org/10.3390/catal11080902
Dimkpa, C. O., & Bindraban, P. S. (2018). Nanofertilizers: New products for the industry? *Journal of Agricultural and Food Chemistry, 66*(26), 6462–6473. https://doi.org/10.1021/acs.jafc.7b02150
Doan, V.-D., Thieu, A. T., Nguyen, T.-D., Nguyen, V.-C., Cao, X.-T., Nguyen, T. L.-H., & Le, V. T. (2020). Biosynthesis of gold nanoparticles using Litsea cubeba fruit extract for catalytic reduction of 4-nitrophenol. *Journal of Nanomaterials, 2020*, 1–10. https://doi.org/10.1155/2020/4548790
Duan, H., Wang, D., & Li, Y. (2015). Green chemistry for nanoparticle synthesis. *Chemical Society Reviews, 44*(16), 5778–5792. https://doi.org/10.1039/C4CS00363B
EL-Moslamy, S. H. (2018). Bioprocessing strategies for cost-effective large-scale biogenic synthesis of nano-MgO from endophytic *Streptomyces coelicolor* strain E72 as an anti-multidrug-resistant pathogens agent. *Scientific Reports, 8*(1), 3820. https://doi.org/10.1038/s41598-018-22134-x
El-Seedi, H. R., El-Shabasy, R. M., Khalifa, S. A. M., Saeed, A., Shah, A., Shah, R., Iftikhar, F. J., Abdel-Daim, M. M., Omri, A., Hajrahand, N. H., Sabir, J. S. M., Zou, X., Halabi, M. F., Sarhan, W., & Guo, W. (2019). Metal nanoparticles fabricated by green chemistry using natural extracts: Biosynthesis, mechanisms, and applications. *RSC Advances, 9*(42), 24539–24559. https://doi.org/10.1039/C9RA02225B
Eroglu, E., Chen, X., Bradshaw, M., Agarwal, V., Zou, J., Stewart, S. G., Duan, X., Lamb, R. N., Smith, S. M., Raston, C. L., & Iyer, K. S. (2013). Biogenic production of palladium nanocrystals using microalgae and their immobilization on chitosan nanofibers for catalytic applications. *RSC Advances, 3*(4), 1009–1012. https://doi.org/10.1039/C2RA22402J

Ezhilarasi, P. N., Karthik, P., Chhanwal, N., & Anandharamakrishnan, C. (2013). Nanoencapsulation techniques for food bioactive components: A review. *Food and Bioprocess Technology, 6*(3), 628–647. https://doi.org/10.1007/s11947-012-0944-0

Feizi, H., Rezvani Moghaddam, P., Shahtahmassebi, N., & Fotovat, A. (2012). Impact of bulk and Nanosized titanium dioxide (TiO_2) on wheat seed germination and seedling growth. *Biological Trace Element Research, 146*(1), 101–106. https://doi.org/10.1007/s12011-011-9222-7

Feng, L., Li, X., Xu, C., & Sadeghzadeh, S. M. (2020). Green synthesis of $Dy_2Ce_2O_7$ Nanoparticles immobilized on fibrous Nano-silica for synthesis of 3-Aryl-2-oxazolidinones from alkenes, amines, and carbon dioxide. *Catalysis Letters, 150*(6), 1729–1740. https://doi.org/10.1007/s10562-019-03073-2

Fouda, A., Hassan, S. E.-D., Abdo, A. M., & El-Gamal, M. S. (2020). Antimicrobial, antioxidant and Larvicidal activities of spherical silver Nanoparticles synthesized by Endophytic *Streptomyces* spp. *Biological Trace Element Research, 195*(2), 707–724. https://doi.org/10.1007/s12011-019-01883-4

Fouda, A., Hassan, S. E.-D., Abdel-Rahman, M. A., Farag, M. M. S., Shehal-deen, A., Mohamed, A. A., Alsharif, S. M., Saied, E., Moghanim, S. A., & Azab, M. S. (2021). Catalytic degradation of wastewater from the textile and tannery industries by green synthesized hematite (α-Fe_2O_3) and magnesium oxide (MgO) nanoparticles. *Current Research in Biotechnology, 3*, 29–41. https://doi.org/10.1016/j.crbiot.2021.01.004

Gericke, M., & Pinches, A. (2006). Microbial production of gold nanoparticles. *Gold Bulletin, 39*(1), 22–28. https://doi.org/10.1007/BF03215529

González-Ballesteros, N., Prado-López, S., Rodríguez-González, J. B., Lastra, M., & Rodríguez-Argüelles, M. C. (2017). Green synthesis of gold nanoparticles using brown algae *Cystoseira baccata*: Its activity in colon cancer cells. *Colloids and Surfaces B: Biointerfaces, 153*, 190–198. https://doi.org/10.1016/j.colsurfb.2017.02.020

Gunalan, S., Sivaraj, R., & Venckatesh, R. (2012). Aloe barbadensis Miller mediated green synthesis of mono-disperse copper oxide nanoparticles: Optical properties. *Spectrochimica Acta Part A: Molecular and Biomolecular Spectroscopy, 97*, 1140–1144. https://doi.org/10.1016/j.saa.2012.07.096

Hassan, S. E.-D., Fouda, A., Radwan, A. A., Salem, S. S., Barghoth, M. G., Awad, M. A., Abdo, A. M., & El-Gamal, M. S. (2019). Endophytic actinomycetes *Streptomyces* spp mediated biosynthesis of copper oxide nanoparticles as a promising tool for biotechnological applications. *JBIC Journal of Biological Inorganic Chemistry, 24*(3), 377–393. https://doi.org/10.1007/s00775-019-01654-5

He, H., Huang, L., Zhong, Z., & Tan, S. (2018). Constructing three-dimensional porous graphene-carbon quantum dots/g-C_3N_4 nanosheet aerogel metal-free photocatalyst with enhanced photocatalytic activity. *Applied Surface Science, 441*, 285–294. https://doi.org/10.1016/j.apsusc.2018.01.298

Henglein, A. (1989). Small-particle research: Physicochemical properties of extremely small colloidal metal and semiconductor particles. *Chemical Reviews, 89*(8), 1861–1873. https://doi.org/10.1021/cr00098a010

Hoseinnejad, M., Jafari, S. M., & Katouzian, I. (2018). Inorganic and metal nanoparticles and their antimicrobial activity in food packaging applications. *Critical Reviews in Microbiology, 44*(2), 161–181. https://doi.org/10.1080/1040841X.2017.1332001

Huang, T., Li, X., Maier, M., O'Brien-Simpson, N. M., Heath, D. E., & O'Connor, A. J. (2023). Using inorganic nanoparticles to fight fungal infections in the antimicrobial resistant era. *Acta Biomaterialia, 158*, 56–79. https://doi.org/10.1016/j.actbio.2023.01.019

Ibrahim, H. M. M. (2015). Green synthesis and characterization of silver nanoparticles using banana peel extract and their antimicrobial activity against representative microorganisms. *Journal of Radiation Research and Applied Sciences, 8*(3), 265–275. https://doi.org/10.1016/j.jrras.2015.01.007

Ijaz, I., Gilani, E., Nazir, A., & Bukhari, A. (2020). Detail review on chemical, physical and green synthesis, classification, characterizations and applications of nanoparticles. *Green Chemistry Letters and Reviews, 13*(3), 223–245. https://doi.org/10.1080/17518253.2020.1802517

Jafarizadeh-Malmiri, H., Sayyar, Z., Anarjan, N., & Berenjian, A. (2019). Nano-additives for food industries. In I. H. Jafarizadeh-Malmiri, Z. Sayyar, N. Anarjan, & A. Berenjian (Eds.), *Nanobiotechnology in food: Concepts, applications and perspectives* (pp. 41–68). Springer International Publishing. https://doi.org/10.1007/978-3-030-05846-3_4

Jameel, M. S., Aziz, A. A., & Dheyab, M. A. (2020). Green synthesis: Proposed mechanism and factors influencing the synthesis of platinum nanoparticles. *Green Processing and Synthesis, 9*(1), 386–398. https://doi.org/10.1515/gps-2020-0041

Jiang, J., Oberdörster, G., & Biswas, P. (2009). Characterization of size, surface charge, and agglomeration state of nanoparticle dispersions for toxicological studies. *Journal of Nanoparticle Research, 11*(1), 77–89. https://doi.org/10.1007/s11051-008-9446-4

John, M. S., Nagoth, J. A., Ramasamy, K. P., Mancini, A., Giuli, G., Natalello, A., Ballarini, P., Miceli, C., & Pucciarelli, S. (2020). Synthesis of bioactive silver nanoparticles by a *Pseudomonas* strain associated with the Antarctic psychrophilic protozoon Euplotes focardii. *Marine Drugs, 18*(1), 38. https://doi.org/10.3390/md18010038

Kalpana, V. N., & Devi Rajeswari, V. (2018). A review on green synthesis, biomedical applications, and toxicity studies of ZnO NPs. *Bioinorganic Chemistry and Applications, 2018*, 1–12. https://doi.org/10.1155/2018/3569758

Kalteh, M., Alipour, Z. T., Ashraf, S., Marashi Aliabadi, M., Falah, A., & Nosratabadi. (2018). Effect of silica nanoparticles on basil (*Ocimum basilicum*) under salinity stress. *Journal of Chemical Health Risks., 4*(3), 49–55.

Kapinder, Dangi, K., & Verma, A. K. (2021). Efficient & eco-friendly smart nano-pesticides: Emerging prospects for agriculture. *Materials Today: Proceedings, 45*, 3819–3824. https://doi.org/10.1016/j.matpr.2021.03.211

Karthik, L., Kumar, G., Keswani, T., Bhattacharyya, A., Reddy, B. P., & Rao, K. V. B. (2013). Marine actinobacterial mediated gold nanoparticles synthesis and their antimalarial activity. *Nanomedicine: Nanotechnology, Biology and Medicine, 9*(7), 951–960. https://doi.org/10.1016/j.nano.2013.02.002

Karthikeyan, P., Mohan, D., Abishek, G., & Priya, R. (2015). Synthesis of silver nanoparticles using Phytoplankton and its characteristics. *International Journal of Fisheries and Aquatic Studies, 2*(6), 398–401. https://www.researchgate.net/publication/281643330

Karim, N. A., Rubinsin, N. J., Burukan, M. A. A., & Kamarudin, S. K. (2019). Sustainable route of synthesis platinum nanoparticles using orange peel extract. *International Journal of Green Energy, 16*(15), 1518–1526. https://doi.org/10.1080/15435075.2019.1671422

Khalafi, T., Buazar, F., & Ghanemi, K. (2019). Phycosynthesis and enhanced photocatalytic activity of zinc oxide Nanoparticles toward organosulfur pollutants. *Scientific Reports, 9*(1), 6866. https://doi.org/10.1038/s41598-019-43368-3

Kharey, P., Dutta, S. B., Manikandan, M., Palani, I. A., Majumder, S. K., & Gupta, S. (2020). Green synthesis of near-infrared absorbing eugenate capped iron oxide nanoparticles for photothermal application. *Nanotechnology, 31*(9), 095705. https://doi.org/10.1088/1361-6528/ab56b6

Kharissova, O. V., Kharisov, B. I., Oliva González, C. M., Méndez, Y. P., & López, I. (2019). Greener synthesis of chemical compounds and materials. *Royal Society Open Science, 6*(11), 191378. https://doi.org/10.1098/rsos.191378

Khati, P., Parul, Bhatt, P., Nisha, Kumar, R., & Sharma, A. (2018). Effect of nanozeolite and plant growth promoting rhizobacteria on maize. *3 Biotech, 8*(3), 141. https://doi.org/10.1007/s13205-018-1142-1

Kim, Y., Lee, J. H., Ha, H., Im, S. W., & Nam, K. T. (2016a). Material science lesson from the biological photosystem. *Nano Convergence, 3*(1), 19. https://doi.org/10.1186/s40580-016-0079-5

Kim, M., Lee, B. Y., Ham, H. C., Han, J., Nam, S. W., Lee, H., Park, J. H., Choi, S., & Shin, Y. (2016b). Facile one-pot synthesis of tungsten oxide (WO_3-x) nanoparticles using sub and supercritical fluids. *The Journal of Supercritical Fluids, 111*, 8–13. https://doi.org/10.1016/j.supflu.2016.01.011

Kumar, C. R., Betageri, V. S., Nagaraju, G., Pujar, G. H., Onkarappa, H. S., & Latha, M. S. (2020). Synthesis of core/shell (ZnO/Ag) nanoparticles using *Calotropis gigantea* and their applications in photocatalytic and antibacterial studies. *Journal of Inorganic and Organometallic Polymers and Materials, 30*(9), 3410–3417. https://doi.org/10.1007/s10904-020-01507-8

Kumari, S., Tehri, N., Gahlaut, A., & Hooda, V. (2021). Actinomycetes mediated synthesis, characterization, and applications of metallic nanoparticles. *Inorganic and Nano-Metal Chemistry, 51*(10), 1386–1395. https://doi.org/10.1080/24701556.2020.1835978

Kunoh, T., Takeda, M., Matsumoto, S., Suzuki, I., Takano, M., Kunoh, H., & Takada, J. (2018). Green synthesis of gold nanoparticles coupled with nucleic acid oxidation. *ACS Sustainable Chemistry & Engineering, 6*(1), 364–373. https://doi.org/10.1021/acssuschemeng.7b02610

Llorens, A., Lloret, E., Picouet, P. A., Trbojevich, R., & Fernandez, A. (2012). Metallic-based micro and nanocomposites in food contact materials and active food packaging. *Trends in Food Science & Technology, 24*(1), 19–29. https://doi.org/10.1016/j.tifs.2011.10.001

Lovley, D. R., Stolz, J. F., Nord, G. L., & Phillips, E. J. P. (1987). Anaerobic production of magnetite by a dissimilatory iron-reducing microorganism. *Nature, 330*(6145), 252–254. https://doi.org/10.1038/330252a0

Lu, S., Zhang, X., & Xue, Y. (2017). Application of calcium peroxide in water and soil treatment: A review. *Journal of Hazardous Materials, 337*, 163–177. https://doi.org/10.1016/j.jhazmat.2017.04.064

Madhura, L., Singh, S., Kanchi, S., Sabela, M., Bisetty, K., & Inamuddin. (2019). Nanotechnology-based water quality management for wastewater treatment. *Environmental Chemistry Letters, 17*(1), 65–121. https://doi.org/10.1007/s10311-018-0778-8

Maiorano, G., Sabella, S., Sorce, B., Brunetti, V., Malvindi, M. A., Cingolani, R., & Pompa, P. P. (2010). Effects of cell culture media on the dynamic formation of protein−nanoparticle complexes and influence on the cellular response. *ACS Nano, 4*(12), 7481–7491. https://doi.org/10.1021/nn101557e

Makarov, V. V., Love, A. J., Sinitsyna, O. V., Makarova, S. S., Yaminsky, I. V., Taliansky, M. E., & Kalinina, N. O. (2014). "Green" nanotechnologies: Synthesis of metal nanoparticles using plants. *Acta Naturae, 6*, 35–45.

Margolis, D. J. A., Hoffman, J. M., Herfkens, R. J., Jeffrey, R. B., Quon, A., & Gambhir, S. S. (2007). Molecular imaging techniques in body imaging. *Radiology, 245*(2), 333–356. https://doi.org/10.1148/radiol.2452061117

Marslin, G., Siram, K., Maqbool, Q., Selvakesavan, R., Kruszka, D., Kachlicki, P., & Franklin, G. (2018). Secondary metabolites in the green synthesis of metallic nanoparticles. *Materials, 11*(6), 940. https://doi.org/10.3390/ma11060940

Mba, I. E., & Nweze, E. I. (2021). Nanoparticles as therapeutic options for treating multidrug-resistant bacteria: Research progress, challenges, and prospects. *World Journal of Microbiology and Biotechnology, 37*(6), 108. https://doi.org/10.1007/s11274-021-03070-x

McGee, C. F., Storey, S., Clipson, N., & Doyle, E. (2017). Soil microbial community responses to contamination with silver, aluminium oxide and silicon dioxide nanoparticles. *Ecotoxicology, 26*(3), 449–458. https://doi.org/10.1007/s10646-017-1776-5

Mirkin, C. A., Letsinger, R. L., Mucic, R. C., & Storhoff, J. J. (2020). *A DNA-based method for rationally assembling nanoparticles into macroscopic materials*. Jenny Stanford Publishing.

Mishra, A., Kumari, M., Pandey, S., Chaudhry, V., Gupta, K. C., & Nautiyal, C. S. (2014). Biocatalytic and antimicrobial activities of gold nanoparticles synthesized by *Trichoderma* sp. *Bioresource Technology, 166*, 235–242. https://doi.org/10.1016/j.biortech.2014.04.085

Møller, P., Jacobsen, N. R., Folkmann, J. K., Danielsen, P. H., Mikkelsen, L., Hemmingsen, J. G., Vesterdal, L. K., Forchhammer, L., Wallin, H., & Loft, S. (2010). Role of oxidative damage in toxicity of particulates. *Free Radical Research, 44*(1), 1–46. https://doi.org/10.3109/10715760903300691

Molnár, Z., Bódai, V., Szakacs, G., Erdélyi, B., Fogarassy, Z., Sáfrán, G., Varga, T., Kónya, Z., Tóth-Szeles, E., Szűcs, R., & Lagzi, I. (2018). Green synthesis of gold nanoparticles by thermophilic filamentous fungi. *Scientific Reports, 8*(1), 3943. https://doi.org/10.1038/s41598-018-22112-3

Monowar, T., Rahman, M. S., Bhore, S. J., & Sathasivam, K. V. (2021). Endophytic bacteria Enterobacter hormaechei fabricated silver nanoparticles and their antimicrobial activity. *Pharmaceutics, 13*(4), 511. https://doi.org/10.3390/pharmaceutics13040511

Moreno, V., Llorent-Martínez, E. J., Zougagh, M., & Ríos, A. (2018). Synthesis of hybrid magnetic carbon nanotubes – C18-modified nano SiO_2 under supercritical carbon dioxide media and their analytical potential for solid-phase extraction of pesticides. *The Journal of Supercritical Fluids, 137*, 66–73. https://doi.org/10.1016/j.supflu.2018.03.007

Mowafi, S., Rehan, M., Mashaly, H. M., Abou El-Kheir, A., & Emam, H. E. (2017). Influence of silver nanoparticles on the fabrics functions prepared by *in-situ* technique. *The Journal of The Textile Institute, 108*(10), 1828–1839. https://doi.org/10.1080/00405000.2017.1292649

Mudunkotuwa, I. A., & Grassian, V. H. (2015). Biological and environmental media control oxide nanoparticle surface composition: The roles of biological components (proteins and amino acids), inorganic oxyanions and humic acid. *Environmental Science: Nano, 2*(5), 429–439. https://doi.org/10.1039/C4EN00215F

Mukherjee, P., Ahmad, A., Mandal, D., Senapati, S., Sainkar, S. R., Khan, M. I., Parishcha, R., Ajaykumar, P. V., Alam, M., Kumar, R., & Sastry, M. (2001). Fungus-mediated synthesis of silver Nanoparticles and their immobilization in the mycelial matrix: A novel biological approach to nanoparticle synthesis. *Nano Letters, 1*(10), 515–519. https://doi.org/10.1021/nl0155274

Mukherjee, A., Sun, Y., Morelius, E., Tamez, C., Bandyopadhyay, S., Niu, G., White, J. C., Peralta-Videa, J. R., & Gardea-Torresdey, J. L. (2016). Differential toxicity of bare and hybrid ZnO Nanoparticles in green pea (*Pisum sativum* L.): A life cycle study. *Frontiers in Plant Science, 6*. https://doi.org/10.3389/fpls.2015.01242

Musarrat, J., Dwivedi, S., Singh, B. R., Al-Khedhairy, A. A., Azam, A., & Naqvi, A. (2010). Production of antimicrobial silver nanoparticles in water extracts of the fungus *Amylomyces rouxii* strain KSU-09. *Bioresource Technology, 101*(22), 8772–8776. https://doi.org/10.1016/j.biortech.2010.06.065

Muthulakshmi, V., & Sundrarajan, M. (2020). Green synthesis of ionic liquid assisted ytterbium oxide nanoparticles by *Couroupita guianensis* abul leaves extract for biological applications. *Journal of Environmental Chemical Engineering, 8*(4), 103992. https://doi.org/10.1016/j.jece.2020.103992

Nagababu, P., & Umamaheswara, R. V. (2016). Cost-effective green synthesis and characterization of silver nanoparticles from *Avicennia alba* blume leaves and their antibacterial activity. *Asian Journal of Pharmaceutical and Clinical Research*. 9, 301–303.

Nabi, G., Raza, W., & Tahir, M. B. (2020). Green synthesis of TiO_2 nanoparticle using cinnamon powder extract and the study of optical properties. *Journal of Inorganic and Organometallic Polymers and Materials, 30*(4), 1425–1429. https://doi.org/10.1007/s10904-019-01248-3

Nel, A., Xia, T., Mädler, L., & Li, N. (2006). Toxic potential of materials at the Nanolevel. *Science, 311*(5761), 622–627. https://doi.org/10.1126/science.1114397

Nongbet, A., Mishra, A. K., Mohanta, Y. K., Mahanta, S., Ray, M. K., Khan, M., Baek, K.-H., & Chakrabartty, I. (2022). Nanofertilizers: A smart and sustainable attribute to modern agriculture. *Plants, 11*(19), 2587. https://doi.org/10.3390/plants11192587

Oliveira, A. C. D. J., Araújo, A. R. D., Quelemes, P. V., Nadvorny, D., Soares-Sobrinho, J. L., Leite, J. R. S. D. A., Da Silva-Filho, E. C., & Silva, D. A. D. (2019). Solvent-free production of phthalated cashew gum for green synthesis of antimicrobial silver nanoparticles. *Carbohydrate Polymers, 213*, 176–183. https://doi.org/10.1016/j.carbpol.2019.02.033

Otari, S. V., Patil, R. M., Nadaf, N. H., Ghosh, S. J., & Pawar, S. H. (2012). Green biosynthesis of silver nanoparticles from an actinobacteria *Rhodococcus* sp. *Materials Letters, 72*, 92–94. https://doi.org/10.1016/j.matlet.2011.12.109

Pal, G., Rai, P., & Pandey, A. (2019). Green synthesis of nanoparticles: A greener approach for a cleaner future. In *Green synthesis, characterization and applications of nanoparticles* (pp. 1–26). Elsevier. https://doi.org/10.1016/B978-0-08-102579-6.00001-0

Pandiyan, N., Murugesan, B., Arumugam, M., Chinnaalagu, D., Samayanan, S., & Mahalingam, S. (2021). Ionic liquid mediated green synthesis of Ag-Au/Y_2O_3 nanoparticles using leaves extracts of *Justicia adhatoda:* Structural characterization and its biological applications. *Advanced Powder Technology, 32*(7), 2213–2225. https://doi.org/10.1016/j.apt.2021.04.030

Patra, J. K., & Baek, K.-H. (2014). Green nanobiotechnology: Factors affecting synthesis and characterization techniques. *Journal of Nanomaterials, 2014*, 1–12. https://doi.org/10.1155/2014/417305

Pelaz, B., Alexiou, C., Alvarez-Puebla, R. A., Alves, F., Andrews, A. M., Ashraf, S., Balogh, L. P., Ballerini, L., Bestetti, A., Brendel, C., Bosi, S., Carril, M., Chan, W. C. W., Chen, C., Chen, X., Chen, X., Cheng, Z., Cui, D., Du, J., . . ., Parak, W. J. (2017). Diverse applications of nanomedicine. *ACS Nano, 11*(3), 2313–2381. https://doi.org/10.1021/acsnano.6b06040

Perera, S., Wijesekara, D., Thiripuranathar, G., & Menaa, F. (2022). The use of nanoparticles to enhance performance in the textile industry – A concise review. *Current Nanoscience, 18*(3), 319–335. https://doi.org/10.2174/1573413717666210715121307

Portela, C. M., Vidyasagar, A., Krödel, S., Weissenbach, T., Yee, D. W., Greer, J. R., & Kochmann, D. M. (2020). Extreme mechanical resilience of self-assembled nanolabyrinthine materials. *Proceedings of the National Academy of Sciences, 117*(11), 5686–5693. https://doi.org/10.1073/pnas.1916817117

Prasad, R., Bhattacharyya, A., & Nguyen, Q. D. (2017). Nanotechnology in sustainable agriculture: Recent developments, challenges, and perspectives. *Frontiers in Microbiology, 8*, 1014. https://doi.org/10.3389/fmicb.2017.01014

Prathna, T. C., Chandrasekaran, N., Raichur, A. M., & Mukherjee, A. (2011). Kinetic evolution studies of silver nanoparticles in a bio-based green synthesis process. *Colloids and Surfaces A: Physicochemical and Engineering Aspects, 377*(1–3), 212–216. https://doi.org/10.1016/j.colsurfa.2010.12.047

Priyadharshini, R. I., Prasannaraj, G., Geetha, N., & Venkatachalam, P. (2014). Microwave-mediated extracellular synthesis of metallic silver and zinc oxide nanoparticles using macro-algae (*Gracilaria edulis*) extracts and its anticancer activity against human PC3 cell lines. *Applied Biochemistry and Biotechnology, 174*(8), 2777–2790. https://doi.org/10.1007/s12010-014-1225-3

Priyadarshini, S. S., Sethi, S., Rout, S., Mishra, P. M., & Pradhan, N. (2023). Green synthesis of microalgal biomass-silver nanoparticle composite showing antimicrobial activity and heterogenous catalysis of nitrophenol reduction. *Biomass Conversion and Biorefinery, 13*(9), 7783–7795. https://doi.org/10.1007/s13399-021-01825-y

Rafique, M., Sadaf, I., Rafique, M. S., & Tahir, M. B. (2017a). A review on green synthesis of silver nanoparticles and their applications. *Artificial Cells, Nanomedicine, and Biotechnology, 45*(7), 1272–1291. https://doi.org/10.1080/21691401.2016.1241792

Rafique, M., Shaikh, A. J., Rasheed, R., Tahir, M. B., Bakhat, H. F., Rafique, M. S., & Rabbani, F. (2017b). A review on synthesis, characterization and applications of copper nanoparticles using green method. *Nano, 12*(04), 1750043. https://doi.org/10.1142/S1793292017500436

Raghunath, A., & Perumal, E. (2017). Metal oxide nanoparticles as antimicrobial agents: A promise for the future. *International Journal of Antimicrobial Agents, 49*(2), 137–152. https://doi.org/10.1016/j.ijantimicag.2016.11.011

Rajabairavi, N., Raju, C. S., Karthikeyan, C., Varutharaju, K., Nethaji, S., Hameed, A. S. H., & Shajahan, A. (2017). Biosynthesis of novel zinc oxide nanoparticles (ZnO NPs) using endophytic bacteria *Sphingobacterium thalpophilum.* In J. Ebenezar (Ed.), Recent Trends in Materials Science and Applications (Vol. 189, pp. 245–254).

Rana, A., Yadav, K., & Jagadevan, S. (2020). A comprehensive review on green synthesis of nature-inspired metal nanoparticles: Mechanism, application and toxicity. *Journal of Cleaner Production, 272*, 122880. https://doi.org/10.1016/j.jclepro.2020.122880

Raheman, F., Deshmukh, S., Ingle, A., Gade, A., & Rai, M. (2011). Silver nanoparticles: Novel antimicrobial agent synthesized from a endophytic fungus *Pestalotia* sp. isolated from leaves of *Syzygium cumini* (L). *Nano Biomedicine and Engineering, 3*(3), 174–178. https://doi.org/10.5101/nbe.v3i3.p174-178

Ramakrishna, M., Rajesh Babu, D., Gengan, R. M., Chandra, S., & Nageswara Rao, G. (2016). Green synthesis of gold nanoparticles using marine algae and evaluation of their catalytic activity. *Journal of Nanostructure in Chemistry, 6*(1), 1–13. https://doi.org/10.1007/s40097-015-0173-y

Raman, J., Reddy, G. R., Lakshmanan, H., Selvaraj, V., Gajendran, B., Nanjian, R., Chinnasamy, A., & Sabaratnam, V. (2015). Mycosynthesis and characterization of silver nanoparticles from *Pleurotus djamor* var. Roseus and their in vitro cytotoxicity effect on PC3 cells. *Process Biochemistry, 50*(1), 140–147. https://doi.org/10.1016/j.procbio.2014.11.003

Rehmanullah, M. Z., Inayat, N., & Majeed, A. (2020). Application of nanoparticles in agriculture as fertilizers and pesticides: Challenges and opportunities. In A. Rakshit, H. B. Singh, A. K. Singh, U. S. Singh, & L. Fraceto (Eds.), *New frontiers in stress management for durable agriculture* (pp. 281–293). Springer Singapore. https://doi.org/10.1007/978-981-15-1322-0_17

Romero, C. M., Alvarez, A., Martínez, M. A., & Chaves, S. (2018). Fungal nanotechnology: A new approach toward efficient biotechnology application. In R. Prasad, V. Kumar, M. Kumar, & S. Wang (Eds.), *Fungal nanobionics: Principles and applications* (pp. 117–143). Springer Singapore. https://doi.org/10.1007/978-981-10-8666-3_5

Rónavári, A., Igaz, N., Gopisetty, M. K., Szerencsés, B., Kovács, D., Papp, C., Vágvölgyi, C., Boros, I., Kónya, Z., Kiricsi, M., & Pfeiffer, I. (2018). Biosynthesized silver and gold nanoparticles are potent antimycotics against opportunistic pathogenic yeasts and dermatophytes. *International Journal of Nanomedicine, 13*, 695–703. https://doi.org/10.2147/IJN.S152010

Rostami-Vartooni, A., Alizadeh, M., & Bagherzadeh, M. (2015). Green synthesis, characterization and catalytic activity of natural bentonite-supported copper nanoparticles for the solvent-free synthesis of 1-substituted 1 *H* -1,2,3,4-tetrazoles and reduction of 4-nitrophenol. *Beilstein Journal of Nanotechnology, 6*, 2300–2309. https://doi.org/10.3762/bjnano.6.236

Sadhasivam, S., Shanmugam, P., Veerapandian, M., Subbiah, R., & Yun, K. (2012). Biogenic synthesis of multidimensional gold nanoparticles assisted by Streptomyces hygroscopicus and its electrochemical and antibacterial properties. *BioMetals, 25*(2), 351–360. https://doi.org/10.1007/s10534-011-9506-6

Sadri, R., Hosseini, M., Kazi, S. N., Bagheri, S., Abdelrazek, A. H., Ahmadi, G., Zubir, N., Ahmad, R., & Abidin, N. I. Z. (2018). A facile, bio-based, novel approach for synthesis of covalently functionalized graphene nanoplatelet nano-coolants toward improved thermo-physical and heat transfer properties. *Journal of Colloid and Interface Science, 509*, 140–152. https://doi.org/10.1016/j.jcis.2017.07.052

Salvadori, M. R., Ando, R. A., Oller Do Nascimento, C. A., & Corrêa, B. (2014). Intracellular biosynthesis and removal of copper nanoparticles by dead biomass of yeast isolated from the wastewater of a mine in the Brazilian Amazonia. *PLoS ONE, 9*(1), e87968. https://doi.org/10.1371/journal.pone.0087968

Salem, S. S., & Fouda, A. (2021). Green synthesis of metallic nanoparticles and their prospective biotechnological applications: An overview. *Biological Trace Element Research, 199*(1), 344–370. https://doi.org/10.1007/s12011-020-02138-3

Sangeetha, M., Thejaswini, G., Shoba, A., Santoshi Gaikwad, S., Amretasre, R. T., & Nivedita, S. (2021). Design and development of a crop quality monitoring and classification system using IoT and blockchain. *Journal of Physics: Conference Series, 1964*(6), 062011. https://doi.org/10.1088/1742-6596/1964/6/062011

Sarno, M. (2020). Nanotechnology in energy storage: The supercapacitors. In *Studies in surface science and catalysis* (Vol. 179, pp. 431–458). Elsevier. https://doi.org/10.1016/B978-0-444-64337-7.00022-7

Sayed, M. T. E., & El-Sayed, A. S. (2020). Biocidal activity of metal nanoparticles synthesized by *Fusarium solani* against multidrug-resistant bacteria and mycotoxigenic fungi. *Journal of Microbiology and Biotechnology, 30*(2), 226–236. https://doi.org/10.4014/jmb.1906.06070

Scarano, G., & Morelli, E. (2003). Properties of phytochelatin-coated CdS nanocrystallites formed in a marine phytoplanktonic alga (*Phaeodactylum tricornutum*, Bohlin) in response to Cd. *Plant Science, 165*(4), 803–810. https://doi.org/10.1016/S0168-9452(03)00274-7

Schröfel, A., Kratošová, G., Šafařík, I., Šafaříková, M., Raška, I., & Shor, L. M. (2014). Applications of biosynthesized metallic nanoparticles—A review. *Acta Biomaterialia, 10*(10), 4023–4042. https://doi.org/10.1016/j.actbio.2014.05.022

Sekhon, B. S., & Kamboj, S. R. (2010). Inorganic nanomedicine—Part 1. *Nanomedicine: Nanotechnology, Biology and Medicine, 6*(4), 516–522. https://doi.org/10.1016/j.nano.2010.04.004

Selvi, A. M., Palanisamy, S., Jeyanthi, S., Vinosha, M., Mohandoss, S., Tabarsa, M., You, S., Kannapiran, E., & Prabhu, N. M. (2020). Synthesis of Tragia involucrata mediated platinum nanoparticles for comprehensive therapeutic applications: Antioxidant antibacterial and mitochondria-associated apoptosis in HeLa cells Process. *Biochemistry, 9821–9833*. https://doi.org/10.1016/j.procbio.2020.07.008

Shah, M., Fawcett, D., Sharma, S., Tripathy, S., & Poinern, G. (2015). Green synthesis of metallic Nanoparticles via biological entities. *Materials, 8*(11), 7278–7308. https://doi.org/10.3390/ma8115377

Shah, S. T., Basit, A., Ullah, I., & Mohamed, H. I. (2021). Cyanobacteria and algae as biocontrol agents against fungal and bacterial plant pathogens. In H. I. Mohamed, H. E.-D. S. El-Beltagi, & K. A. Abd-Elsalam (Eds.), *Plant growth-promoting microbes for sustainable biotic and abiotic stress management* (pp. 1–23). Springer International Publishing. https://doi.org/10.1007/978-3-030-66587-6_1

Sharma, R., Ragavan, K. V., Thakur, M. S., & Raghavarao, K. S. M. S. (2015). Recent advances in nanoparticle based aptasensors for food contaminants. *Biosensors and Bioelectronics, 74*, 612–627. https://doi.org/10.1016/j.bios.2015.07.017

Sharon, M., Choudhary, A. K., & Kumar, R. (2010). Nanotechnology in agricultural diseases and food safety. *Journal of Phytology, 2*(4), 83–92.

Shanmugasundaram, T., Radhakrishnan, M., Gopikrishnan, V., Pazhanimurugan, R., & Balagurunathan, R. (2013). A study of the bactericidal, anti-biofouling, cytotoxic and antioxidant properties of actinobacterially synthesised silver nanoparticles. *Colloids and Surfaces B: Biointerfaces, 111*, 680–687. https://doi.org/10.1016/j.colsurfb.2013.06.045

Sharma, B., Purkayastha, D. D., Hazra, S., Thajamanbi, M., Bhattacharjee, C. R., Ghosh, N. N., & Rout, J. (2014). Biosynthesis of fluorescent gold nanoparticles using an edible freshwater red alga, *Lemanea fluviatilis* (L.) C.Ag. and antioxidant activity of biomatrix loaded nanoparticles. *Bioprocess and Biosystems Engineering, 37*(12), 2559–2565. https://doi.org/10.1007/s00449-014-1233-2

Shenton, W., Douglas, T., Young, M., Stubbs, G., & Mann, S. (1999). Inorganic-organic nanotube composites from template mineralization of tobacco mosaic virus. *Advanced Materials, 11*(3), 253–256. https://doi.org/10.1002/(SICI)1521-4095(199903)11:3<253::AID-ADMA253>3.0.CO;2-7

Shnoudeh, A. J., Hamad, I., Abdo, R. W., Qadumii, L., Jaber, A. Y., Surchi, H. S., & Alkelany, S. Z. (2019). Synthesis, characterization, and applications of metal Nanoparticles. In *Biomaterials and bionanotechnology* (pp. 527–612). Elsevier. https://doi.org/10.1016/B978-0-12-814427-5.00015-9

Sillen, W. M. A., Thijs, S., Abbamondi, G. R., Janssen, J., Weyens, N., White, J. C., & Vangronsveld, J. (2015). Effects of silver nanoparticles on soil microorganisms and maize biomass are linked in the rhizosphere. *Soil Biology and Biochemistry, 91*, 14–22. https://doi.org/10.1016/j.soilbio.2015.08.019

Silvestre, C., Duraccio, D., & Cimmino, S. (2011). Food packaging based on polymer nanomaterials. *Progress in Polymer Science, 36*(12), 1766–1782. https://doi.org/10.1016/j.progpolymsci.2011.02.003

Singh, M. J. (2016). Green nano actinobacteriology—An interdisciplinary study. In D. Dhanasekaran & Y. Jiang (Eds.), *Actinobacteria—basics and biotechnological applications*. InTech. https://doi.org/10.5772/61308

Singh, S., Gupta, D., Anand, R. S., & Kumar, V. (2015). Nonsubsampled shearlet based CT and MR medical image fusion using biologically inspired spiking neural network. *Biomedical Signal Processing and Control, 18*, 91–101. https://doi.org/10.1016/j.bspc.2014.11.009

Singh, P., Kim, Y.-J., Zhang, D., & Yang, D.-C. (2016). Biological synthesis of Nanoparticles from plants and microorganisms. *Trends in Biotechnology, 34*(7), 588–599. https://doi.org/10.1016/j.tibtech.2016.02.006

Singh, J., Dutta, T., Kim, K.-H., Rawat, M., Samddar, P., & Kumar, P. (2018). 'Green' synthesis of metals and their oxide nanoparticles: Applications for environmental remediation. *Journal of Nanobiotechnology, 16*(1), 84. https://doi.org/10.1186/s12951-018-0408-4

Singh, A., Gautam, P. K., Verma, A., Singh, V., Shivapriya, P. M., Shivalkar, S., Sahoo, A. K., & Samanta, S. K. (2020). Green synthesis of metallic nanoparticles as effective alternatives to treat antibiotics resistant bacterial infections: A review. *Biotechnology Reports, 25*, e00427. https://doi.org/10.1016/j.btre.2020.e00427

Siril, P. F., & Türk, M. (2020). Synthesis of metal nanostructures using supercritical carbon dioxide: A green and Upscalable process. *Small, 16*(49), 2001972. https://doi.org/10.1002/smll.202001972

Sivaraj, R., Salam, H. A., Rajiv, P., & Rajendran, V. (2015). Green nanotechnology: The solution to sustainable development of environment. In P. Thangavel & G. Sridevi (Eds.), *Environmental sustainability* (pp. 311–324). Springer India. https://doi.org/10.1007/978-81-322-2056-5_18

Składanowski, M., Wypij, M., Laskowski, D., Golińska, P., Dahm, H., & Rai, M. (2017). Silver and gold nanoparticles synthesized from *Streptomyces* sp. Isolated from acid forest soil with special reference to its antibacterial activity against pathogens. *Journal of Cluster Science, 28*(1), 59–79. https://doi.org/10.1007/s10876-016-1043-6

Soliman, H., Elsayed, A., & Dyaa, A. (2018). Antimicrobial activity of silver nanoparticles biosynthesised by *Rhodotorula* sp. strain ATL72. *Egyptian Journal of Basic and Applied Sciences, 5*(3), 228–233. https://doi.org/10.1016/j.ejbas.2018.05.005

Stampoulis, D., Sinha, S. K., & White, J. C. (2009). Assay-dependent phytotoxicity of nanoparticles to plants. *Environmental Science & Technology, 43*(24), 9473–9479. https://doi.org/10.1021/es901695c

Sudarsan, S., Kumar Shankar, M., Kumar Belagal Motatis, A., Shankar, S., Krishnappa, D., Mohan, C. D., Rangappa, K. S., Gupta, V. K., & Siddaiah, C. N. (2021). Green synthesis of silver nanoparticles by *Cytobacillus firmus* isolated from the stem bark of Terminalia arjuna and their antimicrobial activity. *Biomolecules, 11*(2), 259. https://doi.org/10.3390/biom11020259

Suk, J. S., Xu, Q., Kim, N., Hanes, J., & Ensign, L. M. (2016). PEGylation as a strategy for improving nanoparticle-based drug and gene delivery. *Advanced Drug Delivery Reviews, 99*, 28–51. https://doi.org/10.1016/j.addr.2015.09.012

Teimuri-mofrad, R., Hadi, R., Tahmasebi, B., Farhoudian, S., Mehravar, M., & Nasiri, R. (2017). Green synthesis of gold nanoparticles using plant extract: Mini-review. *Nanochemistry Research, 2*(1), 10.22036/ncr.2017.01.002.

Torres-Chavolla, E., Ranasinghe, R. J., & Alocilja, E. C. (2010). Characterization and functionalization of biogenic gold nanoparticles for biosensing enhancement. *IEEE Transactions on Nanotechnology, 9*(5), 533–538. https://doi.org/10.1109/TNANO.2010.2052926

Tripathy, P., Sethi, S., Panchal, D., Prakash, O., Sharma, A., Mondal, R. B., & Pal, S. (2023). Biogenic synthesis of nanoparticles by amalgamating microbial endophytes: Potential environmental applications and future perspectives. In *Microbial endophytes and plant growth* (pp. 215–231). Elsevier. https://doi.org/10.1016/B978-0-323-90620-3.00003-9

Uma Suganya, K. S., Govindaraju, K., Ganesh Kumar, V., Stalin Dhas, T., Karthick, V., Singaravelu, G., & Elanchezhiyan, M. (2015). Blue green alga mediated synthesis of gold nanoparticles and its antibacterial efficacy against Gram positive organisms. *Materials Science and Engineering: C, 47*, 351–356. https://doi.org/10.1016/j.msec.2014.11.043

Unfried, K., Albrecht, C., Klotz, L.-O., Von Mikecz, A., Grether-Beck, S., & Schins, R. P. F. (2007). Cellular responses to nanoparticles: Target structures and mechanisms. *Nanotoxicology, 1*(1), 52–71. https://doi.org/10.1080/00222930701314932

Vankova, R., Landa, P., Podlipna, R., Dobrev, P. I., Prerostova, S., Langhansova, L., Gaudinova, A., Motkova, K., Knirsch, V., & Vanek, T. (2017). ZnO nanoparticle effects on hormonal pools in *Arabidopsis thaliana*. *Science of the Total Environment, 593–594*, 535–542. https://doi.org/10.1016/j.scitotenv.2017.03.160

Verma, A., & Mehata, M. S. (2016). Controllable synthesis of silver nanoparticles using Neem leaves and their antimicrobial activity. *Journal of Radiation Research and Applied Sciences, 9*(1), 109–115. https://doi.org/10.1016/j.jrras.2015.11.001

Vigneshwaran, N., Ashtaputre, N. M., Varadarajan, P. V., Nachane, R. P., Paralikar, K. M., & Balasubramanya, R. H. (2007). Biological synthesis of silver nanoparticles using the fungus *Aspergillus flavus*. *Materials Letters, 61*(6), 1413–1418. https://doi.org/10.1016/j.matlet.2006.07.042

Vijayan, S. R., Santhiyagu, P., Singamuthu, M., Kumari Ahila, N., Jayaraman, R., & Ethiraj, K. (2014). Synthesis and characterization of silver and gold Nanoparticles using aqueous extract of seaweed, *Turbinaria conoides,* and their Antimicrofouling activity. *The Scientific World Journal, 2014*, 1–10. https://doi.org/10.1155/2014/938272

Vijayaraghavan, K., & Ashokkumar, T. (2017). Plant-mediated biosynthesis of metallic nanoparticles: A review of literature, factors affecting synthesis, characterization techniques and applications. *Journal of Environmental Chemical Engineering, 5*(5), 4866–4883. https://doi.org/10.1016/j.jece.2017.09.026

Vinay Gopal, J., Thenmozhi, M., Kannabiran, K., Rajakumar, G., Velayutham, K., & Rahuman, A. A. (2013). Actinobacteria mediated synthesis of gold nanoparticles using *Streptomyces* sp. VITDDK3 and its antifungal activity. *Materials Letters, 93*, 360–362. https://doi.org/10.1016/j.matlet.2012.11.125

Waghmare, S. R., Mulla, M. N., Marathe, S. R., & Sonawane, K. D. (2015). Ecofriendly production of silver nanoparticles using *Candida utilis* and its mechanistic action against pathogenic microorganisms. *3 Biotech, 5*(1), 33–38. https://doi.org/10.1007/s13205-014-0196-y

Wang, X., Yang, L., Chen, Z., & Shin, D. M. (2008). Application of nanotechnology in cancer therapy and imaging. *CA: a Cancer Journal for Clinicians, 58*(2), 97–110. https://doi.org/10.3322/CA.2007.0003

Wang, S., Wu, X.-M., Liu, C.-H., Shang, J.-Y., Gao, F., & Guo, H.-S. (2020). Verticillium dahliae chromatin remodeling facilitates the DNA damage repair in response to plant ROS stress. *PLoS Pathogens, 16*(4), e1008481. https://doi.org/10.1371/journal.ppat.1008481

Warheit, D. B. (2003). Comparative pulmonary toxicity assessment of single-wall carbon nanotubes in rats. *Toxicological Sciences, 77*(1), 117–125. https://doi.org/10.1093/toxsci/kfg228

Weissleder, R. (2002). Scaling down imaging: Molecular mapping of cancer in mice. *Nature Reviews Cancer, 2*(1), 11–18. https://doi.org/10.1038/nrc701

Yan, A., & Chen, Z. (2019). Impacts of silver nanoparticles on plants: A focus on the phytotoxicity and underlying mechanism. *International Journal of Molecular Sciences, 20*(5), 1003. https://doi.org/10.3390/ijms20051003

Yusuf, A., Almotairy, A. R. Z., Henidi, H., Alshehri, O. Y., & Aldughaim, M. S. (2023). Nanoparticles as drug delivery systems: A review of the implication of nanoparticles' physicochemical properties on responses in biological systems. *Polymers, 15*(7), 1596. https://doi.org/10.3390/polym15071596

Zhang, D., Ma, X.-L., Gu, Y., Huang, H., & Zhang, G.-W. (2020). Green synthesis of metallic nanoparticles and their potential applications to treat cancer. *Frontiers in Chemistry, 8*, 799. https://doi.org/10.3389/fchem.2020.00799

Zhao, L., Sun, Y., Hernandez-Viezcas, J. A., Servin, A. D., Hong, J., Niu, G., Peralta-Videa, J. R., Duarte-Gardea, M., & Gardea-Torresdey, J. L. (2013). Influence of CeO_2 and ZnO nanoparticles on cucumber physiological markers and bioaccumulation of Ce and Zn: A life cycle study. *Journal of Agricultural and Food Chemistry, 61*(49), 11945–11951. https://doi.org/10.1021/jf404328e

Part IV
Recent Trends and Development

Chapter 20
Integrated Environmental Technologies for Waste Management

Amrita Ghosh Majumdar, Biswajit Pany, Sankha Subhra Parua, Satyabrata Si, Junpei Yamanaka, and Priti S. Mohanty

1 Introduction

In the twenty-first century, our world is facing an ever-growing and increasingly complex waste crisis. The rapid expansion of global population, urbanization, and consumerism has led to an unprecedented surge in waste generation. From towering landfills to choking oceans, the repercussions of our unsustainable waste management practices are becoming increasingly evident. Conventional waste management practices, which primarily involve the collection, transportation, and disposal of waste in landfills or through incineration, are gradually becoming obsolete for several compelling reasons (Nazari et al., 2021). The limitations and drawbacks of these traditional methods have become increasingly apparent as society grapples with a growing waste crisis and seeks more sustainable and efficient alternatives.

A. G. Majumdar
School of Biotechnology, Kalinga Institute of Industrial Technology (KIIT), Deemed to be University, Bhubaneswar, India

B. Pany · S. Si · P. S. Mohanty (✉)
School of Biotechnology, Kalinga Institute of Industrial Technology (KIIT), Deemed to be University, Bhubaneswar, India

School of Chemical Technology, Kalinga Institute of Industrial Technology (KIIT), Deemed to be University, Bhubaneswar, India
e-mail: pritisundar.mohanty@kiitbiotech.ac.in

S. S. Parua
School of Chemical Technology, Kalinga Institute of Industrial Technology (KIIT), Deemed to be University, Bhubaneswar, India

J. Yamanaka
Graduate School of Pharmaceutical Sciences, Nagoya City University, Mizuho-ku, Nagoya, Japan

V. Kumar et al. (eds.), *Environmental Engineering and Waste Management*,
https://doi.org/10.1007/978-3-031-58441-1_20

(a) *Environmental Impact:* Conventional waste management practices often lead to significant environmental degradation (Babar et al., 2022). Landfills, for instance, can leach harmful chemicals into the soil and groundwater, emit methane (a potent greenhouse gas), and have a lasting impact on the land's aesthetics and ecosystems (Un et al. 2023). Incineration can release pollutants into the air, contributing to air pollution and posing health risks. Plastic waste has also emerged as a major concern due to its persistence in the environment and severe impacts on the ecosystem. Improper disposal of plastic waste led to widespread plastic pollution in oceans, rivers, and terrestrial environments. It poses a risk to marine life, contributes to the breakdown of ecosystems, and also affects human health. These environmental consequences are no longer acceptable in an era of heightened environmental consciousness.
(b) *Limited Land Availability:* Conventional landfill-based waste disposal relies on the availability of land for waste storage. As urban areas expand and available land diminishes, finding suitable locations for new landfills has become increasingly challenging (Siddiqua et al., 2022). This has led to land scarcity issues, especially in densely populated regions.
(c) *Resource Depletion:* The traditional "take-make-dispose" linear model of production and waste management depletes finite resources at an unsustainable rate (Yang et al., 2022). It fails to tap into the potential value of materials present in waste, such as metals, plastics, and organic matter, which can be recovered, recycled, or repurposed. In a world grappling with resource scarcity, the inefficiency of conventional waste management is becoming increasingly evident.
(d) *Health Concerns:* It can pose health hazards to both workers and nearby communities. Landfill workers are exposed to hazardous waste and gases, while incineration can release toxic substances into the atmosphere (Aziz et al., 2022). The health concerns associated with these methods have prompted calls for safer alternatives.
(e) *Economic Costs:* Traditional waste management practices can be costly in the long run. The expenses associated with landfill site acquisition, maintenance, and monitoring are substantial. Moreover, incineration facilities require significant investments in technology and infrastructure. As economic pressures mount and the focus on cost-effectiveness and sustainability grows, alternatives that offer more economic advantages are gaining favor (Nepal et al., 2022).

In response to these challenges and the recognition of the limitations of conventional waste management, integrated waste management (IWM) technologies have gained a lot of attention in recent years. These innovative approaches prioritize waste reduction, recycling, recovery, and resource efficiency, offering a more sustainable and forward-thinking response to the global waste crisis (Rena et al., 2022). By addressing the environmental, economic, and social challenges associated with waste management, integrated solutions are replacing the obsolete practices of the past, setting a course towards a more sustainable future. It is a comprehensive approach that aims to minimize the generation of waste, maximize resource recovery, and effectively manage the remaining waste in an environmentally sustainable

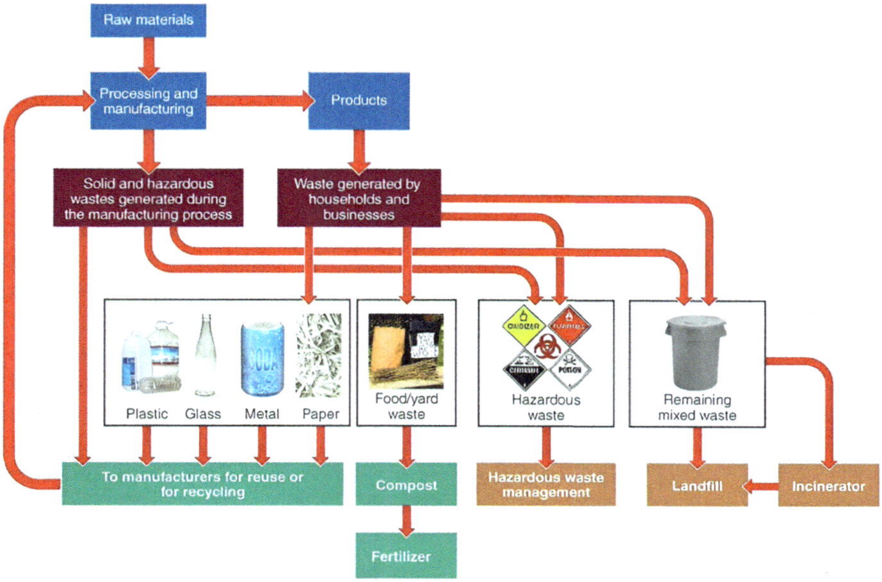

Fig. 20.1 *Schematic representation* of integrated waste management (IWM) practices in a nutshell

manner (Christensen et al., 2020). It involves the coordination and integration of various waste management practices, technologies, and strategies to achieve optimal waste reduction, recycling, and disposal outcomes as shown in Fig. 20.1. IWM technologies are versatile, adaptable, and can be tailored to the specific needs of different regions, urban and rural areas, and industries. They offer a comprehensive approach that aligns with the principles of sustainability and the circular economy. The core principles of integrated waste management technologies include:

(a) *Waste Reduction and Minimization:* The first step in IWM is to reduce waste generation at the source. This involves promoting practices like waste avoidance, source reduction, and product design that prioritize durability, recyclability, and reusability (Mostaghimi et al. 2023). Waste minimization not only conserves resources but also decreases the need for waste disposal facilities.
(b) *Source Separation, Segregation, Recycling, and Recovery:* To facilitate the recycling and recovery of materials, waste should be sorted at the source. Communities and businesses are encouraged to separate their waste into different categories such as organic, recyclable, and hazardous waste. This separation streamlines the recycling process and minimizes contamination (Zhang et al., 2022). This is followed by recycling of materials including paper, plastic, glass, and metals, to create new products. Additionally, organic waste can be composted to produce valuable soil amendments.

(c) *Waste-to-Energy (WtE) Technologies:* In cases where recycling or recovery is not feasible, waste-to-energy technologies offer an alternative (Karmakar et al., 2023). These processes convert nonrecyclable waste into energy through methods like incineration or anaerobic digestion. While some WtE methods generate electricity or heat, others produce biogas or biofuels, contributing to sustainable energy production.
(d) *Safe and Sustainable Disposal:* For waste that cannot be recycled, recovered, or converted to energy, integrated waste management emphasizes the importance of safe disposal (Perkumienė et al., 2023). This means following stringent environmental and health standards when operating landfills and treating hazardous waste.
(e) *Monitoring and Continuous Improvement:* An essential aspect of integrated waste management is monitoring and assessing the effectiveness of waste management practices (Campitelli et al., 2022). By tracking waste generation, recovery rates, and environmental impacts, communities, and organizations can make data-driven decisions to continuously improve their waste management strategies.
(f) *Public Engagement and Education:* The success of integrated waste management relies on the active involvement of communities and individuals. Raising awareness, educating the public about the benefits of recycling, and promoting responsible waste disposal are vital components of this approach (Xiao et al., 2023).

Integrated waste management, while an evolving and comprehensive approach to waste handling, is not without its knowledge gaps and challenges. These knowledge gaps stem from the complexity of waste management systems, the need for ongoing research and development, and the diverse range of issues that must be addressed in waste management. Developing and implementing these technologies is a continuously evolving process. To fully understand the environmental impact of IWM, comprehensive life cycle assessments (LCA) are crucial. These assessments consider the entire life cycle of a product, from raw material extraction to disposal (Rotthong et al., 2022). Another fundamental challenge of IWM is changing human behavior. Encouraging waste reduction, recycling, and responsible disposal requires a deep understanding of human psychology, cultural norms, and socioeconomic factors. Developing standardized data collection methods and systems, as well as improving reporting and data sharing, is also a crucial challenge. While recycling and waste-to-energy technologies have made significant progress, finding innovative ways to reduce waste generation at the source remains a knowledge gap. This includes product design, packaging reduction, and the promotion of a "zero-waste" mindset. Global nature of IWM particularly in the case of e-waste and plastic pollution requires international cooperation and harmonized standards. Bridging gaps in global collaboration and coordinating efforts to address all these issues are the major obstacles yet to be overcome.

This current work offers a comprehensive overview of integrated environmental technologies and their application in the treatment of pollutants found in wastewater from various sources. It covers both applied and newly developed technologies that address major trends in wastewater treatment, including nutrient removal, resource recovery, detoxification of organic and inorganic compounds, bioplastic production, energy conservation, and biofuel/bioenergy production. It majorly divides waste into two major categories biodegradable and nonbiodegradable and then dives into the depth of different integrated waste management technologies by broadly dividing it into biotechnological, nanotechnological, and artificial intelligence (AI) mediated waste management technologies. Furthermore, it provides valuable insights into the limitations, challenges, and potential applications of these integrated technologies. We hope this may serve as a guide for researchers, scientists, and professionals working in the field of bioremediation and phytoremediation, offering information on laboratory and field-scale approaches to remediate pollutants in wastewater. The aim is to achieve sustainable development in wastewater treatment by considering environmental sustainability and future directions for research.

2 Types of Waste Materials

Proper management of waste materials using appropriate technologies depends on the type of waste and the extent of impact to the environment. Broadly, waste can be divided into two categories based on their decomposition property, biodegradable and nonbiodegradable.

2.1 Biodegradable Waste Materials

Biodegradable waste materials refer to organic materials that can be broken down naturally by biological processes (Dhanker et al., 2022). These materials are derived from living organisms and can decompose over time with the help of microorganisms such as bacteria and fungi. Biodegradable waste is often composed of plant-based materials, food scraps, yard waste, and other organic substances. Biodegradable waste materials have environmental advantages over nonbiodegradable materials since they can break down naturally and contribute to the nutrient cycle. Proper management of biodegradable waste includes practices such as composting, anaerobic digestion, and vermicomposting (Saharia et al., 2024). These processes help convert biodegradable waste into valuable compost or biogas, reducing the amount of waste sent to landfills and minimizing greenhouse gas emissions. It is important to separate biodegradable waste from nonbiodegradable waste and ensure proper disposal or recycling methods are followed to maximize the environmental benefits of biodegradable materials. Some common examples of biodegradable waste materials include the following:

(a) *Food Waste:* Biodegradable food waste refers to organic materials from food that can naturally break down over time through biological processes, typically with the help of microorganisms, into substances that are not harmful to the environment. It includes leftover food, fruit and vegetable peels, spoiled food, and other food scraps. These organic materials can be easily decomposed by bacteria and other microorganisms. During this biodegradation, food waste is broken down into simpler substances, such as water, carbon dioxide, and organic matter, which can be used to enrich soil. Many households and communities have implemented composting programs to divert biodegradable food waste from landfills.
(b) *Garden Waste:* Garden waste, also known as green waste, refers to organic materials and debris generated during gardening and yard maintenance activities. These materials are typically biodegradable and can be recycled, composted, or otherwise managed in an environmentally friendly manner. Leaves, grass clippings, branches, and other organic materials, etc., are some of the common garden waste products. They can be composted or mulched to facilitate their decomposition. Proper management of garden waste not only reduces waste but also enriches the soil, reduces the need for chemical fertilizers, and promotes overall garden health. Composting is a sustainable way to recycle these materials, creating nutrient-rich compost that can be used to enhance soil quality and support plant growth.
(c) *Paper Products:* Paper waste refers to discarded or unused paper products that are no longer needed or have reached the end of their useful life. It can encompass various forms of paper materials, including documents, packaging, newspapers, magazines, and cardboard. They can be recycled or composted to prevent unnecessary land filling. Proper management of paper waste is also very essential to reduce its environmental impact, as paper production and disposal can have significant consequences for natural resources, energy consumption, and waste management.
(d) *Wood Waste:* Wood waste refers to various forms of discarded or unused wood materials generated from various sources. Wood chips, sawdust, and other woody materials are major examples of wood waste. Wood is a highly valuable natural resource, and recycling wood waste helps conserve forests and eventually reduces the demand for new timber. They can be used for composting, mulching, or processed into wood chips for various applications.
(e) *Animal Waste:* Animal waste, often referred to as animal manure, is the organic material excreted by animals, including livestock, pets, and wildlife. It consists of faeces and urine, as well as bedding material, such as straw or sawdust in the case of confined animals. Proper management of animal waste is essential for several reasons, including environmental protection and public health. The approach to handling and disposing of animal waste can vary depending on the type of animal and the scale of animal husbandry.

2.2 Nonbiodegradable Waste

Nonbiodegradable waste refers to materials that cannot be broken down naturally by biological processes within a reasonable period of time. These materials persist in the environment for long periods and can have harmful effects on ecosystems if not properly managed. Nonbiodegradable waste typically consists of synthetic or man-made materials that do not easily decompose. Here are some examples:

(a) *Plastics:* Plastic waste is a significant environmental concern worldwide due to its persistence in the environment, impact on ecosystems, and threat to human health. Many types of plastics, such as polyethylene (PE), polypropylene (PP), polystyrene (PS), and polyvinyl chloride (PVC), are nonbiodegradable. Marine life can ingest or become entangled in plastic debris, leading to injury or death (Dey et al., 2023). Plastic waste, including single-use items like bottles, bags, and packaging materials, can persist in the environment for hundreds of years. As plastic items break down, they can become microplastics of less than 5 mm in size. Microplastics are found in water bodies, soil, and even the air potentially affecting human health to a great extent (Gudeta et al., 2023; Sharma et al., 2023). Many countries and regions have implemented regulations to reduce plastic waste, including bans on single-use plastics, restrictions on plastic bag use, and extended producer responsibility (EPR) programs that hold manufacturers responsible for the lifecycle of their products.

(b) *Glass:* Glass is highly resistant to decomposition and does not biodegrade. While it is inert and does not release harmful substances as it breaks down, it can take a very long time to degrade in the environment. But, glass is 100% recyclable and can be recycled indefinitely without a loss in quality. Recycling glass conserves natural resources, such as sand, soda ash, and limestone, which are used in glass production. It requires significantly less energy compared to producing new glass from raw materials. While glass is endlessly recyclable, some glass recycling processes may lead to downcycling, where the recycled glass is used for less demanding applications like aggregate in construction.

(c) *Metals:* Metals such as aluminium, steel, lead, mercury, and cadmium are nonbiodegradable. While metals can corrode and break down over an extended period, they do not undergo biological degradation. They can be toxic to human health as well as the environment. Managing metals and heavy metals waste is very essential to prevent environmental contamination and health risks. The mining and processing of raw metals can have a significant environmental impact, including deforestation, soil erosion, and habitat destruction. Many metals are highly recyclable, and recycling scrap metal reduces the need for virgin materials, conserving natural resources and energy.

(d) *Styrofoam:* Styrofoam waste, also known as expanded polystyrene (EPS) waste, refers to discarded materials made from expanded polystyrene foam, a lightweight and versatile plastic material. Styrofoam is commonly used for packaging, insulation, disposable food containers, and various consumer products because of its lightweight and insulating properties. Improper disposal of

Styrofoam waste, such as littering or inadequate recycling, can lead to environmental pollution. Styrofoam litter is a common sight in many natural settings, where it can harm wildlife and aquatic ecosystems. Recycling Styrofoam can be challenging due to its low density and the need for specialized equipment. In areas where EPS recycling is available, the material can be compacted and processed into pellets for reuse in new products or as a filler material in various applications.

(e) *Synthetic fibers:* Synthetic fibers such as polyester, nylon, acrylic, and spandex, have become increasingly popular due to their durability, affordability, and versatility. However, the disposal of synthetic fiber waste presents environmental challenges. Synthetic textiles can shed microplastic fibers during washing. These microfibers smaller than 5 mm in length, can enter waterways, potentially harming aquatic life and ecosystems. Recycling synthetic fiber waste can be technically challenging due to the variety of synthetic fiber types and the need for specialized processes. Sustainable and eco-friendly alternatives to synthetic fibers, such as natural fibers (cotton, wool, and hemp) and bio-based materials, are gaining popularity as a way to reduce reliance on synthetic materials.

(f) *Rubber:* Rubber waste refers to discarded materials made from natural or synthetic rubber, which is commonly used in various products, including tires, footwear, industrial equipment, and consumer goods. Improper disposal or burning of rubber waste can release harmful pollutants, including toxic chemicals and greenhouse gases. The management of rubber waste is essential for environmental and safety reasons, as improper disposal can lead to pollution and pose health risks. Rubber recycling can be an effective method for managing rubber waste where old tires and other rubber products can be recycled to create new rubber-based materials and products. Old tires can be used as a source of energy in cement kilns, industrial boilers, and waste-to-energy plants. Efforts to find alternative materials for tires and industrial applications, such as eco-friendly materials and sustainable rubber sources, can help reduce the generation of rubber waste in the long run.

(g) *Composite Materials:* Composite material waste refers to discarded materials that are composed of two or more distinct materials bonded together to create a single, stronger, and more durable material. Composite materials, such as fiber glass and carbon fiber composites, are nonbiodegradable materials made from a combination of different substances. These materials are durable and have various applications in industries like construction, aerospace, and automotive. Managing such waste presents unique challenges due to the diverse nature of these materials. They can be categorized into thermoset and thermoplastic composites. Thermoset composites, once cured, cannot be melted and reshaped, making recycling really difficult. On the other hand, thermoplastic composites can be melted and reshaped, offering more recycling potential. Ongoing research and development aim to find innovative recycling technologies for composite materials, such as using solvents to dissolve the matrix and recover the reinforcing fibers.

(h) *Batteries:* Batteries contain toxic and nonbiodegradable components, including heavy metals such as lead, cadmium, and mercury. Improper disposal of batteries can lead to environmental contamination and health risks. Lead-acid batteries, commonly used in vehicles and uninterruptible power supplies (UPS), are one of the most recycled consumer products. Nickel-cadmium (NiCd) batteries contain cadmium, a toxic heavy metal. Lithium-ion batteries are used in electronic devices and electric vehicles. Battery recycling should be considered an essential practice to recover valuable materials and minimize hazardous environmental impact.

(i) *E-waste:* It refers to electronic waste, like discarded electronic or electrical devices. These devices can range from small household appliances to large industrial equipment. E-waste is a growing global concern due to the increasing consumption of electronic devices and the rapid pace of technological advancements. Computers and laptops, mobile phones and tablets, televisions, printers and scanners, batteries, networking equipment, etc., are some of the major E-waste to name. E-waste contains a variety of materials, including metals (such as copper, gold, silver, and palladium), plastics, glass, and hazardous substances (like lead, mercury, cadmium, and flame retardants). The improper disposal and management of e-waste can lead to significant environmental and health risks, as these hazardous substances can contaminate soil, water, and air if not handled properly.

To mitigate the impact of nonbiodegradable waste, various strategies (3R) can be implemented, such as recycling, reusing, and reducing consumption of single-use items. Proper disposal methods, such as land filling and incineration with energy recovery, can help minimize the environmental impact of nonbiodegradable waste. It is crucial to adopt sustainable practices and promote the use of environmentally friendly alternatives to nonbiodegradable materials to reduce the accumulation of waste and protect the environment.

3 Environmental Technologies for Waste Management

Integrated waste management encompasses various environmental technologies that aim to optimize waste handling, treatment, and disposal processes while minimizing their environmental impact. Some of the major environmental technologies commonly employed in integrated waste management are discussed in detail.

3.1 Biotechnological Approaches

Biotechnology-based approaches in integrated waste management (IWM) involve the application of biological processes and techniques to treat and manage various types of waste effectively. These approaches leverage the capabilities of microorganisms, enzymes, and other biological agents to break down, transform, or remove waste materials.

3.1.1 Microbial Fuel Cell

A microbial fuel cell (MFC) is a bioelectrochemical device that uses microorganisms to convert organic matter into electricity (Slate et al., 2019). It combines elements of microbiology, electrochemistry, and environmental engineering to create a sustainable and potentially renewable energy source. The basic structure of an MFC consists of an anode and a cathode compartment, separated by a semi-permeable membrane or a proton exchange membrane (PEM). The anode compartment contains bacteria or microorganisms known as electrogenic bacteria, which oxidize organic matter and release electrons and protons. The electrons flow from the anode through an external circuit to the cathode, creating an electric current. The protons generated in the anode compartment migrate through the membrane to the cathode compartment, where they combine with oxygen (O_2) or another electron acceptor to form water. This creates a closed-loop system that produces electricity as long as there is a supply of organic matter and an electron acceptor. A schematic representation of MFC (Ling et al., 2019) is demonstrated in Fig. 20.2. MFCs hold various potential applications, including wastewater treatment, desalination, remote power generation, and biosensors. They offer advantages such as renewable energy production, low maintenance requirements, and the ability to utilize organic waste as a fuel source. The use of MFCs for wastewater treatment in industrial processes can contribute to a significant reduction in the carbon footprint of these facilities. Microorganisms in MFCs are capable of breaking down a wide range of organic pollutants and contaminants found in industrial wastewater and by harnessing the metabolic activities of microorganisms, MFCs generate a continuous, low-level electrical current that can be used to power onsite equipment or even contribute to the grid. However, there are also challenges associated with MFC technology, such as low power output, slow reaction rates, and the need for further research to optimize performance and improve efficiency. Nonetheless, ongoing research and development aim to overcome these limitations and make microbial fuel cells a more viable and sustainable energy option.

3.1.2 Cyanobacteria and Microalgae-Mediated Waste Removal

Cyanobacteria and microalgae are photosynthetic microorganisms that can play a significant role in waste removal and environmental remediation. These organisms have the ability to remove pollutants from water, air, and soil through various mechanisms. A schematic representation of cyanobacteria and microalgae-mediated sustainable green technology (Gradíssimo et al., 2020) is shown in Fig. 20.3. These

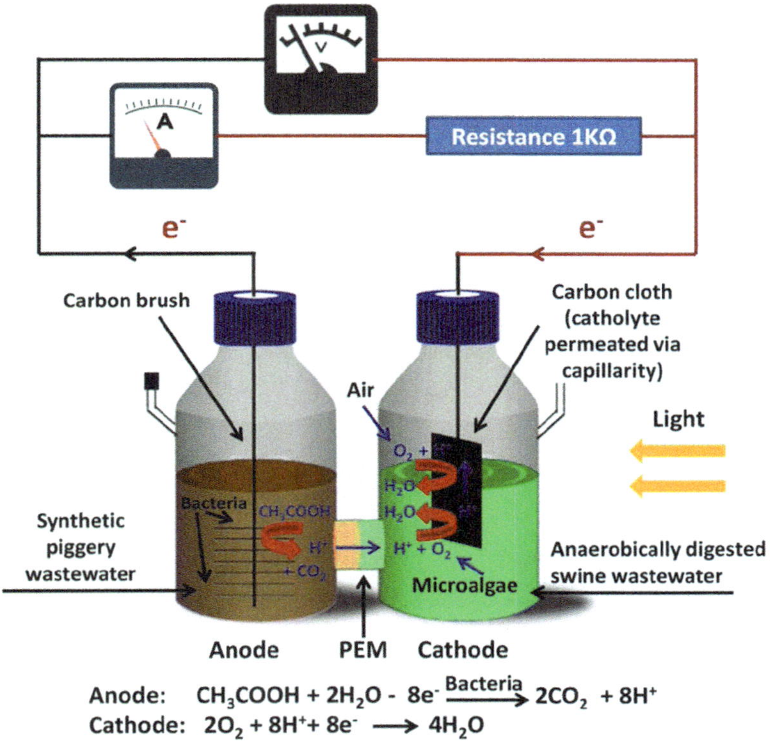

Fig. 20.2 *Schematic* representation of microbial fuel cell (MFC) reactor. (Reprinted with permission from (Ling et al., 2019) under the Creative Commons Attribution License (https://creativecommons.org/licenses/by/4.0/))

organisms can play a pivotal role in waste management via different techniques such as.

(a) *Carbon dioxide (CO_2) Mitigation:* Cyanobacteria and microalgae are capable of photosynthesis, where they utilize CO_2 and release oxygen (Tarafdar et al., 2023). This makes them effective in mitigating CO_2 emissions from various sources such as industrial processes or power plants. These microorganisms can be cultivated in bioreactors or ponds, and their photosynthetic activity helps sequester CO_2 from the atmosphere.
(b) *Bioremediation of Contaminated Water and Soil:* They have also been studied for their ability to remediate polluted water bodies and soil (Priyanka et al., 2020). They can accumulate heavy metals, such as copper, lead, or cadmium, from contaminated environments, thereby reducing their concentrations.
(c) *Nutrient Recovery:* These microorganisms can be employed in nutrient recovery processes (Canizales et al., 2023). When grown in nutrient-rich wastewater or agricultural runoff, they can accumulate nutrients like nitrogen and phosphorus.

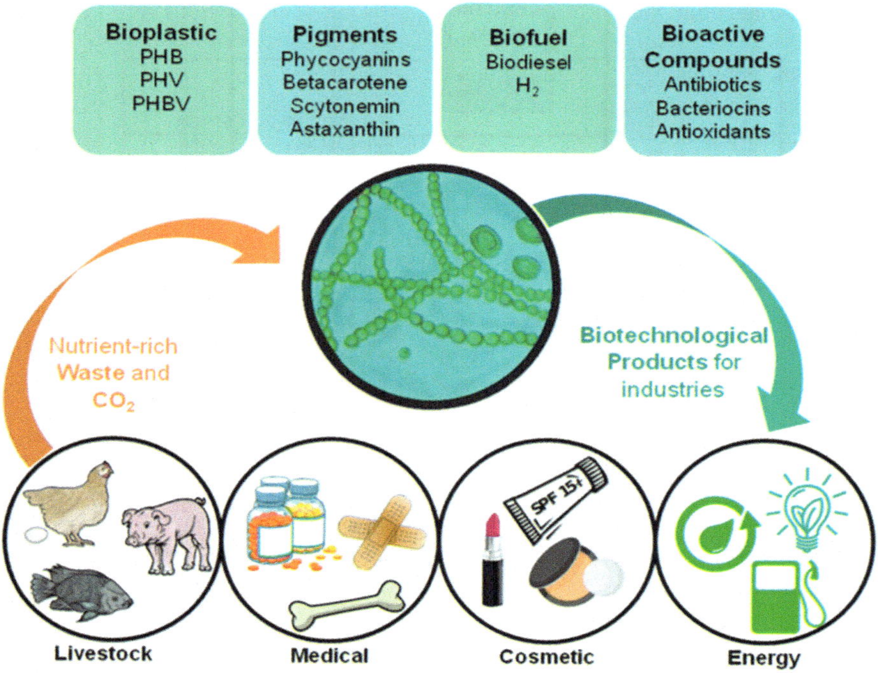

Fig. 20.3 Schematic representation of cyanobacteria and microalgae-mediated sustainable green technology. (Reprinted with permission from (Gradíssimo et al., 2020) under the Creative Commons Attribution License (https://creativecommons.org/licenses/by/4.0/))

After harvesting, the biomass can be processed to extract these nutrients, which can then be used as fertilizers or other agricultural inputs.

(d) *Biofuel Production:* Cyanobacteria and microalgae are also of interest for biofuel production (Sahasrabuddhe et al., 2022). They can accumulate high amounts of lipids (oils) and carbohydrates, which can be converted into biofuels like biodiesel or bioethanol. This offers a potential solution for sustainable and renewable energy generation while utilizing waste resources. Furthermore, the effectiveness of cyanobacteria and microalgae in waste removal depends on various factors, including the specific strains used, cultivation conditions, and the nature of the pollutants. Further research and development are ongoing to optimize these processes and enhance the efficiency of waste removal mediated by these microorganisms.

3.1.3 Anaerobic Ammonium Oxidation (ANAMMOX) Technology

Anaerobic ammonium oxidation (anammox) is a biological process that plays a crucial role in the nitrogen cycle, specifically in the removal of nitrogen compounds from wastewater or other nitrogen-rich environments (Neogi et al., 2021). Anammox technology harnesses this process to treat nitrogen-containing wastewater in a more efficient and sustainable manner compared to conventional methods. Anammox bacteria can convert ammonium (NH_4^+) and nitrite (NO_2^-) ions directly into nitrogen gas (N_2) under anaerobic conditions. This conversion occurs in a specialized compartment within the bacterial cells called the anammoxosome. The overall reaction can be represented as follows:

$$NH_4^+ + NO_2^- \rightarrow N_2 + 2H_2O \quad (20.1)$$

Anammox bacteria have a slower growth rate compared to conventional bacteria used in wastewater treatment. This leads to lower sludge production, reducing the need for sludge handling and disposal. This process is cost effective and requires less oxygen compared to conventional nitrification and denitrification processes, resulting in lower energy demands. The compact nature of anammox reactors, combined with their higher treatment capacity, allows for smaller reactor sizes and a reduced footprint compared to traditional nitrogen removal systems. This technology can be widely used in integrated waste management in different ways. A schematic representation of a two-stage anammox technology is shown in Fig. 20.4 (Liu et al., 2018). Anammox technology is widely used in the treatment of ammonium-rich wastewater (Cho et al., 2019) such as in municipal wastewater treatment plants and industrial settings. It provides an efficient and sustainable

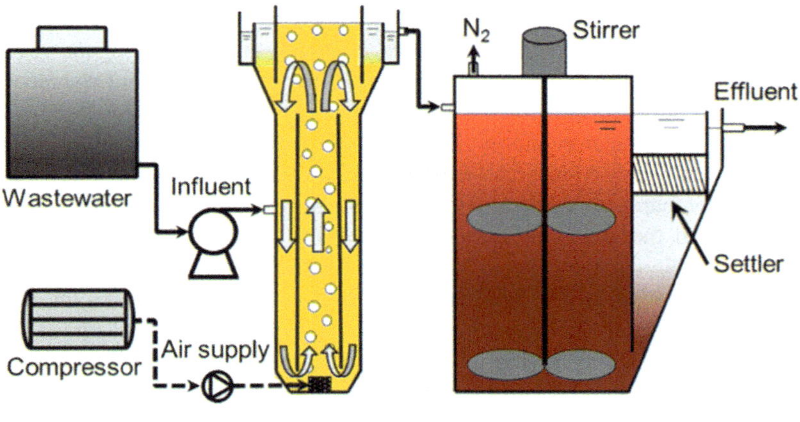

Fig. 20.4 Schematic representation of two-stage PN anammox process used in wastewater management. (Reprinted with permission from (Liu et al., 2018) Copyright 2018 Springer Nature BV (https://creativecommons.org/licenses/by/4.0/))

approach for nitrogen removal. Anammox bacteria can be integrated into anaerobic digestion processes to enhance nitrogen removal and improve the overall efficiency of the digestion process (Cao et al., 2022). This technology has also been explored in the context of bioenergy production. By coupling anammox with anaerobic digestion or other bioenergy processes, the released nitrogen gas can be captured and utilized, while simultaneously treating wastewater. It continues to be an active area of research and development. Challenges include optimizing reactor design, maintaining stable operation, and understanding the microbial community dynamics involved in anammox processes. Further research aims to improve process efficiency, expand its application range, and address any limitations associated with scaling up anammox technology.

3.1.4 Integrated Production of Polyhydroxyalkonate (Bioplastic)

Polyhydroxyalkanoates (PHAs) are a class of bioplastics that are synthesized by microorganisms through the fermentation of renewable carbon sources (Behera et al., 2022). They belong to the broader category of polyesters and are characterized by their ability to accumulate as intracellular granules in the microbial cells. They are considered environmentally friendly alternatives to traditional petroleum-based plastics because they are biodegradable and derived from renewable resources (Kumar et al., 2021). They can be engineered to exhibit a wide range of physical and mechanical properties (Sharma et al., 2021b), making them suitable for various applications. Depending on their composition, PHAs can be rigid or flexible, transparent, or opaque and have different melting points and degradation rates. PHAs can be broken down by naturally occurring microorganisms into carbon dioxide and water under specific environmental conditions (Meereboer et al., 2020) such as composting or anaerobic digestion. This feature reduces their impact on the environment compared to nonbiodegradable plastics. PHAs can be produced from a variety of renewable carbon sources, including plant-based sugars, vegetable oils, and waste streams from agricultural and industrial processes (Mahato et al., 2023). This reduces reliance on fossil fuels and contributes to the development of a sustainable and circular economy. Despite these advantages, they face some challenges that limit their widespread adoption. These challenges include high production costs compared to conventional plastics, limited processing and mechanical properties, and the need for further research and development to optimize their performance. Nonetheless, ongoing efforts are being made to overcome these challenges and enhance the production and application of PHAs. As the demand for sustainable materials increases, PHAs hold promise as a biodegradable and renewable alternative to traditional plastics in various industries. They hold huge potential in becoming a backbone of next-generation integrated waste management systems.

3.2 *Nanotechnological Approach*

Nanotechnological approaches in IWM involve the application of nanotechnology to address various challenges associated with waste treatment, recycling, and disposal. Nanotechnology utilizes nanoscale materials and processes to improve the efficiency and effectiveness of waste management strategies. Here, we focus on the major nanotechnology-based approaches commonly used in IWM.

3.2.1 Adsorption

Adsorption is the most widely used process for efficient removal of biodegradable and nonbiodegradable pollutants from wastewater (Shi et al., 2023). It is based on the principle of selective attachment of the pollutants (adsorbate) onto the surface of a solid material (adsorbent) from an aqueous solution. Different adsorbent nanomaterials such as carbon-based nanomaterials (e.g., activated carbon, grapheme, CNT, etc.), metal oxides (MO) (Grimm et al., 2021), metal organic frameworks (MOF) (Liu et al., 2022), hydrogels (Darban et al., 2022), and polymer nanocomposites (Agboola et al., 2021) have been extensively studied over the years. These nanomaterials have a high surface area and porosity, providing ample sites for adsorption. Surface functionalization also considerably improves adsorption efficiency and selectivity of these nanomaterials. The adsorption capacity is mainly determined by their pore structure and surface area. The contact time between adsorbate and adsorbent, as well as the flow rate of the adsorbate, also impacts the efficiency of the process. This allows for the effective removal of even low concentrations of pollutants from wastewater.

3.2.1.1 Activated Carbon-Based Filtration (ACF)

It is a widely used method for the removal of impurities from wastewater. It involves the adsorption of substances onto the surface of activated carbon, which is a highly porous material with a large internal surface area. The most common application of activated carbon filtration is the removal of organic substances. Organic compounds such as pesticides, herbicides, industrial chemicals, micropollutants including pharmaceuticals such as antibiotics, personal care products, endocrine-disrupting compounds, and other trace organic compounds can be effectively adsorbed by activated carbon. Activated carbon (AC) has a high affinity for these substances and can significantly reduce their concentration in water (Muhammad et al., 2022). It can be in granular or powdered form, depending on the specific treatment requirements and desired water quality. In addition to organic impurities, it can also be used for the removal of inorganic impurities, such as arsenic.

3.2.1.2 Carbon Nanotubes (CNTs)

Carbon nanotubes (CNTs) have indeed been extensively studied for their adsorption properties toward various organic compounds. The unique properties of CNTs make them suitable candidates for adsorption applications. However, it is important to note that the adsorption efficiency of CNTs can be influenced by factors such as CNT type, surface functionalization, solution pH, and the specific characteristics of the target compounds. There are four distinguishable adsorption sites in CNT: (a) external surface, (b) internal surface, (c) interstitial pathways between two individual CNT, and (d) external groove sites (Liu et al., 2021a). SWCNTs are composed of a single cylindrical carbon sheet, whereas MWCNTs consist of multiple concentric carbon layers. These structural differences result in variations in their surface area, pore size distribution, and accessibility of adsorption sites. Extensive research is ongoing to optimize the use of CNTs for efficient adsorption of various organic and inorganic pollutants. CNTs have shown promising adsorption capabilities for the removal of organic dyes from wastewater (Yuan et al., 2021). Organic dyes used in industries such as textile, printing, and dyeing can be effectively adsorbed onto the surface of CNTs, leading to their removal from the water. Cyclic aromatic compounds, including polycyclic aromatic hydrocarbons (PAHs), can also be adsorbed by CNTs (Jin et al., 2020). PAHs are commonly found in industrial wastewater and are known to be hazardous pollutants. Phenolic compounds, which are often found in industrial effluents, can also be effectively adsorbed by CNTs (Peng et al., 2023). Phenols are considered priority pollutants due to their toxicity and environmental persistence. CNTs also hold huge potential for adsorbing pharmaceutical compounds present in wastewater (Kurwadkar et al., 2019) (Taleb et al., 2023). Pharmaceuticals, including antibiotics, hormones, and personal care products, are emerging contaminants of concern. Dissolved organic matter (DOM) consists of complex organic compounds that are naturally present in water bodies. CNTs can also adsorb certain fractions of DOM, contributing to the overall removal of dissolved organic matter from water (Massad et al., 2021; Engel & Chefetz, 2019).

3.2.1.3 Graphene

Graphene is a two-dimensional material consisting of a single layer of sp2 carbon atoms arranged in a hexagonal lattice structure. Since its first use as an adsorbent in 2004, graphene and its derivatives have been extensively studied for their adsorption properties. Graphene can be prepared as a single-layer graphene (SLG) or multiple-layer graphene (MLG) from graphite-based materials. Additionally, graphene oxide (GO) and reduced graphene oxide (RGO) can be produced through chemical oxidation and reduction processes, respectively. These graphene derivatives, such as GO and RGO, exhibit enhanced adsorptive characteristics compared to pristine graphene, further improving their adsorption capabilities. Graphene-based materials have shown great potential for the adsorption of various dyes (Paton-Carrero et al.,

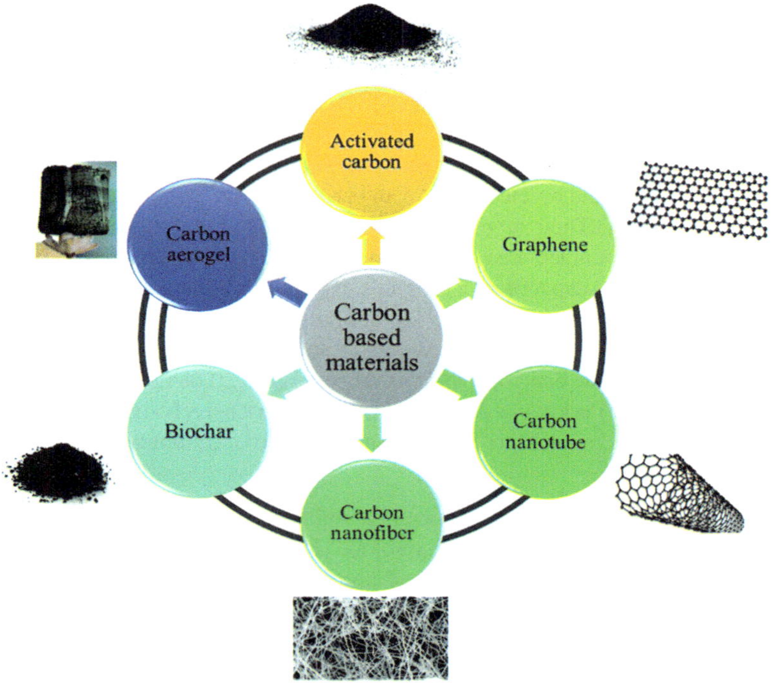

Fig. 20.5 Different carbon-based nanomaterials used in adsorption process. (Reprinted with permission from (Sabzehmeidani et al., 2021) under the Creative Commons Attribution License (https://creativecommons.org/licenses/by-nc/3.0/))

2022). Dyes used in industries such as textile, printing, and dyeing can be effectively adsorbed by graphene sheets due to their large surface area and unique adsorption properties. They have also been investigated for the adsorption of antibiotics present in wastewater (Kern et al., 2022). Antibiotics are emerging contaminants of concern due to their potential impact on ecosystems and the development of antibiotic resistance. Their adsorption capabilities have been further studied for oil and hydrocarbons (Sikiru et al., 2021). They can be utilized for the removal of oil spills or in the treatment of oily waste water. Graphene-based materials have also shown promise in adsorbing pesticides and pharmaceutical compounds from water sources (Casabella-Font et al., 2023). These contaminants, when present in water bodies, can have detrimental effects on aquatic ecosystems. Figure 20.5 schematically demonstrates different types of carbon-based nanomaterials (Sabzehmeidani et al., 2021) used in adsorption processes.

3.2.1.4 Metal Oxides (MO)

MO-based adsorbents are a class of materials that are widely used for adsorption applications in waste management due to their unique properties. They have active surface sites, such as hydroxyl groups (-OH) or Lewis acid/base sites, which enable interactions with target molecules. These surface sites can participate in hydrogen bonding, coordination, or electrostatic interactions with adsorbates, allowing for effective adsorption. Their large surface area provides ample room for adsorption and increases the contact area between the adsorbent and adsorbate, leading to improved adsorption capacities. These adsorbents are typically robust and can be regenerated after adsorption cycles. By applying appropriate desorption methods such as temperature swing, pressure swing, or solvent elution, the adsorbates can be removed from the adsorbent, allowing its reuse. Figure 20.6a demonstrates different MO nanoparticles (Parkinson & Diebold, 2016) extensively used in adsorption-based waste management. Here are some examples of commonly used metal oxide adsorbents:

(a) *Activated Alumina (Al_2O_3):* Activated alumina (Srivastav & Srivastava, 2009) is a highly porous form of aluminium oxide. It possesses a large surface area and high adsorption capacity. Activated alumina is commonly employed for the adsorption of moisture, fluoride (Alhassan et al., 2020), arsenic, and other contaminants in water treatment, air purification (Mohammad et al., 2019), and industrial processes.

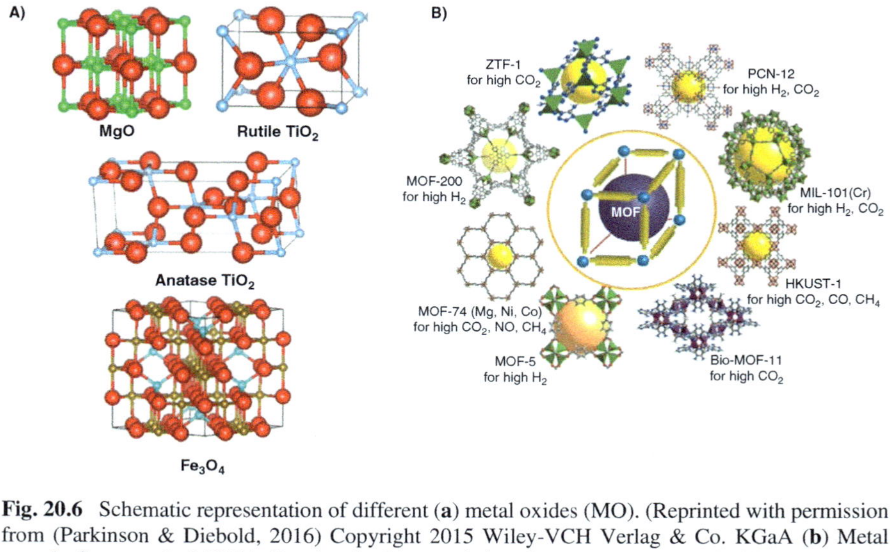

Fig. 20.6 Schematic representation of different (**a**) metal oxides (MO). (Reprinted with permission from (Parkinson & Diebold, 2016) Copyright 2015 Wiley-VCH Verlag & Co. KGaA (**b**) Metal organic framework (MOFs). Reprinted with permission from (Augustus et al., 2017) under the Creative Commons Attribution License (https://creativecommons.org/licenses/by/4.0/))

(b) *Silica gel (SiO_2):* Silica gel (Grande et al., 2020) is a porous form of silicon dioxide that has a high surface area and strong adsorption properties. It is widely used for moisture control, desiccation, and as a drying agent (Soni et al., 2022) in various applications such as electronics, pharmaceuticals, and food packaging.
(c) *Titanium dioxide (TiO_2)*: Titanium dioxide (Liao et al., 2021) is a versatile metal oxide that exhibits excellent photocatalytic properties and adsorption capabilities. It is commonly used in applications such as wastewater treatment, air purification, and the removal of organic pollutants and dyes (Cano et al., 2023). TiO_2 can adsorb pollutants onto its surface and can also be activated by UV light to degrade organic compounds.
(d) *Zirconium oxide (ZrO_2):* Zirconium oxide (Liu et al., 2021b) is a durable and stable metal oxide with high adsorption capacity. It is utilized in various adsorption applications, including gas purification, catalyst support, and removal of heavy metals (Al-Mhyawi et al., 2023) from water and wastewater.
(e) *Iron oxide (Fe_2O_3)*: Iron oxide (Boruah et al., 2023), particularly hematite and magnetite, has shown effective adsorption capabilities for the removal of arsenic, heavy metals, and organic pollutants (Sharma et al., 2021a) from water. It can be used in both aqueous and gas-phase adsorption processes.
(f) *Cerium oxide (CeO2):* Cerium oxide, or ceria, is a versatile metal oxide with excellent redox properties (Mishra & Ahmaruzzaman, 2021). It has been employed as an adsorbent for gas-phase applications, such as catalytic converters and carbon monoxide removal. They have also demonstrated adsorption of organic dyes and pollutants in water treatment (Kurian 2020).

3.2.1.5 Metal–Organic Frameworks (MOF)

Metal-organic frameworks (MOFs) are another class of materials consisting of metal ions or clusters coordinated with organic ligands. MOFs have garnered significant attention as adsorbents due to their exceptional properties such as high porosity, tunable properties, selectivity, specificity, and regenerability. MOFs have found applications in diverse areas, including gas storage and separation, catalysis, drug delivery, and environmental remediation. They have shown great promise in capturing and storing gases such as carbon dioxide (CO_2), hydrogen (H_2), or methane (CH_4), as well as removing harmful pollutants from air or water. It is important to note that while MOFs offer exceptional adsorption properties, their practical application can be influenced by factors such as stability, scalability, and cost-effectiveness. Ongoing research and development efforts are focused on addressing these challenges to make MOFs more viable for industrial-scale adsorption processes. There are numerous metal–organic frameworks (MOFs) that have been used for adsorption applications (Augustus et al., 2017) as shown in Fig. 20.6b.

(a) *MOF-5 (also known as IRMOF-1):* MOF-5 is one of the earliest and most extensively studied MOFs (Liu et al., 2022). It is composed of zinc metal nodes coordinated with benzene dicarboxylate (terephthalate) linkers. MOF-5 exhibits high porosity and has been explored for various adsorption processes, including the capture and storage of gases like hydrogen and methane.

(b) *UiO-66:* UiO-66 is a zirconium-based MOF that consists of zirconium nodes connected by carboxylate linkers (Khudozhitkov et al., 2021). It possesses excellent thermal stability and a large surface area, making it suitable for gas adsorption, including the capture of carbon dioxide and volatile organic compounds (VOCs).

(c) *MIL-101:* MIL-101 (also known as Cr-MIL-101 or MIL-101(Cr)) is a chromium-based MOF that has a large-pore structure and exceptional chemical stability (Zou et al., 2022). MIL-101 has been extensively investigated for gas adsorption, particularly for the capture of carbon dioxide and other greenhouse gases.

(d) *HKUST-1:* HKUST-1 (also known as Cu-BTC or MOF-199) is a copper-based MOF (Nian et al., 2023) composed of copper nodes connected by benzene tricarboxylate linkers. It has been widely used for gas storage and separation, including the capture of carbon dioxide, methane, and hydrogen.

(e) *ZIF-8:* ZIF-8 (zeolitic imidazolate framework-8) is a zinc-based MOF that features zeolite-like structures (Bergaoui et al., 2021). ZIF-8 has been extensively studied for gas adsorption, specifically for applications involving carbon dioxide capture, methane storage, and gas separation.

3.2.1.6 Microgels

Microgels have gained significant attention in the field of adsorption for the removal of pollutants in recent years. Microgels are cross-linked polymer particles that swell in a solvent to form a three-dimensional network. Microgels possess a large surface area due to their porous structure providing ample sites for the adsorption of pollutants (Karg et al., 2019). This high surface area enhances the adsorption capacity of microgels, allowing them to effectively capture and remove pollutants from water or other liquid environments. Microgels can be tailored to have specific properties, such as pore size, surface charge, and functional groups in order to selectively adsorb certain pollutants while ignoring others (Scheffold 2020; Marsili et al., 2021). They exhibit a swelling behavior in response to changes in temperature, pH, or other external stimuli (Haney et al., 2020). This swelling behavior can be utilized to control the release of adsorbed pollutants or to enhance the adsorption capacity. Swelling and deswelling cycles of microgels can facilitate the release of captured pollutants and regenerate the adsorbent for further use. Reusability of microgel particles is another important parameter that make them cost-effective and environmentally friendly adsorbents. After adsorption, the pollutants can be desorbed from the microgel surface using appropriate methods, allowing the microgel to be reused in subsequent adsorption cycles (Pany et al., 2023). Therefore, thorough characterization and optimization of the microgel properties are essential for achieving efficient pollutant removal.

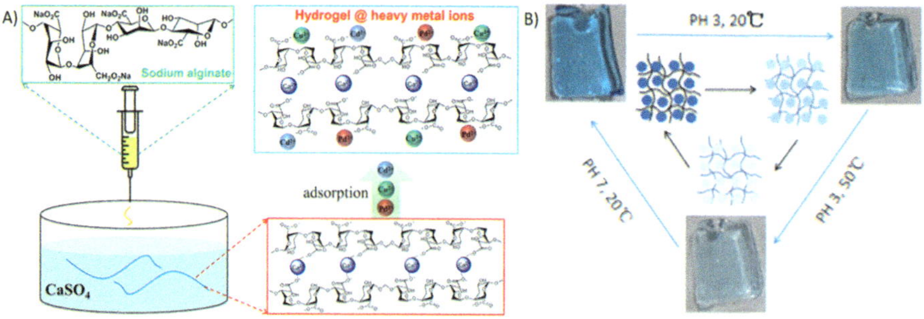

Fig. 20.7 (**a**) Schematic illustration of synthesis of alginate fibroid hydrogel (AFH) and its application in efficient removal of heavy metal ions. (Reprinted with permission from (Kong et al., 2020) under the Creative Commons Attribution License (https://creativecommons.org/licenses/by/4.0/). (**b**) Adsorption and desorption of organic dye via stimuli responsive PNIPAM-co-AAc microgel based PGM. Reprinted with permission from (Pany et al., 2023) Copyright 2023 Elsevier)

It is important to note that the selection of microgel particles depends on the specific pollutant, its physicochemical properties, and the desired adsorption mechanisms. PNIPAM(N-isopropylacrylamide) microgels have temperature-responsive properties, undergoing volume changes in response to temperature variations (Sbeih et al., 2021). This characteristic has been exploited for the removal of organic pollutants from water. At lower temperatures, the microgels swell, enhancing their adsorption capacity. When the temperature is raised, the microgels shrink, facilitating the release of the adsorbed pollutants. Microgels can be synthesized using various crosslinked polymers, such as poly(ethylene glycol) (PEG), poly(vinyl alcohol) (PVA), poly(ethyleneimine) (PEI), and poly acrylic acid (PAA). Poly acrylic acid (PAA) microgels are pH-responsive and can be used for the removal of heavy metal ions from water (Lv et al., 2019). At higher pH values, the microgels swell and form a gel-like structure, allowing for the effective adsorption of different pollutants such as metal ions and organic dyes. Lowering the pH causes the microgels to shrink, releasing the adsorbed material for subsequent regeneration. Adsorption-based efficient removal of heavy metals Cu^{2+}, Cd^{2+}, and Pb^{2+} via sodium alginate fibroid hydrogel (AFH) was showed in the studies of (Kong et al., 2020). The process is shown in a schematic in Fig. 20.7a. One of our recent works demonstrated stimuli-responsive PNIPAM-co-AAc-based polymerized gel matrix (PGM) for efficient removal of methylene blue dye via adsorption (Pany et al., 2023). Reusability of this PGM material is demonstrated in a schematic in Fig. 20.7b. Chitosan, a biopolymer derived from chitin, has been utilized to develop microgels for pollutant removal (De-Paz-Arroyo et al., 2023). Chitosan-based microgels can be functionalized with specific groups to target different types of pollutants. They have been employed for the adsorption of dyes, heavy metals, and organic contaminants from water. Magnetic microgels (Qian et al., 2020) combine the adsorption properties of microgels with the magnetism of iron oxide nanoparticles. These microgels can be easily separated from the solution using an external magnetic field, simplifying the recovery process. These microgels can be tailored to efficiently remove different types of pollutants, including organic compounds, metal ions, and emerging contaminants. Researchers continue to explore and develop new microgel formulations to address specific pollutant removal challenges and optimize their adsorption performance.

3.2.2 Catalysis

Catalysis is another approach to mitigate environmental pollution, in particular water pollution and recovery of valuable chemicals from toxic materials. Catalysis plays a crucial role in waste management by facilitating various processes for the treatment and transformation of waste materials (Martín et al., 2021). Catalytic processes involve the use of catalysts, which are substances that increase the rate of chemical reactions without being consumed in the process. Over the years, catalysis has been extensively used in integrated waste management in many different ways. They are used to convert waste materials into useful energy sources. For example, catalytic pyrolysis involves heating organic waste in the absence of oxygen to produce bio-oil, gas, and char. Catalysts help enhance the efficiency and selectivity of the pyrolysis reactions, improving the yield and quality of the resulting energy products. Catalytic hydrothermal liquefaction is a process that converts wet biomass, such as sewage sludge or algae, into bio-oil. The biomass is reacted with a catalyst in high-temperature and high-pressure water, leading to the conversion of organic components into a liquid product that can be further processed into bio-fuels or specialty chemicals (Lee et al., 2022). They can be employed to mitigate air pollution by treating waste gases and reducing harmful emissions. Catalytic converters in vehicles, for instance, contain catalysts such as platinum, palladium, and rhodium that facilitate the conversion of harmful gases like carbon monoxide, nitrogen oxides, and volatile organic compounds into less harmful substances through selective catalytic reduction (SCR) (Han et al., 2019) or lean-burn catalysis. Catalysis is also used in various wastewater treatment processes to remove contaminants and pollutants. Catalytic oxidation processes, such as advanced oxidation processes (AOPs), employ catalysts like titanium dioxide or metal oxides to break down organic pollutants into nontoxic compounds through oxidation reactions. Catalytic ozonation is a prime example of catalytic oxidation-based degradation of organic pollutants. It combines catalysis with ozonation (An et al., 2023) a process that utilizes ozone (O_3) for water and wastewater treatment. One of the recent works demonstrated catalytic degradation of hazardous nitro aromatic compound 4-NP (Nitrophenol) and their conversion into less hazardous 4-AP (Aminophenol) via Microgel-Metal nanoparticle hybrid nanocomposites (PNIPAM-co-AAc @AuNR and PNIPAM-co-AAc @AgNS) (Satapathy et al., 2017). Steam reforming or catalytic gasification of organic waste can generate hydrogen-rich syngas (Ayodele et al., 2019), which can be further processed for various applications, including fuel cells or industrial processes. In recent years, catalytic processes are being explored for the upcycling or conversion of plastic waste into valuable chemicals or feedstocks. Catalytic depolymerization or cracking of plastics can break down polymers into smaller molecular units (Hou et al., 2021), which can then be used as raw materials for the production of fuels (gasoline and diesel), chemicals, or new plastics. Catalysts facilitate the breakdown of complex biomass components, such as lignocellulosic materials into simpler intermediates that can be further refined into useful products. Methane is a potent greenhouse gas emitted from various sources, including landfills and wastewater treatment plants. It can be catalytically converted into syngas (a mixture of hydrogen and carbon monoxide) (Huang et al., 2023) or methanol (Nandy et al., 2022), which can be utilized as a fuel or chemical feedstock. Can facilitate the conversion of carbon dioxide (CO_2) into value-added chemicals or fuels (Samanta & Srivastava, 2020), enabling the utilization of CO_2 as a feedstock rather than its release into the atmosphere. Catalytic desulfurization processes remove sulfur compounds from fuels and gases, such as natural gas and biogas. Catalysts, often based on metals like nickel or cobalt, facilitate the reaction between the sulfur compounds and hydrogen, resulting in the formation of hydrogen sulfide (H_2S). The hydrogen sulfide can then be separated and further processed or converted into elemental sulfur (Georgiadis et al., 2020). Catalysis in waste management enables more efficient and environmentally friendly processes for waste treatment, energy recovery, and resource utilization. Ongoing research and development in catalytic technologies continue to enhance their effectiveness, selectivity, and sustainability in waste management applications.

3.2.3 Photocatalysis

Photocatalysis (Melchiorre 2022) has emerged as another promising technology for wastewater treatment. Its ability to utilize light energy to drive oxidation reactions makes it an attractive option for addressing environmental challenges. Photocatalysis offers high reaction rates and efficient degradation of wide range of organic pollutants, pesticides, heavy metals, pharmaceutical waste products, etc., in wastewater (Al-Nuaim et al., 2022). Here, we focus on the different steps and principle of photocatalysis which is further explained in Fig. 20.8 (Ren et al., 2021). The activation of photocatalysts by light enables the generation of highly reactive species, such as hydroxyl radicals (·OH), which have strong oxidizing power and can effectively decompose organic contaminants. Contaminants present in water are initially adsorbed onto the surface of the photocatalytic material. This adsorption process enhances the charge mobility of the photocatalyst and improves its redox ability. Subsequently, a series of chemical reactions occur between the adsorbed contaminants and the active species generated by the photocatalyst, leading to the degradation of the contaminants and the formation of degradation products. The redox reactions involved in photocatalysis process are mentioned below in Eqs. (20.2, 20.3, 20.4, 20.5, 20.6, 20.7, 20.8, 20.9 and 20.10).

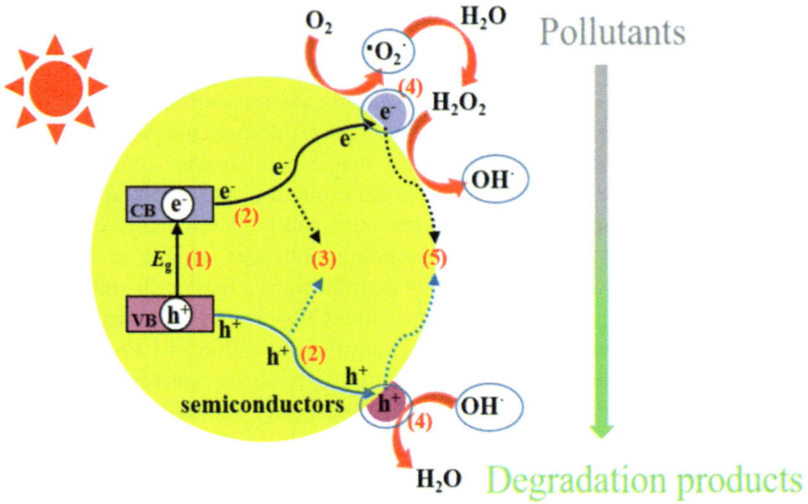

Fig. 20.8 Schematic representation of Photocatalytic degradation of pollutants. (Reprinted with permission from (Ren et al., 2021) under the Creative Commons Attribution License (https://creativecommons.org/licenses/by/4.0/))

$$\text{Semiconductor} + hv \rightarrow h^{+} + e^{-} \tag{20.2}$$

$$e^{-} + \mathrm{O_2} \rightarrow \cdot \mathrm{O_2}^{\cdot} \tag{20.3}$$

$$\mathrm{O_2}^{\cdot} + \mathrm{H}^{+} \rightarrow \cdot \mathrm{OOH} \tag{20.4}$$

$$2 \cdot \mathrm{OOH} \rightarrow \mathrm{O_2} + \mathrm{H_2O_2} \tag{20.5}$$

$$\mathrm{H_2O_2} + \cdot \mathrm{O_2}^{-} \rightarrow \cdot \mathrm{OH} + \mathrm{OH}^{-} + \mathrm{O_2} \tag{20.6}$$

$$\mathrm{H_2O_2} + hv \rightarrow 2 \cdot \mathrm{OH} \tag{20.7}$$

$$h^{+} + \mathrm{H_2O} \rightarrow \cdot \mathrm{OH} + \mathrm{H}^{+} \tag{20.8}$$

$$h^{+} + \mathrm{OH}^{-} \rightarrow \cdot \mathrm{OH} \tag{20.9}$$

$$\text{Pollutants} + (\cdot \mathrm{OH}, h^{+}, e^{-}, \cdot \mathrm{OOH} \text{ or} \cdot \mathrm{O_2}^{-}) \rightarrow \text{Degradation products} \tag{20.10}$$

This efficient degradation process contributes to the overall effectiveness of photocatalysis as a wastewater treatment method. It is a sustainable technology because it operates on renewable energy sources, primarily light. It does not rely on consumables or additives, reducing the need for additional chemicals in the treatment process. Moreover, the photocatalytic materials can be easily regenerated and reused, resulting in a more sustainable and cost-effective approach compared to traditional wastewater treatment methods. Recent research (Yang et al., 2020) involving a visible near infrared (NIR) light responsive hybrid material shows excellent photocatalytic activity, particularly in the removal of petroleum hydrocarbon contaminants. The hybrid material is composed of a g-$C_3N_4H_x^+$ cation and a decatungstate anion. Additionally, the material demonstrated good reusability, indicating that it could be used multiple times for the removal of petroleum hydrocarbons without significant loss of its photocatalytic performance. Photocatalysis has also been extensively studied in removal of toxic heavy metals such as Pb, Hg, and Cd from wastewater. The research conducted by Kanakaraju et al. (2019) focuses on the use of multifunctional TiO_2/Alg/FeNPs (titanium dioxide/alginate/iron nanoparticles) magnetic beads as catalysts for the removal of heavy metal ions. The study specifically investigates the removal of mixed heavy metals, namely chromium (III) (Cr(III)), copper (II) (Cu(II)), and lead (II) (Pb(II)) ions from waste water. Pharmaceutical waste materials such as tetracycline and sulfamethoxazole can be found in ground water, soil, and even in drinking water contributing to potential risk factors regarding human health and aquatic ecosystem (Wang et al., 2020). The study conducted by (Isari et al., 2020) focused on the synthesis and application of N-Cu co-doped TiO_2@CNTs (nitrogen-copper Co-doped titanium dioxide @carbon nanotubes) as a heterogeneous catalyst for sewage treatment. The researchers investigated the catalytic performance of this material under visible light and ultrasonic irradiation. The removal efficiencies of sulfamethoxazole (an antibiotic), chemical oxygen demand (COD), and total organic carbon (TOC) were evaluated under ideal conditions.

3.3 Recycling and Reuse of Industrial and Municipal Waste

Recycling and reuse of industrial and municipal waste are essential components of sustainable waste management strategies. These practices help conserve resources, reduce environmental pollution, and minimize the amount of waste sent to landfills or incineration. Several technologies are involved in the recycling and reuse of municipal and industrial waste.

3.3.1 Material Recovery Facilities (MRFs)

MRFs are specialized facilities equipped with conveyor belts, sorting screens, magnets, and other equipment to separate and sort recyclable materials from mixed waste streams (Citrasari et al., 2019). MRFs play a central role in municipal waste recycling by efficiently processing large volumes of mixed recyclables.

3.3.2 Composting Technology

Composting involves the controlled decomposition of organic waste, such as food scraps and yard trimmings, to produce compost a nutrient-rich soil conditioner. Composting technologies may include aerobic and anaerobic processes, with some systems using mechanical agitation to speed up the composting process (Liu et al., 2022).

3.3.3 Bioremediation

Bioremediation is a sustainable and environmentally friendly approach to managing and mitigating the environmental impact of industrial waste (Yadav et al., 2022). It leverages the metabolic activities of microorganisms, such as bacteria and fungi, to break down, transform, or immobilize contaminants in waste materials like such as petroleum hydrocarbons, solvents, and chlorinated compounds and can be really valuable in cleaning up soil and groundwater contaminated by fuel spills or industrial chemicals. Bioremediation can be broadly divided into in situ and ex situ remediation. In situ remediation involves introducing specific strains of microbes or nutrients to enhance the natural biodegradation process at the contaminated site itself, whereas ex situ bioremediation involves removing contaminated material from the site and treating it elsewhere. Certain bacteria and fungi have the ability to bind or precipitate heavy metals in a process known as biosorption. This method is used to remove heavy metals from industrial wastewaters, preventing their release into the environment. In situations where native microorganisms are unable to effectively degrade specific contaminants, bioaugmentation involves introducing specialized microbial cultures or genetically engineered microorganisms to enhance the bioremediation process. This can significantly improve the treatment of recalcitrant compounds present in industrial waste. Phytoremediation combines plants and microorganisms to remediate contaminated soil or water. Plants known as hyperaccumulators can absorb heavy metals, and the associated rhizosphere microorganisms contribute to their immobilization or transformation. Over the years, bioremediation has been successfully applied in a wide range of industrial settings, including the cleanup of oil spills, the treatment of contaminated groundwater, and the remediation of sites contaminated with industrial chemicals. It offers an eco-friendly and efficient alternative to traditional waste management practices, contributing to the reduction of environmental harm and the restoration of contaminated sites.

3.3.4 Advanced Sorting Technologies

Optical sorters, eddy current separators, and other advanced sorting technologies are used to improve the efficiency and accuracy of waste separation in recycling processes (Taneepanichskul et al., 2022). These technologies can identify and separate different materials based on their optical properties, magnetic properties, or density.

3.3.5 Pyrolysis

Pyrolysis is a thermal decomposition process or thermochemical process that breaks down organic materials at high temperatures in the absence of oxygen (Ławińska et al., 2022). It converts biomass and some types of plastic waste into bio-oil, char, and syngas (Eze et al., 2021). It's important to note that the specific conditions may vary for different types of pyrolysis processes, such as fast pyrolysis, slow pyrolysis, or microwave pyrolysis, and the choice of conditions can significantly impact the yield and quality of the products. Therefore, these conditions should be carefully selected based on the goals and feedstock of the pyrolysis process. Major operating conditions of pyrolysis are as followed.

(a) *Temperature:* It is a critical parameter and can vary depending on the feedstock. Typical pyrolysis temperatures range from 300 °C to 800 °C. The temperature selected can influence the product yield and composition.
(b) *Residence Time:* The time the feedstock spends in the pyrolysis reactor is known as the residence time. It can vary depending on the specific process design and objectives but usually ranges from a few seconds to several minutes.
(c) *Heating Rate:* The rate at which the temperature is increased is also very important to control the pyrolysis process. Slow heating can result in different product distributions compared to rapid heating.
(d) *Pressure:* Most pyrolysis processes are conducted under atmospheric pressure, but some variations, such as vacuum pyrolysis or high-pressure pyrolysis, exist for specific applications.
(e) *Feedstock Preparation:* The size and moisture content of the feedstock can also affect the efficiency of the pyrolysis process. Feedstock is often dried and preprocessed to a suitable size.
(f) *Inert Gas or Vacuum:* The absence of oxygen is crucial in this scenario. Many systems use inert gases (such as nitrogen) or create a vacuum to ensure an oxygen-free environment.

3.3.6 Mechanical Biological Treatment (MBT)

MBT combines mechanical processes, such as shredding and screening, with biological processes like composting to treat mixed municipal waste (Fuss et al., 2020). It helps recover recyclable materials while producing a stabilized organic fraction for composting or energy recovery.

3.3.7 Vermicomposting

It is an eco-friendly and sustainable method of managing organic waste in various industrial settings. It involves the use of earthworms (usually species like *Eisenia fetida* or *Eisenia andrei*) to decompose organic materials, turning them into nutrient-rich compost. It is primarily used to manage organic waste generated in different industrial processes. This can include food waste from cafeterias and restaurants, agricultural residues, paper and cardboard, yard trimmings, and organic sludge from wastewater treatment plants. By diverting organic waste from landfills, vermicomposting reduces the strain on landfill capacity and mitigates the production of harmful greenhouse gases that result from the decomposition of organic matter in anaerobic landfill conditions. The resulting vermicompost is rich in nutrients, making it an excellent soil conditioner and fertilizer. Overall, it reduces the environmental footprint of industrial operations, improves soil fertility, and supports the overall goals of reducing waste and promoting environmental sustainability.

3.3.8 Reverse Vending Machines

Reverse vending machines are automated systems that accept empty beverage containers and provide rewards or refunds in return, encouraging people to recycle (Sambhi and Dahiya, 2020).

3.3.9 Upcycling Technologies

Upcycling involves transforming waste materials into products of higher value or quality. Various creative and innovative technologies are used in upcycling to convert waste into new and desirable products (Wang et al., 2021).

3.3.10 Industrial Symbiosis Platforms

Industrial symbiosis platforms are collaborative systems that facilitate and promote the sharing of resources, materials, energy, and expertise among different industries and businesses. These platforms aim to create a mutually beneficial relationship where waste or by-products from one industry become valuable resources for another, thereby promoting sustainability, reducing environmental impacts, and improving economic efficiency (Akrivou et al., 2021). Different industrial symbiosis platforms include the Kalundborg Symbiosis in Denmark, where several companies collaborate to share resources and reduce environmental impacts, and the Eco-Industrial Parks in the United States, which promote sustainable resource management within industrial zones. As environmental concerns and resource scarcity continue to grow, the adoption of these platforms is expected to become more widespread across various industries and regions.

3.3.11 Microbial Enhanced Oil Recovery (MEOR)

Microbial enhanced oil recovery (MEOR) is a tertiary oil recovery technique that involves the use of microorganisms to improve the extraction of oil from reservoirs and often considered as complementary to the traditional oil recovery techniques. It is considered an environmentally friendly and sustainable method for increasing oil production from mature or depleted oil fields. MEOR techniques harness the metabolic activities of microorganisms to alter the physical and chemical properties of reservoirs and improve oil flow. The success of MEOR depends on a thorough understanding of reservoir conditions, the selection of appropriate microorganisms, and effective injection strategies.

3.3.12 Microbial Desulfurization

The mining and metallurgical industries as well as refinaries often produce waste materials containing sulfur compounds, such as sulfide minerals. These materials can create serious environmental issues, including acid mine drainage. Microbial desulfurization has been applied widely to such industrial waste to meet environmental regulations (Ahmad et al., 2023). This process can help mitigate the environmental impact of industrial waste by employing microbes to remove sulfur from industrial materials, such as coal and petroleum and help in reducing sulfur emissions during combustion.

3.3.13 Methanogenesis

Methanogenesis is the biological process by which microorganisms produce methane (CH_4) gas as a metabolic by-product (Vítězová et al., 2020). This process is used in various industrial waste management applications, primarily for the treatment and utilization of organic waste materials. Methanogenesis can be harnessed to convert organic waste into biogas which can be used for combined heat and power (CHP) systems to simultaneously generate heat and electricity. This renewable energy source can help industries reduce their reliance on fossil fuels and lower their carbon footprint. The optimization of methanogenesis involves factors such as temperature, pH, organic loading rate, and the choice of microorganisms. Research continues to improve the efficiency and effectiveness of these processes.

3.3.14 Remanufacturing Processes

Remanufacturing technologies are used to refurbish and restore used products and equipment, extending their lifespan and reducing the need for new manufacturing (Okorie et al., 2021). The integration of these technologies into waste management systems contributes to a more circular economy, where waste is viewed as a valuable resource rather than a burden on the environment. Implementing these technologies requires collaboration between governments, industries, and waste management stakeholders to create efficient and sustainable recycling and reuse processes.

3.4 Artificial Intelligence (AI) in Integrated Waste Management

Artificial Intelligence (AI) is playing an increasingly important role in waste management, revolutionizing various aspects of the industry (Andeobu et al., 2022). AI is transforming waste management by improving waste sorting and recycling processes, optimizing operations, reducing costs, and supporting data-driven decision-making. The integration of AI technologies in waste management systems contributes to resource efficiency, environmental sustainability, and the transition towards a circular economy by means of different integrated smart sensors like fill level sensors, gas, temperature, moisture, odour sensors, leachate sensors, and flow sensors. AI-based smart waste management technologies are already being used in different smart cities all over the world.

(a) *Waste Sorting and Recycling:* AI-powered image recognition algorithms are used to automatically identify and sort different types of waste materials. By analyzing images or video footage, AI systems can classify waste items into specific categories such as plastics, paper, metals, and organic waste. This technology improves the efficiency and accuracy of waste sorting processes, enabling more effective recycling and resource recovery.
(b) *Predictive Analytics and Optimization:* AI algorithms can analyze large volumes of data from waste management operations, including historical data on waste generation, collection routes, and processing facilities. By identifying patterns and trends, AI systems can generate predictive models that optimize waste collection schedules, route planning, and resource allocation. This helps reduce costs, improve operational efficiency, and minimize environmental impacts.
(c) *Smart Bin Monitoring:* Smart waste bins, also known as intelligent or connected waste bins, are innovative devices that leverage technology to optimize waste management processes (Ahmed et al., 2020). They can monitor the fill level of waste bins in real-time using sensors and data analytics. By analyzing the data, waste management companies can optimize collection routes, reduce unnecessary pickups, and ensure that bins are emptied at the appropriate time, reducing overflowing bins and improving overall waste collection efficiency.
(d) *Contamination Detection:* AI algorithms can detect and identify contamination in waste streams, such as nonrecyclable items mixed with recyclables. By analyzing visual characteristics or using spectroscopic techniques, AI systems can identify contaminants and help improve the quality of recycled materials. This reduces the contamination rate and enhances the value of recycled products.
(e) *Waste Management Decision Support:* AI technologies provide decision support systems for waste management professionals. By analyzing data, AI systems can provide insights and recommendations for waste management strategies, such as optimal landfill site selection, facility planning, and waste treatment options. These decision support systems assist in making informed decisions and developing sustainable waste management practices. Figure 20.9 demonstrates a

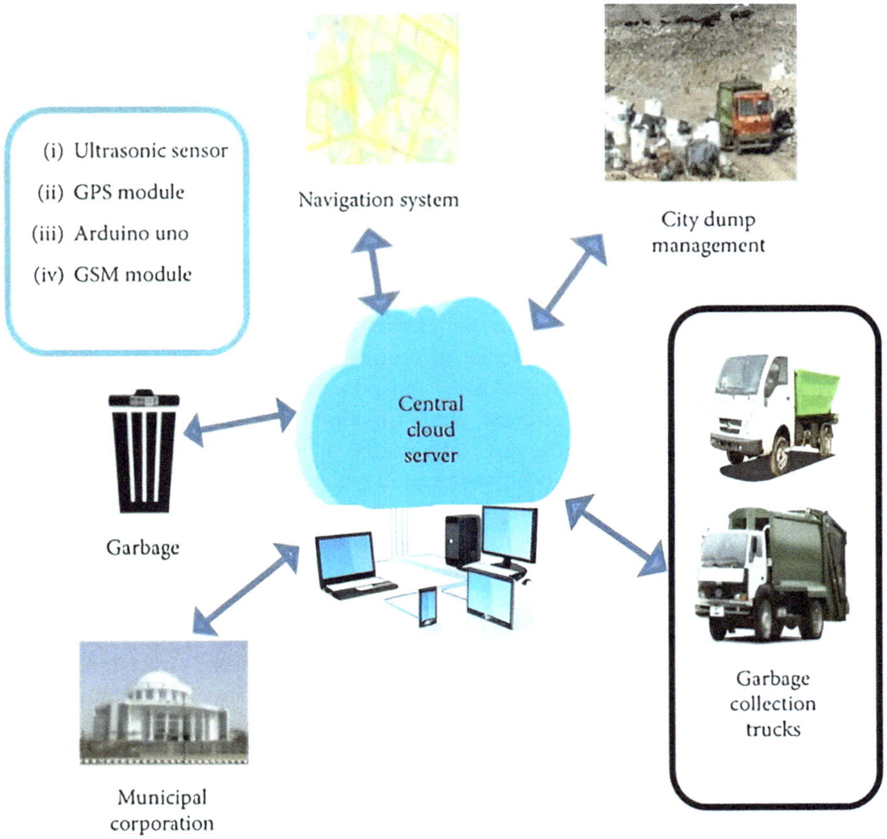

Fig. 20.9 Schematic of AI-based smart waste management system. (Reprinted with permission from (Bano et al., 2020) under the Creative Commons Attribution License (https://creativecommons.org/licenses/by/4.0/))

schematic of AI-based smart waste management system (Bano et al., 2020). AI-powered analytics platforms can process and analyze large amounts of waste-related data, providing real-time insights and generating comprehensive reports. Waste analytics can include waste composition analysis, recycling rates, waste generation patterns, and environmental impact assessments. This information helps waste management companies, policymakers, and stakeholders monitor progress, set targets, and develop strategies for sustainable waste management.

4 Challenges and Future Outlook

The book chapter "Integrated Environmental Technologies for Waste Management" delves into a profound analysis of the challenges and promising future prospects within the realm of waste management technology. It lays the foundation for a transformative future in waste management practices. As the world faces mounting environmental challenges, the integration of advanced technologies promises to reshape the landscape of waste management, steering us toward a more sustainable and resource-efficient era. In addressing the challenges, the chapter confronts the complex nature of waste generation, inadequate waste segregation, and the absence of comprehensive waste management policies in many regions. These challenges are further exacerbated by the persistence of traditional waste disposal methods that contribute to environmental degradation and pose health risks. Additionally, the intricate interplay between different waste fractions and the diversity of waste streams poses hurdles in designing universally applicable integrated solutions. Moreover, the lack of public awareness and engagement hampers the adoption of advanced waste management technologies.

As the chapter peers into the future, it envisions a landscape transformed by innovative technologies that hold the potential to revolutionize waste management practices. The chapter foresees the continued evolution of waste-to-energy conversion technologies. The integration of artificial intelligence and data analytics into waste management processes is expected to revolutionize waste collection, sorting and tracking, optimizing efficiency, and reducing the environmental footprint. Biological treatment techniques are poised to become even more sophisticated, enabling the efficient breakdown of organic waste and producing nutrient-rich soil amendments for agricultural use. Decentralized waste management systems are foreseen to rise, fostering localized resource recovery and diminishing the burden on centralized facilities. Nanotechnology applications could potentially enable enhanced waste separation and recovery at a microscopic level. Additionally, the growing emphasis on sustainability and circular economy principles is likely to drive the development of more efficient recycling and upcycling techniques. The chapter's future outlook encapsulates a shift towards a more harmonious coexistence between technological advancements and environmental conservation, envisioning waste management as a crucial catalyst in the transition towards a greener and more sustainable world.

5 Conclusion

To summarize, integrated environmental technologies play a crucial role in waste management by providing innovative and sustainable solutions for the treatment and handling of waste. These technologies contribute to environmental protection, resource conservation, and the promotion of sustainable development. They are more efficient and effective in waste treatment compared to conventional techniques.

Advanced techniques such as bioremediation, photocatalysis, anaerobic digestion, and membrane filtration enhance the removal of pollutants, reduce the environmental impact of waste, and promote the safe disposal or recycling of waste materials.

These technologies increasingly prioritize energy efficiency and sustainability. Integration of renewable energy sources, optimization of energy consumption in treatment processes, and utilization of waste as a resource for bioenergy production contribute to a more sustainable and environmentally friendly waste management approach. Through the application of AI, IoT, and data analytics, waste management processes can be optimized, resource allocation can be improved, and environmental performance can be monitored in real time. They also play a very important role in raising public awareness and engagement in waste management. Mobile apps, interactive platforms, and educational materials enhance public understanding of waste reduction, recycling, and responsible waste management practices, promoting behavioral change and participation in sustainable waste management initiatives. Continued research, development, and implementation of these technologies are crucial for addressing the global waste management challenges and achieving a more sustainable, green and safe future.

References

Agboola, O., Fayomi, O., Ayodeji, A., Ayeni, A. O., Alagbe, E. E., Sanni, S. E., Okoro, E. E., Moropeng, L., Sadiku, R., Kupolati, K. W., & Oni, B. A. (2021). A review on polymer nanocomposites and their effective applications in membranes and adsorbents for water treatment and gas separation. *Membranes, 11*(2), 139. https://doi.org/10.3390/membranes11020139

Ahmad, A., Zamzami, M. A., Ahmad, V., Al-Thawadi, S., Akhtar, M. S., & Khan, M. J. (2023). Bacterial biological factories intended for the desulfurization of petroleum products in refineries. *Fermentation, 9*(3), 211. https://doi.org/10.3390/fermentation9030211

Ahmed, S., Akkad, M. Z., & Rima, A. (2020). Smart bin monitoring system for smart waste management. *Multidiszciplináris Tudományok*. https://doi.org/10.35925/j.multi.2020.2.45

Akrivou, C., Łękawska-Andrinopoulou, L., Tsimiklis, G., & Amditis, A. (2021). Industrial symbiosis platforms for synergy identification and their most important data points: A systematic review. *Open Research Europe, 1*, 101. https://doi.org/10.12688/openreseurope.13893.2

Alhassan, S. I., Huang, L., He, Y., Yan, L., Wu, B., & Wang, H. (2020). Fluoride removal from water using alumina and aluminum-based composites: A comprehensive review of progress. *Critical Reviews in Environmental Science and Technology, 51*(18), 2051–2085. https://doi.org/10.1080/10643389.2020.1769441

Al-Mhyawi, S. R., Salem, A., Bajaber, M. A., Dayem, S. M. A. E., Ragab, A. H., El-Rahem, K. A. A., Gado, M. A., Atia, B. M., & Cheira, M. F. (2023). Zirconium oxide with graphene oxide anchoring for improved heavy metal ions adsorption: Isotherm and kinetic study. *Journal of Materials Research and Technology, 22*, 3058–3074. https://doi.org/10.1016/j.jmrt.2022.11.121

Al-Nuaim, M. A., Alwasiti, A. A., & Shnain, Z. Y. (2022). The photocatalytic process in the treatment of polluted water. *Chemical Papers, 77*(2), 677–701. https://doi.org/10.1007/s11696-022-02468-7

An, C., Jiang, X., Hong, W., Yu, S., & Zhu, T. (2023). To promote the catalytic ozonation of typical VOCs by modifying NiO with Cetyltrimethylammonium bromide. *PRO, 11*(7), 1893. https://doi.org/10.3390/pr11071893

Andeobu, L., Wibowo, S., & Grandhi, S. (2022). Artificial intelligence applications for sustainable solid waste management practices in Australia: A systematic review. *Science of the Total Environment, 834*, 155389. https://doi.org/10.1016/j.scitotenv.2022.155389

Augustus, E. N., Nimibofa, A., Kesiye, I. A., & Wankasi, D. (2017). Metal-organic frameworks as novel adsorbents: A preview. *American Journal of Environmental Protection, 5*(2), 61–67. https://doi.org/10.12691/env-5-2-5

Ayodele, B. V., Mustapa, S. I., Abdullah, T. A. R. B. T., & Salleh, S. F. (2019). A mini-review on hydrogen-rich syngas production by thermo-catalytic and bioconversion of biomass and its environmental implications. *Frontiers in Energy Research, 7*. https://doi.org/10.3389/fenrg.2019.00118

Aziz, H. A., Kadir, M. O. A., Halim, H. A., & Hung, Y. (2022). *Health-care waste management* (pp. 163–229). Springer eBooks. https://doi.org/10.1007/978-3-030-96989-9_4

Babar, Z. B., Haider, R., & Sattar, H. (2022). *Conventional and emerging practices in hazardous waste management* (pp. 57–93). Springer eBooks. https://doi.org/10.1007/978-3-030-95262-4_3

Bano, A., Din, I. U., & Al-Huqail, A. A. (2020). AIOT-based smart bin for real-time monitoring and management of solid waste. *Scientific Programming, 2020*, 1–13. https://doi.org/10.1155/2020/6613263

Behera, S., Priyadarshanee, M., Vandana, & Das, S. (2022). Polyhydroxyalkanoates, the bioplastics of microbial origin: Properties, biochemical synthesis, and their applications. *Chemosphere, 294*, 133723. https://doi.org/10.1016/j.chemosphere.2022.133723

Bergaoui, M., Khalfaoui, M., Awadallah-F, A., & Al-Muhtaseb, S. A. (2021). A review of the features and applications of ZIF-8 and its derivatives for separating CO_2 and isomers of C_3- and C_4- hydrocarbons. *Journal of Natural Gas Science and Engineering, 96*, 104289. https://doi.org/10.1016/j.jngse.2021.104289

Boruah, H., Tyagi, N., Gupta, S. K., Chabukdhara, M., & Malik, T. (2023). Understanding the adsorption of iron oxide nanomaterials in magnetite and bimetallic form for the removal of arsenic from water. *Frontiers in Environmental Science, 11*. https://doi.org/10.3389/fenvs.2023.1104320

Campitelli, A., Kannengießer, J., & Schebek, L. (2022). Approach to assess the performance of waste management systems towards a circular economy: Waste management system development stage concept (WMS-DSC). *MethodsX, 9*, 101634. https://doi.org/10.1016/j.mex.2022.101634

Canizales, S., Chen, P., Temmink, H., Wijffels, R., & Janssen, M. (2023). Cyanobacteria cultivation on human urine for nutrients recovery. *Algal Research-Biomass Biofuels and Bioproducts, 71*, 103064. https://doi.org/10.1016/j.algal.2023.103064

Cano, F., Reyes-Vallejo, O., Ashok, A., De La, L., Olvera, M., Velumani, S., & Kassiba, A. (2023). Mechanisms of dyes adsorption on titanium oxide– Graphene oxide nanocomposites. *Ceramics International, 49*(13), 21185–21205. https://doi.org/10.1016/j.ceramint.2023.03.249

Cao, S., Koch, K., Du, R., Wells, G. F., Ye, L., & Drewes, J. E. (2022). Toward mainstream anammox by integrating sidestream treatment. *Environmental Science & Technology, 56*(15), 10553–10556. https://doi.org/10.1021/acs.est.2c03256

Casabella-Font, O., Ponzelli, M., Papapanou, M., Balcázar, J. L., Pijuan, M., & Radjenović, J. (2023). Impact of graphene oxide addition on pharmaceuticals removal in anaerobic membrane bioreactor. *Bioresource Technology, 383*, 129252. https://doi.org/10.1016/j.biortech.2023.129252

Cho, S., Kambey, C. S. B., & Nguyen, V. K. (2019). Performance of Anammox processes for wastewater treatment: A critical review on effects of operational conditions and environmental stresses. *Water, 12*(1), 20. https://doi.org/10.3390/w12010020

Christensen, T. H., Damgaard, A., Levis, J. W., Zhao, Y., Björklund, A., Arena, U., Barlaz, M. A., Starostina, V. Y., Boldrin, A., Astrup, T. F., & Bisinella, V. (2020). Application of LCA modelling in integrated waste management. *Waste Management, 118*, 313–322. https://doi.org/10.1016/j.wasman.2020.08.034

Citrasari, N., Sitogasa, P. S. A., Burhan, A. L., & Sari, N. K. (2019). *The design of Material Recovery Facilities (MRF)-based Temporary Disposal Site (TDS) at Universitas Airlangga campus C*. IOP Conference Series. https://doi.org/10.1088/1755-1315/245/1/012010

Darban, Z., Shahabuddin, S., Gaur, R., Ahmad, I., & Sridewi, N. (2022). Hydrogel-based adsorbent material for the effective removal of heavy metals from wastewater: A comprehensive review. *Gels, 8*(5), 263. https://doi.org/10.3390/gels8050263

De-Paz-Arroyo, G., Picos-Corrales, L. A., Pérez-Sicairos, S., & Licea-Claveríe, Á. (2023). Flocculants based on responsive polymers and chitosan for removal of metallic nanoparticles as contaminants of emerging concern present in water. *Colloids and Surfaces A: Physicochemical and Engineering Aspects, 675*, 132045. https://doi.org/10.1016/j.colsurfa.2023.132045

Dey, S., Anand, U., Kumar, V., Kumar, S., Ghoraj, M., Ghosh, A., Kant, N., Suresh, S., Bhattacharya, S., Bontempi, E., Bhat, S. A., & Dey, A. (2023). Microbial strategies for degradation of microplastic generated from COVID-19 healthcare waste. *Environmental Research, 216*, 114438. https://doi.org/10.1016/j.envres.2022.114438

Dhanker, R., Rawat, S., Chandna, V., Deepa, Das, S., Sharma, A., Kumar, V., & Kumar, R. (2022). Recovery of silver nanoparticles and management of food wastes: Obstacles and opportunities. *Environmental Advances, 9*, 100303. https://doi.org/10.1016/j.envadv.2022.100303

Engel, M., & Chefetz, B. (2019). The missing link between carbon nanotubes, dissolved organic matter and organic pollutants. *Advances in Colloid and Interface Science, 271*, 101993. https://doi.org/10.1016/j.cis.2019.101993

Eze, W. U., Umunakwe, R., Obasi, H. C., Ugbaja, M. I., Uche, C. C., & Madufor, I. C. (2021). Plastics waste management: A review of pyrolysis technology. *Clean Technologies and Recycling, 1*(1), 50–69. https://doi.org/10.3934/ctr.2021003

Fuss, M., Vergara-Araya, M., De Vasconcelos Barros, R. T., & Poganietz, W. (2020). Implementing mechanical biological treatment in an emerging waste management system predominated by waste pickers: A Brazilian case study. *Resources Conservation and Recycling, 162*, 105031. https://doi.org/10.1016/j.resconrec.2020.105031

Georgiadis, A. G., Charisiou, N. D., & Goula, M. A. (2020). Removal of hydrogen sulfide from various industrial gases: A review of the most promising adsorbing materials. *Catalysts, 10*(5), 521. https://doi.org/10.3390/catal10050521

Gradíssimo, D. G., Xavier, L. P., & Santos, A. V. (2020). Cyanobacterial Polyhydroxyalkanoates: A sustainable alternative in circular economy. *Molecules, 25*(18), 4331. https://doi.org/10.3390/molecules25184331

Grande, C. A., Morence, D. G. B., Bouzga, A. M., & Andreassen, K. A. (2020). Silica gel as a selective adsorbent for biogas drying and upgrading. *Industrial & Engineering Chemistry Research, 59*(21), 10142–10149. https://doi.org/10.1021/acs.iecr.0c00949

Grimm, O. C., Somaratne, R. M. D. S., Wang, Y., Kim, S., & Whitten, J. E. (2021). Thiol adsorption on metal oxide nanoparticles. *Physical Chemistry Chemical Physics, 23*(14), 8309–8317. https://doi.org/10.1039/d1cp00506e

Gudeta, K., Kumar, V., Bhagat, A., Julka, J. M., Bhat, S. A., Ameen, F., Qaqdri, H., Singh, S., & Amarowicz, R. (2023). Ecological adaptation of earthworms to cope with plant polyphenols, heavy metals, and microplastics in the soil: A critical review. *Heliyon, 9*(3), e14572. https://doi.org/10.1016/j.heliyon.2023.e14572

Han, L., Cai, S., Gao, M., Hasegawa, J., Wang, P., Zhang, J., Shi, L., & Zhang, D. (2019). Selective catalytic reduction of NO_x with NH_3 by using novel catalysts: State of the art and future prospects. *Chemical Reviews, 119*(19), 10916–10976. https://doi.org/10.1021/acs.chemrev.9b00202

Haney, B., Werner, J. G., Weitz, D. A., & Ramakrishnan, S. (2020). Stimuli responsive Janus microgels with convertible hydrophilicity for controlled emulsion destabilization. *Soft Matter, 16*(15), 3613–3620. https://doi.org/10.1039/d0sm00255k

Hou, Q., Zhen, M., Qian, H., Nie, Y., Bai, X., Xia, T., Rehman, M. L. U., Li, Q., & Mei, J. (2021). Upcycling and catalytic degradation of plastic wastes. *Cell Reports Physical Science, 2*(8), 100514. https://doi.org/10.1016/j.xcrp.2021.100514

Huang, E., Rui, N., Rosales, R., Liu, P., & Rodriguez, J. A. (2023). Activation and conversion of methane to syngas over ZrO_2/Cu(111) catalysts near room temperature. *Journal of the American Chemical Society*. https://doi.org/10.1021/jacs.3c01980

Isari, A. A., Hayati, F., Kakavandi, B., Rostami, M., Motevassel, M., & Dehghanifard, E. (2020). N, Cu co-doped TiO2@functionalized SWCNT photocatalyst coupled with ultrasound and visible-light: An effective sono-photocatalysis process for pharmaceutical wastewaters treatment. *Chemical Engineering Journal, 392*, 123685. https://doi.org/10.1016/j.cej.2019.123685

Jin, J., Sun, J., Lv, K., Guo, X., Liu, J., Bai, Y., Huang, X., Liu, J., & Wang, J. (2020). Oxygen-vacancy-rich BiO_{2-x}/Ag_3PO_4/CNT composite for polycyclic aromatic hydrocarbons (PAHs) removal via visible and near-infrared light irradiation. *Industrial & Engineering Chemistry Research, 59*(13), 5725–5735. https://doi.org/10.1021/acs.iecr.0c00232

Kanakaraju, D., Shahdad, N. R. B. M., Lim, Y. C., & Pace, A. (2019). Concurrent removal of Cr (III), Cu(II), and Pb(II) ions from water by multifunctional TiO_2/Alg/FeNPs beads. *Sustainable Chemistry and Pharmacy, 14*, 100176. https://doi.org/10.1016/j.scp.2019.100176

Karg, M., Pich, A., Hellweg, T., Hoare, T., Lyon, L. A., Crassous, J. J., Suzuki, D., Gumerov, R. A., Schneider, S., Potemkin, I. I., & Richtering, W. (2019). Nanogels and microgels: From model colloids to applications, recent developments, and future trends. *Langmuir, 35*(19), 6231–6255. https://doi.org/10.1021/acs.langmuir.8b04304

Karmakar, A., Daftari, T., Sivagami, K., Chandan, M. R., Shaik, A. H., Kiran, B., & Chakraborty, S. (2023). A comprehensive insight into waste to energy conversion strategies in India and its associated air pollution hazard. *Environmental Technology and Innovation, 29*, 103017. https://doi.org/10.1016/j.eti.2023.103017

Kern, M., Škulj, S., & Rožman, M. (2022). Adsorption of a wide variety of antibiotics on graphene-based nanomaterials: A modelling study. *Chemosphere, 296*, 134010. https://doi.org/10.1016/j.chemosphere.2022.134010

Khudozhitkov, A. E., Arzumanov, S. S., Kolokolov, D. I., & Stepanov, A. G. (2021). UiO-66 (Zr) MOF as a promising material for butane isomers separation: Evidence based on the analysis of the adsorbed alkanes mobility by ^{2}H NMR and molecular dynamics simulation. *Journal of Physical Chemistry C, 125*(24), 13391–13400. https://doi.org/10.1021/acs.jpcc.1c02849

Kong, C., Zhao, X., Li, Y., Yang, S., Chen, Y. M., & Yang, Z. (2020). Ion-induced synthesis of alginate fibroid hydrogel for heavy metal ions removal. *Frontiers in Chemistry, 7*. https://doi.org/10.3389/fchem.2019.00905

Kumar, V., Srivastava, S., & Thakur, I. S. (2021). Enhanced recovery of polyhydroxyalkanoates from secondary wastewater sludge of sewage treatment plant: Analysis and process parameters optimization. *Bioresource Technology Reports, 15*, 100783. https://doi.org/10.1016/j.biteb.2021.100783

Kurian, M. (2020). Cerium oxide based materials for water treatment—A review. *Journal of Environmental Chemical Engineering, 8*(5), 104439. https://doi.org/10.1016/j.jece.2020.104439

Kurwadkar, S., Hoang, T. V., Malwade, K. R., Kanel, S. R., Harper, W. F., & Struckhoff, G. C. (2019). Application of carbon nanotubes for removal of emerging contaminants of concern in engineered water and wastewater treatment systems. *Nanotechnology for Environmental Engineering, 4*(1). https://doi.org/10.1007/s41204-019-0059-1

Ławińska, O., Korombel, A., & Zajemska, M. (2022). Pyrolysis-based municipal solid waste management in Poland—SWOT analysis. *Energies, 15*(2), 510. https://doi.org/10.3390/en15020510

Lee, K., Jing, Y., Wang, Y., & Yan, N. (2022). A unified view on catalytic conversion of biomass and waste plastics. *Nature Reviews Chemistry, 6*(9), 635–652. https://doi.org/10.1038/s41570-022-00411-8

Liao, J., Zhang, Y., He, X., Zhang, L., & He, Z. (2021). The synthesis of a novel titanium oxide aerogel with highly enhanced removal of uranium and evaluation of the adsorption mechanism. *Dalton Transactions, 50*(10), 3616–3628. https://doi.org/10.1039/d0dt04320f

Ling, J., Xu, Y., Lu, C., Lai, W., Xie, G., Li, Z., Talawar, M. P., Du, Q., & Li, G. (2019). Enhancing stability of microalgae biocathode by a partially submerged carbon cloth electrode for bioenergy production from wastewater. *Energies, 12*(17), 3229. https://doi.org/10.3390/en121732292019

Liu, W., Yang, D., Shen, Y., & Wang, J. (2018). Two-stage partial nitritation-anammox process for high-rate mainstream deammonification. *Applied Microbiology and Biotechnology, 102*(18), 8079–8091. https://doi.org/10.1007/s00253-018-9207-y

Liu, W., Li, Z., Zhang, S., Jian, W., & Ma, D. (2021a). Adsorption performance of multi-walled carbon nanotube-SiO_2 adsorbent for toluene. *Journal of Fuel Chemistry and Technology, 49*(6), 861–872. https://doi.org/10.1016/s1872-5813(21)60090-7

Liu, Y., Lingyuan, M., Han, K., & Sun, S. (2021b). Synthesis of nano-zirconium-iron oxide supported by activated carbon composite for the removal of Sb(V) in aqueous solution. *RSC Advances, 11*(49), 31131–31141. https://doi.org/10.1039/d1ra06117h

Liu, H., Fu, T., & Mao, Y. (2022). Metal–organic framework-based materials for adsorption and detection of uranium(VI) from aqueous solution. *ACS Omega, 7*(17), 14430–14456. https://doi.org/10.1021/acsomega.2c00597

Lv, Q., Hu, X., Zhang, X., Huang, L., Liu, Z., & Sun, G. (2019). Highly efficient removal of trace metal ions by using poly(acrylic acid) hydrogel adsorbent. *Materials & Design, 181*, 107934. https://doi.org/10.1016/j.matdes.2019.107934

Mahato, R. P., Kumar, S., & Singh, P. (2023). Production of polyhydroxyalkanoates from renewable resources: A review on prospects, challenges and applications. *Archives of Microbiology, 205*(5). https://doi.org/10.1007/s00203-023-03499-8

Marsili, L., Bo, M. D., Berti, F., & Toffoli, G. (2021). Thermoresponsive chitosan-grafted-poly (N-vinylcaprolactam) microgels via ionotropic gelation for oncological applications. *Pharmaceutics, 13*(10), 1654. https://doi.org/10.3390/pharmaceutics13101654

Martín, A. J., Mondelli, C., Jaydev, S. D., & Pérez-Ramírez, J. (2021). Catalytic processing of plastic waste on the rise. *Chem, 7*(6), 1487–1533. https://doi.org/10.1016/j.chempr.2020.12.006

Massad, Y., Hanbali, G., Jodeh, S., Hamed, O., Bzour, M., Dagdag, O., & Samhan, S. (2021). The efficiency of removal of organophosphorus malathion pesticide using functionalized multi-walled carbon nanotube: Impact of dissolved organic matter (DOM). *Separation Science and Technology, 57*(1), 1–12. https://doi.org/10.1080/01496395.2021.1881118

Meereboer, K. W., Misra, M., & Mohanty, A. K. (2020). Review of recent advances in the biodegradability of polyhydroxyalkanoate (PHA) bioplastics and their composites. *Green Chemistry, 22*(17), 5519–5558. https://doi.org/10.1039/d0gc01647k

Melchiorre, P. (2022). Introduction: Photochemical catalytic processes. *Chemical Reviews, 122*(2), 1483–1484. https://doi.org/10.1021/acs.chemrev.1c00993

Mishra, S. R., & Ahmaruzzaman, M. (2021). Cerium oxide and its nanocomposites: Structure, synthesis, and wastewater treatment applications. *Materials Today Communications, 28*, 102562. https://doi.org/10.1016/j.mtcomm.2021.102562

Mohammad, N. K., Ghaemi, A., & Tahvildari, K. (2019). Hydroxide modified activated alumina as an adsorbent for CO_2 adsorption: Experimental and modeling. *International Journal of Greenhouse Gas Control, 88*, 24–37. https://doi.org/10.1016/j.ijggc.2019.05.029

Mostaghimi, K., & Behnamian, J. (2023). Waste minimization towards waste management and cleaner production strategies: A literature review. *Environment, Development and Sustainability, 25*, 12119–12166. https://doi.org/10.1007/s10668-022-02599-7

Muhammad, S., Khalil, H. P. S. A., Hamid, S. A., Albadn, Y. M., Suriani, A. B., Kamaruzzaman, S., Mohamed, A., Allaq, A. A., & Yahya, E. B. (2022). Insights into agricultural-waste-based nano-activated carbon fabrication and modifications for wastewater treatment application. *Agriculture, 12*(10), 1737. https://doi.org/10.3390/agriculture12101737

Nandy, A., Duan, C., Goffinet, C., & Kulik, H. J. (2022). New strategies for direct methane-to-methanol conversion from active learning exploration of 16 million catalysts. *JACS Au, 2*(5), 1200–1213. https://doi.org/10.1021/jacsau.2c00176

Nazari, L., Xu, C., & Ray, M. B. (2021). Conventional approaches for waste management—A Canadian perspective. *In Green chemistry and sustainable technology*, 27–54. https://doi.org/10.1007/978-981-15-9267-6_2

Neogi, S., Sarkar, P., Chaterjee, P. K., Dutta, S., & Dey, A. (2021). *Anammox technology for the food industry nitrogenous wastewater treatment* (pp. 431–450). Elsevier eBooks. https://doi.org/10.1016/b978-0-12-822956-9.00023-4

Nepal, M., Nepal, A. K., Khadayat, M. S., Rajesh, K., Shyamsundar, P., & Somanathan, E. (2022). Low-cost strategies to improve municipal solid waste management in developing countries: Experimental evidence from Nepal. *Environmental and Resource Economics*. https://doi.org/10.1007/s10640-021-00640-3

Nian, L., Wang, M., Zeng, Y., Jiang, J., Cheng, S., & Cao, C. (2023). Modified HKUST-1-based packaging with ethylene adsorption property for food preservation. *Food Hydrocolloids, 135*, 108204. https://doi.org/10.1016/j.foodhyd.2022.108204

Okorie, O., Salonitis, K., & Charnley, F. (2021). *Remanufacturing and refurbishment in the age of industry 4.0: An integrated research agenda* (pp. 87–107). Elsevier eBooks. https://doi.org/10.1016/b978-0-12-818115-7.00001-8

Pany, B., Majundar, A. G., Mohanty, M., Fyis, K., Dey, T., Tripathy, G., Bhat, S., Yamanaka, J., & Mohanty, P. S. (2023). Polymerized stimuli-responsive microgels for the removal of organic dye from water. *Journal of Molecular Liquids, 375*, 121267. https://doi.org/10.1016/j.molliq.2023.121267

Parkinson, G. S., & Diebold, U. (2016). Adsorption on metal oxide surfaces. *Wiley Online Library, 28*, 793–817. https://doi.org/10.1002/9783527680580.ch44

Paton-Carrero, A., Sánchez, P., Sanchez-Silva, L., & Romero, A. (2022). Graphene-based materials behaviour for dyes adsorption. *Materials Today Communications, 30*, 103033. https://doi.org/10.1016/j.mtcomm.2021.103033

Peng, J., Zhou, P., Zhou, H., Huang, B., Sun, M., He, C., Zhang, H., Ao, Z., Liu, W., & Lai, B. (2023). Removal of phenols by highly active periodate on carbon nanotubes: A mechanistic investigation. *Environmental Science & Technology, 57*(29), 10804–10815. https://doi.org/10.1021/acs.est.2c08266

Perkumienė, D., Atalay, A., Perkumienė, D., & Grigienė, J. (2023). Sustainable waste management for clean and safe environments in the recreation and tourism sector: A case study of Lithuania. *Turkey and Morocco. Recycling, 8*(4), 56. https://doi.org/10.3390/recycling8040056

Priyanka, Kumar, C., Chatterjee, A., Wang, W., Yadav, D., & Singh, P. K. (2020). *Cyanobacteria* (pp. 193–202). Elsevier eBooks. https://doi.org/10.1016/b978-0-12-818095-2.00010-2

Qian, J., Zhou, L., Yang, X., Hua, D., & Wu, N. (2020). Prussian blue analogue functionalized magnetic microgels with ionized chitosan for the cleaning of cesium-contaminated clay. *Journal of Hazardous Materials, 386*, 121965. https://doi.org/10.1016/j.jhazmat.2019.121965

Ren, G., Han, H., Wang, Y., Liu, S., Zhao, J., Meng, X., & Li, Z. (2021). Recent advances of photocatalytic application in water treatment: A review. *Nanomaterials, 11*(7), 1804. https://doi.org/10.3390/nano11071804

Rena, Yadav, S., Patel, S., Killedar, D. J., Kumar, S., & Kumar, R. (2022). Eco-innovations and sustainability in solid waste management: An indian upfront in technological, organizational, start-ups and financial framework. *Journal of Environmental Management, 302*, 113953. https://doi.org/10.1016/j.jenvman.2021.113953

Rotthong, M., Takaoka, M., Oshita, K., Rachdawong, P., Gheewala, S. H., & Prapaspongsa, T. (2022). Life cycle assessment of integrated municipal organic waste management systems in Thailand. *Sustainability, 15*(1), 90. https://doi.org/10.3390/su15010090

Sabzehmeidani, M. M., Mahnaee, S., Ghaedi, M., Heidari, H., & Roy, V. A. L. (2021). Carbon based materials: A review of adsorbents for inorganic and organic compounds. *Materials Advances, 2*(2), 598–627. https://doi.org/10.1039/d0ma00087f

Saharia, M., Dey, G., & Kumar, V. (2024). Vermiremediation of plant agro waste to recover residual nutrients and improve crop productivity. In *Earthworm technology in organic waste management: Recent trends and advances* (pp. 79–113). Elsevier.

Sahasrabuddhe, D., Sengupta, A., Sengupta, S., Mishra, V., & Wangikar, P. P. (2022). *Cyanobacteria as a renewable resource for biofuel production* (pp. 475–499). Elsevier eBooks. https://doi.org/10.1016/b978-0-323-88427-3.00006-4

Samanta, S., & Srivastava, R. (2020). Catalytic conversion of CO2 to chemicals and fuels: The collective thermocatalytic/photocatalytic/electrocatalytic approach with graphitic carbon nitride. *Materials Advances, 1*(6), 1506–1545. https://doi.org/10.1039/d0ma00293c

Sambhi, S., & Dahiya, P. (2020). Reverse vending machine for managing plastic waste. *International Journal of Systems Assurance Engineering and Management, 11*, 635–640. https://doi.org/10.1007/s13198-020-00967-y

Satapathy, S., Bhol, P., Chakkarambath, A., Mohanta, J., Samantaray, K., Bhat, S., Panda, S. K., Mohanty, P. S., & Si, S. (2017). Thermo-responsive PNIPAM-metal hybrids: An efficient nanocatalyst for the reduction of 4-nitrophenol. *Applied Surface Science, 420*, 753–763. https://doi.org/10.1016/j.apsusc.2017.05.172

Sbeih, S., Mohanty, P. S., Yethiraj, A., & Morrow, M. R. (2021). 2H NMR study of polymer segmental dynamics at varying cross-linking in poly(N-isopropylacrylamide) microgels. *Langmuir, 37*(46), 13664–13675. https://doi.org/10.1021/acs.langmuir.1c02269

Scheffold, F. (2020). Pathways and challenges towards a complete characterization of microgels. *Nature Communications, 11*(1), 4315. https://doi.org/10.1038/s41467-020-17774-5

Sharma, R. K., Solanki, K., Dixit, R., Sharma, S., & Dutta, S. (2021a). Nanoengineered iron oxide-based sorbents for separation of various water pollutants: Current status, opportunities and future outlook. *Environmental Science, 7*(5), 818–860. https://doi.org/10.1039/d1ew00108f

Sharma, V., Sehgal, R., & Gupta, R. (2021b). Polyhydroxyalkanoate (PHA): Properties and modifications. *Polymer, 212*, 123161. https://doi.org/10.1016/j.polymer.2020.123161

Sharma, U., Sharma, S., Rana, V. S., Rana, N., Kumar, V., Sharma, S., Qadri, H., Kumar, V., & Bhat, S. A. (2023). Assessment of microplastics pollution on soil health and eco-toxicological risk in horticulture. *Soil Systems, 7*(1), 7. https://doi.org/10.3390/soilsystems7010007

Shi, M., Xie, Q., Li, Z., Pan, Y., Yuan, Z., Lin, L., Xu, X., & Li, H. (2023). Adsorption of heavy metals on biodegradable and conventional microplastics in the Pearl River Estuary, China. *Environmental Pollution, 322*, 121158. https://doi.org/10.1016/j.envpol.2023.121158

Siddiqua, A., Hahladakis, J. N., & Al-Attiya, W. A. K. A. (2022). An overview of the environmental pollution and health effects associated with waste landfilling and open dumping. *Environmental Science and Pollution Research, 29*(39), 58514–58536. https://doi.org/10.1007/s11356-022-21578-z

Sikiru, S., Rostami, A., Soleimani, H., Yahya, N., Afeez, Y., Aliu, O., Yusuf, J. Y., & Oladosu, T. L. (2021). Graphene: Outlook in the enhance oil recovery (EOR). *Journal of Molecular Liquids, 321*, 114519. https://doi.org/10.1016/j.molliq.2020.114519

Slate, A. J., Brownson, D. A. C., & Banks, C. E. (2019). Microbial fuel cells: An overview of current technology. *Renewable & Sustainable Energy Reviews, 101*, 60–81. https://doi.org/10.1016/j.rser.2018.09.044

Soni, P., Sur, A., & Gaba, V. K. (2022). Experimental investigation of silica-gel based composite adsorbent for adsorption refrigeration system. *Thermal Science and Engineering Progress, 29*, 101194. https://doi.org/10.1016/j.tsep.2022.101194

Srivastav, A., & Srivastava, V. C. (2009). Adsorptive desulfurization by activated alumina. *Journal of Hazardous Materials*. https://doi.org/10.1016/j.jhazmat.2009.05.088

Taleb, A., Al-Sharif, M. N., Al-Mutair, M. A., Almasoudi, S., & Rahman, M. M. (2023). *Modification and application of carbon nanotubes for the removal of emerging contaminants from wastewater: A review*. In IntechOpen eBooks. https://doi.org/10.5772/intechopen.109781

Taneepanichskul, N., Purkiss, D., & Miodownik, M. (2022). A review of sorting and separating technologies suitable for compostable and biodegradable plastic packaging. *Frontiers in Sustainability, 3*. https://doi.org/10.3389/frsus.2022.901885

Tarafdar, A., Sowmya, G., Yogeshwari, K., Rattu, G., Negi, T., Awasthi, M. K., Hoang, A., & Sirohi, R. (2023). Environmental pollution mitigation through utilization of carbon dioxide by microalgae. *Environmental Pollution, 328*, 121623. https://doi.org/10.1016/j.envpol.2023.121623

Un, C. (2023). A sustainable approach to the conversion of waste into energy: Landfill gas-to-fuel technology. *Sustainability, 15*(20), 14782. https://doi.org/10.3390/su152014782

Vítězová, M., Kohoutová, A., Vítěz, T., Hanišáková, N., & Kushkevych, I. (2020). Methanogenic microorganisms in industrial wastewater anaerobic treatment. *PRO, 8*(12), 1546. https://doi.org/10.3390/pr8121546

Wang, X., Jiang, L., Li, K., Wang, J., Fang, D., Zhang, Y., Tian, D., Zhang, Z., & Dionysiou, D. D. (2020). Fabrication of novel Z-scheme SrTiO3/MnFe2O4 system with double-response activity for simultaneous microwave-induced and photocatalytic degradation of tetracycline and mechanism insight. *Chemical Engineering Journal, 400*, 125981. https://doi.org/10.1016/j.cej.2020.125981

Wang, G., Krzywda, D., Kondrashev, S., & Vorona-Slivinskaya, L. (2021). Recycling and upcycling in the practice of waste management of construction giants. *Sustainability, 13*(2), 640. https://doi.org/10.3390/su13020640

Xiao, L., Fu, B., Lin, T., Meng, L., Zhang, O., & Gao, L. (2023). Promoting and maintaining public participation in waste separation policies—A comparative study in Shanghai, China. *Resources, Environment and Sustainability, 12*, 100112. https://doi.org/10.1016/j.resenv.2023.100112

Yadav, A. N., Suyal, D. C., Kour, D., Rajput, V. D., Rastegari, A., & Singh, J. (2022). Bioremediation and waste management for environmental sustainability. *Journal of Applied Biology and Biotechnology, 1–5*. https://doi.org/10.7324/jabb.2022.10s201

Yang, X., Yu, J., Zhang, Y., Peng, Y., Li, Z., Feng, C., Sun, Z., Yu, X., Cheng, J., & Wang, Y. (2020). Visible-near-infrared-responsive g-C_3N_4H+ reduced decatungstate with excellent performance for photocatalytic removal of petroleum hydrocarbon. *Journal of Hazardous Materials, 381*, 120994. https://doi.org/10.1016/j.jhazmat.2019.120994

Yang, M., Chen, L., Wang, J., Msigwa, G., Osman, A. I., Fawzy, S., Rooney, D., & Yap, P. (2022). Circular economy strategies for combating climate change and other environmental issues. *Environmental Chemistry Letters, 21*(1), 55–80. https://doi.org/10.1007/s10311-022-01499-6

Yuan, L., Huang, Q., Dou, J., Huang, H., Yang, G., Deng, F., Liu, M., Li, X., Zhang, X., & Wei, Y. (2021). Fast adsorptive removal of cationic organic dye by anionic group functionalized carbon nanotubes with high efficiency. *Colloid and Interface Science Communications, 40*, 100328. https://doi.org/10.1016/j.colcom.2020.100328

Zhang, X., Liu, C., Chen, Y., et al. (2022). Source separation, transportation, pretreatment, and valorization of municipal solid waste: A critical review. *Environment, Development and Sustainability, 24*, 11471–11513. https://doi.org/10.1007/s10668-021-01932-w

Zou, M., Dong, M., & Zhao, T. (2022). Advances in metal-organic frameworks MIL-101(CR). *International Journal of Molecular Sciences, 23*(16), 9396. https://doi.org/10.3390/ijms23169396

Chapter 21
Integrated Physicochemical and Biological Treatment Processes and Resource Recovery as Futuristic Approach for the Management of Paper and Pulp Industry Effluent

Om Prakash, Charu Juneja, Purusottam Tripathy, Abhishek Sharma, Deepak Panchal, and Sukdeb Pal

1 Introduction

Over the past few decades, there has been extensive and expeditious development of industries on a global scale, leading to a perturbation of the environment from its earlier state. The pulp/paper industry is one of the chief and entrenched sectors worldwide, and it is well known for its environmental pollution and freshwater utilization issues (Gupta et al., 2014; Negi & Suthar, 2018). This industry usually demands tremendous water throughout several operational phases such as cleaning, pulping, bleaching, and paper manufacturing. It has been reported that paper mills consume 250–300 m^3 of water to manufacture one ton of paper, resulting in the discharge of considerable liquid waste into the surroundings, which contains several deleterious compounds (Chaudhry & Paliwal, 2018). The Ministry of Environment, Forest and Climate Change (MoEF&CC), India, has classified the pulp/paper industries as being in the "Red Category" in a directory of the 17 most polluted industries based on their harmful emissions (Garg, 2022). According to the studies, the paper-making process generates around 250 chemical compounds at various operational steps. These chemicals include xenobiotic compounds such as adsorbable organic halogens (AOXs), chlorinated resin acid, halogens, dioxins, phenols, chlorinated

O. Prakash · P. Tripathy
Wastewater Technology Division, CSIR-National Environmental Engineering Research Institute (CSIR-NEERI), Nagpur, India

C. Juneja · A. Sharma · D. Panchal · S. Pal (✉)
Wastewater Technology Division, CSIR-National Environmental Engineering Research Institute (CSIR-NEERI), Nagpur, India

Academy of Scientific and Innovative Research (AcSIR), Ghaziabad, India
e-mail: s_pal@neeri.res.in

V. Kumar et al. (eds.), *Environmental Engineering and Waste Management*,
https://doi.org/10.1007/978-3-031-58441-1_21

lignin, metal ions, extractable organic acids, chlorinated phenol, chlorophenols, etc., which are formed along with lignin and other biopolymers (Chaudhry & Paliwal, 2018; Dagar et al., 2022; Kumar & Verma, 2023). Due to inadequate wastewater treatment systems, these chemicals are discharged into aquatic environments by the industries. This wastewater contains high levels of toxic chemicals and lignin, resulting in an abnormally high biological oxygen demand (BOD ~5000 mg/L) and chemical oxygen demand (COD ~10,000 mg L^{-1}) (Kumar et al., 2023). This leads to the generation of dark brown colored liquid, which can deteriorate aquatic life by limiting photosynthesis and lowering the level of pH and dissolved oxygen in the water (Raj et al., 2014). Nevertheless, environmental control authorities in developed countries such as North America and Europe have inaugurated wastewater treatment systems and strict regulations that limit the release of chlorinated compounds (Abedinzadeh et al., 2018; Sathya et al., 2022). Consequently, proper wastewater treatment is essential; otherwise, it may cause a serious threat to the environment, such as adverse effects on aquatic life, loss of environmental aesthetics, and increased soil salinity which can eventually economically impact a country (Mahesh et al., 2016; Patel & Dudhagara, 2020). As a consequence, researchers are continually working on several techniques to treat the effluent from the pulp and paper industry. Microbiological organisms, such as fungi and bacteria, represent considerable prospects among potential techniques owing to their high biotransformation and degradation efficiency (Mir-Tutusaus et al., 2018). Large-scale deployment of low-cost reactors and biological sequencing techniques is worthwhile.

Primarily in developing countries like India, the pulp and paper industry is one of the most environmentally conscious industries (Bajpai, 2018). Approximately 10.11 million tons of paper are expected to be produced per year in India, making up 2.6% of the world's total. With yearly per capita consumption growth rates of over 10%, India offers one of the fastest-growing pulp and paper sectors (Yadav et al., 2022). Hardwood, bamboo fibers, and agricultural waste make up 60–70% of the raw materials required for the production of paper in India, while recycled materials make up 30–40% (Ren et al., 2022). Globally, the demand for paper and board is generally rising at a rate of 5–6% per annum (Bhatnagar, 2015). The worldwide significance of the pulp and paper industry is due to its economic profits, but it is also considered one of the major sources of pollution for valuable water resources (Hasan et al., 2023). Millions of tons of polluted wastewater are released by the industry into aquatic resources each year, making it one of the top five contributors to water pollution (Tripathy et al., 2022). Polluting substances are produced by the pulp and paper industry at a rate of about 100 million kg per year (Gopal et al., 2019). Paper manufacturing plants in India generate 11 million tons of paper yearly and account for 40–50 kg of dry sludge for every ton of paper produced. Paper mill sludge (PMS), which is found in large quantities, is either treated as waste or disposed of or burned, resulting in the release of greenhouse gases (Fang et al., 2017). The pulp and paper sector has long-faced challenges when it comes to disposing of or utilizing PMS (Kumar & Verma, 2024). PMS has been investigated for its potential inclusion in cement, ceramics, and construction materials due to its abundance and increased cellulose content, which gives mechanical strength (Kumar & Sharma, 2022). The lignocellulosic makeup of PMS highlights the possibility of extracting valuable components via a biorefinery strategy, which would be a suitable approach for the pulp and paper industry. Manufacturing processes employed during the production of paper are debarking and chipping, separation of the pulp from chipped woods, bleaching of pulp, stock preparation, and finally generation of paper (Li et al., 2019). The main objective is to efficiently remove the lignin seal to reveal the cellulose component for converting and manufacturing valuable materials. For the delignification of lignocellulose biomass, numerous conventional physicochemical processes are used. The amount of freshwater needed by modern paper mills to produce one ton of paper is less than 6 m^3, whereas older mills may need up to 60 m^3 for the same operation (Mahesh et al., 2016). The pulp and paper industry produces more than 42% of the world's wastewater, which poses an alarming threat to the environment and highlights the need for the management of wastewater (Majumdar, Paul, et al., 2020b). Additionally, the large amount of sludge generated during wastewater treatment is a notable source of soil pollution (Sátiro et al., 2022). A small portion of waste paper is utilized in recycling, while the majority is either disposed of or burned without any economic value. Despite contributing significantly to the country's economy, the industry has not implemented an effective waste recycling system. To solve this problem, a biorefinery strategy is required, utilizing wastewater and PMS for the production of valuable products that may improve the economy and lessen environmental risks.

Proper treatment of wastewater is indeed essential to mitigate environmental damage and protect ecosystems. Failing to do so can have a range of adverse effects, including harming the aesthetics of the environment, negatively impacting aquatic life, and causing soil salinity issues due to nutrient imbalances. These problems, in turn, can have economic repercussions for a country.

In response to these concerns, environmental researchers are continuously exploring various methods for treating and recycling pulp and paper wastewater. Microbial groups, such as fungi and bacteria, are being recognized as a promising avenue due to their efficiency in biotransformation and biodegradation. Utilizing low-cost reactors and sequencing processes in biological treatments holds promise for scaling up these approaches in industrial settings. This approach aligns with the goal of sustainable and environmentally friendly wastewater management in the pulp and paper industry.

This chapter highlights the in-depth knowledge about the various existing strategies to treat pulp and paper industry wastewater and focuses on eco-friendly integrated physicochemical and biological treatment processes and also emphasizes on recovery of value-added byproducts as futuristic approach for sustainable management of paper and pulp industry effluent.

1.1 Generation of Waste from Pulp and Paper Industry

As shown in Fig. 21.1, the paper and pulp industry generate waste at several stages of the paper manufacturing process. The preparation of wood chips, pulping, bleaching, and paper manufacturing produce large amounts of effluent. Lignin and hemicelluloses are extracted during the wood chip preparation and pulping process using NaOH or Na_2S treatment under alkaline conditions in the kraft process, producing a lignin-rich effluent referred to as black liquor. Hazardous chemicals such as calcium oxide, ozone, chlorine, hydrochloric acid, hydrogen peroxide, and ozone are employed to treat the pulp during the bleaching process, which increases the toxicity of the effluent (Lewis et al., 2018). Raw materials, the pulping procedure, the quantity of water used, and effluent recirculation all affect the properties of the wastewater produced. (Mir-Tutusaus et al., 2018). High levels of chlorine compounds, BOD, and COD, collectively known as adsorbable organic halides (AOXs), are frequently found in wastewater. These compounds are used in the bleaching process and are known as adsorbable organic halides (AOXs) (Chaudhry & Paliwal, 2018; Kumar et al., 2020). The COD value is crucial for wastewater treatment as it indicates the number of organic pollutants present in the wastewater. The proportion of easily degradable organic compounds is represented by the BOD/COD ratio, which serves as a biodegradability indicator. A BOD/COD ratio of >0.4 signifies the biodegradable nature (Patel & Dudhagara, 2020).

1.2 Characteristics and Associated Potential Hazards of Waste Generated from Paper and Paper Mills

Pulp and paper mill waste is divided into liquid and solid waste categories, each with unique characteristics. The alkaline, dark-colored liquid waste (black liquor) generated by pulp and paper industry is composed of inorganic compounds and is the primary source of environmental pollution from the paper mills due to the presence of kraft lignin (KL) and phenol derivatives (Chandra et al., 2011; Hooda et al., 2018; Raj et al., 2014). Black liquor can be easily distinguished by its color, high levels of BOD and COD (Tiku et al., 2010), and pH ranging from 6.1 to 8.3, which can be attributed to the metabolites of microflora and fungal groups (Kesalkar et al., 2012). The wastewater from black liquor typically contains 395–2500 mg L^{-1} of total dissolved solids, and COD ranges from 480 to 4450 mg L^{-1} (Kumar et al., 2018). Additionally, black liquor contains natural polymers like tannins and lignin, as well as xenobiotic components generated during the manufacturing process, such as phenols, polyaromatic hydrocarbons, resin acids, and, chlorinated lignins (Lewis et al., 2018). The recycled wastewater also includes non-biodegradable organic compounds like plastic components, sand, metal, and glass sand, glass (Buyukkamaci & Koken, 2010), along with a surfactant called 2,4,7,9-tetramethyl-Sdecyne-4,7-diol that is commonly applied in printing inks and paints (Kamali &

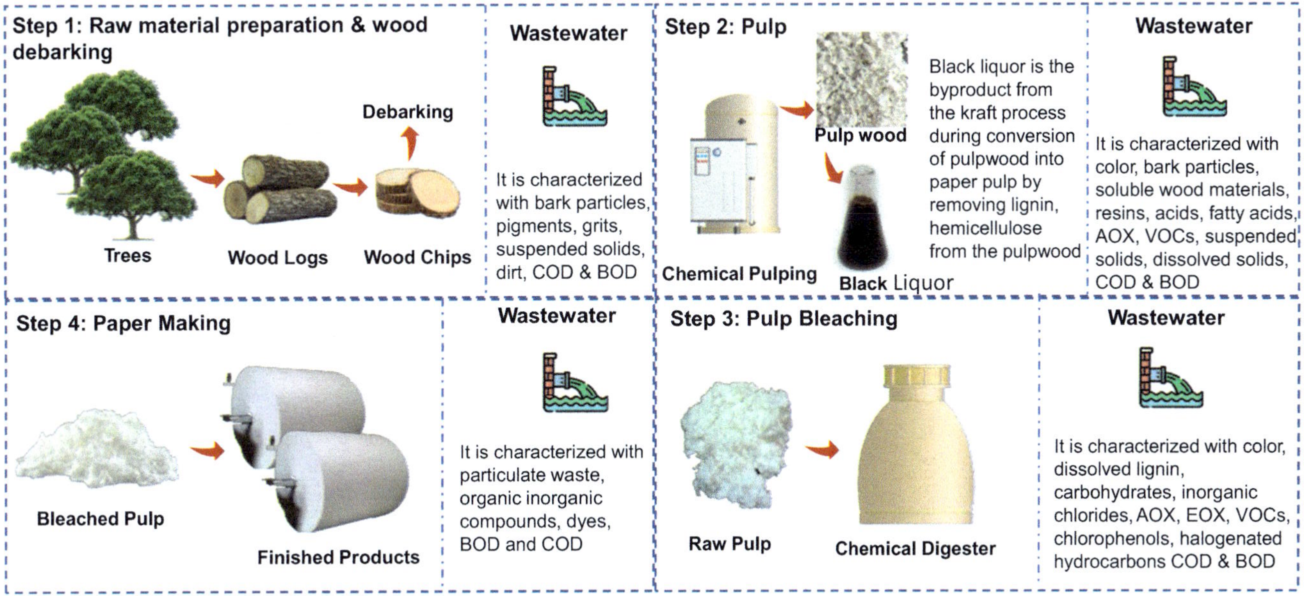

Fig. 21.1 Wastewater generation at different processes of pulp and paper making from the wood source

Khodaparast, 2015). The discharge of untreated wastewater into the environment presents a serious threat to plant, animal, and aquatic life. Pulp and paper mills, which are often located proximal to water bodies, produce alkali-soluble lignin fragments and their derivatives throughout the paper-making process, which can harm aquatic life (Chhonkar et al., 2000). The properties of the wastewater produced in the paper mills have been documented in various studies and are summarized in Table 21.1. A significant concentration of persistent organic pollutants (POPs) is produced when black liquor is discharged as effluent, which can cause respiratory stress, toxicity, genotoxicity, mutagenicity, mixed oxygenase activity, and, liver damage. Exposure to these pollutants can also lead to some health problems such as nausea, vomiting, eye irritation, diarrhea, and headaches in children and workers (Mandal, 1996). The delignification of lignin and its derivatives produces a dark brown color that blocks sunlight, exacerbating the detrimental impacts on aquatic life and eventually causing an adverse effect on the health of human through biomagnification process in the food chain (Karrasch et al., 2006). The production of solid waste from PMS poses potential hazards, with its generation occurring during pulping, deinking, and wastewater treatment processes. The characteristics and quantity of solid waste generated depend on a variety of components, including the raw materials utilized, the production methods used, and the quality of the paper which is being produced. Typically, solid waste is moist and contains a significant amount of ash (Monte et al., 2009). This waste emits various gases, including sodium sulfides, nitrogen oxides, sulfur compounds, methyl mercaptan, hydrogen sulfides, and chlorine dioxides which can lead to chronic health issues such as headaches and nausea. Additionally, the effluents from the sludge release trace amounts of metals into water bodies, which have negative impacts on the environment and contribute to climate change (Krzeminski et al., 2017).

2 Wastewater Treatment

Due to rising concerns about freshwater usage, expanding economic considerations, and stringent environmental regulations, the significance of water management and effective wastewater treatment has been underlined in the pulp and paper industries. The recycling and proper treatment of wastewater can help reduce the use of freshwater, minimize effluent discharges to the surroundings, and advance environmental conservation. Therefore, the primary objective of wastewater treatment plants (WWTPs) is to eliminate pollutants from wastewater by coupling physico-chemical, biological, and integrated treatment processes (Ashrafi et al., 2015) (Fig. 21.2).

Table 21.1 Reported physicochemical characteristics of wastewater from pulp and paper industry. All the values are measured in mg/L except pH and color (HU).

Characteristics								
pH	Color	Total solids	BOD	COD	Lignin	Phenol	Sulfate	Reference
8.1 ± 1.0	1065 ± 89.3	859 ± 45,3	426 ± 30.6	774 ± 43.7	529 ± 20.10	–	866 ± 47	Zainith et al. (2019)
7.7 ± 0.02	1202 ± 6.53	1686 ± 10.58	180.54 ± 4.70	584 ± 3.62	–	–	–	Kumar et al. (2018)
6.4–7.3	660 ±230	354–563	142–221	1170–1510	133–265	–	–	Eskelinen et al. (2010)
8.2 ± 1.0	2242 ± 56	850 ± 30	385 ± 12	792 ± 70	436 ± 18	42 ± 2.5	993 ± 6	Raj et al. (2014)
9.8 ± 1.3	7253 ± 64.2	1753 ± 34.2	2934 ± 38.3	6735 ± 45.3	1863 ± 52.6	–	852.4 ± 34.1	Rajwar et al. (2017)
7 ± 0.20	8942 ± 15.0	1799 ± 17.61	8296 ± 45	22,189 ± 39	1124 ± 12.09	364 ± 20.13	1084 ± 19.67	Chandra et al. (2011)
7.4 ± 0.1	1761 ± 2.3	2359 ± 14.2	185 ± 2.9	2420 ± 17	–	–	–	Garg et al. (2012)
Indian standards for discharge of effluent of large pulp and paper news print/rayon-grade plants of capacity above 24,000 MT per annum								
7–8	–	500	30	350				CPCB (1986)

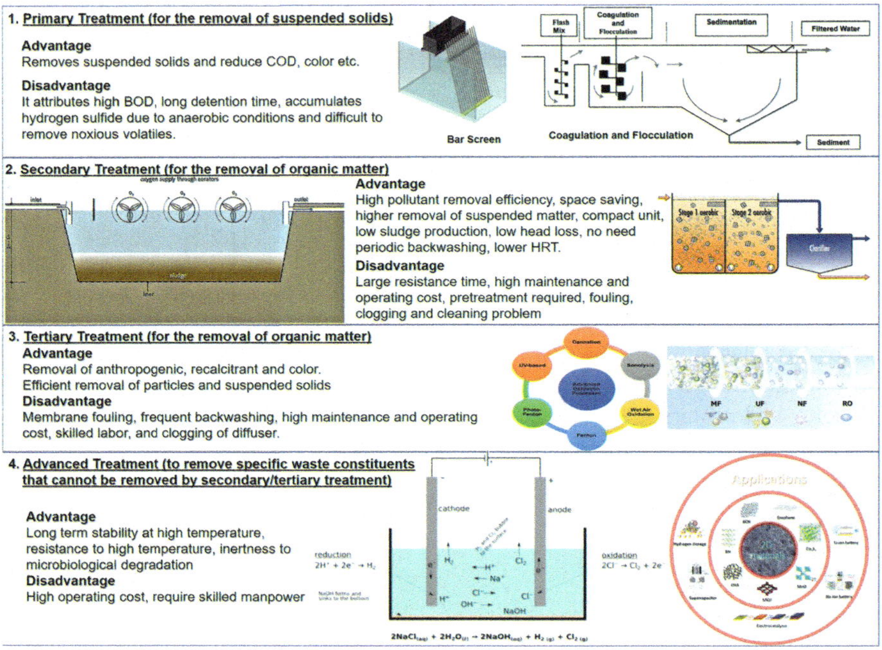

Fig. 21.2 Different primary, secondary, tertiary, and advanced treatment technologies that are being used to treatment the effluents from pulp and paper industries

2.1 *Physicochemical Treatment*

Several physiochemical treatment technologies such as sedimentation, screening, adsorption, coagulation, and ultra- and nanofiltration are attributed to the removal of suspended solids (SSs), colloidal matters, dyes, and hazardous pollutants. Some of the physicochemical processes are explained below.

2.1.1 Sedimentation/Flotation

Generally, bark corpuscle, fiber, fiber litter, resins, and cladding materials are found in the suspended form of pulp and paper industries wastewater. According to Thompson et al. (2001), more than 80% of suspended solids of paper mills wastewater were removed by using a simple sedimentation technique. It also was also found that more than 75% suspended solids were removed by primary clarifier (Rajvaidya & Markandey, 1998). Gubelt and group (Gubelt et al. (2000) reported that more than 90% of total suspended solid (TSS) was removed using air flotation unit (unstable system) (Gubelt et al., 2000). Similarly, the elimination of TSS (> 90%) using dissolved air flotation system was also claimed (Wenta & Hartman, 2002).

2.1.2 Coagulation and Precipitation

In the treatment of pulp and paper mill wastewater, coagulation and flocculation are typically used in the tertiary treatment stage rather than the primary treatment. Flocculants and coagulants have been extensively explored for the treatment of wastewater and water to remove colloidal suspensions and formation of micro-flocs from solutions and suspensions (Sharma et al., 2023). Various coagulants such as metalloenzyme (horseradish peroxidase), aluminum sulfate, $H_2N(CH_2)_6NH_2$ C_3H_5ClO polycondensate (HE), and polyaziridine were used for the removal of AOX, TOC, and dyes. Tong et al. and Ganjidoust et al. were found that the modified chitosan showed effective removal of these contaminants rather than other coagulants (Ganjidoust et al., 1997). In a study, alum was used as a coagulant for the eradication of COD of paper mill, puree, and bleaching-based wastewater. Among them, alum effectively removed 96% of COD of paper machine-based effluents followed by pulping (50%) and bleaching-based (20%) effluent. However, Rohella and group observed that polyelectrolytes removed turbidity, COD, and dyes more effectively rather than alum (as coagulant) (Rohella et al., 2001). In 1989, Sheela and Distidar reported that 63% of dissolved solids was removed from black liquor effluent using precipitation method under CO_2 atmosphere (Sheela & Distidar, 1989). However, Wang and Pan observed that not only the settleability but also the concentration of COD, turbidity, and suspended solids were increased in treated solid while using the polyethylene oxide as a coagulant (Wang et al., 1997). In 1994, Chernoberezhskii and group found that the coagulation with aluminum sulfate effectively removed the SO_4^{2-} and SO_3^{2-} from pulp and paper industries (Chernoberezhskii et al., 1994).

2.1.3 Adsorption

It is one of the most commonly and extensively used technologies accepted by industrial sectors for the treatment of wastewater. It occurs due to the interfacial interaction of pollutants (adsorbates) onto the solid surface (adsorbent) from wastewater. In 1991, Murthy and group reported an effective eradication of dye using activated charcoal (Murthy et al., 1991). Similarly, activated coke significantly removed the dye, DOC, AOX, and COD from bleached effluent. A blast furnace dust (BFD) and slag were used for the eradication of lignin. During the analysis, BFD showed an effective removal efficiency (80.4%) than that of slag (61%). However, AOX (originated from kraft effluents) were effectively remove using PACT™ process (Narbaitz et al., 1997).

2.1.4 Chemical Oxidation

In the chemical oxidation procedure, the oxidizing catalysts initiated the movement of electrons in the reaction medium for the oxidation of chemical moieties. In 1999, Balcioglu and Ferhan found that the efficacy of photocatalytic oxidation for the treatment of kraft pulp wastewater completely depended on the concentration of COD and chloride present in the sample (Akmehmet Balcioğlu & Çeçen, 1999). Horseradish peroxidase effectively decolorized (>50%) the kraft effluent within 3 h. Additionally, a comparative study was performed for the degradation of phenolic compounds using different photocatalysts such as $O_2/TiO_2/UV$, $O_2/ZnO/UV$, O_3, and O_3/UV. Among them, $O_2/TiO_2/UV$ and $O_2/ZnO/UV$ showed the substantial oxidative degradation of phenols. However, photo-Fenton and Fenton reactions were found to be a promising treatment technology for kraft mill effluents. Similarly, Hassan and Hawkyard also worked on a combined treatment technology (ozone coupled Fenton's reagent) for the removal of dyes (Hassan & Hawkyard, 2002). The advanced combined technology effectively removed color (100%) at pH 4–5. In 2000, Dufresne and group catalytically oxidized the total reduced sulfur resulting odor-free products.

2.1.5 Membrane Filtration

Membrane filtration (MF) treatment technology was found to be an appropriate technology for the removal of COD, AOX, and colors from pulp and paper industries. A comparative study was performed between ultrafiltration and ultrafiltration-attached DAF system against TOC, colors, and SS of the effluent. Herein, ultrafiltration plus air flotation method showed effective removal in TOC (65%) and colors (90%) rather than bare ultrafiltration. However, SS of the sample was completely removed by both treatment technology. However, it was also observed that the MF and granular MF are also one of the effective processes for the eradication of heavy metals (Merrill et al., 2001).

2.1.6 Ozonation

The utilization of ozone to eliminate pollutants is known as ozonation, which can be achieved either by direct ozone-contaminant reactions or through the oxidation process of free radicals produced during ozone decomposition in water. Unlike chemical oxidants, these free radicals are highly reactive and less discriminatory. The COD, TOC, and toxicity of pulp mill effluent were efficiently reduced using ozonation process. Chelants such as ethylenediaminetetraacetic acid (EDTA) that can significantly inhibit the performance of an aerobic biological treatment system, and COD from pulp mill effluent were successfully treated using ozonation. Herein, 90% of EDTA and 65% of COD were removed using this effective removal process (Korhonen et al., 2000). Similarly, COD and DOC of pulp mill effluent were effectively removed using ozonation (Oeller et al., 1997). In 2000, Laari and group reported the ozonation was effectively removed the lipophilic wood extract from TMP effluents. The high dose of ozone (100–300 mg/dm^3) was necessitated to half-fold of total concentration of extract. Similarly, 90% of resin acid (RA) was successfully removed using 0.2 mgO_3/initial mg COD of ozone in the reaction system. Heterogenous photocatalysis and ozonation were found to be efficient methods for the removal of TOC, COD, AOX, and dyes from BKME1 wastewater (TORRADES et al., 2001). However, at doses of ozonation for 10 min, dyes were efficiently removed.

2.2 *Biological Treatment*

2.2.1 Aerobic Treatment

2.2.1.1 Activated Sludge Process

A two-stage activated sludge method was utilized to reduce both BOD and COD. Similarly, upgraded activate sludge with sequential floating biological bed enhanced the removal of COD from 51% to 90% and BOD from 70% to 93%, respectively. Citrobacter sp., Pseudomonas putida, and Enterobacter sp. functionalized ASP effectively removed the dyes, COD, BOD, and phenolic and sulfide compounds. 1,1-dichlorodimethyl sulfone (DDS), chlorinated phenols, and chlorinated acetic acids were successfully removed using oxygenated sludge (Singh, 2018). Similarly, AOX was effectively removed (46%) using two activated sludge systems (Fälth et al., 2001). Further, in 1999, Demirbas and group reported the substantial removal of AOX using activated sludge process (Sharma et al., 2023).

2.2.1.2 Aerated Lagoons

An aerated lagoon was found to remove approximately 70% of the AOX, with a short residence time during treatment, when wastewaters generated from mills gets mixed with effluents at chlorinated stage. Simple mixing was found to achieve almost COD removal of 15–46% (Peitz & Xavier, 2019). A study was conducted to check the efficacy of an aerated lagoon at temperatures of about 22–35 °C and pH ~7.3 which achieved ammonia removal of 67% from black liquor spill. In a controlled batch study, an operating lagoon can show significant variations in the removal of TOC and AOX from bleached kraft effluent under both anaerobic and aerobic conditions (Bailón-Salas et al., 2017). A full-scale lagoon was able to remove 30–40% of COD, while a pilot-scale plant achieved 60–70% COD removal (Bailón-Salas et al., 2017). A study reported the AOX removal of 65% from pulp and paper mill effluent using an aerated lagoon, while the removal of 50–75% BOD_7 and

10–50% chlorinated phenolics were achieved with the same treatment (Kumar et al., 2020). Oxidation ponds were able to remove chemical compounds over 50%, whereas aerated lagoon effectively removes BOD, polychlorinated phenolics AOX, and chlorinated (Godini et al., 2021).

2.2.1.3 Aerobic Biological Reactors

Several studies have shown that sequencing batch reactor (SBR) treatment is effective in removing organic pollutants from kraft mill wastewater (Mazhar et al., 2019). One study found that SBR treatment achieved methanol removal of 100% and dissolved COD removal of 90%. Similarly, high-rate compact reactors (HCRs) were able to remove substantial amounts of BOD, COD, TOC, resin, and lignin from TMP wastewater with a retention time of 1.5 h (Patel et al., 2021). A biological compact reactor achieved BOD removal of 93% and COD removal of 65%, while a membrane bioreactor was found to be capable of achieving around 93% removal of TOC (Tawfik et al., 2022). A suspended carrier biofilm reactor (SCBR) is highly effective at removing chronic toxicity from the effluent. A biofilter achieved 76% removal of BOD, 62% removal of COD, 81% removal of SS, and 48% removal of AOX whereas another study revealed that aerobic treatment of white water with high strength was able to achieve BOD removal of 70–80% (Jagaba et al., 2022).

2.2.2 Anaerobic Treatment

It is often preferred to treat high-strength organic wastewater. In the past, pulp industries effluents are less concentrated with a BOD ranging from 300 to 2000 mg L^{-1} draws limited attention for anaerobic treatment (Tawfik et al., 2022). Several anaerobic processes, including up-flow sludge blanket (UASB), anaerobic filter, anaerobic lagoon, fluidized bed, and anaerobic contact reactors are generally used for the treatment of pulp and paper effluents. Continuous anaerobic treatment of black liquor generated from kraft mill was found to be feasible at different loading rates, achieving biodegradable COD removal of 96% (Gupta & Gupta, 2019). Both anaerobic and aerobic treatments were effective in removing COD from TMP mill effluent, achieving 84% and 86% removal, respectively. However, due to the presence of toxic substances and low biodegradability (harm methanogens), effluents bleached with chlorine were not well suited for anaerobic treatment (Chakraborty et al., 2019). A novel treatment approach has been developed that involved a three-step process. The first step involved removing S^{-} and other volatile compounds from the condensate through a process called stripping. The second step is regenerative thermal oxidation of the stripper off-gases. The third and final step involves the adsorption of sulfur oxide to treat foul condensate containing sulfide from the black liquor (Vítězová et al., 2020).

Anaerobic biological granular-activated carbon achieved a removal of 50% of COD and color. A UASB reactor with a hydraulic retention time of 6 h can achieve

COD removal of 66% and sulfate removals of 73% (Sarfraz et al., 2022). In a study conducted by Peerbhoi at the University of Roorkee, India, it was determined that the pollutants in black liquor were not easily degradable and therefore anaerobic biological treatment was not a feasible option (Singh et al., 2022). An evaluation of two anaerobic systems, namely anaerobic filters and fluidized bed, was conducted in laboratory-scale reactors. It was reported that the fluidized bed with porous packing achieved an organic removal efficiency of 81.5%, whereas the anaerobic filters on corrugated plastic tubes achieved a 50% removal. In a study, a fluidized bed reactor was found to reduce the BOD of debarking wastewater by 50% (Patel et al., 2021). Reports suggested that an 80% of COD removal was consistently achievable, but the leftover COD was found around 800 mg L^{-1}, indicating the need for additional treatment. Anaerobic treatment systems were less effective than aerobic systems for the treatment of sulfite-spent liquor (Tawfik et al., 2022).

2.2.3 Fungal Treatment

A fungal species was discovered (*Pencillium* sp.) that removed50% of AOX and color from soft-wood bleachery effluents in just two days. Various authors have also investigated the ability of different fungal species to reduce the color in kraft mill effluent (Sharma & Rath, 2021). A significant reduction in color and COD was also reported by using the white rot fungi *Phanerochaete chrysosporium* and *Trametes versicolor*. A combination of white-rot fungi, including *P. chrysosporium, Pycnoporus sanguineus, Pleurotus ostreatus,* and *Heterobasidion annosum*, along with surfactants, could effectively remove color, COD, and lignin content. *Pleurotus ostreatus*, a fungal species, achieved lignin removal of 77%, BOD removal of 76.8%, COD removal of 60%, and color removal of 80% (Ram et al., 2020). The combined use of enzymes and fungi was investigated for the removal of harmful organics from white water, reporting a >90% reduction in lignin and a 20% reduction in resin and fatty acids. It was also found that fungus *T. versicolor* and fungal culture filtrate (FCF) obtained from these organisms were effective in degrading dissolved and colloidal substances (Kumar et al., 2022a, b).

3 Combined Treatment Processes

An integrated system combines the unique features of two or more processes to improve the overall performance as shown in Tables 21.2 and 21.3. For example, wet oxidation and coagulation combinedly can achieve COD removal efficiency of 51%, color removal of 83%, and lignin removal of 75%. A biofilm reactor with the help of ozone can achieve COD removal of 80%, while chemical oxidation with ozone at 150 °C can achieve COD removal of up to 50% and can remove 90% of wood extract from TMP wastewater (Sathya et al., 2022). Posttreatment methods such as ozonation or electrolysis were recommended for the treatment of effluents

Table 21.2 Combined treatment processes for wastewater from pulp and paper industry

Combined processes	Pollutant removal (%)	References
Coagulation followed by wet oxidation	COD 51%, color 83%, Lignin 75%	Verenich et al. (2000)
Biofilm reactor combined with ozone	COD 80%	Helble et al. (1999)
Chemical oxidation combined with ozone	Wood extractives 90%, COD 50%	Laari et al. (1999)
Activated sludge process combined with ozone	Lignin degraded effectively	Kabdasli et al. (1997)
Dissolved air flotation combined with chemical precipitation	SS 93%, BOD_7 50%, COD 57%, phosphate 92%, nitrogen 52%	Jokela and Keskitalo (1999)
Activated sludge combined with ozonation	COD 97% and BOD 97%	Arslan-Alaton et al. (2008)
Chemical combined with biological methods (BioFerric)	COD 40 to 50% compared to the activated sludge system	Kabdasli et al. (1997)
Kaldnes (anaerobic combined with aerobic) at 55 °C	60% COD (soluble)	Jahren and Ødegaard (1999)
Anaerobic combined with aerobic treatment system	For the treatment of bleached Kraft pulp and paper industries effluents	Duncan et al. (1992)
Aerated lagoon combined with secondary clarifier	Treatment of bleached Kraft effluent 87% SS, 96% BOD_7, 65% COD, 53% AOX, and 22% color	Lescot and Jappinen (1994)
An anoxic selector with an aerated basin and a secondary clarifier in series	77% COD, 98–99% BOD, 72% TN, and 81% TP. The removals of resin and fatty acids: 96% and 98%, respectively. Microtox toxicity reduction: 99%.	Araujo et al. (2014)
Aerobic process called low sludge production (LSP)	Lower the biological sludge by 80–90%.	Krzeminski et al. (2017)
Physicochemical and biological treatment systems combined with aerated lagoon for effluent polishing	BOD 98–99%, COD 91%, SS 97%, and color 90%	Eskelinen et al. (2010)
Biofilm reactor combined with two aerobic and one anaerobic reactor	COD 50%, BOD_7 90%, AOX 50%, and ClO_3 90%.	Shaw et al. (2002)
Aerobic reactor combined with anaerobic reactor	Color 94% and TOC 66%	Schnell et al. (2000)

contaminated with high concentration of NH_4^+–N and COD to comply with certain discharge standards. Activated sludge process coupled with ozone can efficiently degrade lignin (Ahmad et al., 2021). Similarly, dissolved air flotation combined with chemical precipitation can achieve suspended solids (SSs) removal of 93%, BOD_7 removal of 50%, COD removal of 57%, phosphorus removal of 92%, and nitrogen removal of 52%. When activated sludge is combined with tertiary treatment (ozonation), the process can achieve COD and BOD removal of up to 97% (Agrawal et al.,

Table 21.3 Removal in COD of pulp and paper industries wastewater after physiochemical, biological, and aerobic and anaerobic treatment processes

Process	Initial (mg L^{-1})	% removal
Use of aluminum chloride as a coagulant and (starch-g-PAM-g-PDMC) Flocculant	1358	
Electrocoagulation (Al)	2500	~95
Flocculation (polydiallyldimethylammonium)	2900 ± 90	>90
Flocculation (chitosan)	500–1800	>80
Flocculation (polyaluminum chloride)		40–45
Precipitation (CaO)	1510	Up to 90
Electrocoagulation (EC/Fe°) flotation and UV/TiO_2/H_2O_2	1310	88
Electrocoagulation (Al)	426.03	75%
Laccase-polymerized membrane filtration	785	>60
Ozonation	1315	~20
Ozonation+electrolysis	1382	>50
Fungi+solar photo-Fenton (Cryptococcus+Fe^{2+}/H_2O_2)	1802	>90
A consortium of *Merulius aureus*, an unidentified genus and *Fusarium sambucinum*	2013 ± 11.7	89.4
Solar photo-Fenton (Fe^{2+}/H_2O_2/UV)	898.9	~90
Pleurotus sajor-caju	392 ± 2	72
Trametes versicolor	392 ± 2	77
Solar photocatalytic degradation (nTiO_2/UV)	2075	75
Phanerochaete chrysosporium	392 ± 2	81
Rhizopus oryzae	392 ± 2	82
Aspergillus niger	7521	60
Activated sludge	863	76
Hydrogenation with $NaBH_4$ reduction of activated sludge	1231	95
Granular-activated carbon sequencing batch biofilm reactor (GAC-SBBR)	750	97.2
GAC-SBBR	1152 ± 93	92 ± 6
Thermophilic submerged aerobic membrane bioreactor (TSMABR)	3650 ± 289	92.3 ± 0.7
TASMBR—Electrooxidation (EO)	3650 ± 289	98.2 ± 0.3
Upflow anaerobic sludge blanket (UASB)	1410 ± 10	~78
Modified anaerobic baffled reactor (MABR)	1000	71
Submerged anaerobic membrane bioreactor (SAnMBR)	2782–3460	~90

Source: Acquah et al. (2021), Jallouli et al. (2018), Kamali and Khodaparast (2015)

2020). Combining chemical and biological methods resulted in an additional 40–50% of COD removal compared to the activated sludge system. A moving bed biofilm reactor (anaerobic followed by aerobic) operating at 55 °C removed approx. 60% of soluble COD from TMP white water (Liang et al., 2021). A combined anaerobic and aerobic treatment system was suggested as the treatment of bleached kraft pulp and paper industries effluents. A study was conducted in Finland to treat effluent generated from bleached kraft mill using an aerated lagoon combined with a secondary clarifier. The treatment achieved removal efficiency of 87% for SS, 96% for BOD_7, 65% for COD, 53% for AOX, and 22% for color. While upgrading the aerated basin at Monsteras mill resulted in 77% COD removal, 98–99% BOD removal, 72% TN removal, and 81% TP removal (Bhanot et al., 2020). The system includes an aerated basin, an anoxic selector, and a secondary clarifier in series which reduces 99% toxicity with removal efficiency of 96% and 98% for resin and fatty acids, respectively (Dagar et al., 2023). Additionally, an aerobic biological process known as low sludge production (LSP) was investigated to check its effectiveness for reduction in biological sludge by 80–90%. In Brazil, aerated lagoon was investigated through a combination of physicochemical and biological system including an aeration basin, a primary clarifier and a secondary clarifier, achieved BOD, COD, SS, and color removal of 99%, 91%, 97%, and 90% respectively, from the pulp and paper industries effluent (Elakkiya & Niju, 2020). A biofilm reactor combined with two aerobic and one anaerobic reactor was used to remove COD (50%), BOD_7 (80–90%), AOX (50%), and ClO_3 (90%). Combining an aerobic reactor with an anaerobic reactor was able to remove 94% of color and 66% of TOC.

A biological treatment was able to achieve removal rates of 87–95% for BOD, 70–77% for COD, and 80–94% for resin and fatty acids (Elakkiya & Niju, 2020). An experiment was conducted using an SBR and found that at temperatures between 20 and 40 °C, it was able to achieve resin acid (RA) removal of 100%, fatty acid FA removal of 96%, DCOD removal of 76%, and TDS removal of 34%. Meanwhile, an MBR can remove FA and RA, DCOD, TDS up to 100%, 84%, and 37%, respectively, at 40–50 °C (Poojamnong et al., 2020). Several techniques were proposed to enhance the integrated treatment system for the removal of pollutants from paper mill effluent and are as follows:

- Expanding the aeration basin's capacity.
- Incorporating an additional DAF.
- Adding Cl_2 gas to address the sludge stored issue.
- Introducing oxygen to treat BOD during heavy rainfall and flooded conditions.

A study was conducted to check the efficacy of biological treatment systems in degrading chlorinated organic compounds found in wastewater generated from pulp and paper industry. The results revealed that aerated lagoon and activated sludge process can achieve AOX removal of 32% and 10–65%, respectively. Additionally, a bacterial species (*Aeromonas formicans*) was identified to be effective in treating black liquor generated from pulp and paper industries (Sharma et al., 2021).

4 Futuristic Approach: Wastewater Treatment and Recovery of Value-Added Products

In addition, to protect the environment from the unwanted effects of waste, wastewater treatment is a key platform for the utilization of recovered resources. There is necessity for some advanced technologies in wastewater treatment sector so that organic matter, energy, and other products can be recovered as resource to the recovery of value-added by-products instead of being needlessly discarded. Waste recycling and recovery will markedly boost the product diversity scheme and economic growth (Sátiro et al., 2022). As of now, researchers and industrial sectors are focusing on an integrated approach to biorefinery for environmental protection and extraction of costs and benefits from waste. The industrial sectors can implement biorefinery approach in two ways: first, the industry should construct integrated biorefinery in the industrial premises itself, second, the industrial sector should collect all the types of generated effluent and try to extract valuable byproducts from the effluent. This second approach may provide an opportunity for new industries which can set up a resource to recovery facility for effluents received from different sources. The only demerit in the second case is that the collection and transportation costs would increase the total operating and maintenance cost. (Sailwal et al., 2020).

PMS is a waste material which also contains value-added products that could be recovered by various physicochemical and biological treatment technologies. However, the chemical composition of PMS varies to a large extent based on the raw material being used in the paper mills, and the major component in the biomass is lignocellulose (Acquah et al., 2021). Many researches have been carried out to establish the potential usage of PMS and the applications include biogas production, vermicomposting (Badhwar & Singh, 2022), ethanol, (Yoon et al., 2022), pyrolysis, material preparation (Goel & Kalamdhad, 2017), and compost production (Boshoff et al., 2016). On addition, Majumdar et al. reported PMS as a support matrix for bacterial species growth and development in solid-phase fermentation (Majumdar et al., 2019; Majumdar et al., 2020b), as a substrate that produces bacterial pigment (Majumdar et al., 2020a), and could be used as a source of cellulose nanocrystals through bacterial delignification (Majumdar et al., 2020a). Moreover, the presence of lignin makes PMS a valuable lignin-based substance for the production of lignin-based nanoparticles, a good binder for composite or pancaked wood products, a conditioner for soil, and an adhesive for linoleum (Kumar et al., 2022b).

4.1 Lignin: The Unexploited Wealth from Wastewater and Sludge of Pulp and Paper Industries

Pulp and paper industry effluent and PMS are characterized with hemicellulose, lignin, phenol, and few usual xenobiotics (Strassberger et al., 2014). In general, hemicellulose and cellulose are extensively used in sugar, biofuel, and paper production, whereas lignin is the byproduct released during the production process of these materials (Cao et al., 2018). Consequently, lignin has failed to gain attention by the researcher and industrialists, in spite of having high calorific value and could be used as source of energy in industrial sector (Preethi et al., 2022). Commonly, lignin is being used as a source of bioenergy in the pulp and paper sector. Moreover, kraft lignin (KL) is main source of energy in kraft sector, while lignosulfonates are being sold in the market to generate revenue by the industries as they have many applications (Shu et al., 2016). In the present scenario, many researches have been published and many works are being carried out on lignin-based materials and application due to its excellent physicochemical properties as shown in Fig. 21.3. Due to the complex structure and chemical properties like recalcitrant nature of lignin, the researchers face most significant caveat in carrying out lignin-based novel works (Domínguez-Robles et al., 2020).

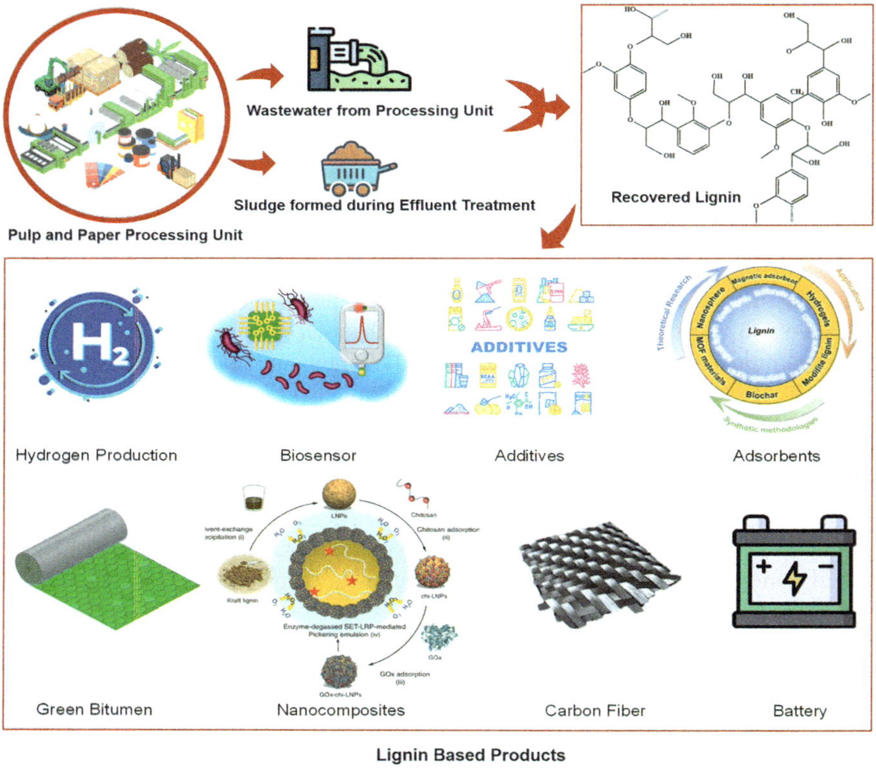

Fig. 21.3 The utilization of lignin that can be recovered from effluent of pulp and paper industry

4.2 Lignin Recovery from Black Liquor

The recovery of lignin is a process which is being carried out in the last twentieth century. Black liquor constitutes of approx. 40% lignin that can be used for energy generation and production of various chemical feedstocks (Wang et al., 2021). It can also be used for the generation of numerous value-added products, but its applications are limited till date. Lignin recovery from the black liquor can be a laborious job. However, the development of various processes like ultrafiltration, acid precipitation, and ion exchange for the recovery of lignin from black liquor have made the job simple (Oliveira et al., 2020).

The value-added products made from purified lignin include phenol, activated carbon, aromatic compounds, bio-oil, and others (Cao et al., 2018). Black liquor is used for a variety of purposes; some researchers reported that it may be transformed into syngas, biofuel, and charcoal when heated between 400 and 900 °C in the absence of oxygen. Both biofuels and syngas may be used to generate electricity. In contrast, biochar may be utilized as sequestering agents, adsorbents, biofertilizers, immobilizing heavy metals like arsenic and cadmium in the environment, and other similar applications (Roy et al., 2019; Sailwal et al., 2020). The bacteria *Paenibacillus glucanolyticus* has the ability to break down the lignin in black liquor to produce succinic, ethanol, lactic, malonic acids, and propanoic among other compounds of added value (Mathews et al., 2016).

5 Challenges and Futuristic Approach

The paper and pulp industries are witnessing daily expansion, and their demands are steadily on the rise. While not of an extreme nature, the environmental impact of these industries is a cause for concern, owing to the substantial volume of waste they generate. These industries produce modified lignin and hemicellulose, which constitute the primary components of their wastewater, thus rendering the local environment susceptible to contamination. Sludge generation is indeed a significant issue associated with the treatment of wastewater in the pulp and paper industry. The moist waste produced in the pulp and paper industry gives rise to a range of harmful gases, including sodium sulfides, nitrogen oxides, sulfur compounds, methyl mercaptan, hydrogen sulfides, and chlorine dioxides. These emissions have the potential to result in chronic health problems, such as headaches and nausea, when individuals are exposed to them. Additionally, the effluents stemming from the sludge contain minute quantities of metals that are released into water bodies. This release of trace metals has adverse environmental consequences, contributing to climate change and negatively impacting ecosystems.

Therefore, the pulp and paper industry confront several notable challenges within the realm of water treatment processes, owing to its considerable water usage and the production of wastewater containing substantial quantities of organic and chemical pollutants. Among the principal challenges encountered in water treatment processes for this sector are the following:

A. Wastewater Volume and Composition:

- Generate large volumes of wastewater, often exceeding their capacity for treatment and disposal.
- The wastewater contains a diverse mix of organic compounds, lignin, pulp fibers, and various chemicals, making treatment complex.

B. Chemical Usage and Residues:

- Recovering and reusing chemicals from the pulping and bleaching processes.
- The presence of harmful chemicals, such as chlorine, chlorine dioxide, and other pulping chemicals, necessitates their removal or conversion to less harmful forms.

C. Biological Treatment:

- High BOD in wastewater requires effective biological treatment systems to reduce the organic load and oxygen demand.
- Nutrient Management: Balancing the nutrient levels in wastewater for optimal microbial activity can be challenging.

D. Advanced Treatment:

- Membrane fouling due to high BOD, COD, TOC, and TS.
- Reduced treatment efficiency.
- High cost for the back cleaning of membrane.

E. Solid Waste Management:

- Managing and disposing of the sludge generated during water treatment processes.

F. Innovative Technologies:

- Development of innovative and efficient technologies is time-consuming process.

Addressing these challenges often requires a combination of advanced water treatment technologies, process optimization, environmental monitoring, and the adoption of sustainable practices. Many pulp and paper mills are investing in research and development to improve their water treatment processes, reduce their environmental impact, and meet stringent regulatory requirements.

It is crucial to employ conventional and advanced treatment methods for managing wastewater in the pulp and paper industry due to its significant environmental threat. Rapidly advancing state-of-the-art treatment technologies such as biosorption, photo-electrolysis, advanced oxidation methods (such as photo-Fenton oxidation), filtration-assisted crystallization technology (FACT), etc., represent promising avenues for future wastewater treatment techniques. In this book chapter, integrated approaches, including combining coagulation with wet oxidation, chemical oxidation with ozone, activated sludge processes with ozone, dissolved air flotation with ozone, chemical methods combined with biological approaches (like BioFerric), Kaldnes (mixing anaerobic and aerobic processes), and aerated lagoon with a secondary clarifier, have been explored. These strategies are effective in maximizing the overall benefits of a comprehensive approach while reducing operational and maintenance costs. In addition to treating black liquor, these hybrid treatment technologies can also be used to recover lignin and value-added products (like ethanol) from paper mill sludge. This promotes the concept of turning waste into wealth in industrial effluent treatment. As the scarcity of clean water becomes an increasing challenge in this century, the reuse of treated wastewater is imperative. To achieve this, various computational, analytical, hydrodynamic, and advanced simulations such as black-box models, fluid dynamics, artificial intelligence, and artificial neural networks, as well as statistical models like IAWQ-activated sludge models (kinetic-based), Takács' models (settling velocity functions), and double exponential models, may be employed in future research to address water demand issues. Pulp mill sludge recovered through various physiochemical and biological treatment technologies can potentially be used in a variety of applications, including the production of ethanol, biogas, compost, and vermicomposting. This opens up opportunities for innovative business strategies and creates favorable circumstances, reinforcing the intersection of energy, the environment, health, and water.

6 Conclusion

The following conclusions are reached in light of the aforementioned prospects:

1. Wastewater from pulp and paper industries can be treated using aerobic and anaerobic treatment systems, with the exception of bleaching kraft wastewaters, as presence of chlorine in wastewater causes degradation of bacterial cells and hampers the anaerobic or aerobic treatment systems.
2. The wastewater characterized with very high pollutant concentration requires step-down treatment facilities after anaerobic treatment due to the presence of residual COD in the waste stream after anaerobic treatment.
3. Anaerobic treatment system followed by aerobic treatment has always proved to be a best treatment option for the effluent from pulp and paper industry.
4. Fungal treatment, coagulation, flocculation, and advanced oxidation processes are all capable of efficient color removal.
5. Adsorption, membrane filtration, and ozonation can be used to remove AOX and chlorinated phenolic chemicals.
6. When more and two physicochemical processes are combined together, then there are chances that they can effectively remove recalcitrant pollutants.

7. On coupling physicochemical and biological treatment processes and optimizing the various influential factors result in a sustainable treatment technology for black liquor.
8. Further, there is requirement for detailed studies on the effective removal of chlorinated phenolic compounds and AOX.

Acknowledgments Director of CSIR-NEERI is thankfully acknowledged for giving the opportunity to pursue the work in CSIR-NEERI, Nagpur, India. CJ and AS acknowledge the University Grant Commission, New Delhi, India, for providing Junior and Senior Research Fellowship, respectively. The article is checked for plagiarism using the iThenticate software and recorded in the Knowledge Resource Center, CSIR-NEERI, Nagpur, for anti-plagiarism.

References

Abedinzadeh, N., Shariat, M., Monavari, S. M., & Pendashteh, A. (2018). Evaluation of color and COD removal by Fenton from biologically (SBR) pre-treated pulp and paper wastewater. *Process Safety and Environmental Protection, 116*, 82–91. https://doi.org/10.1016/j.psep.2018.01.015

Acquah, M. N., Essandoh, H. M. K., Oduro-Kwarteng, S., Appiah-Effah, E., & Owusu, P. A. (2021). Degradation and accumulation rates of fresh human excreta during vermicomposting by Eisenia fetida and Eudrilus eugeniae. *Journal of Environmental Management, 293*, 112817. https://doi.org/10.1016/j.jenvman.2021.112817

Agrawal, K., Bhatt, A., Bhardwaj, N., Kumar, B., & Verma, P. (2020). Integrated approach for the treatment of industrial effluent by physico-chemical and microbiological process for sustainable environment. In M. Shah & A. Banerjee (Eds.), *Combined application of physico-chemical & microbiological processes for industrial effluent treatment plant* (pp. 119–143). Springer. https://doi.org/10.1007/978-981-15-0497-6_7

Ahmad, N. N. R., Ang, W. L., Leo, C. P., Mohammad, A. W., & Hilal, N. (2021). Current advances in membrane technologies for saline wastewater treatment: A comprehensive review. *Desalination, 517*, 115170. https://doi.org/10.1016/j.desal.2021.115170

Akmehmet Balcioğlu, I., & Çeçen, F. (1999). Treatability of Kraft pulp bleaching wastewater by biochemical and photocatalytic oxidation. *Water Science and Technology, 40*(1), 281–288. https://doi.org/10.1016/S0273-1223(99)00396-0

Araujo, J. R., Archanjo, B. S., de Souza, K. R., Kwapinski, W., Falcão, N. P. S., Novotny, E. H., & Achete, C. A. (2014). Selective extraction of humic acids from an anthropogenic Amazonian dark earth and from a chemically oxidized charcoal. *Biology and Fertility of Soils, 50*(8), 1223–1232. https://doi.org/10.1007/s00374-014-0940-9

Arslan-Alaton, I., Gursoy, B. H., & Schmidt, J.-E. (2008). Advanced oxidation of acid and reactive dyes: Effect of Fenton treatment on aerobic, anoxic and anaerobic processes. *Dyes and Pigments, 78*(2), 117–130. https://doi.org/10.1016/j.dyepig.2007.11.001

Ashrafi, O., Yerushalmi, L., & Haghighat, F. (2015). Wastewater treatment in the pulp-and-paper industry: A review of treatment processes and the associated greenhouse gas emission. *Journal of Environmental Management, 158*, 146–157. https://doi.org/10.1016/j.jenvman.2015.05.010

Badhwar, V. K., & Singh, C. (2022). Vermicomposting of textile mill sludge employing *Eisenia fetida*: Role of cow dung and tea waste amendments. *Environmental Science and Pollution Research, 29*(13), 19823–19834. https://doi.org/10.1007/s11356-021-17185-z

Bailón-Salas, A. M., Ordaz-Díaz, L. A., Valle-Cervantes, S., Lopez-Miranda, J., Urtiz-Estrada, N., Páez-Lerma, J. B., de León-Mata, G. D., & Rojas-Contreras, J. A. (2017). Bacterial diversity in two aerated lagoons of a pulp and paper effluent and their interaction with a commercial inoculum using PCR-DGGE. *BioResources, 12*(3), 3.

Bajpai, P. (2018). Biological treatment of pulp and paper mill effluents. In P. Bajpai (Ed.), *Biotechnology for pulp and paper processing* (pp. 313–369). Springer. https://doi.org/10.1007/978-981-10-7853-8_16

Bhanot, P., Celin, S. M., Sreekrishnan, T. R., Kalsi, A., Sahai, S. K., & Sharma, P. (2020). Application of integrated treatment strategies for explosive industry wastewater—A critical review. *Journal of Water Process Engineering, 35*, 101232. https://doi.org/10.1016/j.jwpe.2020.101232

Bhatnagar, A. (2015). Assessment of physico-chemical characteristics of paper industry effluents. *Rasayan Journal of Chemistry, 8*, 143–145.

Boshoff, S., Gottumukkala, L. D., van Rensburg, E., & Görgens, J. (2016). Paper sludge (PS) to bioethanol: Evaluation of virgin and recycle mill sludge for low enzyme, high-solids fermentation. *Bioresource Technology, 203*, 103–111. https://doi.org/10.1016/j.biortech.2015.12.028

Buyukkamaci, N., & Koken, E. (2010). Economic evaluation of alternative wastewater treatment plant options for pulp and paper industry. *Science of the Total Environment, 408*(24), 6070–6078. https://doi.org/10.1016/j.scitotenv.2010.08.045

Cao, L., Yu, I. K. M., Liu, Y., Ruan, X., Tsang, D. C. W., Hunt, A. J., Ok, Y. S., Song, H., & Zhang, S. (2018). Lignin valorization for the production of renewable chemicals: State-of-the-art review and future prospects. *Bioresource Technology, 269*, 465–475. https://doi.org/10.1016/j.biortech.2018.08.065

Central Pollution Control Board (CPCB). (1986). *The environment (protection) rules*. Government of India.

Chakraborty, D., Shelvapulle, S., Reddy, K. R., Kulkarni, R. V., Puttaiahgowda, Y. M., Naveen, S., & Raghu, A. V. (2019). Integration of biological pre-treatment methods for increased energy recovery from paper and pulp biosludge. *Journal of Microbiological Methods, 160*, 93–100. https://doi.org/10.1016/j.mimet.2019.03.015

Chandra, R., Abhishek, A., & Sankhwar, M. (2011). Bacterial decolorization and detoxification of black liquor from rayon grade pulp manufacturing paper industry and detection of their metabolic products. *Bioresource Technology, 102*(11), 6429–6436. https://doi.org/10.1016/j.biortech.2011.03.048

Chaudhry, S., & Paliwal, R. (2018). Techniques for remediation of paper and pulp mill effluents: Processes and constraints. *Handbook of Environmental Materials Management, 12*.

Chernoberezhskii, Y. M., Dyaglieva, A., & Barysheva, I. A. (1994). Coagulation treatment of waste-water from paper and pulp plants. *Russian Journal of Applied Chemistry, 67*(3), 355–359.

Chhonkar, P. K., Datta, S. P., Joshi, H. C., & Pathak, H. (2000). Impact of industrial effluents on soil health and agriculture—Indian experience: Part I—Distillery and paper mill effluents. *JSIR, 59*(5) [May 2000]. http://nopr.niscpr.res.in/handle/123456789/17773

Dagar, S., Singh, S. K., & Gupta, M. K. (2022). Economics of advanced technologies for wastewater treatment: Evidence from pulp and paper industry. *Frontiers in Environmental Science, 10*. https://www.frontiersin.org/articles/10.3389/fenvs.2022.960639

Dagar, S., Singh, S. K., & Gupta, M. K. (2023). Integration of pre-treatment with UF/RO membrane process for waste water recovery and reuse in agro-based pulp and paper industry. *Membranes, 13*(2), 2. https://doi.org/10.3390/membranes13020199

Domínguez-Robles, J., Cárcamo-Martínez, Á., Stewart, S. A., Donnelly, R. F., Larrañeta, E., & Borrega, M. (2020). Lignin for pharmaceutical and biomedical applications—Could this become a reality? *Sustainable Chemistry and Pharmacy, 18*, 100320. https://doi.org/10.1016/j.scp.2020.100320

Duncan, A., Thia, B. C., Donlon, P., & Jelbart, J. (1992). Treatment of pulp and paper mill effluent using anaerobic-aerobic lagoons. https://publications.csiro.au/rpr/pub?list=BRO&pid=procite:33285a4d-3e86-4c8e-931c-2b8bcdf5403a.

Elakkiya, E., & Niju, S. (2020). Application of microbial fuel cells for treatment of paper and pulp industry wastewater: Opportunities and challenges. In K. M. Gothandam, S. Ranjan,

N. Dasgupta, & E. Lichtfouse (Eds.), *Environmental biotechnology* (Vol. 2, pp. 125–149). Springer International Publishing. https://doi.org/10.1007/978-3-030-38196-7_6

Eskelinen, K., Särkkä, H., Kurniawan, T. A., & Sillanpää, M. E. T. (2010). Removal of recalcitrant contaminants from bleaching effluents in pulp and paper mills using ultrasonic irradiation and Fenton-like oxidation, electrochemical treatment, and/or chemical precipitation: A comparative study. *Desalination, 255*(1), 179–187. https://doi.org/10.1016/j.desal.2009.12.024

Fälth, F., Pfromm, P. H., & Bryant, P. S. (2001). Ultrafiltration of E1 stage effluent for partial closure of the bleach plant. *Ultrafiltration of E1 Stage Effluent for Partial Closure of the Bleach Plant, 102*(1), 46–48.

Fang, S., Yu, Z., Lin, Y., Lin, Y., Fan, Y., Liao, Y., & Ma, X. (2017). A study on experimental characteristic of co-pyrolysis of municipal solid waste and paper mill sludge with additives. *Applied Thermal Engineering, 111*, 292–300. https://doi.org/10.1016/j.applthermaleng.2016.09.102

Ganjidoust, H., Tatsumi, K., Yamagishi, T., & Gholian, R. N. (1997). Effect of synthetic and natural coagulant on lignin removal from pulp and paper wastewater. *Water Science and Technology, 35*(2), 291–296. https://doi.org/10.1016/S0273-1223(96)00943-2

Garg, S. (2022). Industrial wastewater: Characteristics, treatment techniques and reclamation of water. In *Advanced industrial wastewater treatment and reclamation of water*. Springer. https://link.springer.com/chapter/10.1007/978-3-030-83811-9_1

Garg, S. K., Tripathi, M., Kumar, S., Singh, S. K., & Singh, S. K. (2012). Microbial dechlorination of chloroorganics and simultaneous decolorization of pulp–paper mill effluent by Pseudomonas putida MTCC 10510 augmentation. *Environmental Monitoring and Assessment, 184*(9), 5533–5544. https://doi.org/10.1007/s10661-011-2359-1

Godini, K., Azarian, G., Kimiaei, A., Dragoi, E. N., & Curteanu, S. (2021). Modeling of a real industrial wastewater treatment plant based on aerated lagoon using a neuro-evolutive technique. *Process Safety and Environmental Protection, 148*, 114–124. https://doi.org/10.1016/j.psep.2020.09.057

Goel, G., & Kalamdhad, A. S. (2017). An investigation on use of paper mill sludge in brick manufacturing. *Construction and Building Materials, 148*, 334–343. https://doi.org/10.1016/j.conbuildmat.2017.05.087

Gopal, P. M., Sivaram, N. M., & Barik, D. (2019). Chapter 7—Paper industry wastes and energy generation from wastes. In D. Barik (Ed.), *Energy from toxic organic waste for heat and power generation* (pp. 83–97). Woodhead Publishing. https://doi.org/10.1016/B978-0-08-102528-4.00007-9

Gubelt, G., Lumpe, C., Verstraeten, E., & Joore, L. (2000). Towards zero liquid effluent at Niederauer Mühle: The validation of two novel separation technologies. *Towards Zero Liquid Effluent at Niederauer Mühle: The Validation of Two Novel Separation Technologies, 41*(8), 41–48.

Gupta, A., & Gupta, R. (2019). Treatment and recycling of wastewater from pulp and paper mill. In R. L. Singh & R. P. Singh (Eds.), *Advances in biological treatment of industrial waste water and their recycling for a sustainable future* (pp. 13–49). Springer. https://doi.org/10.1007/978-981-13-1468-1_2

Gupta, V. K., Nayak, A., Agarwal, S., & Tyagi, I. (2014). Potential of activated carbon from waste rubber tire for the adsorption of phenolics: Effect of pre-treatment conditions. *Journal of Colloid and Interface Science, 417*, 420–430.

Hasan, H. A., Muhamad, M. H., Ji, B., Nazairi, N. A., Jiat, K. W., Sim, S. I. S. W. A., & Poh, A. F. M. S. (2023). Revolutionizing wastewater treatment with microalgae: Unveiling resource recovery, mechanisms, challenges, and future possibilities. *Ecological Engineering, 197*, 107117. https://doi.org/10.1016/j.ecoleng.2023.107117

Hassan, M. M., & Hawkyard, C. J. (2002). Decolourisation of aqueous dyes by sequential oxidation treatment with ozone and Fenton's reagent. *Journal of Chemical Technology & Biotechnology, 77*(7), 834–841. https://doi.org/10.1002/jctb.641

Helble, A., Schlayer, W., Liechti, P.-A., Jenny, R., & Möbiust, C. H. (1999). Advanced effluent treatment in the pulp and paper industry with a combined process of ozonation and fixed bed biofilm reactors. *Water Science and Technology, 40*(11), 343–350. https://doi.org/10.1016/S0273-1223(99)00737-4

Hooda, R., Bhardwaj, N. K., & Singh, P. (2018). Brevibacillus parabrevis MTCC 12105: A potential bacterium for pulp and paper effluent degradation. *World Journal of Microbiology and Biotechnology, 34*(2), 31. https://doi.org/10.1007/s11274-018-2414-y

Jagaba, A. H., Kutty, S. R. M., Baloo, L., Birniwa, A. H., Lawal, I. M., Aliyu, M. K., Yaro, N. S. A., & Usman, A. K. (2022). Combined treatment of domestic and pulp and paper industry wastewater in a rice straw embedded activated sludge bioreactor to achieve sustainable development goals. *Case Studies in Chemical and Environmental Engineering, 6*, 100261. https://doi.org/10.1016/j.cscee.2022.100261

Jahren, S. J., & Ødegaard, H. (1999). Treatment of thermomechanical pulping (TMP) whitewater in thermophilic (55°C) anaerobicaerobic moving bed biofilm reactors. *Water Science and Technology, 40*(8), 81–89. https://doi.org/10.1016/S0273-1223(99)00612-5

Jallouli, N., Pastrana-Martínez, L. M., Ribeiro, A. R., Moreira, N. F. F., Faria, J. L., Hentati, O., Silva, A. M. T., & Ksibi, M. (2018). Heterogeneous photocatalytic degradation of ibuprofen in ultrapure water, municipal and pharmaceutical industry wastewaters using a TiO_2/UV-LED system. *Chemical Engineering Journal, 334*, 976–984. https://doi.org/10.1016/j.cej.2017.10.045

Jokela, P., & Keskitalo, P. (1999). Plywood mill water system closure by dissolved air flotation treatment. *Water Science and Technology, 40*(11), 33–41. https://doi.org/10.1016/S0273-1223(99)00698-8

Kabdasli, I., Tunay, O., & Eldem, N. (1997). *Treatability of wastepaper pulping process wastewater*. Industrial Waste Conference (USA). https://scholar.google.com/scholar_lookup?title=Treatability+of+wastepaper+pulping+process+wastewater&author=Kabdasli%2C+I.+%28Istanbul+Technical+University%2C+Maslak%2C+Istanbul%2C+Turkey.%29&publication_year=1997

Kamali, M., & Khodaparast, Z. (2015). Review on recent developments on pulp and paper mill wastewater treatment. *Ecotoxicology and Environmental Safety, 114*, 326–342. https://doi.org/10.1016/j.ecoenv.2014.05.005

Karrasch, B., Parra, O., Cid, H., Mehrens, M., Pacheco, P., Urrutia, R., Valdovinos, C., & Zaror, C. (2006). Effects of pulp and paper mill effluents on the microplankton and microbial self-purification capabilities of the Biobío River, Chile. *Science of The Total Environment, 359*(1), 194–208. https://doi.org/10.1016/j.scitotenv.2005.03.029

Kesalkar, V. P., Khedikar, I. P., & Sudame, A. M. (2012). Physico-chemical characteristics of wastewater from paper industry. *International Journal of Engineering Research, 2*(4).

Korhonen, M. S., Sirpa, E. M., & Tuula, A. T. (2000) Removal of ethylenediaminetetraacetic acid (EDTA) from pulp mill effluents by ozonation. *Ozone: Science& Engineering, 22,* 279–286.

Krzeminski, P., Leverette, L., Malamis, S., & Katsou, E. (2017). Membrane bioreactors—A review on recent developments in energy reduction, fouling control, novel configurations, LCA and market prospects. *Journal of Membrane Science, 527*, 207–227. https://doi.org/10.1016/j.memsci.2016.12.010

Kumar, D., & Sharma, C. (2022). Paper industry wastewater treatment by electrocoagulation and aspect of sludge management. *Journal of Cleaner Production, 360*, 131970. https://doi.org/10.1016/j.jclepro.2022.131970

Kumar, V., & Verma, P. (2023). A critical review on environmental risk and toxic hazards of refractory pollutants discharged in chlorolignin waste of pulp and paper mills and their remediation approaches for environmental safety. *Environmental Research, 236*, 116728. https://doi.org/10.1016/j.envres.2023.116728

Kumar, V., & Verma, P. (2024). Pulp-paper industry sludge waste biorefinery for sustainable energy and value-added products development: A systematic valorization towards waste management. *Journal of Environmental Management, 352, 120052. https://doi.org/10.1016/j.jenvman.2024.120052*

Kumar, D., Gaurav, V. K., & Sharma, C. (2018). Ecofriendly remediation of pulp and paper industry wastewater by electrocoagulation and its application in agriculture. *American Journal of Plant Sciences, 9*(12), 12. https://doi.org/10.4236/ajps.2018.912178

Kumar, V., Thakur, I. S., & Shah, M. P. (2020). Bioremediation approaches for treatment of pulp and paper industry wastewater: Recent advances and challenges. In M. P. Shah (Ed.), *Microbial Bioremediation & Biodegradation* (pp. 1–48). Springer. https://doi.org/10.1007/978-981-15-1812-6_1

Kumar, A., Prerna, Bilal, M., Singh, A. K., Ratna, S., Talluri Rameshwari, K. R., Ahmed, I., & Iqbal, H. M. N. (2022a). Chapter 14—enzyme cocktail: A greener approach for biobleaching in paper and pulp industry. In R. Bhat, A. Kumar, T. A. Nguyen, & S. Sharma (Eds.), *Nanotechnology in paper and wood engineering* (pp. 303–328). Elsevier. https://doi.org/10.1016/B978-0-323-85835-9.00007-6

Kumar, A., Singh, A. K., Bilal, M., Prasad, S., Rameshwari, K. R. T., & Chandra, R. (2022b). Paper and pulp mill wastewater: Characterization, microbial-mediated degradation, and challenges. In R. Bhat, A. Kumar, T. A. Nguyen, & S. Sharma (Eds.), *Nanotechnology in paper and wood engineering* (pp. 371–387). Elsevier. https://doi.org/10.1016/B978-0-323-85835-9.00011-8

Kumar, R., Maurya, A., & Raj, A. (2023). Emerging technological solutions for the management of paper mill wastewater: Treatment, nutrient recovery and fourth industrial revolution (IR 4.0). *Journal of Water Process Engineering, 53*, 103715. https://doi.org/10.1016/j.jwpe.2023.103715

Laari, A., Korhonen, S., Tuhkanen, T., Verenich, S., & Kallas, I. (1999). Ozonation and wet oxidation in the treatment of thermomechanical pulp (TMP) circulation waters. *Water Science and Technology, 40*(11), 51–58. https://doi.org/10.1016/S0273-1223(99)00700-3

Lescot, J.-C., & Jappinen, H. (1994). Effluent treatment in pulp and paper mills: Present technology and development trends. *APPITA, 47*(4), 330–332.

Lewis, R., Cohen, J., Awad, J., Burger, H., Marzouk, J., Burch, G., Lewis, D. M., & van Leeuwen, J. A. (2018). Study of the impacts of process changes of a pulp and paper mill on aerated stabilization basin (ASB) performance. *Chemosphere, 211*, 767–774. https://doi.org/10.1016/j.chemosphere.2018.07.178

Li, F., Ma, F., Zhao, H., Zhang, S., Wang, L., Zhang, X., & Yu, H. (2019). A lytic polysaccharide monooxygenase from a white-rot fungus drives the degradation of lignin by a versatile peroxidase. *Applied and Environmental Microbiology, 85*(9), e02803–e02818. https://doi.org/10.1128/AEM.02803-18

Liang, J., Mai, W., Wang, J., Li, X., Su, M., Du, J., Wu, Y., Dai, J., Tang, Q., Gao, J., Liu, Y., Tang, J., & Wei, Y. (2021). Performance and microbial communities of a novel integrated industrial-scale pulp and paper wastewater treatment plant. *Journal of Cleaner Production, 278*, 123896. https://doi.org/10.1016/j.jclepro.2020.123896

Mahesh, S., Garg, K. K., Srivastava, V. C., Mishra, I. M., Prasad, B., & Mall, I. D. (2016). Continuous electrocoagulation treatment of pulp and paper mill wastewater: Operating cost and sludge study. *RSC Advances, 6*(20), 16223–16233. https://doi.org/10.1039/C5RA27486A

Majumdar, S., Priyadarshinee, R., Kumar, A., Mandal, T., & Dasgupta Mandal, D. (2019). Exploring *Planococcus* sp. TRC1, a bacterial isolate, for carotenoid pigment production and detoxification of paper mill effluent in immobilized fluidized bed reactor. *Journal of Cleaner Production, 211*, 1389–1402. https://doi.org/10.1016/j.jclepro.2018.11.157

Majumdar, S., Mandal, T., & Dasgupta Mandal, D. (2020a). Production kinetics of β-carotene from *Planococcus* sp. TRC1 with concomitant bioconversion of industrial solid waste into crystalline cellulose rich biomass. *Process Biochemistry, 92*, 202–213. https://doi.org/10.1016/j.procbio.2020.01.012

Majumdar, S., Paul, I., Dey, S., Dutta, S., Mandal, T., & Mandal, D. D. (2020b). Biotransformation of paper mill sludge by *Serratia marcescens* NITDPER1 for prodigiosin and cellulose nanocrystals: A strategic valorization approach. *Biochemical Engineering Journal, 164*, 107766. https://doi.org/10.1016/j.bej.2020.107766

Mandal, T. N. (1996). Studies in physico-chemical and biological characteristics of pulp and paper mill effluent and its impact on human beings. *Journal of Freshwater Biology, 8*(4), 191–196.

Mathews, S. L., Grunden, A. M., & Pawlak, J. (2016). Degradation of lignocellulose and lignin by *Paenibacillus glucanolyticus*. *International Biodeterioration & Biodegradation, 110*, 79–86. https://doi.org/10.1016/j.ibiod.2016.02.012

Mazhar, S., Ditta, A., Bulgariu, L., Ahmad, I., Ahmed, M., & Nadiri, A. A. (2019). Sequential treatment of paper and pulp industrial wastewater: Prediction of water quality parameters by Mamdani fuzzy logic model and phytotoxicity assessment. *Chemosphere, 227*, 256–268. https://doi.org/10.1016/j.chemosphere.2019.04.022

Merrill, D. T., Van Maltby, C., Kahmark, K., Gerhardt, M., & Melcer, H. (2001). Evaluating treatment processes to reduce metals concentrations in pulp and paper mill wastewaters to extremely low values. *Evaluating Treatment Processes to Reduce Metals Concentrations in Pulp and Paper Mill Wastewaters to Extremely Low Values, 84*(4), 63.

Mir-Tutusaus, J. A., Baccar, R., Caminal, G., & Sarrà, M. (2018). Can white-rot fungi be a real wastewater treatment alternative for organic micropollutants removal? A review. *Water Research, 138*, 137–151. https://doi.org/10.1016/j.watres.2018.02.056

Monte, M. C., Fuente, E., Blanco, A., & Negro, C. (2009). Waste management from pulp and paper production in the European Union. *Waste Management, 29*(1), 293–308. https://doi.org/10.1016/j.wasman.2008.02.002

Murthy, B. S. A., Sihorwala, T. A., Tilwankar, H. V., & Killedar, D. J. (1991). Removal of colour from pulp and paper mill effluents by sorption technique—A case study. *Indian Journal of Environmental Protection, 11*(5), 360.

Narbaitz, R. M., Droste, R. L., Fernandes, L., Kennedy, K. J., & Ball, D. (1997). PACT—Process for treatment of Kraft mill effluent. *Water Science and Technology, 35*(2), 283–290. https://doi.org/10.1016/S0273-1223(96)00942-0

Negi, R., & Suthar, S. (2018). Degradation of paper mill wastewater sludge and cow dung by brown-rot fungi *Oligoporus placenta* and earthworm (*Eisenia fetida*) during vermicomposting. *Journal of Cleaner Production, 201*, 842–852. https://doi.org/10.1016/j.jclepro.2018.08.068

Oeller, H.-J., Demel, I., & Weinberger, G. (1997). Reduction in residual COD in biologically treated paper mill effluents by means of combined ozone and ozone/UV reactor stages. *Water Science and Technology, 35*(2), 269–276. https://doi.org/10.1016/S0273-1223(96)00940-7

Oliveira, D. M., Hoshino, É. P., Mota, T. R., Marchiosi, R., Ferrarese-Filho, O., & dos Santos, W. D. (2020). Modulation of cellulase activity by lignin-related compounds. *Bioresource Technology Reports, 10*, 100390. https://doi.org/10.1016/j.biteb.2020.100390

Patel, K., & Dudhagara, P. (2020). Compatibility testing and enhancing the pulp bleaching process by hydrolases of the newly isolated thermophilic *Isoptericola variabilis* strain UD-6. *Biocatalysis and Biotransformation, 38*(2), 144–160. https://doi.org/10.1080/10242422.2019.1711067

Patel, K., Patel, N., Vaghamshi, N., Shah, K., Duggirala, S. M., & Dudhagara, P. (2021). Trends and strategies in the effluent treatment of pulp and paper industries: A review highlighting reactor options. *Current Research in Microbial Sciences, 2*, 100077. https://doi.org/10.1016/j.crmicr.2021.100077

Peitz, C., & Xavier, C. R. (2019). Evaluation of aerated lagoon modified with spongy support medium treating Kraft pulp mill effluent. *Revista Facultad de Ingeniería Universidad de Antioquia, 92*, Article 92. https://doi.org/10.17533/udea.redin.20190725

Poojamnong, K., Tungsudjawong, K., Khongnakorn, W., & Jutaporn, P. (2020). Characterization of reversible and irreversible foulants in membrane bioreactor (MBR) for eucalyptus pulp and paper mill wastewater treatment using fluorescence regional integration. *Journal of Environmental Chemical Engineering, 8*(5), 104231. https://doi.org/10.1016/j.jece.2020.104231

Preethi, Rajesh Banu, J., Kumar, G., Tyagi, V. K., Bajhaiya, A. K., Gugulothu, P., & Gunasekaran, M. (2022). Biohydrogen production from waste activated sludge through thermochemical mechanical pretreatment. *Bioresource Technology, 358*, 127301. https://doi.org/10.1016/j.biortech.2022.127301

Raj, A., Kumar, S., Haq, I., & Singh, S. K. (2014). Bioremediation and toxicity reduction in pulp and paper mill effluent by newly isolated ligninolytic *Paenibacillus* sp. *Ecological Engineering, 71*, 355–362. https://doi.org/10.1016/j.ecoleng.2014.07.002

Rajvaidya, N., & Markandey, D. K. (1998). *Advances in environmental science and technology: Treatment of pulp and paper industrial effluent*. APH Publishing.

Rajwar, D., Paliwal, R., & Rai, J. P. N. (2017). Biodegradation of pulp and paper mill effluent by co-culturing ascomycetous fungi in repeated batch process. *Environmental Monitoring and Assessment, 189*(9), 482. https://doi.org/10.1007/s10661-017-6198-6

Ram, C., Rani, P., Gebru, K. A., & Abrha, M. G. M. (2020). Pulp and paper industry wastewater treatment: Use of microbes and their enzymes. *Physical Sciences Reviews, 5*(10). https://doi.org/10.1515/psr-2019-0050

Ren, J., Li, X., Zhang, W., Li, Z., Wang, Q., Li, S., Wang, S., & Li, H. (2022). Evaluation of application potential of dye-decolorizing peroxidase from *Bacillus amyloliquefaciens* in bioremediation of paper and pulp mill effluent. *Frontiers in Microbiology, 13*. https://www.frontiersin.org/articles/10.3389/fmicb.2022.1031853

Rohella, R. S., Choudhury, S., Manthan, M., & Murty, J. S. (2001). Removal of colour and turbidity in pulp and paper mill effluents using polyelectrolytes. *Indian Journal of Environmental Health, 43*(4), 159–163.

Roy, S., Kumar, U., & Bhattacharyya, P. (2019). Synthesis and characterization of exfoliated biochar from four agricultural feedstock. *Environmental Science and Pollution Research, 26*(7), 7272–7276. https://doi.org/10.1007/s11356-018-04117-7

Sailwal, M., Banerjee, A., Bhaskar, T., & Ghosh, D. (2020). Chapter 23—Integrated biorefinery concept for Indian paper and pulp industry. In T. Bhaskar, A. Pandey, E. R. Rene, & D. C. W. Tsang (Eds.), *Waste biorefinery* (pp. 631–658). Elsevier. https://doi.org/10.1016/B978-0-12-818228-4.00023-X

Sarfraz, U., Ali Shah, S. I., Muhammad Awais, M., & Nawaz, S. (2022). Automatic pH control for anaerobic bacteria of wastewater treatment of paper and pulp industry. *2022 19th international Bhurban conference on applied sciences and technology (IBCAST)*, 464–475. https://doi.org/10.1109/IBCAST54850.2022.9990224.

Sathya, K., Nagarajan, K., Carlin Geor Malar, G., Rajalakshmi, S., & Raja Lakshmi, P. (2022). A comprehensive review on comparison among effluent treatment methods and modern methods of treatment of industrial wastewater effluent from different sources. *Applied Water Science, 12*(4), 70. https://doi.org/10.1007/s13201-022-01594-7

Sátiro, J., Cunha, A., Gomes, A. P., Simões, R., & Albuquerque, A. (2022). Optimization of microalgae–bacteria consortium in the treatment of paper pulp wastewater. *Applied Sciences, 12*(12), Article 12. https://doi.org/10.3390/app12125799

Schnell, A., Steel, P., Melcer, H., Hodson, P. V., & Carey, J. H. (2000). Enhanced biological treatment of bleached Kraft mill effluents—I. Removal of chlorinated organic compounds and toxicity. *Water Research, 34*(2), 493–500. https://doi.org/10.1016/S0043-1354(99)00160-8

Sharma, P., & Rath, S. K. (2021). Chapter 10—Potential applications of fungi in the remediation of toxic effluents from pulp and paper industries. In V. K. Sharma, M. P. Shah, S. Parmar, & A. Kumar (Eds.), *Fungi bio-prospects in sustainable agriculture, environment and Nano-technology* (pp. 193–211). Academic Press. https://doi.org/10.1016/B978-0-12-821925-6.00010-1

Sharma, P., Tripathi, S., Purchase, D., & Chandra, R. (2021). Integrating phytoremediation into treatment of pulp and paper industry wastewater: Field observations of native plants for the detoxification of metals and their potential as part of a multidisciplinary strategy. *Journal of Environmental Chemical Engineering, 9*(4), 105547. https://doi.org/10.1016/j.jece.2021.105547

Sharma, A., Paul, S., Charu, Kaur, R., Bhaumik, J., & Pal, S. (2023). Chapter 25—Lignin-based hybrid materials in wastewater cleanup. In V. Kumar, M. Bilal, S. K. Shahi, & V. K. Garg

(Eds.), *Metagenomics to bioremediation* (pp. 619–640). Academic Press. https://doi.org/10.1016/B978-0-323-96113-4.00019-6

Shaw, C. B., Carliell, C. M., & Wheatley, A. D. (2002). Anaerobic/aerobic treatment of coloured textile effluents using sequencing batch reactors. *Water Research, 36*(8), 1993–2001. https://doi.org/10.1016/S0043-1354(01)00392-X

Sheela, V., & Distidar, M. G. (1989). Treatment of black liquor wastes from small paper mills. *Indian Journal of Environmental Protection, 9*(9), 661–666.

Shu, H.-Y., Huang, S.-W., & Tsai, M.-K. (2016). Comparative study of acid blue 113 wastewater degradation and mineralization by UV/persulfate and UV/Oxone processes. *Desalination and Water Treatment, 57*(60), 29517–29530. https://doi.org/10.1080/19443994.2016.1172031

Singh, S. (2018). White-rot fungal Xylanases for applications in pulp and paper industry. In S. Kumar, P. Dheeran, M. Taherzadeh, & S. Khanal (Eds.), *Fungal biorefineries* (pp. 47–63). Springer International Publishing. https://doi.org/10.1007/978-3-319-90379-8_3

Singh, A. K., Kumar, A., & Chandra, R. (2022). Environmental pollutants of paper industry wastewater and their toxic effects on human health and ecosystem. *Bioresource Technology Reports, 20*, 101250. https://doi.org/10.1016/j.biteb.2022.101250

Strassberger, Z., Tanase, S., & Rothenberg, G. (2014). The pros and cons of lignin valorisation in an integrated biorefinery. *RSC Advances, 4*(48), 25310–25318. https://doi.org/10.1039/C4RA04747H

Tawfik, A., Bakr, M. H., Nasr, M., Haider, J., Mesfer, M. K. A., Lim, H., Qyyum, M. A., & Lam, S. S. (2022). Economic and environmental sustainability for anaerobic biological treatment of wastewater from paper and cardboard manufacturing industry. *Chemosphere, 289*, 133166. https://doi.org/10.1016/j.chemosphere.2021.133166

Tiku, D. K., Kumar, A., Chaturvedi, R., Makhijani, S. D., Manoharan, A., & Kumar, R. (2010). Holistic bioremediation of pulp mill effluents using autochthonous bacteria. *International Biodeterioration & Biodegradation, 64*(3), 173–183. https://doi.org/10.1016/j.ibiod.2010.01.001

Torrades, F., Peral, J., Perez, M., Domenech, X., Garcia Hortal, J. A., & Riva, M. C. (2001). Removal of organic contaminants in bleached Kraft effluents using heterogeneous photocatalysis and ozone. *Removal of Organic Contaminants in Bleached Kraft Effluents Using Heterogeneous Photocatalysis and Ozone, 84*(6), 63.

Tripathy, A. P., Dixit, P. K., & Panigrahi, A. K. (2022). Impact of effluent of Pulp & Paper industry on the flora of river basin at Jaykaypur, Odisha, India and its ecological implications. *Environmental Research, 204*, 111769. https://doi.org/10.1016/j.envres.2021.111769

Verenich, S., Laari, A., & Kallas, J. (2000). Wet oxidation of concentrated wastewaters of paper mills for water cycle closing. *Waste Management, 20*(4), 287–293. https://doi.org/10.1016/S0956-053X(99)00308-6

Vítězová, M., Kohoutová, A., Vítěz, T., Hanišáková, N., & Kushkevych, I. (2020). Methanogenic microorganisms in industrial wastewater anaerobic treatment. *PRO, 8*(12), Article 12. https://doi.org/10.3390/pr8121546

Wang, X., Mize, T. H., Saunders, F. M., & Baker, S. A. (1997). Biotreatab ility test of bleach wastewaters from pulp and paper mills. *Water Science and Technology, 35*(2), 101–108. https://doi.org/10.1016/S0273-1223(97)84672-0

Wang, C., Chen, Z., Li, Y., Feng, K., Peng, Z., Zhu, Y., & Yang, X. (2021). Refinery wastewater treatment via a multistage enhanced biochemical process. *Scientific Reports, 11*(1), 10282. https://doi.org/10.1038/s41598-021-89665-8

Wenta, B., & Hartman, B. (2002). Dissolved air flotation system improves wastewater treatment at Glatfelter. *Pulp & Paper, 76*(3) https://elibrary.ru/item.asp?id=4370245

Yadav, V., Sarker, A., Yadav, A., Miftah, A. O., Bilal, M., & Iqbal, H. M. N. (2022). Integrated biorefinery approach to valorize citrus waste: A sustainable solution for resource recovery and environmental management. *Chemosphere, 293*, 133459. https://doi.org/10.1016/j.chemosphere.2021.133459

Yoon, K., Cho, D.-W., Wang, H., & Song, H. (2022). Co-pyrolysis route of chlorella sp. and bauxite tailings to fabricate metal-biochar as persulfate activator. *Chemical Engineering Journal, 428*, 132578. https://doi.org/10.1016/j.cej.2021.132578

Zainith, S., Purchase, D., Saratale, G. D., Ferreira, L. F. R., Bilal, M., & Bharagava, R. N. (2019). Isolation and characterization of lignin-degrading bacterium *Bacillus aryabhattai* from pulp and paper mill wastewater and evaluation of its lignin-degrading potential. *3 Biotech, 9*, (3), 92. https://doi.org/10.1007/s13205-019-1631-x

Chapter 22
Circular Waste Management System and Greenhouse Gases Reduction

Akshay Botle, Sayli Salgaonkar, Rahul Tiwari, Gayatri Barabde, and Mihir Herlekar

1 Introduction

In our atmosphere, nitrogen (N_2), oxygen (O_2), water vapor, argon, and carbon dioxide (CO_2) are present in 78%, 21%, 0–4%, 1%, and 0.04% concentrations, respectively, followed by methane (CH_4), ozone (O_3), nitrous oxide (N_2O), and halocarbons as trace gases. These gases support life on Earth. Thus, every ecology relies on these greenhouse gases (GHGs). Although GHGs are essential, their recent growth has harmed humans and other life forms (Audi et al., 2020). Throughout the past century, the planet's temperature has increased due to economic expansion and fossil fuel use resulting in global warming driven by anthropogenic GHGs (Zhang et al., 2011). Global warming has become a significant issue, as the industrial boom in the early twentieth century caused the Earth's yearly temperature to rise. The surface temperature in 2022 exhibited an increase of 1.55 °F (0.86 °C) from the average temperature observed during the twentieth century (Climate.gov). Though numerous natural forces cause climate change, human intervention has emerged as a significant climatic driver in the past two centuries, and its influence is described as

A. Botle · S. Salgaonkar · M. Herlekar (✉)
Department of Environmental Science, The Institute of Science, Dr. Homi Bhabha State University, Mumbai, India
e-mail: tevs.mihir@iscm.ac.in

R. Tiwari
CSIR-National Environmental and Engineering Research Institute (CSIR-NEERI), Nagpur, India

G. Barabde
Department of Environmental Science, The Institute of Science, Dr. Homi Bhabha State University, Mumbai, India

Department of Analytical Chemistry, The Institute of Science, Dr. Homi Bhabha State University, Mumbai, India

V. Kumar et al. (eds.), *Environmental Engineering and Waste Management*,
https://doi.org/10.1007/978-3-031-58441-1_22

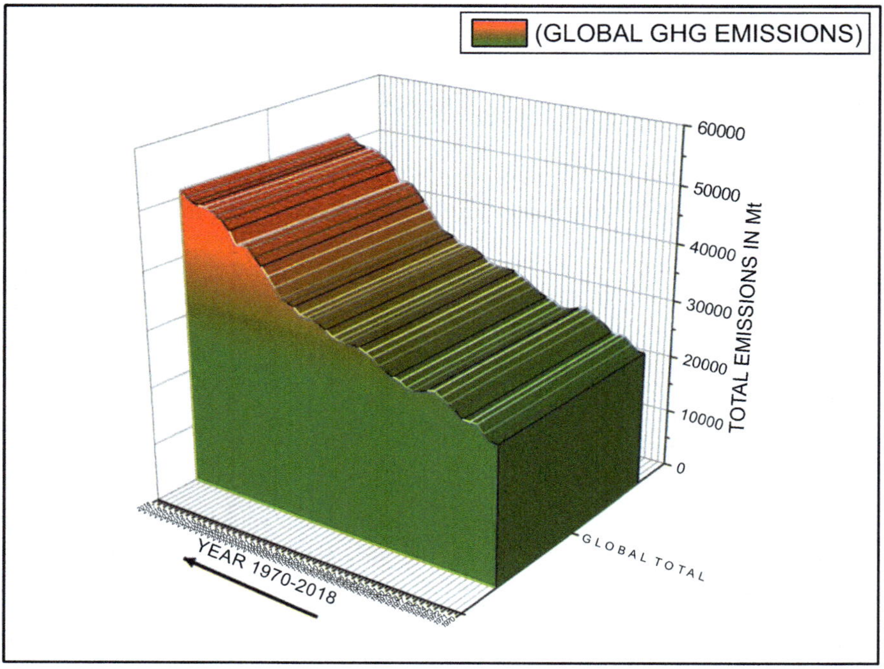

Fig. 22.1 European commission report data indicating elevated concentrations in total GHG emission around the globe (1970–2018)

the greenhouse effect (GE) (Mikhaylov et al., 2020). GHGs heat the lower atmosphere, raising the Earth's surface temperature where air temperature rises, causing eternal changes to the climate and rising temperatures (Huang et al., 2016). Discussing the phenomena started in 1827, when Jean Baptiste Joseph Fourier's "A Note on the Temperatures of the Globe and other planets" article explained the GE mechanics and origins (Inamdar & Ramanathan, 1994).

Global energy dependence on fossil fuels causes environmental, social, and economic instability. Burning fossil fuels for energy releases large volumes of GHGs, including CO_2, CH_4, and NO_2, which deteriorate the climate (Kumar et al., 2022). These alterations may be natural; however, since the 1800s, human actions like using non-replenishable fuels have been the prominent causes of climate change (Hazarika, 2023). From observations, industrialization and usage of such fuel-based vehicles are generating a severe upsurge in GHG discharges, where freight contributes for one-fifth of cumulative CO_2 discharges, whereas automobiles contribute 75% (Siddique et al., 2021). From the data published by the European Commission (2021) report, it was found that there was a constant increase in total GHG emissions around the globe (Fig. 22.1). The World Bank Group (2018) reported that worldwide CO_2 emissions rose from 22,150 megaton (Mt) in 1990 to 36,140 Mt. in 2014. CH_4 discharge rose from 6600 to 8010 Mt. CO_2 equivalent. Thus, the global CO_2, CH_4, and N_2O emission studies are crucial (Liu et al., 2019a, b). Thus, addressing climate change requires lowering human GHG emissions or enhancing the absorption of pollutants (Liu et al., 2019a, b). The Paris Agreement was proposed at the United Nations Framework Convention on Climate Change's (UNFCCC) 21st Conference of Parties (COP21) on December 12, 2020, by leaders of 197 nations to combat climate change and adjust to its massive impact on humanity. The Paris Agreement aims to limit the rise in worldwide temperatures to 247.65 K (Horowitz, 2016). The Intergovernmental Panel on Climate Change (IPCC), set by the World Meteorological Organization and the United Nations Environment Programme, indicated 2011–2020 to be the warmest decade on record (Singh, 2022) when the Earth's temperature increased by 1.1 °C. Thus, this chapter aims to investigate the notion of circular waste management and its role in mitigating GHG emissions, the causes and sources of GHGs, emphasizing the need for effective reduction strategies, waste management processes and their link to GHG emissions, highlighting the importance of adopting circular approaches, and finally explores circular waste management strategies, technological innovations, and relevant policies and regulations followed by challenges and barriers faced during the implementation.

2 Greenhouse Gases and Climate Change

2.1 Causes and Sources of Greenhouse Gases

GHGs like CH_4, CO_2, and N_2O absorb and reemit energy in the lower atmosphere, thus enhancing the GE (Montzka et al., 2011). Since the early twentieth century, GHGs have been studied because they impact climate change (Kristanto & Koven, 2019). Emissions of GHGs will change the planet's climate and damage ecological systems, causing enormous socioeconomic impacts (Li et al., 2021). If GHG emissions continue to rise, the planet's average temperature will rise by 6.4 °C by the twenty-first century, according to the IPCC (IPCC, 2013a). CO_2 is the most common and widely distributed GHG, followed by CH_4, which is 34 times stronger than CO_2 over 100 years (IPCC, 2013b). CH_4 releases have increased for decades, contributing to increased temperatures (Saunois et al., 2016). N_2O has 298 times the 100-year global warming potential of CO_2 (IPCC, 2013b), although its modest emissions have garnered less focus (Kumar et al., 2019). These prominent GHGs have varying global warming potentials and emission extents, affecting the GE (Li et al., 2021). CO_2, CH_4, and N_2O emissions have increased by 142%, 253%, and 121% since the beginning of industrialization (Bhattacharyya et al., 2013).

CO_2 is emitted by burning fossil fuels, waste, plants, and various organic materials (Yoro et al., 2018). Burning nonrenewable fuels for electricity and freight is the principal anthropogenic activity that emits CO_2, although land use changes like deforestation are also prominent contributors (Amaral et al., 2019). CH_4 emissions are likely to rise beyond 2030; rice and livestock breeding, mining for coal, crude oil and natural gas extraction and transportation, biomass burning, and garbage dumps have all contributed to increased CH_4 concentrations in the environment (Flores-Jiménez et al., 2019; Turner et al., 2019). CH_4 has a 12-year atmospheric residence time and (×25) times the global warming potential of CO_2 over 100 years (Solomon, 2007). Potential sources of CH_4 discharges are mining for coal, petroleum-based products, crude oil, and stationary and portable emission facilities. These activities released 303 tonnes of CH_4 in 2009 (Baillie et al., 2019). N_2O, like CO_2 and CH_4, may impart to the GE. Biogas production emits N_2O, which increases the overall warming constraint (Oshita et al., 2014; Paolini et al., 2018). Wastewater comprising organic-based nitrogen compounds, including human or animal waste, may release nitrous oxide (Yoro & Daramola, 2020). Moreover, humans emit nitrous oxide primarily via fertilizers from nitrogen enrichment of soils (Uchida & von Rein, 2018). It is estimated that N_2O emissions have increased by 290–330 ppb from 1940 to 2017 (Thompson et al., 2019). Sulfur dioxide (SO_2) is emitted when coal, diesel, and other sulfur-containing fuels emit when burned. Power plants and industrial fossil fuel burning release the most SO_2 into the atmosphere (Sadare et al., 2018). Smaller SO_2 producers include metal manufacturing, foundry operations, volcanic eruptions, engines (such as vessels and various other big vehicles), and sulfur-containing combustion appliances. High SO_2 concentrations around extensive industrial operations have many health and environmental impacts (Rogers et al., 2000).

2.2 *Impacts of Greenhouse Gases on Climate Change*

GHGs have raised both precipitation and temperature rates. Recent years have seen the melting of Antarctica's ice caps, glaciers, and other snowclad mountains worldwide due to rising temperatures, which has decreased water quantity in the mountain's glacier-fed rivers and lakes (Majumder, 2016). Moreover, saline water intrusion in coastal areas has grown owing to Antarctica's melting (Chen et al., 2016; Robins et al., 2016), followed by a decrease in inland water resulting in increased groundwater salinity (Xia et al., 2016; Hoover et al., 2017). The temperature rise has lowered precipitation in some locations and increased it in others, but snowfall has dramatically decreased (Majumder, 2016) in all regions where snowfall is frequently monitored. Consequently, soil moisture has also reduced due to rainwater runoff in increased precipitation areas. Other consequences, like poor water quality, are witnessed due to low water quantity levels in low precipitation areas (Barry et al., 2016; Dirnböck et al., 2016; Willems et al., 2016). In addition, pollutants are rising in certain regions, followed by population growth with reduced water availability per person. (Majumder, 2016). Water allocation disputes across sectors are also rising and witnessed (Oki, 2015). The sources and effects of GHGs are described in Fig. 22.2.

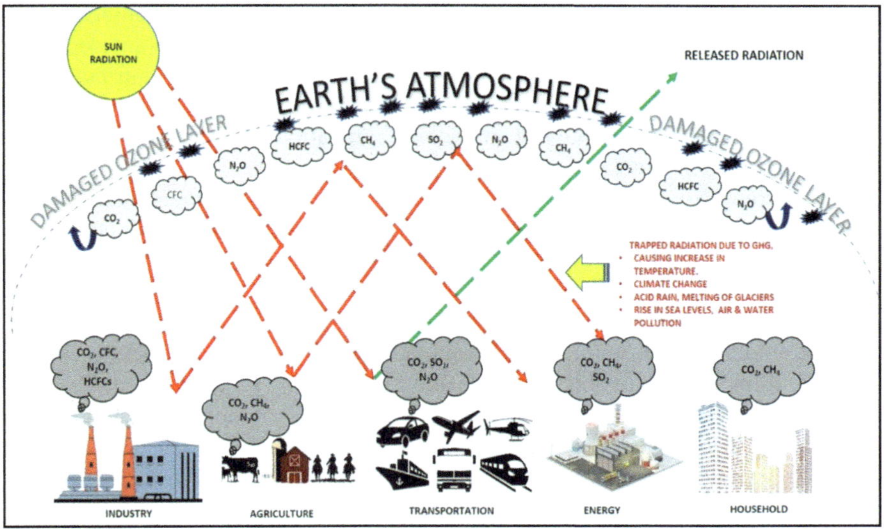

Fig. 22.2 Various anthropogenic sources of GHGs and their detrimental effects

2.3 Need for GHG Reduction Strategies

GHG reduction methods are critical for minimizing the destructive consequences of climate change and protecting the ecosystem. We can control global warming, conserve ecosystems, and maintain biodiversity by reducing CO_2, CH_4, and other GHGs. These initiatives improve resource sustainability and energy reliability by encouraging innovation in clean technology and renewable energy sources. They also enhance air quality, which benefits the general public's well-being and standard of life. By sticking to international agreements and incorporating climate action into sustainable development objectives, nations can pave the road for a more resilient and equitable future. Adopting these measures prepares us for future problems while ensuring a healthy world for current and future generations to flourish. Collective actions from governments, companies, communities, and people are required to create genuine change and successfully tackle climate change.

3 Link Between Waste Management and GHGs

3.1 Waste Generation and GHG Emissions

The dump site was a popular solid waste management (SWM) option in the early twentieth century. The notion of a "sanitary landfill" evolved in the 1930s, originating in America (Wang et al., 2015a, b). Dumpsites emit anthropogenic GHGs and contribute to global warming. Around 3–19% of global anthropogenic CH_4 emissions come from landfills (IPCC, 1996), whereas landfills contribute to 5% of worldwide GHG emissions (Stocker et al., 2014). Biodegradation of municipal solid waste (MSW) by aerobic and anaerobic degradation pathways produces GHGs such as CH_4, N_2O, and CO_2.

MSW comprises food waste, waste paper, textile scraps, inorganic trash, bamboo and timber trash, glassware trash, waste plastic, metallic junk, and latex trash. Recyclables include paper, textiles, plastic, glass, rubber, and metal (Du et al., 2006), whereas kitchen waste, textile, wood, and bamboo waste are degradable (Bo-Feng et al., 2014). India emits 29% of its GHGs as CH_4, double the world average (Siddiqui & Khan, 2011). Biodegradable organic materials make up most MSW and undergo anaerobic decomposition in landfills, producing landfill gas. It contains CH_4 (60%), CO_2 (40%), and modest levels of non-methane chemical molecules and other trace gases (Singh et al., 2016). Incineration is a treatment technology that converts waste energy into power while emitting GHGs and reduces trash volumes effectively (Chen, 2018). Japan and China use it, but China still considers it as a secondary option due to economic and technical reasons (Zhao et al., 2012). Incineration has drawbacks as it generates fly ash and toxic fumes polluting the environment.

3.2 *Recycling and GHG Reduction*

Recycling is critical to lowering GHG emissions and supporting environmentally sound waste management practices. Recycling reduces the need for energy-intensive manufacturing procedures and raw material extraction by transforming waste materials into new goods, thus cutting GHG emissions. Recycling also helps in the fight against deforestation by reducing the need for new paper manufacturing, which contributes to reducing carbon emissions from land use changes. Recycling contributes to a circular economy (CE) by extending product life cycles and lowering GHG emissions associated with manufacturing and transportation.

Over a span of two decades, the garbarge concerns about upstream GHG emissions from consumer goods production have renewed interest in recycling. Recycling is encouraged to lessen this upstream externality as recycled inputs produce items with lower emissions than virgin inputs (Calcott & Walls, 2000). Walls and Palmer (2001) demonstrated that in Pigovian circumstances, the best approach is to impose an emission tariff equivalent to the social cost that is marginal to the upstream externality and a downstream refund equivalent to waste disposal expenses. Finally, recycling and other sustainable practices are essential to climate action.

4 Understanding Circular Waste Management

4.1 *Definition and Concept*

The circular economy is an economic model characterized by the perpetual utilization of materials, thereby preventing their disposal as waste, and the restoration of natural resources (Ellen Macarthur Foundation). CE, which includes the top tiers of the waste structure, contributes straight to clean manufacturing by producing and discarding less waste from production and raw material operations. Therefore, the notion of a CE is a robust bridging concept that may help to establish fundamental linkages between resource usage, emissions, and waste contribution to the integration of environmental and financial guidelines (Mayer et al., 2019). The plan encourages CE at different stages, viz. economic structure, from manufacturing to usage, maintenance and production, dealing with waste, and the reuse of residual raw materials (European Commission, 2018).

4.2 *Principles of Circular Waste Management*

"Anything that does not create value" is considered waste (Business for Social Responsibility, 2010). According to 2016 data, 201 crore tonnes of MSW were created worldwide and is predicted to surge to 340 crore tonnes by 2050 (World Bank, 2018). The Central Pollution Control Board (CPCB) also observed a rampant increase in garbage creation in India (CPCB India, 2018).

Moreover, our manufacturing system is linear "take-make-use-dispose." This linear model overuses scarce natural resources and degrades the environment (Gaur et al., 2022). Thus, the "Zero waste"approach is founded on the principles of "3R" "Reduce:Reuse:Recycle" (Fig. 22.3). This represents a paradigm change away from the cradle-to-grave economic model, which concentrates on dumps and incineration, toward an environmentally friendly cyclical economical model, which uses refuse from one cycle of production as input for the second. These 3R principles promote ecological balance to save resources and energy and help the environment. As products reach the end of their practical lifespans, they are repurposed, repaired, traded, or reallocated (Gaur et al., 2022).

4.3 *Benefits and Importance*

March 2020 saw the European Commission's CE action plan (D'Adamo et al., 2022). This strategy covers the whole product life cycle, from design to waste prevention, to maximize resource usage in the economic cycle. CE principles show managers how new business models might improve waste management

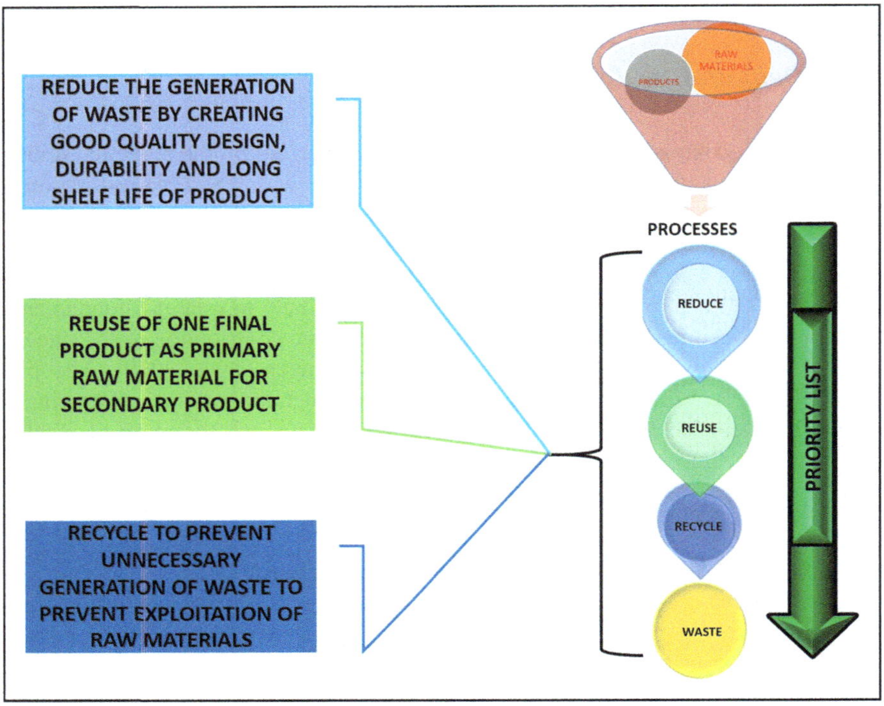

Fig. 22.3 The Zero waste approach based on 3R (Reduce, Reuse, and Recycle) Principles of circular waste management

(Rosa & Terzi, 2018). CE may profit as precious resources are reclaimed (Favot & Massarutto, 2019). However, it requires integrating many tools, including digitization and intelligent robotics, into the waste management value chain (Sarc et al., 2019). This could eliminate financial, organizational, cultural, and social constraints and meet the needs of Generation Z and Teen's rising demands for environmental protection (Salmenperä et al., 2021) through enhancing logistics resiliency (Appolloni et al., 2021).

If the government drives CE adoption, it may reap several advantages, for example, 0.65 million recycling operators create 7% of Japan's gross domestic product (GDP). Ducks and fishes both consume insects, thus reducing insecticide use; similarly, paddy weeds provide nitrogen, reducing fertilizer use and yielding 20–50% more rice compared with industrial farming practices used in India, China, Japan, and several additional major rice producers (Ghosh & Agamuthu, 2018). Cows and buffaloes in India eat leftover veggies from marketplaces, shifting them from lower-value inputs to composting. Most unsold veggies given to cattle become cow manure in 24 h. Cow manure is used as fertilizer or in small biogas plants to increase the carbon:nitrogen ratio of biofertilizers and produce cooking gas. By

2015, Austria, Denmark, South Korea, Switzerland, Netherlands, Norway, and Sweden had under 6% of MSW disposed of in dumps (Ghosh & Agamuthu, 2018).

5 Circular Waste Management Approaches

5.1 *Waste Reduction and Prevention Strategies*

"Source reduction and reuse" minimizes waste (Muzenda, 2014). Source minimization and reusing can also comprise preventive actions to optimize energy and resource consumption in upstream materials manufacturing. "Light weighting" beverage cans could reduce aluminum usage, while keeping functionality or recycling half-printed sheets in printing might reduce tree harvesting and increase forest carbon sequestration (EPA, 2015a, b). "Source reduction and reuse" might conserve natural resources, be energy efficient, minimize upstream contamination, can minimize waste toxicity, and save consumers' and manufacturers' capital (EPA, 2015b).

"Designing long-life products" is one technique for reducing trash output. It is focused on ensuring that items have a long useful life. (Chapman, 2015). "Design for durability" refers to physical durability, such as creating items that can withstand wear and tear without failing. The choice of materials for long life is a crucial aspect of design development. When maintained, following the manufacturer's instructions, "design for reliability" means designing for the probability that a material would function for a given duration without encountering a chargeable failure (Bocken et al., 2016; Moss, 1985).

5.2 *Waste Segregation and Recycling*

Solid waste management (SWM) involves generation, collection, transportation, and disposal. Source reduction solutions begin with waste generation. Waste segregation is needed to reduce waste via reuse and recycling (Oberlin, 2013). Source waste segregation may minimize trash quantities, improving assortment and increasing removal efficiency. Sorting garbage at the source streamlines management, processing, retrieval of resources, recycling, and reuse while also lowering operational expenses. Waste segregation has been shown to improve recycling programs. Stoeva and Alriksson (2017) emphasize that residential trash segmentation promotes recycling and reuse. In addition, waste segregation is crucial to effective recycling programs (Troschinetz & Mihelcic, 2009). Researchers say waste segregation is crucial to a CE (Maletz et al., 2018).

Recycling, defined as any process of recovery through which recyclables can be transformed into goods, components, or materials for original or additional uses, has substantial financial and environmental advantages and is more efficient and viable than incineration and landfilling (Omran et al., 2009). The recycling business is growing as Indians become more aware of the benefits of recycling to reduce waste and natural resource use. Thus, segregation is necessary to redirect more garbage to recycling companies (Miranda Carreño & Blanco Suárez, 2010). Waste reuse and recycling need source segregation. Garbage segregation at the source reduces contamination of recyclable materials and the quantity of garbage sent to sanitary landfills (Matter et al., 2013). Segregation also minimizes contaminated waste volume (Jin et al., 2017; Kumbhar et al., 2013), which reduces waste recovery time, cost, and occupational dangers (Banga, 2011; Sin-Yee & Sheau-Ting, 2016).

5.3 Composting and Organic Waste Management

Composting is a safe waste management strategy where anaerobic or aerobic processes convert waste to organic waste and produce compost (Lasaridi et al., 2018). Microorganisms decompose complex degradable constituents into inorganic and organic byproducts in aerobic composting (Toledo et al., 2018). Globally, 46% of MSW generated is organic, and 64% is generated in low-income countries (Lu et al., 2020). Since past few years, Australia has been adopting waste composting (Zaman, 2014), where The Australian National Garbage Report estimated that 60% of recyclable garbage was composted (Pickin & Randell, 2017). Compost has significant agricultural value and may be used to preserve soil humus and moisture-holding capacity (Hubbe et al., 2010).

MSW management values composting for its resilience and ability to provide a beneficial soil supplement as it recycles mineral nutrients (N, P, and K) from organic waste to compost (Wang et al., 2015a, b). Moreover, selectively collected biowaste compost has surplus organic matter and low bulk density. As per an investigation, the average MSW compost included carbon: residual (71%), organic (20%), and carbonate (8%). As a result, due to its surplus organic matter, selectively collected biowaste compost is known to elevate the ability of water retention of the soil (Soumare et al., 2003).

5.4 Energy Recovery from Waste

"Waste-to-energy" (WtE) technologies have been proposed to promote SWM while fostering healthier and cleaner urban environments via waste recovery and energy production. It is predicted that around 800 thermal WtE facilities are active in almost 40 nations throughout the globe and that 11% of processed MSW generates up to 429 TWh of energy (Dlamini et al., 2019). According to Tan et al. (2015), with the correct engagement in technological and institutional improvements, garbage might turn into a supply that subsidizes the socioeconomic growth of cities and countries. WtE systems use thermal and biological methods to extract useable energy contained in the organic fraction of MSW and generate heat or electricity (Brunner & Rechberger, 2015). This includes retrieving the biodegradable portion of waste by incineration, anaerobic digestion (AD), and landfill gas recovery. (Al-Salem et al., 2014). The technologies have been further discussed in this chapter.

6 Technological Innovations for Circular Waste Management

6.1 Advanced Recycling Technologies

With advanced recycling technology, modern waste management systems can handle rising trash quantities and environmental issues. These cutting-edge methods include recycling and reusing diverse waste streams, decreasing landfill trash, and preserving resources. These also help to reduce GHG emissions and promote sustainability. As the world grapples with the effects of fast industrialization and consumerism, adopting and investing in sophisticated recycling technology is crucial to attaining a CE and a cleaner, greener planet for future generations.

Although waste electrical and electronic equipment (WEEE) only makes up about 3–5% of solid garbage, it contains more precious metals than typical municipal waste (Li & Xu, 2015). WEEE is an urban mine; thus, recycling metals from WEEE reduces secondary waste and energy use, according to a US Environmental Protection Agency assessment (He et al., 2006). Bioleaching is considered as one of the prominent technologies where microorganisms bioleach insoluble metal complexes into water-soluble ones (Sand et al., 2001). Metal ions undergo mobilization through oxidation of biological compounds and complexing (Rohwerder et al., 2003), where microbes perform numerous roles. At first, they create Fe^{3+} ions and protons and later saturate them in the sample and bacterial cell, thus increasing sample leaching. This results in iron oxidation, which destroys the matrix and allows other metals to dissolve. Bioleaching recovers 10–20% of copper globally (Schippers et al., 2013) and is primarily used in industry to recover Cu, U, and Au from low-grade ores. Bioleaching of secondary raw materials has grown in popularity due to growing demand and pricing. According to Glombitza et al. (2013), bioleaching of secondary materials may be promising for industrial use. Secondary materials like fluorescent phosphors, electronic gadgets, batteries, catalysts, fly ashes, and slags from industries are tested in labs for bioleaching (Auerbach et al., 2019).

Pyro metallurgical recycling and pollution remediation have improved where crushed waste comprising metals is burnt in a furnace. In this process, contaminants are converted into slags, whereas the metals are vaporized by reacting with chemicals or heat (Zhang & Xu, 2016). Due to the amount of pollution generated and resources consumed, this method is less preferred. On the contrary, biosorption is a physicochemical and metabolism-independent process that removes compounds from solutions using biological compounds (Gadd, 2009). Metal ions are trapped on dead or inactive material from biomass of microbes generated from industrial wastewater and aqueous solutions. Biosorbents are manufactured for precious metal from biowaste or by biosorption made from algae, fungi, actinomycetes, biowaste, bacteria, and yeasts (Zhang & Xu, 2016). Recent research focuses on mild extracting, modified, and novel hydrometallurgical processes as they are better for metal extraction or pretreatment, more straightforward to manage chemical processes, and less polluting than the pyrometallurgical techniques (Ghosh et al., 2015).

The electrochemical approach is promising for recovering precious and base metals from WEEE owing to its eco-friendliness, low chemical use, and energy competence. Researchers used this approach to extract metals from WEEE. Using electrogenerated chlorine, Kim et al. (2011a) leached metals from waste-printed circuit boards. Thus, high oxidation potential allows electrogenerated chlorine to leach valuable metals (Kim et al., 2011b). Moreover, supercritical technology is used due to its high mass transfer coefficient, low viscosities, greater diffusivity, and greater organic solubility to disintegrate organic polymers and recycle metals in an environmentally friendly way (Zhang & Xu, 2016).

6.2 Anaerobic Digestion and Biogas Production

AD is a complicated biochemical technique by which a consortium of microbes breaks down organic materials without oxygen to produce methane-rich biogas (Lim et al., 2018). AD involves (a) hydrolysis, where fermentative bacteria hydrolyze high molecular weight compounds into water-soluble organic intermediates, including amino acids, glucose, and fatty acids; (b) acidogenesis, which produces ammonia (NH_3), volatile fatty acids (VFAs), CO_2, and hydrogen sulfide (H_2S); (c) acetogenesis converts VFAs into acetate, hydrogen (H_2), and CO_2; and (d) the last, methanogenesis, which produces biomethane from acetate (Uçkun Kiran et al., 2016).

Biogas is 50–75% CH_4 with 35,846 kJ/m^3 energy and 25–40% CO_2 (Khanal & Li, 2017). Biogas is a renewable energy resource used for cooking, lighting, and producing heat and electricity via combined power units and heat, natural gas substitutes, and transportation fuel. Thus, a biorefinery might make AD financially robust to the market mandate. For example, the biogas sector of Germany produced 40,000 employment jobs and supplied 37,470 m^3/h biogas for natural gas pipeline use (Zhang et al., 2019a, b). Similarly, 557 UK plants transformed food waste, agricultural waste, and wastewater into biogas for power generation and waste management, thus creating 3500 employments considering the CE (Torrijos, 2016). Before 2015, Denmark developed centralized biogas facilities to transform energy crops, organic industrial wastes, and manure into biogas (Zhang et al., 2019a, b). Singapore's biogas businesses use sewage sludge, horticultural waste, food waste, and animal manure to generate electricity (Ng et al., 2017). In rural

China, nearly 30,000 large and medium-sized biogas facilities handling agricultural waste have been developed (Chen et al., 2012).

6.3 "Waste-to-Energy" Technologies

"Waste-to-energy" technologies (WtE-T) offer the potential for underdeveloped nations to convert garbage into electricity. WtE-T is part of integrated solid waste management systems in industrialized countries to produce derivatives and combat climate change and global warming (Moya et al., 2017). WtE-T help manage trash and mitigate environmental challenges worldwide (Brunner & Rechberger, 2015; Di Matteo et al., 2017). Hydrothermal liquefaction, composting, incineration, landfilling, fermentation, pyrolysis, AD, and gasification can convert MSW to energy-dense products (Fig. 22.4). Thermochemical conversion processes have become more prevalent in recent decades due to their enhanced rate of conversion, zero-waste principle, lower time resistance, improved cost-effectiveness, and integration with an array of substrates (Varjani et al., 2022).

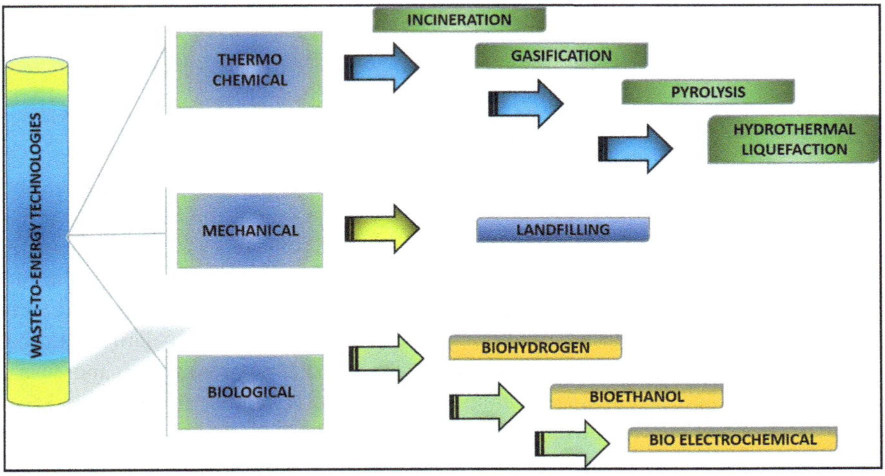

Fig. 22.4 Various waste-to-energy technologies utilized for the conversion of MSW to energy-dense product

6.3.1 Thermochemical Technologies

6.3.1.1 Incineration

Incineration is a well-known process for obtaining energy at a low cost. At temperatures over 800 °C, the flammable constituent combines with oxygen in the incineration procedure, leading to thermoelectricity, fumes, and ashes. The primary goal is to generate vapor while producing energy by managing waste incineration with the recovery of heat. This method is desirable considering its small assembly footprint, excellent efficiency, and enhanced waste volume minimization. The net energy production from this technique is computed based on the waste's compact structure, level of moisture, and fraction of inert materials. As an inert residue of solid waste burning, ash may be used to make cement and construction materials, among other things. The exhaust gases are tracked and treated in incineration facilities at a high expense and with substantial drawbacks (Khan et al., 2022).

6.3.1.2 Gasification

It transforms biomass into gaseous substances, such as CO_2, H_2, CO, and CH_4, in the presence of steam, oxygen, or supercritical water inside a gasifier at 500–1200 °C. Because it can utilize a broad range of substrates and create several valuable services, gasification has the ability to substitute traditional waste processing procedures such as landfills and incineration (Wu et al., 2022). Gasification of organic wastes consists of various steps; step I: polysaccharides such as pectin, cellulose, lignin, fat, starch, hemicellulose, proteins, and lipids release water vapor and volatile chemicals when heated to 500 °C. Step II: disintegrating components generate aliphatics, aromatics, olefins, aldehydes, and phenolics. Step III: fractures intermediate substrates into gases like CO and CO_2 and tertiary chemicals like benzene, pyrene, and naphthalene (Varjani et al., 2022).

6.3.1.3 Pyrolysis

Pyrolysis disintegrates MSW at 300–1000 °C without oxygen, where syngas, bio-oil, and char are produced. Pyrolysis temperature, retention duration, thermal rate, size of particles, and waste input composition affect product quality and yield (Beyene et al., 2018). Increased temperature reactions promote volatile splitting, water loss, decarboxylation, and additional biochar disintegration, thus increasing syngas output but decreasing biochar and bio-oil yields (Chintala, 2018; Hasan et al., 2021). Prolonged retention time accelerates secondary breakdown, recondensation, and repolymerization, thus increasing biochar output. Thus, pyrolysis, which recovers oil, gas, and carbon black from tires, plastics, wood, straw, rubber, and other waste (Foong et al., 2020), has garnered interest lately in the research community.

6.3.1.4 Hydrothermal Liquefaction

Hydrothermal liquefaction depolymerizes biomass that is wet into biochemicals and bio-oil at 250–370 °C at 35 MPa. It generates higher yield and higher-quality bio-oil having improved physicochemical characteristics. Mixed solvents like water-alcohol, hot water (compressed), and various organic solvents are used in a high-pressure continuous or batch reactor to avoid the procedure of drying for wet substrate, thus conserving both capital and energy. Hydrothermal liquefaction depolymerizes, decomposes, and recombines biomass where it is hydrolytically and thermally depolymerized to oligomers and cracked units. Furthermore, such fractured units are chemically and thermally destroyed to yield intermediates. In the next process, reactive intermediates repolymerize, condense, and cyclize to generate gas, biocrude, and char (Wu et al., 2022).

6.3.2 Landfilling Equipped with Biogas Production

The majority of MSW is still disposed through landfilling. India dumps 90% of its MSW in unsanitary landfills. Experts recommend disposing 10–15% of garbage in landfills as a last option in metropolitan settings with limited land. However, landfills produce 30–70 crore tons of CH_4 annually, adding to climate change. Although it is considered as the least unfavorable waste management approach, attempts are undertaken to turn it into a WtE-T (Varjani et al., 2022).

6.3.3 Biohydrogen

H_2 is the most abundant element present. The world produces over 1 billion (m^3/day) of H_2 from natural gas, oil, coal, and electrolysis. Due to its sustainability and waste-to-fuel potential, it is considered the future fuel (Mohan, 2009). Microorganisms convert MSW into biohydrogen (Lui et al., 2020). Biologically produced H_2 requires less energy and investment (Brandon & Kurban, 2017; Hanley et al., 2018; Staffell et al., 2019).

6.3.4 Bioethanol

Bioethanol is a good substitute for petrol due to its high flashpoint (13 vs. –23 °C) and octane number (113 vs. 85–95). Mixing bioethanol with petrol reduces GHG emissions and other pollutants, proving its sustainability (Jeswani et al., 2020). Biochemical routes can produce bioethanol from wastes like grape pomace, cafeteria garbage, banana peels, domestic waste, and potato peels. Proper substrate pretreatment, hydrolysis with enzymes, and microbial fermentation to generate ethanol are the main steps in ethanol production. Finally, to produce superior-quality bioethanol, the fermented broth requires distillation (Varjani et al., 2022).

6.3.5 Bio-electrochemical Process

Microbial fuel cells (MFCs) combine organic matter elimination with the retrieval of energy by converting chemical energy in the matrix directly into electricity. MFCs are identified as a viable bio-electrochemical technique for a wide range of trash. Electrochemically active microbes degrade organic material in an anodic chamber, emitting protons and electrons. Several parameters may influence MFC performance, including substrate form and quantity, material of the electrode and surface area, ion concentration, pH, and cell structure. The benefits of bio-electrochemical technology include minimal power requirement and the ability to regain direct energy, both of which are highly reliant on system design and functioning circumstances (Bolognesi et al., 2021).

7 Policies and Regulations for Circular Waste Management

7.1 International Agreements

Waste management regulations evolved in the late 20th and early 21st centuries due to global environmental issues and resource depletion. The Resource Conservation and Recovery Act (RCRA, significantly amended in 1984), Pollution Prevention Act (amended in 2002), and Resource Conservation Challenge (2004) have shaped US waste management policy (Sakai et al., 2011). The RCRA sets the standards for solid waste treatment and harmful waste reduction and management; however, each state manages MSW. European Union (EU) Member States have enacted waste management rules under the Waste Framework Directive (2008) (Directive, 2008). The Directive on the Incineration of Waste (2000) principally provides requirements for using the best available techniques/best environmental practice for environmental conservation, which has reduced dioxin discharges. The 1999 Directive on Waste Landfilling requires a phased decrease in organic waste disposal at landfills. The regulation encourages 3R plans to decrease the ultimate disposal of waste (European Environment Agency, 2009; Fischer, 2009). The 2002 Directives on Waste Electrical and Electronic Equipment and Restriction of Dangerous Substances (RoHS) promoted material recycling and prevented contamination from dangerous substances in such wastes.

In 1992, Germany was the pioneering EU country to implement extended producer responsibility (EPR) for packaging trash (Unterseh, 2009), followed by laws on aggregating and reprocessing end-of-life vehicles (ELVs) and trash batteries. In

Italy, 15% of local governments, or 29% of people, use a unit-based charge model to incentivize better garbage management (Douvan, 2009). The core legislation for developing a Material Cycles Society (2000) in Japan identifies recyclable materials and outlines their use (Ministry of the Environment, Japan, 2009). The legislation promotes the 3R and ecologically sound waste management to decrease natural resource use and environmental burden. The Waste Disposal and Public Cleansing Law (2010) governs waste management in Japan, whereas the Law for the Promotion of Effective Utilities of Resources (2001) governs recycling. Korea has implemented The Waste Management Act (2007) and the Act on Promotion of Resources Saving and Recycling (2008) from Korea's waste management framework (Sakai et al., 2011). The 1991 Waste Management Act full-text amendment included trash reuse, a collection system, and regulations for burning and dumping (Ministry of Environment, 2007a, b). The Act on Promotion of Resources Saving and Recycling (2002) provides the general strategy for resource recycling, the waste processing charge system, one-way packaging and products rules, and EPR (Ministry of Environment, 2008a, b).

The Environmental Protection Law of the People's Republic of China (1989) is the foundation of China's environmental regulations. The 1996 Solid Waste Environmental Pollution Prevention and Control Law was revised in 2005. It used 3R policies to solidify local, commercial, and toxic waste and reduce the waste quantity and danger (Choate, 2008). China's primary waste and material cycle regulation is the 2008 Circular Economy Promotion regulation (The Sino-Japan Friendship Centre for Environmental Protection, 2008). The CE Promotion Law was enacted due to a scarcity of resources, inadequate recycling, and a national policy to address resource depletion and promote sustainable economic development. Thus, China utilized more recycled materials due to a resource shortage and a fast-growing economy. In Vietnam, The Environmental Protection Law (2005) is Vietnam's primary waste management law. The Decree on Solid Waste Management (2007) provided ecological protections for solid waste treatment (Chi, 2009). The National Strategy on Integrated SWM for 2025–2050 (Vietnam Government, 2009) sets waste management goals for Vietnam. MSW separation is the 2025 midterm strategy goal. 100% of urban business and industrial toxic and nontoxic wastes must be handled environmentally, and 90% of urban construction and suburban MSW must be collected. Major international agreements have been described in Table 22.1. The Kyoto Protocol mandates 37 industrialized nations to cut their GHG emissions to a significant percentage level by 2008–2012. All signatory nations must inventory their GHG emissions (Shrotriya & Smout, 2010).

7.2 National Waste Management Policies

India has waste management laws and regulations (Chintan, 2013). The Ministry of Environment, Forest and Climate Change (MoEF&CC) oversees municipal solid waste management (MSWM) in India. In addition, CPHEEO (Central Public Health

Table 22.1 International policies for waste management

Sr. No	Country	Policy	Year	Reference
1.	USA	EPA Strategic Plan Resource Conservation & Recovery Act Resource Conservation Challenge Pollution Prevention Act	2006–2011 1976, **(1984)*** 2004–05 2002.	Sakai et al. (2011)
2.	EU	Community Waste Strategies Waste Framework Directive Regulation on Shipment of Waste Incineration of Waste Landfill of Waste Packaging End-of-Life Vehicles WEEE	1996, **(2005)*** 2008 2006 2000 1999 2004 2000 2002	Fischer (2009)
3.	Japan	Basic Environment Law Basic Environmental Plan Basic Law for Establishing a Recycling Based Society Law for Promotion of Effective Utilities of Resources Containers & Packaging Recycling Law Home Appliance Recycling Law Food Recycling Law Construction Waste Recycling Law End-of-Life Vehicles Recycling Law Law on Promoting Green Purchasing	1994 2006 2001, **(2003)*** 2001 2000, **(2006)*** 2001 2001, **(2007)*** 2002 2005 2001	Yoshida (2009)
4.	Korea	Act on Installation of Waste Disposal Facilities & Assistance to Adjacent Areas Act on the Management & Use of Livestock Manure Act on the Disposal of Sewage, Excreta & Livestock Wastewater Act on Promotion of Saving & Recycling of Resources Act on Promotion of Construction Waste Recycling Act on Resource Recycling of Electrical & Electronic Equipment and Vehicles	1995 2006 2006 1992, **(2008)*** 2003 2007	Ministry of Environment (2007a, b, 2008a, b) and (Oh 2009)
5	China	The National Eleventh Five Year Plan for Environmental Protection Environmental Protection Law of The People's Republic of China Law of The People's Republic of China on the Prevention and Control of Environmental Pollution by Solid	2008 1989 1996, **(2005)*** 2001 2006 & 2009	Li (2009)

(continued)

Table 22.1 (continued)

Sr. No	Country	Policy	Year	Reference
		Waste Law of The People's Republic of China on the End-of-Life-Vehicles Management Ordinance on RoHS & WEEE Circular Economy Promotion Law of The People's Republic of China Law of The People's Republic of China on Promoting Clean Production Law of The People's Republic of China on regenerable energies	2007 2003 2003	
6	Vietnam	The National Strategy for Environmental Protection Towards 2010 and Orientations Towards 2020 The National Strategy on Integrated SWM to 2025 and Vision 2050 Law on Environmental Protection The Decree on SWM Hazardous Waste Medical Waste	2003 2009 **(2005)*** 2007 2006 2007	Chi (2009) and Vietnam Government (2009)

Note: (*Amended). *WEEE* Waste electricals and electronic equipment, *RoHS* Restriction of Dangerous Substances, *SWM* Solid waste management

Environmental Engineering Organization) under the Ministry of Urban Development (MoUD) issued the MSW Manual in 2000 to help implement these guidelines (Dube, 2015). Municipal, plastic, and biomedical waste acquisition, segregation, transportation, storage, treatment, and management laws are enforced at the state level by municipalities and urban local bodies (ULBs) (Gupta et al., 2016). The Ministry of Law also prepares state government legal frameworks and policies. Committee on Urban Waste (1972), The Ministry of Health and Family Welfare (MoHFW), established one of the first solid waste management committees. The 1975 study recommended SWM acquisition, segregation, transportation, and management. It also compared municipal administrations in different states and stressed the necessity for a legislative enacting model (Madhav, 2010). Furthermore, The Central Government established a "High Power Committee on urban solid waste management in India" (Bajaj Committee) in 1994 to address the growing urban waste issue and the delayed development of viable SWM methods. The study recommended source segregation, primary collecting units, composting, recycling, and waste transportation equipment and vehicles (Mahadevia & Wolfe, 2008). National agreements have been described in Table 22.2.

Table 22.2 National policies for waste management

Sr. No.	Policy	Year	Issuer	Implementing agencies
1.	National Environmental Policy	2000	MoEF&CC	Govt. of India (GOI)
2.	National Action Plan for Climate Change	2009	Prime Ministers Council on Climate Change	MoEF&CC
3.	National Waste Management Council	1990	MoEF&CC	MoEF&CC
4.	Bajaj Committee.	1994	Planning Commission GOI	MoHFW Ministry of Health and Family Welfare
5.	Asim Burman Committee.	1999	Hon. Supreme Court of India.	MoEF&CC
6.	CAG Audit on MSW in India.	2008	MoEF&CC, CPCB	MoEF&CC
7.	Solid Waste Management Parliamentary Standing Committee (5th Report)	2010	MoUD, Ministry of Urban Development.	MoEF&CC
8.	The Waste (Prevention & Control of Pollution) Act	1974, **(1988, 2003)***	Parliament of India	CPCB
9.	The Air (Prevention & Control of Pollution) Act	1981, **(1987)***	Parliament of India	CPCB
10.	Environment Protection Act	1986, **(1991)***	MoEF&CC	MoEF&CC
11.	Hazardous Waste (Management & Handling)	1989	MoEF&CC	MoEF&CC
12.	Biomedical Waste (Management & Handling)	1989, **(2003)***	MoEF&CC	CPCB
13.	Municipal Solid Waste (Management & Handling)	2000, **(2013, 2015)***	MoEF&CC	-
14.	The Recycled Plastics Manufacture & Usage Rule	1989, **(1999)***	MoEF&CC	CPCB

Note: (*Amended). *MoEF&CC* Ministry of Environment, Forest and Climate Change, *CPCB* Central Pollution Control Board

Municipal Solid Waste (Management and Handling) Rule, 2000: The Government of India (GOI) and the MoEF&CC announced the MSW rule among several policies and committees. The regulation sought scientific methods for municipal solid waste disposal. It also required every local authority to execute all regional regulations to create an efficient waste acquisition, segregation, transport, storage, treatment, and management system (Gupta et al., 2016). The National Environment Policy (2006) conserves critical natural resources and turns them into social and economic development policies, programs, and initiatives. It also encourages responsibility for the environment for effective management and the optimal use of natural resources to reduce ecological damage (MoEF, 2006). The National Urban

Sanitation Policy (2008) urban sanitation strategy aims to end open defecation. In the National Action Plan for Climate Change (2009), eight operations executed the climate change plan. The National Mission in Sustainable Habitat (one of the operations) stresses building energy efficiency, public transit, and SWM. The action plan also included the development of energy, sewage, and recycling technology (Byravan & Rajan, 2012). The Swachh Bharat Mission (SBM, 2014) was to end open defecation, manual scavenging, modern and scientific municipal solid waste management, and good sanitation habits. The SBM emphasized scientific MSWM at all levels with contemporary technology involvement, which is seen to have changed the Indian MSWM movement. The 2016 Solid Waste Management Rules were established to meet SBM's goals. This statute was based on 14 years of MSW rule experience. Thus, the guidelines require ULB to use a planning process to create, revise, and execute MSWM plans. CPHEEO (2015) developed a seven-step MSWM planning process for ULBs that emphasizes community or stakeholder input and local authority interdepartmental collaboration to achieve success.

7.3 Extended Producer Responsibility

Extended producer responsibility (EPR), a European concept and prominent environmental policy, was adopted to safeguard human health and the environment (Hammami et al., 2018). Organization for Economic Co-operation and Development (OECD) calls EPR "an environmentally friendly strategy that involves a producer's duty to protect a product and extend it to the postconsumer phase of a product's life cycle involving its ultimate disposal." EPR policies often transfer responsibilities from upstream producers to downstream channel members and incentivize companies to create products with environmental concerns (Gupt & Sahay, 2015). EPR has different meanings, but it usually means the manufacturer should pay for "end-of-pipe" pollution and be accountable for the whole product life cycle (Maxwell, 2001). In recent decades, consumer awareness and environmental laws have changed enterprises' operations to decrease waste (Aydinliyim & Pangburn, 2012). Many companies use reverse and closed-loop supply chain (CLSC) activities to be "green" and profitable (Ferguson et al., 2011; Clottey et al., 2012; Wu & Zhou, 2016).

7.4 Incentives and Support Programs for Circular Economy

The CE is strongly tied to sustainable development practices, such as enhanced energy conservation, regenerative energy and waste recovery, and recyclable or sustainable supply of raw materials (Lieder & Rashid, 2016; Moreno et al., 2016). Thus, CE incorporates environmental management and generates new avenues and sustainable practices by utilizing trash as resources and providing collective output (Baumgartner, 2018). CE practices in developing economies need government intervention. Braun et al. (2018) advised stakeholders to collaborate and agree on CE to produce a profit. By implementing environmental standards for each sector, the government may encourage enterprises to adopt CE practices (Scupola, 2003). The government also raises public awareness and understanding of CE concepts (Mathiyazhagan et al., 2013; van Buren et al., 2016). Thus, the authorities should include CE laws in their environmental regulations and laws. This legislation and accompanying mechanisms must involve financial efforts, technical assistance, incentive programs, cooperation platforms, and guidelines to monitor and audit supply chain transparency (Lewandowski, 2016).

Governments may help companies move to CE by providing technology, infrastructure, new laws, regulations, and eco-labels (Geng & Doberstein, 2008; Manninen et al., 2018). Governments may support circular public procurement to encourage circular business practices and rules (Yuan et al., 2008). The Dutch Government promoted circular manufacturing with the Waste to Resources (Gr Van Afval Naar Grondstof) paradigm (Leendertse, 2016). Jia et al. (2018) suggested that ISO 14001-certified enterprises may get tax breaks and subsidies for implementing circular practices. Government green procurement and supply programs may encourage firms to use CE (Jia et al., 2018). Public procurement may boost recycled material demand. According to Mangla et al. (2018), government purchase and logistics regulations would help establish circular procurement models. The government coordinates, mediates, and facilitates. The government must also create laws, regulations, economic policies, public awareness, and supply chain collaboration (Lau & Wang, 2009). The lack of legislative grants and incentives for finance prevents enterprises from having eco-labels, environmental certifications, and drills (Jia et al., 2018). These authorizations help CE implementation and assess business environmental skills (Jaeger & Upadhyay, 2020; Scarpellini et al., 2020).

8 Challenges and Barriers to Circular Waste Management and GHG Reduction

8.1 Economic Challenges

The financial capital shortage affects individual investors who wish to participate in businesses and government incentives. Companies looking to create new enterprises or discover new revenue streams may need extra capital to spend on amenities, new technology, and overheads, particularly the costs of implementation. This may be expensive when moving to a new business model. As a result, local enterprises commonly emphasize expensive initial or scaling-up investment expenses, followed by comments about "lack of sufficient funds for the implementation of our project" and "high investment costs." Moreover, transport concerns economic feasibility stating logistics costs may damage the economy; thus, all stakeholders should aim to enhance logistics, especially the local economy (Salvador et al., 2022).

8.2 Technological and Infrastructural Limitations

Lack of modern technology and technical skills are technological impediments. Materials are becoming more complicated to meet ergonomic requirements, which limits their regeneration and recovery. Regeneration and repair necessitate aggressive technological innovation (Zhang et al., 2019a, b). In addition, green technologies need trained and experienced labor (Rizos et al., 2016). However, innovative enabling technologies have immense promise and several technical hurdles where technology integration is a significant obstacle (Hannan et al., 2015). When a supply chain or company employs incompatible technological platforms, it becomes more challenging, followed by technology's limited lifespan, which is another issue.

A scarcity of facilities for reverse logistics and disassembly, as well as an inconsistent supply chain design for a circular system, are operational impediments and increase functioning decision-making complexity. The current supply chain architecture supports forward material and service movement but cannot support material circularity's extensive reverse logistics and disassembly (Govindan & Hasanagic, 2018). According to the linear model (Pagell & Wu, 2009), the current supply chain architecture emphasizes the efficiency (low cost and fast speed). Integration of CE facilities and supply chain—e.g., sharing restorable materials—are slow to build (Ritzén & Sandström, 2017). Thus, circular resource flows alter supply chain planning and operational decision factors, necessitating a more complicated production scheduling and oversight system (Govindan et al., 2015).

8.3 Behavioral and Cultural Factors

A company's culture may prohibit or make it difficult to shift to a new or modified business that meets the circular economy. Culture may sometimes hinder innovation. Due to a lack of information, consumers may see nontraditional items and practices as unusual and doubt their advantages. "People see including insects in their diets as taboo," which may reduce demand-pull. Workforce problems, such as a shortage of workers wanting to conduct hard labor such as waste sorting, are also addressed. COVID-19-related obstacles have also occurred. COVID-19 has halted business activities. "COVID-19 caused companies the halt manufacturing for an unidentified amount of duration, owing to the risk of dealing polluted waste." At the same time, enterprises sat idle with no idea when to resume operations.

8.4 Policy and Regulatory Barriers

Waste reduction strategies will generally drive up the price of raw materials in subsequent years of CE. Furthermore, most nations with bioeconomy regulations have concentrated on the slightest favored biofuels and bioenergy, which deliver the lowest value-added methods. Due to a lack of official backing, there are few to no particular legislation or policies promoting ecologically sound product creation via a bioeconomy. The European Commission, on the other hand, has formed a joint venture between businesses and the public sector for biobased industries (Salvador et al., 2022).

Regulatory impediments impact policy, the regulatory environment, and specific problematic laws and regulations. Fiscal incentives are one kind of incentive. The regulatory discrepancy is often emphasized. This covers the need for global CE policy support and objectives beyond waste diversion. Trash disposal and categorization standards are mentioned as being restrictive by many researchers. Furthermore, there are no incentives for CE. The ranking is topped by public procurement, fiscal incentives, and producer responsibility (Hart et al., 2019).

8.5 Challenges and Future Perspectives

Implementing a Circular Waste Management System for GHG reduction is packed with several challenges and complications. The primary challenge faced is economic restrictions that need significant expenditures from both the public and private sectors. Overcoming technical and infrastructure constraints is critical for effective waste circularity is a major challenge. Targeted programs are required to shift society's behavior and culture toward less trash creation and more recycling. Furthermore, it is critical to ensure a steady supply of recyclable materials and to undertake extensive environmental effect evaluations. It is critical to include stakeholders and the general public in decision-making processes. Nonetheless, collaborative, multidisciplinary initiatives offer great potential for the effective construction of a long-term circular waste management system.

Prospects for circular waste management and GHG reduction include advancing new technologies, introducing additional circular business models, and enacting more stringent legislation and regulations. These technologies, business models, and legislation will assist in minimizing the quantity of waste created in the initial place and making recycling and reusing resources more superficial and more cost-effective. Nevertheless, several hurdles and impediments exist to develop an improved circular waste management system. The challenges include economic constraints, technology and infrastructure limits, behavioral and cultural reasons, and legislative and regulatory hurdles. Despite these obstacles, there is a rising movement favoring circular waste management. As the advantages of this method become more evident, we expect to see much more advancement in the upcoming years.

9 Conclusion

The linear "take-make-dispose" manufacturing and consumption approach is unsustainable, generating waste and emitting GHGs. Switching to a CE provides a solution by increasing resource efficiency and decreasing waste. Coordination among organizations, governments, and people may help accomplish this. Circular waste management, which includes recycling, reusing, and remanufacturing, has the potential to drastically decrease the load on landfills and natural resources, resulting in a more sustainable and healthier environment. It may also have a significant influence on lowering GHG emissions. One can help to mitigate environmental consequences, reduce trash, and save resources by using circular waste management processes. Switching to a CE and adopting circular waste management practices is a societal obligation and a revolutionary opportunity. Thus, we can construct the road to a sustainable future by cooperating and adopting circular methods at all levels.

References

Al-Salem, S. M., Evangelisti, S., & Lettieri, P. (2014). Life cycle assessment of alternative technologies for municipal solid waste and plastic solid waste management in the Greater London area. *Chemical Engineering Journal, 244*, 391–402. https://doi.org/10.1016/j.cej.2014.01.066

Amaral, S. S., Costa, M. A. M., Soares Neto, T. G., Costa, M. P., Dias, F. F., Anselmo, E., Santos, J. C. D., & Carvalho, J. A. D. (2019). CO2, CO, hydrocarbon gases and PM2.5 emissions on dry season by deforestation fires in the Brazilian Amazonia. *Environmental Pollution, 249*, 311–320. https://doi.org/10.1016/j.envpol.2019.03.023

Appolloni, A., D'Adamo, I., Gastaldi, M., Santibanez-Gonzalez, E. D., & Settembre-Blundo, D. (2021). Growing e-waste management risk awareness points towards new recycling scenarios: The view of the Big Four's youngest consultants. *Environmental Technology & Innovation, 23*, 101716. https://doi.org/10.1016/j.eti.2021.101716

Audi, M., Ali, A., & Kassem, M. (2020). Greenhouse gases: A review of losses and benefits. *International Journal of Energy Economics and Policy, 10*(1), 403–418. https://doi.org/10.32479/ijeep.8416

Auerbach, R., Bokelmann, K., Stauber, R., Gutfleisch, O., Schnell, S., & Ratering, S. (2019). Critical raw materials–Advanced recycling technologies and processes: Recycling of rare earth metals out of end of life magnets by bioleaching with various bacteria as an example of an intelligent recycling strategy. *Minerals Engineering, 134*, 104–117. https://doi.org/10.1016/j.mineng.2018.12.022

Aydinliyim, T., & Pangburn, M. S. (2012). Reducing packaging waste and cost via consumer price discounts. *Decision Sciences, 43*(6), 1063–1089. https://doi.org/10.1111/j.1540-5915.2012.00385.x

Baillie, J., Risk, D., Atherton, E., O'Connell, E., Fougère, C., Bourlon, E., & MacKay, K. (2019). Methane emissions from conventional and unconventional oil and gas production sites in

southeastern Saskatchewan. *Canada. Environmental Research Communications, 1*(1), 011003. https://doi.org/10.1088/2515-7620/ab01f2
Banga, M. (2011). Household knowledge, attitudes and practices in solid waste segregation and recycling: The case of urban Kampala. *Zambia Social Science Journal, 2*(1), 4.
Barry, M., Chiu, C. A., & Westerhoff, P. (2016). Severe weather effects on water quality in Central Arizona. *Journal-American Water Works Association, 108*(4), E221–E231. https://doi.org/10.5942/jawwa.2016.108.0027
Baumgartner, R. J. (2018). *Circular economy: Sustainability implications and corporate management challenges*. https://www.cec4europe.eu/wp-content/uploads/2018/09/Chapter-4.6._Baumgartner_Circular-Economy-Sustainability-Implications-and-corporate-managementchallenges.Pdf
Beyene, H. D., Werkneh, A. A., & Ambaye, T. G. (2018). Current updates on waste to energy (WtE) technologies: A review. *Renewable Energy Focus, 24*, 1–11. https://doi.org/10.1016/j.ref.2017.11.001
Bhattacharyya, P., Nayak, A. K., Mohanty, S., Tripathi, R., Shahid, M., Kumar, A., Raja, R., Panda, B. B., Roy, K. S., Neogi, S., Dash, P. K., Shukla, A. K., & Rao, K. S. (2013). Greenhouse gas emission in relation to labile soil C, N pools and functional microbial diversity as influenced by 39 years long-term fertilizer management in tropical rice. *Soil and Tillage Research, 129*, 93–105. https://doi.org/10.1016/j.still.2013.01.014
Bocken, N. M., De Pauw, I., Bakker, C., & Van Der Grinten, B. (2016). Product design and business model strategies for a circular economy. *Journal of Industrial and Production Engineering, 33*(5), 308–320. https://doi.org/10.1080/21681015.2016.1172124
Bo-Feng, C., Jian-Guo, L., Qing-Xian, G., Xiao-Qin, N., Dong, C., Lan-Cui, L., Ying, Z., & Zhan-Sheng, Z. (2014). Estimation of methane emissions from municipal solid waste landfills in China based on point emission sources. *Advances in Climate Change Research, 5*(2), 81–91. https://doi.org/10.3724/SP.J.1248.2014.081
Bolognesi, S., Cecconet, D., Callegari, A., & Capodaglio, A. G. (2021). Bioelectrochemical treatment of municipal solid waste landfill mature leachate and dairy wastewater as co-substrates. *Environmental Science and Pollution Research, 28*(19), 24639–24649. https://doi.org/10.1007/s11356-020-10167-7
Brandon, N. P., & Kurban, Z. (2017). Clean energy and the hydrogen economy. *Philosophical Transactions of the Royal Society A: Mathematical, Physical and Engineering Sciences, 375*(2098), 20160400. https://doi.org/10.1098/rsta.2016.0400
Braun, A. T., Kleine-Möllhoff, P., Reichenberger, V., & Seiter, S. (2018). *Survey concerning enablers for material efficiency activities in manufacturing, their supply chains and the transformation towards circular economy*. Reutlinger Diskussionsbeiträge zu Marketing & Management. No. 2018-3. ESB Business School. https://doi.org/10.15496/publikation-23180
Brunner, P. H., & Rechberger, H. (2015). Waste to energy–key element for sustainable waste management. *Waste Management, 37*, 3–12. https://doi.org/10.1016/j.wasman.2014.02.003
Business for Social Responsibility (BSR). (2010). *The new frontier in sustainability, The business opportunity in tackling sustainable consumption*.
Byravan, S., & Rajan, S. C. (2012). *National Action Plan on climate change* (pp. 5–7). Centre for Development Finance: IIT Madras.
Calcott, P., & Walls, M. (2000). Can downstream waste disposal policies encourage upstream "design for environment"? *American Economic Review, 90*(2), 233–237. https://doi.org/10.1257/aer.90.2.233
Chapman, J. (2015). *Emotionally durable design: Objects, experiences and empathy*. Routledge. https://doi.org/10.4324/9781315738802
Chen, Y. C. (2018). Evaluating greenhouse gas emissions and energy recovery from municipal and industrial solid waste using waste-to-energy technology. *Journal of Cleaner Production, 192*, 262–269. https://doi.org/10.1016/j.jclepro.2018.04.260
Chen, L., Zhao, L., Ren, C., & Wang, F. (2012). The progress and prospects of rural biogas production in China. *Energy Policy, 51*, 58–63. https://doi.org/10.1016/j.enpol.2012.05.052

Chen, W., Chen, K., Kuang, C., Zhu, D. Z., He, L., Mao, X., Liang, H., & Song, H. (2016). Influence of sea level rise on saline water intrusion in the Yangtze River Estuary, China. *Applied Ocean research, 54*, 12–25. https://doi.org/10.1016/j.apor.2015.11.002

Chi, N. K. (2009). 3R policy development and solid waste management status in Vietnam. In: *International workshop on 3R and waste management, Kyoto City, 2009.*

Chintala, V. (2018). Production, upgradation and utilization of solar assisted pyrolysis fuels from biomass–A technical review. *Renewable and Sustainable Energy Reviews, 90*, 120–130. https://doi.org/10.1016/j.rser.2018.03.066

Chintan (Environmental research and Action Plan). (2013). Waste rules in India—Booklet; India (pp. 1–6).

Choate, A. (2008). *Landfill analysis overview.* In EPA organics workshop.

Climate.gov. https://www.climate.gov/news-features/understanding-climate/climate-change-globaltemperature#:~:text=The%202022%20surface%20temperature%20was,period%20(1880%2D1900). Accessed 26 Aug 2023.

Clottey, T., Benton, W. C., Jr., & Srivastava, R. (2012). Forecasting product returns for remanufacturing operations. *Decision Sciences, 43*(4), 589–614. https://doi.org/10.1111/j.1540-5915.2012.00362.x

CPCB India. (2018). *MSW – Generation & composition CPCB 2018.* https://cpcb.nic.in/uploads/MSW/Waste_generation_Composition.pdf

D'Adamo, I., Mazzanti, M., Morone, P., & Rosa, P. (2022). Assessing the relation between waste management policies and circular economy goals. *Waste Management, 154*, 27–35. https://doi.org/10.1016/j.wasman.2022.09.031

Di Matteo, U., Nastasi, B., Albo, A., & Astiaso Garcia, D. (2017). Energy contribution of OFMSW (Organic Fraction of Municipal Solid Waste) to energy-environmental sustainability in urban areas at small scale. *Energies, 10*(2), 229. https://doi.org/10.3390/en10020229

Directive 2008/98/EC of the European Parliament and of the Council on Waste and Repealing Certain Directives. (2008). 19 November 2008. https://eur-lex.europa.eu/legal-content/EN/TXT/PDF/?uri=CELEX:02008L0098-20180705&from=SV. Accessed 29 July 2023.

Dirnböck, T., Kobler, J., Kraus, D., Grote, R., & Kiese, R. (2016). Impacts of management and climate change on nitrate leaching in a forested karst area. *Journal of Environmental Management, 165*, 243–252. https://doi.org/10.1016/j.jenvman.2015.09.039

Dlamini, S., Simatele, M. D., & Serge Kubanza, N. (2019). Municipal solid waste management in South Africa: From waste to energy recovery through waste-to-energy technologies in Johannesburg. *Local Environment, 24*(3), 249–257. https://doi.org/10.1080/13549839.2018.1561656

Douvan, A. R. (2009). Italian approaches on 3R developments. In *International workshop on 3R and waste management, 2009, Kyoto City.*

Du, W., Gao, Q., Zhang, E. C., Miao, Q., & Wu, J. (2006). The emission status and composition analysis of municipal solid waste in China. *Journal of Research of Environmental Sciences, 19*(2006a), 85–90.

Dube, R. (2015). *Municipal solid waste management.* Indo-German Environment Partnership (IGEP) Programme.

Ellen Macarthur Foundation. https://ellenmacarthurfoundation.org/topics/circular-economy-introduction/overview. Accessed 26 Aug 2023.

Environmental Protection Agency (EPA). (2015a). *Documentation for greenhouse gas emission and energy factors used in the waste reduction model (WARM).*

Environmental Protection Agency (EPA). (2015b). *Sustainable materials management: Non-hazardous materials and waste management hierarchy.*

European Commission. (2018). *The early warning report for Croatia, accompanying the report on the implementation of EU waste legislation, including the early warning report for Member States at risk of missing the 2020 preparation for re-use/recycling target on municipal waste.* SWD(2018) 414 final. eur-lex.europa.eu/legalcontent/EN/TXT/PDF/?uri=CELEX:52018SC0414&from=EN

European commission. 2021 report. https://edgar.jrc.ec.europa.eu/report_2021. Accessed 10 June 2023.

European Environment Agency. (2009). *Diverting waste from landfill—Effectiveness of waste-management policies in the European Union*. http://www.eea.europa.eu/publications/divertingwaste-from-landfill-effectiveness-of-waste-management-policiesin-the-european-union/at_download/file(En). Accessed 29 July 2023.

Favot, M., & Massarutto, A. (2019). Rare-earth elements in the circular economy: The case of yttrium. *Journal of Environmental Management, 240*, 504–510. https://doi.org/10.1016/j.jenvman.2019.04.002

Ferguson, M. E., Fleischmann, M., & Souza, G. C. (2011). A profit-maximizing approach to disposition decisions for product returns. *Decision Sciences, 42*(3), 773–798. https://doi.org/10.1111/j.1540-5915.2011.00330.x

Fischer, C. (2009). EU waste policy: Development and achievements. In *International workshop on 3R and waste management, Kyoto City, 2009*.

Flores-Jiménez, D. E., Carbajal, N., Algara-Siller, M., Aguilar-Rivera, N., Álvarez-Fuentes, G., Ávila-Galarza, A., & García, A. R. (2019). Atmospheric dispersion of methane emissions from sugarcane burning in Mexico. *Environmental Pollution, 250*, 922–933. https://doi.org/10.1016/j.envpol.2019.04.025

Foong, S. Y., Liew, R. K., Yang, Y., Cheng, Y. W., Yek, P. N. Y., Wan Mahari, W. A., Lee, X. Y., Han, C. S., Vo, D.-V. N., Van Le, Q., Aghbashlo, M., Tabatabaei, M., Sonne, C., Peng, W., & Lam, S. S. (2020). Valorization of biomass waste to engineered activated biochar by microwave pyrolysis: Progress, challenges, and future directions. *Chemical Engineering Journal, 389*, 124401. https://doi.org/10.1016/j.cej.2020.124401

Gadd, G. M. (2009). Biosorption: Critical review of scientific rationale, environmental importance and significance for pollution treatment. *Journal of Chemical Technology & Biotechnology: International Research in Process, Environmental & Clean Technology, 84*(1), 13–28. https://doi.org/10.1002/jctb.1999

Gaur, A., Gurjar, S. K., & Chaudhary, S. (2022). Circular system of resource recovery and reverse logistics approach: Key to zero waste and zero landfill. In *Advanced organic waste management* (pp. 365–381). Elsevier. https://doi.org/10.1016/B978-0-323-85792-5.00008-3

Geng, Y., & Doberstein, B. (2008). Developing the circular economy in China: Challenges and opportunities for achieving leapfrog development. *The International Journal of Sustainable Development & World Ecology, 15*(3), 231–239. https://doi.org/10.3843/SusDev.15.3:6

Ghosh, S. K., & Agamuthu, P. (2018). Circular economy: The way forward. *Waste Management & Research, 36*(6), 481–482. https://doi.org/10.1177/0734242X18778444

Ghosh, B., Ghosh, M. K., Parhi, P., Mukherjee, P. S., & Mishra, B. K. (2015). Waste printed circuit boards recycling: An extensive assessment of current status. *Journal of Cleaner Production, 94*, 5–19. https://doi.org/10.1016/j.jclepro.2015.02.024

Glombitza, F., Berger, W., Kothe, E., Ondruschka, J., Pinka, J., Schippers, A., Willscher, S., Zehnsdorf, A., & Brandl, H. (2013). 4 Rückstände aus Industrie und Umwelt–Aufbereitung mittels Geobiotechnologie? *Geobiotechnologie, 23*.

Govindan, K., & Hasanagic, M. (2018). A systematic review on drivers, barriers, and practices towards circular economy: A supply chain perspective. *International Journal of Production Research, 56*(1–2), 278–311. https://doi.org/10.1080/00207543.2017.1402141

Govindan, K., Soleimani, H., & Kannan, D. (2015). Reverse logistics and closed-loop supply chain: A comprehensive review to explore the future. *European Journal of Operational Research, 240*(3), 603–626. https://doi.org/10.1016/j.ejor.2014.07.012

Gupt, Y., & Sahay, S. (2015). Review of extended producer responsibility: A case study approach. *Waste Management & Research, 33*(7), 595–611. https://doi.org/10.1177/0734242X15592275

Gupta, V., Goel, S., & Rupa, T. G. (2016). Good governance and solid waste management: An overview of legislative regulations in India. *Journal of Business and Management Studies, 2*(1), 1–9.

Hammami, R., Nouira, I., & Frein, Y. (2018). Effects of customers' environmental awareness and environmental regulations on the emission intensity and price of a product. *Decision Sciences, 49*(6), 1116–1155. https://doi.org/10.1111/deci.12302

Hanley, E. S., Deane, J. P., & Gallachóir, B. Ó. (2018). The role of hydrogen in low carbon energy futures–a review of existing perspectives. *Renewable and Sustainable Energy Reviews, 82*, 3027–3045. https://doi.org/10.1016/j.rser.2017.10.034

Hannan, M. A., Al Mamun, M. A., Hussain, A., Basri, H., & Begum, R. A. (2015). A review on technologies and their usage in solid waste monitoring and management systems: Issues and challenges. *Waste Management, 43*, 509–523. https://doi.org/10.1016/j.wasman.2015.05.033

Hart, J., Adams, K., Giesekam, J., Tingley, D. D., & Pomponi, F. (2019). Barriers and drivers in a circular economy: The case of the built environment. *Procedia CIRP, 80*, 619–624. https://doi.org/10.1016/j.procir.2018.12.015

Hasan, M. M., Rasul, M. G., Khan, M. M. K., Ashwath, N., & Jahirul, M. I. (2021). Energy recovery from municipal solid waste using pyrolysis technology: A review on current status and developments. *Renewable and Sustainable Energy Reviews, 145*, 111073. https://doi.org/10.1016/j.rser.2021.111073

Hazarika, L. (2023). Rising temperatures, rising risks: A review of the impact of climate change on human cardiac health. *Journal of Environment Pollution and Human Health, 11*(2), 33–36. https://doi.org/10.12691/jephh-11-2-2

He, Y., Duan, Z., Wang, H., & Song, S. (2006). *The recycling of the electric and electronic waste*. Chemical Industry Press.

Hoover, D. J., Odigie, K. O., Swarzenski, P. W., & Barnard, P. (2017). Sea-level rise and coastal groundwater inundation and shoaling at select sites in California, USA. *Journal of Hydrology: Regional Studies, 11*, 234–249. https://doi.org/10.1016/j.ejrh.2015.12.055

Horowitz, C. A. (2016). Paris agreement. *International Legal Materials, 55*(4), 740–755. https://doi.org/10.1017/S0020782900004253

Huang, S. K., Kuo, L., & Chou, K. L. (2016). The applicability of marginal abatement cost approach: A comprehensive review. *Journal of Cleaner Production, 127*, 59–71. https://doi.org/10.1016/j.jclepro.2016.04.013

Hubbe, M. A., Nazhad, M., & Sánchez, C. (2010). Composting as a way to convert cellulosic biomass and organic waste into high-value soil amendments: A review. *BioResources, 5*(4), 2808–2854.

Inamdar, A. K., & Ramanathan, V. (1994). Physics of greenhouse effect and convection in warm oceans. *Journal of Climate, 7*(5), 715–731. https://doi.org/10.1175/1520-0442(1994)007%3C0715:POGEAC%3E2.0.CO;2

Intergovernmental Panel on Climate Change (IPCC). (1996). *Report of the twelfth season of the Intergovernmental Panel on Climate Change, Mexico City*.

Intergovernmental Panel on Climate Change (IPCC). (2013a). *Climate change: The physical science basis. Working Group I contribution to the IPCC fifth assessment report*. Cambridge University Press.

Intergovernmental Panel on Climate Change (IPCC). (2013b). In T. F. Stocker, D. Qin, G.-K. Plattner, M. Tignor, S. K. Allen, J. Boschung, A. Nauels, Y. Xia, V. Bex, & P. M. Midgley (Eds.), *Climate change 2013: The physical science basis*. Contribution of Working Group I to the fifth assessment report of the Intergovernmental Panel on Climate Change. Cambridge University Press.

Jaeger, B., & Upadhyay, A. (2020). Understanding barriers to circular economy: Cases from the manufacturing industry. *Journal of Enterprise Information Management, 33*, 729–745. https://doi.org/10.1108/JEIM-02-2019-0047

Jeswani, H. K., Chilvers, A., & Azapagic, A. (2020). Environmental sustainability of biofuels: A review. *Proceedings of the Royal Society A, 476*(2243), 20200351. https://doi.org/10.1098/rspa.2020.0351

Jia, F., Zuluaga-Cardona, L., Bailey, A., & Rueda, X. (2018). Sustainable supply chain management in developing countries: An analysis of the literature. *Journal of Cleaner Production, 189*, 263–278. https://doi.org/10.1016/j.jclepro.2018.03.248

Jin, R., Li, B., Zhou, T., Wanatowski, D., & Piroozfar, P. (2017). An empirical study of perceptions towards construction and demolition waste recycling and reuse in China. *Resources, Conservation and Recycling, 126*, 86–98. https://doi.org/10.1016/j.resconrec.2017.07.034

Khan, A. H., López-Maldonado, E. A., Khan, N. A., Villarreal-Gómez, L. J., Munshi, F. M., Alsabhan, A. H., & Perveen, K. (2022). Current solid waste management strategies and energy recovery in developing countries-state of art review. *Chemosphere, 291*, 133088. https://doi.org/10.1016/j.chemosphere.2021.133088

Khanal, S. K., & Li, Y. (Eds.). (2017). Bioenergy: Principles and applications. Wiley.

Kim, E. Y., Kim, M. S., Lee, J. C., & Pandey, B. D. (2011a). Leaching kinetics of copper from waste printed circuit boards by electro-generated chlorine in HCl solution. *Hydrometallurgy, 107*(3–4), 124–132. https://doi.org/10.1016/j.hydromet.2011.02.009

Kim, E. Y., Kim, M. S., Lee, J. C., & Pandey, B. D. (2011b). Selective recovery of gold from waste mobile phone PCBs by hydrometallurgical process. *Journal of Hazardous Materials, 198*, 206–215. https://doi.org/10.1016/j.jhazmat.2011.10.034

Kristanto, G. A., & Koven, W. (2019). Estimating greenhouse gas emissions from municipal solid waste management in Depok, Indonesia. *City and Environment Interactions, 4*, 100027. https://doi.org/10.1016/j.cacint.2020.100027

Kumar, A., Yang, T., & Sharma, M. P. (2019). Greenhouse gas measurement from Chinese freshwater bodies: A review. *Journal of Cleaner Production, 233*, 368–378. https://doi.org/10.1016/j.jclepro.2019.06.052

Kumar, A., Singh, P., Raizada, P., & Hussain, C. M. (2022). Impact of COVID-19 on greenhouse gases emissions: A critical review. *Science of the Total Environment, 806*, 150349. https://doi.org/10.1016/j.scitotenv.2021.150349

Kumbhar, S., Gupta, A., & Desai, D. (2013). Recycling and reuse of construction and demolition waste for sustainable development. *OIDA International Journal of Sustainable Development, 6*(7), 83–92. https://ssrn.com/abstract=2383436

Lasaridi, K. E., Manios, T., Stamatiadis, S., Chroni, C., & Kyriacou, A. (2018). The evaluation of hazards to man and the environment during the composting of sewage sludge. *Sustainability, 10*(8), 2618. https://doi.org/10.3390/su10082618

Lau, K. H., & Wang, Y. (2009). Reverse logistics in the electronic industry of China: A case study. *Supply Chain Management: An International Journal, 14*(6), 447–465. https://doi.org/10.1108/13598540910995228

Leendertse, P. W. (2016). *The transition from a linear to circular economy: An innovation system analysis of the composites industry* (Master's thesis), MSc Sustainable Business & Innovation, University Supervisors: Dr. Simona Negro.

Lewandowski, M. (2016). Designing the business models for circular economy—Towards the conceptual framework. *Sustainability, 8*(1), 43–70. https://doi.org/10.3390/su8010043

Li, J. (2009). Summary policy and legislative tools in promoting waste management under circular economic model in China. In *International workshop on 3R and waste management, Kyoto City, 2009.*

Li, K., & Xu, Z. (2015). Application of supercritical water to decompose brominated epoxy resin and environmental friendly recovery of metals from waste memory module. *Environmental Science & Technology, 49*(3), 1761–1767. https://doi.org/10.1021/es504644b

Li, Y., Shang, J., Zhang, C., Zhang, W., Niu, L., Wang, L., & Zhang, H. (2021). The role of freshwater eutrophication in greenhouse gas emissions: A review. *Science of the Total Environment, 768*, 144582. https://doi.org/10.1016/j.scitotenv.2020.144582

Lieder, M., & Rashid, A. (2016). Towards circular economy implementation: A comprehensive review in context of manufacturing industry. *Journal of Cleaner Production, 115*, 36–51. https://doi.org/10.1016/j.jclepro.2015.12.042

Lim, J. W., Ge, T., & Tong, Y. W. (2018). Monitoring of microbial communities in anaerobic digestion sludge for biogas optimisation. *Waste Management, 71*, 334–341. https://doi.org/10.1016/j.wasman.2017.10.007

Liu, D., Guo, X., & Xiao, B. (2019a). What causes growth of global greenhouse gas emissions? Evidence from 40 countries. *Science of the Total Environment, 661*, 750–766. https://doi.org/10.1016/j.scitotenv.2019.01.197

Liu, Y., Tang, H., Muhammad, A., & Huang, G. (2019b). Emission mechanism and reduction countermeasures of agricultural greenhouse gases – A review. *Greenhouse Gases: Science and Technology, 9*(2), 160–174. https://doi.org/10.1002/ghg.1848

Lu, H. R., Qu, X., & El Hanandeh, A. (2020). Towards a better environment-the municipal organic waste management in Brisbane: Environmental life cycle and cost perspective. *Journal of Cleaner Production, 258*, 120756. https://doi.org/10.1016/j.jclepro.2020.120756

Lui, J., Chen, W. H., Tsang, D. C., & You, S. (2020). A critical review on the principles, applications, and challenges of waste-to-hydrogen technologies. *Renewable and Sustainable Energy Reviews, 134*, 110365. https://doi.org/10.1016/j.rser.2020.110365

Madhav, R. (2010). *Untapped potential: Securing livelihoods dependant on 'waste'. A Review of Law and Policy in India*. WIEGO Final Report.

Mahadevia, D., & Wolfe, J. M. (2008). *Solid waste management and Indian cities: Status and emerging practices* (pp. 19–22). Concept Publication Company.

Majumder, M. (2016). *Minimization of climatic vulnerabilities on mini-hydro power plants: Fuzzy AHP, fuzzy ANP techniques and neuro-genetic model approach*. Springer.

Maletz, R., Dornack, C., Ziyang, L., & (Eds.). (2018). *Source separation and recycling*. Springer International Publishing. https://doi.org/10.1007/978-3-319-69072-8

Mangla, S. K., Luthra, S., Mishra, N., Singh, A., Rana, N. P., Dora, M., & Dwivedi, Y. (2018). Barriers to effective circular supply chain management in a developing country context. *Production Planning and Control, 29*(6), 551–569. https://doi.org/10.1080/09537287.2018.1449265

Manninen, K., Koskela, S., Antikainen, R., Bocken, N., Dahlbo, H., & Aminoff, A. (2018). Do circular economy business models capture intended environmental value propositions? *Journal of Cleaner Production, 171*, 413–422. https://doi.org/10.1016/j.jclepro.2017.10.003

Mathiyazhagan, K., Govindan, K., Haq, A. N., & Geng, Y. (2013). An ISM approach for the barrier analysis in implementing green supply chain management. *Journal of Cleaner Production, 47*, 283–297. https://doi.org/10.1016/j.jclepro.2012.10.042

Matter, A., Dietschi, M., & Zurbrügg, C. (2013). Improving the informal recycling sector through segregation of waste in the household–The case of Dhaka Bangladesh. *Habitat International, 38*, 150–156. https://doi.org/10.1016/j.habitatint.2012.06.001

Maxwell, D. (2001). Products and the environment–Extended producer responsibility for manufacturers. Environmental Energy Management September/October, 11-5.

Mayer, A., Haas, W., Wiedenhofer, D., Krausmann, F., Nuss, P., & Blengini, G. A. (2019). Measuring progress towards a circular economy: A monitoring framework for economy-wide material loop closing in the EU28. *Journal of Industrial Ecology, 23*(1), 62–76. https://doi.org/10.1111/jiec.12809

Mikhaylov, A., Moiseev, N., Aleshin, K., & Burkhardt, T. (2020). Global climate change and greenhouse effect. *Entrepreneurship and Sustainability Issues, 7*(4), 2897. https://doi.org/10.9770/jesi.2020.7.4(21)

Ministry of Environment. (2007a). *Resource recirculation policy of Korea*.

Ministry of Environment. (2007b). *Wastes control act*. http://eng.me.go.kr/file.do?method=fileDownloader&attachSeq=984(En). Accessed 29 July 2023.

Ministry of Environment. (2008a). *Korea's 3R policy*.

Ministry of Environment. (2008b). *Act on the promotion of saving and recycling of resources*. http://eng.me.go.kr/file.do?method=fileDownloader&attachSeq=1608(En). Accessed 29 July 2023.

Ministry of Environment and Forest (MoEF). 2006 National Environment Policy (NEP) https://ibkp.dbtindia.gov.in/DBT_Content_Test/CMS/Guidelines/20190411103521431_National%20Environment%20Policy,%202006.pdf. Accessed 29 July 2023.

Ministry of the Environment, Japan. (2009). *White paper on the environment*. Ministry of Environment, Japan HP, Enactment and Amendment. http://eng.me.go.kr/content.do?method=moveContent&menuCode=law_law_his_amendment(En)

Miranda Carreño, R., & Blanco Suárez, Á. (2010). Environmental awareness and paper recycling.

Mohan, S. V. (2009). Harnessing of biohydrogen from wastewater treatment using mixed fermentative consortia: Process evaluation towards optimization. *International Journal of Hydrogen Energy, 34*(17), 7460–7474. https://doi.org/10.1016/j.ijhydene.2009.05.062

Montzka, S. A., Dlugokencky, E. J., & Butler, J. H. (2011). Non-CO2 greenhouse gases and climate change. *Nature, 476*(7358), 43–50. https://doi.org/10.1038/nature10322

Moreno, M., De los Rios, C., Rowe, Z., & Charnley, F. (2016). A conceptual framework for circular design. *Sustainability, 8*(9), 937–951. https://doi.org/10.3390/su8090937

Moss, M. A. (1985). *Designing for minimal maintenance expense: The practical application of reliability and maintainability* (Vol. 1). CRC Press.

Moya, D., Aldás, C., López, G., & Kaparaju, P. (2017). Municipal solid waste as a valuable renewable energy resource: A worldwide opportunity of energy recovery by using waste-to-energy technologies. *Energy Procedia, 134*, 286–295. https://doi.org/10.1016/j.egypro.2017.09.618

Muzenda, E. (2014). A discussion on waste generation and management trends in South Africa.

Ng, B. J. H., Mao, Y., Chen, C. L., Rajagopal, R., & Wang, J. Y. (2017). Municipal food waste management in Singapore: Practices, challenges and recommendations. *Journal of Material Cycles and Waste Management, 19*, 560–569. https://doi.org/10.1007/s10163-015-0405-8

Oberlin, A. S. (2013). Resource recovery potential: A case study of household waste in Kinondoni municipality, Dar es Salaam. *TaJONAS: Tanzania Journal of Natural and Applied Sciences, 4*(1), 563–574. https://doi.org/10.5901/ajis.2013.v2n13p35

Oh, G. J. (2009). 3R and waste management policy and outcome in Korea. In *International workshop on 3R and waste management, Kyoto City, 2009*.

Oki, T. (2015). Water resources management and adaptation to climate change. In *Water security, climate change and sustainable development* (pp. 27–40). Springer Singapore. https://doi.org/10.1007/978-981-287-976-9_3

Omran, A., Mahmoud, A., ABDUL, A. H., & Robinson, G. M. (2009). Investigating households attitude toward recycling of solid waste in Malaysia: A case study.

Oshita, K., Okumura, T., Takaoka, M., Fujimori, T., Appels, L., & Dewil, R. (2014). Methane and nitrous oxide emissions following anaerobic digestion of sludge in Japanese sewage treatment facilities. *Bioresource Technology, 171*, 175–181. https://doi.org/10.1016/j.biortech.2014.08.081

Pagell, M., & Wu, Z. (2009). Building a more complete theory of sustainable supply chain management using case studies of 10 exemplars. *Journal of Supply Chain Management, 45*(2), 37–56. https://doi.org/10.1111/j.1745-493X.2009.03162.x

Paolini, V., Petracchini, F., Segreto, M., Tomassetti, L., Naja, N., & Cecinato, A. (2018). Environmental impact of biogas: A short review of current knowledge. *Journal of Environmental Science and Health, Part A, 53*(10), 899–906. https://doi.org/10.1080/10934529.2018.1459076

Pickin, J., & Randell, P. (2017). *Australian National Waste Report 2016*. Department of the Environment and Energy & Blue Environment Pty Ltd. https://www.environment.gov.au/system/files/resources/d075c9bc-45b3-4ac0-a8f26494c7d1fa0d/files/national-waste-report-2016.pdf. Accessed 03 July 2023.

Ritzén, S., & Sandström, G. Ö. (2017). Barriers to the circular economy–Integration of perspectives and domains. *Procedia CIRP, 64*, 7–12. https://doi.org/10.1016/j.procir.2017.03.005

Rizos, V., Behrens, A., Van Der Gaast, W., Hofman, E., Ioannou, A., Kafyeke, T., Flamos, A., Rinaldi, R., Papadelis, S., Hirschnitz-Garbers, M., & Topi, C. (2016). Implementation of circular economy business models by small and medium-sized enterprises (SMEs): Barriers and enablers. *Sustainability, 8*(11), 1212. https://doi.org/10.3390/su8111212

Robins, P. E., Skov, M. W., Lewis, M. J., Giménez, L., Davies, A. G., Malham, S. K., Neill, S. P., McDonald, J. E., Whitton, T. A., Jackson, S. E., & Jago, C. F. (2016). Impact of climate change on UK estuaries: A review of past trends and potential projections. *Estuarine, Coastal and Shelf Science, 169*, 119–135. https://doi.org/10.1016/j.ecss.2015.12.016

Rogers, J. F., Thompson, S. J., Addy, C. L., McKeown, R. E., Cowen, D. J., & Decoufle, P. (2000). Association of very low birth weight with exposures to environmental sulfur dioxide and total suspended particulates. *American Journal of Epidemiology, 151*(6), 602–613. https://doi.org/10.1093/oxfordjournals.aje.a010248

Rohwerder, T., Gehrke, T., Kinzler, K., & Sand, W. (2003). Bioleaching review part A: Progress in bioleaching: Fundamentals and mechanisms of bacterial metal sulfide oxidation. *Applied Microbiology and Biotechnology, 63*, 239–248. https://doi.org/10.1007/s00253-003-1448-7

Rosa, P., & Terzi, S. (2018). Improving end of life vehicle's management practices: An economic assessment through system dynamics. *Journal of Cleaner Production, 184*, 520–536. https://doi.org/10.1016/j.jclepro.2018.02.264

Sadare, O. O., Masitha, M., Yoro, K. O., & Daramola, M. O. (2018). Removal of sulfur (eg DBT) from petroleum distillates using activated carbon in a continuous packed-bed adsorption column. In *Proceedings of the World Congress on Engineering and Computer Science* (Vol. 2).

Sakai, S., Yoshida, H., Hirai, Y., Asari, M., Takigami, H., Takahashi, S., Tomoda, K., Peeler, M. V., Wejchert, J., Schmid-Unterseh, T., Douvan, A. R., Hathaway, R., Hylander, L. D., Fischer, C., Oh, G. J., Jinhui, L., & Chi, N. K. (2011). International comparative study of 3R and waste management policy developments. *Journal of Material Cycles and Waste Management, 13*(2), 86–102. https://doi.org/10.1007/s10163-011-0009-x

Salmenperä, H., Pitkänen, K., Kautto, P., & Saikku, L. (2021). Critical factors for enhancing the circular economy in waste management. *Journal of Cleaner Production, 280*, 124339. https://doi.org/10.1016/j.jclepro.2020.124339

Salvador, R., Barros, M. V., Donner, M., Brito, P., Halog, A., & Antonio, C. (2022). How to advance regional circular bioeconomy systems? Identifying barriers, challenges, drivers, and opportunities. *Sustainable Production and Consumption, 32*, 248–269. https://doi.org/10.1016/j.spc.2022.04.025

Sand, W., Gehrke, T., Jozsa, P. G., & Schippers, A. (2001). (bio) chemistry of bacterial leaching—Direct vs. indirect bioleaching. *Hydrometallurgy, 59*(2–3), 159–175. https://doi.org/10.1016/S0304-386X(00)00180-8

Sarc, R., Curtis, A., Kandlbauer, L., Khodier, K., Lorber, K. E., & Pomberger, R. (2019). Digitalisation and intelligent robotics in value chain of circular economy oriented waste management – A review. *Waste Management, 95*, 476–492. https://doi.org/10.1016/j.wasman.2019.06.035

Saunois, M., Jackson, R. B., Bousquet, P., Poulter, B., & Canadell, J. G. (2016). The growing role of methane in anthropogenic climate change. *Environmental Research Letters, 11*(12), 120207. https://doi.org/10.1088/1748-9326/11/12/120207

Scarpellini, S., Valero-Gil, J., Moneva, J. M., & Andreaus, M. (2020). Environmental management capabilities for a "circular eco-innovation". *Business Strategy and the Environment, 29*, 1850–1864. https://doi.org/10.1002/bse.2472

Schippers, A., Dold, B., Drobe, M., Glombitzsa, F., Gutzmer, J., Ondruschka, J., Sand, W., Vasters, J., & Willscher, S. (2013). Biomining – Bergbau mit Bakterien. In: DECHEMA Gesellschaft für Chemische Technik und Biotechnologie e.V. (Ed.), *Geobiotechnologie: Stand und Perspektiven*, Frankfurt am Main. Science 35, 881–889.

Scupola, A. (2003). The adoption of internet commerce by SMEs in the south of Italy: An environmental, technological and organizational perspective. *Journal of Global Information Technology Management, 6*(1), 52–71. https://doi.org/10.1080/1097198X.2003.10856343

Shrotriya, V., & Smout, I. (2010). GHG emissions: An assessment at municipal solid waste disposal site in Indore, India. *International Journal of Environmental Technology and Management, 13*(3–4), 362–371. https://doi.org/10.1504/IJETM.2010.038014

Siddique, A., Shahzad, A., Lawler, J., Mahmoud, K. A., Lee, D. S., Ali, N., Bilal, M., & Rasool, K. (2021). Unprecedented environmental and energy impacts and challenges of COVID-19 pandemic. *Environmental Research, 193*, 110443. https://doi.org/10.1016/j.envres.2020.110443

Siddiqui, F. Z., & Khan, M. E. (2011). Landfill gas recovery and its utilization in India: Current status, potential prospects and policy implications. *Journal of Chemical and Pharmaceutical Research, 3*(5), 174–183.

Singh, I. K. (2022). *IPCC climate change report 2022 and its implications for India > WCS-India*. Wildlife Conservation Society of India, 2022. https://india.wcs.org/Newsroom/Blog/ID/17734/IPCC-Climate-Change-Report-2022-and-its-Implications-for India?gclid=Cj0KCQiAx6ugBhCcARIsAGNmMbjK1hNXlHBOK

Singh, S. K., Anunay, G., Rohit, G., Shivangi, G., & Vipul, V. (2016). Greenhouse gas emissions from landfills: A case of NCT of Delhi, India. *Journal of Climatology & Weather Forecasting, 4*(1), 1–6. https://doi.org/10.4172/2332-2594.1000157

Sin-Yee, T., & Sheau-Ting, L. (2016). Attributes in fostering waste segregation behaviour. *International Journal of Environmental Science and Development, 7*(9), 672. https://doi.org/10.18178/ijesd.2016.7.9.860

Solomon, S. (Ed.). (2007). *Climate change 2007-the physical science basis: Working group I contribution to the fourth assessment report of the IPCC* (Vol. 4). Cambridge University Press.

Soumare, M., Tack, F. M. G., & Verloo, M. G. (2003). Characterisation of Malian and Belgian solid waste composts with respect to fertility and suitability for land application. *Waste Management, 23*(6), 517–522. https://doi.org/10.1016/S0956-053X(03)00067-9

Staffell, I., Scamman, D., Velazquez Abad, A., Balcombe, P., Dodds, P. E., Ekins, P., Shah, N., & Ward, K. R. (2019). The role of hydrogen and fuel cells in the global energy system. *Energy & Environmental Science, 12*(2), 463–491. https://doi.org/10.1039/C8EE01157E

Stocker, T. F., Qin, D., Plattner, G.-K., Tignor, M. M. B., Allen, S. K., Boschung, J., Nauels, A., Xia, Y., Bex, V., & Midgley, P. M. (Eds.). (2014). *Climate change 2013: The physical science basis. Contribution of working group I to the fifth assessment report of IPCC the intergovernmental panel on climate change*. Cambridge University Press. https://doi.org/10.1017/CBO9781107415324

Stoeva, K., & Alriksson, S. (2017). Influence of recycling programmes on waste separation behaviour. *Waste Management, 68*, 732–741. https://doi.org/10.1016/j.wasman.2017.06.005

Tan, S. T., Ho, W. S., Hashim, H., Lee, C. T., Taib, M. R., & Ho, C. S. (2015). Energy, economic and environmental (3E) analysis of waste-to-energy (WTE) strategies for municipal solid waste (MSW) management in Malaysia. *Energy Conversion and Management, 102*, 111–120. https://doi.org/10.1016/j.enconman.2015.02.010

The Sino-Japan Friendship Centre for Environmental Protection. (2008). *The recycling economy promotion law*. http://www.zhb.gov.cn/japan/CNE/CNE04_78.htm (Jpn)

The World Bank Group. (2018). *World development indicators*. https://data.worldbank.org/

Thompson, R. L., Lassaletta, L., Patra, P. K., Wilson, C., Wells, K. C., Gressent, A., Koffi, E. N., Chipperfield, M. P., Winiwarter, W., Davidson, E. A., Tian, H., & Canadell, J. G. (2019). Acceleration of global N2O emissions seen from two decades of atmospheric inversion. *Nature Climate Change, 9*(12), 993–998. https://doi.org/10.1038/s41558-019-0613-7

Toledo, M., Siles, J. A., Gutiérrez, M. C., & Martín, M. A. (2018). Monitoring of the composting process of different agroindustrial waste: Influence of the operational variables on the odorous impact. *Waste Management, 76*, 266–274. https://doi.org/10.1016/j.wasman.2018.03.042

Torrijos, M. (2016). State of development of biogas production in Europe. *Procedia Environmental Sciences, 35*, 881–889. https://doi.org/10.1016/j.proenv.2016.07.043

Troschinetz, A. M., & Mihelcic, J. R. (2009). Sustainable recycling of municipal solid waste in developing countries. *Waste Management, 29*(2), 915–923. https://doi.org/10.1016/j.wasman.2008.04.016

Turner, A. J., Frankenberg, C., & Kort, E. A. (2019). Interpreting contemporary trends in atmospheric methane. *Proceedings of the National Academy of Sciences, 116*(8), 2805–2813. https://doi.org/10.1073/pnas.1814297116

Uchida, Y., & von Rein, I. (2018). Mitigation of nitrous oxide emissions during nitrification and denitrification processes in agricultural soils using enhanced efficiency fertilizers. In *Soil contamination and alternatives for sustainable development*. IntechOpen.

Uçkun Kiran, E., Stamatelatou, K., Antonopoulou, G., & Lyberatos, G. (2016). Production of biogas via anaerobic digestion. In R. Luque, C. S. K. Lin, K. Wilson, & J. Clark (Eds.), *Handbook of biofuels production: Processes and technologies* (pp. 259–301). Woodhead Publishing. https://doi.org/10.1016/B978-0-08-100455-5.00010-2

Unterseh, T. S. (2009). Extended producer responsibility—The German experience. In: *International workshop on 3R and waste management, Kyoto City, 2009.*

van Buren, N., Demmers, M., van der Heijden, R., & Witlox, F. (2016). Towards a circular economy: The role of Dutch logistics industries and governments. *Sustainability, 8*(7), 647–663. https://doi.org/10.3390/su8070647

Varjani, S., Shahbeig, H., Popat, K., Patel, Z., Vyas, S., Shah, A. V., Barceló, D., Hao Ngo, H., Sonne, C., Shiung Lam, S., Aghbashlo, M., & Tabatabaei, M. (2022). Sustainable management of municipal solid waste through waste-to-energy technologies. *Bioresource Technology, 355*, 127247. https://doi.org/10.1016/j.biortech.2022.127247

Vietnam Government. (2009). National strategy on integrated SWM to 2025 and vision to 2050 Vietnam (2149/QD-TTg).

Walls, M., & Palmer, K. (2001). Upstream pollution, downstream waste disposal, and the design of comprehensive environmental policies. *Journal of Environmental Economics and Management, 41*(1), 94–108. https://doi.org/10.1006/jeem.2000.1135

Wang, J., Chen, Z., Yang, L., & Xi, S. (2015a). Study on trends and performance of landfill research from 1999 to 2013 by using bibliometric analysis. *Environmental Progress & Sustainable Energy, 34*(5), 1349–1355. https://doi.org/10.1002/ep.12126

Wang, K., He, C., You, S., Liu, W., Wang, W., Zhang, R., Qi, H., & Ren, N. (2015b). Transformation of organic matters in animal wastes during composting. *Journal of Hazardous Materials, 300*, 745–753. https://doi.org/10.1016/j.jhazmat.2015.08.016

Willems, W., Kasper, G., Klotz, P., Stricker, K., & Zimmermann, A. (2016). Surface water: Discharge rate and water quality. In *Regional assessment of global change impacts: The project GLOWA-Danube* (pp. 229–235). https://doi.org/10.1007/978-3-319-16751-0_29

World Bank. (2018). *What a waste, 2.0: A global snapshot of solid waste management to 2050*. Urban Development. Available online: https://elibrary.worldbank.org/doi/pdf/10.1596/978-1-4648-1329-0. Accessed on 26 Aug 2023.

Wu, X., & Zhou, Y. (2016). Does the entry of third-party remanufacturers always hurt original equipment manufacturers? *Decision Sciences, 47*(4), 762–780. https://doi.org/10.1111/deci.12194

Wu, Y., Wang, H., Li, H., Han, X., Zhang, M., Sun, Y., Fan, X., Tu, R., Zeng, Y., Xu, C. C., & Xu, X. (2022). Applications of catalysts in thermochemical conversion of biomass (pyrolysis, hydrothermal liquefaction and gasification): A critical review. *Renewable Energy*. https://doi.org/10.1016/j.renene.2022.07.031

Xia, J., Zhao, X., Chen, Y., Fang, Y., & Zhao, Z. (2016). Responses of water and salt parameters to groundwater levels for soil columns planted with *Tamarix chinensis*. *PLoS One, 11*(1), e0145828. https://doi.org/10.1371/journal.pone.0145828

Yoro, K. O., & Daramola, M. O. (2020). CO2 emission sources, greenhouse gases, and the global warming effect. In *Advances in carbon capture* (pp. 3–28). Woodhead Publishing. https://doi.org/10.1016/B978-0-12-819657-1.00001-3

Yoro, K. O., Isafiade, A. J., & Daramola, M. O. (2018). Sequential synthesis of mass exchanger networks for CO_2 capture. In *Proceedings of The World Congress on Engineering and Computer Science* (pp. 503–508).

Yoshida, H. (2009). Japan's experiences of policy developments on 3R and waste management. In: *International workshop on 3R and waste management, Kyoto City, 2009*.

Yuan, Z., Bi, J., & Moriguichi, Y. (2008). The circular economy: A new development strategy in China. *Journal of Industrial Ecology, 10*(1–2), 4–8.

Zaman, A. U. (2014). Measuring waste management performance using the 'Zero Waste Index': The case of Adelaide, Australia. *Journal of Cleaner Production, 66*, 407–419. https://doi.org/10.1016/j.jclepro.2013.10.032

Zhang, L., & Xu, Z. (2016). A review of current progress of recycling technologies for metals from waste electrical and electronic equipment. *Journal of Cleaner Production, 127*, 19–36. https://doi.org/10.1016/j.jclepro.2016.04.004

Zhang, Y.-M., Hu, C.-S., Zhang, J.-B., Dong, W.-X., Wang, Y.-Y., & Song, L.-N. (2011). Research advances on source/sink intensities and greenhouse effects of CO_2, CH_4 and N_2O in agricultural soils. *Chinese Journal of Eco-Agriculture, 19*(4), 966–975. https://doi.org/10.3724/SP.J.1011.2011.00966

Zhang, A., Venkatesh, V. G., Liu, Y., Wan, M., Qu, T., & Huisingh, D. (2019a). Barriers to smart waste management for a circular economy in China. *Journal of Cleaner Production, 240*, 118198. https://doi.org/10.1016/j.jclepro.2019.118198

Zhang, L., Loh, K. C., & Zhang, J. (2019b). Enhanced biogas production from anaerobic digestion of solid organic wastes: Current status and prospects. *Bioresource Technology Reports, 5*, 280–296. https://doi.org/10.1016/j.biteb.2018.07.005

Zhao, Y., Xing, W., Lu, W., Zhang, X., & Christensen, T. H. (2012). Environmental impact assessment of the incineration of municipal solid waste with auxiliary coal in China. *Waste Management, 32*(10), 1989–1998. https://doi.org/10.1016/j.wasman.2012.05.012

Chapter 23
Industry 4.0: A New Window for the Future Waste Management Systems

W. J. Pech-Rodríguez, Nihat Ege Sahin, G. G. Suarez-Velázquez, and E. Rocha-Rangel

1 Introduction

The increasing population leads to an imminent environmental crisis due to excessive waste generation giving rise to health problems. Thus, waste management systems have been a global goal to obtain sustainable development, which is crucial for the well-being of individuals and ecosystems (Roy et al., 2023). Nevertheless, it should be mentioned that the government and stakeholders have not prioritized waste management as energy and water because the overall pathway is expensive, and new technologies and logistics are required (Gatto, 2023). According to Vasconcelos et al., the participatory approach of all sectors can be adopted as a natural way for resource management and prevention, including urban waste, to obtain innovative and sustainable opportunities (Vasconcelos et al., 2022). In this sense, Fitchel and coworkers studied solid waste management in Ghana, an African country whose population is growing rapidly (Fichtel & Duram, 2022). They observed that although communities actively manage residues, education and

W. J. Pech-Rodríguez (✉)
Department of Mechatronics, Polytechnic University of Victoria, Tamaulipas, Mexico
e-mail: wpechr@upv.edu.mx

N. E. Sahin
Department of Biological and Chemical Engineering – Process and Materials Engineering, Aarhus University, Aarhus, Denmark

G. G. Suarez-Velázquez
Universidad Politécnica de Altamira, Departamento de Ingeniería en energía, Altamira, Tamaulipas, Mexico

E. Rocha-Rangel
Department of Master Engineering program, Polytechnic University of Victoria, Tamaulipas, Mexico

V. Kumar et al. (eds.), *Environmental Engineering and Waste Management*,
https://doi.org/10.1007/978-3-031-58441-1_23

enforcement measures are the main limitations due to political indifference and insufficient resources.

Electrification of the primary sectors globally is the most viable strategy for sustainable development. However, a lot of undesired items are generated, such as batteries, plastic, hazardous chemicals, and electronic waste (Neumann et al., 2022; Pražanová et al., 2022). Malinauskaite et al. reported that the EU ends up with approximately 190,000 tonnes of batteries each year, which are not properly collected and recycled (Malinauskaite et al., 2021). The latter results in economic issues (considering batteries are constructed with limited chemical elements) and contributes to environmental risk due to hazardous compounds. The same effect is observed for solar photovoltaic panels with a lifetime of 25 years that eventually will become hazardous waste due to the harmful chemicals in the panel (Chowdhury et al., 2020).

I4.0 revolution is a digital transformation of the industrial processes proposed to improve life quality, increase income, and decrease adverse environmental impact. This revolution is based on the implementation of smart technologies such as the Internet of Things (IoT), machine learning (ML), data mining, big data, sensors, and so on. Applying the circular economy under the I4.0 paradigm could be a feasible strategy to face the waste issue by making a visual pathway for the flow of virgin materials and recovery while reducing energy consumption and the carbon footprint (Albertsen et al., 2021; Birkel & Müller, 2021). Some authors bet on the complete digitalization of the manufacturing process to obtain affordable and sustainable production. Digitalization has tremendous advantages because it can simulate a whole manufacturing chain, and then it can be possible to prevent unnecessary costs related to the trial-and-error process. The latter can easily be achieved by adopting the I4.0 approach, where digital twinning, artificial intelligence (AI), and sensors can be used to competitive production lines. A practical application of this paradigm is the concept of a smart city that uses all kinds of novel technology to digitalize the process and exchange information with the public to obtain higher-quality services (Xiangru Chen, 2022). This includes the smart waste management system in which sensors interconnected with IoT infrastructure allow real-time waste bin monitoring (Sheng et al., 2020).

For example, an interesting study was conducted based on a case study for waste management 4.0 in Perth, Western Australia (Zaman, 2022). The case study reveals that household waste selection (based on YOLOv4 real-time object detection platform) was achieved with acceptable precision, more than 70% for selected items. However, the author claimed that this novel management system's main obstacles are the initial investment in technologies, platforms, and data acquisition. The study concluded that waste management through I4.0 technologies reduces cost, improves supply chain efficiency, and eliminates waste. Meanwhile, Fatimah et al. reported a case study conducted in Indonesia where I4.0 technologies were proposed for waste management and, in this form, developed a sustainable and smart country (Fatimah et al., 2020). The goal was achieved by adopting the sustainable circular economy approach in which real-time, smart, and flexible waste management was performed considering the umbrella and social dimensions. Another interesting case study was published by Kurniawah et al., where the authors evaluated the existing waste management situation in St. Petersburg (Kurniawan et al., 2022). Digitalization was the primary tool proposed by the author to meet the city's needs. Nonetheless, they found some drawbacks, such as a lack of proper infrastructure, low public interest, technological issues, and poor coordination between institutions.

According to the information discussed above, this study highlights the potential of I4.0 as a feasible tool to obtain benefits from an economic, environmental, and social point of view. It should be mentioned that the application of cutting-edge technologies in waste management is still limited but is emerging (Abdallah et al., 2020). Thus, it is imperative to create legislation and incentives that motivate the adoption of I4.0 technologies to reach an effective recycling process and reduce harmful materials out of landfills.

2 Recycling Batteries Under the I4.0 Paradigm

Nothing batteries have been utilized as an innovative strategy to support transport electrification worldwide as well as support intermittent renewable sources like solar cells and wind turbines (Islam & Iyer-Raniga, 2022). Now, there is exponential growth in the implementation of electric vehicles (EVs) as a green alternative to reduce environmental pollution. By 2040, at least 70% of passenger vehicles are expected to be electric (Ayerbe et al., 2022). Batteries are identified as primary (non-rechargeable) and secondary (rechargeable) (Kurzweil, 2023). During the practical operation of the battery, two latent stages are observed: the decrease in charge retention and the end-of-life of this device. Here, it should be mentioned that the battery is fabricated by incorporating several chemical elements such as Li, Zn, Cu, Ni, Mg, and others. Most of the cited chemical elements easily dissolve in water, so we need to be aware that batteries might be a contaminant source if inadequate waste management is implemented. Actually, many discarded batteries are directly landfilled without adequate selection for reuse.

It is anticipated that in the near future, EVs will replace the existing combustion engine, and this would generate an enormous number of waste battery packs due to the short lifetime of batteries. Moreover, world reserves for battery fabrication are limited, and this task is usually labor-intensive so waste management will be a critical point (Kaunda, 2020). The global battery recycling market is estimated at USD 10.21 billion in 2021, with a projection of 18.96 billion by 2023 (Pregowska et al., 2022).

Battery digitalization is proposed as an efficient approach to deal with battery supply chains, life cycle, end-life, and recycling. Nonetheless, an economic issue in battery fabrication is that the sources of the used chemicals are concentrated in Congo, Australia, and China, increasing the cost of production (Gulley, 2022). The former is relevant in the battery manufacturing industry, considering the exponential demand and high cost of raw materials for battery fabrication. Battery technology aims not only to reach high energy density and fast charge/discharge cycle but also to guarantee high-quality standards with affordable prices. Thus, the I4.0 technologies might be a pivotal function in improving battery cell performance while reducing cost (Wanner et al., 2021). Furthermore, this revolution can help in the environmentally benign production of these storage devices by embracing a circular economy that contemplates recycling and its social implications. In this sense, many authors proposed adopting the I4.0 paradigm to improve the battery cycle, including manufacturing, reusing, and minimizing waste generation and raw materials.

The reverse supply chain (from supplier to customer and vice versa) has a critical role in managing operations during battery recovery, and I4.0 technologies like IoT, ML, and so on can be implemented to pursue this goal. For instance, a reverse logistic study was conducted in Canada to determine the best places for lithium-ion battery (LIB) dismantling using a developed software that considers geographical information systems (Gonzales-Calienes et al., 2022). On the other hand, system dynamics modeling was also used to study the reverse supply chain of EV batteries in terms of cost, revenue, and strategy (Alamerew & Brissaud, 2020). From these studies, the authors concluded that battery remanufacturing and repurposing have market potential.

The traditional waste management system uses a large task force to recycle and dispose of waste material. This is a tough time-consuming process because there is a lot of material diversity with the potential to reuse. In the case of the battery, this is most difficult considering that discharge, disarming, selection, and recovery, such as pyrometallurgy and hydrometallurgy processes or a combination of both, are required (Fan et al., 2020). Moreover, this traditional recycling procedure exposes active chemical elements and leakage of hazardous liquid waste contaminating soil and water. Thus, a new strategy for battery recycling is necessary to decrease inherent costs with environmental benefits.

Kintscher et al. propose the adoption of I4.0 in order to fulfill the requirements of a sustainable recycling process (Kintscher et al., 2020). Recycling I4.0 considers the use of advanced technologies such as ML, the IoT, and big data that make the information exchange between each part of the smart recycling process possible. Furthermore, the battery's State of Health (SoH) can be determined in real time, and this information will be analyzed to propose battery change and determine the path for the second life of recycling. Figure 23.1 shows the pictorial representation of the information marketplace that considers the I4.0 paradigm under a secure environment, especially data privacy. The final stakeholder can be a machine or human, which is very important because some information can be shared without decoding.

Recently, our research group reported a methodology based on electrochemical impedance spectroscopy (EIS) to determine the parameters in batteries by implementing genetic algorithms (GAs) (Pech-Rodríguez et al., 2023). The document used experimental EIS for LIB data to optimize the battery parameters according to a proposed electronic circuit model. The results demonstrate that GA is a versatile tool to accurately determine each electrical element's values in the circuit model. It should be mentioned that the EIS measurement can be achieved in real time, but a lot of work and regulations need to be developed.

The technologies of I4.0 not only can help to measure the EIS of batteries in mobile vehicles but also can be used to share information through an interconnected

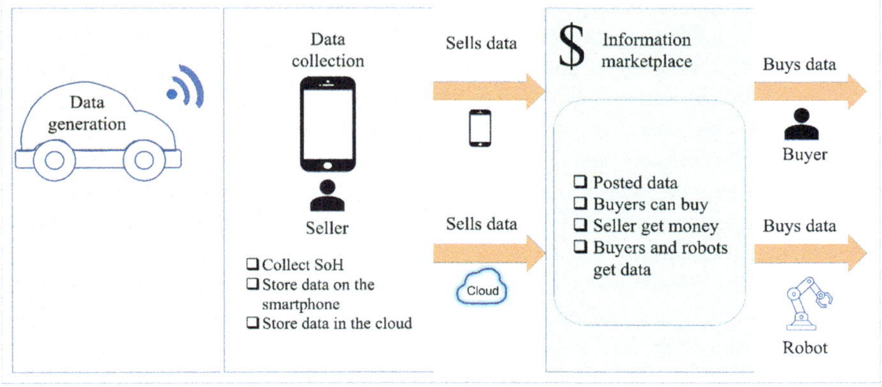

Fig. 23.1 Pictorial representation of information marketplace in the I4.0 paradigm. (Adapted from Kintscher et al., 2020)

platform where users, sellers, and workshops can consult de charge status, SoH, and the trace battery for reuse. In fact, it has been demonstrated that the SoH can be indirectly determined through the EIS parameters. For instance, Haram and coworkers (Haram et al., 2021) use the measured impedance to calculate the SoH according to the following equation:

$$\mathrm{SoH}(Z) = \left(2 - \frac{Z_{\mathrm{m}}}{Z_{\mathrm{n}}}\right) * 100 \tag{23.1}$$

where Z_{m} is the experimental impedance and Z_{n} is the nominal impedance of a new battery. They recommended that an increment of 200% in impedance or internal resistance indicates battery replacement. In addition, this group states that batteries that lost 20–30% of their initial capacity will have to be replaced because the vehicle does not have enough autonomy, and these devices are not more useful for electric cars.

The battery state of charge (SoC) is another relevant parameter. It is defined as the ratio of the remaining charge in the battery and the maximum charge that can be obtained from the battery. For battery second-life use, measuring this parameter to determine the best configuration for the re-ensembled battery will be convenient. Galiounas et al. reported an interesting approach based on machine learning to estimate the SoC (Galiounas et al., 2022). This group used actuation pulse signals in conjunction with acoustic measurements to obtain the acoustic data point that was further analyzed using a feed-forward neural network. Similarly, Hannan et al. developed a deep-learning approach to estimate the SoC in LiB using a self-supervised model (Hannan et al., 2021). The result shows that this model is accurate, achieving a root mean square error of 0.9%.

Second-life batteries may play a vital role as an alternative strategy to face the great pressure of battery recycling. Recycled batteries can be used for smart grid storage or residential applications bringing economic value and, at the same time, environmental benefits. Nevertheless, retired batteries must be selected before applying for a second use, which is one of the main obstacles to overcome. Garg et al. conducted an in-depth study about the primary technical issues related to battery recovery for second use (Garg et al., 2020). They found that traditional screening of battery recovery is visual inspection, voltage measurement, and cycle performance evaluation.

Meanwhile, EIS measurements have been achieved to obtain accurate capacity, internal resistance, and diffusion coefficients. This group proposed implementing a Self-Organizing Map (SOM) neuronal network ML algorithm for battery classification and regrouping. The algorithm simultaneously evaluates the capacity, voltage, temperature, and internal resistance and gives the optimal combination for the second-life battery assembly. This methodology based on machine learning demonstrates promising results in reducing time and cost in second-life batteries.

In recent work, Tripathy et al. developed a systematic study to understand the most critical drivers of Li-ion battery recycling under the I.40 umbrella (Tripathy et al., 2023). They used the concept of the triple bottom line of the I4.0 to determine the main drives for the circular economy of LIB based on the I4.0 technologies. First, they focus on the economic aspect, where they found that the main motivation for LIB recycling is institutional incentives, reducing production costs in LIB by reuse of materials, and reducing material imports. They also mention that LIB recycling will bring coordination between each production chain and give importance to landfill bans. The second scenario was the social drivers, where they highlight the creation of employment opportunities due to the expected massive generation of LIB.

Furthermore, they also consider that human health risk is one of the main drivers that motivate the adoption of battery recycling. Last but not least, this group found that the main driver in the environmental dimension is the reduction of CO_2 emissions. They concluded that policy planners are critical in successfully implementing sustainable LIB recycling. Besides, they highlight the importance of incentives as persuasive means for stakeholders and users to encourage the recovery of batteries, which depends on the availability of I4.0 technologies.

Stena Recycling is a company leading the development of sustainable circular solutions in Europe (Skärin et al., 2022). Currently, they are building seven new battery recycling facilities across Europe for black mass processing in conjunction with some expert partners, and their goal is to recycle 95% of LiBs (Metall, 2023). To do that, three phases are visualized that include the following tasks: collection, discharge, and dismantling the first phase. Then, phase two is recycling batteries from smart devices, power tools, etc. Finally, the last phase is the introduction of recycled materials back into the loop. Stena Nordic Recycling Centre is a novel facility that incorporates the latest technology of the I4.0 such as Artificial Intelligence (AI), ML, and vision to reuse and recycle complex products that deliver a reliable stream of recycled raw material. Figure 23.2 depicts the usual mechanical

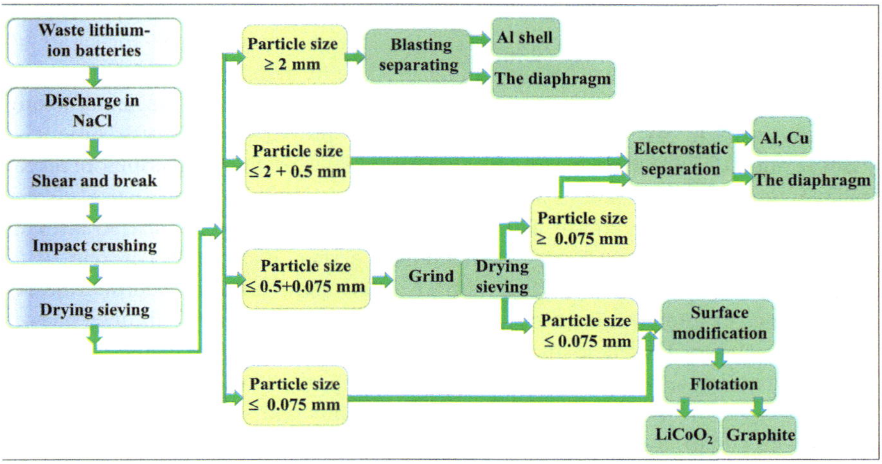

Fig. 23.2 Typical flow chart for mechanical battery separation. (Adapted from Li et al., 2019)

Table 23.1 Estimated cost for disassembling a Smart ForFour 17.6 kWh battery (Rallo et al., 2020)

Activity	Battery item	Module item	Cell item
Removing the battery	117 €	117 €	117 €
Post-battery assessment	442 €	442 €	442 €
Disassembly of battery to modules	–	500 €	500 €
Disassembly module to cells	–	–	275 €
Total	558 €	1058 €	1333 €

whole-battery screening separation route (Li et al., 2019). The first step is battery discharge in NaCl solution, and then shear, crushing, and drying are achieved to classify the part by particle size. Then, chemical elements or compounds are recovered using blasting, electrostatic separation, or flotation.

The widespread introduction of EVs in mobile service has been limited by the high cost of batteries representing 30–40% of the final EV price (Picatoste et al., 2022). In this context, Rallo et al. investigate the economic aspects of battery recycling and reuse as key elements in reducing electric vehicle prices (Rallo et al., 2020). It should be highlighted that EV batteries consist of multiple interconnected cells forming battery packs. This group analyzed the current state of the art for disassembly and remanufacturing activities for module or cell level and provided an estimated cost for each operation. Table 23.1 summarizes the cost of adapting the batteries from the first to second life; the process was done manually because these devices have different packing configurations, making it challenging to apply the automatic process. It can be seen that cell and module dismantling is the most expensive process. Schmied et al. conducted a study where the I4.0 framework was used for LiB cell production. This research group claimed that the only way that batteries become competitive is if the manufacturers adopt a network and digitalization production (Schmied et al., 2022).

The IoT can benefit the LIB recycling process by reducing the time for waste battery collection and facilitating effective communication between manufacturers and customers. In this context, Lu and coworkers propose using Radio Frequency Identification (RFID) to trace battery information such as chemical components and manufacturers, and this would aim for real-time traceability for adequate classification (Riexinger et al., 2020). The obtained information can be shared with the blockchain by IoT technology. As mentioned earlier, disassembly is one of the most expensive tasks in battery recycling because some authors proposed the I4.0 technologies to improve this process (Hossain et al., 2023). Lu et al. designed and developed a novel automatic disassembly platform for end-of-life batteries based on machine learning techniques (Lu et al., 2022). They use computer vision to classify different types of batteries according to their brands and sizes. Thermal cameras, data-driven, smart control, and neural network models were used to measure temperature and control the quality of cutting defects. The results demonstrate that machine learning is a feasible strategy to enhance battery disassembly.

DemoSens project has been proposed to appropriately label and automatically disarm batteries using ML methods (Zorn et al., 2022). Full computer vision was implemented using a high-resolution camera, and the analyzed data was sent to another next stage, like loosening screws. Pregowska et al. researched the progress of AI in battery waste management as a tool for sustainable development, and they claimed that there is still a gap or missing link between laboratory studies and the real world (Pregowska et al., 2022). They proposed ML and computational intelligence to close the breach and obtain an effective battery life cycle. Spent zinc-manganese-based alkaline batteries contain valuable chemical elements crucial to recovery and keeping into the production chain. It has been reported that this type of storage device has 12–28% of zinc and 26–45% of manganese that can be extracted from several treatments (Hu et al., 2021). Ruhatiya et al. developed an optimized model for maximum metal extraction in this type of battery. Their studies used statistical neural automated network techniques to determine the better input parameter values for biorecovery of Zn and Mn (Ruhatiya et al., 2020). They studied the influence of four parameters, energy substrate concentration, pH, temperature, and pulp density of the solution used for the bioleaching finding that this model can recover 10% of Zn and 8% of Mn using the optimal values. In this sense, Priyadarshini et al. adopt the ML method to predict the maximum metal recovery from zinc-manganese batteries using the experimental data of the Ruhatiya research group (Priyadarshini et al., 2022). The same input was also studied, but this group makes a detailed comparison between five ML approaches: linear regression, random forest regression, AdaBoost regression, gradient boosting regression, and XG boost regression. The results demonstrated that XG boost regression is more efficient in predicting the metal recovery from the bioleaching reactor.

3 Hazardous Waste Management Under the I4.0 Paradigm

Every day, humankind generates a large amount of hazardous waste, which represents a potential health threat to the ecosystem if this material is not properly managed. This trash can be found in different forms, such as solid, liquid, and even gas, and is present in almost all industrial processes. Waste management of hazardous substances is very complex because few people are interested in these activities. The latter can be observed in household dumped toxic waste where no appropriate selection and disposal are achieved.

Contrary to battery waste management, this topic has not been thoroughly studied, and only a few documents can be found in the literature. Zhang et al. unveil that small and micro enterprises face high collection costs of hazardous waste, 5–10 times compared to large enterprises, which provokes the illegal dumping of these materials (Zhang et al., 2023). So, this research group analyzed the potential of a platform called hazardous waste collection (third-party service) in conjunction with the dual-channel reverse supply chain model that the government regulates. The results show that using these approaches makes it possible to considerably reduce environmental risk because waste can be canalized to these centres at a low cost under this paradigm. The industrialization process results in large amounts of chemical waste production, which is vital for disposal safely. Waste management companies help dispose and recycle, but this task is risky, and many challenges must be overcome. AAEON is a technology manufacturer working with waste management companies to find affordable, practical devices to trace waste (AAEON, 2023). For example, they developed an SRG-ADIO edge IoT gateway device connected to truck sensors, and this data is ready to share in the cloud. This technology can be monitored in real time, and the company can be sure that waste is correctly handled. In this sense, Samieifard identified and classified several facilities in hazardous waste management and proposed using IoT technology to monitor data such as the fullness of storage tanks and select optimal routes ("Use of Internet of Things (IoT) in Hazardous Waste Management System," 2021).

Medical waste is a particular pathogenic residue that needs special handling because it is categorized as hazardous and toxic waste (B3). It usually contains infections, sharp objects, chemical, cytotoxic, and radioactive materials (Candra Nugraha Deni & Shabirah, 2022). However, incineration is the most recurred modal method for treating this harmful waste because new novel tactics are required to minimize negative ecosystem impact. Chauhan et al. studied seven criteria for the smart healthcare disposal system under the circular economy paradigm, and the analysis was achieved using a decision-making trial and evaluation laboratory method (Chauhan et al., 2021). They claimed that RFID labeling, GPS (global positioning system) tracking, vision, digitalization, and App development could help maximize resources' circularity. Similar work was reported by Yu et al., where a stochastic network was designed for hazardous waste management (Yu et al., 2020). Actually, this model was applied in a third-level grade-A hospital in Wuhan. Still, due to the highly infectious nature of residues, recycling was not conducted in that city. Instead, five possible incineration centers and three landfills were considered variables and solved by the Monte Carlo simulation-based SSA-GP heuristic approach. Conversely, Saeidi and coworkers proposed a real-world application for hazardous waste management based on a bilevel and robust optimization framework (Saeidi-Mobarakeh et al., 2020). The case study was conducted in Isfahan, Iran, in a hospital that generates medical waste management. The algorithm was applied to construct an integral facility with infrastructure for treating and disposing of this waste. The data fed into the model were the investment, truck capacity, and transportation costs. Computational experiments solved the problem, and from this, several scenarios were obtained, considering environmental and economic outputs.

The COVID-19 pandemic has led to extra hazardous medical waste that represents several human health treatments if not correctly managed. Moreover, this pandemic exponentially increases the demand for personal protective equipment. Mosallanezhad et al. reported a fascinating scientific article where they analyzed in detail the supply chain impact during the COVID-19 pandemic in Puebla, Mexico (Mosallanezhad et al., 2023). IoT enabled the reverse supply chain to retrieve real-time data, and the model was solved by a metaheuristic algorithm considering seven evaluation indicators. In this direction, Erdem designed a sustainable logistic network based on Adaptative Large Neighborhood Search (ALNS) algorithms and local search heuristics (Erdem, 2022). The author proposed a small-size logistic for waste management consisting of eight medical centers, two charging stations, and a landfill. Figure 23.3 presents the diagram for this logistic, and it can be observed that the author used electric vehicles to collect medical waste in each hospital.

Big data is another technology of I4.0 that has been proposed as an alternative to developing sustainable strategies for waste management. Zhan et al. constructed a spatial metabolic path for managing organic solvent waste in the city of Chengdu using big data (Zhan et al., 2022). The results suggest that the automobile and electronic devices manufacturing industry is the main generator of organic solvents. Furthermore, the automotive sector observed a decrease in the quantity of this waste compared to electronic device fabricants. Finally, they proposed as a future study the data acquisition from multi-source such as satellite remote sensing, aerial vehicle, and radio frequency identification.

4 Electronic Waste Management in the Context of I4.0

In today's digital era, electronic waste (e-waste) has become a significant environmental and health concern. According to Seif et al., while 44.4 million metric tons of e-waste are generated over the next 5 years annually, only less than 17.4% is recycled (Seif et al., 2023). This not only poses a threat to the environment but also to human health, as e-waste often contains several hazardous heavy metals such as lead, chromium, mercury, and cadmium. Recycling e-waste has many environmental benefits. It allows the recovery of elements (such as glass, plastic, and metals)

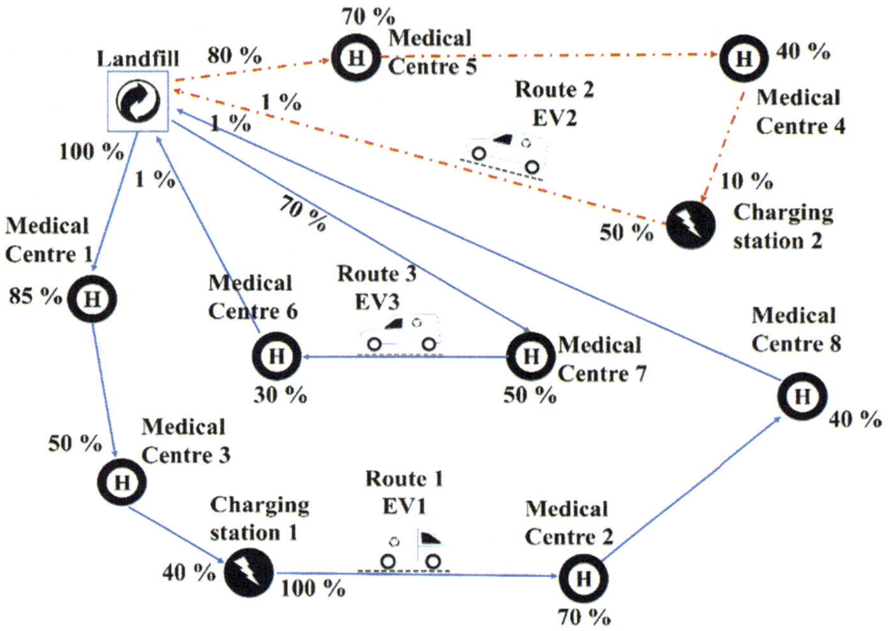

Fig. 23.3 The schematic diagram for the logistics of electric vehicles, medical centers, and charging stations. (Adapted from Erdem, 2022)

that return to the production cycle. This reduces the extraction of raw materials, protects natural resources, and reduces air, water, and soil pollution. Moreover, it produces social benefits, such as the prevention of health risks to people, because it prevents toxic materials from entering the atmosphere or waterways. It also has economic advantages, insofar as it generates job opportunities, for example, separating materials (Shahabuddin et al., 2023).

The ongoing industrial revolution represents a potential for substantial benefits in recycling and waste management, as with new cost-effective and smart sensors, many industries will have the ability to closely track the entire life cycle of any item sold. This will enable product designers to optimize design, minimize materials, avoid damage, improve maintenance, and ultimately develop clean materials. The "zero waste"approach for specific supply chains will become more realistic than ever.

In this sense, some technologies of I4.0 such as ML, AI, RFID, and wireless communications enable the traceability of materials to better understand the opportunities in circular economy. Furthermore, the interconnectivity of systems and automatic transfer of data permit the creation of a more complex system that houses a larger solution space such as cloud computing to reduce the cost by removing the individual computing systems. For example, Oscar Gómez (Gómez Borzdynski, 2020) used several AI algorithms, such as vector machine regression, gradient boosting regression, or Gaussian processes to predict the waste generated in some place and a certain period. Deep learning was tested to see if it could help in the management of electronic waste recycling, using a sequential neural network and a layered neural network. From this work, it was determined that the use of artificial intelligence technologies including deep learning and neural networks can successfully support e-waste management, based on the amount of data generated by different countries. Another way in which I4.0 can contribute to e-waste management is through the integration of systems and data, in other words, by automating the processes. With automation, it is possible to reduce costs, time, and waste, increase productivity, minimize failures, and control, in real-time, all processes. In addition, I4.0 can enable the creation of a more circular economy for electronics (Stevels, 2023). The circular economy has been conceptualized as an industrial system, which is regenerative by intention and aims to improve waste management through the generation of optimized cycles of components and materials, keeping them at their highest utility and value. In the framework of e-waste management, this could include creating closed-loop supply chains in which electronics are reused or repurposed rather than discarded.

Puppala Ramya et al. (2023) in their work designed an e-waste sorting approach, called convolutional neural network (CNN) Shepherd, to sort e-waste using an IoT platform and the cloud. So, they collected images of e-waste using IoT nodes and then stored them in the cloud data storage which then performs the routing process using different algorithms. In addition, they extracted effective features to perform the e-waste classification process. The e-waste classification is performed using FrHHGO, which is a combination of Fractional Henry Gas Optimization (FHGO) and Horse Herd Optimization Algorithm (HOA). In conclusion, the developed e-waste classification system enhances social, environmental, and economic sustainability using FrHHGO and Shepherd Convolutional Neural Network (ShCNN). I4.0 presents both opportunities and challenges for e-waste management. While new technologies can help to improve the efficiency and sustainability of e-waste management, a holistic approach to e-waste management is necessary to ensure a sustainable future. This will require collaboration between stakeholders in the e-waste value chain, as well as investment in the necessary infrastructure to enable a more circular economy for electronics.

5 Metals Waste Management in the Context of I4.0

Although the recycling of metals has been carried out for many years, more attention has been paid in recent years due to the generation of a large amount of electronic waste which contains a considerable amount of heavy metals. There has been a need to recycle metals through more efficient processes, given the danger that these metals can pose both for the environment and for the health of living beings [9–10]. In an I4.0 framework, new technologies and processes are developing that could transform the way metals are recycled, creating a more sustainable and efficient metals recycling industry. These technologies have the potential to create more efficient and sustainable production methods, which in turn could lead to better metal recycling management.

Currently, there are two main methods for metal recycling: pyrometallurgy, which uses traditional processes for metal recovery, including incineration, smelting in electric furnaces and casting in molds and hydrometallurgy, where metal recovery includes two stages: chemical leaching (solubilization of metals with aqueous chemical reagents in leachates) and selective recovery of metals dissolved in the leachates (Baum et al., 2022). In both cases, the processes can be fully automated to ensure good control of each stage of the process and thus ensure the highest efficiency in recovering metals from waste. Thus, with the advent of I4.0 automation technologies, a good recovery of light and heavy metals from e-waste can be achieved. Both pyrometallurgy and hydrometallurgy are highly polluting and dangerous industries due to the high working temperatures and the evaporation that emanates from the process. Thus, by using robots that can disassemble the components and carry out the riskiest operations in each process, the recovery of valuable metals from these products will be easier and safer.

Another way in which I4.0 can contribute to metal recycling is through the use of advanced sensors. In metal recovery processes, there is a crushing stage to reduce the size of the material and facilitate its subsequent classification. A magnet recovers the ferrous fraction and then the material is sieved to ensure, at a later stage of the process, efficient sorting supported by sensors and computer vision, the plant can increase product yield, generate material fractions of significantly higher purity, and detect and sort metals and other materials in the e-waste. The metals recycling industry is part of a complex global supply chain that includes mining, manufacturing, and distribution. The integration of I4.0 technologies will require collaboration and integration across the entire supply chain to create a more efficient system. All this will only be possible if one or more of the technologies that are part of I4.0 are used. Although I4.0 technologies present many opportunities for metals recycling management, numerous challenges must be addressed to fully reach their potential. To achieve these challenges, it is necessary to collaborate with all the participants in the metals recycling value chain, including regulators, manufacturers, recyclers, and technology providers. Just working together, the industry can create a more sustainable and efficient system for metal recycling management.

One case study of metal recycling in the context of Industry 4.0 was reported in the metal work-copper industry (Aivaliotis et al., 2021). The project aims to facilitate the optimization of resource management, using I4.0 technologies, and reduce copper scrap. The project includes the implementation of sensors, data acquisition, data analytics, and some algorithms to monitor and improve the production process in real time. The digital information coming from the shop-floor sensors allows the recreation of different scenarios and the prediction of the behavior of the production process, permitting more efficient and sustainable production. The project aims to increase the proportion of recycled copper in the production process, reducing the demand for virgin materials and the environmental impact of copper production. The aforementioned project has confirmed the potential of Industry 4.0 technologies to improve the copper production process and endorse sustainability in this industry.

6 Plastic Waste Management Using I4.0 Technologies

The application of I4.0 in managing plastic waste has been analyzed from the social, environmental, and economic points of view to impulse the circular economy. A systematic literature review has been made, highlighting the use of cloud computing, AI, and the IoT (da Silva et al., 2022). The work of Ortega-Gras focuses on a desk-research about policies in Europe for the twin digital and green transition and cases where the circular economy uses Key Enabling Technologies (KET), supported by I4.0 technologies (Ortega-Gras et al., 2021). In this review, legislative instruments were identified in the diverse processes such as product design and production and use or consumption for the management of plastic. Meanwhile, others focus on the management of plastics by conducting real data analysis. Bollegraaf was built for plastic recycling. AEVAE created a system of recycling of packing based in (HDDE y LPDE) for the Agricola sector, and RecySmart made by Recircula Solutions S.L identifies plastic bottles by IoT and AI. Radical Innovations Group studied the incorporation of AI into multi-sensors in plastic segregation. Two smart contracts were also incorporated, driving public and private information. The study concludes that this combination helps to make decisions of the involved reliably (Chidepatil et al., 2020).

ML in plastic waste has been used mainly to identify and sort. Zaman et al. adopted an ML model to identify dangerous waste and patterns of incorrect recycling by using a Database of Labelled Images in real time in Perth, Western Australia (Zaman, 2022). The case study application shows that waste management under the I4.0 framework reduces costs and improves efficiency in the supply chain while eliminating waste.

Besides, a combination of supervised and no supervised algorithms has been used to help a satellite called PRISMA, by the Italian Space Agence, recognizing the plastic waste in coastal and ocean (Taggio et al., 2022). ML plays a vital role here because of its potential to find hidden patterns. For sorting purposes, a system based on a convolutional neural network in solid waste using two image resolutions was developed, and the results indicate that the lighter model has an accuracy of 80.88% (Nnamoko et al., 2022). Moreover, the models such as K-nearest neighbor (KNN), support vector machine (SVM), and back propagation neural network (BPNN) have shown excellent results in the classification of commercial plastics (Yang et al., 2020). For the effects of quantification, algorithms of CNN have been proposed to develop the system named APLASTIC–Q able to detect and quantify plastic waste (Wolf et al., 2020). This system can be easily adapted to detect and quantify plastic litter from aerial and space missions.

Also, the use of learning machine-based decision trees, random forests, SVM, partial least squares, CNN, and infrared spectroscopy has been applied to get over the problems of the classification of plastic bottles for recycling (Xiaozheng Chen et al., 2023). On the other hand, Malik and coworkers adopted deep learning neural network models to design the EfficientNet-b3 architecture that was trained with 34 classes of waste images. This group reached an 85% image classification accuracy in sorting solid waste (Malik et al., 2022). The ML-based decision tree and the artificial neural network have been used in the prediction process as the composition of liquid and gas processes in plastic waste pyrolysis (Cheng et al., 2023). The I4.0 has demonstrated its versatility and innovative way of handling waste in unimaginable scenarios. For example, Mikula et al. review filament-based recycled polymers such as polypropylene (PP), polyvinyl chloride (PVC), polyethylene terephthalate (PET), and so on for application in 3D printers (Mikula et al., 2021). The author discusses the complete extraction process and the composite modification with additives such as crat lignin, carbon fiber reinforced (CFR), polydopamine (PDA), tropolone, and p-benzoquinone hydroquinone. Besides, commercial filaments of recycled polymers are obtained such as B-PET, filament, fila-cycle, refill, innofil3D, fishy filaments, portcurno, tride, and CREAMELT. Ruiz et al. reviewed related works to plastic waste from the vehicle industry applied to 3D printing, recycling, and circular economy. Still, they did not present many details about their relationship (Ruiz et al., 2022). In addition, aspects such as degradation, the cons of recycling with catalyst/solvent, direct reuse, and processing and reuse of different 3D filaments has been already studied (Zhu et al., 2021).

In the application of plastic waste in 3D, case studies have found where the waste plastic is recycled for a specific application or to analyze its properties. Garrido et al. studied the waste plastic from the ocean for reutilization in 3D printers in the Vigo Port and other ports from España (Garrido et al., 2020). In another study, the same authors proposed the variability of plastic filaments based on maritime plastic waste for their integration into 3D technologies (Julio Garrido et al., 2021). Table 23.2 presents examples of the I4.0 technologies applied to plastic waste management.

7 Paper Waste Management Using I4.0 Technologies

Paper and cardboard waste is another severe problem that many industries face, which represents a direct environmental impact considering that trees are the raw material for paper fabrication. Furthermore, it must be regarded that different chemicals are required during paper production, which in most cases, are not environmentally friendly (Qian, 2021). In this regard, society has been applying the recycling process to reduce paper waste and, at the same time, obtain economic revenue. Nonetheless, paper selection and management are time-consuming and lower income to the industrial recycling sector.

Contrary to plastic and metal waste management under the I4.0 paradigm, there are only a few papers exclusively applied to waste paper. The major reported manuscripts deal with waste paper joint to glass and plastic, food or other-oriented to identify, sort, and recycle. In this sense, ML algorithms have been used to predict,

Table 23.2 Examples of I4.0 technologies in plastic waste

Technology I 4.0	Type of plastic	Work description	Reference
3D	Polymers	Formation of filaments	Mikula et al. (2021)
3D	Maritime plastic waste recycled	Adaptation of 3D printer for work with a wide range of types or qualities of plastic	Julio Garrido et al. (2021)
3D	Polyethylene terephthalate	Comparison of new and recycled material	Oussai et al. (2021)
AI	Recycled plastics	First-of-its-kind blockchain smart contracts powered by multi-sensor	Chidepatil et al. (2020)
AI Deep learning	Paper, plastic, glass, and organic material	Use of convolutional neural networks and image identification for the automatic recycling	Nañez Alonso et al. (2021)
Machine learning	Plastic, glass, metal, paper	Use image recognition with deep learning neural network models	Malik et al. (2022)
Machine learning	PP, PS, polyethylene-high density (PE-HD), and PET	Use machine learning for the separation of garbage into the four mentioned categories of plastic	Bobulski and Kubanek (2021)
Machine learning	Metal, plastic, paper, and general waste	Segregation by the use of LoRa communication protocol and TensorFlow-based deep learning model	Sheng et al. (2020)
Machine learning	Aquatic plastic litter	Proposed a detection and classification system named APLASTIC-Q based on convolutional neural networks (CNNs)	Wolf et al. (2020)
Machine learning	Plastic bottles	Recycling by a decision tree, random forest, support vector machine, partial least squares, and convolutional neural network	Xiaozheng Chen et al. (2023)

classify, and optimize solid waste such as paper, glass, and plastic (Xia et al., 2021). An interesting case study was conducted by Rahman et al., where I4.0 technologies were adopted to develop a system capable of sorting digestible and indigestible waste using a CNN (Rahman et al., 2022). To this end, a camera was used to obtain images from waste materials processed by a Rasberry pi which ultimately sent the motor drive signal to put the waste in the corresponding box. In the case of paper sort, the developed system shows an accuracy of 93.61% from 53 measured items. In the same direction, Gupta and coworkers design and fabricate hardware called "smartBin" for waste segregation based on AI (Gupta et al., 2022). This group highlighted that this task is arduous because garbage shares similar characteristics. One relevant point was the use of infrared and standard low-light sensitivity cameras.

Other relevant works highlight the implementation of Keras framework that included Visual Geometric Group 16 (VGG16), Visual Geometric Group 19 (VGG19), and ResNet15V2 applied to paper, plastic, glass, and organic material sorting in events, office, and airports (Nañez Alonso et al., 2021). In separating paper, cans, glass, and other recyclable materials, an ML-based waste recycling framework based on IoT and ML for smart cities has been proposed. Other work proposed recycling glass, paper, and plastic by a methodology that tests the reliability of convolutional neural networks and image identification (Nañez Alonso et al., 2021).

Classification techniques such as the Artificial Hummingbird Algorithm (AHA) coupled with Opposition-based learning (OBL) have been proposed for recognizing cardboard, glass, metal, paper, plastic, and rubbish (Ali et al., 2022). In a recent project called "ZRR2," artificial intelligence in conjunction with sensors, cameras, and others was used to develop a robot to sort municipal garbage in Barcelona. This system sorts 13 different materials between cardboard, paper, and Tetra Pak (Wilts et al., 2021).

8 Food Waste Management Under the I4.0 Umbrella

Food production is considered a key contributor to the negative environmental impact worldwide because this activity requires the use of fuel, water, land, and some chemical materials. For this reason, the generation of waste food is undesired. Approximately 1.2 billion of the global farm-stage food produced worldwide is wasted annually (Parfitt et al., 2021). To have an idea, Europe generates 88 million tons of food waste (FW) per year (Ioannou et al., 2022). Thus, countries have sought to reduce FW or convert these residues into value-added products such as livestock feed, biomass, and biofertilizers(Liu et al., 2021). An innovative approach to deal with this problem is the adoption of I4.0 technologies; see Table 23.3. Various literature reviews about incorporating Industry 4.0 for managing FW have been reported, and they analyze the current state and challenges (Ada et al., 2021; Derakhti et al., 2023; Furstenau et al., 2020).

Under the umbrella of I4.0, an important approach to food waste has been the circular economy. Kerdlap et al. researched the transition to a circular economy in the urban zone of Singapore. They proposed using IoT technology in smart waste collection and waste valorization (Arora et al., 2021). Other cases found in the literature focused on the impact of the I4.0 on a triple helix structure formed by governments, academics, and organizations in developed countries (Reis et al., 2021).

On the other hand, the IoT has been applied in the management and security of food by video location and sensors of humidity, pollution, and gases. At the same time, communication technologies have allowed the exchange of information at all levels and have been proposed in food manufacturing and waste reduction (Said et al., 2023). In this sense, IoT was used for traceability and quality monitoring of postharvest carrots involving farmers, suppliers, and exporting firms (Kailaku & Djatna, 2022). From those mentioned earlier, it can be inferred that the best control of quality allows the reduction of waste. So, IoT and other tools have been applied in

Table 23.3 Examples of I4.0 technologies applied to food waste across the world

Machine learning used in the world				
Continent	Country	Algorithm ML	Aim	Reference
Europe		Long and short-term memory networks Feedforward neural networks Support vector regression Random forests Holt–winters statistical model	Supply chains by proposing daily fresh fish demand forecasting models	Miguéis et al. (2022)
	Spain	Linear model with stepwise selection Regression tree Bagged tree Random forest Gradient boosting Ridge regression Lasso regression Elastic net Spline regression	Production planning of a food industry	Garre et al. (2020)
	Swedish public	Random forest	Planification of attendance for food waste	Malefors et al. (2022)
	UK	Support vector machines	Monitor the stability of tomato fruit during storage	Emsley et al. (2022)
Asia	Japan	Random forest, naive Bayes, extreme gradient boosting (Xgboost), and decision tree	Garbage content estimation using internet of things and machine learning	Likotiko et al. (2023)
African	Audi Arabia	Fu logic	Analysis of attitudes toward food waste	Malibari et al. (2023)
	Iran	Support vector machine	Waste recycling of fruits	Farjami et al. (2020)
Oceania	Australia	ML algorithm/full YOLOv4	Identify and measure household waste contamination	Zaman (2022)

packaging various types of food, such as meat and fruit. Agri-food producers of two Spanish firms used the IoT to adequately manage meat, fruit, and vegetables (Oltra-Mestre et al., 2021). The author concluded that adopting I4.0 is a process innovation which enables competitive outcomes. Besides, incorporating IoT and using nano-technology in pH and oxygen sensors can also help detect microbial growth and oxidation and share information with food suppliers(Atoche et al., 2023).

It is essential to highlight that IoT, combined with other technologies in the manufacturing process, can reduce food waste, water, and energy. As shown in the work of Jagtap et al., where the process data were collected, monitored by IoT and saved by gateways and internet networks to further decision-making analysis (Jagtap et al., 2021). Meanwhile, other authors reviewed challenges and technological applications on the supply chain during the COVID-19 pandemic to improve food storage and reduce waste (Abideen et al., 2021).

Another less common application of I4.0 is using FW to create filaments for 3D printing. The company Krill Design RET uses orange peel waste and dry to obtain a powder compound mixed with a plant-derived biopolymer. The composite was extruded to form a filament and then implemented in a 3D printer to create the lamp's body (Ortega-Gras et al., 2021). This idea was further discussed by Wong et al., who studied the feasibility of using jackfruit seeds as printable fruit ink (Hooi Chuan Wong et al., 2022). The study shows that although the waste fruit is printable, it can only be extracted with a 4-mm nozzle, resulting in a low-quality shape.

The cloud computing paradigm is the new evolution of computing that allows the development of new internet services, such as ubiquitous and on-demand networks capable of accessing configurable computing interfaces (Soni & Kumar, 2022). Consequently, this model can be utilized to enable waste reduction. For example, using cloud computing with other technologies, such as machine learning and QR code, a SmartNoshWaste framework has been proposed to reduce food waste in smart cities (Dey et al., 2022). The experimental evaluation of this platform reveals that potato waste was reduced by 9.46%. Another proposed model is Agritech Business 4.0, which integrates food production and distribution (Eashwar & Chawla, 2021). This transparent model includes the Internet of things, smart DLT, big data, cloud computing, and sensor-based automation to help optimize, reduce wastage, and increase the profit margin.

ML is a kind of artificial intelligence capable of data decisions that has been used in diverse areas of food to predict and reduce waste (Garre et al., 2020). For example, forward neural networks, support vector regression, random forest, long- and short-term memory networks, and statistical methods were tested in demand and reduction of fish food where better accuracy results were obtained by long- and short-term memories (Miguéis et al., 2022). Other ML algorithms have been used in specific prediction cases in food waste composition. The neural network used for this aim in municipal solid waste (paper, plastic, organic, and textile) in different seasons (Adeleke et al., 2021).

9 The Proposed Approaches to Incentivize Waste Management

As it is well known, there is a lot of work to do to motivate institutions and society to actively participate in managing vast volumes of plastic, metals, food waste, etc. One proposed pathway is a persuasive incentives scheme where participants are encouraged to recycle and separate by vouchers or payments. For example, in Hong Kong, the Environmental Protection Department released a pilot project called "Green Community" (Department, 2022). To encourage the public to separate and recycle waste, they use smart cards to earn points and exchange them with gifts such as daily necessities and groceries. In this way, this department promotes an environmental awareness culture in the community. Besides, awareness campaigns for citizens, forums, workshops, seminars for waste experts, and job opportunities are other alternatives to promote waste management.

On the other hand, the collection approach might play a critical role in waste management. Today, it was envisaged two alternatives: door-to-door collection and drop-off collection. The first is advantageous for users but is operationally costly, and in the second, users lift the waste at centralized stations. In fact, there are a lot of potential schemes that have been proposed to engage waste management.

Furthermore, policy and regulations need to be developed and applied, and this might be achieved by considering the most problematic residues, geography, demography, and technological capacity. It also should be considered an instrument for measuring the applied policies and how they can be improved. For instance, e-waste management policies are constantly analyzed using the SWOT (Strengths, Weaknesses, Opportunities, and Threats) (Bhardwaj et al., 2023). There are no unified regulations for waste treatment; some countries pass responsibility to consumers, and others share commitment between stakeholders, manufacturers, and consumers (Castro et al., 2023).

10 Challenges for Waste Management

The generation of a huge quantity of waste because of the rapid increase in population and urbanization creates a disposal problem in rural and urban areas. Most of the developing countries have been facing difficulties managing waste classification and quantification because of the inadequate practice of collection and separation, treatment, and disposal of a variety of waste. Particularly, solid waste management shows up more challenging task in developing countries as compared to developed ones. The main reason for this is the lack of documentation for the proper assessment and disposal of waste processes gives rise to vulnerable conditions in the ecosystem such as environmental pollution and exposure to harmful substances and therefore triggers health issues. Besides, there is a lack of investment in advanced waste treatment processes and a lack of financial input for waste disposal. Hence, life cycle assessment (LCA), which is very effective for evaluating environmental impacts, should be periodically documented by municipalities. No doubt that evaluation and follow-up of LCA reports make easier waste collection, classification, and disposal or valorization. Even though there is growing support in countries for waste management systems, international conventions and agreements are required under constructive government policies.

On the other hand, it is highly crucial to classify types of wastes with physical, chemical, and biological characteristics such as garbage, combustible, biodegradable, agricultural, hazardous, construction and demolition, and sewage. Waste management also suffers from a lack of awareness and lack of proper knowledge regarding waste generation, classification, and collection. It is not sufficient to have an educated homeowner who correctly classifies their trash if there are not adequate facilities and trained personnel for the management of the residues. For example, it is normal to see cans for organic and inorganic trash, but when it is collected, they are dumped at the same site. This is evidence of the lack of policies and the suitable application of coercive or persuasive measurements. Decision-makers in the municipality should consider providing an efficient system to the inhabitants.

Transportation is another strong challenge for waste based on the content and type of waste. It should offer real-time monitoring and online mapping optimal route design. In addition, a lack of technical skills among personnel within municipalities, poor vehicles and roads, and a lack of equipment for the collection of sorted materials also influence effective transportation. The municipality authorities should push forward their organizational capacities, training, and professional knowledge.

The lack of practical waste management plants is another issue to be addressed. No doubt that investments in collecting, sorting, recycling, valorization, and disposal of waste improve achieving environmental sustainability. To achieve efficient waste management, enough large spaces to enable disposal and valorization of the distinguished waste should be introduced by considering the population of each city. Moreover, today most landfills lack advanced technologies for adequate waste management. So, it is mandatory to design facilities by incorporating the I4.0 technologies which can not only help in the management of residues but also control odors, methane gas. The imminent solution is the creation of policies where the participation of stakeholders, homeowners, industrial representatives, and civil associations is necessary.

11 Future Perspectives and Opportunities

Recycling and Valorization: The development of efficient and sustainable waste management processes plays a crucial role in the recycling and valorization of solid wastes. Recycling solid waste into renewable form enables the reduction of a huge volume of residues from land disposal. Developed countries are adopting the circular economy transition as it contributes to the reduction of environmental pollution. Providing accurate knowledge of source-based waste (residential, industrial, municipal, hazardous, agricultural, and commercial, etc.) and type-based waste (farm, biodegradable, ashes, combustible, garbage, bulky, construction, sewage, etc.), classification waste is essential for efficient valorization of wastes into value-added chemicals and fuel. Among the advanced valorization techniques, biotechnological processes are the key to sustainable waste management systems to convert the wastes into useful chemicals. Therefore, high investments are required to reinforce the link between advanced processes and innovation systems.

Reduce waste generation: This can be achieved by continuous awareness campaigns and economic incentives. On the industrial scale, the adoption of cutting-edge technologies can play a critical role by improving the use of materials.

One pending task is the optimization of waste management activities which can attract the attention of the private industry due to the inherent monetary incomes (Khan et al., 2022). Furthermore, the adoption of regulations might result in new job opportunities.

Another pending point is the standardized source separation. The former is very important because it is a critical component of the current waste management system, and a lot of resources are invested in separation and classification. Thus, waste separation can drastically change the perception of investors who are willing to have higher profitability ratios. Waste separation is not an easy task because it depends on different members such as homeowners, consumers, and stakeholders, so guidelines for each one need to be developed.

Technological innovation enables providing effective implementation of sustainable waste treatment and hence constructing efficient business models using IoT, cloud computing, and artificial intelligence.

On the other hand, it should be mentioned that the creation of opportunities is through the development and implementation of novel initiatives.

12 Conclusion

This chapter discusses waste management systems and current strategies involving innovations such as the I4.0 technologies in detail. From the literature, it can be inferred that waste generation is a big problem to solve, considering that inadequate management negatively affects human health and ecosystems. Developing routes under the I4.0 umbrella might be one answer to alleviate and reduce negative environmental impacts before they develop into more complex problems.

According to the information discussed earlier, metal and plastic recycling is typically the most recurrent form of waste management because the recycling industry reaches positive net earnings. The latter showed that waste management had been mainly implemented as an economic strategy, not for environmental and social concerns. The I4.0 cutting-edge technologies could play a significant role in the widespread adoption of waste management where economic, environmental, and social dimensions are considered. It should be highlighted that incorporating these technologies is very expensive, and an educated workforce will be required to develop and manage this infrastructure.

Furthermore, to promote waste management at all levels, policymakers and governments must implement persuasion economic incentives and coercion measurements and encourage education as essential to waste management implementation to reach sustainable development goals. It should be mentioned that currently, some countries apply economic incentives, and users receive vouchers or discounts. At the same time, waste management is an obligation for consumers without financial incentives in other regions, such as Europe. Finally, it has been observed that countries with any policy for waste management have a low collection, which reveals the necessity of friendly strategies where consumers feel invited to adopt sustainable behavior.

Acknowledgments The author would like to thank the Polytechnic University of Victoria for the time provided to develop this research.

Declaration of Competing Interest The authors declare that there is no competing financial interest or personal relationship that influences this work.

References

AAEON. (2023). *Driving chemical waste management with IoT solutions*. Application story. Retrieved from https://www.aaeon.com/jp/ai/application-story-srg-adio-chem-waste.

Abdallah, M., Abu Talib, M., Feroz, S., Nasir, Q., Abdalla, H., & Mahfood, B. (2020). Artificial intelligence applications in solid waste management: A systematic research review. *Waste Management, 109*, 231–246. https://doi.org/10.1016/j.wasman.2020.04.057

Abideen, A. Z., Sundram, V. P., Pyeman, J., Othman, A. K., & Sorooshian, S. (2021). Food supply chain transformation through technology and future research directions – A systematic review. *Logistics, 5*(4), 1–24. https://doi.org/10.3390/logistics5040083

Ada, N., Kazancoglu, Y., Sezer, M. D., Ede-Senturk, C., Ozer, I., & Ram, M. (2021). Analyzing barriers of circular food supply chains and proposing Industry 4.0 solutions. *Sustainability, 13*(12), 1–29. https://doi.org/10.3390/su13126812

Adeleke, O., Akinlabi, S. A., Jen, T.-C., & Dunmade, I. (2021). Application of artificial neural networks for predicting the physical composition of municipal solid waste: An assessment of the impact of seasonal variation. *Waste Management & Research, 39*(8), 1058–1068. https://doi.org/10.1177/0734242X21991642

Aivaliotis, P., Anagiannis, I., Nikolakis, N., Alexopoulos, K., & Makris, S. (2021). Intelligent waste management system for metalwork-copper industry. *Procedia CIRP, 104*, 1571–1576. https://doi.org/10.1016/j.procir.2021.11.265

Alamerew, Y. A., & Brissaud, D. (2020). Modelling reverse supply chain through system dynamics for realizing the transition towards the circular economy: A case study on electric vehicle batteries. *Journal of Cleaner Production, 254*, 120025. https://doi.org/10.1016/j.jclepro.2020.120025

Albertsen, L., Richter, J. L., Peck, P., Dalhammar, C., & Plepys, A. (2021). Circular business models for electric vehicle lithium-ion batteries: An analysis of current practices of vehicle manufacturers and policies in the EU. *Resources, Conservation and Recycling, 172*, 105658. https://doi.org/10.1016/j.resconrec.2021.105658

Ali, M. A. S., P. P, F. R, & Salama Abd Elminaam, D. (2022). A feature selection based on improved artificial hummingbird algorithm using random opposition-based learning for solving waste classification problem. *Mathematics, 10*(15), 1–34. https://doi.org/10.3390/math10152675

Arora, S., Jung, J., Liu, M., Li, X., Goel, A., Chen, J., et al. (2021). Gasification biochar from horticultural waste: An exemplar of the circular economy in Singapore. *Science of the Total Environment, 781*, 146573. https://doi.org/10.1016/j.scitotenv.2021.146573

Atoche, A. C., García, N. C., Estrada-López, J. J., Vázquez-Castillo, J., Heredia-Lozano, J., Burgos-Reyes, M., et al. (2023). A study on the renewable power generation capacity of microalgae microbial fuel cells for powering GIoT sensor nodes. *Journal of Power Sources, 580*, 233185. https://doi.org/10.1016/j.jpowsour.2023.233185

Ayerbe, E., Berecibar, M., Clark, S., Franco, A. A., & Ruhland, J. (2022). Digitalization of battery manufacturing: Current status, challenges, and opportunities. *Advanced Energy Materials, 12*(17), 2102696. https://doi.org/10.1002/aenm.202102696

Baum, Z. J., Bird, R. E., Yu, X., & Ma, J. (2022). Lithium-ion battery recycling—Overview of techniques and trends. *ACS Energy Letters, 7*(2), 712–719. https://doi.org/10.1021/acsenergylett.1c02602

Bhardwaj, S., Khurana, S., & Kumar, A. (2023). Chapter 7 – Electronic (E-waste) conduct: Chemical assessment and treatment methods. In P. Singh, P. Verma, R. Singh, A. Ahamad, & A. C. S. Batalhão (Eds.), *Waste management and resource recycling in the developing world* (pp. 143–161). Elsevier.

Birkel, H., & Müller, J. M. (2021). Potentials of Industry 4.0 for supply chain management within the triple bottom line of sustainability—A systematic literature review. *Journal of Cleaner Production, 289*, 125612. https://doi.org/10.1016/j.jclepro.2020.125612

Bobulski, J., & Kubanek, M. (2021). Deep learning for plastic waste classification system. *Applied Computational Intelligence and Soft Computing, 2021*, 6626948. https://doi.org/10.1155/2021/6626948

Candra Nugraha Deni, M., & Shabirah, A. (2022). Planning and implementation of hazardous waste management emergency response program—A case study of PT. X medical hazardous waste treatment. *IOP Conference Series: Earth and Environmental Science, 1111*(1), 012041. https://doi.org/10.1088/1755-1315/1111/1/012041

Castro, F. D., Botelho Júnior, A. B., Bassin, J. P., Tenório, J., Cutaia, L., Vaccari, M., & Espinosa, D. (2023). Chapter 13 – E-waste policies and implementation: A global perspective. In P. Singh, P. Verma, R. Singh, A. Ahamad, & A. C. S. Batalhão (Eds.), *Waste management and resource recycling in the developing world* (pp. 271–307). Elsevier.

Chauhan, A., Jakhar, S. K., & Chauhan, C. (2021). The interplay of circular economy with Industry 4.0 enabled smart city drivers of healthcare waste disposal. *Journal of Cleaner Production, 279*, 123854. https://doi.org/10.1016/j.jclepro.2020.123854

Chen, X. (2022). Machine learning approach for a circular economy with waste recycling in smart cities. *Energy Reports, 8*, 3127–3140. https://doi.org/10.1016/j.egyr.2022.01.193

Chen, X., Kroell, N., Althaus, M., Pretz, T., Pomberger, R., & Greiff, K. (2023). Enabling mechanical recycling of plastic bottles with shrink sleeves through near-infrared spectroscopy and machine learning algorithms. *Resources, Conservation and Recycling, 188*, 106719. https://doi.org/10.1016/j.resconrec.2022.106719

Cheng, Y., Ekici, E., Yildiz, G., Yang, Y., Coward, B., & Wang, J. (2023). Applied machine learning for prediction of waste plastic pyrolysis towards valuable fuel and chemicals production. *Journal of Analytical and Applied Pyrolysis, 169*, 105857. https://doi.org/10.1016/j.jaap.2023.105857

Chidepatil, A., Bindra, P., Kulkarni, D., Qazi, M., Kshirsagar, M., & Sankaran, K. (2020). From trash to cash: How Blockchain and multi-sensor-driven artificial intelligence can transform circular economy of plastic waste? *Administrative Sciences, 10*(2), 1–16. https://doi.org/10.3390/admsci10020023

Chowdhury, M. S., Rahman, K. S., Chowdhury, T., Nuthammachot, N., Techato, K., Akhtaruzzaman, M., et al. (2020). An overview of solar photovoltaic panels' end-of-life material recycling. *Energy Strategy Reviews, 27*, 100431. https://doi.org/10.1016/j.esr.2019.100431

da Silva, L. F., Resnitzkyd, M. H., Santibanez Gonzalez, E. D., de Melo Conti, D., & da Costa, P. R. (2022). Management of Plastic Waste and a circular economy at the end of the supply chain: A systematic literature review. *Energies, 15*(3), 1–15. https://doi.org/10.3390/en15030976

Department, E. P. (2022). Green@community. Retrieved from https://www.wastereduction.gov.hk/en/community/crn_intro.htm

Derakhti, A., Santibanez Gonzalez, E. D. R., & Mardani, A. (2023). Industry 4.0 and beyond: A review of the literature on the challenges and barriers facing the Agri-food supply chain. *Sustainability, 15*(6), 1–22. https://doi.org/10.3390/su15065078

Dey, S., Saha, S., Singh, A. K., & McDonald-Maier, K. (2022). SmartNoshWaste: Using blockchain, machine learning, cloud computing and QR code to reduce food waste in decentralized web 3.0 enabled smart cities. *Smart Cities, 5*(1), 162–176. https://doi.org/10.3390/smartcities5010011

Eashwar, S., & Chawla, P. (2021). Evolution of agritech Business 4.0—Architecture and future research directions. *IOP Conference Series: Earth and Environmental Science, 775*(1), 012011. https://doi.org/10.1088/1755-1315/775/1/012011

Emsley, N. E. M., Holden, C. A., Guo, S., Bevan, R. S., Rees, C., McAinsh, M. R., et al. (2022). Machine learning approach using a handheld near-infrared (NIR) device to predict the effect of storage conditions on tomato biomarkers. *ACS Food Science & Technology, 2*(1), 187–194. https://doi.org/10.1021/acsfoodscitech.1c00420

Erdem, M. (2022). Designing a sustainable logistics network for hazardous medical waste collection a case study in COVID-19 pandemic. *Journal of Cleaner Production, 376*, 134192. https://doi.org/10.1016/j.jclepro.2022.134192

Fan, E., Li, L., Wang, Z., Lin, J., Huang, Y., Yao, Y., et al. (2020). Sustainable recycling Technology for Li-Ion batteries and beyond: Challenges and future prospects. *Chemical Reviews, 120*(14), 7020–7063. https://doi.org/10.1021/acs.chemrev.9b00535

Farjami, J., Dehyouri, S., & Mohamadi, M. (2020). Evaluation of waste recycling of fruits based on Support Vector Machine (SVM). *Cogent Environmental Science, 6*(1), 1712146. https://doi.org/10.1080/23311843.2020.1712146

Fatimah, Y. A., Govindan, K., Murniningsih, R., & Setiawan, A. (2020). Industry 4.0 based sustainable circular economy approach for smart waste management system to achieve sustainable development goals: A case study of Indonesia. *Journal of Cleaner Production, 269*, 122263. https://doi.org/10.1016/j.jclepro.2020.122263

Fichtel, P. S., & Duram, L. A. (2022). The perceived role of communities and government officials in solid waste Management in Ghana, West Africa. *International Journal of Applied Geospatial Research (IJAGR), 13*(1), 1–18. https://doi.org/10.4018/IJAGR.295863

Furstenau, L. B., Sott, M. K., Kipper, L. M., Machado, Ê. L., López-robles, J. R., Dohan, M. S., et al. (2020). Link between sustainability and Industry 4.0: Trends, challenges and new perspectives. *IEEE Access, 8*, 140079–140096. https://doi.org/10.1109/ACCESS.2020.3012812

Galiounas, E., Tranter, T. G., Owen, R. E., Robinson, J. B., Shearing, P. R., & Brett, D. J. L. (2022). Battery state-of-charge estimation using machine learning analysis of ultrasonic signatures. *Energy and AI, 10*, 100188. https://doi.org/10.1016/j.egyai.2022.100188

Garg, A., Yun, L., Gao, L., & Putungan, D. B. (2020). Development of recycling strategy for large stacked systems: Experimental and machine learning approach to form reuse battery packs for secondary applications. *Journal of Cleaner Production, 275*, 124152. https://doi.org/10.1016/j.jclepro.2020.124152

Garre, A., Ruiz, M. C., & Hontoria, E. (2020). Application of machine learning to support production planning of a food industry in the context of waste generation under uncertainty. *Operations Research Perspectives, 7*, 100147. https://doi.org/10.1016/j.orp.2020.100147

Garrido, J., Sáez, J., Armesto, J. I., Espada, A. M., Silva, D., Goikoetxea, J., et al. (2020). 3D printing as an enabling technology to implement maritime plastic circular economy. *Procedia Manufacturing, 51*, 635–641. https://doi.org/10.1016/j.promfg.2020.10.089

Garrido, J., Silva, D., Portela, B., & Lekube, B. (2021). Integrated user-oriented service for 3D printing environments with recycled material from maritime plastic waste. *Applied Sciences, 11*(9), 1–20. https://doi.org/10.3390/app11093787

Gatto, A. (2023). Quantifying management efficiency of energy recovery from waste for the circular economy transition in Europe. *Journal of Cleaner Production, 414*, 136948. https://doi.org/10.1016/j.jclepro.2023.136948

Gómez Borzdynski, Ó. (2020). *Modelado y predicción de generación de residuos sólidos usando enfoque de aprendizaje automático*. (PhD), Universidad Autonoma de Madrid, Madrid. Retrieved from https://repositorio.uam.es/handle/10486/693636

Gonzales-Calienes, G., Yu, B., & Bensebaa, F. (2022). Development of a reverse logistics modeling for end-of-life lithium-ion batteries and its impact on recycling viability—A case study to support end-of-life electric vehicle battery strategy in Canada. *Sustainability, 14*(22), 1–23. https://doi.org/10.3390/su142215321

Gulley, A. L. (2022). One hundred years of cobalt production in The Democratic Republic of the Congo. *Resources Policy, 79*, 103007. https://doi.org/10.1016/j.resourpol.2022.103007

Gupta, T., Joshi, R., Mukhopadhyay, D., Sachdeva, K., Jain, N., Virmani, D., & Garcia-Hernandez, L. (2022). A deep learning approach based hardware solution to categorise garbage in environment. *Complex & Intelligent Systems, 8*(2), 1129–1152. https://doi.org/10.1007/s40747-021-00529-0

Hannan, M. A., How, D. N. T., Lipu, M. S. H., Mansor, M., Ker, P. J., Dong, Z. Y., et al. (2021). Deep learning approach towards accurate state of charge estimation for lithium-ion batteries using self-supervised transformer model. *Scientific Reports, 11*(1), 19541. https://doi.org/10.1038/s41598-021-98915-8

Haram, M. H. S. M., Lee, J. W., Ramasamy, G., Ngu, E. E., Thiagarajah, S. P., & Lee, Y. H. (2021). Feasibility of utilising second life EV batteries: Applications, lifespan, economics, environmental impact, assessment, and challenges. *Alexandria Engineering Journal, 60*(5), 4517–4536. https://doi.org/10.1016/j.aej.2021.03.021

Hooi Chuan Wong, G., Pant, A., Zhang, Y., Kai Chua, C., Hashimoto, M., Huei Leo, C., & Tan, U. X. (2022). 3D food printing– Sustainability through food waste upcycling. *Materials Today: Proceedings, 70*, 627–630. https://doi.org/10.1016/j.matpr.2022.08.565

Hossain, R., Sarkar, M., & Sahajwalla, V. (2023). Technological options and design evolution for recycling spent lithium-ion batteries: Impact, challenges, and opportunities. *WIREs Energy and Environment, 12*(5), e481. https://doi.org/10.1002/wene.481

Hu, X., Robles, A., Vikström, T., Väänänen, P., Zackrisson, M., & Ye, G. (2021). A novel process on the recovery of zinc and manganese from spent alkaline and zinc-carbon batteries. *Journal of Hazardous Materials, 411*, 124928. https://doi.org/10.1016/j.jhazmat.2020.124928

Ioannou, A., Georgali, P.-Z., & Fokaides, P. A. (2022). Quantification of food waste in an insular Island state for all stages of the food supply chain. *Resources, Conservation and Recycling, 185*, 106486. https://doi.org/10.1016/j.resconrec.2022.106486

Islam, M. T., & Iyer-Raniga, U. (2022). Lithium-Ion battery recycling in the circular economy: A review. *Recycling, 7*(3), 1–40. https://doi.org/10.3390/recycling7030033

Jagtap, S., Garcia-Garcia, G., & Rahimifard, S. (2021). Optimisation of the resource efficiency of food manufacturing via the internet of things. *Computers in Industry, 127*, 103397. https://doi.org/10.1016/j.compind.2021.103397

Kailaku, S. I., & Djatna, T. (2022). Traceability and quality monitoring improvement throughout carrot supply chain with the implementation of internet-of-things. *IOP Conference Series: Earth and Environmental Science, 1024*(1), 012079. https://doi.org/10.1088/1755-1315/1024/1/012079

Kaunda, R. B. (2020). Potential environmental impacts of lithium mining. *Journal of Energy & Natural Resources Law, 38*(3), 237–244. https://doi.org/10.1080/02646811.2020.1754596

Khan, S., Anjum, R., Raza, S. T., Ahmed Bazai, N., & Ihtisham, M. (2022). Technologies for municipal solid waste management: Current status, challenges, and future perspectives. *Chemosphere, 288*, 132403. https://doi.org/10.1016/j.chemosphere.2021.132403

Kintscher, L., Lawrenz, S., Poschmann, H., & Sharma, P. (2020). Recycling 4.0 – Digitalization as a key for the advanced circular economy. *Journal of Communications, 15*(1), 652–660. https://doi.org/10.12720/jcm.15.9.652-660

Kurniawan, T. A., Maiurova, A., Kustikova, M., Bykovskaia, E., Othman, M. H. D., & Goh, H. H. (2022). Accelerating sustainability transition in St. Petersburg (Russia) through digitalization-based circular economy in waste recycling industry: A strategy to promote carbon neutrality in era of Industry 4.0. *Journal of Cleaner Production, 363*, 132452. https://doi.org/10.1016/j.jclepro.2022.132452

Kurzweil, P. (2023). Electrochemical power sources: Primary and secondary batteries. In *Reference module in chemistry, molecular sciences and chemical engineering*. Elsevier.

Li, H., Dai, J., Wang, A., Zhao, S., Ye, H., & Zhang, J. (2019). Recycling and treatment of waste batteries. *IOP Conference Series: Materials Science and Engineering, 612*(5), 052020. https://doi.org/10.1088/1757-899X/612/5/052020

Likotiko, E., Matsuda, Y., & Yasumoto, K. (2023). Garbage content estimation using internet of things and machine learning. *IEEE Access, 11*, 13000–13012. https://doi.org/10.1109/ACCESS.2023.3242547

Liu, Y., Ma, X., Shu, L., Hancke, G. P., & Abu-Mahfouz, A. M. (2021). From Industry 4.0 to Agriculture 4.0: Current status, enabling technologies, and research challenges. *IEEE Transactions on Industrial Informatics, 17*(6), 4322–4334. https://doi.org/10.1109/TII.2020.3003910

Lu, Y., Maftouni, M., Yang, T., Zheng, P., Young, D., Kong, Z. J., & Li, Z. (2022). A novel disassembly process of end-of-life lithium-ion batteries enhanced by online sensing and machine learning techniques. *Journal of Intelligent Manufacturing, 34*(1), 2463–2475. https://doi.org/10.1007/s10845-022-01936-x

Malefors, C., Secondi, L., Marchetti, S., & Eriksson, M. (2022). Food waste reduction and economic savings in times of crisis: The potential of machine learning methods to plan guest attendance in Swedish public catering during the Covid-19 pandemic. *Socio-Economic Planning Sciences, 82*, 101041. https://doi.org/10.1016/j.seps.2021.101041

Malibari, A., Alsawah, G., Saleh, W., & Lashin, M. M. A. (2023). Analysis of attitudes towards food waste in the Kingdom of Saudi Arabia using fuzzy logic. *Sustainability, 15*(4). https://doi.org/10.3390/su15043668

Malik, M., Sharma, S., Uddin, M., Chen, C.-L., Wu, C.-M., Soni, P., & Chaudhary, S. (2022). Waste classification for sustainable development using image recognition with deep learning neural network models. *Sustainability, 14*(12), 1–12. https://doi.org/10.3390/su14127222

Malinauskaite, J., Anguilano, L., & Rivera, X. S. (2021). Circular waste management of electric vehicle batteries: Legal and technical perspectives from the EU and the UK post Brexit. *International Journal of Thermofluids, 10*, 100078. https://doi.org/10.1016/j.ijft.2021.100078

Metall, S. (2023). New battery recycling facility close to customers. *NEWS & INSIGHTS*. Retrieved from https://www.stenarecycling.com/news-insights/insights-inspiration/guides-articles/new-battery-recycling-facility-close-to-customers/

Miguéis, V. L., Pereira, A., Pereira, J., & Figueira, G. (2022). Reducing fresh fish waste while ensuring availability: Demand forecast using censored data and machine learning. *Journal of Cleaner Production, 359*, 131852. https://doi.org/10.1016/j.jclepro.2022.131852

Mikula, K., Skrzypczak, D., Izydorczyk, G., Warchoł, J., Moustakas, K., Chojnacka, K., & Witek-Krowiak, A. (2021). 3D printing filament as a second life of waste plastics—A review. *Environmental Science and Pollution Research, 28*(10), 12321–12333. https://doi.org/10.1007/s11356-020-10657-8

Mosallanezhad, B., Gholian-Jouybari, F., Cárdenas-Barrón, L. E., & Hajiaghaei-Keshteli, M. (2023). The IoT-enabled sustainable reverse supply chain for COVID-19 Pandemic Wastes (CPW). *Engineering Applications of Artificial Intelligence, 120*, 105903. https://doi.org/10.1016/j.engappai.2023.105903

Nañez Alonso, S. L., Reier Forradellas, R. F., Pi Morell, O., & Jorge-Vazquez, J. (2021). Digitalization, circular economy and environmental sustainability: The application of artificial intelligence in the efficient self-management of waste. *Sustainability, 13*(4), 1–19. https://doi.org/10.3390/su13042092

Neumann, J., Petranikova, M., Meeus, M., Gamarra, J. D., Younesi, R., Winter, M., & Nowak, S. (2022). Recycling of lithium-ion batteries—Current state of the art, circular economy, and next generation recycling. *Advanced Energy Materials, 12*(17), 2102917. https://doi.org/10.1002/aenm.202102917

Nnamoko, N., Barrowclough, J., & Procter, J. (2022). Solid waste image classification using deep convolutional neural network. *Infrastructures, 7*(4). https://doi.org/10.3390/infrastructures7040047

Oltra-Mestre, M. J., Hargaden, V., Coughlan, P., & Segura-García del Río, B. (2021). Innovation in the agri-food sector: Exploiting opportunities for Industry 4.0. *Creativity and Innovation Management, 30*(1), 198–210. https://doi.org/10.1111/caim.12418

Ortega-Gras, J.-J., Bueno-Delgado, M.-V., Cañavate-Cruzado, G., & Garrido-Lova, J. (2021). Twin transition through the implementation of Industry 4.0 technologies: Desk-research analysis and practical use cases in Europe. *Sustainability, 13*(24), 1–18. https://doi.org/10.3390/su132413601

Oussai, A., Bártfai, Z., & Kátai, L. (2021). Development of 3D printing raw materials from plastic waste. A case study on recycled polyethylene terephthalate. *Applied Sciences, 11*(16), 1–10. https://doi.org/10.3390/app11167338

Parfitt, J., Croker, T., & Brockhaus, A. (2021). Global food loss and waste in primary production: A reassessment of its scale and significance. *Sustainability, 13*(21), 1–18. https://doi.org/10.3390/su132112087

Pech-Rodríguez, W. J., Suarez-Velázquez, G. G., Armendáriz-Mireles, E. N., Calles-Arriaga, C. A., & Rocha-Rangel, E. (2023). Performance assessment of heuristic genetic algorithm (HGA) for electrochemical impedance spectroscopy parameter estimation. *Axioms, 12*(1), 1–14. https://doi.org/10.3390/axioms12010084

Picatoste, A., Justel, D., & Mendoza, J. M. F. (2022). Exploring the applicability of circular design criteria for electric vehicle batteries. *Procedia CIRP, 109*, 107–112. https://doi.org/10.1016/j.procir.2022.05.222

Pražanová, A., Knap, V., & Stroe, D.-I. (2022). Literature review, recycling of lithium-ion batteries from electric vehicles, Part I: Recycling technology. *Energies, 15*(3), 1–29. https://doi.org/10.3390/en15031086

Pregowska, A., Osial, M., & Urbańska, W. (2022). The application of artificial intelligence in the effective battery life cycle in the closed circular economy model – A perspective. *Recycling, 7*(6), 1–17. https://doi.org/10.3390/recycling7060081

Priyadarshini, J., Elangovan, M., Mahdal, M., & Jayasudha, M. (2022). Machine-learning-assisted prediction of maximum metal recovery from spent zinc-manganese batteries. *PRO, 10*(5), 1–14. https://doi.org/10.3390/pr10051034

Qian, K. (2021). Environmental concerns of the pulp and paper industry: Focusing on household and sanitary paper products. In T. Yahara (Ed.), *Decision science for future earth: Theory and practice* (pp. 181–195). Springer Singapore.

Rahman, M. W., Islam, R., Hasan, A., Bithi, N. I., Hasan, M. M., & Rahman, M. M. (2022). Intelligent waste management system using deep learning with IoT. *Journal of King Saud University – Computer and Information Sciences, 34*(5), 2072–2087. https://doi.org/10.1016/j.jksuci.2020.08.016

Rallo, H., Benveniste, G., Gestoso, I., & Amante, B. (2020). Economic analysis of the disassembling activities to the reuse of electric vehicles Li-ion batteries. *Resources, Conservation and Recycling, 159*, 104785. https://doi.org/10.1016/j.resconrec.2020.104785

Ramya, P., Ramya, V., & Abau Rao, M. (2023). E-waste management using hybrid optimization-enabled deep learning in IoT-cloud platform. *Advances in Engineering Software, 176*, 103353. https://doi.org/10.1016/j.advengsoft.2022.103353

Reis, J. S., Espuny, M., Nunhes, T. V., Sampaio, N. A., Isaksson, R., Campos, F. C., & Oliveira, O. J. (2021). Striding towards sustainability: A framework to overcome challenges and explore opportunities through Industry 4.0. *Sustainability, 13*(9), 1–28. https://doi.org/10.3390/su13095232

Riexinger, G., Doppler, J. P., Haar, C., Trierweiler, M., Buss, A., Schöbel, K., et al. (2020). Integration of traceability Systems in Battery Production. *Procedia CIRP, 93*, 125–130. https://doi.org/10.1016/j.procir.2020.04.002

Roy, S., Rautela, R., & Kumar, S. (2023). Towards a sustainable future: Nexus between the sustainable development goals and waste management in the built environment. *Journal of Cleaner Production, 415*, 137865. https://doi.org/10.1016/j.jclepro.2023.137865

Ruhatiya, C., Shaosen, S., Wang, C.-T., Jishnu, A. K., & Bhalerao, Y. (2020). Optimization of process conditions for maximum metal recovery from spent zinc-manganese batteries: Illustration of statistical based automated neural network approach. *Energy Storage, 2*(3), e111. https://doi.org/10.1002/est2.111

Ruiz, L. E., Pinho, A. C., & Resende, D. N. (2022). 3D printing as a disruptive technology for the circular economy of plastic components of end-of-life vehicles: A systematic review. *Sustainability, 14*(20), 1–15. https://doi.org/10.3390/su142013256

Saeidi-Mobarakeh, Z., Tavakkoli-Moghaddam, R., Navabakhsh, M., & Amoozad-Khalili, H. (2020). A bi-level and robust optimization-based framework for a hazardous waste management problem: A real-world application. *Journal of Cleaner Production, 252*, 119830. https://doi.org/10.1016/j.jclepro.2019.119830

Said, Z., Sharma, P., Thi Bich Nhuong, Q., Bora, B. J., Lichtfouse, E., Khalid, H. M., et al. (2023). Intelligent approaches for sustainable management and valorisation of food waste. *Bioresource Technology, 377*, 128952. https://doi.org/10.1016/j.biortech.2023.128952

Schmied, J., Puchta, A., Scharmann, T., Többer, H.-C., Kampker, A., Jürgen, F., et al. (2022). *Framework for the application of Industry 4.0 in lithium-ion battery cell production*. Paper presented at the Proceedings of the Conference on Production Systems and Logistics, CPSL 17th – 20th May, Vancouver 2022, Vancouver, Kanada.

Seif, R., Salem, F. Z., & Allam, N. K. (2023). E-waste recycled materials as efficient catalysts for renewable energy technologies and better environmental sustainability. *Environment, Development and Sustainability, 1*, 5473. https://doi.org/10.1007/s10668-023-02925-7

Shahabuddin, M., Uddin, M. N., Chowdhury, J. I., Ahmed, S. F., Uddin, M. N., Mofijur, M., & Uddin, M. A. (2023). A review of the recent development, challenges, and opportunities of electronic waste (e-waste). *International journal of Environmental Science and Technology, 20*(4), 4513–4520. https://doi.org/10.1007/s13762-022-04274-w

Sheng, T. J., Islam, M. S., Misran, N., Baharuddin, M. H., Arshad, H., Islam, M. R., et al. (2020). An internet of things based smart waste management system using LoRa and Tensorflow deep learning model. *IEEE Access, 8*, 148793–148811. https://doi.org/10.1109/ACCESS.2020.3016255

Skärin, F., Rösiö, C., & Andersen, A.-L. (2022). An explorative study of circularity practices in Swedish manufacturing companies. *Sustainability, 14*(12), 1–16. https://doi.org/10.3390/su14127246

Soni, D., & Kumar, N. (2022). Machine learning techniques in emerging cloud computing integrated paradigms: A survey and taxonomy. *Journal of Network and Computer Applications, 205*, 103419. https://doi.org/10.1016/j.jnca.2022.103419

Stevels, A. (2023). The challenge of introducing design for the circular economy in the electronics industry: A proposal for metrics. *Circular Economy, 2*(3), 100051. https://doi.org/10.1016/j.cec.2023.100051

Taggio, N., Aiello, A., Ceriola, G., Kremezi, M., Kristollari, V., Kolokoussis, P., Karathanassi, V. B., & Enrico;. (2022). A combination of machine learning algorithms for marine plastic litter detection exploiting hyperspectral PRISMA data. *Remote Sensing, 14*(15), 1–17. https://doi.org/10.3390/rs14153606

Tripathy, A., Bhuyan, A., Padhy, R. K., Kumar Mangla, S., & Roopak, R. (2023). Drivers of lithium-ion batteries recycling industry toward circular economy in industry 4.0. *Computers & Industrial Engineering, 179*, 109157. https://doi.org/10.1016/j.cie.2023.109157

Use of Internet of Things (IoT) in Hazardous Waste Management System. (2021). *Future Generation of Communication and Internet of Things, 1*(2), 50–55.

Vasconcelos, L. T., Silva, F. Z., Ferreira, F. G., Martinho, G., Pires, A., & Ferreira, J. C. (2022). Collaborative process design for waste management: Co-constructing strategies with stakeholders. *Environment, Development and Sustainability, 24*(7), 9243–9259. https://doi.org/10.1007/s10668-021-01822-1

Wanner, J., Bahr, J., Full, J., Weeber, M., Birke, K. P., & Sauer, A. (2021). Technology assessment for digitalization in battery cell manufacturing. *Procedia CIRP, 99*, 520–525. https://doi.org/10.1016/j.procir.2021.03.110

Wilts, H., Garcia, B. R., Garlito, R. G., Gómez, L. S., & Prieto, E. G. (2021). Artificial intelligence in the sorting of municipal waste as an enabler of the circular economy. *Resources, 10*(4), 1–9. https://doi.org/10.3390/resources10040028

Wolf, M., van den Berg, K., Garaba, S. P., Gnann, N., Sattler, K., Stahl, F., & Zielinski, O. (2020). Machine learning for aquatic plastic litter detection, classification and quantification (APLASTIC-Q). *Environmental Research Letters, 15*(11), 114042. https://doi.org/10.1088/1748-9326/abbd01

Xia, W., Jiang, Y., Chen, X., & Zhao, R. (2021). Application of machine learning algorithms in municipal solid waste management: A mini review. *Waste Management & Research, 40*(6), 609–624. https://doi.org/10.1177/0734242X211033716

Yang, Y., Zhang, X., Yin, J., & Yu, X. (2020). Rapid and nondestructive on-site classification method for consumer-grade plastics based on portable NIR spectrometer and machine learning. *Journal of Spectroscopy, 2020*, 6631234. https://doi.org/10.1155/2020/6631234

Yu, H., Sun, X., Solvang, W. D., Laporte, G., & Lee, C. K. M. (2020). A stochastic network design problem for hazardous waste management. *Journal of Cleaner Production, 277*, 123566. https://doi.org/10.1016/j.jclepro.2020.123566

Zaman, A. (2022). Waste Management 4.0: An application of a machine learning model to identify and measure household waste contamination-A case study in Australia. *Sustainability, 14*(5), 1–17. https://doi.org/10.3390/su14053061

Zhan, L., Zhao, R., Wu, Y., Zeng, S., & Yuan, Y. (2022). Construction of a spatial–temporal metabolic path for hazardous waste management based on the fusion of reported data and web text data. *Environmental Technology & Innovation, 28*, 102541. https://doi.org/10.1016/j.eti.2022.102541

Zhang, M., Wang, Y., Qian, X., Zhao, J., Nie, Y., & Qian, G. (2023). Competition and price strategies of hazardous waste collection for small and micro enterprises based on dual-channel reverse supply chain. *Journal of Cleaner Production, 386*, 135714. https://doi.org/10.1016/j.jclepro.2022.135714

Zhu, C., Li, T., Mohideen, M. M., Hu, P., Gupta, R., Ramakrishna, S., & Liu, Y. (2021). Realization of circular economy of 3D printed plastics: A review. *Polymers, 13*(5), 1–16. https://doi.org/10.3390/polym13050744

Zorn, M., Ionescu, C., Klohs, D., Zähl, K., Kisseler, N., Daldrup, A., et al. (2022). An approach for automated disassembly of lithium-ion battery packs and high-quality recycling using computer vision, labeling, and material characterization. *Recycling, 7*(4), 1–18. https://doi.org/10.3390/recycling7040048

Chapter 24
Life Cycle and Cost Estimation of Solid Waste Management Practices: A Conceptual Framework

Gulzar Alam Khan and Parag Shil

1 Introduction

The term municipal solid waste (MSW) describes the waste that is disposed of in urban areas, including commercial waste, household waste, and small-scale industrial waste that is collected and disposed of by municipalities (Cheng & Hu, 2010). Household waste consists of papers, wood, cotton, leather, etc., which are mostly biodegradable. Industrial solid waste is waste of an industrial nature, which may include debris from the demolition of residential buildings, fossil fuel waste, plastic waste, rubber waste, and textile waste.

To ensure environmental and human health protection, MSW must be properly disposed of and treated (Eggers et al., 2008). As a result of rapid socioeconomic growth, there is a disconnection between the growth in waste generation rates and the decline in waste disposal capacity, which is only becoming more pressing as time goes on (Lu et al., 2009). The growing awareness of environmental issues among the general public has led governments, municipalities, and waste management utilities to search for new technical and organizational solutions for waste management systems in the future (Sundberg et al., 1994). Waste management is the process of collection of waste, transportation, treatment, recycling, and disposal procedure of waste and keeping track of it (Wikipedia, 2010). In 1997, on a global scale, MSW produced in developing countries amounted to 0.49 billion tons and is expected to grow at a rate of 2–3% per annum (Suocheng et al., 2001). It is estimated that around two billion tons of MSW were produced around the world in 2007 (UNEP, 2009).

Despite numerous attempts to deal with the problem of municipal solid waste management (MSWM), the situation in most urban areas in developing countries remains unsatisfactory. Municipal solid waste management (MSWM) is not a one-stop shop but rather a multistep process that includes the collection and

G. A. Khan · P. Shil (✉)
Department of Commerce, Assam University, Silchar, India

V. Kumar et al. (eds.), *Environmental Engineering and Waste Management*,
https://doi.org/10.1007/978-3-031-58441-1_24

transportation of waste, the recovery of usable materials, and the final disposal of waste. These actions could eventually have a negative effect on the environment, like releasing more greenhouse gases. In the context of India, a huge amount of MSW is produced daily, and proper management of MSW is one of the top priorities of local municipalities and is getting more attention due to its economic and environmental implications.

Due to the difficulty of MSW treatment systems, it can be challenging to properly evaluate their performance. LCA (Life cycle assessment) has been proven to be a reliable and consistent way to measure how waste management is doing for the environment (Lazarevic et al., 2012). However, it disregards economic performance, which is an important element in waste management decisions (De Menna et al., 2018; Lavee & Nardiya, 2013). Cost projection is an important factor to be considered when deciding how to properly allocate funds for MSW management (Lavee & Khatib, 2010). The economic aspect must therefore be analyzed systematically as much as the environmental aspect to ensure that they are complementary in the decision-making process. In recent years, the economic performance of the MSW treatment system has been analyzed based on life cycle thinking.

A life cycle approach is a term used to describe a set of tools used to evaluate the economic and environmental performance of a product, service, or process. Generally, LCA and Life Cycle Cost Analysis (LCCA) are used to evaluate environmental and economic performance. These two methods help assess and measure environmental and economic impacts over the entire life cycle of a product, process, or service. According to IS 13174, 1991, "life cycle cost (LCC) analysis is an economic evaluation technique that covers all the costs from project conception to final scrapping and disposal and includes all costs of operation, repairs, maintenance, and energy consumption, in addition to the initial costs of development, and all the costs are discounted to the same point in time." Although LCC analysis is relatively uncommon in the literature, it has become increasingly important in recent years in the context of MSW management. The research on LCC analysis has looked at how the economy is influenced by many types of MSW management systems, and it is a potent tool for providing a thorough economic study of waste management systems (Woon & Lo, 2016; Li et al., 2016; Elagroudy et al., 2011; Reich, 2005).

Reich was one of the first to use the LCC technique to evaluate MSW management solutions by contrasting the four scenarios using materials recycling, incineration, anerobic digestion, composting, and disposal in Uppsala, Sweden, based on the 0.09 million metric tons of garbage produced yearly. The results of the study show that landfilling is the most expensive scenario, costing SEK 54 million per year, while composting is the cheapest option, costing SEK 43 million per year. The study makes the point that future studies for an expanded waste management system should also take the environmental cost into account to have a thorough analysis. In a similar study using the LCC analysis method, the researcher in Iraq analyzed three potential MSW management scenarios as alternatives to the open dumping depending on their economic success (Elagroudy et al., 2011).

LCC technique has been successfully employed for an economic evaluation of the three MSW treatment systems currently in use in Hangzhou, China, viz. landfill,

landfill with biogas conversion to power, and fluidized bed incinerator with energy recovery (Dong et al., 2014). The cost information was gathered from landfill and incinerator facilities that are currently in operation and have daily waste capacities of 2400 and 1200 metric tons, respectively. The findings of the analysis indicate that landfill with biogas conversion to electricity fared the best of the three scenarios, with a net LCC of 16.57 Yuan per ton of MSW, compared to 34.78 Yuan and 45.87 Yuan for incineration and landfill, respectively. A similar LCC analysis was carried out to estimate the cost of a 600-ton-per-day MSW incineration power plant project in China (Li et al., 2016). This study developed a model based on the time value of acquisition costs, operating and maintenance costs, and fault costs derived from the LCC analysis. In the MSW project referred to above, the LCC and the total revenue were calculated as 588 × 106 Yuan and 746 × 106 Yuan, respectively, for the waste treatment of 600 ton over 20 years.

LCC analysis is not often applied in India for the economic evaluation of waste management systems. Using LCC to manage solid waste has only been the subject of a few studies so far. Recently, researchers have conducted a study in Bangalore, India, in which they contrasted four disposal options, viz. open dumps, landfills with and without gas recovery, and landfills with bioreactors (Babu et al., 2014). With a net cost of $47 million USD, the landfill with a bioreactor was judged to be the most popular choice among these. The least preferred dump, however, was energy-free and would have cost USD 28 million to process 28 million tons of waste over 50 years. As advised by the integrated solid waste management hierarchy, this study did not focus much on the combination of various treatment technologies for extracting economic advantages from a variety of waste fractions. Additionally, their study focused mostly on landfill treatment and ignored alternative treatment approaches. The review of the literature reveals that the cost of the life cycle differs by nation. Therefore, life cycle cost analysis research for MSW in India is necessary. So far, few studies have been conducted in India, and none of the waste management strategies have been properly analyzed.

Despite the fact that studies on the life cycle costing (LCC) of solid waste management have grown significantly over the past two decades, different researchers and organizations have used diverse frameworks and approaches as a result of the absence of a widely used life cycle costing approach. It has been noted throughout the past few decades that innovative methods and strategies for carrying out life cycle cost analysis are changing, progressively moving away from typical life cycle costing and toward social and environmental life cycle costing. Despite this, there is still a lack of comprehensive, systematic research into life cycle costing techniques and approaches in the municipal solid waste management sector. An extensive review of the literature on municipal waste life cycle cost analysis, with a major focus on the methodology, approach, strategy, integration with life cycle projection, cost computation, and handling uncertainty, may offer an understanding of MSW and its management.

In this case, a comprehensive analysis of the life cycle and cost forecasting of solid waste management practices has been conducted. This study concentrates on life cycle costing (LCC) approaches, techniques, models, and a conceptual framework. The broad objectives of the study are as follows:

- To explore the life cycle trends and cost estimates for municipal solid waste by conducting a systematic review.
- To identify and report the environmental and economic consequences of MSWM practices.
- To explore the ways and means of ensuring the best solid waste management practices.

2 Conceptual Framework

2.1 *Municipal Solid Waste Management: An Introduction*

The primary sources of urban solid waste are household refuse (including food waste, plastics, and other household waste), commercial, industrial, and institutional solid waste (including hospitals), garden waste, street sweepings, and market waste, which are generated either in the municipalities or localities that have been notified. Management is the systematic goal setting procedure, creating long-term plans, planning, budgeting, implementing, operating and maintaining, monitoring and evaluating, controlling costs, reviewing objectives and plans, etc. Municipal solid waste management (MSWM) is the process of collecting solid waste, transporting it, treating it, recycling it, recovering it, and finally disposing of it in urban areas by the highest standards of public health, environmental protection, aesthetics, and other considerations.

2.2 *Importance of Municipal Solid Waste Management*

Municipal solid waste management is a fundamental component of environmental protection and is the primary responsibility of local municipalities, which typically account for approximately 20–50% of municipal expenditure in developing countries. Water pollution, air pollution, and soil contamination can result from an ineffective waste management system, which can range from an inadequate or nonexistent collection system to inefficient disposal. Municipal solid waste has created grave environmental hazards as a result of population growth and lifestyle advancements. The successful disposal of waste is a multifaceted endeavor that relies heavily on collaboration between households, municipalities, private businesses, and municipal entities. In addition, it is contingent upon the development and implementation of suitable technical solutions for the collection, transportation, recycling, and removal of waste. Waste management has always been linked to the environment's quality, which has a substantial impact on people's health and well-being, as well as the city's sustainability, lifestyle standards, productivity, and efficiency of the urban economy. Municipal solid waste may be recycled, and items like paper, plastic, glass, etc., can be separated from one another and converted into valuable goods, conserving natural resources.

2.3 Goals and Principles of Municipal Solid Waste Management (MSWM)

Municipal solid waste management (MSWM) aims to promote environmental sustainability, public health, and resource conservation and often includes several major objectives in these areas. The first objective of MSWM is to implement waste management techniques that safeguard public health, especially that of low-income populations, who are most negatively impacted by inadequate waste management. This is done by reducing exposure to hazardous waste, managing pests, and limiting the spread of disease. The second objective of MSWM is to promote and improve environmental conditions by reducing pollutant sources (air, soil, cross-media pollution, and water pollution) and to ensure the sustainability of the ecosystems in the urban area by reducing the impact of waste on the ecosystems, water sources, and air quality. The third objective of MSWM is to promote the economic growth of urban areas by providing appropriate waste disposal services and the efficient utilization and preservation of valuable resources and materials. The purpose of recycling and resource recovery is to reduce the use of natural resources, decrease energy consumption, and reduce the requirement for raw material extraction to ensure economic growth. The fourth objective of MSWM is to reduce waste in the initial stages of its development, thereby decreasing the burden of landfilling and the use of resources. Finally, MSWM seeks to create employment and revenue through its implementation.

To accomplish the objectives outlined above, it is essential to encourage sustainable solid waste disposal methods that reduce the environmental impact, protect public health and safety, and optimize the use of resources that can accommodate the requirements of the entire city, including the underprivileged. The process of waste management encompasses the production, distribution, and consumption of trash, the gathering of waste, and the proper disposal of trash. The strategy must focus on reducing waste generation through public education initiatives, the encouragement of reuse and recycling, the promotion of composting, and the implementation of waste regulations and policies. It is necessary to foster collaborations with stakeholders, including businesses, community organizations, and waste management service providers, to develop and implement effective waste management strategies. Continuous monitoring of waste generation rates, recycling rates, and other key performance indicators must be done to evaluate the effectiveness of waste management programs.

2.4 *Scope of Municipal Solid Waste Management (MSWM)*

Within the scope of MSWM, there are several activities and considerations to be taken into account to ensure efficient and sustainable waste management. The following tasks and techniques are under the purview of MSWM within the larger context of urban management:

$$\text{Collection of Garbage} \rightarrow \text{Separation of Garbage} \\ \rightarrow \text{Treatment and Processing of Garbage} \rightarrow \text{Disposal of Garbage}$$

1. Management and planning
   - Strategic plan of action
   - Legal and regulatory structure
   - Public participation
   - Financial management (budgeting, accounting, cost recovery, etc.)
   - Institutional framework (including participation of private sector)
   - Disposal or dumping facility setting
2. Generation of waste
   - Characterization of waste (source, composition, rates, etc.)
   - Minimization of waste and separation of source
3. Handling of waste
   - Gathering of waste
   - Transfer of waste, treatment, and dumping or disposal of garbage
   - Special kind of wastes (wastes from small industries, medical, etc.)

The effectiveness of the process of municipal solid waste management depends on a systematic method of trash management. The method of managing solid waste is shown in Fig. 24.1.

Education and Public Awareness Public involvement and awareness are crucial for the success of MSWM. Informing and educating the community about waste reduction, segregation, and responsible disposal practices can help promote behavioral changes and improve waste management outcomes.

2.5 *Financial Aspects*

The purpose of MSWM is to create practical budgetary and cost accounting systems in order to ensure transparency regarding the actual costs of waste management, which may cause a rise in operational effectiveness. The financial aspects of waste management that need to be closely monitored are proper investment in waste

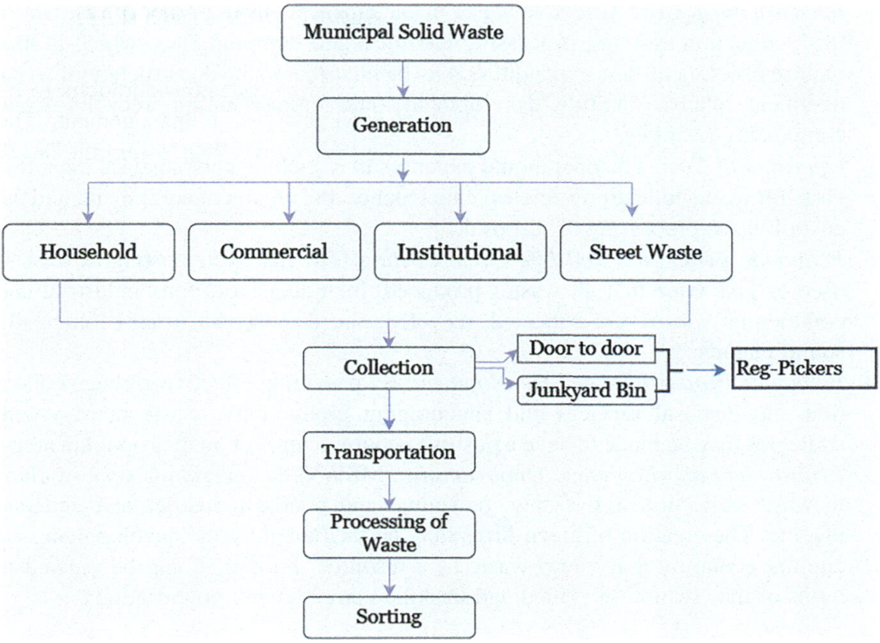

Fig. 24.1 Schematic diagram of different steps of municipal solid waste management (MSWM)

facilities through resource mobilization, collection, and cost-oriented revenue generation for waste management activities, as well as operational cost reduction with improved efficiency.

The financial components of MSWM are as follows:

- Systems for cost accounting and budgeting
- Re-allocation of resources for investments in capital
- Cost recovery and operational funding
- Cost reduction and cost control

2.6 Economic Aspects

MSWM has several economic aspects that play a significant role in its proper implementation and long-term viability. The main economic considerations associated with MSWM include:

1. *Infrastructural Cost*: Effective waste management requires proper infrastructure trash collection systems, processing facilities, and dumping sites, which in turn require investment. The expenditure may be incurred on the construction of waste treatment plants, landfill development, and implementing recycling and composting facilities.
2. *Operational Cost*: The operational expenses may include personnel salaries, fuel costs for waste collection vehicles, maintenance and repairs of equipment, and the cost of waste processing or disposal.
3. *Economic Evaluation and Life Cycle Costing*: The life cycle costing of trash is used to guarantee that all wastes produced, including hazardous industrial and commercial wastes, are collected, recycled, and disposed of in an ecologically sound manner.
4. *Economic Productivity and Development*: By providing effective garbage collection and disposal services and encouraging productivity, waste management strategies may be made to have a positive economic impact on metropolitan areas.
5. *Employment and Economic Opportunities*: MSWM can create job opportunities in waste collection, processing, recycling, and related industries and generate income. The creation of green firms may be facilitated by the development of a circular economy that views waste as a resource. Further, it can be viewed in terms of the creation of skilled and unskilled employment opportunities.

2.7 Hazardous and Special Waste Management

Hazardous waste poses a serious threat to both aquatic and terrestrial life. In order to prevent hazardous wastes from entering the municipal system, including landfills, sewers, and drains, it is necessary to make concerted efforts to implement and strengthen environmental monitoring and controls. The most essential thing is to identify, record, and target possible sources of hazardous elements in industrial wastes for appropriate treatment, whether they are handled by private or public trash gathering.

Establishing emergency response plans and protocols is prime requirement for effective hazardous waste management. This involves training personnel in emergency procedures, having appropriate safety equipment and spill containment measures in place, and keeping coordinating with local emergency response agencies to address any untoward incidents or accidents that may occur during handling, transportation, or treatment of hazardous waste. Proper maintenance of accurate records of hazardous and special waste management activities is essential for regulatory compliance and monitoring.

3 LCA of Municipal Solid Waste Management

Rapid population expansion, development, lifestyle shifts, and a lack of available land resources in the vicinity of metropolitan centers are limiting trash handling efforts in cities with poor planning regulations (Arena et al., 2003). In developing countries such as India, where landfills remain the preferred option mainly due to economic constraints or a lack of technical knowledge in alternative methods such as biological or thermal treatment, solid waste is becoming a major source of challenges and environmental threats due to inadequate management. It is therefore desirable to select appropriate waste processing technologies through a comprehensive, sustainable trash handling solution with the least environmental impact and lesser amount of emissions.

The life cycle assessment (LCA) technique has been acknowledged in this context as a useful tool that has been widely accepted as a globally standardized technique (IOS, 2006a, b). It is efficient at assessing the environmental effects of management alternatives, simplifying the decision-making process by contrasting the many solutions that are now accessible.

3.1 Concept of Life Cycle Assessment (LCA)

Life cycle assessment (LCA) is a "cradle-to-grave" technique for assessing the environmental effects of materials, processes, commercial goods, industrial systems, and services. The procedure starts with the removal of raw minerals mined from the earth to produce a product, and it is completed when all materials have been returned to the earth in some way. In 1990s, life cycle analysis (LCA) of municipal solid trash was initiated to investigate its environmental impacts (Barton et al., 1996).

A methodology used to study the environmental effects connected with all phases of a commercial product, activity, or service is called the life cycle assessment (LCA) approach. It is a strategy for evaluating all environmental exchanges that occur during the life cycle of an activity, including resource, energy, emissions, and waste exchanges. LCA is widely used to evaluate how different trash handling techniques may affect the environment. The different phases and typical inflows and outflows in the life cycle analysis of trash are displayed in Fig. 24.2.

Different guidelines have been published by international organization for standardization (ISO, 2006a, b) time to time to ease the process and its implementation. Nowadays, LCA is considered as an important assessment criterion of MSW. Following the establishment of an important evaluation criterion, a variety of approaches and techniques were employed in the LCA of MSW. LCA has gained a lot of attention in recent years, and accounting methods based on LCA have been created to estimate emissions from various systems (Laurent et al., 2014a, b). Recently, LCA applications have been focused on developed economies when solid waste management is evaluated from an environmental perspective (Di Maria

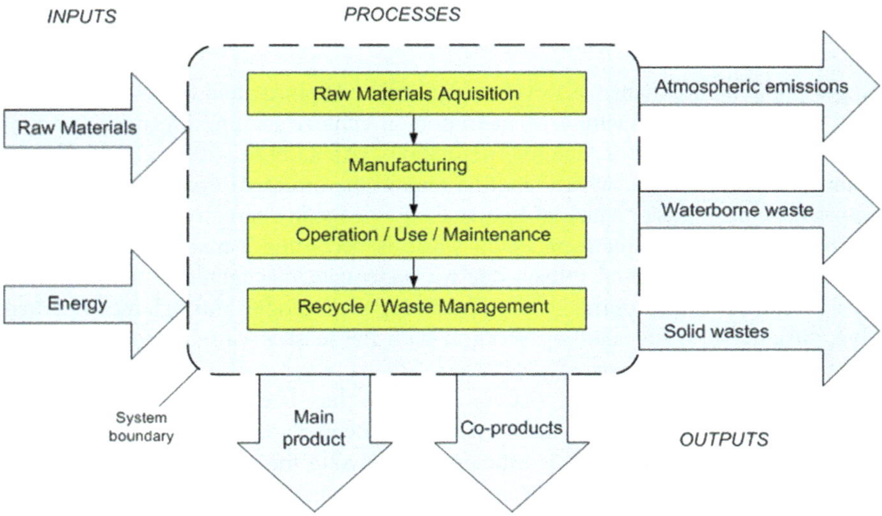

Fig. 24.2 Distinct phases and typical inflows and outflows in life cycle assessment (LCA). (Credit: Mark Fedkin) (https://www.e-education.psu.edu/eme807/node/690)

et al., 2016; Di Maria & Sisani, 2017). In case of developing countries, the nature of trash composition differs from location to location and accordingly the management strategies mainly focus on open dumping, composting, or landfilling. The studies indicate that these developing economies still receive poor attention (Liu et al., 2017a, b; Laurent et al., 2014a, b; Noya et al., 2018).

3.2 *Life Cycle Assessment Technique*

The normal standard procedure to carry out LCA study involves several steps (Fig. 24.3). The main steps are given as follows:

1. *Identification of scope and objective*: First step is identification of a product, setting the system boundaries, LCA mapping, and material flow chart development.
2. *Analysis of Inventory*: This step involves collection of data and raw materials categorization, water and energy under inputs, and emissions and side products as outputs.
3. *Impact assessment*: This involves impact categorization, metrics development, and comparative analysis through proper calculations.
4. *Data interpretation and result analysis*: Finally, the data interpretation is done to assess the human and environmental effects. The result is processed to deliver information to target audience and issue recommendations.

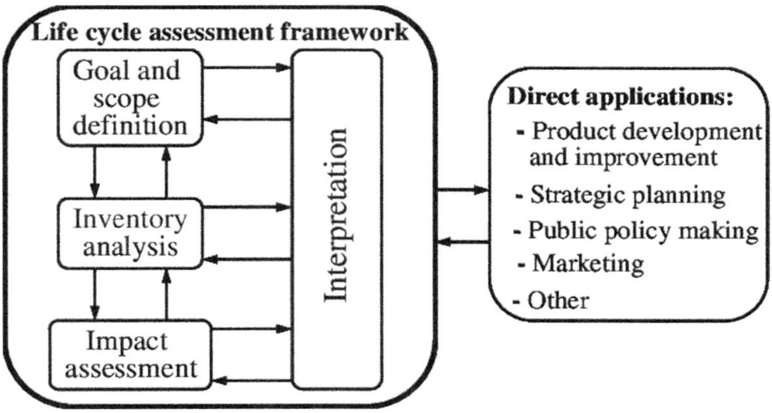

Fig. 24.3 Different stages of life cycle assessment (LCA). (ISO, 2006a, b)

3.3 Limitations of LCA

Data are necessary for LCA to be accurate and comprehensive, yet acquiring such data might be difficult. The uncertainty and assumptions must be well understood. LCA, in contrast to traditional risk assessment, is a broad evaluation of environmental consequences rather than necessarily attempting to quantify any specific real repercussions. The LCA is a broad undertaking that necessitates multidisciplinary expertise in both the technical and economic fields, even for a tiny system. Since it is difficult for a single individual to complete an LCA with adequate precision, it is often given to a team of specialists.

4 Cost Estimation of Solid Waste Management

Globally, municipalities are constrained by a lack of funding, which is perhaps their biggest solid waste management (SWM) difficulty. According to research done in India, institutional and financial problems are the biggest obstacles in further advancements of solid waste management practices (Hanrahan et al., 2006). Specifically, to create and implement realistic investment estimates, there is a critical need for greatly improved medium-term planning at the state and local levels.

Cost estimation is a strategy used to assess the need for resources while also being conscious of the risks involved (Ostwald & McLaren, 2004). As a rule of thumb, the municipal budget is made up of three parts: operating expenses, capital expenditures, and recurrent revenue (Schaeffer, 2000). However, a projection itself is made up of different solid waste management elements, including wages, the cost of equipment, and the expenses of routine maintenance. When striving to increase the level of service and implement effective waste management, precise and excellent cost estimations for solid waste management may assist in setting a suitable budget and help in defending the budget.

Cost planning encompasses both quantitative and qualitative aspects. Some research has predominantly employed quantitative techniques, such as cost–benefit analysis, programming, optimization, and statistical methodologies (Huang et al., 2001; Clark et al., 1971). While past studies have mostly focused on qualitative assessments of the costs of certain processes such as privatization, waste reduction, collection, and disposal (Strathman et al., 1995; Palmer & Walls, 1997; McDavid, 1985; Jenkins, 1991). From time to time, various models have been created to facilitate the efficient management and decision-making of solid waste, as well as the empirical choices that are useful in selecting MSWM policies. The Seattle public utilities created the "Recycling Potential Assessment and System Analysis Model (RPA/SAM)" to support several policy and planning activities (Bagby et al., 1998). To anticipate the entire system costs related to SWM in Seattle, the model specifically incorporates prior cost estimations.

SWM is a significant issue for local governments in India as well, and access to funding is a barrier to successful SWM. A grant of USD 550 million to municipalities of India for the years 2005–2010 was suggested by the 12th Finance Commission, with at least 50% of that amount designated for SWM (Appasamy & Nelliyat, 2007).

4.1 Unit Cost Method (UCM)

The unit cost method (UCM) is a method for estimating expenses in which each action, from collection to treatment and disposal, is broken down into individual expenditures like consumables, fuel, maintenance, and labor. In the subsequent stage, the required amount of each item is reported, together with the price per item or unit cost (created using preexisting data or derived from price quotations) is multiplied by this amount to determine the overall cost per item. The overall cost of providing the service is then determined by adding the expenses for each item. The cost-estimating approach may be used to prepare a budget for a certain task, build up a new facility, purchase more resources, and more. The benefits of using UCM to estimate SWM costs are that it is straightforward to create, dependable owing to its top-down methodology, and simple to comprehend. There is not much space for speculation in this method, since it takes a deterministic approach to cost estimating and the independent variable(s) are essentially measures of the object being evaluated (Christensen & Dysert, 2003).

The approach is simple and easy to understand in theory, but it takes work to implement it correctly. It requires an accurate estimate of the quantity of units needed, for instance, if it is used for forecasting. Due to inadequate accounting practices and changing circumstances in a city or municipality region, several other issues, such as accurate documentation, may be challenging. Additionally, unit prices may vary because of differing standards within a framework (e.g., alternate day collection in residential zones and daily collection in commercial zones), and these changes are required to be taken into account when producing cost estimates.

Thus, the cost model's accuracy determines how reliable the procedure as a whole is. Due to the potential complexity involved in modeling huge systems, different techniques may be able to offer more easily accessible recommendations on cost estimation. It should be mentioned that the unit cost technique is a commonly used approach for estimating SWM expenses globally due to its simplicity and obvious assumptions.

4.2 Benchmarking

Another accurate approach to cost assessment is benchmarking. In essence, it is an efficient approach for determining an acceptable cost forecast utilizing actual cost information from a comparable firm that has implemented the sort of modification being considered. The Department of Urban Services, Canberra, Australia, employed benchmarking analysis in their 1999–2000 budgets to predict the costs of trash handling and recycling. Comparative data have been gathered using the 1998–1999 budget data from a similar region to estimate the landfill expenses in the 1999–2000 budgets (www.treasury.act.gov.au, date of citation- 23/03/2011). The Vermont Department of Environmental Conservation's Solid Waste Program (DSM, 2005) similarly used the information from the results of the residential and business pricing survey results of 1999 to assess the overall expenses for collecting, recycling, and disposing of solid waste for planning purposes in 2005. The 1999 statistics were valuable because they were used to compare SWM pricing across the state and by area and served as a baseline cost. It is also anticipated that it will act as a standard for comparison in the future.

Benchmark costs must encompass all expenditures since they are inclusive by nature. The UNEP's (2004) "Introductory Guide for Decision-Makers" notes that since waste gathering, landfills, equipment, and other installations that are invariably associated with integrated trash handling systems have various depreciation periods and lifetimes, the total yearly costs, i.e., operating costs plus the annual payback for capital investments, must be calculated. The report highlighted the importance of separately estimating the costs associated with general administrative initiatives like the issuance of permits and legislation, as well as specific trash handling activities, such as the composting and recycling of waste streams. Furthermore, cost estimation may enable the comparison of the current trash handling system costs with the future waste management plan costs (UNEP, 2004).

Despite being a useful tool, benchmarking may not always allow for decent comparisons. Sufficient full-cost accounting is one possible restriction, and certain benchmark costs could not account for capital expenses. The employment of this strategy may be deterred by database deficiencies (such as the absence of cost data by year). Additionally, the range or caliber of the services offered may differ substantially, restricting their applicability. It is possible for a dataset to have bias, which would reduce the utility of using it. A budget may have been overstated by management to subsequently use it as evidence that they did well by making long-term cost cuts, or it may have been understated to allow for the quick approval of money. When used as benchmarks to estimate costs elsewhere, such biased costs might result in major shortcomings in long-term planning.

4.3 Cost Modeling

4.3.1 Production and Cost Functions

Typically, the production function is defined by economists as the connection between a process' output and its required input resources (Wohl & Hendrickson, 1984; Fullerton & Kinnaman, 1995). The phrase "cost function" is generally used to refer to the connection between costs and variables. Solid waste management costs are correlated with production costs or with other factors, such as the density of the population or the nature of the service being offered. A variety of input variables, including trucks, personnel, frequency of gathering, total tonnage collected, etc., may be used to represent the parameters of cost and production functions. These functions can either be linear or nonlinear. When a scale variable, such as tons per year, is the only input variable taken into account, the function then provides the economy-of-scale impact for that cost. Regression techniques are frequently used to experimentally determine the coefficients in production functions and costs using the datasets that are already available.

The most current research on cost-estimating approaches employing cost and production functions was published by De Jaeger and colleagues (2011) and by Weng and colleagues (2011). The authors propose that, to manage the increased complexity and uncertainty in contemporary waste management systems, the data envelopment analysis approach and the econometric modeling technique respectively should be applied.

4.3.2 System Models

Several models that concentrate on employing linear programming or other optimization approaches to minimize expenses have been used to analyze the economic aspects. By combining current linear programming and optimization techniques, the advanced optimization modeling framework created by Xu et al. can balance out the uncertain components of the decision-making process for waste management Xu et al., 2010). To develop a long-term cost planning model, the applicability of their methodology was shown using a hypothetical SWM example with three municipalities and two treatment alternatives (landfilling and incineration).

The COSEPRE (costs of urban cleaning services) application created by Sandoval et al. (PAHO, 2001) allows for the cost evaluation of scenarios and calculates the yearly and unit costs per service based on data given by the user. Only when the user has easy access to a comprehensive, full-cost accounting system are the costs of each service calculated. The difficulty of generalizing system models to different contexts is one of the main difficulties encountered while employing them. The underlying cost functions can be challenging to evaluate, and it can be much more challenging to understand how they were created and how they could be used in various contexts.

5 Challenges Associated with Municipal Solid Waste Management (MSWM)

Although municipal facilities are improving day by day, yet the most significant challenges associated with municipal solid waste management in India lies with the poor organizational setup, use of unscientific methods, financial constraints, shortage of suitable land for landfilling, and lack of public awareness regarding hazardous effects of solid waste. Most of the landfill sites in metropolitan cities have already reached their design life and are overflowing. Moreover, developing and operating new landfill sites in those cities using the scientific disposal method are challenging and needs considerable attention.

Developing countries like India faces challenges related to identification and segregation of waste, waste policy, selection of waste technology, and the availability of well-trained people for engagement in the waste management sector. Until and unless these minimum requirements are met, India will continue to suffer from poor waste management and its associated impacts on public health and the environment. Non-availability of qualified and well-trained waste management professionals greatly hampers the solid waste management practices. Lack of accountability of the system at every level throughout India is also responsible for poor management and is a challenge in effective disposal of the task. Insufficient allocation of budget to municipal authorities to cover the costs associated with the development of proper waste collection, storage, treatment, and disposal are responsible for poor MSW in India and is great challenge that needs to be addressed. The lack of strategic MSW plans, waste collection/segregation methodology, and government finance regulatory framework are major barriers in achieving a sustainable solid waste management goal in India.

6 Solutions and Future Prospective

1. Waste collection and transport: Proper waste collection from the source and improvement in the both waste collection and transport infrastructure will create jobs and improve public health and scope of tourism in the country.
2. Engineered landfills in India: Engineered landfill will allow the safe disposal of residual MSW on land and in turn will protect the ground and surrounding environment from pollution and avoids air emissions, wind-blown litter, odor, etc. Exhaustive planning can be done to develop engineered landfills in the country.
3. Waste-to-energy: The problems associated with non-scientific and improper waste disposal could be significantly mitigated through material recovery. Source separation of inert and high moisture content fractions and other treatment options need to be developed for sustainable growth.
4. Setup of a strong and independent regulating authority is needed to regulate waste management practices. Without clear regulation and enforcement, attainment of success in MSW management is impossible. Government and other local bodies should work on it.
5. Finance for waste management companies and funding for infrastructure development must be raised from waste producers through a waste tax and recycling of valuable materials from waste.
6. More progress can be made through proper training and capacity building at every level. The school children and youth should understand the importance of waste management and ill effects of poor waste management on the environment and public health. The needs to be aware of their role and responsibilities toward waste management practices.
7. With increasing size of the communities, the problems of waste become more acute, and this provides opportunities for decentralized waste management by self-help groups and NGOs.
8. It is essential to segregate recyclable solid waste components from garbage bins and landfills [131]. Authorities have implemented policy measures to include the unorganized sector in solid waste management.
9. The matters related to awareness can be addressed by assessing public awareness and involvement through knowledge gap analysis, which promotes informed decision-making. This will ensure better policy formulation, clear laws, and fee methodology.
10. The study definitely will be great help in analyzing the current waste management scenario, progress so far, and steps that needs to be taken further for sustainable growth.

7 Conclusion

The appropriate management and treatment of municipal solid waste are of the utmost significance to safeguard the environment's quality and the public's health across the world. The theoretical and conceptual underpinnings of solid waste management techniques were examined in this chapter. It has been noted that solid waste management only receives a very small portion of the total municipal budget and receives little attention since municipal authorities give this activity a low priority. The financial and economic aspects of municipal solid waste management must be taken into consideration when selecting a particular method of solid waste treatment. The life cycle assessment (LCA) technique has been acknowledged as an important instrument that has been extensively used as a globally standardized technique that is efficient in measuring the environmental consequences of solid waste when comparing various scenarios. LCA may be a crucial tool for municipalities as they develop integrated waste management strategies aimed at reducing the harmful environmental effects of discarded solid waste. By using an integrated strategy, LCA might significantly aid in enhancing law enforcement and supporting legislative activities for the transition to more sustainable municipal solid waste management plans.

References

Appasamy, P. P., & Nelliyat, P. (2007). *Financing solid waste management: Issues and options*. International conference on sustainable solid waste management. Chennai, India, pp. 537–542.

Arena, U., Mastellone, M. L., & Perugini, F. (2003). Life cycle assessment of a plastic packaging recycling system. *International Journal of Life Cycle Assessment, 8*(2), 92–98.

Babu, G. L. S., Lakshmikanthan, P., & Santhosh, L. G. (2014). *Life cycle analysis of municipal solid waste (MSW) land disposal options in Bangalore City*. International conference on sustainable infrastructure, Long Beach, pp. 6–8.

Bagby, J., Ernsdorff, S., Kipperberg, G., & Perrin, L. (1998). *Seattle's solid waste plan: On the path to sustainability. City of Seattle's Recycling Potential Assessment/System Analysis Model.* Seattle Public Utilities.

Barton, J. R., Dalley, D., & Patel, V. S. (1996). Life cycle assessment for waste management. *Waste Management, 16*(1–3), 35–50.

Cheng, H., & Hu, Y. (2010). Municipal solid waste as a renewable source of energy: Current and future practices in China. *Bioresource Technology, 101*, 3816–3824.

Christensen, P., & Dysert, L. R. (2003). *AACE International Recommended Practice No. 17R-97 Cost Estimate Classification System*. AACE International.

Clark, R., Grupenhoff, B., Garland, G., & Klee, A. (1971). Cost of residential solid waste collection. *Journal of the Sanitary Engineering Division, 97*, 563–568.

De Jaeger, S., Eyckmans, J., Rogge, N., & Puyenbroeck, T. V. (2011). Wasteful waste-reducing policies? The impact of waste reduction policy instruments on collection and processing costs of municipal solid waste. *Waste Management, 31*, 1429–1440.

De Menna, F., Dietershagen, J., Loubiere, M., & Vittuari, M. (2018). Life cycle costing of food waste: A review of methodological approaches. *Waste Management, 73*, 1–13.

Di Maria, F., Micale, C., & Contini, S. (2016). A novel approach for uncertainty propagation applied to two different bio-waste management options. *International Journal of Life Cycle Assessment, 21*, 1529–1537.

Di Maria, F., & Sisani, F. (2017). A life cycle assessment of conventional technologies for landfill leachate treatment. *Environmental Technology and Innovation, 8*, 411–422.

Dong, J., Chi, Y., Zou, D., Chao, F., Huang, Q., & Ni, M. (2014). Energy–environment–economy assessment of waste management systems from a life cycle perspective: Model development and case study. *Journal of Applied Energy, 114*(C), 400–408.

DSM. (2005). *The price of solid waste management services in Vermont Waterbury: Vermont Department of Environmental Conservation*. DSM Environmental Services, Inc.

Eggers, J., Lindner, M., Zudin, S., Zaehle, S., & Liski, J. (2008). Impact of changing wood demand, climate and land use on European forest resources and carbon stocks during the 21st century. *Global Change Biology, 14*, 2288–2303.

Elagroudy, S., Elkady, T., & Ghobrial, F. (2011). Comparative cost benefit analysis of different solid waste management scenarios in Basrah, Iraq. *Journal of Environmental Protection, 2*(5), 555–563.

Fedkin, M. Published on *EME 807: Technologies for sustainability systems*. https://www.e-education.psu.edu/eme807

Fullerton, D., & Kinnaman, T. C. (1995). Garbage, recycling, and illicit burning or dumping. *Journal of Environmental Economics and Management, 29*, 78–91.

Hanrahan, D., Srivastava, S., & Ramakrishna, A. (2006). *Improving management of municipal solid waste in India: Overview and challenges*. Published by Environment and Social Development Unit, South Asia Region, World Bank.

Huang, G. H., Sae-Lim, N., Liu, L., & Chen, Z. (2001). An interval-parameter fuzzy-stochastic programming approach for municipal solid waste management and planning. *Environmental Modelling & Assessment, 6*, 271–283.

ISO. (2006a). *ISO 14040:2006. Environmental management — Life cycle assessment — Principles and framework*. International Organization for Standardization. Retrieved from https://www.iso.org/obp/ui/en/#iso:std:iso:14040:ed-2:v1:en

ISO. (2006b). *ISO 14044:2006. Environmental management – Life cycle assessment – Requirements and guidelines*. International Organization for Standardization. Retrieved from https://www.iso.org/standard/38498.html

Jenkins, R. (1991). *Municipal demand for solid-waste-disposal services: The impact of user fees*. Maryland University.

Laurent, A., Bakas, I., Clavreul, J., Bernstad, A., Niero, M., Gentil, E., Hauschild, M. Z., & Christensen, T. H. (2014a). Review of LCA studies of solid waste management systems–Part I: Lessons learned and perspectives. *Waste Management, 34*(3), 573–588.

Laurent, A., Clavreul, J., Bernstad, A., Bakas, I., Niero, M., Gentil, E., Christensen, T. H., & Hauschild, M. Z. (2014b). Review of LCA studies of solid waste management systems–Part II: Methodological guidance for a better practice. *Waste Management, 34*(3), 589–606.

Lavee, D., & Nardiya, S. (2013). A cost evaluation method for transferring municipalities to solid waste source-separated system. *Waste Management, 33*(5), 1064–1072.

Lavee, D., & Khatib, M. (2010). Benchmarking in municipal solid waste recycling. *Waste Management, 30*(11), 2204–2208.

Lazarevic, D., Buclet, N., & Brandt, N. (2012). The application of life cycle thinking in the context of European waste policy. *Journal of Cleaner Production, 29*, 199–207.

Liu, Y., Ni, Z., Kong, X., & Liu, J. (2017a). Greenhouse gas emissions from municipal solid waste with a high organic fraction under different management scenarios. *Journal of Cleaner Production, 147*, 451–457.

Liu, Y., Sun, W., & Liu, J. (2017b). Greenhouse gas emissions from different municipal solid waste management scenarios in China: Based on carbon and energy flow analysis. *Waste Management, 68*, 653–661.

Li, W. C., Tse, H. F., & Fok, L. (2016). Plastic waste in the marine environment: A review of sources, occurrence and effects. *Science of the total environment, 566*, 333–349.
Lu, H. W., Huang, G. H., He, L., & Zeng, G. M. (2009). An inexact dynamic optimization model for municipal solid waste management in association with greenhouse gas emission control. *Journal of Environmental Management, 90*, 396–409.
McDavid, J. (1985). The Canadian experience with privatizing residential solid waste collection services. *Public Adminstration Review, 45*(5), 602–608.
Noya, I., Inglezakis, V., González-García, S., Katsou, E., Feijoo, G., & Moreira, M. T. (2018). Comparative environmental assessment of alternative waste management strategies in developing regions: A case study in Kazakhstan. *Waste Management & Research, 36*(8), 689–697.
Ostwald, P., & McLaren, T. (2004). *Cost analysis and estimating for engineering andmanagement*. Prentice Hall Inc..
PAHO. (2001). *COSEPRE: Costs of urban cleaning services*. Working Paper series no. 9, PAHO/WHO, World Bank, Washington DC. Computer program software stored with CD roms.
Palmer, K., & Walls, M. (1997). Optimal policies for solid waste disposal taxes, subsidies, and standards. *Journal of Public Economics, 65*, 193–205.
Reich, M. C. (2005). Economic assessment of municipal waste management systems—Case studies using a combination of life cycle assessment (LCA) and life cycle costing (LCC). *Journal of Cleaner Production, 13*(3), 253–263.
Schaeffer, M. (2000). *Municipal budgeting toolkit*. World Bank.
Strathman, J., Rufolo, A., & Mildner, G. (1995). The demand for solid waste disposal. *Land Economics, 71*, 57–64.
Sundberg, J., Gipperth, P., & Wene, C. O. (1994). A systems approach to municipal solid waste management: A pilot study of Göteborg. *Waste Management & Research, 12*, 73–91.
Suocheng, D., Tong, K. W., & Yuping, W. (2001). Municipal solid waste management in China: Using commercial management to solve a growing problem. *Utilities Policy, 10*(1), 7–11.
UNEP. (2009). *Developing integrated solid waste management plan, Training manual* (p. 176). United Nations Environmental Programme, Division of Technology, Industry and Economics, International Environmental Technology Centre.
UNEP-IETC. (2004). *Waste management planning, an environmentally sound approach for sustainable urban waste management – An introductory guide for decision-makers*. United Nations Environment Programme.
Weng, Y. C., & Fujiwara, T. (2011). Examining the effectiveness of municipal solid waste management systems: An integrated cost–benefit analysis perspective with a financial cost modeling in Taiwan. *Waste Management, 31*, 1393–1406.
Wikipedia. (2010). *Waste management*. Available from: http://en.wikipedia.org/wiki/Waste_management
Wohl, M., & Hendrickson, C. (1984). *Transportation investment and pricing principles: anintroduction for engineers, planners, and economists*. Wiley.
Woon, K. S., & Lo, I. M. C. (2016). An integrated life cycle costing and human health impact analysis of municipal solid waste management options in Hong Kong using modified eco-efficiency indicator. *Resource Conservation and Recycling, 107*, 104–114.
Xu, Y., Huang, G. H., Qin, X. S., Cao, M. F., & Sun, Y. (2010). An interval-parameter stochastic robust optimization model for supporting municipal solid waste management under uncertainty. *Waste Management, 30*(2), 316–327.

Index

V. Kumar et al. (eds.), *Environmental Engineering and Waste Management*, https://doi.org/10.1007/978-3-031-58441-1

M

N

O

P

R

S

SPRINGER NATURE

GPSR Compliance

The European Union's (EU) General Product Safety Regulation (GPSR) is a set of rules that requires consumer products to be safe and our obligations to ensure this.

If you have any concerns about our products, you can contact us on ProductSafety@springernature.com

In case Publisher is established outside the EU, the EU authorized representative is:

Springer Nature Customer Service Center GmbH
Europaplatz 3
69115 Heidelberg, Germany

The manufacturer's authorised representative in the EU is Springer Nature Customer Service Centre GmbH, Europaplatz 3, 69115 Heidelberg, Germany. If you have any concerns regarding our products, please contact ProductSafety@springernature.com

Printed and bound by CPI Group (UK) Ltd, Croydon, CR0 4YY
30/09/2026
02225586-0020